# METAL-ENHANCED FLUORESCENCE

# METAL-ENHANCED FLUORESCENCE

*Edited By*

**Chris D. Geddes**

WILEY

A JOHN WILEY & SONS, INC., PUBLICATION

Published by John Wiley & Sons, Inc., Hoboken, New Jersey
Published simultaneously in Canada



***Library of Congress Cataloging-in-Publication Data:***

Metal-enhanced fluorescence / edited by Chris D. Geddes.
p. cm.
Includes index.
ISBN 978-0-470-22838-8 (cloth)
1. Fluorescence spectroscopy. 2. Nanoparticles. 3. Radioactive decay. 4. Plasmons (Physics)
I. Geddes, Chris D.
QC477.M475 2010
543′.56–dc22

2010004247

Printed in the United States of America

10 9 8 7 6 5 4 3 2 1

# CONTENTS

# PREFACE

Today conventional fluorescence spectroscopy, i.e. far-field fluorescence, is entrenched in the Biosciences. For over 150 years we have both observed and enjoyed an era of the growth of both the theory and applications of fluorescence. In nearly all examples of classical far-field fluorescence, a fluorophore is excited from the far field, optics enable the collection of light followed by detection. In these simple examples, the fundamental properties of the fluorophore, such as extinction coefficient or oscillator strength of a transition as well as the spatial distribution of emission remain for the most part unchanged. However, over the last 10 years, there has been intense research on the influences of near-field (less than 1 wavelength of light away) metallic nanoparticles on traditional far-field fluorescence properties. In contrast to traditional far-field fluorescence we know relatively little today concerning near-field fluorescence, although its potential applications look set to fuel yet another era in fluorescence, a likely paradigm shift in the ways we both think and use fluorescence spectroscopy today.

In about 2000, my laboratory started to study the interactions of fluorophores with metallic nanoparticles, both solution-based and surface-immobilized. Our findings agreed with other workers whom had observed increases in fluorescence emission coupled with a decrease in the fluorophores' radiative lifetime. Subsequently, we applied classical far-field fluorescence descriptions to these experimental observations, which ultimately suggested a modification in the fluorophores's intrinsic radiative decay rate, a rate thought to be mostly unchanged and only weakly dependent on external environmental factors. This simple description, coupled with what seemed like a limitless amount of applications led to a paper published by our laboratory in 2001 entitled "Metal-Enhanced Fluorescence", or MEF, a term now widely used today almost a decade later.

In 2005 and onwards we reported in numerous papers that the underlying mechanisms for MEF were potentially much more elegant than the simplistic radiative rate modifications, demonstrated by our group and others, which was approximated by a collection of modified far-field fluorescence rate equations on the emission side and an enhanced near-field absorption. In this new mechanistic description of MEF, a close-proximity excited fluorophore induces a mirror dipole in the metallic nanoparticle, which itself radiates with high efficiency the coupled quanta, resulting in significantly enhanced fluorescence. Today, near-field interactions have been extended to include enhanced chemiluminescence, phosphorescence as well as enhanced Bioluminescence signatures. The coupled fluorophore lifetime is also observed to be considerably shorter than the free-space or far-field fluorescence lifetime, reflecting the "surface plasmon" lifetime of the coupled emission. While this thinking is somewhat different than earlier descriptions, it has enabled new plasmonic discoveries, such as Plasmonic Electricity, by the Institute of Fluorescence, a digital form of metal-enhanced fluorescence and indeed far-field fluorescence with a plethora of potential applications. The enhanced absorption of fluorophores in close-proximity to plasmonics nanoparticles is also now further understood, with the wavelength dependence and distance from a metallic nanoparticle both modeled and experimentally verified.

While the mechanism/s of MEF is/are most certainly likely to be debated in the coming years, it is evident that the applications of MEF are forever growing. Subsequently, I have invited a collection of works from outstanding scientists around the world which have collectively contributed to our advancement of this new concept of near-field fluorescence, namely Metal-Enhanced Fluorescence, MEF, from both a theoretical and experimental perspective. At this time, I would like to thank the authors for their most timely cutting edge contributions and their willingness to share them with us, thank you. I would also like to thank members of my lab for helping me collate this volume, including, Caroleann Aitken, Rodd Pribik, Jeff Roeser, and Micah Weisenberg whom have spent many an hour typesetting the entire volume. Thanks also to Dr Jennie Hunter-Cevera, President of UMBI, for her support of the Institute of Fluorescence's MEF programs. In closing, I hope you find this volume a valuable resource which attempts to reflect current thinking and trends in this new era of fluorescence spectroscopy.

*Institute of Fluorescence*
*Baltimore, Maryland*
*18th August 2009*

PROFESSOR CHRIS D. GEDDES

# CONTRIBUTORS

DR. DAMIAN AHERNE, School of Chemistry, Trinity College, Dublin, Ireland

DR. JON P. ANDERSON, LI-COR Biosciences Inc., Lincoln, USA

DR. RENATO E. DE ARAUJO, Department of Electronic and Systems, Universidade Federal de Pernambuco, Recife, Brazil

DR. R. F. AROCA, Laser Dynamics Laboratory, School of Chemistry and Biochemistry, Georgia Materials & Surface Science Group, Faculty of Science, University of Windsor, Windsor, Canada

DR. KADIR ASLAN, The Institute of Fluorescence, University of Maryland Biotechnology Institute, Baltimore, USA

DR. J. AUBARD, The MacDiarmid Institute for Advanced Materials and Nanotechnology, School of Chemical and Physical Sciences, Victoria University of Wellington, Wellington, New Zealand

DR. E. BLACKIE, The MacDiarmid Institute for Advanced Materials and Nanotechnology, School of Chemical and Physical Sciences, Victoria University of Wellington, Wellington, New Zealand

DR. STEVE BLAIR, Department of Electrical and Computer Engineering, University of Utah, Salt Lake City, USA

DR. YING-FENG CHANG, Institute of Biophotonics, National Yang Ming University, Taipei, Taiwan

DR. CHII-CHANG CHEN, Department of Optics and Photonics, National Central University, Jhung-li, Taiwan

DR. SI-HAN CHEN, Department of Chemistry, National Tsing Hua University, Hsin-chu, Taiwan

DR. YEECHI CHEN, Department of Chemistry, University of Washington, Seattle, WA

DR. NAN-FU CHIU, National Taiwan University, Taipei, Taiwan

DR. CHIEN CHOU, Department of Biomedical Imaging and Radiological Sciences, National Yang Ming University, Taipei, Taiwan

DR. P. G. ETCHEGOIN, The MacDiarmid Institute for Advanced Materials and Nanotechnology, School of Chemical and Physical Sciences, Victoria University of Wellington, Wellington, New Zealand

DR. IVAN H. EL-SAYED, Department of Otolaryngology-Head and Neck Surgery, Comprehensive Cancer Center, University of California at San Francisco, San Francisco, USA

DR. MOSTAFA A. EL-SAYED, Laser Dynamics Laboratory, School of Chemistry and Biochemistry, Georgia Institute of Technology, Atlanta, USA

Dr. N. Félidj, The MacDiarmid Institute for Advanced Materials and Nanotechnology, School of Chemical and Physical Sciences, Victoria University of Wellington, Wellington, New Zealand

Dr. Chen Fu, Department of Chemistry, National Tsing Hua University, Hsin-chu, Taiwan

Dr. Chris D. Geddes, The Institute of Fluorescence, University of Maryland Biotechnology Institute, Baltimore, USA

Dr. David S. Ginger, Department of Chemistry, University of Washington, Seattle, WA

Dr. Anderson S. L. Gomes, Department of Physics, Universidade Federal de Pernambuco, Recife, Brazil

Dr. J. Grand, The MacDiarmid Institute for Advanced Materials and Nanotechnology, School of Chemical and Physical Sciences, Victoria University of Wellington, Wellington, New Zealand

Dr. Stephen K. Gray, Chemistry Sciences and Engineering Division Argonne National Laboratory, Argonne, USA

Dr. Mark Griffiths, LI-COR Biosciences Inc., Lincoln, USA

Dr. Daniel L. Grone, LI-COR Biosciences Inc., Lincoln, USA

Dr. Jong-in Hahm, Department of Chemical Engineering, The Pennsylvania State University, University Park, USA.

Dr. Ja-An Annie Ho, Department of Chemistry, National Tsing Hua University, Hsin-chu, Taiwan.

Dr. A. Hohenau, The MacDiarmid Institute for Advanced Materials and Nanotechnology, School of Chemical and Physical Sciences, Victoria University of Wellington, Wellington, New Zealand.

Dr. Xiaohua Huang, Laser Dynamics Laboratory, School of Chemistry and Biochemistry, Georgia Institute of Technology, Atlanta, USA.

Dr. John M. Kelly, School of Chemistry, Trinity College, Dublin, Ireland.

Dr. J. R. Krenn, The MacDiarmid Institute for Advanced Materials and Nanotechnology, School of Chemical and Physical Sciences, Victoria University of Wellington, Wellington, New Zealand.

Dr. Ting-Yang Kuo, Department of Chemistry, National Tsing Hua University, Hsin-chu, Taiwan.

Dr. Deirdre M. Ledwith, School of Chemistry, Trinity College, Dublin, Ireland.

Dr. Chih-Kung Lee, National Taiwan University, Taipei, Taiwan.

Dr. Jiun-Haw Lee, National Taiwan University, Taipei, Taiwan.

Dr. D.Y. Lei, Department of Physics, The Chinese University of Hong Kong, Hong Kong, People's Republic of China.

Dr. G. Lévi, The MacDiarmid Institute for Advanced Materials and Nanotechnology, School of Chemical and Physical Sciences, Victoria University of Wellington, Wellington, New Zealand.

Dr. J. Li, Department of Physics, The Chinese University of Hong Kong, Hong Kong, People's Republic of China.

DR. CHII-WANN LIN, National Taiwan University, Taipei, Taiwan.

DR. WEI-CHIH LIU, Department of Physics, National Taiwan Normal University, Taipei, Taiwan.

DR. BRIAN D. MACCRAITH, Biomedical Diagnostics Institute, Dublin City University, Dublin, Ireland.

DR. LUIGI MARTIRADONNA, National Nanotechnology Laboratory of CNR-INFM, IIT Research Unit, ISUFI, University of Salento, Lecce, Italy.

DR. COLETTE MCDONAGH, Biomedical Diagnostics Institute, Dublin City University, Dublin, Ireland.

DR. LYLE M. MIDDENDORF, LI-COR Biosciences Inc., Lincoln, USA.

DR. KEIKO MUNECHIKA, Department of Chemistry, University of Washington, Seattle, WA.

DR. MING-YAW NG, Department of Physics, National Taiwan Normal University, Taipei, Taiwan.

DR. ROBERT NOONEY, Biomedical Diagnostics Institute, Dublin City University, Dublin, Ireland.

DR. H.C. ONG, Department of Physics, The Chinese University of Hong Kong, Hong Kong, People's Republic of China.

DR. SHANLIN PAN, University of Alabama, Department of Chemistry, Tuscaloosa, USA.

DR. NICHOLAS P.W. PIECZONKA, Laser Dynamics Laboratory, School of Chemistry and Biochemistry, Georgia Materials & Surface Science Group, Faculty of Science, University of Windsor, Windsor, Canada.

DR. PIER PAOLO POMPA, National Nanotechnology Laboratory of CNR-INFM, IIT Research Unit, ISUFI, University of Salento, Lecce, Italy.

DR. DIEGO RATIVA, Department of Physics, Universidade Federal de Pernambuco, Recife, Brazil.

DR. DANIEL J. ROSS, LASER DYNAMICS LABORATORY, School of Chemistry and Biochemistry, Georgia Materials & Surface Science Group, Faculty of Science, University of Windsor, Windsor, Canada.

DR. LEWIS ROTHBERG, University of Rochester, Department of Chemistry, Rochester, USA.

DR. E. C. LE RU, The MacDiarmid Institute for Advanced Materials and Nanotechnology, School of Chemical and Physical Sciences, Victoria University of Wellington, Wellington, New Zealand.

DR. S. SHIV SHANKAR, National Nanotechnology Laboratory of CNR-INFM, IIT Research Unit, ISUFI, University of Salento, Lecce, Italy.

DR. JESSICA M. SMITH, Department of Chemistry, University of Washington, Seattle, WA.

DR. DAVE L. STEFFENS, LI-COR Biosciences Inc., Lincoln, USA.

DR. ONDREJ STRANIK, Biomedical Diagnostics Institute, Dublin City University, Dublin, Ireland.

DR. JÉRÔME WENGER, Institut Fresnel, CNRS, Domaine Universitaire de Saint-JÉrôme, Marseille, France.

DR. JOHN G. WILLIAMS, LI-COR Biosciences Inc., Lincoln, USA.

Dr. J.B. Xu, Department of Electronic Engineering, The Chinese University of Hong Kong, Hong Kong, People's Republic of China.

Dr. Yongxia Zhang, The Institute of Fluorescence, University of Maryland Biotechnology Institute, Baltimore, USA.

# 1

# Metal-Enhanced Fluorescence: Progress Towards a Unified Plasmon-Fluorophore Description

Kadir Aslan and Chris D. Geddes*
The Institute of Fluorescence,
University of Maryland Biotechnology Institute, 701 East Pratt St., Baltimore, MD, 21202, USA

## 1.1. INTRODUCTION

Since the first experimental demonstration of the spontaneous emission rate of fluorescent species could be modified by changing the local photonic mode density (PMD) by metal surfaces by Drexhage and coworkers, [1, 2] there has been numerous studies to describe the interactions of fluorophores with metals placed in close proximity. [3, 4] Barnes provided an excellent detailed summary of fluorescence near interfaces in a review article in 1998.[5] In recent years our laboratory has also both described and demonstrated many new favorable photophysical effects of electronically excited states (fluorophores / luminophores) in close-proximity to plasmon resonant particles.[6-8] In these examples of Metal-Enhanced Fluorescence (MEF), it has become apparent that the true mechanism for MEF is far more elegant than the simple radiative decay rate description postulated by others. [9] In fact MEF appears to be underpinned by the formation of a mirror dipole on the metallic nanoparticle, which itself radiates highly efficiently, effectively resulting in enhanced luminescence (Figure 1.1).[8, 10-12] In this regard, our research laboratory has been working a "Unified Fluorophore-Plasmon Description", which is summarized in Figure 1.1 and which describes the effects of fluorophores in a broad range of fluorescence disciplines. The coupled fluorophore lifetime is also observed to be much shorter than the "free-space lifetime", reflecting the very fast "plasmon" lifetime of the coupled quanta.[5-7] Further, it is thought that the MEF effect is a consequence of an excited state coupling with the scattering mode of nanoparticles, which is thought to account for very fast MEF lifetimes, i.e. a coupled elastic scattering event.[8, 13] For metallic nanoparticles, the extinction spectrum is comprised of both an absorption and scattering component.[14] For smaller particles (< 25 nm), then the extinction is for the most part dominated by absorption, while for larger particles, by scattering. For a fixed wavelength, this scales as the radius cubed and to the sixth power respectively.[14] Given that MEF is thought to be the ability of a nanoparticle to plasmon scatter coupled quanta (i.e. fluorescence, phosphorescence etc),[8] then concordantly larger particles have subsequently been shown to be ideal for MEF.[8, 13] It is this train of thought which readily accounts for the numerous reports of luminescence quenching by small metallic nanoparticles. While our mechanism for MEF is fairly new,[8] we have reported compelling data in the recent past. In these reports, i) MEF is seen to be more pronounced for larger particles, ii) the wavelength dependence of MEF suggests a correlation between the scattering spectra of the nanoparticles and the emission spectra of the fluorophores, iii) MEF is angular

*MEF: Progress towards a Unified Plasmon Fluorophore Description*
Edited by Chris D. Geddes.

dependent,[15] both from an observation and excitation perspective, similar to scattering by nanoparticles themselves,[16] iv) metals ideal for MEF are those with high free electron densities[17] and v) continuous metallic films have been shown to couple fluorophore emission, when fluorophores are positioned less than 100 nm from the surface. In addition, the coupled emission has been shown to be completely *p*-polarized, strongly indicating that the coupled-plasmon system is radiating.[18, 19]

In the opening chapter of this new exciting book, we aim to provide a broad overview of our interpretation of metal-fluorophore interactions, which exhibit the MEF concept with numerous experimental data. We hope this chapter will be useful in introducing the MEF concept to the newcomers to field, as well as interest to other researchers whom have already made contributions to this field.

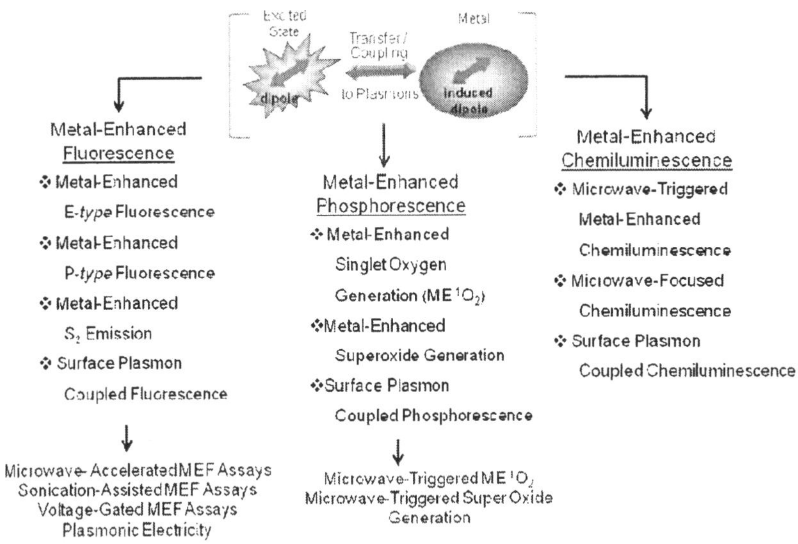


**Figure 1.1** A schematic diagram of the unified plasmon/fluorophore description. Fluorophores induce surface plasmons in metals and energy is effectively transferred in a non-radiative fashion. This interaction of excited states with surface plasmons leads to a wealth of new fluorescence, chemiluminescence and phosphorescence phenomena and technologies we describe as a *Unified Description.*

## 1.2. METAL-ENHANCED FLUORESCENCE (MEF)

As briefly mentioned in the Introduction the MEF phenomenon is a result of the interactions between the excited states of the fluorophores and the induced surface plasmons of metal nanoparticles or roughened surfaces. Figure 1.2A depicts our laboratory's interpretation of the processes involved in the interactions of fluorophores with metal nanoparticles in close proximity to one another. There are two main processes thought responsible in MEF 1) non-radiative coupling from the excited state of the fluorescent species to surface plasmons of the metallic nanoparticles; which is subsequently radiated by the nanoparticles themselves and 2) enhanced absorption of light by fluorophores due to the increased electric fields between and around the metal nanoparticles.[20] Fluorophores located in between the metal nanoparticles, where the electric fields are larger than the electric fields around the individual metal nanoparticles can contribute significantly to the MEF phenomenon.[20] Figure 1.2B shows emission spectra of fluorescein isothiocyanate (FITC) placed on silver island films (SIFs) and on a blank glass slide. The emission peak at 520 nm of FITC on SIFs is ≈ 5-fold larger than that of FITC on glass. Real-color photographs (Figure 1.2B-inset) provides a visual evidence for the observation of increased fluorescence emission. In addition, it has also been reported that the decay time of the excited states is affected as a result of the interactions of fluorophores with metal nanoparticles.[7] Figure 1.2C shows the fluorescence emission intensity decays for FITC on both SIFs and glass. The intensity decay for FITC on glass follows a single-exponential decay, corresponding to a fluorescence lifetime of ≈ 3 ns, and follows a multi-exponential decay on SIFs, corresponding to a much shorter lifetime ≈ 1.5 ns. The fact that one observes a multi-exponential decay on SIFs indicates the presence of multiple processes involved in the decay.

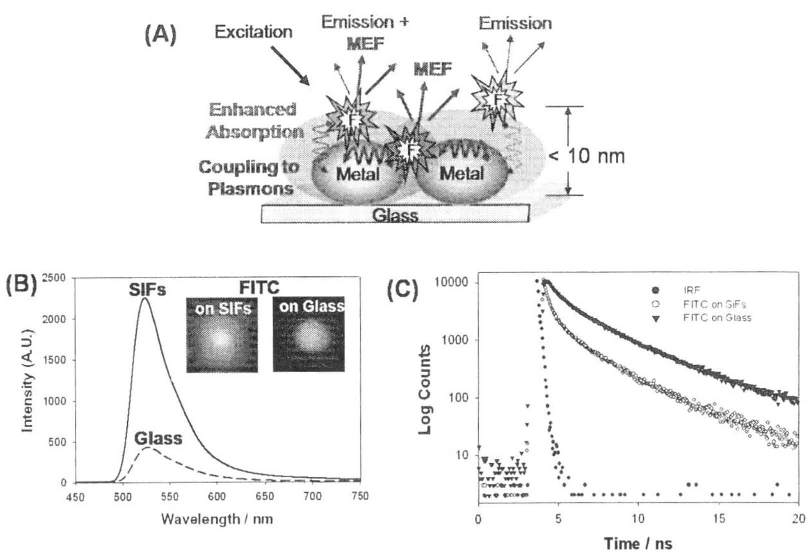


**Figure 1.2.** (A) A schematic diagram depicting the processes in close proximity to metals (< 10 nm) involved in Metal-Enhanced Fluorescence: enhanced absorption and coupling to surface plasmons. (B) Emission spectra of FITC deposited onto SIFs and glass. The inset shows the real-color photographs of FITC emission from these surfaces. (C) Intensity decays for FITC on both glass and SiFs. IRF: Instrument Response Function.

In the publications on MEF to-date,[7, 21, 22] silver has been the most studied metal as compared to other metals. Figure 1.3A shows the photographs of SIFs deposited onto glass and plastic supports, where the left-half of these supports was left intentionally blank to visually demonstrate the deposition of silver nanoparticles and utilize them as control surfaces in MEF experiments. A typical absorption spectrum of SIFs is shown in Figure 1.3B. Our research laboratory has also shown MEF from copper,[23] gold[24] and zinc nanostructures.[25] The reader is referred to the specific publications for detailed information regarding MEF from metals other than silver. As shown in Figure 1.3B, the surface plasmon resonance (SPR) peak for SIFs occur at ≈ 420 nm; the SPR peak depends on the type, size and the shape of the metal deposited onto the solid support. The surface morphology of SIFs on glass and plastic supports studied by atomic force microscopy, Figures 1.3C and 1.3D, show SIFs were deposited onto glass as individual particles of size ≈100 nm in a homogeneous fashion. On the other hand, AFM images of SIFs on plastic support reveal that the deposition was more heterogeneous, which stems from the fact the surface of the plastic was chemically modified and was not smooth as the glass surfaces.

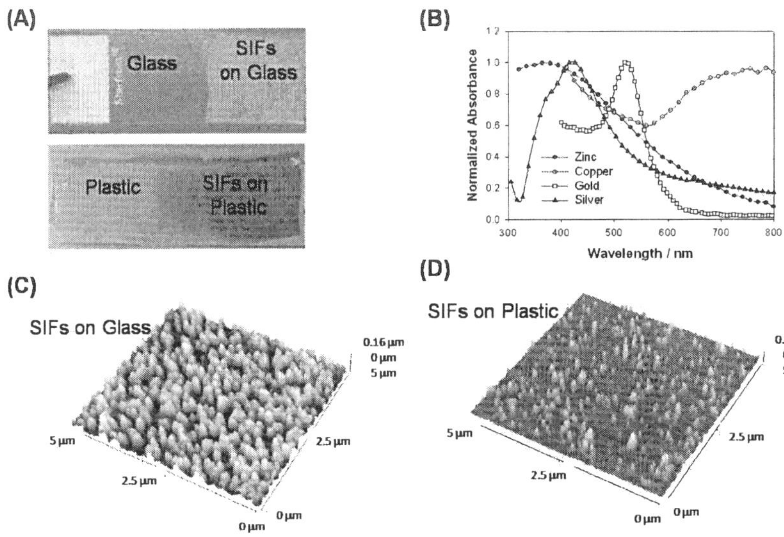


**Figure 1.3.** (A) Photographs of silver island films (SIFs) deposited onto glass and plastic supports. (B) Normalized absorbance of zinc, copper, gold and silver nanostructured particles on a glass support. Atomic force microscope images of SIFs on (C) glass (D) plastic support.

It is well known that several different modes of surface plasmons can be created in a single metal nanoparticle by changing the shape of the nanoparticle itself.[26, 27] That is, while a single surface plasmon mode typically exists in a spherical nanoparticle for $\lambda/10$ since all the electrons in the metal oscillate with the same energy, the presence of a secondary path or an axis on the metal, e.g,. a longer longitudinal axis in rod-shape nanoparticle, a secondary surface plasmon mode occurs. In a rod-shaped nanoparticle, electrons can oscillate at two different frequencies giving rise to an additional SPR peak which is red-shifted with respect to the SPR peak as a result of the electron oscillations on the shorter axis. One can also observe additional SPR peaks for a triangular shaped nanoparticle, etc. The presence of multiple SPRs in a single nanoparticle results in a greater probability of coupling excited states of fluorophores, e.g. fluorophores randomly oriented near nanoparticles have greater probability of coupling to one or multiple SPR modes. In addition, electric fields in between these anisotropic nanoparticles are significantly stronger as compared to the electric fields in between isotropic (e.g., spherical) nanoparticles, which could result in further increases in the extent of light absorbed by the fluorophores. In this regard, our research laboratory has recently demonstrated MEF from anisotropic silver structures.[26, 27] Figures 1.4A and 1.4B show the AFM images of silver nanorods and triangles deposited onto glass supports, respectively. Figures 1.4C and 1.4D show the emission spectra of indocyanine green (ICG) from these surfaces. Figures 1.4C and 1.4D also show that emission intensities from anisotropic silver structures are significantly larger than from control surfaces and also depends on the extent of loading of anisotropic silver structures on the surfaces.

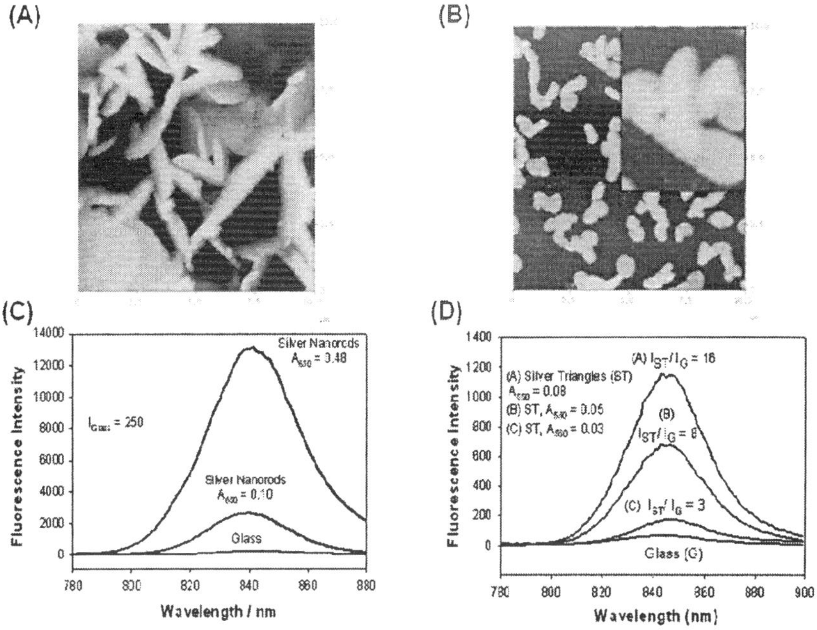


**Figure 1.4.** Atomic Force Microscope Images of (A) silver rods and (B) triangles deposited onto glass substrates. (C) and (D) Fluorescence emission spectra of Indocyanine Green (ICG) deposited onto both surfaces, respectively.

In nearly all the applications of MEF, the excitation and emission observation have been on the same side of the assay platform as the incoming excitation. It is well known that when fluorophores are near interfaces with different refractive indices (such as air and glass), a significant part of the fluorescence can be coupled into the medium of higher refractive index,[28] with a unique angular dependence peaking at the critical angle.[29] Moreover, the fluorescence emission through the high refractive index medium (and in air) could further be increased, when metallic nanoparticles are placed between the fluorophores and the interface. This phenomenon arises from the fact that metallic nanoparticles (especially silver, gold and copper) are known to scatter light efficiently and in an angular-dependent fashion.[14] In fact, 30 nm silver nanoparticles scatter light (at 530 nm) ≈ 9 and 124 times more efficiently than 30 nm gold and polystyrene nanoparticles, respectively.[14] Subsequently, light scattering by silver and gold nanoparticles can be detected at concentrations as low as $10^{-16}$ M.[14] It was previously shown that scattered light by metallic nanoparticles is highest at observation angles of 0 and 180° with respect to the incident light, i.e., backward and forward scatter. Thus, when silver (also gold or copper) nanoparticles are placed between the fluorophores and the interface (glass substrate), coupled light originated from the fluorophore will be preferentially scattered in backward and forward directions by the silver nanoparticles. The forward scattered light then couples to glass and is subsequently emitted from the back of the glass substrate, while the backward scattered light is emitted into free space. In addition, the angular-dependent scattered light from metallic nanoparticles depends on the size, shape and composition of the nanoparticles, as well as the refractive index of the suspending medium and is significantly different than the light scattered by planar glass.[14]

In a recent paper, our research laboratory has demonstrated the angular-dependent nature of MEF from silver nanoparticles deposited onto glass support.[15, 24, 30] Figure 1.5A shows the experimental setup, where the excitation was normal to the surface, a common angle in fluorescence spectroscopy, and the fluorescence emission was collected at all angles (0-360 degrees). Real-color photographs of FITC on SIFs and glass, collected separately at 225 degrees, show the emission is more intense from SIFs than from glass at the same angle. The quantitative interpretation of the observed angular-dependent MEF can be accomplished by MEF enhancement factors (emission intensity at 517 nm (I517) on silver colloids divided by I517 on glass) were calculated at all observation angles and are shown in Figure 1.5B. MEF enhancement factors of 3-4 and 7-9 can be seen for FITC emission on silver colloids with low and medium loading, and high loading, respectively. One can see that when silver colloids were present between the fluorophores and the interface, the fluorescence emission from the air and back side, increased to a similar extent. Clearly, the enhanced fluorescence was significantly greater from surfaces with a higher degree of colloid loading (OD = 0.4). The utility of this observation lies in the fact that one can construct a bioassay on the silvered surfaces and measure a concentration-dependent response at a chosen angle, i.e angular-ratiometric assays.[15] The angular-ratiometric intensity measurements are independent of initial intensity and fluctuations in the excitation light, colloid loading and also the loading of the fluorophore. Thus, angular-ratiometric measurements combined with the use of more sensitive detectors should further enhance the capability of MEF-based surface assays.[15]

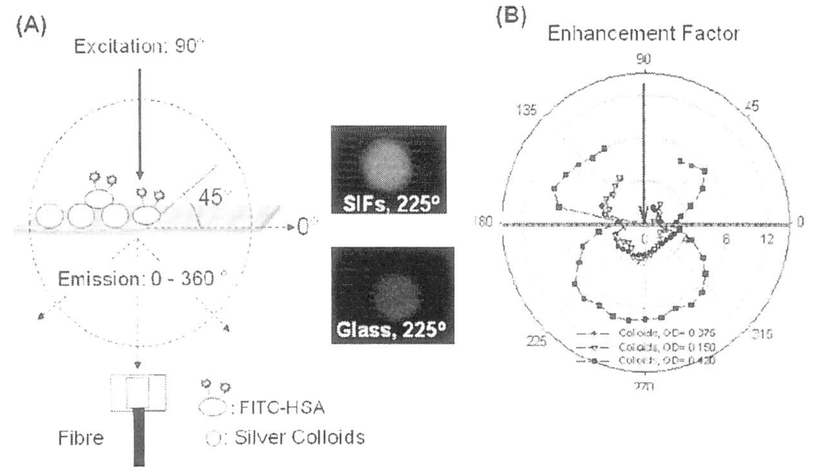


**Figure 1.5**. (A) A schematic diagram depicting the angular-dependent nature of Metal-Enhanced Fluorescence. The real-color photographs show the fluorescence emission taken at an angle of 225 degrees (back of the surface) from SIFs and glass. (B) Angular-dependent Metal-Enhanced Fluorescence (MEF) enhancement factor (emission intensity at 517 nm ($I_{517}$) on silver colloids divided by $I_{517}$ on glass (control sample) for all angles).

In addition to the benefits of MEF from metal nanostructures deposited onto solid supports that are very useful in surface bioassays, MEF can also be observed from individual nanostructures in bioassays carried out in solution. In this regard, fluorophores and metal nanostructures can be assembled in core-shell architecture and can be used as fluorescent nanoparticles as indicators in biological applications such as imaging of cellular activity or single-molecule sensing.[22, 31]

In a recent paper,[31] our research laboratory have developed core-shell (silver core-silica shell) nanoparticles with various shell thicknesses featuring a variety of fluorophores, to show the versatility of the core-shell architecture, and have demonstrated their applicability for MEF and single nanoparticle sensing.[31] The broad potential applications of these fluorescent nanostructures were demonstrated by employing near-infra red emitting probes (Rh800) for potential applications in cellular imaging and the use of highly photostable long lifetime (μS) lanthanide probes, probes suitable for off-gating biological auto fluorescence. In addition, a commonly used long-wavelength fluorophore, Alexa 647, was also incorporated into to these core-shell nanostructures to demonstrate that fluorophores can be readily covalently linked to the core-shell particles also, for MEF applications.[31]

Figures 1.6A and 1.6B show the fluorescence emission intensity from Eu-TDPA-doped Ag@$SiO_2$ and Rh800-doped Ag@$SiO_2$ and from the corresponding fluorescent nanobubbles (control samples), Eu-TDPA-doped $SiO_2$ and Rh800-doped $SiO_2$. The emission intensity was approximately *8-fold* and *20-fold* higher for Eu-TDPA-doped Ag@$SiO_2$ and Rh800-doped Ag@$SiO_2$ than Eu-TDPA-doped $SiO_2$ and Rh800-doped $SiO_2$, respectively. Figure 1.6C shows representative scanning confocal images of individual fluorescent core-shell nanoparticles, Alexa 647 Ag@$SiO_2$, and the corresponding nanobubbles, Alexa 647@$SiO_2$. The bright spots in Figure 1.6C-1 represent fluorescence emission from the single fluorescent core-shell nanoparticles, while the dimmer spots in Figure 1.6C-2 represent the single nanobubbles. The significant differences in the peak intensities of the two images are immediately evident from Figure 1.6C. For fluorescent core-shell nanoparticles the average value of the peak intensity was approximately 10-fold higher than that of the nanobubbles.

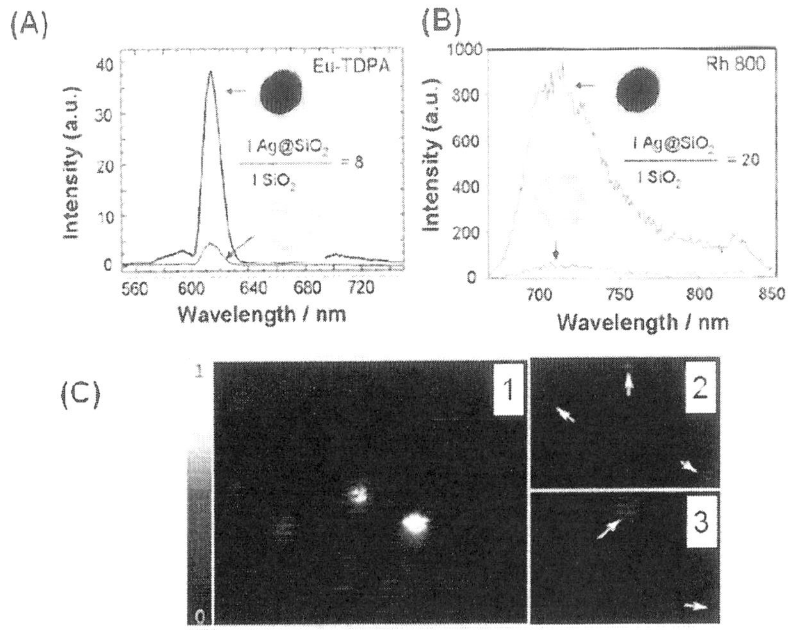


**Figure 1.6.** Fluorescence emission intensity of (A) Eu-TDPA-doped $Ag@SiO_2$ and (B) Rh800-doped $Ag@SiO_2$ and from the corresponding fluorescent nanobubbles (control samples), Eu-TDPA-doped $SiO_2$ and Rh800-doped $SiO_2$. The diameter of the Ag is 130±10 nm and the thickness of the shell is 11±1 nm for all the samples. (C) Scanning confocal images (20x20 mm) of (1) Alexa 647 $Ag@SiO_2$, (2) Alexa 647@$SiO_2$, (3) zoomed in version of (B). Intensity counts in the scale were normalized to 1.

In a recent publication, our research laboratory has studied another interesting phenomenon observed in fluorescence: emission from a second excited state ($S_2$) level. It is well-known that the azulene molecule emits fluorescence from $S_2$ upon excitation, instead of emitting from the normal $S_1$ that is observed for almost all fluorophores. Since its discovery azulene has been used in tracking cellular activity due to its affinity towards binding DNA and tubulin of the cytoplasm. On the other hand, azulene's $S_2$ emission is still weak. In this regard, our research laboratory has demonstrated that azulene's $S_2$ emission can be further increased by placing azulene in close proximity to silver nanoparticles. Figure 1.7A shows the graphical representation of metal-enhanced $S_2$ emission, where the interactions of azulene's second excited state level of with surface plasmons of silver nanoparticles are enhanced. Figure 1.7B shows the $S_2$ fluorescence emission spectra of azulene from SIFs and from glass at room temperature, where a ≈1.5-fold increase is observed. It is important to note that the true enhancement factor is ≈37-fold. This is because the MEF phenomenon is distance dependent, where with a sample thickness of ≈1 μm, and an enhanced interaction region <20 nm, then only 4% of the sample is within the MEF enhancement region. In addition, the effect of temperature on $S_2$ emission was also studied, where > 2-fold enhancement of fluorescence from SIFs at 77K was observed, Figure 1.7C.

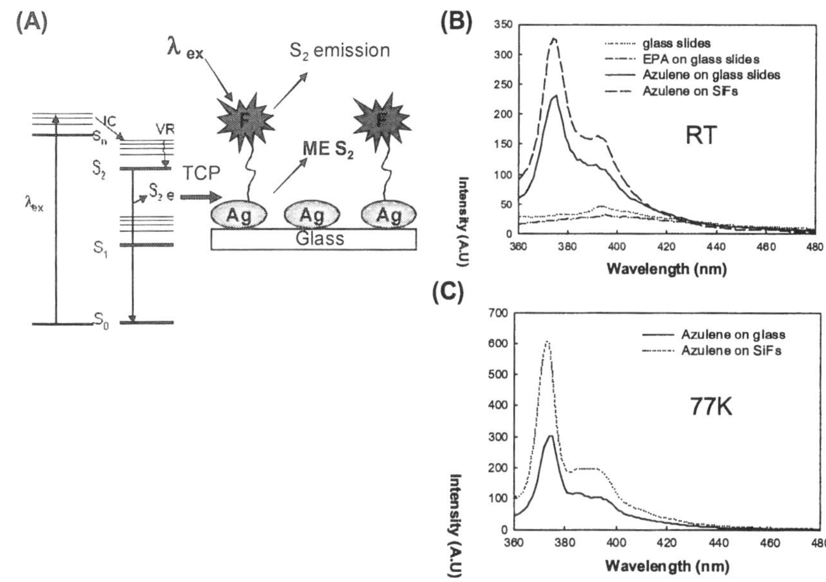


**Figure 1.7.** (A) Graphical representation of our laboratory's current interpretation of Metal-Enhanced $S_2$ emission (**Bottom**). IC-Internal Conversion, VR-Vibrational energy relaxation. Ag-Silver nanoparticle (SIFs), TCP-Transfer/coupling to Plasmons, $MES_2$ –Metal Enhanced $S_2$ Emission. Energy level spacing not drawn to scale. Fluorescence emission spectra, lex = 338 nm, of Azulene sandwiched between two SiFs and unsilvered slides at room temperature (B) Room Temperature, RT and (C) at 77K.

In all the examples summarized above, fluorescence emission of fluorophores was unstructured, where excitation of fluorophores with a single or wide range of wavelengths generates an entire range of allowed transitions that populate the vibrational energy levels, Figure 1.8A-top. There exist some fluorophores with several vibrational structures emitted over a range of wavelengths, that is, the transitions from the lowest vibrational level of an excited state to several vibrational levels of the ground state usually have a much higher degree of probability than others, and when combined, constitute the structured emission spectrum of that molecule. In this regard, amplification of vibrational structures may yield better analysis of intra- and intermolecular interactions as well as very detailed finger print for the discrimination of chemical substances with subtle differences, or in slightly difference physical states and surroundings can be achieved. In a recent publication,[12] enhancement of structured emission of perylene, a fluorophore with structured absorption and emission bands (Figures 1.8B and 1.8C) in close proximity to SIFs was reported. It is thought that the vibrational bands of perylene can couple to surface plasmons of silver nanoparticles and subsequently be emitted at a increased intensity, Figure 1.8A-bottom. Since the fluorophores placed close proximity to SIFs experience increased electric fields and they absorb more light than as shown in Figure 1.8B. Figure 1.8C shows the structured fluorescence emission spectra of perylene on SIFs and glass. Perylene shows significant structure due to transitions from the lowest vibrational level of the excited $S_1$ state to several vibrational levels of the ground state, those having a much higher degree of probability than others. Figure 1.8C also shows the enhanced structured fluorescence intensity was ≈ 2 -fold brighter from the silver, as compared to glass. An additional experiment was undertaken to verify that the structured emission is plasmon enhanced: SIFs were replaced by a continuous silver film (50 nm thick), where the fluorescence emission was heavily quenched. Since the generation of surface plasmons in continuous metal films occurs only under certain unique conditions, as compared to metallic nanoparticles, the observation of quenching supports the notion that the structured emission on SIFs is indeed plasmon-enhanced.[12]

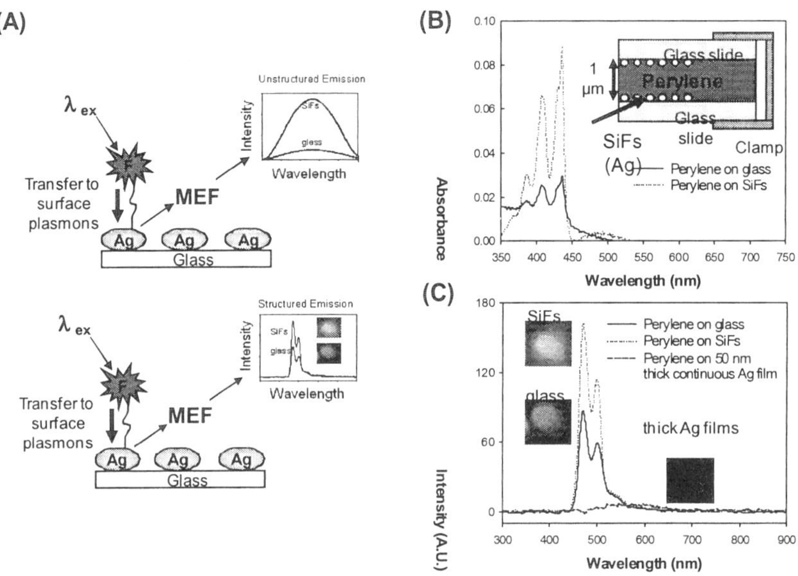


**Figure 1.8**. (A) Current interpretation Metal-Enhanced unstructured (Top) and structured emission (Bottom). F-Fluorophore. MEF- Metal Enhanced Fluorescence. (B) Absorption spectra of Perylene sandwiched between two silvered and unsilvered slides respectively (Sandwich experimental geometry (Top Insert). (C) Fluorescence emission spectra of Perylene sandwiched between two silvered, unsilvered slides, 50 nm thick continuous Ag films, respectively.

Previous MEF studies were exclusively focused on *prompt* fluorescence emission,[7] where excited singlet state populations rapidly decayed to ground states. E-type fluorescence is well-known to arise from thermal activation of the $S_1$ state from the $T_1$ state (reverse intersystem crossing), and hence its decay time (excited state lifetime) is underpinned by the decay of the $T_1$ state, which typically has a long decay time in the millisecond range.[32] Because of its long decay time, e-type delayed fluorescence has been widely used to investigate the rotational diffusion time of biological macro-molecules in membranes on micro – millisecond time scales.[33] Our research laboratory has also studied whether E-type fluorescence can be enhanced by SIFs when placed in close proximity.[34] Figure 1.9A shows e-type fluorescence and phosphorescence spectra after Argon purging in a cuvette at different temperatures. One can observe not only e-type delayed fluorescence at ≈ 550 nm, which is spectrally identical to *prompt* $S_1$ fluorescence, but also the phosphorescence emission peak is evident at ≈ 690 nm. The delayed emission band readily disappears if the solution is not deoxygenated. The e-type fluorescence peak readily increases with increased temperature, while the phosphorescence peak subsequently decreases. On SIFs we have observed different extents of e-type fluorescence and phosphorescence from both glass slides and SIFs at different temperatures, with Argon purging. Figure 9-insert shows e-type fluorescence is ≈2.5-fold brighter as compared to phosphorescence, ≈ 1.9-fold brighter on SIFs. Visual evidence for metal-enhanced e-type fluorescence is provided in Figure 1.9C; eosin fluorescence emission intensity on glass is a light greenish; on SIFs it is brighter than on glass due to the MEF effect. After heating the sample for 2 minutes, the eosin fluorescence emission on glass is much brighter than the pre-heating emission. This indicates more e-type fluorescence with increased temperature. The eosin emission is even more pronounced on SIFs after heating, indicating that e-type fluorescence emission was enhanced by surface plasmons.

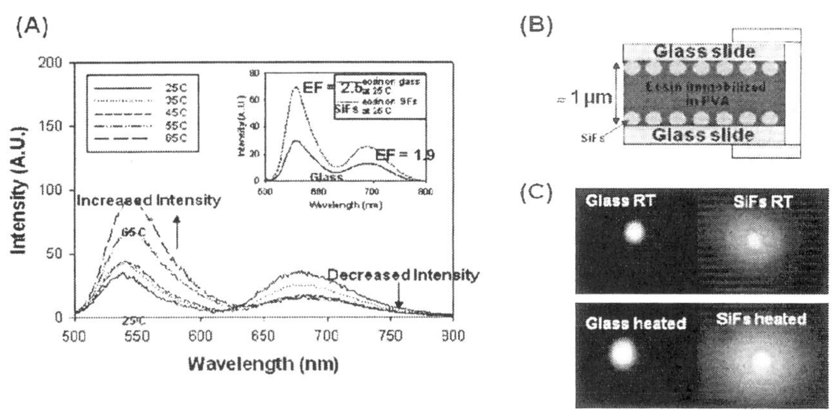


**Figure 1.9**. (A) E-type Fluorescence and phosphorescence emission spectra, $\lambda_{ex}$ = 473 nm, of Eosin in a cuvette at different temperatures. **Insert** Eosin immobilized in PVA sandwiched between two silvered and unsilvered slides at 25C EF - Enhancement Factor. RT – Room Temperature. (B) experimental sample geometry. (C) Real-color photographs of Eosin emission from glass and SIFs, before and after 2 mins heating. $\lambda_{ex}$ = 473 nm. SIFs - Silver Island Films. The real- color photographs were taken through an emission filter (488 nm razor edge).

Fluorescence can also be used to track intermolecular interactions of molecules.[32] These interactions can result in the formation of an "excimer" abbreviation for an excited state dimmer, which is formed between the ground state (M) and lowest electronically excited singlet state (M*) of a species, Figure 1.10A. The aromatic molecule pyrene is well-known for forming an excimer.[32] Pyrene has been used to probe RNA folding and DNA duplex formation by monitoring the monomers and excimer emission fluctuations that arise from local base stacking and the quenching effect.[35] Pyrene has also been used for the selective detection of cellular mRNA by yielding a strong excimer emission at 485 nm in the presence of the target. However, the sensitivity of these approaches of the singly labeled pyrene probe is typically more dependent on the excimer emission, due to the inherently high biological autofluorescence in the pyrene monomer emission region. In a recent publication, our research laboratory has studied MEF from pyrene when placed in close proximity to SIFs (Figure 1.10B and 1.10C). Due to transitions from the lowest vibrational level of the monomer excited state to several vibrational levels of the ground state, the fluorescence emission spectrum of the monomer of pyrene from quartz and SIFs has fine structured bands at 370 nm, while the broad excimer emission yields an unstructured band ≈ 470 nm.[36] Enhanced monomer emission (≈1.5 fold brighter) and excimer fluorescence emission (≈ 2.5 fold brighter) was typically observed from the SIFs as compared to a quartz control substrate (Figure 1.10C), containing no silver nanoparticles. These findings of metal-enhanced monomer and excimer fluorescence of pyrene are consistent with our research laboratory's previous reported findings for structured and unstructured $S_1$ emission for fluorophores sandwiched between silver nanostructures.[12] Metal-enhanced excimer fluorescence can also be seen visually (Figure 1.10C-inset) from real-color photographs taken through a 450 nm long-pass filter.

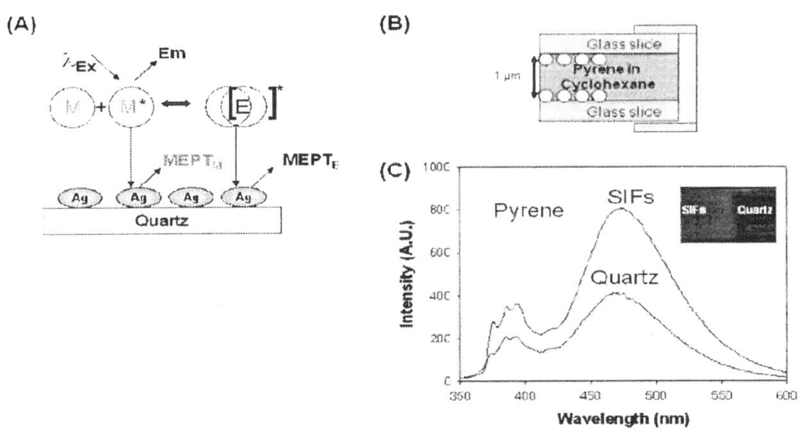


**Figure 1.10**. (A) Graphical representation of current interpretation for Metal-Enhanced *P-type* Fluorescence. (B) Experimental sample geometry. (C) Fluorescence emission spectra and photographs of pyrene emission from quartz (control sample) and SIFs.

## 1.3. METAL-ENHANCED CHEMILUMINESCENCE (MEC)

In recent years, our research laboratory has also studied the interactions of chemically excited states (i.e., chemiluminescence) with metal nanoparticles extensively, metal-enhanced chemiluminescence.[11, 37, 38] The reader is referred to Chapter 15 of this book for a detailed summary of MEC. The authors intend to provide a short introduction to MEC within the context of the current Chapter. Unlike fluorescence excited states which are induced by an external excitation source, the excited states in chemiluminescence are created as a result of chemical reactions. Once the optically inactive chemicals are activated (creation of excited states) by chemical reactions, these excited states can interact with surface plasmons in metal nanoparticles in a similar fashion to their fluorescence counter parts. It is also important to note that, in addition to enhancement of chemiluminescence, metal nanoparticles are thought to act as catalyst of the chemical reactions that produce light. Figure 1.11A depicts the current interpretation of the MEC phenomenon, where chemically excited states couple to surface plasmons of the metals and chemiluminescence is in turn enhanced by the metals. A simple experimental geometry as shown in Figure 1.11B can be used to study MEC. Here the chemiluminescent solution is placed in between two metal-coated glass slides and the emission is measured. In this experimental geometry, the gap between the slides was estimated to be ≈1 μm. Figure 1.11C shows the emission spectra of green chemiluminescence material placed in between two SIFs. Left-half of each silvered glass slide is left intentionally blank to demonstrate the benefits of using silver. Figure 1.11C shows the emission from SIFs is significantly brighter than from glass, as also can be seen visually from the real-color photographs of chemiluminescence material placed in between two SIFs.

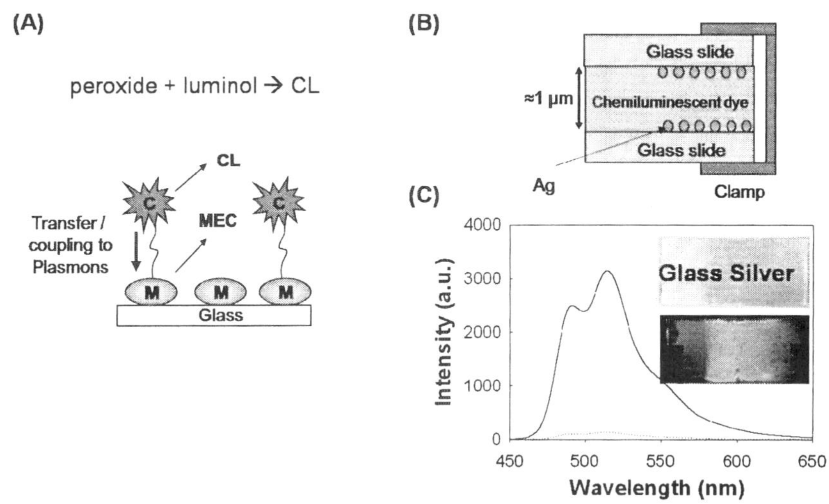


**Figure 1.11.** (A) Graphical representation of our current interpretation for Metal-Enhanced Chemiluminescence. (B) Experimental sample geometry. (C)

Chemiluminescence emission spectra and real-color photographs of green chemiluminescence emission from glass and SIFs. M- Metal.

Based on our research laboratory's observations of MEF from metals other than silver, MEC from chromium, copper, nickel and zinc was also studied. Figure 1. 12 shows the enhancement factors for green chemiluminescence from these metals with various thicknesses. A typical enhancement factor of 2-3-fold is observed from all metal surfaces, which implies that chemically excited states can couple to these plasmon resonant metal particles. It is interesting to note that the chemiluminescence emission is dependent on the amount of reactants in the solution and diminishes once one of the reactants is used up. To test whether the remainder of the inactive chemiluminescent dye can be excited with an external light source and still emit luminescence, additional experiments were undertaken where the chemiluminescence solution was excited with a laser at 473 nm. Interestingly, the inactive chemiluminescence dye can be optically excited and still emit luminescence with enhancement factors similar to the chemically excited conditions being observed. A detailed investigation of MEC from different metals is currently underway and will be reported in due course.

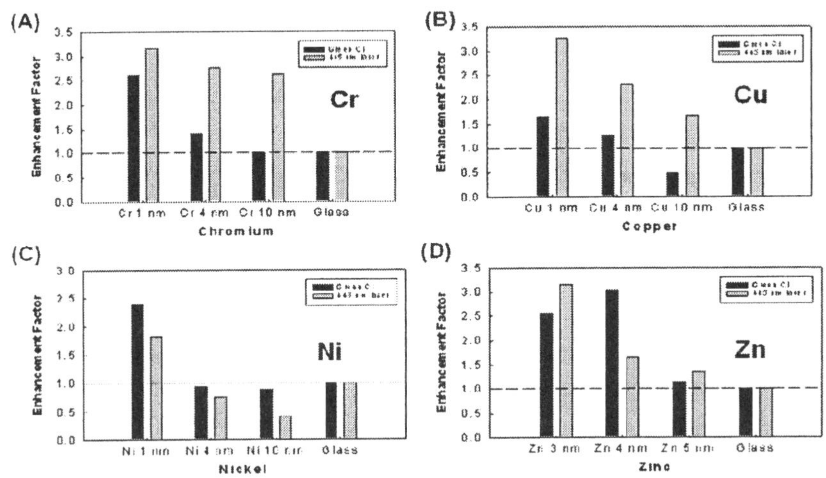


**Figure 1.12** – Enhancement factor versus metal deposition thickness for (A) chromium (Cr), (B) copper (Cu), (C) nickel (Ni), (D) zinc (Zn) for both a green chemiluminescence solution and the same solution after reaction completion optically excited at $\lambda_{ex} = 473$ nm.

## 1.4. METAL-ENHANCED PHOSPHORESCENCE (MEP)

Photodynamic therapy (PDT) has potential in the diagnosis and treatment of several diseases such as diabetes, peripheral vascular diseases, cerebrovascular and cardiovascular. [39, 40] In all PDT treatments, the main focus is on inducing cell death when a suitable dye (lumophore), light and oxygen are combined. Three primary processes are known to be involved in the photodynamic effect: Firstly, the ground state of the dye ($S_0$) is optically excited to an excited singlet state, Then a population from the excited stated is typically transferred to the dye's lowest triplet state by intersystem crossing. Finally, collisional energy transfer from the triplet dye to ground-state molecular oxygen ($^3O_2$) produces highly reactive singlet oxygen, returning the dye to its original ground state. The singlet oxygen can react readily with many biological targets and destroy a wide variety of cells. However, modern PDT is limited by an insufficient quantity of singlet oxygen, while reacting with biological targets. Since singlet oxygen plays a very important role for cell damage, an abundant supply of oxygen is very important.[40] If the consumption of oxygen by the photodynamic process is faster than it can be resupplied, it causes oxygen depletion. An alternative method to resolve this problem is to populate high-lying triplet excited states of several dyes which produce oxygen-independent damage. If these excited triplet states obtain sufficient energies to allow for the cleavage of one of the molecular bonds, then radicals that are even more reactive than singlet oxygen can be produced. The production of these radicals does not require the presence of oxygen. But both mechanisms, i.e. $^1O_2$ and radical, do require high phosphorescence (triplet) quantum yields.

In recent years, our research laboratory has published several papers on the increase of phosphorescence,[41] singlet oxygen[42, 43] and superoxide generation[44] within close proximity to metal nanoparticles. A summary of these reports are provided in this section. The current interpretation of metal-enhanced phosphorescence (MEP) is depicted in Figure 1.13A, where the phosphorescence emission is thought to couple to surface plasmons of metal nanoparticles and is subsequently increased. Figure 1.13B shows the experimental geometry used in MEP experiments. In this regard, a solution of rose bengal was placed in between two SIFs and the phosphorescence emission was measured. It is important to note that phosphorescence is not readily observed at room temperature because of collisional deactivation by oxygen and the presence of quenching impurities. However, phosphorescence emission can be observed at low temperatures and in media where the diffusion of both the luminophore and oxygen is negligible. Figure 1.13C shows the phosphorescence emission spectra and real-color photographs of rose bengal emission from glass and SIFs. From Figure 1.13C the enhanced phosphorescence intensity was ≈5-fold brighter from the SIFs, as compared to glass (497 vs 110 A.U.). The photograph inserts of Figure 1.13C also show the enhanced phosphorescence visibly from both glass and SiFs. The emission intensity is clearly brighter from between the SiFs, but much weaker from the glass control slide. Since fluorescence and phosphorescence are competitive processes, one would not expect to observe MEP. We have presented two possible explanations for the occurrence of both MEF and MEP in the same system. 1) enhanced net system absorbance, which would increase the net singlet and triplet yields, and 2) reverse intersystem crossing, i.e. $T_n \rightarrow S_n$, where the triplet state is directly populated and the back intersystem crossing facilitates MEF. Further information can be found in the corresponding publication.[45]

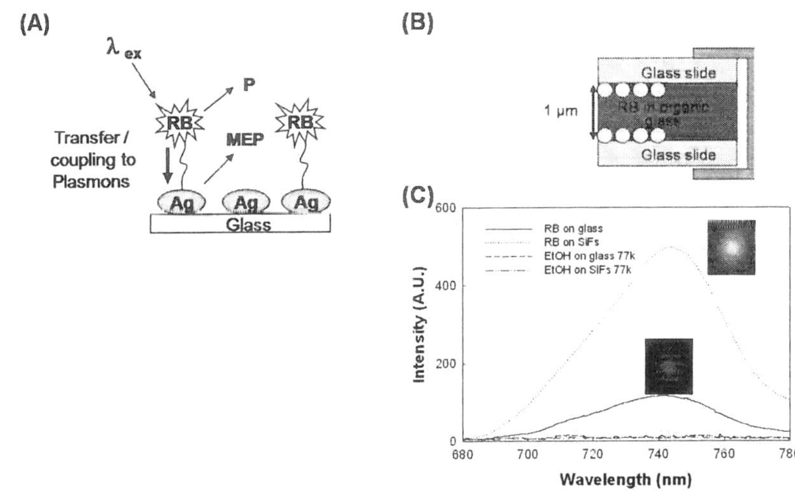


**Figure 1.13**. (A) Graphical representation of current interpretation for Metal-Enhanced Phosphorescence (MEP). (B) Experimental sample geometry. (C) Phosphorescence emission spectra and real-color photographs of Rose Bengal (RB) emission from glass and SIFs.

Molecular oxygen has a unique electronic configuration characterized by a partially filled set of antibonding π* orbitals. As predicted by Hund's rule, the lowest energy state of the molecular oxygen has maximum multiplicity, i.e. is a triplet ground state. Molecular oxygen has two excited singlet states, $^1\Sigma_g^+$ and $^1\Delta_g$, whose electronic energies are 158 and 95 kJ/mol higher than that of the triplet ground state $^3\Sigma_g^-$, respectively. The electronic configuration of $^1\Sigma_g^+$ is very similar to that of the ground state, except the fact that the last two electrons in $^1\Sigma_g^+$ have antiparallel spins while these electrons in the ground state have parallel spins. The other excited singlet state, $^1\Delta_g$, is commonly called singlet oxygen ($^1O_2$) and is more important than $^1\Sigma_g^+$ because of its longer lifetime (lifetime of $^1\Delta_g$ is 45 minutes and of $^1\Sigma_g^+$ is 7-12 sec). Since singlet oxygen can readily react with many biological targets and destroy a wide variety of cells, the photosensitized production of singlet oxygen has significance in a range of areas, especially in PDT.

It is widely accepted that singlet oxygen is the primary cytotoxic agent responsible for photobiological activity. Since singlet oxygen plays a very important role in cell damage, an abundant supply of oxygen is required. In addition, photodynamic therapy is currently limited by the insufficient generation of singlet oxygen while reacting with biological targets. A method to possibly resolve these problems is to increase the triplet yield of sensitizers, by coupling to surface plasmons, which invariably results in more singlet oxygen generation. Our research laboratory has reported the first observation of metal–enhanced singlet oxygen generation ($ME^1O_2$). A photosensitizer (rose bengal) sandwiched between SIFs enables more singlet oxygen to be produced due to the enhanced triplet excited state yield of the sensitizer, Figure 1.14A.

Figure 1.14B and 1.14C show the fluorescence emission spectra of a mixture of GR and Rose Bengal solutions on glass and SiFs, before and after UV light exposure, respectively. While the peak at 588 nm is the Rose Bengal fluorescence emission and not related with singlet oxygen, the peak at 525 nm originates from the GR sensor dye emission and directly correlates with singlet oxygen yield. It can be seen that the fluorescence emission intensity at 525 nm is increased after UV light exposure, which is indicative of singlet oxygen generation. The fluorescence intensity at 525 nm was increased ≈3.3 fold, as compared to that on glass. The increased intensity signifies that more singlet oxygen was generated from the Silver-Rose Bengal system after light exposure, as also evidenced by real-color photographs (Figures 1.14B and 1.14C-inset). The spectra are corrected for the increased intensity of the GR sensor by the MEF effect, so that the ME $^1O_2$ can be solely observed.

Since enhanced electromagnetic fields in proximity to metal nanoparticles are the basis for the increased system absorption, various computational methods are available to predict the extent of the net system absorption and therefore potentially model the relative increase in singlet oxygen generation from photosensitizers.[46-48] In comparison to traditional Mie theory, more accurate computational methods, such as discrete dipole approximation (DDA)[49] or finite difference time domain (FDTD) methods, [48, 50, 51] are often implemented to more accurately approximate field distributions for larger particles with quadruple plasmon resonances, plasmon frequencies of silver nanoparticles, or non-spherical nanoparticles in complex media or arrangements.[49, 50, 52, 53]

FDTD methods were used to demonstrate direct evidence for the relationship of electric field enhancements around nanoparticles and the increase in triplet yields for a photosensitizer and the subsequent increase in singlet oxygen generation. Several photosensitizers with singlet oxygen yields ranging from 0.08 to 1.00 were also studied on SIFs for ME $^1O_2$. An inverse relationship between singlet oxygen enhancement factor and the free-space singlet oxygen quantum yield was observed. In addition, distance dependence for the generation of ME $^1O_2$ from photosensitizers on SIFs substrates using $SiO_x$ layers 0.5, 2, 5, 10 nm thick was also observed. These observations are consistent with numerous distance dependence measurements for MEF and MEP themselves.

The authors offered the following explanation: The mechanism of singlet oxygen generation from the photo-sensitizer is an energy transfer process during a collision of the excited lumophore with $^3O_2$.[43] Since the lifetime of the triplet excited $T_1$ state (micro or milli-second range) is much longer than that of the singlet excited $S_1$ state (nanosecond), it is long enough for the triplet excited state to react with (ground-state triplet oxygen) $^3O_2$ to form $^1O_2$, in essence competing with the more favorable phosphorescence emission.

Metal-enhanced superoxide anion radical generation using acridine as a photosensitizer and a second probe specific to superoxide dihydroethidium (DHE) in close-proximity to SIFs was also studied. In this regard, acridine and DHE was placed on top of the SIFs which had varying thickness of $SiO_2$ layer to investigate the effect of distance to superoxide generation, c.f., Figure 1.16A. DHE is a well known superoxide probe, which is highly selective for superoxide. In the presence of superoxide, it emits an orange fluorescence (excitation/emission: 473 / 586 nm, respectively). Figure 1.16B shows the real-color photographs of DHE and Acridine emission on glass and SIFs before and after 2 minutes of light exposure. On glass, the DHE fluorescence emission was not observed before and after light, Figure 1.16B-top left and 1.16B-bottom left, respectively. However on SIFs, the DHE fluorescence emission was much more intense after light exposure in the presence of the acridine photosensitizer, indicating that more superoxide anion radical is generated on SIFs than on the glass slide, c.f. Figure 1.16B-top right panel and 1.16B-bottom right panel. It is important to note that the photographs were taken through an emission filter and the intensities observed are not due to backscattering of the excitation light by silver. The middle panel shows a photograph of the SIFs, coated on only half of the glass slide.

The presence of MEF, MEP and Metal-Enhanced superoxide anion radical generation in the same system seems surprising at first, as these processes are effectively competitive and ultimately provide a route for deactivation of electronic excited states. As recently shown by the authors, simultaneous photophysical mechanisms can be present within the same system when enhanced absorption effects of the fluorophore near to silver are present (i.e. an enhanced excitation rate). In this case, enhanced absorption of Acridine near-to the plasmon resonant particles facilitates MEF,[7] MEP,[41, 45] ME$^1O_2$ [44] and also Metal-Enhanced superoxide generation *simultaneously* within the same system. Acridine showed an enhanced absorption spectra near-to silver, similar to other probes reported by the authors, in essence acridine absorbs more light.[41]

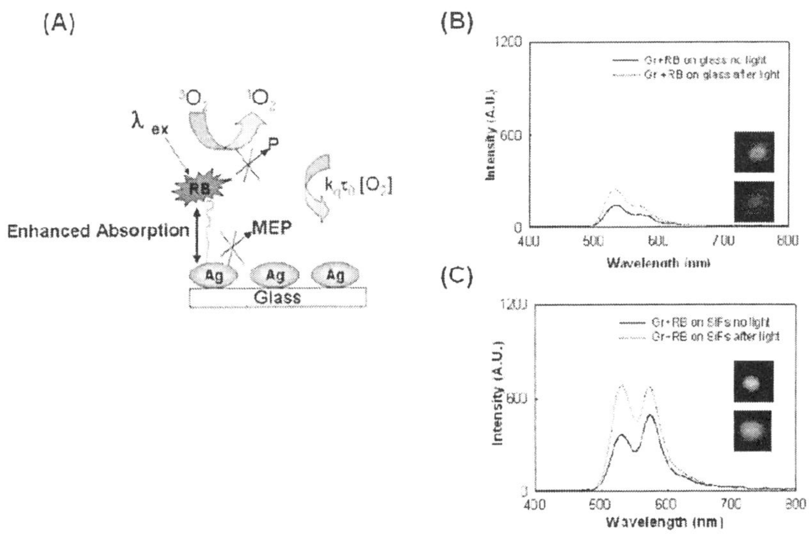


**Figure 1.14.** (A) Graphical representation of the generation of singlet oxygen. F – Fluorophore, RB – Rose Bengal, P - Phosphorescence and MEP – Metal-Enhanced Phosphorescence, $^3O_2$ – triplet ground state oxygen. $^1O_2$ – singlet oxygen. Fluorescence emission spectra of a mixture of green sensor (GR) and Rose Bengal (RB) on (B) glass and (C) SIFs, before and after light exposure (2 min) at room temperature. Light source is 100 W mercury lamp. $\lambda_{ex}$ = 473nm.

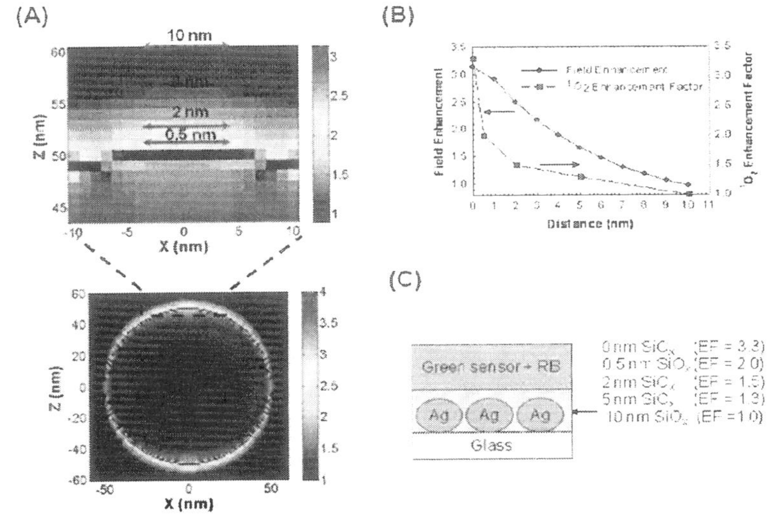


**Figure 1.15.** (A) FDTD calculations for field enhancements around a silver sphere. Zoomed image of 10 nm above the surface of the silver sphere maximum field intensity at z = 10 nm to correlate increased field enhancements in proximity to sphere surface with increased singlet oxygen generation. (B) Distance dependence relationship between for electric field enhancements and singlet oxygen on 100 nm Ag nanoparticles. (C) Distance dependence of singlet oxygen Enhancement Factor of Rose Bengal on SiFs. Top layer is mixed solution of Green Sensor and Rose Bengal. SiOx layer was deposited using thermal vapor deposition.

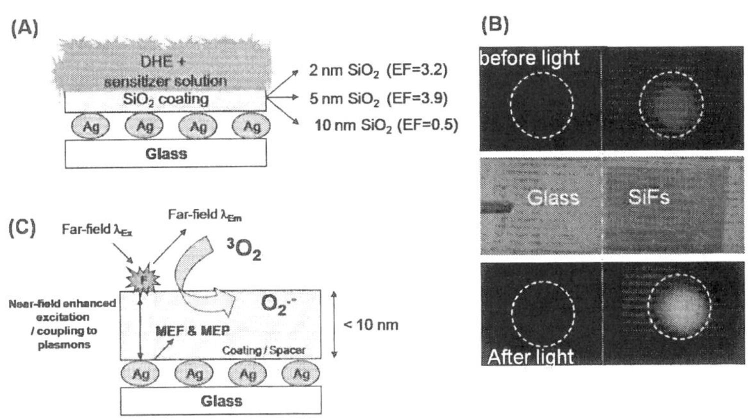


**Figure 1.16.** (A) Sample architecture for the distance dependence of metal-enhanced Superoxide generation. (B) Real color photographs of dihydroethidium (DHE) and Acridine emission from glass and SiFs, before and after 2 mins light exposure (sensitization). (C) Graphical representation of the interpretation of metal-enhanced superoxide generation with an enhanced and distance dependent excitation rate. Light exposure source was a 100 W mercury lamp. $\lambda_{ex}$ = 473 nm.

## 1.5. CLOSING REMARKS

In this review chapter we have summarized some recent advances and understandings of the interactions of excited states with close-proximity metallic nanoparticles. Remarkably, close proximity metallic nanoparticles alter the free space properties of traditional far-field fluorescence concepts, in ways which look certain to change the way we both *use* and *think* about fluorescence spectroscopy today. For example, fluorescence, chemiluminescence and phosphorescence signatures can be significantly enhanced, the excited coupled molecules becoming significantly more photostable with the spatial distribution of the metal also altered. Interestingly, in addition to enhanced luminescence signatures, other properties also benefit from an enhanced excited state yield, such as singlet oxygen and superoxide generation, to name but a few. While metal-enhanced fluorescence is still not fully understood today, we are beginning to develop a *unified understanding* of how far-field luminescence concepts can be applied and understood to molecules in the near-field, i,e. at distances less than 1 wavelength of light away.

## 1.6. ACKNOWLEDGMENTS

The authors acknowledge the Middle Atlantic Regional Center of Excellence for Biodefense and Emerging Infectious Diseases Research (NIH NIAID - U54 AI057168), the National Institute of Neurological Disorders & Stroke (NIH NS055187) and the Institute of Fluorescence, UMBI, for financial support.

## 1.7. REFERENCES

1. K. H. Drexhage. Variation of Fluorescence Decay Time of a Molecule in Front of a Mirror *Berichte Der Bunsen-Gesellschaft Fur Physikaliche Chemie*, 1968, 72, 329.
2. K. H. Drexhage. Influence of a dielectric interface on fluorescence decay time *J. Luminesc*, 1970, 1/2, 693-701.
3. K. H. Drexhage (1974) in E. Wolfe (Ed.), *Progress in Optics*, North-Holland, Amsterdam, pp. 165-232.
4. W. H. Weber and C. F. Eagen. Energy-Transfer from an Excited Dye Molecule to the Surface-Plasmons of an Adjacent Metal *Optics Letters*, 1979, 4, 236-238.
5. W. L. Barnes. Fluorescence near interfaces: the role of photonic mode density *Journal of Modern Optics*, 1998, 45, 661-699.
6. C. D. Geddes and J. R. Lakowicz. Metal-enhanced fluorescence *Journal of fluorescence*, 2002, 12, 121-129.
7. K. Aslan, I. Gryczynski, J. Malicka, E. Matveeva, J. R. Lakowicz, and C. D. Geddes. Metal-enhanced fluorescence: an emerging tool in biotechnology *Current Opinion in Biotechnology* 2005, 16, 55-62.

8. K. Aslan, M. J. R. Previte, Y. X. Zhang, and C. D. Geddes. Metal-enhanced fluorescence (MEF): Progress towards a unified plasmon-fluorophore theory *Biophysical journal*, 2007, 371A-371A.
9. J. R. Lakowicz. Radiative decay engineering: Biophysical and biomedical applications *Analytical biochemistry*, 2001, 298, 1-24.
10. Y. Zhang, K. Aslan, M. J. Previte, and C. D. Geddes. Metal-enhanced $S_2$ fluorescence from azulene *Chemical Physics Letters* 2006, 432, 528-532.
11. M. H. Chowdhury, K. Aslan, S. N. Malyn, J. R. Lakowicz, and C. D. Geddes. Metal-enhanced chemiluminescence: Radiating plasmons generated from chemically induced electronic excited states *Applied Physics Letters*, 2006, 88, 173104.
12. Y. Zhang, K. Aslan, M. J. Previte, and C. D. Geddes. Metal-Enhanced Fluorescence: Surface Plasmons can Radiate a Fluorophores Structured Emission *Applied Physics Letters*, 2007, 90, 053107.
13. K. Aslan, Z. Leonenko, J. R. Lakowicz, and C. D. Geddes. Annealed silver-island films for applications in metal-enhanced fluorescence: Interpretation in terms of radiating plasmons *Journal of fluorescence*, 2005, 15, 643-654.
14. J. Yguerabide and E. E. Yguerabide. Light-scattering submicroscopic particles as highly fluorescent analogs and their use as tracer labels in clinical and biological applications - I. Theory *Analytical Biochemistry*, 1998, 262, 137-156.
15. K. Aslan, S. N. Malyn, and C. D. Geddes. Angular-dependent metal-enhanced fluorescence from silver colloid-deposited films: opportunity for angular-ratiometric surface assays *The Analyst*, 2007, 132, 1112-1121.
16. K. Aslan, P. Holley, L. Davies, J. R. Lakowicz, and C. D. Geddes. Angular-ratiometric plasmon-resonance based light scattering for bioaffinity sensing *Journal of the American Chemical Society*, 2005, 127, 12115-12121.
17. Y. Zhang, K. Aslan, M. J. R. Previte, and C. D. Geddes, Metal-enhanced fluorescence from copper substrates, vol. 90: AIP, 2007, pp. 173116.
18. M. H. Chowdhury, S. N. Malyn, K. Aslan, J. R. Lakowicz, and C. D. Geddes. Multicolor directional surface plasmon-coupled chemiluminescence *Journal of Physical Chemistry B*, 2006, 110, 22644-22651.
19. M. J. R. Previte, K. Aslan, Y. X. Zhang, and C. D. Geddes. Metal-enhanced surface plasmon-coupled phosphorescence *Journal of Physical Chemistry C*, 2007, 111, 6051-6059.
20. K. Aslan, M. J. Previte, Y. Zhang, and C. D. Geddes. Surface plasmon coupled fluorescence in the ultraviolet and visible spectral regions using zinc thin films *Analytical chemistry*, 2008, 80, 7304-7312.
21. K. Aslan, P. Holley, and C. D. Geddes. Metal-enhanced fluorescence from silver nanoparticle-deposited polycarbonate substrates *Journal of Materials Chemistry*, 2006, 16, 2846-2852.
22. K. Aslan, J. R. Lakowicz, H. Szmacinski, and C. D. Geddes. Metal-enhanced fluorescence solution-based sensing platform *Journal of Fluorescence*, 2004, 14, 677-679.
23. Y. Zhang, K. Aslan, M. J. R. Previte, and C. D. Geddes. Metal-enhanced fluorescence from copper substrates *Applied Physics Letters*, 2007, 90, 173116.
24. K. Aslan, S. N. Malyn, and C. D. Geddes. Metal-enhanced fluorescence from gold surfaces: angular dependent emission *J Fluoresc*, 2007, 17, 7-13.

25. K. Aslan, M. J. Previte, Y. Zhang, and C. D. Geddes. Metal-Enhanced Fluorescence from Zinc Particulate Films *The Journal of Physical Chemistry C*, In Press,
26. K. Aslan, J. R. Lakowicz, and C. D. Geddes. Rapid deposition of triangular silver nanoplates on planar surfaces: Application to metal-enhanced fluorescence *Journal of Physical Chemistry B*, 2005, 109, 6247-6251.
27. K. Aslan, Z. Leonenko, J. R. Lakowicz, and C. D. Geddes. Fast and slow deposition of silver nanorods on planar surfaces: Application to metal-enhanced fluorescence *Journal of Physical Chemistry B*, 2005, 109, 3157-3162.
28. J. Enderlein, T. Ruckstuhl, and S. Seeger. Highly efficient optical detection of surface-generated fluorescence *Applied Optics*, 1999, 38, 724-732.
29. J. I. Gersten and A. Nitzan. Accelerated Energy-Transfer between Molecules near a Solid Particle *Chemical physics letters*, 1984, 104, 31-37.
30. K. Aslan, S. N. Malyn, and C. D. Geddes. Angular-Dependent Metal-Enhanced Fluorescence from Silver Island Films *Chemical physics letters*, 2008, 453, 222-228.
31. K. Aslan, M. Wu, J. R. Lakowicz, and C. D. Geddes. Fluorescent core-shell Ag@SiO2 nanocomposites for metal-enhanced fluorescence and single nanoparticle sensing platforms *J Am Chem Soc*, 2007, 129, 1524-1525.
32. J. R. Lakowicz *Titl* 2nd ed. Kluwer Academic, New York.1999
33. L.Brand and B.Witholt. *Methods in Enzymology*, 1967, 11, 776-856.
34. Y. Zhang, K. Aslan, M. J. R. Previte, and C. D. Geddes. Metal-enhanced e-type fluorescence *Applied Physics Letters*, 2008, 92,
35. N. S. patrick Conlon: Chaoyong Jame Yang;Yanrong Wu;Yan Chen;karen Martinez; Youngmi Kim, Angel A. Marti, Steffen Jockusch, Nicholas J. Turro; Weihong Tan. Pyren Excimer Singaling Molecular Beacons for Probing Nucleic Acids *J am. Chem. Soc.* 2008, 130, 336-342.
36. P. C. J. H. W. Offen. Pyrene Fluorescence in ethanol and cyclohexane under pressure *The journal of Chemical Physics*, 1973, 59, 801-806.
37. M. H. Chowdhury, K. Aslan, S. N. Malyn, J. R. Lakowicz, and C. D. Geddes. Metal-enhanced chemiluminescence *Journal of Fluorescence*, 2006, 16, 295-299.
38. M. H. Chowdhury, S. N. Malyn, K. Aslan, J. R. Lakowicz, and C. D. Geddes. First observation of surface plasmon-coupled chemiluminescence (SPCC) *Chemical physics letters*, 2007, 435, 114-118.
39. J. Moan, Q. Peng, R. Sorensen, V. Iani, and J. M. Nesland. The biophysical foundations of photodynamic therapy *Endoscopy*, 1998, 30, 387-391.
40. M. S. Thompson, A. Johansson, T. Johansson, S. Andersson-Engels, S. Svanberg, N. Bendsoe, and K. Svanberg. Clinical system for interstitial photodynamic therapy with combined on-line dosimetry measurements *Appl Opt*, 2005, 44, 4023-4031.
41. Y. Zhang, K. Aslan, S. N. Malyn, and C. D. Geddes. Metal-enhanced phosphorescence (MEP) *Chemical physics letters*, 2006, 427, 432-437.
42. Y. Zhang, K. Aslan, M. J. R. Previte, and C. D. Geddes. Plasmonic engineering of singlet oxygen generation *Proceedings of the National Academy of Sciences of the United States of America*, 2008, 105, 1798-1802.

43. Y. X. Zhang, K. Aslan, M. J. R. Previte, and C. D. Geddes. Plasmon-enhanced singlet oxygen generation *Biophysical Journal*, 2007, 519A-519A.
44. Y. Zhang, K. Aslan, M. J. Previte, and C. D. Geddes. Metal-enhanced Singlet Oxygen Generation: A Consequence of Plasmon Enhanced Triplet Yields *Journal of Fluorescence*, 2007, 17, 345-349.
45. Y. Zhang, K. Aslan, M. J. R. Previte, S. N. Malyn, and C. D. Geddes. Metal-enhanced phosphorescence: Interpretation in terms of triplet-coupled radiating plasmons *Journal of Physical Chemistry B*, 2006, 110, 25108-25114.
46. P. W. Barber, R. K. Chang, and H. Massoudi. Electrodynamic Calculations of the Surface-Enhanced Electric Intensities on Large Ag Spheroids *Physical Review B*, 1983, 27, 7251-7261.
47. W. H. Yang, G. C. Schatz, and R. P. Vanduyne. Discrete Dipole Approximation for Calculating Extinction and Raman Intensities for Small Particles with Arbitrary Shapes *Journal of Chemical Physics*, 1995, 103, 869-875.
48. K. S. Yee and J. S. Chen. Impedance boundary condition simulation in the FDTD/FVTD hybrid *Ieee Transactions on Antennas and Propagation*, 1997, 45, 921-925.
49. K. L. Kelly, E. Coronado, L. L. Zhao, and G. C. Schatz. The optical properties of metal nanoparticles: The influence of size, shape, and dielectric environment *Journal of Physical Chemistry B*, 2003, 107, 668-677.
50. W. A. Challener, I. K. Sendur, and C. Peng. Scattered field formulation of finite difference time domain for a focused light beam in dense media with lossy materials *Optics Express*, 2003, 11, 3160-3170.
51. K. S. Yee. Numerical solution of inital boundary value problems involving maxwell's equations in isotropic media *Antennas and Propagation, IEEE Transactions on [legacy, pre - 1988]*, 1966, 14, 302-307.
52. S. Foteinopoulou, J. P. Vigneron, and C. Vandenbem. Optical near-field excitations on plasmonic nanoparticle-based structures *Optics Express*, 2007, 15, 4253-4267.
53. E. Hao and G. C. Schatz. Electromagnetic fields around silver nanoparticles and dimers *Journal of Chemical Physics*, 2004, 120, 357-366.

# 2

# Spectral Profile Modifications In Metal-Enhanced Fluorescence

E. C. Le Ru[a]*, J. Grand[b], N. Félidj[b], J. Aubard[b], G. Lévi[b], A. Hohenau[c], J. R. Krenn[c], E. Blackie[a], P. G. Etchegoin[a]

[a]The MacDiarmid Institute for Advanced Materials and Nanotechnology, School of Chemical and Physical Sciences, Victoria University of Wellington, PO Box 600, Wellington, New Zealand

[b]Laboratoire ITODYS, Université Paris 7 – Denis Diderot, CNRS UMR 7086, 1 rue Guy de la Brosse, F - 75005 Paris, France

[c]Institute of Physics, Karl Franzens University, Universit¨atsplatz 5, A-8010 Graz, Austria

## 2.1 INTRODUCTION

In this chapter, we provide a tutorial review of an often-overlooked aspect of Metal Enhanced Fluorescence (MEF), namely the modification of the spectral profile of emission [1]. Decay rates (radiative and total) of a fluorophore are strongly modified in MEF conditions, as a result of electromagnetic (EM) coupling to localized surface plasmon (LSP) resonances in the metallic substrate [2,3,4,5,6,7,8,9,10,11,12]. These decay rate modifications are, as the underlying LSP resonances, wavelength-dependent, a fact that has recently been discussed in detail [13]. These well-accepted facts should naturally result in a modification of the spectral profile of fluorescence in MEF conditions, in addition to the "standard" modifications of the fluorescence intensity (enhancement or quenching). Despite this, spectral profile modifications (SPMs) in MEF conditions have hardly been discussed in the literature, neither theoretically nor experimentally. Up until recently, there had only been isolated reports of small SPMs, typically mentioned in passing in studies dedicated to other aspects of MEF [11, 14]. We have presented in Ref. [1] a theoretical and experimental study dedicated entirely to this particular aspect of MEF (called Surface Enhanced Fluorescence (SEF) in Ref. [1]; MEF and SEF will be used with the same meaning in this chapter). The content of this chapter is therefore largely based on an extension of the concepts and results of Ref. [1], where several of the mechanisms of Spectral Profile Modification (SPM) discussed below were first proposed and evidenced.

This chapter is organized as follows. In Sec. 2.2, we review some basic aspects of the free-space fluorescence process and of its modification close to metal surfaces (MEF). These concepts are then extended in Sec. 2.3 to discuss the various mechanisms of spectral profile modification that can be expected in MEF conditions. Experimental demonstration of some of these concepts is

*Spectral Profile Modifications in MEF*
Edited by Chris D. Geddes.

provided in Sec. 2.4. Finally, we discuss in Sec. 2.5 some extensions of these concepts and experiments along with possible future research directions.

## 2.2 PRELIMINARIES

### 2.2. 1. *Notations And Assumptions*

Let us briefly go through some of the notations and assumptions that will be used throughout this chapter. Units are indicated in brackets [...] when deemed useful.

The fluorophore (emitter) will always be assumed to be embedded in a non-absorbing dielectric medium of refractive index $n_M$ (or relative dielectric constant $\varepsilon_M = (n_M)^2$ real). The excitation is assumed to be monochromatic (e.g. a laser) at angular frequency $\omega_L$ [rad s$^{-1}$]. The excitation density is denoted $S_{\mathrm{Inc}}$ [W m$^{-2}$] ($S$ refers to the Poynting vector). Alternatively, the number of photons per unit area per unit time $N_L = S_{\mathrm{Inc}} / (\hbar\omega_L)$ [m$^{-2}$ s$^{-1}$] may be used.

The signal (e.g. fluorescence) intensities will be characterized by their powers $P$ [W]. The spectral profile is characterized by the spectral density $n(\omega)$ [rad$^{-1}$]. By definition, the number of photons per unit time at a frequency between $\omega$ and $\omega + d\omega$ is $n(\omega)d\omega$.

The total power is then given by:

$$P = \int \hbar\omega n(\omega)d\omega. \qquad (1)$$

Experimental spectra will be denoted $I(\omega)$ [counts per second, or arb. units] and correspond to the number of photons per unit time, i.e. they are proportional to the spectral density $n(\omega)$.

To avoid unnecessary complications in the presentation, we will ignore any vectorial, tensorial, or molecule orientation effects, unless specifically stated. This does not affect any of the conclusions, and they can always be included at a later stage if judged necessary.

### 2.2. 2. *Spectral Profile Of Fluorescence: Free-space Case*

In order to discuss any modification to the fluorescence spectral profile, it is important to first understand its origin for a free fluorophore. The model of fluorescence that we discuss below has been simplified as much as possible and retains only the features that are necessary for the generalization to the MEF case. In particular, most of the parameters of this model are introduced *phenomenologically*. The connection to the underlying molecular electronic and vibrational properties is only discussed qualitatively. More detailed descriptions of the fluorescence process can be found in molecular spectroscopy textbooks [15, 16]. Note that the superscript "0" will be used for all free-space properties, to distinguish them from their modified counterparts in MEF conditions (without superscript).

## 2.2. 2.a *The Fluorescence Process*

Let us consider a typical fluorophore, with ground ($S_0$ ) and excited ($S_1$) singlet electronic states. The energy states in the vibrational (or vibronic) substructure will be denoted $S_0(\omega_v)$ $S_1(\omega_v)$, where $\omega_v$ is the vibrational frequency. The lowest energy state in $S_0$ ($S_1$) is accordingly denoted $S_0(0)$ $S_1(0)$. This vibrational substructure is quasi-continuous for most fluorophores (with say, more than $\approx 50$ atoms) and can be characterized by a vibrational (or vibronic) density of state $\rho_0(\omega_v)$ in $S_0$ and $\rho_0(\omega_v)$ in $S_1$.

The transition energy between $S_0(0)$ and $S_1(0)$ is denoted $\hbar\omega_0$. This fluorophore is excited by monochromatic light (a laser) at angular frequency $\omega_L > \omega_0$. The fluorescence process is represented schematically in the simplified Jablonski diagram of Fig. 2.1(a) (see Sec. 2.3) and consists of three separate steps.

### *Absorption:*

The first step corresponds to the absorption of a photon (of energy $\hbar\omega_L$), resulting in the excitation of an electron from the ground ($S_0$) to the excited electronic state ($S_1$). More precisely, the transition occurs from a state $S_0(\omega_{v0})$in the substructure of $S_0$ to a state $S_1(\omega_{v1})$ in the substructure of $S_1$ and energy conservation therefore requires $\omega_{v1} - \omega_{v0} + \omega_0 = \omega_L$. Moreover, at thermal equilibrium at a temperature $T$, the probability of occupation for the initial electron in $S_0$ can in principle be determined from $\rho_0(\omega_v)$ (and is typically within a range $\sim k_B T$ from $S_0(0)$). At room temperature the energy $k_B T$ is of the order of 200 cm$^{-1}$ and we will assume for simplicity that it is negligible compared to the other energy scales in the system. This is equivalent to assuming $T = 0$ K and neglecting any thermal excitation effects. In this case the initial state for the transition is $S_0(0)$.

Optical absorption can be described in terms of the absorption cross section $\sigma^0_{\text{Abs}}$ [m$^2$] of the molecule. For excitation at $\omega_L$, the power absorbed is then given by:

$$P^0_{\text{Abs}} = \sigma^0_{\text{Abs}}(\omega_L) S_{\text{Inc}} \tag{2}$$

For fluorophores, $\sigma^0_{\text{Abs}}(\omega_L)$ can easily be measured experimentally in UV / V is absorption experiments. Note that the spectral profile of absorption is intricately related to the vibrational density of state $\rho_1(\omega_v)$ (and more precisely to overlap integrals between vibrational states within the Franck-Condon approximation). It can therefore be used as an experimental probe of the vibrational density of state.

### *Relaxation:*

The second step is then that of energy relaxation within the excited state $S_1$. This typically occurs through vibrational relaxation (or possibly through interactions with the environment) on picosecond time-scales; sometimes even hundreds of femtoseconds for large molecules. This step is probably the least understood, but is in most cases irrelevant because it is much faster than the other processes. It results, in fact, in fast thermalization of the excited electron in $S_1$. Neglecting thermal excitation, the electron reaches the lowest state $S_1(0)$, typically within picoseconds [15].

### *Decay:*

The final step corresponds to the decay of the electron from $S_1(0)$ to a state $S_0(\omega_v)$ in the substructure of the electronic ground state. This decay may be radiative, i.e. it results in emission of a photon at angular frequency $\omega_S = \omega_0 - \omega_v$; this is what is detected as fluorescence. It may also be non-radiative and the energy is then lost to other degrees of freedom in the system (for example to internal rotations, to the solvent, or to the triplet state). All decay mechanisms are stochastic processes and are therefore characterized by their probability of occurring per unit time or rate; in this case the radiative decay rate $\Gamma^0_{\text{Rad}}$ and the non-radiative decay rate $\Gamma^0_{\text{NR}}$. The total decay rate is then $\Gamma^0_{\text{Tot}} = \Gamma^0_{\text{Rad}} + \Gamma^0_{\text{NR}}$ and $(\Gamma^0_{\text{Tot}})^{-1}$ is the excited state lifetime. Moreover all possible decay paths compete with each other. The quantum yield $Q^0$ – corresponding to the proportion of electrons decaying radiatively– is then defined as $Q^0 = \Gamma^0_{\text{Rad}} / \Gamma^0_{\text{Tot}}$. Following decay to $S_0$, the absorption / relaxation / emission cycle can then start again.

In this simple description, we have not considered possible additional effects such as: excitation to a higher excited state, saturation effects, or inter-system crossing (transition to the triplet state). These will be ignored throughout for simplicity.

In most fluorescence models where the spectral profile is ignored, the emitted fluorescence power is then simply given as $P^0_{\text{Fluo}} = \sigma^0_{\text{Abs}} Q^0 S_{\text{Inc}}$. This can alternatively be characterized by a fluorescence cross-section: $\sigma^0_{\text{Fluo}} = \sigma^0_{\text{Abs}} Q^0$. We now adapt this standard argument to account for the spectral dispersion.

## 2.2. 2.b *A Model For The Spectral Profile*

The spectral profile of fluorescence is characterized by the $\omega_S$-dependence of the fluorescence intensity. Within the simple description given above, it is therefore governed by the final step in the process, i.e. decay from $S_1(0)$ to $S_0(\omega_v)$, which we now study in more detail.

We first need to define the radiative decay rate at frequency $\omega_S$. Because of the continuous nature of $\omega_S$, we have to consider all transitions from $S_1(0)$ to $S_0(\omega_v = \omega_0 - \omega)$ at frequencies $\omega$ within a small frequency range: $\omega_S \leq \omega \leq \omega_S + d\omega_S$. Their radiative decay rate is by definition $\gamma^0_{\text{Rad}}(\omega_S) d\omega_S$. $\gamma^0_{\text{Rad}}(\omega_S)$ [rad$^{-1}$] can be viewed as the spectral density of the radiative decay rate. This will in most cases be a parameter determined by experiments, but is in principle directly related to the ground state vibrational density of state $\rho_0(\omega_v)$ (and to overlap integrals within the Franck-Condon approximation). The (total) radiative decay rate is then obtained by integration of its spectral density over the entire fluorescence spectrum, i.e.

$$\Gamma^0_{\text{Rad}} = \int \gamma^0_{\text{Rad}}(\omega_S) d\omega_S. \tag{3}$$

The non-radiative decay rate is $\Gamma^0_{\text{NR}}$ as before (its spectral density is irrelevant here) and the total decay rate is therefore as before $\Gamma^0_{\text{Tot}} = \Gamma^0_{\text{Rad}} + \Gamma^0_{\text{NR}}$. The spectral density of the radiated (fluorescence) signal is then:

$$n^0_{\text{Rad}}(\omega_S) = \frac{\gamma^0_{\text{Rad}}(\omega_S)}{\Gamma^0_{\text{Tot}}} \sigma^0_{\text{Abs}} N_L. \tag{4}$$

The spectral profile of fluorescence is therefore entirely determined (phenomenologically) by the $\omega_S$-dependence of $\gamma^0_{\text{Rad}}(\omega_S)$. This is linked, microscopically, to the density of states $\rho_0(\omega_0 - \omega_S)$ in the substructure of $S_0$ and the matrix elements of the respective transitions (or overlap integrals of the vibrational states).

The total fluorescence power is then obtained using Eqs. 1 and 4 as

$$P^0_{\text{Fluo}} = Q^{0-\text{Eff}} \sigma^0_{\text{Abs}} S_{\text{Inc}} \quad \text{with} \quad Q^{0-\text{Eff}} = \frac{\Gamma^{0-\text{Eff}}_{\text{Rad}}}{\Gamma^0_{\text{Tot}}}, \tag{5}$$

where $Q^{0-\text{Eff}}$ is an effective quantum yield and the effective radiative decay rate has been defined as:

$$\Gamma^{0-\text{Eff}}_{\text{Rad}} = \int \frac{\omega_S}{\omega_L} \gamma^0_{\text{Rad}}(\omega_S) d\omega_S. \tag{6}$$

It is interesting to highlight the fact that, strictly speaking, the effective radiative decay $\Gamma^{0-\text{Eff}}_{\text{Rad}}$ that enters this expression is different to the real radiative decay rate $\Gamma^0_{\text{Rad}}$ that may be measured in a time-resolved experiment (if $\Gamma^0_{\text{NR}}$ is known). However, in most practical cases, it is easy to see that $\Gamma^{0-\text{Eff}}_{\text{Rad}} \approx \Gamma^0_{\text{Rad}}$ to a very good approximation (since $\omega_L - \omega_S \ll \omega_S$). We will therefore ignore this distinction in the following.

We have here kept the description of the radiative decay rates fairly general and did not discuss their exact relations to the vibronic states (in particular within the Franck-Condon approximation). This description is sufficient for our purpose here and more adapted to the generalization to the case of MEF. It is nevertheless worth mentioning that there is an approximate symmetry between the transitions for absorption: $S_0(0) \rightarrow S_1(\omega_v)$ and those for fluorescence emission: $S_1(0) \rightarrow S_0(\omega_v)$. This results in the well-known "mirror symmetry" around the $\omega = \omega_0$ line between the absorption and fluorescence spectra, i.e. $P^0_{\text{Fluo}}(\omega) \propto P^0_{\text{Abs}}(2\omega_0 - \omega)$.

## 2.2. 3. *Plasmonic Effects In MEF Conditions*

We are now looking at extending the simple model for free-space fluorescence to the case of Metal Enhanced Fluorescence (MEF). We therefore consider a fluorophore with known free-space properties in the presence of optically-active objects (in particular metallic objects).

The EM theory of Metal Enhanced Fluorescence (MEF) was studied and developed extensively in the 70-80's [2,3,4,5,6]. All the EM mechanims involved in MEF can be understood within classical EM theory [3,4,5,6] as confirmed in the simplest cases by quantum studies [2,12,17,18]. In most of these models, the emitter is depicted as a simple two- (or three-) level system, i.e. only one emission wavelength is considered. This is appropriate in general to understand modifications of absorption or emission rates, but it entirely ignores the spectral profile of the fluorescence emission. We will first review these concepts here before extending and generalizing them in the next section (to specifically incorporate the description of the fluorescence spectral profile in MEF conditions).

When a fluorophore is located in close proximity to a metal surface, both its absorption and emission properties may be affected dramatically. This in turn affects its fluorescence properties and may result in either a quenching or an enhancement of the fluorescence signal. This latter situation is obviously of interest for many applications using fluorophores. Let us discuss these steps in more detail.

### 2.2. 3.a *The Local Field Enhancement*

We focus first on the absorption. It is well known that the electric field may change dramatically in the vicinity of metallic surfaces, both in magnitude and polarization. The modified field at the molecule position, which we call the local field $\mathbf{E}_{\mathrm{Loc}}(\omega_L)$, can in principle be deduced and calculated at every point by solving the EM problem with appropriate boundary conditions. The field intensity felt by the molecule is modified by the local field intensity enhancement factor (EF):

$$M_{\mathrm{Loc}}(\omega_L) = \frac{|\mathbf{E}_{\mathrm{Loc}}(\omega_L)|^2}{|E_{\mathrm{Inc}}|^2}, \qquad (7)$$

where $E_{\mathrm{Inc}}$ is the electric field amplitude of the incident excitation. One can then show that the absorption cross-section (or power absorbed) is modified by the same factor, i.e.:

$$\sigma_{\mathrm{Abs}}(\omega_L) = M_{\mathrm{Loc}}(\omega_L)\sigma^0_{\mathrm{Abs}}(\omega_L). \qquad (8)$$

### 2.2. 3.b *Radiative And Non-Radiative Enhancements*

Similar modifications arise for the emission properties of an emitter and affect its radiative decay. We will assume that the intrinsic non-radiative decay path is unchanged (a reasonable assumption since it does not involve photons) and is therefore characterized by the same rate $\Gamma^0_{\mathrm{NR}}$. The radiative decay path, characterized by $\Gamma^0_{\mathrm{Rad}}$ in free-space, is modified in at least three ways:

❖ Firstly, the decay rate through photon emission is modified and we will call the modified rate the total EM decay rate $\Gamma^{\mathrm{EM}}_{\mathrm{Tot}}$. The EM superscript indicates the EM nature of the decay (through emission of a photon).

❖ Secondly, part of the photons emitted during this decay may be absorbed in the environment (for example the metal substrate). These photons are not radiated in the far-field; they correspond to non-radiative emission. Such a non-radiative decay is very different in nature to the intrinsic non-radiative channel, since it does involve exchange of a photon. From an experimental point of view, however, the end result is the same, i.e. a decay without production of a fluorescence photon. To distinguish the two possible EM decay channels, we can define the radiative $\Gamma_{\text{Rad}}$ and non-radiative $\Gamma_{\text{NR}}^{\text{EM}}$ EM decay rates, the sum of which gives $\Gamma_{\text{Tot}}^{\text{EM}}$. One can alternatively define the EM radiative efficiency $\eta_{\text{Rad}}^{\text{EM}} = \Gamma_{\text{Rad}} / \Gamma_{\text{Tot}}^{\text{EM}}$. All these decay rates can instead be characterized with respect to the free-space EM decay rate (i.e. the free-space radiative decay rate) by defining the total EM enhancement factor, the non-radiative EM enhancement factor, and the radiative enhancement factor as:

$$M_{\text{Tot}} = \frac{\Gamma_{\text{Tot}}^{\text{EM}}}{\Gamma_{\text{Rad}}^{0}} \quad M_{\text{NR}} = \frac{\Gamma_{\text{NR}}^{\text{EM}}}{\Gamma_{\text{Rad}}^{0}} \quad M_{\text{Rad}} = \frac{\Gamma_{\text{Rad}}}{\Gamma_{\text{Rad}}^{0}}. \tag{9}$$

The EM radiative efficiency is then given as

$$\eta_{\text{Rad}}^{\text{EM}} = \frac{M_{\text{Rad}}}{M_{\text{Tot}}}. \tag{10}$$

❖ Thirdly, the radiation pattern (angular and polarization dependence of the radiated power) may also be modified. We will ignore this aspect, which is in most cases secondary.

The total decay rate of the emitter is the sum of the decay rates for all possible decay channels (which compete with each other) and is therefore:

$$\Gamma_{\text{Tot}} = \Gamma_{\text{Tot}}^{\text{EM}} + \Gamma_{\text{NR}}^{0} = M_{\text{Tot}}\Gamma_{\text{Rad}}^{0} + \Gamma_{\text{NR}}^{0}. \tag{11}$$

This is what would be observable in time-resolved experiments; the lifetime being simply $(\Gamma_{\text{Tot}})^{-1}$ [7].

## 2.2. 3.c *Fluorescence Enhancement Factor*

The modified quantum yield can therefore be expressed as:

$$Q = \frac{\Gamma_{\text{Rad}}}{\Gamma_{\text{Tot}}} = \frac{M_{\text{Rad}}}{M_{\text{Tot}} + \Gamma_{\text{NR}}^{0} / \Gamma_{\text{Rad}}^{0}} = \frac{M_{\text{Rad}}}{M_{\text{Tot}} + (Q^{0})^{-1} - 1} \approx \eta_{\text{Rad}}^{\text{EM}}, \tag{12}$$

where the latter approximation is valid when $M_{\text{Tot}} \gg (Q^{0})^{-1}$, which is the case in most situations of interests. Note that this means that the modified quantum yield is determined entirely by the EM radiative efficiency, a property of the metallic substrate. All fluorophores should therefore have the same modified quantum yield, irrespective of their intrinsic quantum yield $Q^0$ (see for example Ref. [19] for a discussion of the EM factors affecting the modified quantum yield, along with examples of MEF predictions on spherical metallic nanoparticles).

The modified fluorescence power can be expressed, as in the free space case, as $P_{\text{Fluo}} = Q\sigma_{\text{Abs}} S_{\text{Inc}}$. This may conveniently be expressed as a fluorescence enhancement factor:

$$M_{\text{Fluo}} = \frac{P_{\text{Fluo}}}{P^0_{\text{Fluo}}} = M_{\text{Loc}} \frac{M_{\text{Rad}}}{Q^0 M_{\text{Tot}} + 1 - Q^0} \approx M_{\text{Loc}} \frac{M_{\text{Rad}}}{Q^0 M_{\text{Tot}}} = \frac{M_{\text{Loc}} \eta^{\text{EM}}_{\text{Rad}}}{Q^0}. \tag{13}$$

The fluorescence enhancement is therefore a result of the interplay of three factors:

- The local field enhancement factor $M_{\text{Loc}}$, characterizing the modification of absorption. This results, in most cases of interest, in an enhancement, and possibly a very large enhancement up to $\approx 10^5$.

- The EM radiative efficiency $\eta^{\text{EM}}_{\text{Rad}} = M_{\text{Rad}} / M_{\text{Tot}}$. This is always smaller (and possibly much smaller) than 1, and contributes to a quenching of the fluorescence because of non-radiative emission (fluorescence photons absorbed in the metallic substrate).

- The factor $(Q^0)^{-1}$, an intrinsic property of the fluorophore under consideration. For a good fluorophore with $Q^0 \approx 1$, this factor has no effect. Only for a bad fluorophore with $Q^0 \ll 1$ can this factor lead to a substantial additional fluorescence enhancement. This is, in a way, a consequence of the fact that the modified quantum yield is dictated by the metallic substrate and therefore the same for all fluorophores (good or bad). If one uses a bad fluorophore, greater enhancements can be obtained simply because the free-space fluorescence was poor in the first place.

The exact balance among these three factors determines whether fluorescence quenching or enhancement is observed [10].

### 2.2. 3.d *Link Between The Local Field And Radiative Enhancement Factors*

The local field and decay rate enhancements have the same physical origin: EM coupling to the plasmon resonances of the metallic substrate. This will be discussed in more detail in Sec. 3.7. All these enhancement factors therefore exhibit a strong wavelength- (or equivalently $\omega$-) dependence. This fact will actually be the main reason for the spectral profile modifications predicted in the MEF signal.

Moreover, it is possible to show that the local field enhancement factor $M_{\text{Loc}}(\omega)$ and the radiative enhancement factor $M_{\text{Rad}}(\omega)$ are in many cases of the same order of magnitude. The exact relation between these and the range of validity of the statement above are not trivial and require the use of the optical reciprocity theorem; see for example Refs. [20, 21] for more details. The approximation $M_{\text{Loc}}(\omega) \approx M_{\text{Rad}}(\omega)$ is nevertheless very useful when discussing (semi-quantitatively) the magnitude of the fluorescence enhancement or quenching.

## 2.3 SPECTRAL PROFILE IN MEF CONDITIONS

Due to the intrinsic spectral dependence of localized surgace Plasmon resonances, all the enhancement factors defined so far for emission depend on $\omega_S$ (the emission frequency). This should, accordingly, affect the spectral profile of MEF

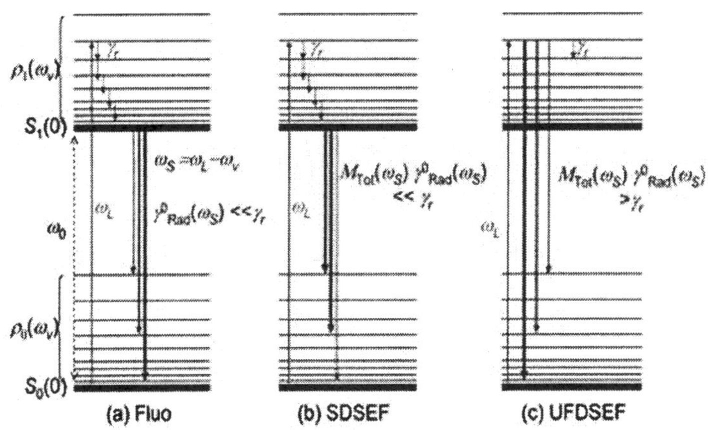


**Figure 2.1:** Adapted and reproduced with permission from *J. Phys. Chem. C* 2007, *111*, 16076–16079. Copyright 2007 American Chemical Society. Simplified Jablonski diagrams depicting the processes involved for a fluorophore in three possible situations discussed in the text: (a) Standard fluorescence, (b) Surface Enhanced Fluorescence in the slow-dynamics regime (SDMEF), and (c) MEF in the ultra-fast-dynamics regime (UFDMEF).

### 2.3. 1. *The Regimes Of Spectral Profile Modification*

We have assumed so far that the electron has sufficient time to relax within the substructure of $S_1$ and therefore reach the lowest state $S_1(0)$ before any decay to the ground state occurs. This is entirely justified if the total decay rate from $S_1$ is much smaller than the typical internal relaxation rate in $S_1$, $\gamma_r$. In MEF conditions however, the factor $M_{\text{Tot}}$ (and therefore the total decay rate $\Gamma_{\text{Tot}}$) may be large enough for this common assumption to become no longer justified. This possibility has in fact been occasionally suggested in the past [22, 23], but never actually pursued. If this is the case, the dynamics of the electron in the substructure of $S_1$ then becomes important. It is therefore necessary to distinguish between three regimes for MEF.

Following the notations of Ref. [1], these are:

- <u>Slow-dynamics MEF (SDMEF)</u>, for which $\Gamma_{\text{Tot}} \ll \gamma_r$. This is the same situation as for standard MEF. The electron relaxes to $S_1(0)$ from which it can then decay to $S_0$.

- <u>Fast-dynamics MEF (FDMEF)</u>, for which $\Gamma_{\text{Tot}}$ is of the order of –and possibly larger than– the internal relaxation rate $\gamma_r$. In this case, decay to the ground state may occur from any intermediate levels $S_1(\omega_v)$ between the one reached after absorption $S_1(\omega_L - \omega_0)$ and the lowest one $S_1(0)$.

- <u>Ultra-fast-dynamics MEF (UFDMEF)</u>, for which $\Gamma_{\text{Tot}} \gg \gamma_r$. In this extreme case of FDMEF, the decay rate is so fast that no energy relaxation occurs in $S_1$. All decay transitions therefore occur from $S_1(\omega_L - \omega_0)$.

Note that the distinction between SDMEF and FDMEF is irrelevant in cases where the excitation energy $\omega_L$ is close to or below the transition energy $\omega_0$. The fluorescence process in the SDMEF and UFDMEF regimes are represented schematically in simplified Jablonski diagrams in Fig. 2.1 along with the standard fluorescence process for comparison. We will now extend the models described so far to these three regimes of spectral profile modification of the MEF.

### 2.3. 2. *Slow-Dynamics Regime*

In the slow-dynamics regime (see Fig. 2.1(b)), the situation is very similar to standard fluorescence (Fig. 2.1(a)), except for the rate modifications introduced by the metallic substrate.

### 2.3. 2.a *Modified Decay Rates*

Let us consider again the transitions between $S_1(0)$ to $S_0(\omega_0 - \omega_S)$ at an energy $\hbar\omega_S$. The modified spectral density of the radiative decay rate is

$$\gamma_{\text{Rad}}(\omega_S) = M_{\text{Rad}}(\omega_S)\gamma^0_{\text{Rad}}(\omega_S), \tag{14}$$

and there is a corresponding spectral density for non-radiative EM emission:

$$\gamma_{\text{NR}}(\omega_S) = M_{\text{NR}}(\omega_S)\gamma^0_{\text{Rad}}(\omega_S). \tag{15}$$

The total EM decay rate is accordingly:

$$\Gamma^{\text{EM}}_{\text{Tot}} = \int(\gamma_{\text{Rad}}(\omega_S) + \gamma_{\text{NR}}(\omega_S))d\omega_S = \int M_{\text{Tot}}(\omega_S)\gamma^0_{\text{Rad}}(\omega_S)d\omega_S. \tag{16}$$

It is convenient here to define the *spectrally averaged total EM enhancement factor* as:

$$\bar{M}_{\text{Tot}} = \int M_{\text{Tot}}(\omega_S)\frac{\gamma^0_{\text{Rad}}(\omega_S)}{\Gamma^0_{\text{Rad}}}d\omega_S. \tag{17}$$

In a similar fashion, the *spectrally averaged radiative decay rate enhancement factor* is:

$$\bar{M}_{\text{Rad}} = \int M_{\text{Rad}}(\omega_S)\frac{\gamma^0_{\text{Rad}}(\omega_S)}{\Gamma^0_{\text{Rad}}}d\omega_S. \tag{18}$$

Note that the weighing factor $\gamma^0_{\text{Rad}}(\omega_S)$ in these integrals is proportional to the free-space fluorescence spectral density $n^0_{\text{Rad}}(\omega_S)$, which can be measured easily experimentally.

Thus, we have simply:

$$\Gamma^{\text{EM}}_{\text{Tot}} = \bar{M}_{\text{Tot}}\Gamma^0_{\text{Rad}}, \text{ and } \Gamma_{\text{Rad}} = \bar{M}_{\text{Rad}}\Gamma^0_{\text{Rad}}. \tag{19}$$

$\bar{M}_{\text{Tot}}$ and $\bar{M}_{\text{Rad}}$ then play the same role that $M_{\text{Tot}}(\omega_S)$ and $M_{\text{Rad}}(\omega_S)$ did in the one-wavelength fluorescence model. The modified total decay rate is then

$$\Gamma_{\text{Tot}} = \Gamma_{\text{Tot}}^{\text{EM}} + \Gamma_{\text{NR}}^{0} = \left(\bar{M}_{\text{Tot}} + (Q^0)^{-1} - 1\right)\Gamma_{\text{Rad}}^{0}. \tag{20}$$

## 2.3. 2.b *Modified Spectral Profile*

The modified spectral density is obtained (as for the free-space case) from:

$$n_{\text{Rad}}^{\text{SDSEF}}(\omega_S) = \frac{\gamma_{\text{Rad}}(\omega_S)}{\Gamma_{\text{Tot}}}\sigma_{\text{Abs}}N_L = \frac{M_{\text{Rad}}(\omega_S)\gamma_{\text{Rad}}^{0}(\omega_S)}{\Gamma_{\text{Tot}}}\sigma_{\text{Abs}}N_L.$$

(21)

From this expression, the MEF spectral profile is simply determined through the product of the free-space spectral profile, proportional to $\gamma_{\text{Rad}}^{0}(\omega_S)$ and the radiative enhancement factor profile $M_{\text{Rad}}(\omega_S)$. This can, in fact, be expressed as a modification factor with respect to the free-space case as:

$$\frac{n_{\text{Rad}}^{\text{SDSEF}}(\omega_S)}{n_{\text{Rad}}^{0}(\omega_S)} = \frac{M_{\text{Rad}}(\omega_S)}{\bar{M}_{\text{Tot}} + (Q^0)^{-1} - 1} M_{\text{Loc}}(\omega_L). \tag{22}$$

The $\omega_S$-dependence in this modification factor is entirely dictated by $M_{\text{Rad}}(\omega_S)$.

## 2.3. 2.c *Fluorescence Intensity*

This latest expression provides in addition the magnitude of the fluorescence intensity enhancement factor at a given frequency $\omega_S$. In some instances and also to make a connection with the standard one-wavelength model, it may be more convenient to consider the total fluorescence intensity.

Following the description above, the *modified quantum yield* is simply obtained as:

$$Q = \frac{\Gamma_{\text{Rad}}}{\Gamma_{\text{Tot}}} = \frac{\bar{M}_{\text{Rad}}}{\bar{M}_{\text{Tot}} + (Q^0)^{-1} - 1}. \tag{23}$$

Similarly, the *total fluorescence power enhancement factor* is:

$$M_{\text{Fluo}} = \frac{P_{\text{Fluo}}}{P^0_{\text{Fluo}}} = \frac{\bar{M}^{\text{Eff}}_{\text{Rad}}}{\bar{M}_{\text{Tot}} + (Q^0)^{-1} - 1} M_{\text{Loc}}(\omega_L), \tag{24}$$

where

$$\bar{M}^{\text{Eff}}_{\text{Rad}} = \int \frac{\omega_S}{\omega_L} M_{\text{Rad}}(\omega_S) \frac{\gamma^0_{\text{Rad}}(\omega_S)}{\Gamma^0_{\text{Rad}}} d\omega_S. \tag{25}$$

In a similar manner to the free-space case, $\bar{M}^{\text{Eff}}_{\text{Rad}}$ (for the power enhancement) is strictly speaking different to $\bar{M}_{\text{Rad}}$ (for the rate enhancement) because of the Stokes shift, but for most practical purposes, $\bar{M}^{\text{Eff}}_{\text{Rad}} \approx \bar{M}_{\text{Rad}}$ is a very good approximation. These expressions for the modified quantum yield and the fluorescence enhancement factor are therefore essentially the same as those obtained in the one-wavelength model, only replacing $M_{\text{Rad}}(\omega_S)$ and $M_{\text{Tot}}(\omega_S)$ by $\bar{M}_{\text{Rad}}$ and $\bar{M}_{\text{Tot}}$. This is an important result, since it means that, as far as the total fluorescence intensity is concerned, any spectral modification can in a first approximation be ignored, and a one-wavelength model (at the frequency of the fluorescence peak $\omega_F$) can be used. Hence, the discussion of the various mechanisms of fluorescence enhancement and quenching given in Sec. 2.2 3 remains valid.

## 2.3. 3. *Ultra-Fast-Dynamics Regime*

### 2.3. 3.a *Model*

The ultra-fast-dynamics regime is, in some ways, similar to the slow-dynamics regime; the only difference being that the decay transitions originate from $S_1(\omega_L - \omega_0)$ rather than $S_1(0)$. Let us denote $\Delta\omega_L = \omega_L - \omega_0$ and assume here that $\Delta\omega_L > 0$. If $\omega_L$ is close to or smaller than $\omega_0$, then internal relaxation in $S_1$ is irrelevant and the treatment given for the SDMEF applies independently of the actual decay rates.

In order to adapt the previous description, one therefore only needs to replace the free-space spectral decay rates from $S_1(0)$, $\gamma^0_{\text{Rad}}(\omega_S)$, by those from $S_1(\Delta\omega_L)$, denoted $\gamma^1_{\text{Rad}}(\omega_S)$ in the following. In a first approximation, one may assume that the dipole moment (or overlap integrals) is the same for both transitions. Therefore, what determines their decay rate is primarily the density of states in the final state in $S_0$ (and the secondary $\omega^3$ dependence of decay rates, which we also neglect here). Accordingly, we can assume that the two transitions to the same final state, $S_1(\Delta\omega_L) \rightarrow S_0(\omega_L - \omega_S)$ (at frequency $\omega_S$ ) and $S_1(0) \rightarrow S_0(\omega_L - \omega_S)$ (at frequency $\omega_S - \Delta\omega_L$), have (in a first approximation) the same decay rate, i.e.:

$$\gamma^1_{\text{Rad}}(\omega_S) \approx \gamma^0_{\text{Rad}}(\omega_S - \Delta\omega_L). \tag{26}$$

From there, all the formula of the previous section apply, simply replacing $\gamma^0_{\text{Rad}}(\omega_S)$ by $\gamma^0_{\text{Rad}}(\omega_S - \Delta\omega_L)$.

In particular, we have:

$$\overline{M}^{\text{UFDSEF}}_{\text{Tot}} = \int M_{\text{Tot}}(\omega_S)\frac{\gamma^0_{\text{Rad}}(\omega_S - \Delta\omega_L)}{\Gamma^0_{\text{Rad}}}d\omega_S. \tag{27}$$

Note that $\overline{M}^{\text{UFDSEF}}_{\text{Tot}}$ is of the same order of magnitude as that obtained for SDMEF (Eq. 17). The spectral density of fluorescence in the UFDMEF regime is then:

$$n^{\text{UFDSEF}}_{\text{Rad}}(\omega_S) = \frac{M_{\text{Rad}}(\omega_S)\gamma^0_{\text{Rad}}(\omega_S - \Delta\omega_L)}{\Gamma^{\text{UFDSEF}}_{\text{Tot}}}\sigma_{\text{Abs}}N_L. \tag{28}$$

This expression is similar to that obtained for the SDMEF case (Eq. 21), except that the radiative enhancement factor $M_{\text{Rad}}(\omega_S)$ is now applied to the *blue-shifted free-space fluorescence profile* $\gamma^0_{\text{Rad}}(\omega_S - \Delta\omega_L)$. It can, in fact, be expressed as a modification factor with respect to the blue-shifted free-space case as:

$$\frac{n^{\text{UFDSEF}}_{\text{Rad}}(\omega_S)}{n^0_{\text{Rad}}(\omega_S - \Delta\omega_L)} = \frac{M_{\text{Rad}}(\omega_S)}{\overline{M}^{\text{UFDSEF}}_{\text{Tot}} + (Q^0)^{-1} - 1}M_{\text{Loc}}(\omega_L). \tag{29}$$

It is possible to also write expressions for the modified quantum yield and fluorescence intensity enhancement factor in the UFDMEF regime. These are similar to those obtained for the SDMEF regime.

### 2.3. 3.b *Comparison With The SDMEF Regime*

From these results, it is clear that the spectral profile in both the SDMEF and UFDMEF regimes is predominantly dictated by the radiative enhancement factor profile $M_{\text{Rad}}(\omega_S)$. Distinguishing between the two regimes is in principle possible by direct comparison of the respective spectral profile of fluorescence.

They are related by the expression:

$$\frac{n_{\mathrm{Rad}}^{\mathrm{UFDSEF}}(\omega_S)}{n_{\mathrm{Rad}}^{\mathrm{SDSEF}}(\omega_S)} = \frac{n_{\mathrm{Rad}}^{0}(\omega_S - \Delta\omega_L)}{n_{\mathrm{Rad}}^{0}(\omega_S)}. \quad (30)$$

Such a distinction is facilitated if $\Delta\omega_L$ is large, and in any case requires an accurate knowledge of $n_{\mathrm{Rad}}^{0}(\omega)$ (the difficulties associated with this are discussed in Sec. 2.4).

### 2.3 4. *Intermediate Regime*

The situation in the intermediate regime is substantially more complicated: both decay to the ground state and energy relaxation can occur from any of the intermediate levels in $S_1$. The exact dynamics therefore not only depends on the modified decay rates, but also on the details of the relaxation dynamics within $S_1$; a process that is not understood in detail and is not easily measurable. We will, therefore, not attempt here a detailed description of the intermediate FDMEF regime. Nevertheless, one should qualitatively expect results that reside in between the two limiting cases of SDMEF and UFDMEF.

### 2.3 5. *Discussion*

Let us summarize the predictions of these simple models in terms of the spectral profile ($\omega_S$-dependence) of the fluorescence:

- The free-space fluorescence spectrum follows $n_{\mathrm{Rad}}^{0}(\omega_S)$.
- The modified spectrum in the SDMEF regime follows $M_{\mathrm{Rad}}(\omega_S)n_{\mathrm{Rad}}^{0}(\omega_S)$.
- The modified spectrum in the UFDMEF regime follows $M_{\mathrm{Rad}}(\omega_S)n_{\mathrm{Rad}}^{0}(\omega_S - \Delta\omega_L)$.

Moreover, for the MEF signal to be observable, it is also necessary that the local field enhancement at the laser frequency $M_{\mathrm{Loc}}(\omega_L)$ is sufficiently large. Since $M_{\mathrm{Loc}}$ and $M_{\mathrm{Rad}}$ follow approximately the same resonant profile, $M_{\mathrm{Rad}}(\omega_L)$ is a good indication of the magnitude of $M_{\mathrm{Loc}}(\omega_L)$.

For many fluorophores, $n_{\mathrm{Rad}}^{0}(\omega_S)$ is simply a "bell-shaped" curve (for example a Lorentzian); possibly slightly asymmetric or with a shoulder. Its full-width at half-maximum (FWHM) is typically of the order of

$\sim 700-1200$ cm$^{-1}$. As discussed already, $M_{\mathrm{Rad}}(\omega_S)$ exhibits a resonance profile at frequencies corresponding to the localized surface plasmon (LSP) resonances of the metallic substrate. If a single resonance dominates, its FWHM is typically of the same order as that of the fluorescence spectrum, i.e. $700-1200$ cm$^{-1}$, for typical Ag or Au substrates in the visible. To understand how such a resonance may affect the MEF spectral profile, it is interesting to consider a few model cases. For the sake of argument, we focus on the cases of Methylene Blue (MB) [24] and Rhodamine 6G (Rh6G) [25] excited at 633 nm, for which the free-space (in water) fluorescence profiles have been measured (dashed lines in Fig. 2.2(a,b,d) for MB and (c) for Rh6G). Note that these free-space fluorescence spectra directly reflect the spectral profile of fluorescence $n^0_{\mathrm{Rad}}(\omega_S)$ as given in Eq. 4. These two examples provide model fluorescence spectra for two typical situations: $(i)$ excitation below (but close to) the absorption maximum (here with $\Delta\omega_L = \omega_L - \omega_0 \approx 900$ cm$^{-1}$ for MB). The fluorescence spectrum then exhibits a strong and broad peak at a smaller energy (Stokes shift), and $(ii)$ excitation at longer wavelength in the absorption tail, $\omega_L < \omega_0$ (the case of Rh6G at 633 nm excitation), where the fluorescence spectrum is much weaker and exhibits a monotonous decay with wavelength. Only residual fluorescence is then observed and, since there is no internal relaxation in $S_1$ even in the free-space case, the SDMEF and FDMEF regimes are equivalent ($\Delta\omega_L = 0$).

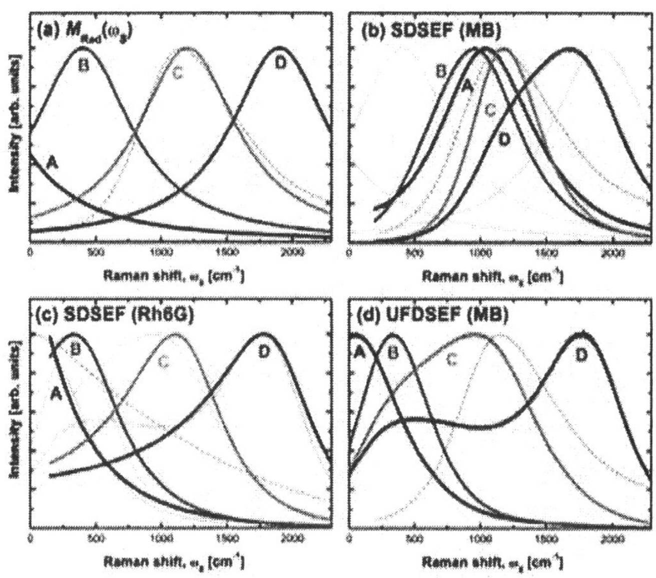

**Figure 2.2:** Predictions of spectral profile modifications in MEF for a few model cases. Spectra are shown in terms of Raman shift (equivalent to $\omega_S$), assuming a laser excitation at 633 nm. All spectra are normalized to 1. (a) Four model cases of radiation enhancement spectral profile, $M_{\text{Rad}}(\omega_S)$, with a resonance peaking at increasing wavelengths; A: $-500$ cm$^{-1}$ (614 nm), B: 400 cm$^{-1}$ (650 nm), C: 1200 cm$^{-1}$ (685 nm), and D: 1900 cm$^{-1}$ (720 nm). These are modelled here by Lorenztian profiles with a FWHM of 900 cm$^{-1}$. The free-space fluorescence spectrum of MB in water ($n^0_{\text{Rad}}(\omega_S)$) is also shown as a dashed line here and in (b) and (d). (b) Predicted SDMEF spectrum of MB, $M_{\text{Rad}}(\omega_S)n^0_{\text{Rad}}(\omega_S)$ for the four cases of (a). The Lorentzian profiles of (a) are repeated as dotted lines for easier comparison. (c) Predicted SDMEF spectrum of Rh6G. Note that this is also the predicted FDMEF spectrum of Rh6G since excitation at 633 nm is below $\omega_0$ (i.e. $\Delta\omega_L = 0$). The free-space spectrum of Rh6G in water [21] is shown as a dashed line (the Raman peaks, observable here because of the weakness of the fluorescence spectrum have been removed). (d) Predicted UFDMEF spectrum of MB, $M_{\text{Rad}}(\omega_S)n^0_{\text{Rad}}(\omega_S - \Delta\omega_L)$ with $\Delta\omega_L = 900$ cm$^{-1}$. These spectra are repeated as dotted lines in (c) for easier comparison.

The predicted modifications to the MEF spectrum are illustrated in Fig.2. 2 for a few representative radiation enhancement factor profiles $M_{\text{Rad}}(\omega_S)$. These results provide a qualitative overview of what to theoretically expect in terms of Spectral Profile Modification (SPM):

- The case of MB in the SDMEF regime (Fig. 2.2(b)) is typical of many MEF experiments. Moreover, to maximize the MEF intensity, the LSP resonance is typically chosen in between the laser excitation and the peak free-space fluorescence emission, i.e. cases B and C in Fig. 2.2(b). Only small SPMs are predicted in these conditions: a small shift for B and a small narrowing for C. These situations are in fact representative of typical MEF conditions and small SPMs should therefore (in general) be expected. Such small modifications have indeed been reported [11], but it is likely that they remain unnoticed or have not been emphasized as such in many cases.

- A larger modification is predicted when the resonance occurs beyond the free-space fluorescence peak (case D of Fig. 2.2(b)). The MEF signal is however much smaller because of the small local field enhancement factor at $\omega_L$.

- Much more spectacular SPMs are predicted in the case of Rh6G (Fig. 2.2(c)). This can be understood simply: because only the fluorescence tail is excited, the emission is almost flat (featureless) in free-space and the MEF profile is therefore imposed by the underlying LSP resonance $M_{\mathrm{Rad}}(\omega_S)$ in MEF conditions. Although less common for fluorescence experiments –which will always use excitation close to the absorption maximum of the molecule under use– this situation can be useful, for example, to study the mechanisms of SPMs in MEF and can even be used as a tool to probe the LSP resonances in the substrate.

❖ Another situation where large SPMs are predicted is the case of UFDMEF for MB as shown in Fig. 2.2(d). The predictions are then similar to those of Rh6G (compare solid and dotted lines in 2.2(c)) and this can again be understood simply: in UFDMEF the radiation enhancement factor $M_{\mathrm{Rad}}(\omega_S)$ applies to the blue-shifted free-space fluorescence $n^0_{\mathrm{Rad}}(\omega_S - \Delta\omega_L)$, which typically peaks around $\omega_L$. It therefore applies to the long-wavelength tail of the fluorescence spectrum, which resembles closely that of Rh6G. In fact, this can be generalized to most fluorophores in the UFDMEF regimes: the UFDMEF spectrum is mostly dominated by $M_{\mathrm{Rad}}(\omega_S)$, i.e. by the LSP resonance spectral profile.

## 2.3 6. *Effect Of Averaging*

In many typical experimental conditions, the fluorescence signal originates from an ensemble of molecules randomly adsorbed on (or close to) the metal substrate. Each of these molecules is therefore subject to different EM enhancement factors, $M_{\mathrm{Loc}}$, $M_{\mathrm{Rad}}$, $M_{\mathrm{Tot}}$, etc...The measured signal should simply be the average signal produced by these individual molecules. Nevertheless, these factors are typically highly non-uniform (spatially) and may exhibit large variations in magnitude, even over relatively small distances [26]. For such distributions, the effect of averaging on complex expressions like Eq. 21 or 24 can be fairly complicated and is beyond the scope of this chapter. This makes predictions of the average absolute fluorescence enhancement a difficult undertaking in general. Fortunately, as far as the spectral profile is concerned, all molecules in the ensemble are subjected to EM enhancement factors originating from the same underlying LSP resonance of the substrate. The modified spectral profile of fluorescence is therefore retained upon spatial averaging, only its absolute magnitude may be affected in a non-trivial way. Nevertheless, there remains the possibility that depending on the spatial distribution of $\overline{M}_{\mathrm{Tot}}$ (and therefore $\Gamma_{\mathrm{Tot}}$), part of the molecules may be in a different MEF regime than others.

## 2.3 7. *Fluorescence Intensities In SDMEF And FDMEF Regimes*

### 2.3 7.a *Qualitative Discussion*

Before moving on to the experimental results, it is worth discussing briefly the conditions (if any) for which FDMEF (or UFDMEF) may be observed. By definition, this requires the decay rate $\Gamma_{\text{Tot}}$ to be sufficiently large (with respect to internal relaxation rates) or equivalently $\bar{M}_{\text{Tot}}$ to be sufficiently large (say, larger than $\sim 10^4$ for the sake of argument). From Eq. 13 –or its equivalent in FDMEF conditions– this may result in a very poor fluorescence enhancement factor. One may then wonder whether the FDMEF spectrum would be at all observable.

Let us consider two model cases to discuss this issue:

❖ *Radiative-Emission-Dominated Substrate*

Let us take for example the case of a metallic nanoparticle coated with a $10$ nm dielectric spacer. The non-radiative emission is then typically negligible and $\bar{M}_{\text{Tot}}$ is of the order of $\bar{M}_{\text{Rad}}$. $\bar{M}_{\text{Rad}}$ may exhibit a strong spatial distribution over the NP surface, but remains below, say $\sim 10^2 - 10^3$. All the molecules are then in the SDMEF regime. Moreover, since $\bar{M}_{\text{Tot}} \approx \bar{M}_{\text{Rad}}$, the fluorescence enhancement factor is of the order of $M_{\text{Fluo}} \approx M_{\text{Loc}}(\omega_L)$. Assuming $M_{\text{Loc}}(\omega_L) \approx M_{\text{Rad}}(\omega_L)$, its maximum value is of the order of $\sim 10^2 - 10^3$, and this results in a fluorescence enhancement factor of the order of $\sim 10 - 100$ after spatial averaging. This case corresponds to a "typical" situation for many MEF studies.

❖ *Non-Radiative-Emission-Dominated Substrate*

Let us now consider the same metallic NP without a dielectric spacer, i.e. the molecules adsorb directly onto the metal substrate. The non-radiative emission is then typically very large, say $M_{\text{NR}} \sim 10^5$ and dominates $\bar{M}_{\text{Tot}}$. Moreover, since it mostly depends on distance from the metal surface, it is of the same order for all molecules. All the molecules are then in the FDMEF (possibly UFDMEF) regime. $\bar{M}_{\text{Rad}}$ does not vary as much with distance from the surface and may as before have a maximum value of $\sim 10^3$. From Eq. 13, the fluorescence enhancement factor is reduced by a factor $\sim 100$ compared to the previous case. Overall, there is therefore no enhancement, but also *no large quenching*, as is often wrongly assumed for molecules directly adsorbed on the surface. The quenching effect of a large $\bar{M}_{\text{Tot}}$ is simply compensated by the local field enhancement factor $M_{\text{Loc}}$ in Eq. 13. The resulting signal is still well within observation capabilities. We conclude that although the FDMEF signal intensity is typically much less enhanced (and possibly even slightly quenched) than its SDMEF counterpart, it should remain easily observable with standard techniques.

Finally, intermediate cases may obviously exist, for example at intermediate dielectric spacers or for substrates with localized regions of large radiative enhancement factors (EM hot-spots).

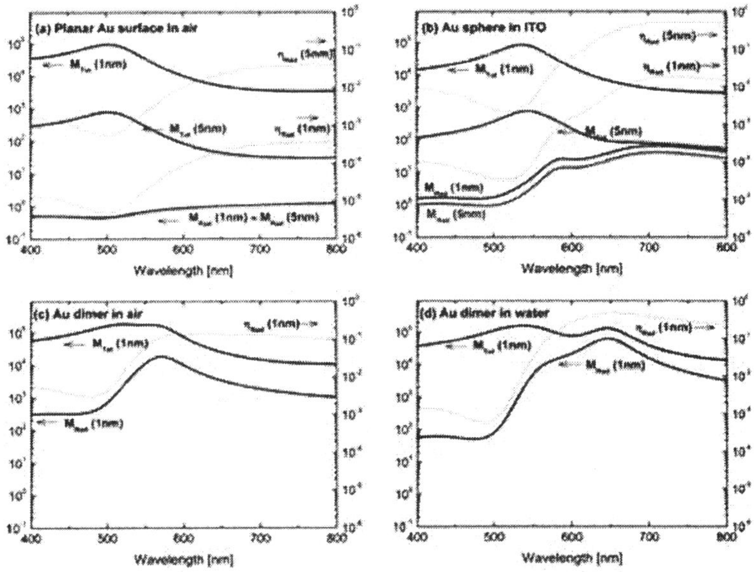

**Figure 2.3:** Adapted and reproduced with permission from *J. Phys. Chem. C* 2007, *111*, 16076–16079. Copyright 2007 American Chemical Society. Predictions of the frequency dependence of $M_{\text{Tot}}$ (blue), $M_{\text{Rad}}$ (red) (solid lines, both scale on the left axis), and the EM radiative efficiency $\eta_{\text{Rad}}^{\text{EM}} = M_{\text{Rad}} / M_{\text{Tot}}$ (dotted lines, scale on the right axis) for a dipole perpendicular to a gold metal surface and at a distance $d = 1$ nm (dark colours) or $d = 5$ nm (light colours) from the surface. Three examples are given: (a) a semi-infinite gold plane in air; (b) a gold nano-sphere of radius $a = 40$ nm in ITO (with $\varepsilon_M = 4.0$); (c-d) a dimer of two gold nano-spheres of radius $a = 30$ nm, and separated by a gap of 2 nm in either air (c) or water (d). The dipole is placed in the middle of the gap in these latter cases (i.e. $d = 1$ nm). Vertical scales have been kept the same for all three plots to facilitate the comparison of relative orders of magnitude. It is clear in these plots that $M_{\text{Tot}}$ can be very large at short distances ($d = 1$ nm), but this does not necessarily imply a small EM radiative efficiency (and therefore a fluorescence quenching), as is the case for the planar surface.

## 2.3. 7.b *Examples Of EM Predictions*

To emphasize these qualitative arguments, we show in Fig. 2.3 some EM predictions for a few examples of representative geometries. $M_{\text{Tot}}$ and

$M_{\mathrm{Rad}}$ can be calculated within the classical EM theory of dipolar emission [19] and we focus here on three cases for which exact results are obtainable. These are for fluorophores close to:

- a planar metal surface, [4,12]
- a metallic sphere [27]; using Mie theory,
- or a pair of metallic spheres (dimer); using Generalized Mie Theory [28,29].

Figure 2.3 shows the results of such predictions for a few representative cases (see the caption for details). The metal is gold and its dielectric function is taken from Ref. [30]. Only the case of a dipole perpendicular to the surface, at a distance $d$ , is shown for clarity. For an emitter very close to the metal surface (typically less than 5 nm), we conclude from these model examples the following:

- For the plane and the sphere, $M_{\mathrm{Tot}} \gg M_{\mathrm{Rad}}$, i.e. the lifetime is dominated by non-radiative emission, and the EM radiative efficiency (approximate modified quantum yield) is small. We also note that $M_{\mathrm{Tot}}$ is then mostly determined by the distance between the metal surface and the emitter, independently of the actual geometry –and therefore of the Localized Surface Plasmon (LSP) resonances– of the metallic object. Its frequency dependence is therefore essentially governed by the intrinsic optical properties of the metal, and its magnitude depends on the distance as $d^{-3}$ [13].

- $M_{\mathrm{Tot}}$ can be quite large, $\approx 10^5$ for $d = 1$ nm at the non-radiative resonance around 500 nm for gold in air (corresponding to $\varepsilon = -1$). Around 650 nm (the region of interest for the experimental results shown later in Sec. 2.4), $M_{\mathrm{Tot}}$ is still of the order of $\sim 10^3 - 10^4$. $M_{\mathrm{Tot}}$ would be even larger for a dipole closer to the surface, although a non-local dielectric function should then be used for quantitative predictions.

- For the gold dimer, $M_{\mathrm{Rad}}$ becomes comparable to the geometry-independent non-radiative contribution to $M_{\mathrm{Tot}}$. $M_{\mathrm{Rad}}$ then contributes significantly to $M_{\mathrm{Tot}}$. In particular, resonances of $M_{\mathrm{Rad}}$ are then observed in $M_{\mathrm{Tot}}$, as seen in Fig. 2.3(c-d). $M_{\mathrm{Tot}}$ then has a relatively large radiative component, and the EM radiative efficiency or modified quantum yield can be reasonably large, $Q > 0.1$.

These predictions confirm that, in typical situations: $M_{\mathrm{Tot}}$ can easily be large enough for FDMEF to exist, and even possibly UFDMEF. This is particularly true for the first few monolayers, i.e. at distances of the order of $d \approx 1$ nm or less from the surface.

## 2.3 7.c *Predicted intensity of the FDMEF signal*

Still, we must now consider whether in such situations any detectable fluorescence is expected. It is indeed often assumed that at such short distances the fluorescence signal would be *completely quenched* because of the dominance of non-radiative decay. This dominance is indeed a fact in many situations, and results in very poor modified quantum yields, $Q$, as evidenced in Fig. 2.3(a-b), but it does not necessarily imply a complete quenching of the MEF signal.

To understand this, we focus on the expression for the fluorescence EF given in Eq. 13 and recall that $M_{\text{Loc}}$ is of the same order as $M_{\text{Rad}}$ in a first approximation. We consider an emitter very close to the surface, say $d = 1$ nm for the sake of argument. Let us first consider the case of a planar surface (see Fig. 3(a)). $M_{\text{Rad}}$ and $M_{\text{Loc}}$ are of the order of 1, while $Q < 10^{-3}$, so that $M_{\text{Fluo}} \approx 10^{-3} - 10^{-4}$. It is clear in this case that a large fluorescence quenching occurs, and this is partly the origin of the common view that fluorescence near metal surfaces is always quenched. Even if FDMEF exists in such a case, its intensity would not be detectable.

The situation is similar in the case of a gold sphere, because the radiative (and local field) enhancements remain small compared to $M_{\text{Tot}}$ for $d = 1$ nm. It is interesting to note at this stage that this is not true for a silver sphere, for which the LSP resonances are much stronger (and the local field and radiative enhancements accordingly much larger); see Ref. [19] for more details. This highlights one important aspect of the optical properties of gold. Because of the large absorption for $\lambda < 600$ nm, the LSP resonances are strongly damped. At longer wavelengths however, the optical absorption of gold becomes comparable with that of silver, and much stronger LSP resonances (and therefore EFs) are then possible. To profit from these, one must have a gold structure where the LSP resonances are red-shifted to this region. This can for example be achieved by using non-spherical structures (like the ones used later), by placing the structures on a dielectric substrate (such as ITO), by making interacting structures, or by embedding the structures in a dielectric of larger dielectric constant (like water or ITO).

We have presented results in Fig. 2.3(c-d) where the resonances have been red-shifted by using interacting objects (a dimer) either in air or water. As illustrated in Fig. 2.3(c-d), it is then possible to have $M_{\text{Rad}}$ almost comparable to $M_{\text{Tot}}$, and therefore a modified quantum yield of the order of $Q \sim 0.1 - 1$, even for an emitter as close as $1$ nm to the surface. No quenching of the fluorescence signal is therefore predicted, and even enhancements are possible (depending on how close the dipole is to the surface) thanks to the local field enhancement effect. It should therefore be possible to observe the fluorescence signal, even in FDMEF or UFDMEF conditions.

Finally, the SERS Enhancement Factor is predicted [1] to be of the order of $M_{\text{Tot}}$ larger than $M_{\text{Fluo}}$; say larger by a factor of $\sim 10^5 - 10^6$. But for typical fluorophores, the non-modified fluorescence cross-section is $\sim 10^{10}$ larger than the non-modified Raman cross-section [31]. The (integrated) MEF signal should therefore be $\sim 10^4$ times larger than the SERS signal, but it is also spread out over a much larger ($\sim 10^2 - 10^3$) spectral range than a typical SERS peak. These considerations, although mostly qualitative, show that for molecules adsorbed on the metal, the MEF signal (SDMEF or FDMEF) is expected to be at least of the order of the SERS peaks, the relative intensity depending on the exact distance of the emitter from the surface.

In conclusion, *the fluorescence signal is not necessarily completely quenched for molecules directly adsorbed on the metal surface*, but it is rather much less enhanced than the Raman signal. As a consequence, if SERS peaks can be observed for a fluorophore, they should in most cases be accompanied by a MEF signal.

## 2.4 EXPERIMENTAL STUDIES

The most direct approach to study the SPMs in MEF conditions is to measure the spectral profile of fluorescence under MEF conditions $I(\omega_S)$ and show that it is different to the free-space profile $n^0_{\text{Rad}}(\omega_S)$. Moreover, to confirm that the observed signal is indeed modified MEF, it is desirable to show in addition that (i) it follows the predicted profile: for example, $M_{\text{Rad}}(\omega_S)n^0_{\text{Rad}}(\omega_S)$ for SDMEF, and (ii) its intensity is compatible with the expected fluorescence enhancement.

We will present results following this approach in this section. These will be largely based on the work presented in Ref. [1]. There have not been, to our knowledge, any other dedicated experimental studies of this aspect of MEF to date in the literature. Small spectral profile modifications were however occasionally reported as part of other studies [11, 14], but have not been analyzed in depth.

### 2.4 1. *Experimental Considerations*

In order to carry out such a study, one must first assess carefully how the various quantities $I(\omega_S)$, $n^0_{\text{Rad}}(\omega_S)$, and $M_{\text{Rad}}(\omega_S)$ can be inferred experimentally. Let us look at each of these three aspects separately.

#### 2.4 1.a *Free-Space Fluorescence Spectrum*

Any experimental verifications should rely on studying the modification of the fluorescence profile with respect to the free-space case. It is therefore necessary to measure the free-space fluorescence (at the same excitation wavelength) first. Ideally, this should be carried out for the fluorophore in conditions as close as possible to the one used for MEF, i.e. in the same dielectric environment (air or water), adsorbed vs. dissolved in a solvent, etc... In any case, there is always a possibility that the intrinsic fluorescence properties of the probe are affected by adsorption onto the metallic substrate (or any dielectric spacer that may be used). For example, its electronic levels could be affected by covalent bonding to a surface or fluorophore / fluorophore interactions at large coverage. The importance of these effects should be assessed on a case-by-case basis.

In order to demonstrate large SPMs, an approximate free-space fluorescence profile is usually sufficient since the profile of the radiation enhancement factor $M_{\text{Rad}}(\omega_S)$ usually dominates. For more subtle effects, like differentiating between SDMEF and FDMEF, a more accurate free-space fluorescence profile may be necessary.

### 2.4 1.b *Radiation enhancement and extinction profile*

Arguably the most difficult quantity to measure experimentally is the radiation enhancement factor profile $M_{\text{Rad}}(\omega_S)$. It could in principle be predicted from EM theory, provided the substrate geometry is well-defined, but an experimental estimate is obviously preferred.

The simplest approach to this problem is to approximate it by the (far-field) extinction profile $Q_e(\omega_S)$, much easier to measure, i.e. $M_{\text{Rad}}(\omega_S) \propto Q_e(\omega_S)$. This is justified qualitatively since both quantities are affected by coupling to the substrate localized surface plasmon resonances and should therefore exhibit the same resonances. Moreover, this approximation can be shown theoretically to be valid for simple structures like spherical nano-particles [32]. This approximation has also been used many times in the context of SERS to study the validity of the SERS EM model of the enhancement factors [21, 33, 34, 35, 36].

For a nano-particle-based substrate, additional complications may arise because of polydispersity, i.e variations in size and / or shape of the particles. This introduces changes in the LSP resonance wavelength from one particle to another and can broaden dramatically the extinction profile $Q_e(\omega_S)$. To avoid such issues, highly uniform NPs must be used. In the examples shown here, the substrates consist of arrays of gold NPs fabricated by e-beam nano-lithography [37, 38]. The high uniformity of these arrays is readily seen in the SEM images in the inset of Figs. 2.5 and 2.8 and is further confirmed by measuring the extinction $Q_e(\omega_S)$ at several positions on the array with a high magnification objective. For our purpose here, the size and shape of the particles is varied from one array to another in order to change the position of the LSP resonances, i.e. $Q_e(\omega_S)$ and therefore $M_{\text{Rad}}(\omega_S)$.

### 2.4 1.c *MEF Spectrum And Background Subtraction*

In the experiments presented here, the fluorophores are adsorbed directly onto the metallic (gold) NP surface. The fluorescence enhancement (if any) is therefore small and the MEF signal may accordingly be weak, especially for weak fluorophores like Rh6G at 633 nm. In fact, the MEF spectrum is accompanied by Raman peaks of comparable intensity (themselves enhanced through Surface Enhanced Raman Scattering, SERS).

When considering the spectral profile of the MEF spectrum, it is extremely important to subtract any background signal not related to the NPs themselves, or at least make sure that it is negligible. The apparent profile can indeed be completely changed by the presence of a background (with its own spectral profile) of comparable intensity, as illustrated in Fig. 2.4.

Such a background may come from the environment itself (glass slide or solvent) or from fluorophores that are not adsorbed on the NPs (and therefore not under MEF conditions). In our case here, the background signal is entirely dominated by the ITO substrate supporting the gold NPs. No change in this background is detected in the presence of the fluorophores, meaning that they either do not adsorb on ITO, or that their non-enhanced fluorescence is negligible.

One can easily measure the ITO background signal $I_{\mathrm{ITO}}(\omega_S)$ by taking a spectrum under identical conditions on ITO only, for example in between two NP arrays. Subtracting directly this background from the MEF signal $I(\omega_S)$ however leads to unphysical results (negative signals in some instances). This is because, in the MEF spectrum, the ITO signal is viewed *through* the NP array and is accordingly modified because of the intrinsic extinction of the NP arrays. This can however be corrected as follows (see also Fig. 2.4).

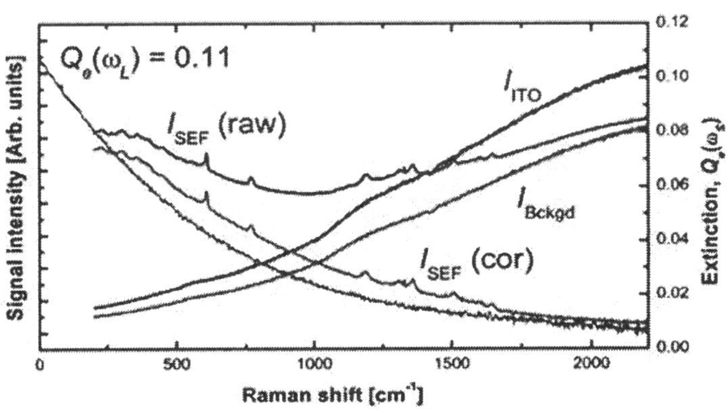


**Figure 2.4:** Illustration of the background subtraction for NP arrays on ITO. $I_{\mathrm{SEF}}$ (raw) and $I_{\mathrm{ITO}}$ are the raw spectra for MEF (with Rh6G adsorbed on the arrays) and ITO. Simple subtraction would result in a negative signal.

One must correct the ITO signal according to Eq. 32 to obtain the real ITO contribution, $I_{\text{Bckgd}}$, which once subtracted gives the corrected MEF spectrum, $I_{\text{SEF}}$, (cor). This step is clearly necessary here to reveal the true spectral profile of the MEF spectrum. The extinction spectrum $Q_e(\omega_S)$ for this array is also shown in the figure.

Simple subtraction would result in a negative signal. One must correct the ITO signal according to Eq. 32 to obtain the real ITO contribution, , which once subtracted gives the corrected MEF spectrum, $I_{\text{SEF}}$ (cor). This step is clearly necessary here to reveal the true spectral profile of the MEF spectrum. The extinction spectrum $Q_e(\omega_S)$ for this array is also shown in the figure.

The extinction spectrum $Q_e(\omega_S)$ is first measured experimentally and by definition characterizes the transmission of the array at frequency $\omega_S$:

$$Q_e = -\log_{10}\frac{I_{\text{Transmitted}}}{I_{\text{Incident}}}. \qquad (31)$$

For the MEF experiment, the incident laser at $\omega_L$ must go through the array to excite the ITO substrate. Only a fraction $10^{-Q_e(\omega_L)}$ of the power is transmitted. The ITO substrate then produces a background signal $10^{-Q_e(\omega_L)}\times I_{\text{ITO}}(\omega_S)$ (where $I_{\text{ITO}}(\omega_S)$ is the reference spectrum taken without the array), which must again go through the array before we detect it. The real ITO contribution to the MEF spectrum is therefore

$$I_{\text{Bckgd}}(\omega_S) = 10^{-Q_e(\omega_L)}\times 10^{-Q_e(\omega_S)}\times I_{\text{ITO}}(\omega_S). \qquad (32)$$

This background should be removed before any interpretation of the MEF spectrum. The effect could be dramatic in some situations as illustrated in Fig. 2.4. This is not as critical when the MEF intensity is much larger than the ITO background (which is generally the case for Methylene Blue (MB) or Crystal Violet (CV) for example).

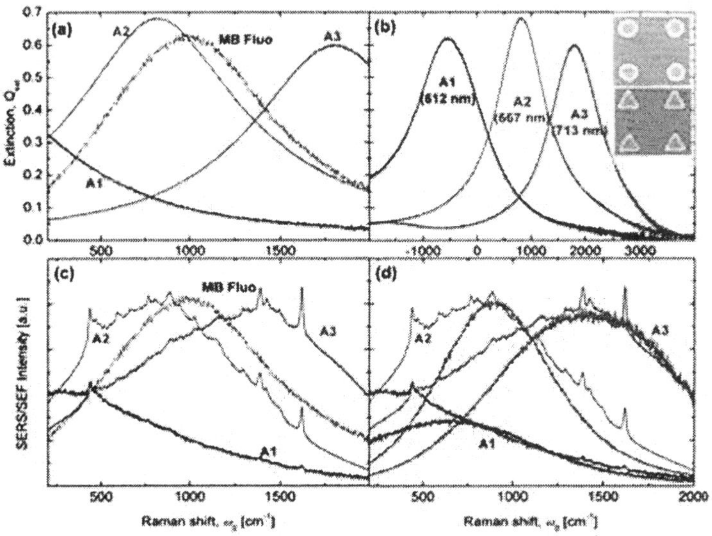


**Figure 2.5:** Reproduced with permission from *J. Phys. Chem. C* 2007, *111*, 16076–16079. Copyright 2007 American Chemical Society. (a-b) Extinction spectra ($Q_{\text{ext}}(\omega_S)$) for three gold nanoparticle arrays: A1 (100 nm edge triangles), A2 (100 nm diameter dots), and A3 (150 nm edge triangles). Resonance wavelengths for each are indicated in brackets in (b); representative SEM images are shown in the inset. The data in (a) are the same as in (b), but on the same energy scale as used in (c) and (d) for easier comparison. (c) SERS / SMF spectra at 633 nm excitation of the same arrays, covered with methylene blue (MB). The standard MB fluorescence (adsorbed on ITO), $P^0_{\text{Fluo}}(\omega_S)$, is shown in (a) and (c). (d) Comparison of the SMF spectra of (c) with the approximated predicted profile: $Q_{\text{ext}}(\omega_S)P^0_{\text{Fluo}}(\omega_S)$.

## 2.4 2. *Evidence Of Spectral Profile Modifications*

One possible approach to evidence strong spectral profile modifications under MEF conditions is to take a given fluorophore (fixed $n^0_{\text{Rad}}(\omega_S)$), and vary the radiation enhancement factor profile $M_{\text{Rad}}(\omega_S)$. The MEF spectral profile should then change in the same direction as $M_{\text{Rad}}(\omega_S)$. The change in $M_{\text{Rad}}(\omega_S)$ can, for example, be obtained by changing the size and / or shape of the gold NPs and is easily monitored by measuring their extinction spectrum $Q_e(\omega_S)$.

An example of the results for such an approach is shown in Fig. 2.5 (from Ref. [1]), which we now discuss. Methylene Blue (MB) is used here as a probe molecule and is transferred onto gold NP arrays by dipping them into a 10 $\mu$M MB solution for 5 minutes. The arrays are plasma-cleaned before any dipping, to ensure that no contaminants are previously adsorbed on the NPs. The MB molecules are therefore adsorbed directly onto the gold surfaces, with possibly some molecules further away from the surface if several monolayers are present. Three NP arrays are used with distinct LSP resonances at 612 nm (array A1), 667 nm (A2) and 713 nm (A3), as evidenced in the extinction spectra of Fig. 2.5(b). The resonance of A2 is close to the peak absorption and fluorescence of MB and would be considered as the "standard" situation for most MEF experiments (except for the direct adsorption onto the metal). A1 and A3 have resonances much further away on either side of the MB fluorescence spectrum. The MEF spectra (corrected for the ITO background), shown in Fig. 2.5(c), exhibit a broad spectrum underneath the SERS (Raman) peaks. The SERS peaks clearly confirm the presence of MB on the NPs surface. The accompanying broad signal is attributed to the modified MB fluorescence (MEF), initially for two reasons:

❖ The signal is not observed in the absence of adsorbed molecules (MB or otherwise). Note that a thorough cleaning (plasma cleaning in this case) is necessary to ensure that no spurious fluorescent species remain on the sample.

❖ The broad signal (and the SERS peaks) decays as a result of photobleaching, and the signal decay rates increase with excitation density.

This interpretation is further confirmed by comparison with the predicted modified MEF spectral profile. For this, we make the assumption discussed earlier that $M_{\text{Rad}}(\omega_S) \propto Q_e(\omega_S)$, and $n^0_{\text{Rad}}(\omega_S)$ can be approximated by the free-space fluorescence profile $P^0_{\text{fluo}}(\omega_S)$ of MB dried on ITO (at high concentration). Hence, the MEF profile (for SDMEF) should follow the product $Q_e(\omega_S)P^0_{\text{fluo}}(\omega_S)$, which is indeed the case to a good approximation as shown in Fig. 2.5(d). The results of Fig. 2.5 demonstrate clearly the possibility of a large spectral profile modification under MEF conditions.

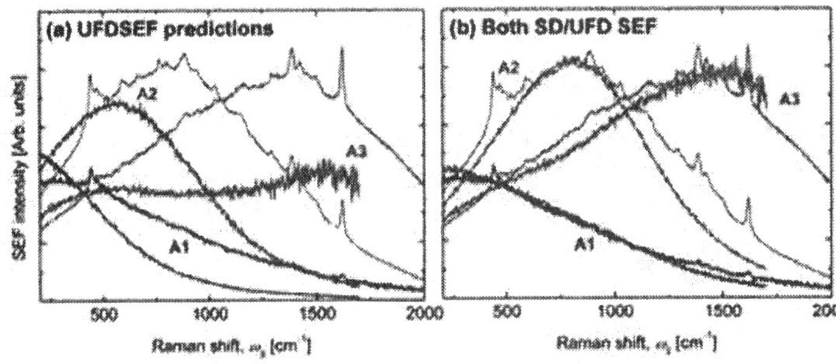


**Figure 2.6:** (a) Comparison of the experimental results of Fig. 5 for MB on arrays A1, A2, and A3 with theoretical predictions in the UFDMEF regime with $\Delta\omega_L = 700$ cm$^{-1}$, i.e. $Q_e(\omega_S)n^0_{\text{Rad}}(\omega_S - \Delta\omega_L)$. (b) Same as (a) using a weighted superposition of the theoretical predictions in the SDMEF regime (shown in Fig. 2.5(d)) and in the UFDMEF regime (shown in (a)).

## 2.4 3. *Evidence For Fast-Dynamics MEF*

The comparison between experiments and SDMEF predictions in Fig. 2.5(d) also validates the simple model presented earlier. One may notice however, that the SDMEF predictions tend to underestimate the short-wavelength side of the modified MEF spectra. This could be a consequence of some molecules being in the FDMEF (or UFDMEF) regimes. It would qualitatively enhance more the short-wavelength side of the spectrum at the expense of the other side as predicted theoretically in Fig. 2.2(d).

To be more quantitative, it is possible to compare the experimental MEF results to the UFDMEF predictions, i.e. to the product $Q_e(\omega_S)P^0_{\text{fluo}}(\omega_S - \Delta\omega_L)$. This is illustrated in Fig. 2.6(a). It is clear that the UFDMEF spectra can explain the discrepancies observed in the SDMEF spectra, but then the long-wavelength side of the spectrum is no longer correctly predicted. Only a weighted superposition of both the SDMEF and UFDMEF predictions results in a better account of the overall spectral profile under MEF conditions, as shown in Fig. 2.6(b). This result can be interpreted in two ways: $(i)$ The majority of molecules are in the intermediate FDMEF regime, where internal relaxation rates are comparable to decay rates. As discussed in Sec. 2.3, an intermediate spectrum between the SDMEF and UFDMEF cases is then expected. Unfortunately, the details of such a spectrum are difficult to predict theoretically. $(ii)$ The total decay rate enhancement factors are non-uniform on the surface, resulting in some molecules being in SDMEF conditions, while other are in FDMEF or even UFDMEF conditions. This could for example be the case if some molecules are further away from the surface (i.e. for multilayer coverage). It is difficult from the sole evaluation of the optical spectra to distinguish between these two scenarios.

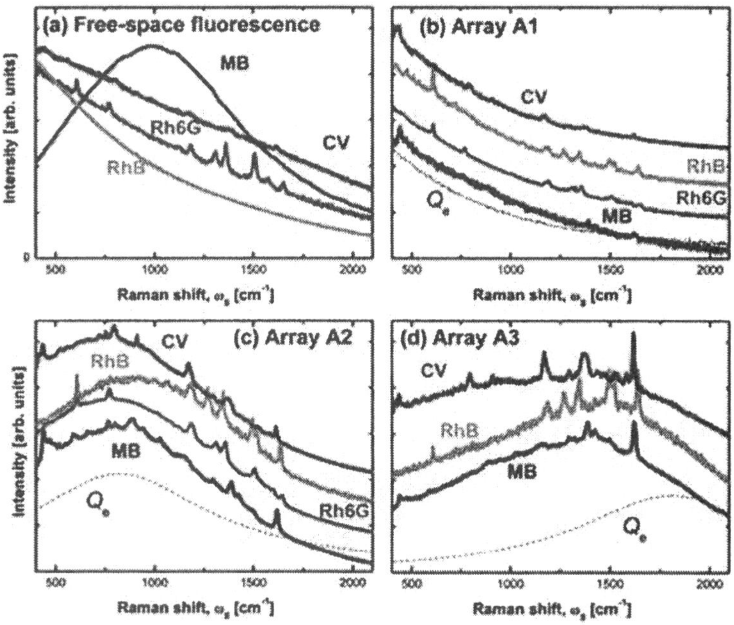


**Figure 2.7:** (a) Free-space fluorescence spectra ($P^0_{\text{Fluo}}(\omega_S)$) for four fluorophores: Methylene Blue (MB), Rhodamine 6G (Rh6G), Rhodamine B (RhB), and Crystal Violet (CV). (b-d) MEF spectra obtained from arrays A1 (b), A2 (c), and A3 (d) after dipping in a 10 $\mu$M solution for each of the four fluorophores in (a). All spectra are obtained for excitation at 633 nm and are shown in terms of the Raman shift (equivalent to $\omega_S$). All spectra intensities have been rescaled and are displaced vertically in (b-d) for convenience. The extinction profile $Q_{\text{ext}}(\omega_S)$ for each array is also shown in (b-d) as a dashed line. Note that the spectrum for Rh6G on array A3 was not measured.

## 2.4 4. *Comparison Between Different Fluorophores*

Another complementary approach to that presented so far is to study different fluorophores (with different free-space fluorescence spectral profiles) on the same NP arrays. This is illustrated in Fig. 2.7. Four fluorophores are used there: Methylene Blue (MB) as before, Rhodamine 6G (Rh6G), Rhodamine B (RhB), and Crystal Violet (CV). All experiments are carried out at 633 nm excitation again, and the corresponding free-space fluorescence spectra are shown in Fig. 2.7(a). Rh6G, RhB, and CV all exhibit a tail-like fluorescence spectrum, very similar to each other. This simply reflects the fact that the excitation (at 633 nm) is close or below (in energy) the main absorption band. In this case, internal relaxation in $S_1$ is expected to play a negligible role, even in free-space conditions. The SDMEF and FDMEF regimes should therefore be equivalent for these three molecules. It should also be similar to the UFDMEF spectrum of any molecules as discussed in Sec. 2.3 5. In fact, similar MEF spectral profiles are obtained for the four molecules on each of the three NP arrays as shown in Fig. 2.7(b-d). The fact that the results for MB are similar to the other three molecules –despite their dissimilar free-space fluorescence spectra– is further indication that the MB molecules are in the FDMEF regimes (possibly UFDMEF for some).

The spectra in Fig. 2.7 also further confirm the possibility of large spectral profile modifications in MEF conditions.

## 2.4 5. *Discussion Of The Experimental Results*

The SPMs observed in Figs 2.5, 2.6, 2.7 are sufficiently large and unambiguous to render a clear-cut identification of spectral modifications of the emission. The agreement between experimental and predicted spectra is also sufficiently satisfactory to support the interpretations and validate the models (despite their simplicity). Nonetheless, a more accurate agreement would be desirable for subtler effects, such as a clear distinction between SDMEF and (U)FDMEF. This is, however, prevented by a number of issues; which we highlight here as a prerequisite towards improved future experiments.

❖ As already mentioned, the "free-space" fluorescence profile may be modified upon molecular adsorption, even before any modifications associated with the EM response of the substrate. This may arise from small alterations in their electronic structure or from molecule / molecule interactions (at high coverage). Unfortunately, there is no simple experimental solution to this issue. One approach is to use several different fluorophores; any strong modifications for one of them will result in anomalous behaviour with respect to the others.

❖ The approximation of the radiation enhancement factor profile $M_{\text{Rad}}(\omega_S)$ by the extinction spectrum $Q_e(\omega_S)$ may not always be valid. EM calculations on the NPs may be used to confirm it.

❖ In all the results shown here, the extinction spectra were measured on bare substrates (not covered with molecules). However, it has been shown that molecular adsorption may affect the underlying LSP resonances and therefore change slightly the extinction profiles [39]. This aspect can be improved upon by measuring the extinction spectrum of the arrays for each molecule.

❖ The theoretical description of the fluorescence process, even for free-space fluorescence, could be improved. For example, the relevance (if any) of the internal relaxation in $S_1$ for excitation in the fluorescence tail below the adsorption band is not well understood. Moreover, the description of the intermediate FDMEF regime requires a detailed understanding of the internal relaxation mechanisms in $S_1$. Further improvements to account for temperature-related effects in the effective populations of $S_0$ and $S_1$ can also be envisioned.

These secondary aspects do not affect the main conclusions of this section, but would certainly contribute to a deeper understanding of the SPMs mechanisms in MEF.

## 2.5 ADVANCED ASPECTS AND OUTLOOK

### 2.5 1. *The SERS Continuum*

An interesting aspect of the large spectral profile modifications that may arise in MEF conditions is the fact that the resulting spectrum is modified to an extent that it may not be easily recognized as a fluorescence spectrum. This means that it may be observed in many standard experimental situations but incorrectly interpreted. Along these lines, we propose that this modified MEF (either SDMEF with a large SPM or FDMEF) may be the origin of the so-called SERS continuum in many common situations (and in particular for dyes, even weakly fluorescent). Such an interpretation is compatible with many well-known features of the SERS continuum [14, 40, 41, 42, 43, 44] which, following Ref. [1], are listed below along with other important predictions of this proposal:

❖ The SDMEF / FDMEF follows the same intensity fluctuations and polarization properties of the SERS signal through the factor $M_{\text{Rad}}(\omega_S)$, as observed for the SERS continuum.

❖ There is no MEF in the absence of adsorbed molecules, as for the SERS continuum. We note that this is still a contentious point in relation to the SERS continuum itself, since there are conflicting reports of whether the SERS continuum is observed or not in the absence of adsorbed molecules. These have been tentatively attributed to problems of surface cleanliness and impurities [41]. Our experiments support this view: the MEF / SERS background only disappears upon thorough plasma cleaning.

❖ The issue of surface cleanliness and impurities is even more problematic for non-resonant molecules, since the SERS signals are much smaller [31] and therefore potentially more affected by contaminations. The MEF from organic impurities [41] could therefore account for the SERS background of non-resonant molecules.

❖ Contrary to what is generally thought, MEF is possible even for molecules in direct contact with the metal, as is often the case in SERS. This was shown and discussed extensively in Sec. 2.3 7.

❖ The underlying resonance that should characterize the SPM of the MEF (and therefore the SERS background within our proposal) is not seen in most SERS experiments because of its large inhomogeneous broadening for most substrates and the typically small spectral window usually probed with high dispersion gratings. Our ability to identify it as such in Figs. 2.5, 2.6, 2.7 is a direct consequence of the high uniformity of the resonances in our structures and the wide spectral window.

❖ As mentioned, in UFDMEF conditions, the spectral profile is mostly independent of the molecule, and could therefore account for the lower energy peak of the SERS continuum observed in Ref. [14] (previously attributed to electronic Raman scattering or to LSP luminescence).

❖ One interesting consequence of the MEF interpretation concerns the behaviour of the SERS background when changing laser excitation. Depending on the specific SERS conditions, the background may consist of either SDMEF, FDMEF (possibly UFDMEF), or both. When changing excitation wavelength, the SDMEF signals should retain their spectral profile, but the FDMEF should be shifted together with the laser in a similar fashion as a Raman signal, thus potentially explaining conflicting reports regarding the actual nature (Raman vs. luminescence) of the SERS background.

This proposal obviously requires further investigation. It is clear, however, that the possibility of a MEF background (possibly strongly modified spectrally) should be considered in any interpretations of the SERS continuum. It is likely that this could account for many (if not all) observations of the SERS continuum, in particular for SERS of resonant or pre-resonant analytes such as dyes. In fact, in all the spectra presented in this chapter, the SERS signal is clearly observed and the underlying background could have been ascribed from a SERS perspective to a "SERS continuum" of unknown origin.

## 2.5 2. *Polarization Effects*

In the theoretical and experimental discussion presented so far, we have not discussed any polarization effects such as: $(i)$ the dependence of the LSP resonances upon the incident polarization, or $(ii)$ the nature of the polarization of the emitted fluorescence signal. This choice simplified the presentation of the principles of SPM in MEF and did not affect the interpretations of the experimental results, which were obtained from NP arrays with no (or only a weak) polarization dependence.

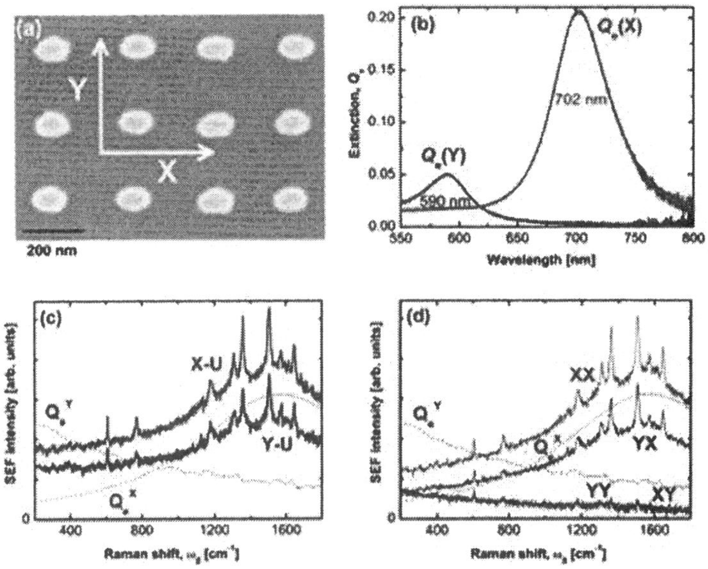


**Figure 2.8:** Adapted and reproduced with permission from *J. Phys. Chem. C* 2008, *112*, 8117-8121. Copyright 2008 American Chemical Society. (a) Representative SEM image of gold prolate NPs. Height is $\approx 50$ nm and dimensions in the plane are 140 nm ($X$) by 90 nm ($Y$) (b) Extinction spectrum, $Q_e^X$ and $Q_e^Y$, for $X$- and $Y$-polarized excitation. (c) MEF spectra at 633 nm excitation after dipping the array in a $10^{-5}$ M Rh6G solution. SERS peaks can be seen on top of the modified MEF spectrum. Detection is unpolarized and excitation is polarized along $X$ ($X-U$) or $Y$ ($Y-U$). Also shown (dashed lines) are the extinction spectra $Q_e^X$ and $Q_e^Y$, for $X$- and $Y$-polarized excitation. The spectral profile of MEF follows $Q_e^X(\omega_S)$ in both cases. (d) MEF spectra for $E$-polarized excitation and $D$-polarized detection (i.e. four configurations, $ED = XX$, $XY$, $YX$, $YY$). It is clear that the MEF spectral profile follows $Q_e^D$, i.e. it is *determined by the detection polarization.* The excitation polarization only affects the overall MEF intensity. As a consequence, the MEF signal for $Y$-polarized excitation is almost entirely $X$-polarized ($YX$ is much larger than $YY$).

It is however possible to design specifically NP arrays with a strong polarization dependence in order to investigate polarization effects. This has recently been reported [21] using arrays of "elongated" gold NPs (with elliptic cross-section), which exhibit two distinct LSP resonances, each associated with a polarization along the two main axis of the ellipse. The spectral response for each resonance, $Q_e^X(\omega)$ and $Q_e^Y(\omega)$, can be measured by polarized extinction measurements. Some of the results of this study are illustrated in Fig. 2.8, which is explained in the caption. We mention here briefly the main conclusions:

- The incident polarization only affects the coupling to the LSP resonances at the laser frequency $\omega_L$, i.e. the local field enhancement factor at $\omega_L$. It has no effect on the spectral profile of fluorescence.

- The MEF spectrum is in general affected by both resonance profiles $Q_e^X(\omega_S)$ and $Q_e^Y(\omega_S)$.

- The respective contributions of these two resonances can be separated experimentally by analyzing the polarization of the MEF signal. The MEF signal for polarized detection along $X$ simply follows a spectral profile $\approx Q_e^X(\omega_S)n_{\text{Rad}}^0(\omega_S)$ (and similarly for $Y$).

- A rather spectacular consequence of this is that the spectral profile of MEF for a given fluorophore adsorbed on a given array may be dramatically different when analyzed with one polarization or the other. Along the same line, the fluorescence depolarization ratio is predominantly dictated by the underlying LSP resonances and is therefore $\omega_S$-dependent and can, under appropriate conditions, be as large as $\approx 10$, i.e. there is an almost complete *polarization rotation* of the MEF spectrum with respect to the incident polarization.

These experimental results confirm and extend the interpretations given earlier. We refer the reader to Ref. [21] for further details and discussion of these results in the context of SERS and in relation to surface selection rules [45, 46].

## 2.5 3. *Further Experiments*

The experimental results presented so far demonstrate clearly the existence of a strong SPM in MEF conditions. However, it remains difficult to distinguish between the Slow-Dynamics and Fast-Dynamics regime of MEF from these results. Additional experiments can be envisaged to provide more direct evidence for FDMEF and study the transition between these two regimes. The easiest approach (possibly) would be to carry out a distance-dependence study of the MEF; for example using dielectric spacers. If non-radiative emission is important for molecules directly adsorbed on the metal, its proportion should decrease dramatically as the molecule moves away from the surface (even by a few nm), i.e. the decay rate $\Gamma_{\text{Tot}}$ becomes slower and SDMEF becomes more likely. In the meantime, the spectral profile of the radiative enhancement factor $M_{\text{Rad}}(\omega_S)$ is not expected to vary much. Any spectral profile changes as a function of distance would therefore be a strong indication of a transition from a FDMEF- (or UFDMEF-) dominated regime to a SDMEF regime.

Another interesting extension of the experiments presented here would be to study the SPM of the MEF in *single molecule* (SM) conditions. The SERS signal of single fluorescing molecules are now routinely observed [47, 48], and one would then only need to study the accompanying fluorescence background. Unfortunately, single molecule SERS has not been yet achieved *in highly controlled conditions* and many problems may therefore arise in the interpretation of the results. It is in particular difficult to $(i)$ measure the underlying LSP resonance associated with a given SM event, and $(ii)$ measure and subtract the background signal not associated with the SM event (extrinsic fluorescence, etc...).

## 2.5 4. *Conclusion And Outlook*

With the benefit of hindsight, it is possible to claim that spectral modifications of the fluorescence profile have been to a large degree overlooked in the literature, although occasionally hinted at [11, 22, 23]. But this oversight can be easily understood in many ways. For a start, most studies in the past have focused on the seemingly more "urgent" matter of total (spectrally integrated) properties of fluorophores on surfaces; and in particular on the very important question of whether the dyes will emit *more* or *less* overall fluorescence through the interaction. The effects described here, however, are a step forward in our basic understanding of fluorescence emission in close proximity of surfaces. By including explicitly in the treatment the *spectral modification* of the emission, a new dimension is added to the problem and new directions are immediately evident. From a wider standpoint, it is to some extent paradoxical that one of the clearest effects revealed by the spectral modification of the fluorescence arises from optical excitation in the low energy *tail* of the absorption of a dye (as in the case of Rh6G at 633 nm). This situation produces a fairly featureless background fluorescence emission that is more susceptible then to reveal the spectral profile of the underlying plasmon resonances. Indeed, it is worth noting that this is a situation that would be deemed as *not very interesting* in the realms of conventional fluorescence spectroscopy (where one would try to excite as much fluorescence as possible by using an excitation at much shorter wavelengths tuned to the dye absorption). This latter point is, in fact, partially (but not completely) responsible for the delay hitherto in the published literature to single out the phenomenon of *spectral modification of the fluorescence emission* as an "effect in its own right".

Undoubtedly –together with an improved understanding of the fluorescence process itself– many possible new paths are now open for further research and the results highlighted in this chapter have much wider implications. The striking connection with the "elusive" nature of SERS backgrounds mentioned in Sec. 2.5, for example, is certainly one such promising area of likely future developments. As it stands at the moment, we would argue the spectral profile modification of the MEF signal provides the most natural explanation to all the known phenomenology of SERS backgrounds. If this latter connection is confirmed by future research, a new layer of understanding will emerge in the related technique of SERS. This would also tighten the link between the two techniques, SERS and MEF, which despite their common "plamonics" basis, have been largely studied so far as two independent entities.

## 2.6 NOTE

A number of recent papers dealing with similar concepts as those discussed in this chapter have appeared in the literature [49,50,51] after the writing of this manuscript.

Some of the basic electromagnetic models mentioned here are also extensively discussed in Ref. [52].

## 2.7 ACKNOWLEDGMENT

This work was supported by the Dumont d'Urville France/NZ exchange programme.

## 2.7 REFERENCES

1 Le Ru, E. C., Etchegoin, P. G., Grand, J., Félidj, N., Aubard, J., and Lévi, G. (2007). Mechanisms of Spectral Profile Modification in Surface-Enhanced Fluorescence. *J. Phys. Chem. C* **111**:16076–16079; see also Supplementary info.

2 Philpott, M. R. (1975). Effect of surface plasmons on transitions in molecules. *J. Chem. Phys.* **62**:1812–1817.

3 Morawitz, H., and Philpott, M. R. (1974). Coupling of an excited molecule to surface plasmons. *Phys. Rev. B* **10**:4863–4868.

4 Chance, R. R., Prock, A., and Silbey, R. (1978). Molecular fluorescence and energy transfer near interfaces. *Adv. Chem. Phys.* **37**:1–65.

5 Metiu, H. (1984). Surface enhanced spectroscopy. *Progress. Surf. Sci.* **17**:153–320; and references therein.

6 Ford, G. W., and Weber, W. H. (1984). Electromagnetic interactions of molecules with metal surfaces. *Phys. Rep.* **113**:195–287; and references therein.

7 Dulkeith, E., Morteani, A. C., Niedereichholz, T., Klar, T. A., Feldmann, J., Levi, S. A., van Veggel, F. C. J. M., Reinhoudt, D. N., Möller, M., and Gittins, D. I. (2002). Fluorescence quenching of dye molecules near gold nanoparticles: Radiative and nonradiative effects. *Phys. Rev. Lett.* **89**:203002-1–4.

8 Geddes, C. D., and Lakowicz, J. R. (2002). Metal-enhanced fluorescence. *J. Fluoresc.* **12**:121–129.

9 Lakowicz J. R., Geddes, C. D., Gryczynski, I., Malicka, J., Gryczynski, Z., Aslan, K., Lukomska, J., Matveeva, E., Zhang, J., Badugu, R., and Huang, J. (2004). Advances in surface-enhanced fluorescence. *J. Fluoresc.* **14**:425–441.

10 Anger, P., Bharadwaj, P., Novotny, L. (2006). Enhancement and Quenching of Single-Molecule Fluorescence. *Phys. Rev. Lett.* **96**:113002-1–4.

11 Kühn, S., Hakanson, U., Rogobete, L., and Sandoghdar, V. (2006). Enhancement of ingle-Molecule Fluorescence Using a Gold Nanoparticle as an Optical Nanoantenna. *Phys. Rev. Lett.* **97**:017402-1–4; see also Supplementary info.

12 Novotny, L., and Hecht, B. (2006). *Principles of Nano-Optics*. Cambridge University Press, Cambridge.

13 Bharadwaj, P., and Novotny, L. (2007). Spectral dependence of single molecule fluorescence enhancement. *Optics Express* **15**:14266–14274.

14 Maruyama, Y., and Futamata, M. (2005). Elastic scattering and emission correlated with single-molecule SERS. *J. Raman Spectrosc.* **36**:581–592.

15 Valeur, B. (2002). *Molecular Fluorescence. Principles and Applications*. Wiley-VCH, Weinheim.

16 Lakowicz, J. R. (2006). *Principles of Fluorescence Spectroscopy*, 3[rd] ed. Springer, New York.

17 Das, P., and Metiu, H. (1985). Enhancement of molecular fluorescenceand photochemistry by small metal particles. *J. Phys. Chem.* **89**:4680–4687.

18 Barnett S. M., Huttner B., and Loudon R. (1992). Spontaneous emission in absorbing dielectric media. *Phys. Rev. Lett.* **68**:3698–3701.

19 Le Ru, E. C., and Etchegoin, P. G. (2005). Surface-Enhanced Raman Scattering (SERS) and Surface-Enhanced Fluorescence (SEF) in the context of modified spontaneous emission. arXiv:physics/0509154v1 pp.1–14.

20 Le Ru, E. C., and Etchegoin, P. G. (2006). Rigorous justification of the $|E|^4$ enhancement factor in Surface Enhanced Raman Spectroscopy *Chem. Phys. Lett.* **423**:63–66.

21 Le Ru, E. C., Grand, J., Félidj, N., Aubard, J., Lévi, G., Hohenau, A., Krenn, J. R., Blackie E., Etchegoin, P. G. (2008). Experimental verification of the SERS electromagnetic model beyond the $|E|^4$-approximation: polarization effects. *J. Phys. Chem. C* in press.

22 Ritchie, G., and Burstein, E. (1981). Luminescence of dye molecules adsorbed at a Ag surface. *Phys. Rev. B* **24**:4843–4846.

23 Leitner, A., Lippitsch, M. E., and Aussenegg, F. R. (1983). in *Surface Studies with Lasers"*. Springer, Berlin.

24 Atherton, S. J., and Harriman, A. (1993). Photochemistry of intercalated methylene blue: photoinduced hydrogen atom abstraction from guanine and adenine. *J. Am. Chem. Soc.* **115**:1816–1822.

25 Eggeling, C., Widengren, J., Rigler, R., and M. Seidel, C. A. (1998). Photobleaching of fluorescent dyes under conditions used for single-molecule detection: evidence of two-step photolysis. *Anal. Chem.* **70**:2651–2659.

26 Le Ru, E. C., Etchegoin, P. G., and Meyer, M. (2006). Enhancement factor distribution around a single SERS hot-spot and its relation to single molecule detection. *J. Chem. Phys.* **125**:204701-1–13.

27 Chew, H. (1987). Transition rates of atoms near spherical surfaces. *J. Chem. Phys.* **87**:1355–1360.

28 Gérardy, J. M., and Ausloos, M. (1982). Absorption spectrum of clusters of spheres from the general solution of Maxwell's equations. II. Optical properties of aggregated metal spheres. *Phys. Rev. B* **25**:4204–4229.

29 Johansson, P., Xu, H., and Käll, M. (2005). Surface-enhanced Raman scattering and fluorescence near metal nanoparticles. *Phys. Rev. B* **72**:035427-1–17.

30 Etchegoin, P. G., Le Ru, E. C., and Meyer, M. (2006). An analytic model for the optical properties of gold. *J. Chem. Phys.*, **125**:164705-1–3.

31 Le Ru, E. C., Blackie, E., Meyer, M., and Etchegoin, P. G. (2007). Surface Enhanced Raman Scattering Enhancement Factors: A Comprehensive Study. *J. Phys. Chem. C* **111**:13794–13803.

32 Messinger, B. J., von Raben, K. U., Chang, R. K., and Barber, P. W. (1981). Local fields at the surface of noble-metal microspheres. *Phys. Rev. B* **24**:649–657.

33 Félidj, N., Aubard, J., Lévi, G., Krenn, J. R., Salerno, M., Schider, G., Lamprecht, B., Leitner, A., and Aussenegg, F. R. (2002). Controlling the optical response of regular arrays of gold particles for surface-enhanced Raman scattering. *Phys. Rev. B* **65**:075419-1–9.

34 Haynes, C. L., and Van Duyne, R. P. (2003). Plasmon-Sampled Surface-Enhanced Raman Excitation Spectroscopy. *J. Phys. Chem. B*, **107**:7426–7433.

35 McFarland, A. D., Young, M. A., Dieringer, J. A., and Van Duyne, R. P. (2005). Wavelength-Scanned Surface-Enhanced Raman Excitation Spectroscopy. *J. Phys. Chem. B* **109**:11279–11285.

36 Le Ru, E. C., Etchegoin, P. G., Grand, J., Félidj, N., Aubard, J., Lévi, G., Hohenau, A., Krenn, J. R. (2008). Surface Enhanced Raman Spectroscopy on Nano-lithography-prepared substrates. *Cur. Appl. Phys.* **8**:467–470.

37 McCord, M. A., and Rooks, M. J. (1997). Electron Beam Lithography. In *Handbook of Microlithography, Micro matching and Micro fabrication. Volume 1: Microlithography*, P. Rai-Choudhury (Ed.), SPIE press book, Washington, pp. 139–250.

38 Gold arrays are prepared by electron beam lithography (EBL) according to the following method. The bare substrate is a glass plate (10 mm × 10 mm) on which a thin indium tin oxide (ITO) substrate is evaporated. The ITO substrate is then spin coated (60 s, at 5000 rpm) with a 100 nm thick Poly methyl-methacrylate (PMMA) polymer, baked at 170 °C for 8 hours and exposed in an electron-beam writing

equipment consisting of a JEOL 6400 scanning electron microscope (SEM) and a RAITH Quantum pattern generator. After exposure, a chemical development is performed by a 1:2 mixture of developer (PMMA-developer, Allresist Comp.) and isopropanol. The resulting PMMA film is then used as a mask for the thermally deposition of a 40 nm thick gold layer. Removing the remaining PMMA layer by a chemical lift off process leaves gold particles on the ITO substrate at the position defined by the e-beam exposure. The array size is a square of 100 $\mu$m side length.

39 Zhao, J., Jensen, L., Sung, J., Zou, S., Schatz, G. C., and Van Duyne, R. P. (2007). Interaction of Plasmon and Molecular Resonances for Rhodamine 6G Adsorbed on Silver Nanoparticles. *J. Am. Chem. Soc.* **129**:7647–7656.

40 Jiang, J., Bosnick, K., Maillard, M., Brus, L. (2003). Single Molecule Raman Spectroscopy at the Junctions of Large Ag Nanocrystals. *J. Phys. Chem. B* **107**:9964–9972.

41 Maruyama, Y., Futamata M. (2005). Inelastic scattering and emission correlated with enormous SERS of dye adsorbed on Ag nanoparticles. *Chem. Phys. Lett.* **412**:65–70.

42 Moskovits, M. (2005). Surface-enhanced Raman spectroscopy: a brief retrospective. *J. Raman Spectrosc.* **36**:485–496.

43 Itoh, T., Biju, V., Ishikawa, M., Kikkawa, Y., Hashimoto, K., Ikehata, A., Ozaki, Y. (2006). Surface-enhanced resonance Raman scattering and background light emission coupled with plasmon of single Ag nanoaggregates. *J. Chem. Phys.* **124**:134708-1–6.

44 Itoh, T., Kikkawa, Y., Biju, V., Ishikawa, M., Ikehata, A., Ozaki, Y. (2006). Variations in Steady-State and Time-Resolved Background Luminescence from Surface-Enhanced Resonance Raman Scattering-Active Single Ag Nanoaggregates. *J.Phys. Chem. B* **110**:21536–21544.

45 Moskovits, M. (1982). Surface Selection Rules. *J. Chem. Phys.* **77**:4408–4416.

46 Le Ru, E. C., Meyer, M., Blackie, E., and Etchegoin, P. G. (2008). Advanced aspects of electromagnetic SERS enhancement factors at a hot spot. *J. Raman Spectrosc.* in press.

47 Le Ru, E. C., Meyer, M., and Etchegoin, P. G. (2006). Proof of Single-Molecule Sensitivity in Surface Enhanced Raman Scattering (SERS) by Means of a Two-Analyte Technique. *J. Phys. Chem. B* **110**:1944–1948.

48 Pieczonka, N. P. W., Aroca, R. F. (2008). Single molecule analysis by surfaced-enhanced Raman scattering. *Chem. Soc. Rev.*, in press, doi:10.1039/b709739p.

# 3

# The Role Of Plasmonic Engineering In Potential Surface-Enhanced Fluorescence

Daniel J. Ross, Nicholas P.W. Pieczonka, R. F. Aroca
[1]Laser Dynamics Laboratory, School of Chemistry and Biochemistry, Georgia
Materials & Surface Science Group, Faculty of Science, University of Windsor, N9B 3P4, Windsor, ON, Canada.

## 3.1 INTRODUCTION

The fluorescence emission from molecules, commonly described in terms of an oscillating electric dipole moment, is characterized by its efficiency or quantum yield (ratio of photons absorbed and photons created) and its fluorescence lifetime, i.e., the average time the molecule stays in its excited state before emitting a photon. The fluorescence cross section is one of the highest in molecular spectroscopy with values in the order of $10^{-17}$ $cm^2$ / molecule. The emission properties of this dipole can be strongly modified when surrounded by materials of different composition. In particular, the coupling of an excited molecule and the surface states of a metal has been and is at the centre of many new developments. For instance, Philpot (1975), in an early paper [*1*], already discusses the "effect of surface plasmons on transitions in molecules", exploring the idea of using fluorescence to probe the surface-plasmon polaritons in metals. The excitations termed surface-plasmon polaritons were observed at the beginning of the $20^{th}$ century (Wood's anomalies); however, it is after Otto's work [*2*] (followed by Kretschmann and Raether) that the field of controlled excitation of surface modes rapidly developed. The most common recent convention is to rename the surface-plasmon polaritons observed in nanostructures as *localized surface plasmon resonances* (LSPR) [3], and this nomenclature is adopted here. For molecules on a metal surface, the coupling to surface plasmons has given rise to two very different branches of spectroscopy. a) Techniques for the study of molecules located at the surface, or in the vicinity of, a flat metal surfaces sustaining surface plasmon polaritons [*4*] termed "surface-plasmon spectroscopy", including the use of the evanescent optical field for the excitation of chromophores, and b) plasmon-enhanced fluorescence (See Figure 3.1) on metallic nanostructures that support localized surface plasmon resonances [*5, 6*]. The spectroscopic properties affected by the presence of the metal are the absorption cross section, the lifetime of the excited state, the radiative decay and the non-radiative energy transfer to the metal, and the corresponding quantum yield for emission. Since, we are here interested in the role of Plasmonics [*7*] in the emission properties of molecules, the discussion is restricted to plasmon enhanced fluorescence using LSPR in plasmon supporting nanostructures. [*3, 8*] Gersten and Nitzan [*5*] provided the first complete electromagnetic (EM) study for a molecule-particle system in terms of a modified local electromagnetic field and plasmon resonance effects. The dimension of the particles was small compared with the wavelength of excitation (static

*Plasmonic Engineering in SEF*
Edited by Chris D. Geddes.

approximation) and they concluded that both the radiationless and radiative lifetimes for the dipole interacting with the particle were quite different. The results of this work were then used as a reference for the interpretation of experimental results. Weitz *et al.* [*9*] reported that fluorescence lifetimes of molecules on silver island films were several orders of magnitude shorter than on a silica surface; however the radiation was not substantially quenched. The latter was in contrast to smooth metal surfaces where the decrease in lifetime was directly related to nonradiative decay leading to fluorescence quenching [*10*]. The role of the plasmons induced in the silver island, leading to radiation, was the experimental evidence of what we call now surface-enhanced fluorescence (Figure 3.1) which has grown with two family names: surface-enhanced fluorescence (SEF) [*11,12*] and metal-enhanced fluorescence (MEF) [*13*]. The experimental results provided further support for the role of the LSPR in the enhanced emission rate of fluorescence at the molecular frequency. It was shown that the experimental fluorescence decay can be fitted by the EM model, and the expressions provided by the model also describe de-enhanced emission defined as the ratio of the emission dipole of the system to the molecular dipole. Soon after, Wokaun *et al.*[*14*] demonstrated a key feature of the EM model, the distance dependence of the enhanced emission of basic fuchsin on silver island film using $SiO_x$ spacer layers. A maximum enhancement of ca. 200 was reported at an optimum silver-molecule separation of 25 Å. The apparent yield for the observed enhanced emission may be simplified as the factor of three terms [*5, 14*]:

$$Y_{app} = \left|A\left(\omega_{exc}\right)\right|^2 \times \left|A\left(\omega_{fluo}\right)\right|^2 \left(\frac{\Gamma^f}{\Gamma}\right)^2 \qquad (1)$$

Where the first term corresponds to the enhanced absorption rate at the excitation frequency, the second is the enhanced emission by the LSPR. Both terms contain the dielectric function of the metal and the metal-particle distance. The last factor is the decay rate of the free molecule to that of the adsorbed molecule coupled to the surface plasmon. A good agreement was found between the calculated and observed distance dependence of the emission, [*14*] i.e., the electromagnetic coupling between the fluorescence molecule and the silver nanostructure is a function of the distance between them. The distance dependence was later also demonstrated using Langmuir-Blodgett (LB) monolayers as spacers on silver [*15, 16*], or by the use of Layer-by-Layer (LbL) films [*17*]. The excited molecule-plasmon coupling was clearly demonstrated in energy transfer experiments using ultrathin metal films, where energy transfer between a donor and acceptor molecule was mediated by surface plasmon polaritons [*18*]. It was shown that discontinuous silver films (30 nm and 60 nm thick ness with respect to 120 nm Ag thickness) were more efficient in the mediated energy transfer from donor to acceptor. A very important point in the physics of SEF is the fact that the excitation frequency and the enhanced emission can be far apart in the electromagnetic spectrum. In simple terms, it means that the model of the perfect sphere cannot be used to explain SEF, since at least two surface modes are needed. In practice, the most common nanostructures used for SEF, are island films or colloidal metals aggregates that can support multiple surface modes, since they are exactly what you need to observe the enhancement. Thereby, engineering nanostructures with multiple resonances is the required pathway for control experiments and optimization of the substrate. There are many SEF examples in the literature with fabricated nanostructures that can support multiple plasmon resonances [*5,19-22*], including, for example, the fabrication of uniform arrays of split gold rings structures [*22*].

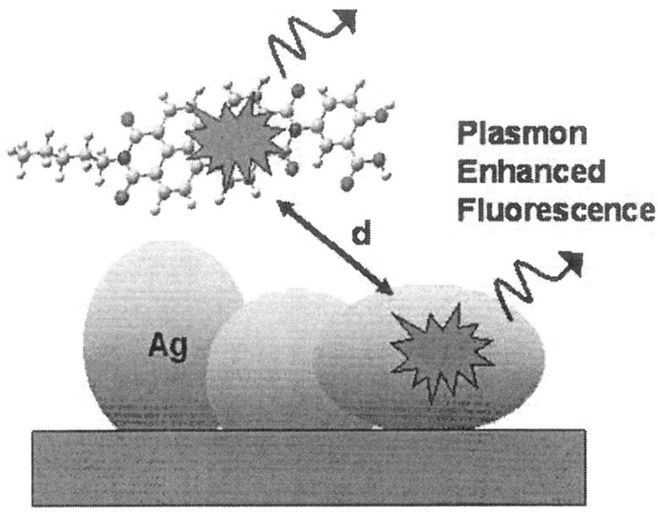


**Figure 3.1:** Cartoon for an excited PTCDA molecule and the silver island dipolar response that results in significant radiation.

In addition to enhanced intensity, the observed spectrum of LSPR mediated SEF also shows the fingerprints of the role of plasmonics, as has been elegantly pointed out recently by Le Ru *et al.* [*23*]. They coined the name "spectral profile modification (SPM)" for these characteristic features of fluorescence close to metal surfaces: "*We argue theoretically and demonstrate experimentally that under SEF conditions the plasmon resonances affect not only the fluorescence intensity (enhancement or quenching) but also its spectral profile, sometimes to a point where the original fluorescence spectrum may no longer be recognizable as such*". It should be pointed out that the SPM described in this work is completely independent of "chemical effects", i.e., perturbation of the electronic states due to metal-molecule strong (chemical) interactions. Experimentally, the need for different resonances is achieved in this work using particles with various shapes and sizes fabricated by nanolithography. The nanolithographic array substrates are advantageous for studies aiming to separate different contributions to the observed enhanced fluorescence signal.

In our group we have studied a series of perylene tetracarboxylic (PTCD) derivatives using Langmuir-Blodgett monolayers deposited onto silver, gold and silver / gold island films [*24*]. These materials are dyes that absorb strongly in the visible (~500 nm) and correspondingly there is a strong Stokes shifted mirror image fluorescence of the absorption seen in dilute solutions [*25*]. In addition, PTCD aggregates, in LB films or solid samples produce a strong structureless excimer fluorescence that is observed more than 100 nm away from the main absorption of the dye [*26*]. A typical example of these results is shown in Figure 3.2, that illustrate spectra containing surface-enhanced excimer fluorescence and the corresponding SERRS (surface-enhanced resonance Raman scattering) [*26*] obtained when the excimer fluorescence is completely quenched in an LB sandwiched between silver island films.

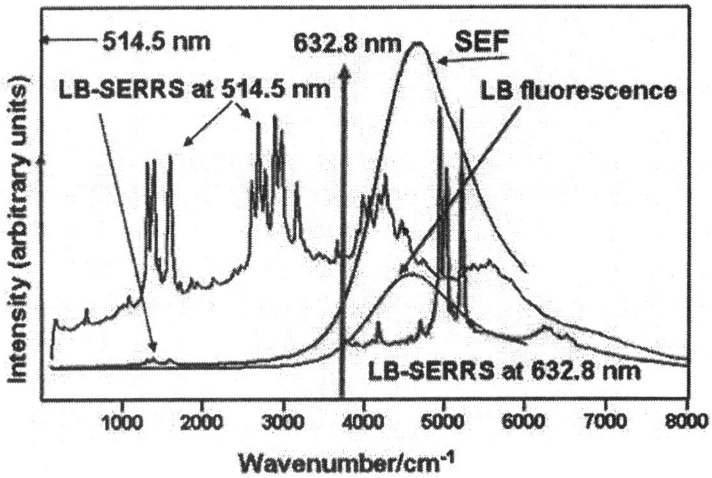


**Figure 3.2:** SEF at 514.5 nm are illustrated using Langmuir-Blodgett monolayer of a PTCD derivative on silver islands with a plasmon resonance in resonance with the molecular absorption. After quenching the Fluorescence with Ag overlayers LB-SERRS, with its overtone series, is clearly seen. The pre-resonant SERS excited at 632.8 nm is also shown to highlight the differences in the intensities of fundamental and overtones bands compared to SERRS.

The LB-SERRS spectrum showing the strong excimer SEF is that of a LB monolayer of mixed monolayers of PTCD with arachidic acid excited in resonance with the absorption of the dye at 514.5 nm. The plasmon absorption of the film is also in the same region and there is a double resonance with the green laser line. Given the strong surface-enhanced excimer fluorescence observed, it can be deduced that the island film provides nanostructures able to support plasmons in the region more than 100 nm away from the main plasmon absorption. These nanostructures can also provide SERRS as demonstrated by the LB-SERRS at 632.8 nm. We are using here the nomenclature of SERRS, when in fact this region corresponds to a pre-resonant effect. The fact that full resonance is not the case is highlighted by the very low relative intensity of the overtone and combinations. In order to bring the full SERRS of the monolayer, silver islands were evaporated onto the original LB on silver (silver overlayer) where the series of overtones and combinations are seen up to the third order. The excimer fluorescence of the LB on glass has also been included for comparison. For practical applications the island films can provide plasmons resonances over a large region of the electromagnetic spectrum, however this property inherits a wide distribution of enhancement factors on the surface of the film [*27*]. The role of plasmonic engineering could be to narrow the region of surface resonances as may be required for the specific application. The lithographic arrays are clearly one way to achieved exactly this objective.

## 3.2 MOLECULE-PLASMON COUPLING

The molecule-single particle model has been very successful in explaining experimental results. However, to gain a better understanding of the physical phenomenon it is necessary to deal with nanostructure aggregates that support multiple surface plasmons and produce spatial locations with extremely high local field (hot spots) [*28*]. A more general model including two metal spheres is now available [*29, 30*] and it will be discussed in detail, vide infra. Computational and experimental developments have brought and are bringing to light the plasmonic properties of different nanostructures [*31, 32*] It can be said that the basic parameters guiding plasmonic engineering are; the response function to polarization (dielectric function) that confines the materials to a few metals (mainly silver and gold), the shape (particles, shells, rings, etc.), size and the spatial geometry of nanostructure arrays or aggregates [*3, 8, 33*].

The prediction of the absorption and scattering by a single particle has been known since the time of Maxwell-Garnett [*34*] and were explicitly calculated for an sphere by Mie [*35*]. Since then, there have been many successful extensions to non-spherical shapes using the dipolar model [*36, 37*]. As the calculation of absorption and scattering relies on the spatial integration of the Poynting vector, explicit calculation of field intensities, and thus, field enhancements, is also known exactly for spheres, and approximated for many other shapes in dipolar models. Although the dipole model is always and approximation of the exact solution, it is appropriate when the particle is much smaller than the wavelength of light, which is often the case for small nanoparticles in the visible region of the electromagnetic spectrum.

The coupled dipole equations (CDE) have been used in calculating the optical properties of composite media, including larger particles, where the dipoles are arranged to mimic a more complicated system, such as those used in DDA [*38*], [*39*], as well as fractal structures [*40*], which could be applied to model aggregation, surface composition, or percolation. The general nature of the solution allows for calculation of optical properties, as well as enhanced Raman and electric fields at any point in space.

In this work, the CDE are used to calculate the enhanced field between 2 particles using polarized light. The addition of the field created by an oscillating dipole also allows for the calculation of enhancement in decay rate, a necessary component in considerations of SEF.

The particles are arranged along the x axis, for simplicity. The light is incident from the z direction, with polarization chosen to be along the axis of the particles. This will excite only the coupled mode of the interacting particles, which will cause a shift in the resonance to lower energies compared to an isolated particle. The particles are chosen to have a radius of 20 nm, so the centre to centre distance is

40 nm. They are separated by a space of d. A schematic of this arrangement is shown in Figure 3.3.

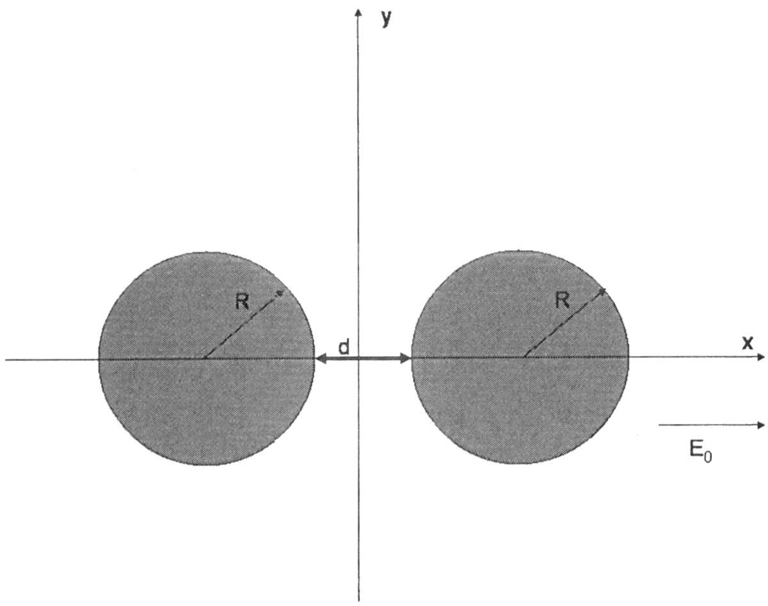


**Figure 3.3:** Schematic of particle arrangement for calculation of enhancement. The particles of radius R are separated by a distance d.

Following the development of Markel, Shalaev, and co-workers [40], an arrangement of N monomers interacts with a plane wave of the form $\vec{E}(\vec{r},t) = \vec{E}_0 e^{i(\vec{k}\cdot\vec{r}-\omega t)}$. The local field induces a dipole moment for each monomer of $\vec{d}_i = \alpha_0 \vec{E}_l$, where α0 is the isotropic polarizability, El the local field, and di the dipole moment of monomer i. Each monomer is simultaneously interacts with all other monomers, and the external field. The total dipole moment each monomer is given by:

$$d_{i\gamma} = \alpha_0 \left( E_{0\gamma} e^{i\vec{k}\cdot\vec{r}_i} + \sum_{j=1}^{N}{}' G_{\gamma\beta}(r_{ij}) d_{j\beta} \right) \qquad (2)$$

Where the greek indicies γ,β are the Cartesian coordinates, over which summation is implied. The primed summation indicates the j=i is excluded from the sum. Also, $\mathbf{r}_{ij} = \mathbf{r}_i - \mathbf{r}_j$. G is a tensor defining the interaction between particles. The elements of this tensor are given by

$$G_{\gamma\beta}(\vec{r}) = k^3\left( A(kr)\delta_{\gamma\beta} + B(kr)\frac{r_\gamma r_\beta}{r^2}\right)$$

$$A(kr) = \left(\frac{1}{kr} + \frac{i}{(kr)^2} - \frac{1}{(kr)^3}\right)e^{ikr} \tag{3}$$

$$B(kr) = \left(-\frac{1}{kr} - \frac{3i}{(kr)^2} + \frac{3}{(kr)^3}\right)e^{ikr}$$

This system of equations is easiest to solve in complex vector space of 3N dimensions, with basis vectors of |iγ >. In this 3N space, the equation for the 3N dimensional dipole moment vector is

$$|d\rangle = \alpha_0|E\rangle + \alpha_0\hat{G}|d\rangle \tag{4}$$

$\hat{G}$ is a 3N x 3N tensor with matrix elements given above, and the elements of E, the incident field, are given by

$$\langle i\gamma|E\rangle = E_{0\alpha}e^{i\vec{k}\cdot\vec{r}} \tag{5}$$

0n

$$Z\hat{1}|d\rangle - \hat{G}|d\rangle = |E\rangle \tag{6}$$

These equations are a matrix equation of the form AX=Y, with complex elements where A is a known matrix, Y is a known vector. The unknown vector X is solved through the conjugate gradient method [*38*].

The polarizability of each particle is given by:

$$\alpha_0 = V\frac{\varepsilon - 1}{\varepsilon + 2} \tag{7}$$

Where α is the polarizability, ε is the frequency dependent dielectric function [*41*], and V the volume of the dipole. The radius of each sphere is calculated using a/R=1.612 [*42*], where a is the spacing between the particles, 40 nm in this case, and R the radius. This value is less than the expected a/R=2, as the dipole response will not be homogeneous: rather, it will be stronger in the region where the particles are in contact. This value of a/R causes the dipoles to overlap, and interact more strongly. It is this overlap between the individual dipolar modes of each particle that causes the coupling.

The electric field is calculated through the addition of a non-interacting particle at the desired position, and the sum of the dipole moment vectors divided by the polarizability of each particle is taken at this point. The enhancement of the electric field, M, is calculated by:

$$M = \left| \frac{\vec{E}_{inc} + \sum_i \vec{E}_i}{\vec{E}_{inc}} \right| \quad (8)$$

For the 2 particle model described above, the field between at a point symmetrically between them (located at a distance d/2 from each particle) is shown in Figure 3.4.

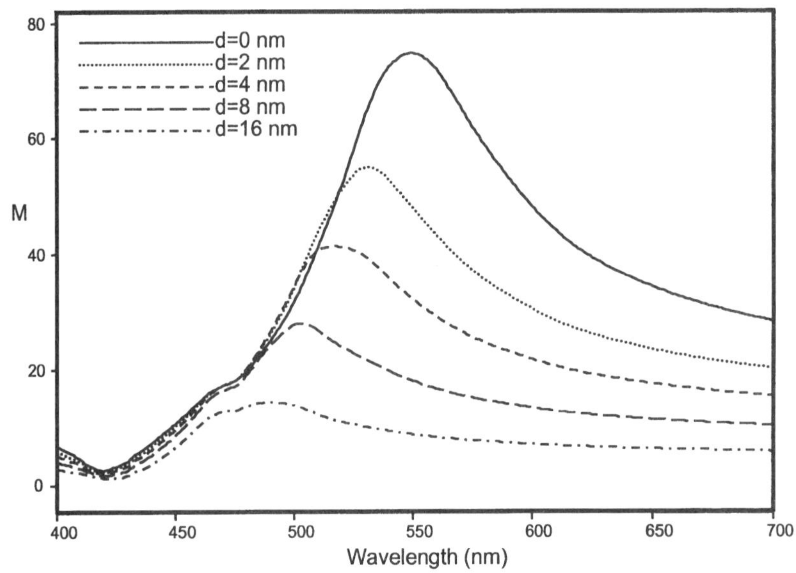


**Figure 3.4:** Field enhancement as a function of wavelength for 2 particle system.

The field enhancements show a clear trend. When the particles are well separated, the wavelength dependence is similar to that obtained for a single particle, with enhancement peaking at 490 nm. As the particles are moved closer together, the peak enhancement increases and red shifts, due to the increased coupling between the particles, as well as the decreased particle-molecule distance.

Finally, the enhancement to the decay rate is obtained by the addition of an oscillating dipole. This dipole emits power as [*43*]

$$P_{free} = \frac{\omega^4 |\vec{p}|^2}{12\pi\varepsilon_0 c^3} \tag{9}$$

Rather than a plane wave as the incident field, the $E_0$ treated as the field of the dipole acting on the particles only. This system of 3 monomers is solved to find the new dipole moment at the particle. The enhancement of the power emission of the oscillating dipole is then

$$|M_d|^2 = \frac{P_{CDE}}{P_{free}} \tag{10}$$

This treatment gives $M_d/M$ values approximately unity as d is increased. As the molecule approaches the particles, however, $M_d/M$ increases.

## 3.3 UNIFIED Model Of SEF and SERS.

Most models for SEF focus on the plasmonics, and treat the molecule as a classical dipole. While the plasmonics models increasingly give more realistic results for the plasmon observed in the system, the treatment of the molecule, and thus the molecule-metal system, is not always as well developed. In their 2005 paper, Johansson, Xu, and Kall [*44*] present a unified model of enhanced Raman scattering and enhanced fluorescence within the context of quantum optics. This model is easily modified to include the field enhancement (M) and decay enhancement ($M_d$), which may be calculated through plasmonics methodology.

This model assumes only 2 electronic states, with the ground, |g>, and excited, |e>, states having energies $\hbar\omega_g$ and $\hbar\omega_e$ respectively. Also, a dipole moment is defined, where $p_0=l_{dip}e$, where e is the elementary charge, and $l_{dip}$ the dipole length. The energies and the dipole moment are all that is required to characterize the electronic states in this model. Finally, there is assumed to be a single vibrational mode of energy $\hbar\omega_{vib}$, which has $N_{vib}$ states in each of the electronic states. Thus there is a total of $N=2\ N_{vib}$ states, $|g,n=1..N_{vib}>$ and $|e,m=1..N_{vib}>$. The equation of motion of the NxN density matrix ρ is:

$$i\frac{d\rho}{dt} = \frac{1}{\hbar}[H_{mol} + H', \rho] + L_{tr}\rho + L_{ph}\rho \tag{11}$$

The Hamiltonian is made up of 2 terms, the molecular ($H_{mol}$) and the molecule-field interaction (H'). $L_{tr}\rho$ is a damping term in the density matrix due to transitions caused by interaction with the vacuum state [*45*], but also includes non-radiative processes and vibrational damping. $L_{ph}\rho$ represents damping due to phase relaxation. This parameter allows broadening of the bands which may be experimentally due to other vibrational states, or temperature broadening.

The diagonal molecular Hamiltonian is a sum of electronic and vibrational energies:

$$H_{mol} = \sum_{n=0}^{N_{vib}-1} \sum_{l=ge} |l;n\rangle(\hbar\omega_l + n\hbar\omega_{vib})\langle l;n| \qquad (12)$$

The interaction Hamiltonian, H', is off diagonal.

$$H' = -\left(\frac{pE_0M(\omega_L)}{2}\right)\sum_{n,m=0}^{N_{vib}-1}\left(|e;m\rangle e^{-i\omega_L t} f(n,m)\langle g;n| + |g;n\rangle e^{i\omega_L t} f(n,m)\langle e;m|\right)$$

(13)

Where $p_0$ is the dipole moment, $E_0$ the incident electric field, and $M(\omega_L)$ the field enhancement at the frequency of the laser line. f(n,m) is the Frank-Condon factor, defined as:

$$f(n,m) = \sqrt{n!m!}e^{-\alpha^2/2}\sum_{k=0}^{\min(n,m)} \frac{(-1)^{(m+k)}\alpha^{n+m-2k}}{k!(n-k!)(m-k)!} \qquad (14)$$

Where $\alpha$ is the dimensionless displacement between states.

The $L_{tr}\rho$ and $L_{ph}\rho$ terms are evaluated using an NxN operator $\sigma_{ij}$, in which all matrix elements are zero except the ij element. The $L_{tr}\rho$ term is then [*45*]:

$$L_{tr}\rho = -\sum_{jk}\frac{i\Gamma_{kj}}{2}\left[\sigma_{jk}\sigma_{kj}\rho + \rho\sigma_{jk}\sigma_{kj} - 2\sigma_{kj}\rho\sigma_{jk}\right] \qquad (15)$$

$\Gamma_{kj}$ is the damping rate due to spontaneous transitions from state j to state k. This is calculated using Fermi's golden rule:

$$\Gamma_{kj} = \frac{|M_d(\omega)|^2\omega^3}{3\pi\hbar\varepsilon_0 c^3}|p_0|^2|f(n,m)|^2 \qquad (16)$$

Where $\omega$ is the frequency of the transition, $\omega = \omega_{ge} + (m-n)\omega_{vib}$, defining $\omega_{ge} = \omega_e - \omega_g$, and $M_d$ the enhancement of the decay rate due to interaction between the molecule and the metal particle. It should be noted that $L_{tr}\rho$ as written does not include any temperature effects, such as thermal excitation, and is therefore a zero temperature case.

Lastly, $L_{ph}\rho$ is:

$$L_{ph}\rho = -\gamma_{ph}\sum_{i\in g}^{j\in e}\left[\sigma_{ij}\rho_{ij} + \sigma_{ji}\rho_{ji}\right] \tag{17}$$

This problem is made more amenable by representing the density matrix as a vector of $N^2$ dimensions, such as

$$\vec{\rho} = \left(\rho_{11}, \rho_{21}, \ldots, \rho_{N1}, \rho_{12}, \rho_{22}, ..\rho_{NN}\right) \tag{18}$$

Equation 10, once solved, can be expressed as:

$$i\hbar\frac{d\vec{\rho}}{dt} = \overleftrightarrow{L}\vec{\rho} \tag{19}$$

Where $\overleftrightarrow{L}$ is an $N^2$x $N^2$ tensor that couples the time evolution of a matrix element with all the matrix elements at time t. The time independent form of $\overleftrightarrow{L}$ is:

$$\overleftrightarrow{L}' = e^{i\overleftrightarrow{\Omega}t}\overleftrightarrow{L}e^{-i\overleftrightarrow{\Omega}t} \tag{20}$$

Where $\overleftrightarrow{\Omega}$ is a diagonal tensor with elements referring to populations or intraband coherences are zero. Upwards coherences have a value of $\Omega_L$, while downwards coherences are - $\Omega_L$.

The differential cross section is given by

$$\frac{d^2\sigma}{d\Omega d(\hbar\omega)} = \frac{\omega^4\left|M(\omega,\theta)\right|^2}{I_{in}8\pi^3c^3\varepsilon_0\hbar}\left|p_0\right|^2\sum_{b,k\in g}^{a,j\in e} f(b,a)f(k,j)\sum_r \mathrm{Re}\left[\rho_{0,ra}G_{0\,jk,rb}(\omega-\Omega_L)\right]$$

(21)

where:

$$I_{in} = \frac{c\varepsilon_0 E_0^2}{2} \tag{22}$$

$G_0$ and $\rho_0$ represent a stationary Green's tensor and initial density matrix, respectively. The initial density matrix can be obtained by solving the system of N equations given by:

$$\left(\overleftrightarrow{L}' - \overleftrightarrow{\Omega}\right)\rho_0 = 0 \tag{23}$$

It is also required that the trace of the initial density matrix is unity.

The Green's tensor is defined as

$$\rho_{jk} = \sum_{rs} G_{jk,rs}\rho(0)_{rs} \tag{24}$$

While each matrix element is given by:

$$\hat{G}_0(\omega) = i\left[(\omega + i\delta)\hat{1} + \hat{\Omega} - \hat{L}'\right]^{-1} \tag{25}$$

To obtain spectra, equation 21 is used, with the elements of the Green's tensor obtained from equation 19 and 25, and the initial density matrix elements from equation 23. Enhanced spectra are calculated using the results from CDE, as shown in the previous section. The results are seen in Figure 3.5.

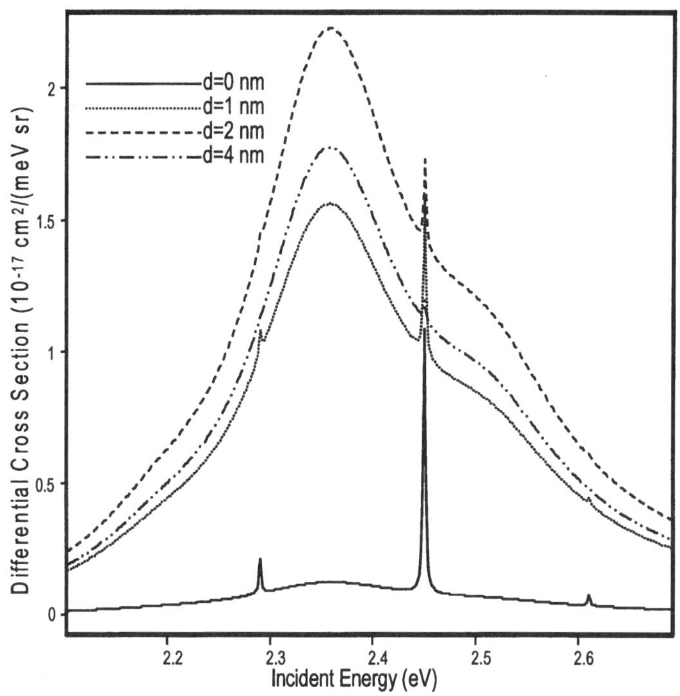


**Figure 3.5:** Scattering and Fluorescence cross sections for d=0, 1, 2, and 4 nm. The parameters used are: $\hbar\omega_L$=2.45 eV, : $\hbar\omega_{ge}$=2.35 eV, $\hbar\omega_{vib}$=0.16 eV, $\gamma_{ph}$=1x10$^{14}$ Hz, $\alpha$=0.5, and $E_0$=1x10$^4$ N/C.

The enhanced Raman scattering and fluorescence at close particle-molecule distances show an almost complete quenching of the fluorescence, with a strong

Raman enhancement. The fluorescence reaches a maximum around d=2 nm, while at intermediate distances the quenching effect is reduced, resulting in fairly strong fluorescence with moderate Raman. Past the optimal distance, both the fluorescence and Raman enhancement decay.

The combination of plasmonics calculations to model the enhancing particles, and rigorous molecular treatment allow for a comprehensive model. The quantum optic model of Kall and co-workers [*30, 44*] gives a unified treatment of a model molecule with two electronic states, and an arbitrary number of vibrational levels. Although more complete methods are possible, by adding different vibrational bands, for example, this includes the most important elements, and allows coupling to calculated plasmonic results.

## 3.4 THE EXPERIMENTAL SEF

In the previous sections we have discussed the historical development of enhanced fluorescence as well as giving a detailed theoretical basis for the phenomena. In this section we will highlight some of the experimental results that illustrate how LSPR supporting nanostructures have been used for enhanced luminescence.

The presence of a surface plasmon strongly modifies the photophysics of a fluorophore thus the optimization of SEF requires a detailed understanding both of the targeted fluorophore but also of the plasmonics of the nanostructure to be used.

In fluorescence experiments the spectral parameters are fluorescence intensity, lifetime and band profile. The measured fluorescence intensity can be given by:

$$F \propto I_o \varepsilon(\lambda_o) \phi N \tag{26}$$

Where $I_o$ is the incident power, $\varepsilon(\lambda_0)$ is the molar absorptivity, N the number density of fluorophores ($m^{-3}$), and $\varphi$ is the quantum yield that is defined in term of rates of processes competing for deactivation:

$$\phi = \frac{k_{rad}}{k_{rad} + k_{non-rad}} \tag{27}$$

What emerges from the rigorous theoretical treatment in the previous section is that the Plasmon / metal interface plays several roles in the enhancement. Thus the measured fluorescence intensity in the presence of plasmon supporting metal nanoparticle is modified to:

$$SEF \propto I_o M(\lambda_o) \varepsilon(\lambda_o) \phi' M(\lambda_{emmisson}) N \tag{28}$$

Here M are the field enhancement factors and $\varphi'$ is the modified quantum yield given by

$$\phi' = \frac{k'_{rad}}{k'_{rad} + k'_{non-rad}} \tag{29}$$

and the complete SEF profile enhancement can be seen as

$$SEF_{Total} \propto I_o M(\lambda_o)\varepsilon(\lambda_o)\phi' N \int_{\lambda_a}^{\lambda_b} s(\lambda_x) M(\lambda_x) d\lambda \tag{30}$$

where $s(\lambda_x)$ is a relative emission function whose integrated area is normalized to one.

There are implicit distance and orientation dependence to SEF that can be expressed as

$$SEF(d,\theta) \propto I_o M(\lambda_o, d, \theta)\varepsilon(\lambda_o, \theta)\phi'(d,\theta) M(\lambda_{emisson}, d, \theta) N \tag{31}$$

The angular dependence is a direct result of the dipole nature of the plasmon polarization, the excitation polarization, and the directionality of the transition dipole moment. This is the most difficult parameter to predict and control, especially in multimode plasmon supporting substrates. In addition, as shown in the previous section, there is competition between the rate of nonradiative energy transfer and the effective field enhancement. Hence, for each fluorophore-nanostructure combination there is a delicate balance to be found. Therefore, finding the optimal distance from the enhancing surface is essential for maximizing SEF.

Within this context then the goal of the experimental design is to engineer a molecule nanoparticle interaction pairing that is tailored to the desired maximized efficiency. The inherent challenge of engineering substrates for SEF is demonstrated by looking at a small sampling of enhancement factors reported over the years for a variety of molecules and substrates as seen in Table 3.1. Similarly, to other plasmon enhanced phenomena, notably SERS, there is not a universal nanostructure, but each case requires the tuning of the plasmonics to attain an effective coupling.

Experimentally, plasmonic engineering of SEF substrates requires then the consideration of the following variable: The response function of the metal to polarization (dispersion of the dielectric function) plasmon resonances can be tuned using different shapes (such as triangles, squares, spheroids, rods), nanowires, shells, rings or holes.

Size and the spatial geometry of nanostructured arrays or aggregates which also allow for plasmonic manipulation of the substrate optimization of the fluorophore-nanostructure distance orientation of the fluorophore on the metallic surface.

**Table 3.1: Reported Enhancement Factors**

| Substrate | Fluorophore | Enhance-ment Factor | ef |
|---|---|---|---|

| | | | |
|---|---|---|---|
| Silver island film using $SiO_x$ spacer layers | Basic fuchsin | 200 | *14*] |
| Silver island film with 15 nm fatty acid spacer layer | Phthalocyanine | 400 | *15*] |
| Silver island film | bis(phenethylimido)-perylene | 400 | *46*] |
| Silver island film | azopolymer films | 99 | *47*] |
| Silver island film | peridinin–chlorophyll–protein | 18 | 51] |
| Silver nanoparticles/ Polymers – LBL | R123 | 10 | 52] |
| Aluminium nanostructured Surface | 2-aminopurine | 9 | 53] |
| Nanoholes in gold films | Oaxine 720 | 82 | 54] |
| Pairs of elliptical gold nanoparticles | Rhodamine 800 | 100 | 19] |
| Silver island film | Rose Bengal | 10 | *48*] |
| Silver Island film | fluorescein-labeled immunoglobulin G (Fl-IgG) | 40 | *49*] |
| Silver particle Film | Platinum Octaethyl Porphyrin | 200 | *50*] |
| Gold nanoparticles | Indocyanine green (ICG) | 50 | *20*] |

It is appropriate here to present a few examples of the various approaches that have successfully applied to the task of tailoring LSPR for enhanced luminescence.

The necessity of multi-plasmon mode substrates has been evident since the first recorded SEF experiments. In particular, our work on enhanced excimer fluorescence, an emission which is strongly red shifted from the excitation wavelength, provides a simple illustration of the need for the nanostructures to support plasmon resonances over a wide spectral region. Island films meet this requirement due the complex plasmonics generated by the interaction of large number of nanostructures that comprise the surface. This is one reason why island films are so effective for SEF and why they have been the workhorse of SEF studies [*6, 14-16, 46, 47, 55, 56*]. Island films are fabricated by various methods of metal evaporation onto solid substrates. By changing parameters such as metal concentration and temperature, various sizes and distributions of nanoparticles can be generated. The following examples will illustrate their effectiveness as SEF substrates as well as providing a chance to discuss issues common to all SEF studies.

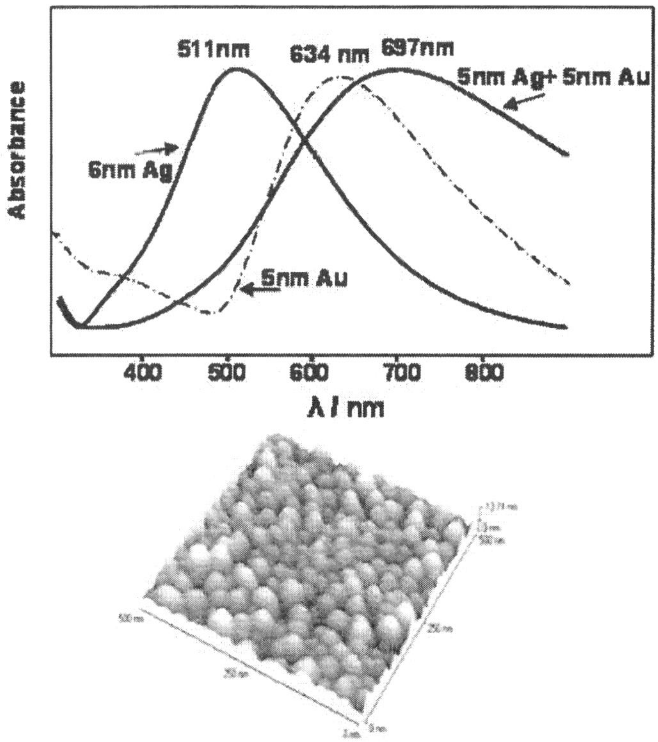


**Figure 3.6:** LSPR of several island films. The bottom image is the AFM of mixed AG/Au island film.

The PTCDA moiety is a class of molecule that readily exhibits excimer fluorescence. A direct method to attain the enhancement factor for a substrate is to deposit a LB monolayer onto a glass slide that contains an area of a silver island film. By comparing the fluorescence intensity measured on glass and the island film under identical conditions, an excellent assessment of the EF can be attained.

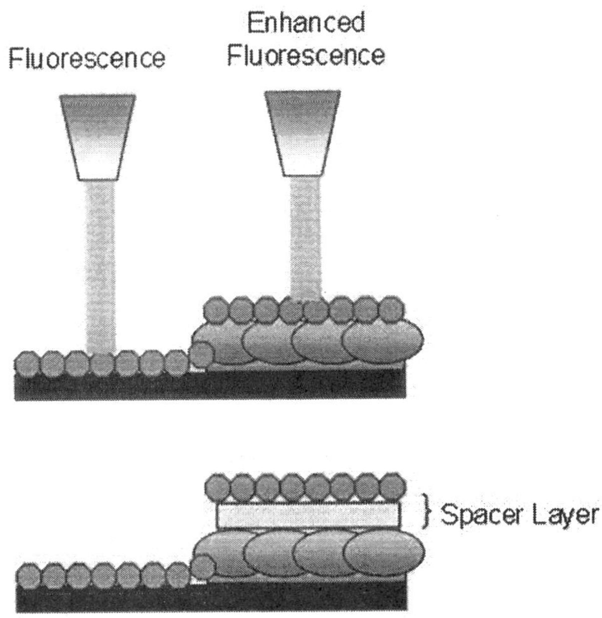


**Figure 3.7:** Schematic of the LB approach to SEF measurements.

In addition to ascertaining the EF the use of monomolecular films allows the use of SEF to be an extremely sensitive probe of molecular properties. For instance, depending on the functional groups present on the PTCD chromophore, the EF can either be at a maximum when the LB is deposited directly onto the SIF (recalling that there exists a 10-15 Å oxide-water layer), while for other cases a spacer layer of which separates the molecule from the surface an additional 10-20 Å is needed for the maximum EF (See Figure 3.4). The results can be explained in terms of the effect molecular organization attained in LB films that fixed a preferential surface orientation of the transition dipole determining the efficiency of the energy transfer to the surface. The latter can be seen in Figure 3.9, where an LB monolayer of a planar molecule deposited directly onto a silver island film gives SEF, suggesting an edge-on orientation on the metal surface

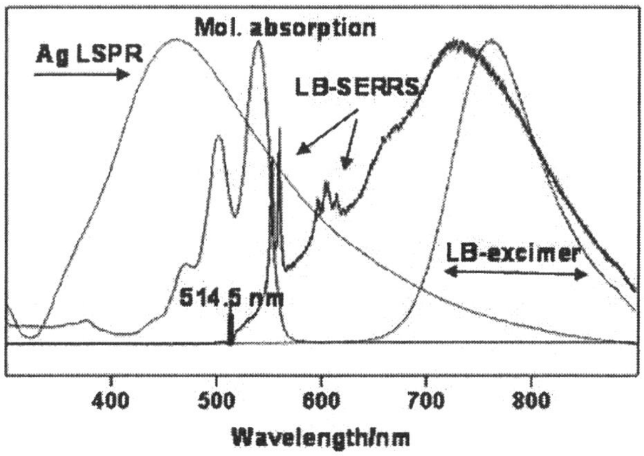


**Figure 3.8:** Molecular absorption of a PTCD derivative, surface plasmon of the evaporated silver film. Excimer fluorescence of an LB monolayer on glass and SERRS with the background SEF of the LB monolayer on the silver island film.

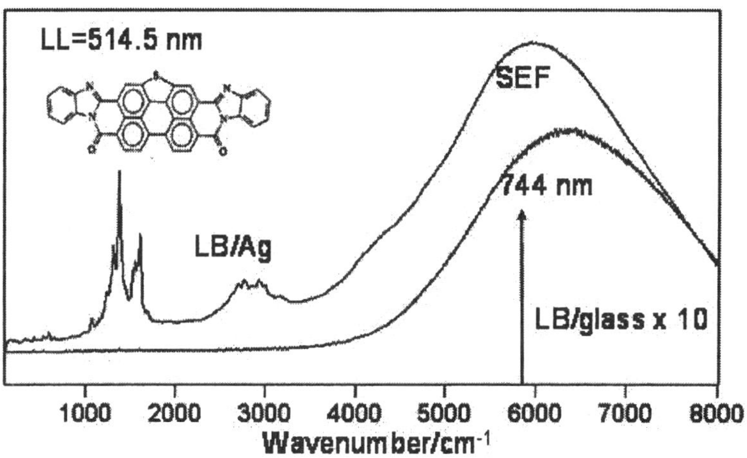


**Figure 3.9:** Example of a molecule where SEF is observed for an LB monolayer directly deposited onto the SIF due to the edge-on molecular orientation.

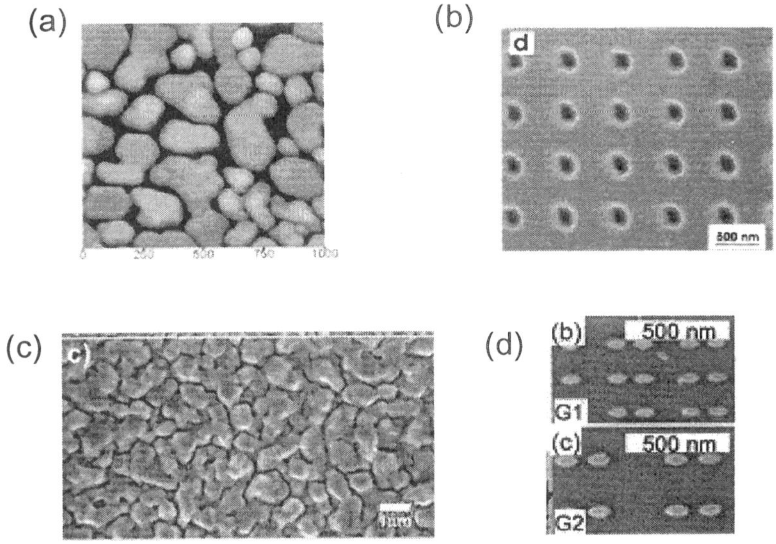


**Figure 3.10:** Various SEF Substrates; (a) pseudotabular nanoparticles [*57*]. (b) a nanohole gold film [*54*]. (c) a aluminium island film [*53*]. (d) Pairs of elliptical gold nanoparticles [*19*].

In recent years, a large variety of structures have shown to be efficient SEF enhancers: a small sampling of these is given in Figure 3.10. All these substrates, while composed of different materials and are of different shapes and morphologies, have a common denominator in that all support LSPR. It is this property that provides the field enhancement factors M(λ) (see Eq. 28). These enhanced fields essentially provide the same role regardless to the nanostructure of origin. In fabricating substrates it is a matter of choosing the nanostructures that have LSPR in the region that overlap with both the excitation wavelength and the emission wavelength.m The implementation of these basic principles is readily found in recent reports, where the plasmon responses are optimized for the target fluorophores [*58*]. This has been demonstrated with single nanoparticles of different shapes [*59*], arrays of various sized nanoparticles and separations [*60*], or even by variation in the thicknesses of nanoshells [*20*]. The LSPR of the materials provide the enhancement for emission regardless of the photoprocess that generated it: be it monomer fluorescence, phosphorescence, chemiluminescence or excimer emission.

## 3.5 CONCLUSION

Plasmonic engineering provides the fundamentals for nanostructure fabrication exploiting the unique optical properties of certain metals (mainly silver and gold) and advancing the development and applications of plasmon enhanced luminescence. Nanostructures can not only be fabricated to provide enhancement throughout the UV-Vis spectrum but also for near infrared excitation and emissions.

The electromagnetic field enhancement provided by nanostructure plasmonics is the key factor to manipulate the quantum efficiency. However, as it is illustrated in the unified theory of enhancement, since both the radiative and non-radiative rates of the molecular systems are affected by proximity of the nanostructure, the tuning has to be done on a case by case basis. In addition, there are factors due to molecule-metal interactions and molecular orientation at the surface causing effects that are much more molecule dependent and as are much more difficult to predict. Given the fact that fluorescence cross sections are the one of the highest in optical spectroscopy the analytical horizon of SEF or MEF is enormous, in particular in the expanding field of nano-bio science.

## 3.6 REFERENCES


1. Philpott, M. R. (1975). Effect of surface plasmons on transitions in molecules. *The Journal of Chemical Physics* 62: 1812-1817.
2. Otto, A. (1968). Excitation of nonradiative surface plasma waves in silver by the method of frustrated total reflection. *Zeitschrift fur Physik* 216: 398-410.
3. Willets, K. A., and Van Duyne, R. P. (2007). Localized surface plasmon resonance spectroscopy and sensing. *Annual Review of Physical Chemistry* 58: 267-297.
4. Liebermann, T., and Knoll, W. (2000). Surface-plasmon field-enhanced fluorescence spectroscopy. *Colloid and Surfaces A. Physicochemical and Engineering Aspects* 171: 115-130.
5. Gersten, J., and Nitzan, A. (1981). Spectroscopic properties of molecules interacting with small dielectric particles. *Journal of Chemical Physics* 75: 1139-1152.
6. Lakowicz, J. R. (2001). Radiative decay engineering: Biophysical and biomedical applications. *Analytical Biochemistry* 298: 1-24.
7. Brongersma, M. L., Zia, R., and Schuller, J. A. (2007). Plasmonics - the missing link between nanoelectronics and microphotonics. *Applied Physics A: Materials Science & Processing* 89: 221-223.
8. Eustis, S., and El-Sayed, M. A. (2006). Why gold nanoparticles are more precious than pretty gold: Noble metal surface plasmon resonance and its enhancement of the radiative and nonradiative properties of nanocrystals of different shapes. *Chemical Society Reviews* 35: 209-217.
9. Weitz, D. A., Garoff, S., Hanson, C. D., and Gramila, T. J. (1982). Fluorescence lifetimes of molecules on silver island films. *Optics Letters* 7: 89-91.
10. Chance, R. R., Prock, A., and Silbey, R. (1978). *Adv. Chem. Phys.* 37: 1.
11. Yamaguchi, T., Kaya, T., and Takei, H. (2007). Characterization of cap-shaped silver particles for surface-enhanced fluorescence effects. *Analytical Biochemistry* 364: 171-179.
12. Zhang, J., Malicka, J., Gryczynski, I., and Lakowicz, J. R. (2005). Surface-enhanced fluorescence of fluorescein-labeled oligonucleotides capped on silver nanoparticles. *Journal of Physical Chemistry B* 109: 7643-7648.

13. Zhang, Y., Aslan, K., Previte, M. J. R., and Geddes, C. D. (2007). Metal-enhanced fluorescence: Surface plasmons can radiate a fluorophore's structured emission. *Applied Physics Letters* 90: 053107/053101-053107/053103.
14. Wokaun, A., Lutz, H. P., King, A. P., Wild, U. P., and Ernst, R. R. (1983). Energy transfer in surface enhanced luminescence. *Journal of Chemical Physics* 79: 509-514.
15. Aroca, R., Kovacs, G. J., Jennings, C. A., Loutfy, R. O., and Vincett, P. S. (1988). Fluorescence enhancement from Langmuir-Blodgett monolayers on silver island films. *Langmuir* 4: 518-521.
16. Ray, K., Badugu, R., and Lakowicz, J. R. (2007). Sulforhodamine Adsorbed Langmuir-Blodgett Layers on Silver Island Films: Effect of Probe Distance on the Metal-Enhanced Fluorescence. *Journal of Physical Chemistry C* 111: 7091-7097.
17. Ray, K., Badugu, R., and Lakowicz, J. R. (2007). Polyelectrolyte Layer-by-Layer Assembly To Control the Distance between Fluorophores and Plasmonic Nanostructures. *Chemistry of Materials* 19: 5902-5909.
18. Andrew, P., and Barnes, W. L. (2004). Energy transfer across a metal film mediated by surface plasmon polaritons. *Science* 396.
19. Bakker, R. M., Yuan, H.-K., Liu, Z., Drachev, V. P., Kildishev, A. V., Shalaev, V. M., Pedersen, R. H., Gresillon, S., and Boltasseva, A. (2008). Enhanced localized fluorescence in plasmonic nanoantennae. *Applied Physics Letters* 92 043101-043103.
20. Tam, F., Goodrich, G. P., Johnson, B. R., and Halas, N. J. (2007). Plasmonic Enhancement of Molecular Fluorescence. *Nano letters* Vol. 7 496-501.
21. Stranik, O., Nooney, R., McDonagh, C., and MacCraith, B. D. (2007). Optimization of nanoparticle size for plasmonic enhancement of fluorescence. *Plasmonics* 2: 15-22.
22. Sheridan, A. K., Clark, A. W., Glidle, A., Cooper, J. M., and Cumming, D. R. S. (2007). Multiple plasmon resonances from gold nanostructures. *Applied Physics Letters* 90: 143105-143103.
23. Le Ru, E. C., Etchegoin, P. G., Grand, J., Felidj, N., Aubard, J., and Levi, G. (2007). Mechanisms of Spectral Profile Modification in Surface-Enhanced Fluorescence. *Journal of Physical Chemistry C* 111: 16076-16079.
24. Geddes, C. D., and Lakowicz, J. R., (Eds.) (2005). *Surface-enhancement of fluorescence near noble metal nanostructures (Goulet, P. J. G. and Aroca, R. F.)* Vol. 8.
25. Goulet, P. J. G., Pieczonka, N. P. W., and Aroca, R. F. (2005). Mapping single-molecule SERRS from Langmuir-Blodgett monolayers on nanostructured silver island films. *Journal of Raman Spectroscopy* 36: 574-580.
26. Aroca, R. (2006). *Surface-enhanced Vibrational Spectroscopy*, John Wiley & Sons, Chichester.
27. Le Ru, E. C., Etchegoin, P. G., and Meyer, M. (2006). Enhancement factor distribution around a single SERS hot-spot and its relation to single molecule detection. *Los Alamos National Laboratory, Preprint Archive, Physics*: 1-13, arXiv:physics/0608139.

28. Stockman, M. I., Faleev, S. V., and Bergman, D. J. (2001). Localization versus delocalization of surface plasmons in nanosystems: can one state have both characteristics? *Physical Review Letters* 87: 167401-167404.
29. Inoue, M., and Ohtaka, K. (1983). Surface enhanced Raman scattering by metal spheres. I. Cluster effect. *Journal of the Physical Society of Japan* 52: 3853-3864.
30. Xu, H., Wang, X.-H., Persson, M. P., Xu, H. Q., Kall, M., and Johansson, P. (2004). Unified Treatment of Fluorescence and Raman Scattering Processes near Metal Surfaces. *Physical Review Letters* 93: 243002/243001-243002/243004.
31. Blanco, L. A., and Garcıa de Abajo, F. J. (2004). Spontaneous light emission in complex nanostructures. *Physical Review B* 69: 205414-205411, 205414-205412.
32. Schmid, G., (Ed.) (2005). *Nanoparticles. From theory to Applications*, Wiley-VCH Verlag, Essen.
33. Wiley, B., Sun, Y., and Xia, Y. (2007). Synthesis of Silver Nanostructures with Controlled Shapes and Properties. *Acc. Chem. Res.* 40: 1067-1076.
34. Maxwell-Garnet, J. C. (1904). Colours in metal glasses and in metallic films. *Philos. Trans. R. Soc. London, Ser. A.* 203: 385-420.
35. Mie, G. (1908). Considerations on the Optics of Turbid Media, Especially Colloidal Metal Sols. *Ann. Physik* 25: 377-445.
36. Mulvaney, P. (1996). Surface Plasmon Spectroscopy of Nanosized Metal Particles. *Langmuir* 12: 788-800.
37. Bohren, C. F., and Huffman, D. R. (1983). *Absorption and Scattering of Light by Small Particles*, Wiley Interscience, New York.
38. Draine, T. (1988). *Astropyhys. J.* 333: 848.
39. Podolski, V. A., Sarychev, A. K., Narimanov, E. E., and Shalaev, V. M. (2005). *Opt. A: Pure Appl. Opt.* 7: S32-S37.
40. Markel, V. A., Shalaev, V. M., Stechel, E. B., Kim, W., and Armstrong, R. L. (1996). Small-particle composites. I. Linear optical properties. *Physical Review B* 53: 2425-2436.
41. Palik, E. D., (Ed.) (1985). *Handbook of Optical Constants of Solids*, Academic, New Work.
42. Markel, V. A., and Shalaev, V. M. (1999). Computational Approaches In Optics Of Fractal Clusters, in *Computational Studies of New Materials* (Jelski, D. A., and George, T. F., Eds.), World Scientific Publishing, Singapore.
43. Jackson, J. D. (1999). *Classical Electrodynamics*, 3rd ed., John Wiley and Sons, New York.
44. Johansson, P., Xu, H., and Kall, M. (2005). Surface-enhanced Raman scattering and fluorescence near metal nanoparticles. *Physical Review B: Condensed Matter and Materials Physics* 72: 035427/035421-035427/035417.
45. Scully, M. O., and Zubaire, M. S. (1997). *Quantum Optics*, Cambridge University Press, New York.
46. Aroca, R. F., Constantino, C. J. L., and Duff, J. (2000). Surface-enhanced Raman scattering and imaging of Langmuir-Blodgett monolayers of bis(phenethylimido)perylene on silver island films. *Applied Spectroscopy* 54: 1120-1125.

47. Constantino, C. J. L., Aroca, R. F., Mendonca, C. R., Mello, S. V., Balogh, D. T., and Oliveira, O. N. (2001). Surface enhanced fluorescence and Raman imaging of Langmuir-Blodgett azopolymer films. *Spectrochimica Acta, Part A: Molecular and Biomolecular Spectroscopy* 57A: 281-289.
48. Lakowicz, J. R., Shen, Y. B., D'Auria, S., Malicka, J., Fang, J. Y., Gryczynski, Z., and Gryczynski, I. (2002). Radiative decay engineering 2. Effects of silver island films on fluorescence intensity, lifetimes, and resonance energy transfer. *Analytical Biochemistry* 301: 261-277.
49. Lakowicz, J. R., Geddes, C. D., Gryczynski, I., Malicka, J., Gryczynski, Z., Aslan, K., Lukomska, J., Matveeva, E., Zhang, J. A., Badugu, R., and Huang, J. (2004). Advances in surface-enhanced fluorescence. *Journal of Fluorescence* 14: 425-441.
50. Pan, S. L., and Rothberg, L. J. (2005). Enhancement of platinum octaethyl porphyrin phosphorescence near nanotextured silver surfaces. *Journal of the American Chemical Society* 127: 6087-6094.
51. Mackowski, S., Woermke, S., Maier, A. J., Brotosudarmo, T. H. P., Harutyunyan, H., Hartschuh, A., Govorov, A. O., Scheer, H., and Braeuchle, C. (2007). Metal-Enhanced Fluorescence of Chlorophylls in Single Light-Harvesting Complexes. *Nano Letters*: ACS ASAP.
52. dos Santos, D. S., Jr., and Aroca, R. F. (2007). Selective surface-enhanced fluorescence and dye aggregation with layer-by-layer film substrates. *Analyst (Cambridge, United Kingdom)* 132: 450-454.
53. Ray, K., Chowdhury, M. H., and Lakowicz, J. R. (2007). Aluminum Nanostructured Films as Substrates for Enhanced Fluorescence in the Ultraviolet-Blue Spectral Region. *Analytical Chemistry (Washington, DC, United States)* 79: 6480-6487.
54. Brolo, A. G., Kwok, S. C., Moffitt, M. G., Gordon, R., Riordon, J., and Kavanagh, K. L. (2005). Enhanced Fluorescence from Arrays of Nanoholes in a Gold Film. *Journal of the American Chemical Society* 127 14936-14941.
55. DeSaja-Gonzalez, J., Aroca, R., Nagao, Y., and DeSaja, J. A. (1997). Surface-enhanced fluorescence and SERRS spectra of N-octadecyl-3,4:9,10-perylenetetracarboxylic monoanhydride on silver island films. *Spectrochimica Acta, Part A: Molecular and Biomolecular Spectroscopy* 53A: 173-181.
56. Lakowicz, J. R., Geddes, C. D., Gryczynski, I., Malicka, J., Gryczynski, Z., Aslan, K., Lukomska, J., Matveeva, E., Zhang, J., Badugu, R., and Huang, J. (2004). Advances in Surface-Enhanced Fluorescence. *Journal of Fluorescence* 14: 425-441.
57. Kawasaki, M., and Mine, S. (2005). Enhanced Molecular Fluorescence near Thick Ag Island Film of Large Pseudotabular Nanoparticles. *Journal of Physical Chemistry B* 109: 17254-17261.
58. Gerber, S., Reil, F., Hohenester, U., Schlagenhaufen, T., Krenn, J. R., and Leitner, A. (2007). Tailoring light emission properties of fluorophores by coupling to resonance-tuned metallic nanostructures. *Physical Review B* 75: 073404-073404.
59. Chen, Y., Munechika, K., and Ginger, D. S. (2007). Dependence of Fluorescence Intensity on the Spectral Overlap between Fluorophores and Plasmon Resonant Single Silver Nanoparticles. *Nano Letters* 7: 690-696.

60. Lakowicz, J. R., Malicka, J., Gryczynski, I., Gryczynski, Z., and Geddes, C. D. (2003). Radiative decay engineering: The role of photonic mode density in biotechnology. *Journal of Physics D: Applied Physics* 36: R240-R249.

4

# Importance of Spectral Overlap: Fluorescence Enhancement by Single Metal Nanoparticles

Keiko Munechika, Yeechi Chen, Jessica M. Smith and David S. Ginger

Department of Chemistry, University of Washington, Seattle, WA 98195-1700

## 4.1 INTRODUCTION

Metal nanoparticles and nanostructured metal films possess localized surface plasmon resonances (LSPRs) that give these materials a number of unique optical properties. In particular, the enhanced and confined local electromagnetic fields that occur near the surface of metal nanoparticles when excited at an LSPR can lead to significantly enhanced fluorescence from nearby fluorophores,[1-11] despite the fact that planar metal films generally quench fluorescence.[12,13] Different aspects and applications of this phenomena, which is known variously as plasmon-enhanced fluorescence, near-field enhanced fluorescence or *metal-enhanced fluorescence*, are discussed through this volume. In this chapter we focus on the wavelength dependence of the observed fluorescence enhancement effects, and summarize our existing understanding in convenient rules for optimizing structures for specific applications.

To understand the importance of spectral overlap to metal-enhanced fluorescence, it is useful to review the basics of metal-enhanced fluorescence. Metal nanostructures can alter the apparent fluorescence from nearby fluorophores in two ways. First, metal nanoparticles can enhance the excitation rate of the nearby fluorophore, as the excitation rate is proportional to the electric field intensity that is increased by the local-field enhancement. Fluorophores in such "hot spots" absorb more light than in the absence of the metal nanoparticle. Second, metal nanoparticles can alter the radiative decay rate and nonradiative decay rate of the nearby fluorophore, thus changing both quantum yield and the lifetime of the emitting species. We can summarize the various effects of a nanoparticle on the apparent fluorescence intensity, $Y_{APP}$, of a nearby fluorophore as:

$$Y_{APP} = \gamma_{ex}(\omega_{ex})Q_{EM}(\omega_{em})\eta_{coll}(\omega_{em})\sigma \qquad (1)$$

where $\gamma_{ex}(\omega_{ex})$ is the excitation rate of the fluorophores in the particle near-field at the excitation frequency, $\omega_{ex}$; $Q_{EM}(\omega_{em})$ is the quantum yield for far-field emission at the emission frequency, $\omega_{em}$; $\eta_{coll}(\omega_{em})$ is the collection efficiency per unit area of the far-field light in the experimental geometry (accounting for any modification of the free-space spatial emission profile and the fixed acceptance of the detector) and $\sigma$ is a normalization factor accounting for the fluorophore attachment density and total area

*The Role Of Plasmonic Engineering In Surface-Enhanced Fluorescence.*
Edited by Chris D. Geddes.

excited. The $\gamma_{ex}(\omega_{ex})$ and $Q_{EM}(\omega_{em})$ terms are both sensitive to frequency, distance, and orientation of the fluorophore relative to the metal. Generally, $\gamma_{ex}(\omega_{ex})$ depends on both the absorption coefficient of the fluorophore and the local (nanoparticle-enhanced) field intensity. Since the field intensity increases closer to the nanoparticle surface, $\gamma_{ex}(\omega_{ex})$ should be maximized closer to the surface. The behavior of $Q_{EM}(\omega_{em})$ is more complicated, as the quantum yield of the fluorophore is a ratio of the radiative decay rate to the sum of all possible decay rates. Not only can the metal-altered local photonic mode density (the number of states into which the photon can be emitted) lead to changes in the radiative decay rate of the fluorophore, but the presence of the metal also opens up new non-radiative decay pathways via energy transfer to the metal.[13-17] In addition, energy transferred to the excited metal plasmon modes can be re-scattered back into the far-field by nanoparticles[10,13] or the energy of the excited fluorophore can be quenched by loss to non-radiative decay pathways in the metal. Thus, a metal nanostructure can lead to either an increase or a decrease in the fluorescence quantum efficiency of a nearby fluorophore, depending on the relative magnitudes of the enhancement and quenching terms.

Although it has been difficult to separate the effects of excitation and emission enhancement, both of these effects should be extremely sensitive functions of the shape of the metal particle, the orientation of the fluorophore, and the distance between the fluorophore and the metal,[7,15,18-21] because the local-field effects depend strongly on these parameters. Many groups have studied variations in fluorescence intensity as a function of the distance between a layer of fluorophores and a number of nanostructured metal surfaces,[22-24] adsorbed colloidal particles[1,25,26] or suspended colloidal particles.[17,27,28] Single-molecule experiments have even provided strong evidence for the existence of a local maximum in the fluorescence intensity versus distance curve.[4-6]

In addition to this distance dependence, local-field enhancements surrounding metal nanostructures are strongly wavelength dependent.[29] For example, Figure 4.1 shows Finite Difference Time Domain (FDTD) calculations[30] of the magnitude of the electric field around a Ag nanoprism with slightly rounded corners both on (508 nm) and off (608 nm) resonance. The electric field intensity (square of the magnitude plotted) is significantly larger on resonance, and is consistent with previous calculations.[29] Given the wavelength dependence of all the materials involved, it seems apparent that the strongest interaction will occur when the nanoparticle plasmon resonance has good overlap with the excitation and/or emission spectra of the fluorophore. However, since both enhancement and quenching factors are wavelength dependent,[7,20] different degrees of spectral offset are required depending on whether maximum fluorescence (e.g. for sensing applications), maximum excitation enhancement (e.g. for photochemistry), maximum acceleration of the fluorophore decay rate (e.g. improved quantum yield/photostability), or maximum overall brightness is desired in specific application.

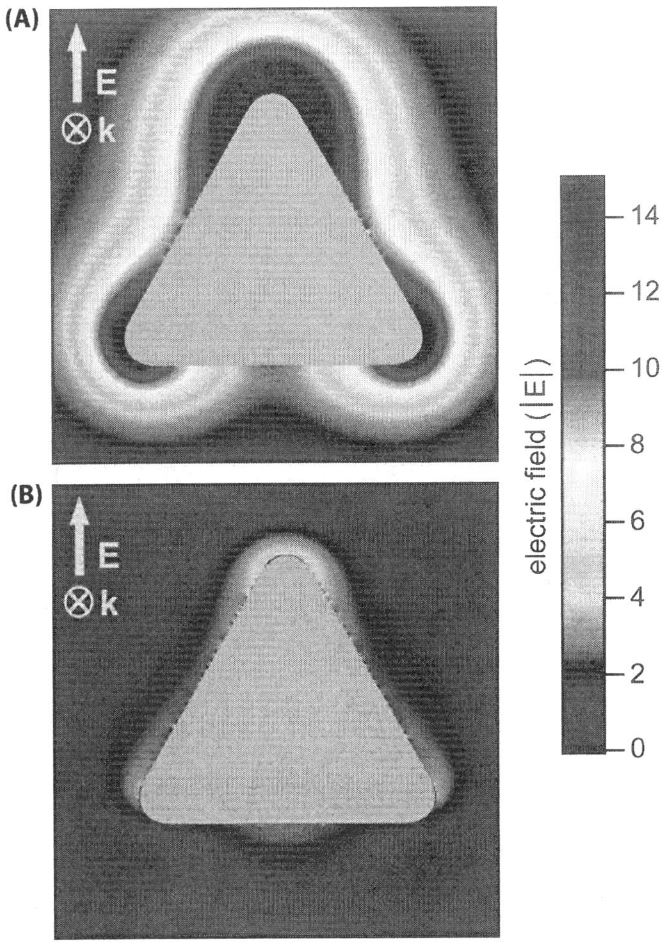


**Figure 4.1:** Finite Difference Time Domain (FDTD) calculations of the local electric field around a Ag nanoprism with slightly rounded corners (100 nm edge length, 12 nm thick) both (A) on (508 nm) and (B) off (608 nm) resonance.

We begin our discussion of the effects of spectral overlap on metal-enhanced fluorescence in Section 4.2 with a discussion of the preparation and characterization of our metal nanoparticle of choice: the Ag nanoprism. We discuss the synthesis, optical and physical characterization of chemically synthesized, single Ag nanoprisms and explain the features that make them particularly well-suited for fluorescence enhancement studies. We also briefly discuss the plasmon linewidths of Ag nanoprisms and explain factors which contribute to the homogeneous linewidth broadening in single particle studies. Next, in Section 4.3, we examine the one-photon fluorescence enhancement of fluorophores near Ag nanoparticles. We show the fluorescence intensity of fluorophores near metal nanoparticle is highly dependent on the spectral overlap between a nanoparticle's LSPR scattering peak and the fluorophore's absorption and emission spectra. Finally, in Section 4.4, we describe the measurement of excitation enhancement using the system of Ag nanoparticles placed near a layer of CdSe nanocrystals in order to separate the effects of excitation and emission.

## 4.2 OVERVIEW OF Ag NANOPRISMS FOR FLUORESCENCE ENHANCEMENT: SYNTHESIS AND OPTICAL PROPERTIES

Although many types of metal nanostructures including fractal aggregates and rough metal island films,[1,3,10] spheres,[4-7,11] spherical shells,[8] periodic arrays,[2] prisms,[9] and nanoholes[31-33] can lead to enhanced fluorescence, we choose to study metal-enhanced fluorescence using chemically synthesized Ag nanoprisms. These nanoprisms are attractive for their large scattering cross sections and the plasmon resonances can be tunable across the visible spectrum.[34-36] Furthermore, as seen in Figure 4.1, calculations show that Ag nanoprisms exhibit strong local-field enhancement due to their sharp tips.[29] We synthesize Ag nanoprisms via sodium borohydride reduction of Ag precursor salts following literature procedures.[36-38] The striking and tunable optical properties of these particles can be seen in Figure 4.2. Figure 4.2A shows a series of Ag nanoparticle solutions that only vary in size and shape. Figure 2B shows the extinction spectra of the same nanoparticle solutions taken using a UV-Vis spectrometer. The nanoparticle plasmon resonances can be tuned across the entire visible spectrum, however, in some cases the full width half maximum (FWHM) of the extinction spectrum spreads well above 300 nm. Although synthetic methods are continually being improved,[39-41] much of the broad solution linewidth in our samples results from inhomogeneous broadening due to the distribution of nanoparticle shapes and sizes. For example, Figure 4.3 shows several scanning electron microscope (SEM) images of chemically synthesized Ag nanoprisms showing the typical distributions in sizes and shapes. In our hands, the nanoprisms typically show various degrees of truncation at the tips.

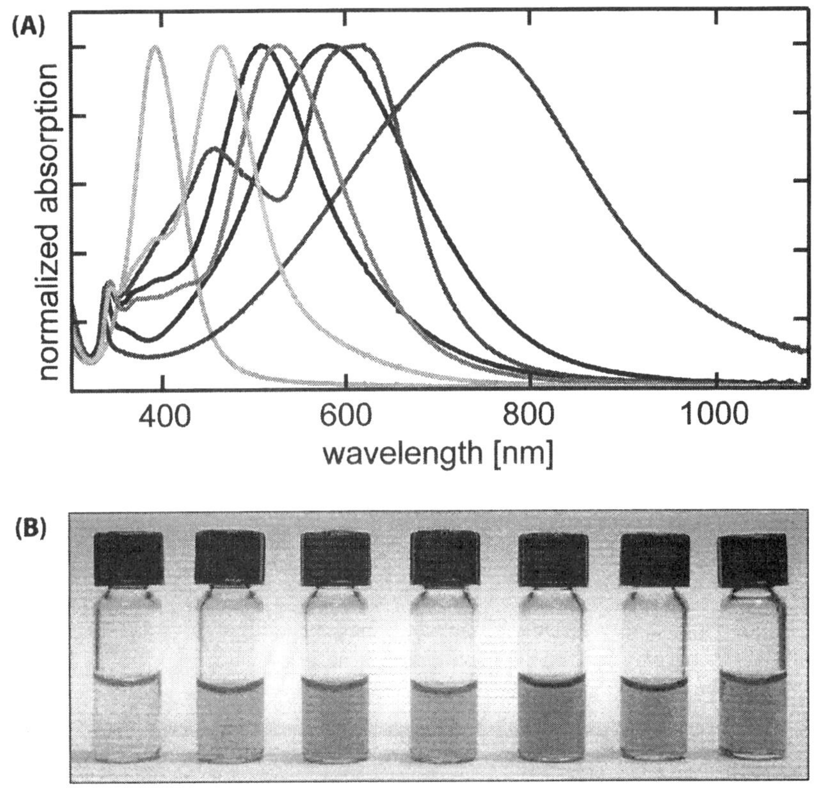


**Figure 4.2:** (A) Series of normalized extinction spectra of chemically synthesized colloidal Ag nanoparticle solutions, showing the tuning of the plasmon resonance across the visible region. (B) Images of colloidal nanoparticle solutions. The differences in color are due to variations in the size and shape of the nanoparticles within each solution.

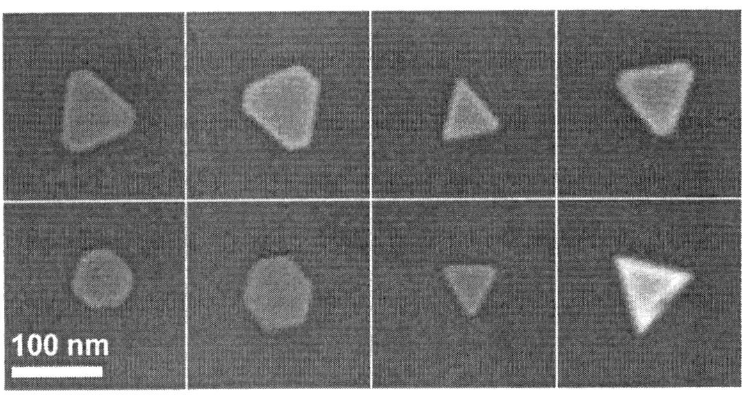

**Figure 4.3:** Collage of SEM images of chemically-synthesized Ag nanoprisms showing the typical size and shape variations. The corners of the nanoprisms also exhibit varying degrees of truncation.

Such broad extinction peaks are undesirable for precise studies of the impact of the spectral overlap between fluorophore and nanoparticle on the apparent fluorescence intensity.[9,42,43] In order to remove this inhomogeneous broadening, we use single particle spectroscopy to collect individual nanoparticle LSPR scattering spectra. Figure 4 shows a schematic illustration of the experimental setup. Optical microscopy and spectroscopy are performed using an inverted microscope fitted with a transmitted darkfield condenser and an objective with a total effective magnification of 75X. The microscope output is either directed to a thermoelectrically-cooled color CCD camera or a fiber optic cable (diameter = 100 μm) coupled to a small CCD spectrometer. A standard tungsten halogen lamp is used for transmitted light darkfield illumination, and a metal halide lamp is used for epi-fluorescence illumination. Figure 4.5A shows a typical distribution of Ag nanoparticles immobilized onto a silanized glass cover slip. Each of the colored particles is a single Ag nanoparticle with a different plasmon resonance. Figure 4.5B shows the single particle scattering spectra of the labeled nanoparticles in the darkfield scattering image in Figure 4.5A. The individual LSPR scattering linewidths we observe range from 30-80 nm, consistent with the other reports.[34,35,44] These spectra are considerably narrower than the ensemble colloidal solution extinction spectra (shown in shaded/dashed line in Figure 4.5B).

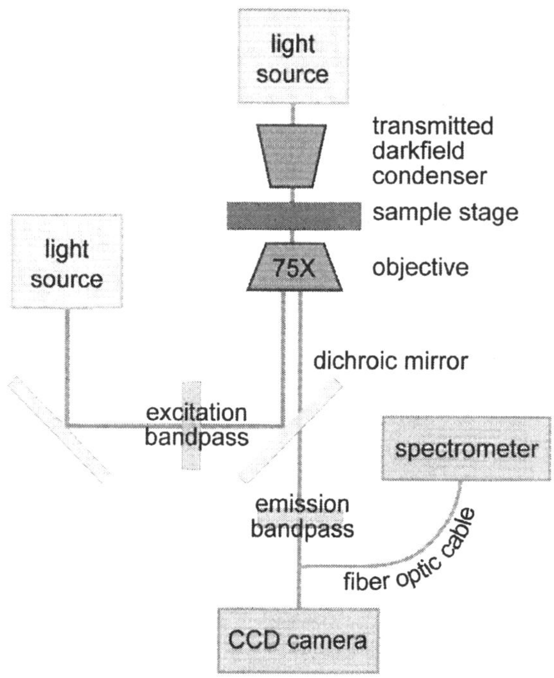

**Figure 4.4:** Schematic of the experimental apparatus used for single-particle darkfield scattering spectroscopy.

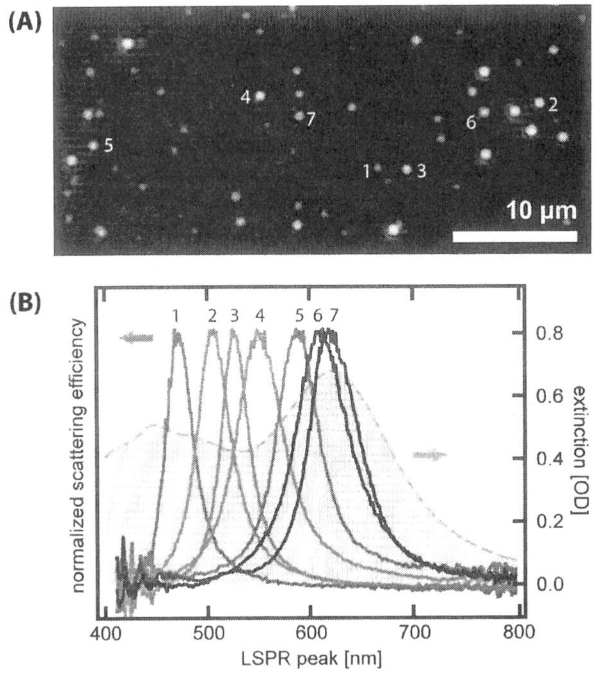


**Figure 4.5** (A) Darkfield optical micrograph of a typical distribution of single Ag nanoparticles immobilized on a glass cover slip. (B) Single-particle darkfield scattering spectra corresponding to the individual Ag nanoprisms labeled in (A). The ensemble solution extinction spectrum is shown as the shaded, dashed curve for comparison. Reprinted with permission from reference 9.

Although single particle spectroscopy can remove the inhomogeneous broadening seen in ensemble colloidal spectra, single particle scattering spectra still show variations in the individual particle linewidths. Since these linewidths are related to the quality of the plasmon resonance, their study is also of interest for applications in metal-enhanced fluorescence. The scattering linewidth of a metal nanoparticle is generally limited by lifetime broadening.[45-47] The linewidth (Γ) of the nanoparticle plasmon resonance is related to the total plasmon dephasing time by $\Gamma_{tot} = 2\hbar / T_{total}$, where $T_{total}$ is total dephasing time of the plasmon due to all possible decay pathways. The primary plasmon decay pathways are radiation into the far field (radiation damping) and dephasing of the electron gas (non-radiative decay). Thus, we can view the total linewidth as containing contributions from both radiative, $\Gamma_{rad}$, and non-radiative, $\Gamma_{non\text{-}rad}$, contributions: $\Gamma_{total} = \Gamma_{rad} + \Gamma_{non\text{-}rad}$.[45,48] Qualitatively, we expect that $\Gamma_{non\text{-}rad}$ should be sensitive to the plasmon peak position due to the frequency-dependent dielectric properties of the metal,[49] and that $\Gamma_{rad}$ should depend both on particle size and plasmon peak position due to the dependence of radiation damping on these two parameters.[15,50]

In order to investigate the factors which contribute to linewidth broadening, we correlated number of single particle scattering spectra with individual SEM images. Figure 6A-B shows the measured single-nanoprism linewidths plotted as a function of both the nanoparticle volume (binned by 10,000 $nm^3$ increments) and the resonance energy (binned by 0.1 eV increments). We found that the linewidths increased linearly as a function of increasing volume and also increased as a function of the plasmon resonance energy. Theoretical calculations indicate that the primary scattering resonance for triangular Ag nanoprisms is an in-plane dipole resonance,[35,51] so we consider the prediction of simple dipole-limit approximations for the changes in our particle linewidths. In the perfect dipole limit, the radiation damping term is directly proportional to particle volume, $\Gamma_{rad} = 2\hbar\kappa V$, where $\kappa$ is the proportionality constant and $V$ is the volume of the particle.[15,52,53] Figure 4.6B shows the plot of the average linewidth as a function of energy (open squares), showing the linear increase in the linewidths as a function of resonance energy. The common expression for the nonradiative contribution to the linewidth, $\Gamma_{non-rad} \approx \frac{2\varepsilon_2}{\varepsilon_1'}$, shows that we should expect the total linewidth to be frequency dependent. In this expression, $\varepsilon_1$ and $\varepsilon_2$ are the real and imaginary parts of the wavelength dependent complex dielectric constant of the metal, and $\varepsilon_1'$ is the first derivative of the real part of the dielectric constant with respect to energy.[49,54] Our observed trends, namely the linear increase in linewidths as a function of both energy and volume of the nanoparticles are consistent with the expectations from the dipole-limit approximations due to the strong dipolar character of the lowest energy scattering peak for Ag nanoprisms with edge lengths less than ~100 nm. The discussion of the detailed analysis is omitted from this chapter, but can be found elsewhere.[42] In summary, we found quantitative agreement with theoretical predictions that larger nanoprisms should have larger linewidths due to increased radiation damping, and that bluer nanoprisms have larger linewidths due to increased energy loss in the metal.

However, the agreement between experiment and theory could only be obtained when using optical constants for Ag as tabulated by Palik,[55] rather than by Johnson & Christy.[56] Figure 4.6B also shows the calculated total linewidths from the dipole-limit approximations using the optical constants values from Palik (solid line) and Johnson and Christy (dashed line). We observe that the measured linewidths (open squares) are in good agreement with the linewidth values calculated from Palik's optical constants in terms of both the magnitude and the energy dependence. On the contrary, the optical constants from Johnson and Christy lead to underestimation of the linewidths, mainly due to the smaller amount of non-radiative loss in the optical constants. This is an important insight to consider when predicting the field enhancement factor for these types of nanoparticles, since underestimation of nonradiative decay may result in accidental overestimation of the local-field enhancement that can be achievable. In the next section, we discuss fluorescence enhancement from fluorophores adsorbed onto these Ag nanoprisms.

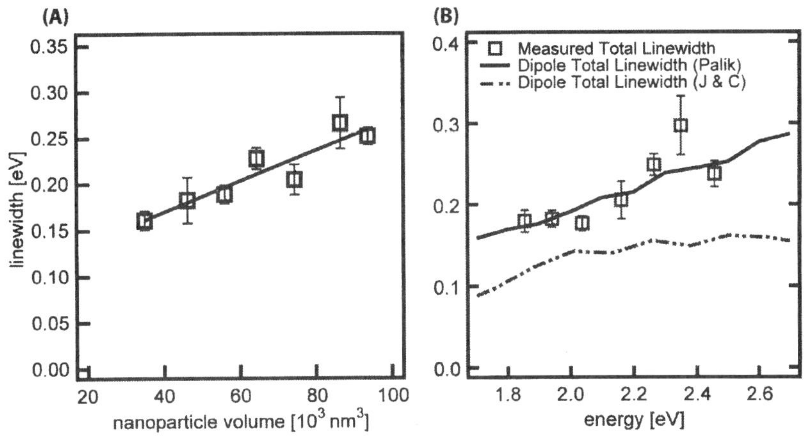


**Figure 4.6:** (A) Plot of linewidths vs. nanoparticle volume binned by particle volume (as determined by SEM). The Y error bars represent the standard deviation of the mean linewidth of the particle within each bin. The solid line is a linear fit to the data showing that the linewidths increase linearly as a function of increasing volume. (B) Plot of the average linewidth vs. plasmon resonance peak energy binned together by 0.1 eV intervals along the X-axis (empty squares). Y-error bars represent the standard deviation of the mean linewidth of the particles within each bin. The black solid line is the linewidth using the dipole-limit approximations with Palik's optical constants (ref 55), showing good agreement with our measured linewidths. The red dashed line is the linewidth using the dipole-limit approximations with Johnson & Christy's optical constants (ref 56) as input. Adapted from reference 42.

## 4.3 FLUORESCENCE ENHANCEMENT BY SINGLE AG NANOPRISMS AND CORRELATION WITH PLASMON PEAK POSITION

While a number of groups have studied metal-enhanced fluorescence from clustered films,[1,3,10,57,58] metal-enhanced fluorescence can also be observed near isolated metal nanoparticles.[2,4-9,11] Figure 4.7A depicts a simple experimental scheme shown by the Rothberg group,[3] that can be used to observe metal-enhanced fluorescence near Ag nanoprisms. Here, a layer of organic fluorophores is adsorbed onto a 3-mercaptopropyltrimethoxysilane functionalized indium tin oxide (ITO) slide. On top of the layer of organic fluorophores, a dilute solution of Ag nanoparticles is drop-coated onto the sample. The nanoparticle density is controlled by concentration and exposure time to ensure that adsorbed nanoparticles are optically isolated and resolvable using a conventional far-field optical microscope. Figure 4.7B shows a darkfield scattering image taken from an area of the substrate. As in the previous darkfield images each of the colored spots corresponds to a single Ag nanoparticle. Figure 4.7C shows the corresponding fluorescence image of the same area shown in the darkfield scattering image in Figure 4.7B. Although the fluorophore density is uniform underneath the adsorbed nanoparticles, there are clearly spots of brightly enhanced fluorescence near Ag nanoprisms. In fact, the bright fluorescence spots correspond directly to the locations of the Ag nanoprisms shown in the darkfield image, showing that these nanoprisms can dramatically enhance the fluorescence of nearby fluorophores. This relatively simple experiment shows that fluorescence enhancement using near-field effects is achievable. One can easily notice that the enhanced fluorescence intensity varies significantly from one nanoparticle to the next. The question arises however: why do some nanoparticles greatly enhance fluorescence from the nearby fluorophores while others do not? The answer, of course, lies in the different terms in Equation 1. Here, however, we wish to focus on the variations that correlate with the different spectral properties of the metal nanoparticles. While differences in distance also affect the resulting fluorescence, we can minimize these effects by using well-defined nanoparticle-fluorophore structures.

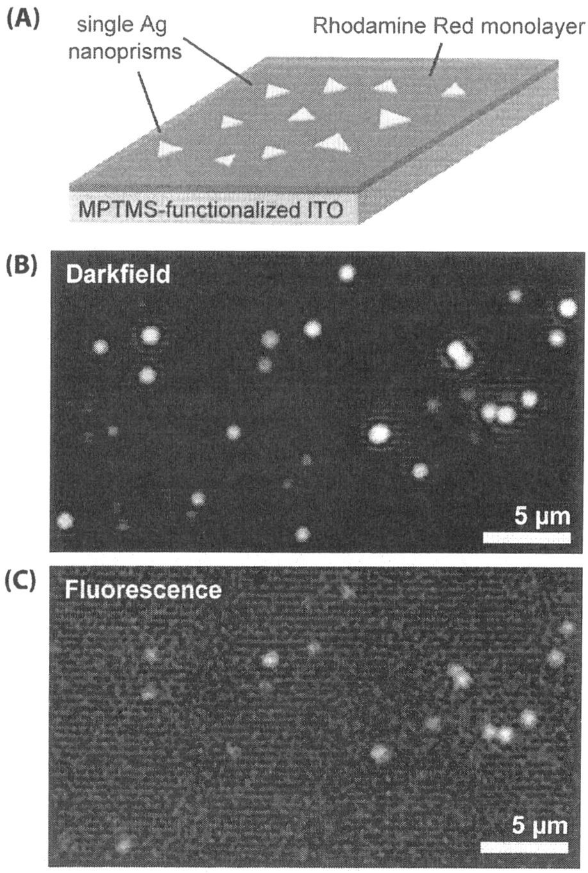


**Figure 4.7** (A) Schematic illustration of fluorescence enhancement experiment; Ag nanoprisms are adsorbed on top of monolayer of Rhodamine red on glass slide. (B) Darkfield scattering image of an area of the substrate. Each of colored

spots indicates single Ag nanoprisms. (C) Fluorescence image of the substrate taken from the same area shown in the darkfield scattering image in (B), showing enhanced fluorescence spots that directly correspond to the locations of Ag nanoprisms.

To this end, we have coupled fluorophores to Ag nanoprisms using double-stranded DNA as a biological linker.[17,59,60] This system allows us to specifically attach fluorophores to nanoparticle surface at a more consistent distance than is achievable with a continuous fluorophore film. A schematic depiction of the assembly of the fluorophore-coupled nanoparticle structure is shown in Scheme 1. More details can be found in the literature.[9] Figure 4.8A shows a darkfield micrograph of a field of Ag nanoparticles just prior to being functionalized with ssDNA: *Seq1* (5' HS AAA AAA AAA ACG CAT TCA GGA TTC TCA ACT CGT A 3'). Figure 4.8B shows a fluorescence micrograph of the same region after attachment of the thiolated-DNA monolayer and incubation of the substrate with non-complementary Rhodamine Red-conjugated DNA (*Seq 1*+Rhodamine Red) as a control. Only very low levels of background fluorescence are detectable either in the background or on the nanoparticles themselves. In contrast, Figure 4.8C shows an identical exposure of the same region of the substrate after it has been incubated with *complementary* fluorophore-conjugated DNA (*Seq 2* + Rhodamine: 5' HS AAA AAA AAA ATA CGA GTT GAG AAT CCT GAA TGC G 3' + Rhodamine Red) The background fluorescence remains low, but significant fluorescence is observed from the fluorophore-functionalized nanoparticles, confirming the specific DNA-mediated attachment of the fluorophore. Although the hybridization is specific, there are still clear variations in the fluorescence intensity from the fluorophores attached to each of the individual metal nanoparticles, as were seen in the case for the nanoparticles adsorbed on top of a layer of Rhodamine Red (Figure 4.7C). We believe a major component of this variation is due to the size and shape dependence of the LSPR scattering peak position of the nanoparticles, which leads to frequency dependent local-field enhancement factors at the absorption and emission frequencies of the fluorophore. Using these nanoparticle-fluorophore assemblies we were able to study these effects of spectral overlap in more detail, as discussed below.

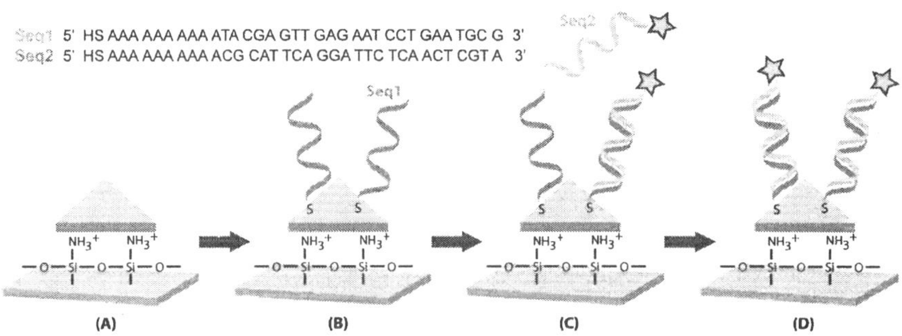


**Scheme 4.1:** Schematic illustration of DNA-dye-nanoprism cluster assembly. (A) Ag nanoprisms are immobilized on a silanized glass slide. (B) A ssDNA monolayer is attached to the nanoprisms via a 5' thiol group. (C) Complementary DNA conjugated with a fluorescent dye is hybridized to the DNA-functionalized nanoprisms, resulting in (D) specific attachment of the dye at a finite distance from the nanoprism surface. Reprinted from reference 9.

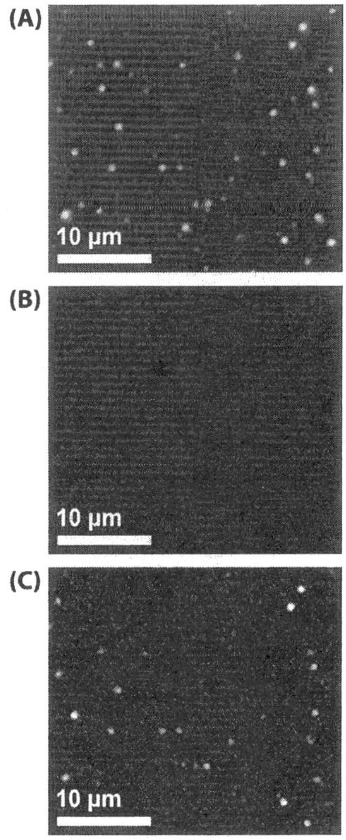


**Figure 4.8** Specific DNA-directed coupling of fluorescent dyes to Ag nanoprisms. (A) Darkfield optical micrograph showing a field of isolated Ag nanoprisms. (B) Incubation of the DNA-functionalized particle field in with non-complementary dye-labeled DNA results in little detectable fluorescence. (C) Subsequent hybridization of the same sample with complementary Rhodamine Red-labeled DNA leads to attachment of the dye and visible fluorescence from the functionalized nanoparticles. Reprinted from reference 9.

Figure 4.9A shows a darkfield scattering image of 4 different Ag nanoparticles which have been incubated with a mixture of two complementary fluorophore-functionalized DNA sequences (Alexa Fluor 488 and Rhodamine Red). Each of the fluorophores was conjugated to the same complementary DNA sequence and then hybridized to the surface in order to tag the Ag nanoparticles with equal amounts of both fluorophores. Figure 4.9B shows the fluorescence image from Alexa Fluor 488, and Figure 4.9C shows the fluorescence image from Rhodamine Red. The two fluorescence images look very different. For example, the yellow particle (#3) in the darkfield image with LSPR scattering peak at 564 nm (Figure 4.8D) shows the most Rhodamine Red fluorescence (red-orange fluorophore), but it shows very little fluorescence from Alexa Fluor 488 (green fluorophore). In contrast, the aqua-blue particle (#1) in the darkfield image with LSPR scattering peak at 517 nm (Figure

4.8D) is by far the brightest particle in the Alexa Fluor 488 fluorophore fluorescence image, while the yellow particle (#3) which is the brightest particle in the Rhodamine Red fluorescence image, has barely visible fluorescence. These results provide strong qualitative evidence that the relative overlap of the LSPR scattering peak and the absorption and emission peaks of the fluorophore play a major role in determining the apparent fluorescence from the nearby fluorophore on a single nanoparticle level.

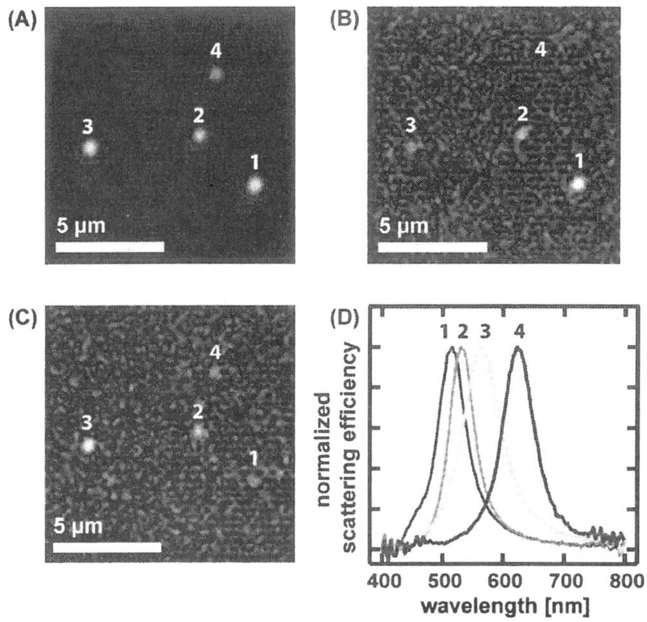


**Figure 4.9** (A) Darkfield optical micrograph of four individual Ag nanoparticles that have been hybridized with a 1:1 mixture of the dyes Alexa Fluor 488 and Rhodamine Red. (B) Fluorescence micrograph of the same area collected using Alexa Fluor 488 excitation and emission. (C) Fluorescence micrograph of the same area collected using Rhodamine Red excitation and emission. (D) Single particle scattering spectra show the LSPR for each particle in (A). Reprinted from reference 9.

Utilizing these nanoparticle-fluorophore assemblies, we have studied the impact of spectral overlap in a more quantitative fashion. We prepared DNA functionalized nanoparticles hybridized with one of the three fluorophores (Alexa Fluor 488, Alexa Fluor 532, and Rhodamine Red). We then correlated the single particle LSPR scattering peak of the nanoparticles with the fluorescence intensity of the fluorophores attached to the nanoparticles. We obtained correlated single particle LSPR scattering spectra with fluorescence intensity measurements for total 457 single nanoparticle-fluorophore clusters. These numbers only include the particles with dominant single spectral features (fwhm < 80 nm). "Particles" with broad and multi-peaked plasmon spectra were excluded from our analysis to eliminate the possibility of including nanoparticle dimers or aggregates.

Figure 4.10 A-C shows our key experimental finding from this set of experiments. Each of the plots shows the fluorescence intensity from a different fluorophore plotted as a function of the nanoparticle LSPR scattering peak (for the nanoparticle that fluorophore is attached to) grouped in 20 nm bins. The maximum fluorescence from Alexa Fluor 488 is observed when coupled to the nanoparticle LSPR scattering peaks at around 505 nm. For Alexa Fluor 532, the maximum fluorescence is observed when coupled to nanoparticles with LSPR scattering peaks at around 525 nm. For Rhodamine Red, the maximum fluorescence is observed when coupled to nanoparticles with LSPR scattering peaks at around 575 nm. For each fluorophore, the excitation and the emission spectrum are overlaid as the shaded curves in Figure 4.10 A-C for comparison. These data indicate that the fluorescence intensity of the fluorophore adsorbed onto the nanoparticle is indeed strongly dependent on the spectral overlap between the nanoparticle LSPR scattering peak and spectral features of the fluorophore. For each case, we estimated the relative "average" brightness ratio "on" and "off" the fluorescence intensity maxima to be about 5-7. We also estimated the absolute fluorescence enhancement factor (versus an equal number of free fluorophore molecules) to be approximately 9-30 for the brightest individual particles, which was obtained by determining the ratio of the fluorescence intensities of the fluorophore-labeled nanoparticles with the intensity of a single fluorophore molecule and correcting for the number of fluorophores on each nanoparticles.

Importantly, our data allow us to experimentally determine the optimal LSPR scattering position for maximum fluorescence with these organic fluorophores. For all three fluorophores, the brightest fluorescence is observed when the fluorophore emission peak is at lower energy (40-120 meV: Alexa Fluor 488, ~70 meV, Alexa Fluor 532, ~ 120 meV, Rhodamine Red, ~ 40 meV) relative to the LSPR scattering peak of the nanoparticles. This offset of the peak is in accord with theoretical calculations that predict the highest fluorescence from fluorophores with emission peaks slightly lower energy from the LSPR scattering peak. One interpretation of this result is that on resonance, both the radiative and nonradiative decay rates are enhanced, which is disadvantageous in terms of quantum yield. However, when the fluorophore emission peak is slightly red-shifted from the plasmon resonance peak of the nanoparticle the non-radiative decay rates decrease faster than the radiative decay rates, and one can thus expect to find the most fluorescence enhancement to the red of the plasmon resonance.[20]

The Novotny group has produced similar theoretical results with calculations on spherical nanoparticles, showing that fluorescence enhancement due to metal near field effects is strongly frequency dependent and that florescence enhancement is maximized when the fluorophore emits red-shifted to the plasmon resonance peak of the nanoparticle. They also explained this result as a consequence of the slight offset of the frequency dependence of the quenching term and enhancement term.[7]

We point out that another partial explanation for the experimentally observed red-shift is the combination of excitation and emission factors that enter the total enhancement expression. Since a fluorophore's fluorescence peak is red-shifted from its absorption peak, it is likely that the point of maximum brightness will also require some compromise between excitation and emission enhancement. This explanation has been given by Rothberg[3] for fluorescence enhancements from random Ag colloidal films and is reminiscent of the optimum position of the LSPR peak in a surface-enhanced Raman experiment.[61] In the next section we consider how to better separate excitation and emission factors.

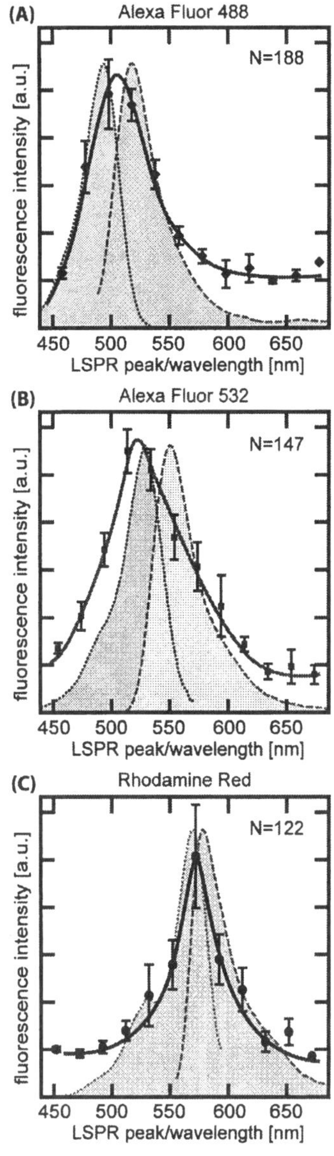


**Figure 4.10:** Summary of 457 individual particle fluorescence vs. LSPR peak position measurements with three different fluorescent dyes. The LSPR peak positions are binned in 20 nm intervals along the x-axis. The average fluorescence intensity observed from particles within each bin is then plotted as a function of the LSPR position for Ag nanoprisms functionalized with (A) Alexa Fluor 488, (B) Alexa Fluor 532 and (C) Rhodamine Red dyes. The absorption spectra (dotted lines) and emission spectra (dashed lines) are plotted for reference for each dye. The solid

line is a guide to the eye. Y-error bars represent the standard deviation of the mean fluorescence intensity observed from particles with LSPR peaks falling within each 20 nm bin. Reprinted from reference 9.

## 4.4 WAVELENGTH DEPENDENCE OF EXCITATION ENHANCEMENT

In order to decouple excitation and emission factors from each other, we switch from organic fluorophores to semiconductor quantum dots. Organic fluorophores, such as the ones used in the previous section, exhibit fairly narrow absorption peaks that partially overlap the emission peak due to the small Stokes shift. This makes it difficult to spectrally separate excitation and emission effects. In contrast, semiconductor quantum dots are attractive fluorophores since the quantum dots have a broad spectral range of absorption while retaining a narrow emission band, which enables us to excite them at many wavelengths to spectrally distinguish the excitation effects. We use CdSe quantum dots to compare photoluminescence intensities over many different excitation wavelengths while having a constant emission wavelength on the same sample. In this section, we discuss the photoluminescence excitation spectrum of CdSe quantum dots near single metal nanoprisms in order to spectrally differentiate excitation and emission effects.

We use a similar approach to the one previously introduced in Figure 4.7 in order to observe fluorescence enhancement of CdSe quantum dots by the metal near-field effects. Although we sacrifice the distance control of the experiments discussed above, the excitation enhancement factors are large enough that we can still achieve acceptable signal to noise. Figure 4.11A depicts the experiment. As before, layer of sparse Ag nanoprisms is first fixed on a silane-treated glass slide as described in the previous section. The Ag nanoprisms are then overcoated with a 5-nm film of CdSe/CdS/CdZnS/ZnS core/multi-shell quantum dots suspended in a poly(methylmethacrylate) (PMMA) matrix. The quantum dots were synthesized following adapted literature methods, then washed to remove excess ligands from the synthesis.[62] Figure 4.11B shows the darkfield scattering image of isolated Ag nanoprisms and Figure 4.11C shows the single particle scattering spectra for the labeled nanoprisms shown in the darkfield scattering image. Figure 4.11 D-I show a series of photoluminescence images of the quantum dots which were excited at different wavelengths (as labeled) while observing the emission wavelength at 625 nm: (D) 410 nm, (E) 440 nm, (F) 470 nm, (G) 490 nm, (H) 570 nm, (I) 590 nm.

Reminiscent of the image shown in Figure 4.7C, which showed varying degrees of fluorescence enhancement from a monolayer of organic fluorophores by nearby Ag nanoprisms, we also observe significantly brighter spots (compared to the background) in the photoluminescence images which correspond to the locations of the Ag nanoprisms in the darkfield scattering image with different excitation wavelength, demonstrating metal-enhanced photoluminescence of the quantum dots.

In order to investigate the wavelength-dependent excitation enhancement factor, we carefully examine each of the photoluminescence images of the different excitation wavelengths (Figure 4.10 D-I). We observe that the emission intensity of the quantum dots near the nanoprisms changes as a function of the excitation wavelength. For example, the photoluminescence of the quantum dots near Nanoprisms #1 and #5 are comparable to the background quantum dots when excited by 410 nm and 440 nm light (Figure 4.11 D, E). However, under 470 nm and 490 nm excitation (Figure 4.11 F, G) the quantum dots near Nanoprism #1 show more photoluminescence than those near Nanoprism #5. Furthermore, the relative emission intensities are reversed when the sample is excited by 570 nm and 590 nm light (Figure 4.11 H, I).

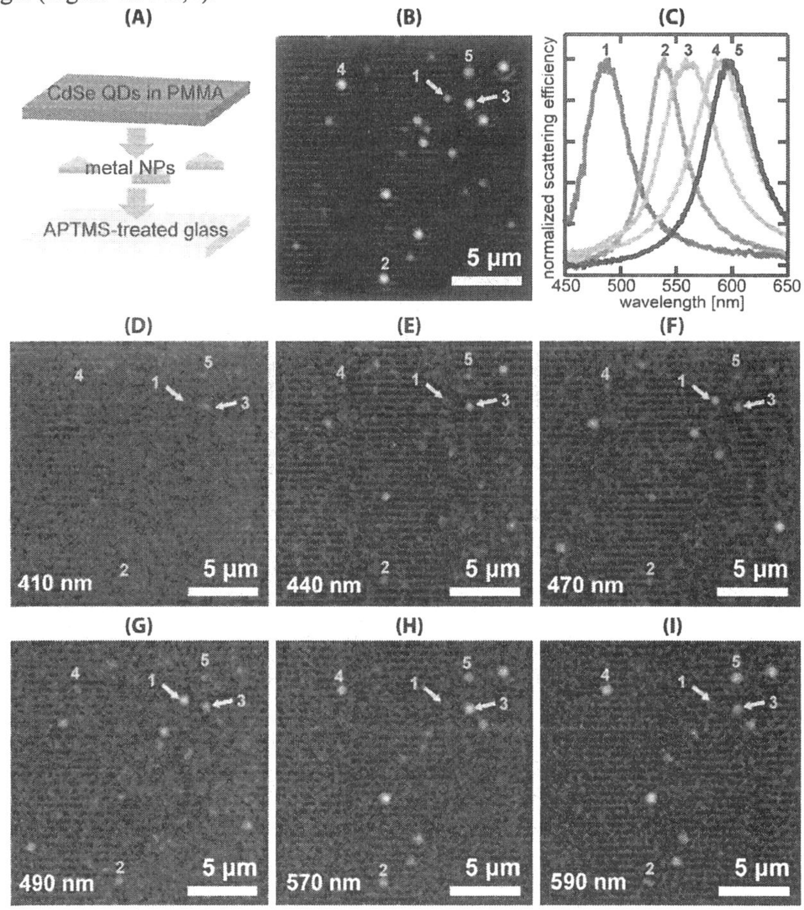


**Figure 4.11:** Photoluminescence of quantum dots near Ag nanoprisms with a series of excitation wavelengths. (A) Schematic diagram of sample preparation: Ag nanoprisms (NP) are attached to a 3-aminopropyltrimethoxysilane-treated glass coverslip and overcoated with a layer of quantum dot-doped PMMA; (B) Darkfield image showing locations of Ag nanoprisms; (C) scattering spectra of each of the labeled nanoprisms in (B). Correlated quantum dot photoluminescence images excited with (D) 410 nm, (E) 440 nm, (F) 470 nm, (G) 490 nm, (H) 570 nm, and (I) 590 nm light. Adapted from reference 43.

Again, the question arises: why do we see different degree of photoluminescence enhancements from the same particle as we vary the excitation wavelength? Unlike the previous experiments with variety of organic fluorophores shown in Figure 4.10, here we only use one kind of fluorophore; CdSe quantum dots which emit at 625 nm. The answer to this question lies in the differences in the excitation rate $\gamma_{ex}(\omega_{ex})$, of the fluorophores in the particle near-field at the excitation frequency (see Equation 1). Since the $\gamma_{ex}(\omega_{ex})$ term is frequency dependent, we expect that the highest field intensity may be achieved near the nanoparticle plasmon resonance, leading to spectral variation in excitation enhancements. This can be seen by examining the LSPR scattering spectra of the nanoprisms, shown in Figure 4.11C. We generally observe the most emission from quantum dots near a metal nanoparticle when those dots are excited with light near the metal nanoparticle's LSPR scattering peak, where the local-field enhancement is the most pronounced.[29] For example, Nanoprism #1 has a LSPR scattering peak at 488 nm (Figure 4.11C) and the nearby quantum dots looks the brightest when illuminated by 470 and 490 nm light (Figure 4.11 F, G). However, when excited off-resonance (from Nanoprism #1 scattering resonance) with 570 and 590 nm light (Figure 4.11 H, I), the same quantum dots appear much dimmer, almost comparable to the background photoluminescence intensity from quantum dots located far from the nanoprisms. Similarly, quantum dots near Nanoprism #5 (which has an LSPR scattering peak at 600 nm) show little near field enhancement when excited off-resonance with 410 – 490 nm light (Figures 4.11 D-G), thus the quantum dots near Nanoprism #5 appear very dim. In contrast, when the same quantum dots are excited near resonance at 570 and 590 nm (Figures 4.11 H, I) they appear extremely bright.

We note that although the distribution of the quantum dots in the film may be locally non-uniform (which would also show increased fluorescence), the lack of contrast in Figure 4.11D (excitation at 410 nm) shows that most of the increased photoluminescence near the nanoprism stems from wavelength-dependent excitation enhancement. This result again (as was seen in Figure 4.7) demonstrates that near-field enhancement by optically isolated, *single* metal nanoparticles (as opposed to clusters or aggregates) is readily achievable, as both our group and others have previously reported.[4-9,11,63]

In order to examine the variations in quantum dot emission intensity as a function of excitation wavelength more systematically, we collected discrete, spatially-resolved photoluminescence excitation spectra for our quantum dot/metal nanoparticle samples over the wavelength range from 400-600 nm. We accomplished this by collecting photoluminescence images of the sample excited with 21 different bandpass (10 nm fwhm) excitation filters. Next, we extracted photoluminescence excitation (PLE) spectra for the quantum dots from the image pixel intensity values, correcting for the excitation flux at each wavelength and the diffractive spread of the enhanced emission near the nanoprisms. See the literature for the detailed image analysis protocol.[43] Figure 4.12A shows two representative PLE spectra for background quantum dots located far from nanoprisms (squares) and near a sample Ag nanoprism (diamonds). Both spectra are normalized for comparison. Figure 4.12B also shows the photoluminescence images of the sample near the nanoprism at selected excitation wavelengths. We observe that the photoluminescence intensity contrast between the background quantum dots and the quantum dots near the nanoprism increases as the excitation wavelength approaches 550 nm, then decreases at longer wavelengths. This change in the photoluminescence intensity is also seen clearly in the divergence of the two PLE spectra.

In addition to the PLE spectra, we calculated the total fluorescence enhancement factor as the ratio of the photoluminescence near the nanoprism to the background photoluminescence. Figure 4.13 shows representative plots of the total fluorescence enhancement factor versus excitation wavelength for quantum dots near different shapes of metal nanoparticles: (A) Ag nanoprism featured in Figure 4.12, (B-D) Ag nanoprisms; average side length=100 nm, (E-F) Ag nanocubes;[64] average side length=40 nm, (G) Au sphere; diameter = 80 nm, and (H) Au sphere, diameter = 100 nm. The LSPR scattering spectrum of each metal nanoparticle is plotted for comparison. In each case, the enhancement factor mirrors the shape of the LSPR scattering spectrum, further demonstrating that excitation effects play a significant role in the observed enhancement. We note that the total fluorescence enhancement factors of the nanoparticles shown in Figure 4.13 do not go to one off-resonance. We believe this residual off-resonance enhancement is due to either plasmonic emission enhancement effects, small differences in the distribution of quantum dots near the nanoparticles, or both. Nevertheless, since both of these effects should be independent of excitation wavelength, we can separate them from excitation enhancement by taking the *ratio* of the total enhancement factor on and off the metal nanoparticle plasmon resonance. This ratio effectively cancels out any contribution from emission enhancement as well as any variations in local quantum dot density, leaving only the change in excitation enhancement factor on and off the metal nanoparticle plasmon resonance. If we assume that the off-resonance excitation enhancement factor is ~1, this on/off ratio provides a conservative measure of the on-resonance excitation enhancement factor.

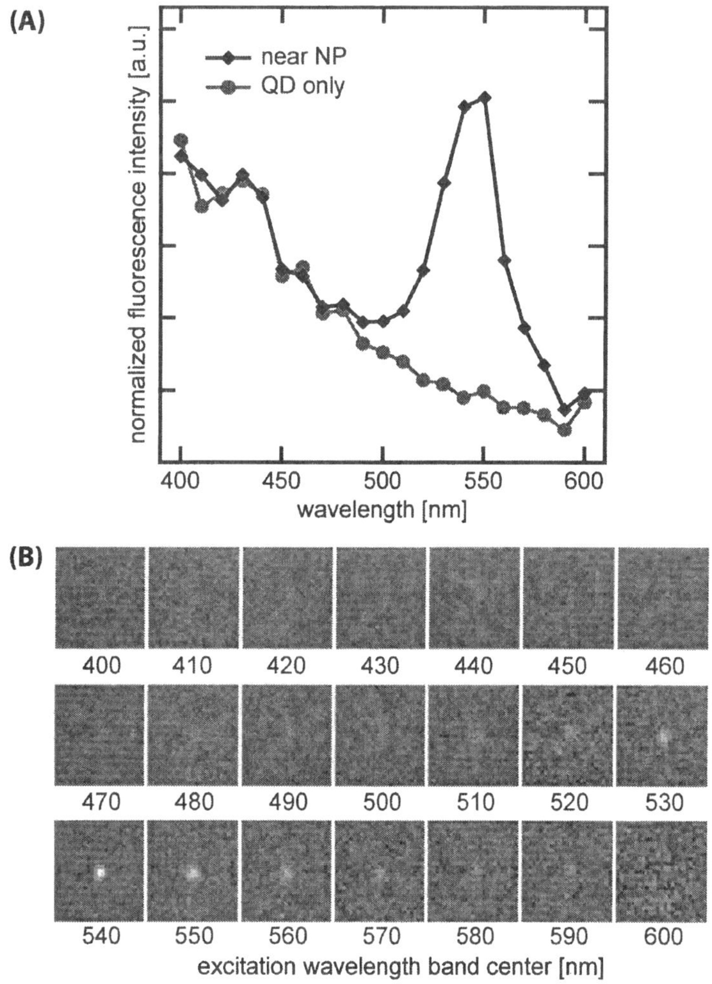


**Figure 4.12:** (A) Comparison of PLE spectra of quantum dots near a single Ag nanoprism (NP) (diamonds) and quantum dots alone (blue circles). Intensities have been normalized to compare the line shape of the two spectra. (B) Series of images of quantum dot photoluminescence near the Ag nanoprism in (A). Adapted from reference 43.

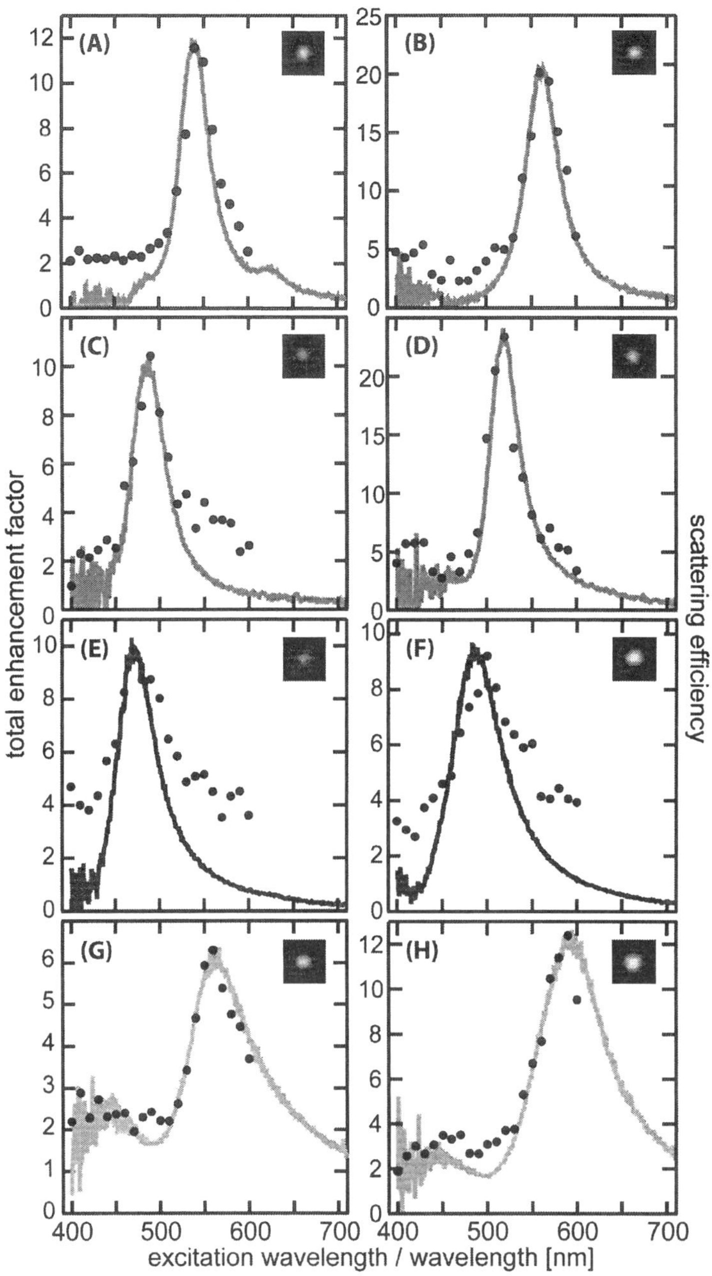


**Figure 4.13:** Enhancement of quantum dots photoluminescence near a nanoparticle as a function of excitation wavelength for several single metal nanoparticles. Black dots are the ratio of the photoluminescence near a nanoparticle to photoluminescence far from a nanoparticle. The trace in each spectra is the LSPR

scattering spectra of (A-D) Ag nanoprisms (E-F) Ag cubes (s~50 nm), and (G) Au spheres (d~80 nm) (H) Au sphere (d~100 nm). The darkfield image of each nanoparticle is also shown. Adapted from reference 43.

We observed different maximum excitation enhancement factors for each type of nanoparticle shape: Ag nanoprisms ~ 10, Au sphere (diameter ~ 80 nm) ~3, Au sphere (diameter ~ 100 nm) ~3.5, and Ag cubes (side ~ 50 nm) ~3. Furthermore, we find that there is variation in the excitation enhancement factors between different individual nanoprisms (Figure 4.13 A-D), even for nanoprisms with similar LSPR scattering peaks. Why do we see these variations from nanoprisms with similar spectral properties? We believe that the difference in the excitation enhancement comes from two factors. One is from the variations in size and sharpness of the tips of nanoprisms under investigation. As we discussed in the previous sections, the distribution and magnitude of the local-field intensity is highly dependent on the shape, size and sharpness of the nanoprisms. As a result, the maximum local-field enhancement at the local "hot spots" may also vary despite having similar LSPR scattering peaks[34,42] since we did not independently determine the shapes of the individual metal nanoparticles in this study. Second, the local density of quantum dots near the "hot-spots" may vary from nanoparticle to nanoparticle. We have observed phase separation of the quantum dots from the PMMA matrix into smaller aggregates which are spaced ~50 nm apart in both AFM and SEM measurements. In other words, despite the fact the photoluminescence images appear uniform far away from the metal nanoparticles, there are still local variations in quantum dot density in the film. In most cases, we likely have observed less than the maximum possible excitation enhancement factor for some nanoprisms. If the quantum dots could be precisely placed in the nanoparticle hotspots, we would expect to see overall higher excitation enhancement factors.

In summary, the photoluminescence of CdSe quantum dots can be strongly enhanced by nearby metal nanoparticles, where most of the enhancement results from excitation effects. We observed that the shape of the PLE spectra of the quantum dots near a metal nanoparticle is significantly altered for both gold and Ag nanoparticles, and shows a new PLE peak coincident with the LSPR peak of the metal nanoparticle. Although the absolute enhancement factor varies from one metal nanoparticle to another, the wavelength dependence of the total enhancement factor still mirrors the line shape of the metal nanoparticle's scattering spectrum. There may be a small offset in the maximum *excitation* enhancement from the nanoparticle's scattering peak (as was described for the total fluorescence in Section 4.3 above), but at present our experiments have not had sufficient spectral resolution to identify any such shift.

## 4.5 CONCLUSIONS

Metal nanostructures can act as small antennas that aid in the reception and broadcasting (absorption and emission) of light from nearby fluorophores. Whether fluorescence enhancement or quenching is observed in a given system is determined by the relative extent of excitation enhancement (increased light absorption), emission enhancement (increased radiative decay), and quenching (increased non-radiative decay) effects. These effects are sensitive both to distance, as discussed elsewhere,[6,7,15,18-21] and wavelength, as discussed in this chapter. To study these effects as a function of the plasmon resonance energy of the metal nanoparticle we have used Ag nanoprisms with narrow scattering linewidths, large scattering cross-sections, and sharp edges. Using single particle spectroscopy, we have correlated the apparent fluorescence intensity of fluorophores attached to Ag nanoprisms via DNA linkage with the LSPR scattering spectra of the nanoprisms. These results and others[7,8] show that the apparent brightness of a fluorophore near a metal nanoparticle is strongly dependent on the spectral overlap between the nanoparticle plasmon resonance scattering peak and the spectral properties of the fluorophore. More quantitatively, we have determined that the maximum total fluorescence intensity is achieved when the LSPR scattering peak is at a slightly higher energy (40-120 meV) than the fluorophore emission peaks for high intrinsic quantum yield organic fluorophores. At present, we attribute this to a combination of both excitation and emission effects leading to the observed enhancement.

In an attempt to better quantify the relative contributions of excitation and emission enhancement to the total change in fluorescence, we have also used CdSe quantum dots as fluorophores since they can be excited at any energy above their band gap to emit very narrow photoluminescence at their band edge. Using quantum dots allowed us to compare photoluminescence intensities over many different excitation wavelengths with a single emission spectrum and the same plasmonic near-field effects. We correlated the scattering spectra of single metal nanoparticles with the photoluminescence excitation (PLE) spectra of nearby quantum dots. Our results show that proximity to the metal nanoparticle changes the shape of the quantum dots' PLE spectra, creating large new peaks that correlate with the plasmon scattering spectra of the metal nanoparticles. For the particles we observed, we were able to measure near-field excitation enhancement factors that eliminated contributions from emission enhancement and quantum dot spatial variations. The excitation enhancement factor seems to be maximized when the excitation light coincides with the nanoparticle's scattering peak wavelength. We compared the maximum observed excitation enhancement factor between Ag nanoprisms, Ag nanocubes, and Au spheres, and found that Ag nanoprisms can increase the excitation of the quantum dots by at least a factor of 10, while the other shapes averaged an excitation enhancement of about 3 times. In addition, we find that in this sample geometry, excitation effects had a bigger effect than potential emission effects.

While it is clear that the degree of spectral overlap between the plasmon and fluorophore has an effect on the ultimate behavior of the fluorophore, some details warrant further experimental study. As research moves to harness larger and more complex metal nanostructures, the scattering and absorption components of the overall plasmon extinction become distinct not only in magnitude, but in spectral dependence as well.[65,66] The excitation enhancement and the modifications to radiative and non-radiative decay rates must be examined as a function of the spectral overlap between the fluorophore absorption and emission with the plasmon spectra. The more we can separate the contributions from excitation and emission modifications, the more we can improve fluorescence in specific applications, depending on whether maximized fluorescence output (e.g. for sensing applications), fluorophore excitation (e.g. for photochemistry), reduced lifetime (e.g. improved photostability) or efficient emission (e.g. for light-producing devices) are required. Indeed, once the "rules" are known, the optimal configuration for a specific application may come from the precise spatial assembly of the desired fluorophore and a metal nanostructure with a tailored plasmon lineshape.

## 4.6 ACKNOWLEDGEMENTS

This chapter is based on work supported by the NSF (DMR 0520567, DMR 0449422), the Air Force Office of Scientific Research, the National Science Foundation Materials Research Science and Engineering Center (MRSEC) program through the Genetically Engineered Materials Science & Engineering Center (GEMSEC) and the American Chemical Society Petroleum research fund. D.S.G thanks the Camille Dreyfus Teacher-Scholar Awards Program for program support. D.S.G. is a Cottrell Scholar of the Research Corporation, and an Alfred P. Sloan Foundation Research Fellow. J.M.S. thanks the NASA Space Grant Program for Summer Undergraduate Fellowship. Scanning electron microscopy was performed at the NanoTech User Facility at the UW, a member of the NSF National Nanotechnology Infrastructure Network. FDTD calculations were performed on the UW chemistry computing cluster supported in part by the NSF (CHE 0342956). The authors thank Andrea Munro and Ilan Jen-La Plante for the CdSe quantum dots and Sara Skrabalak and Younan Xia for the Ag nanocubes.

## 4.7 REFERENCES

1. Kulakovich, O., Strekal, N., Yaroshevich, A., Maskevich, S., Gaponenko, S., Nabiev, I., Woggon, U. and Artemyev, M. (2002). Enhanced Luminescence of CdSe Quantum Dots on Gold Colloids. *Nano Lett.* **2**: 1449-1452.
2. Pompa, P. P., Martiradonna, L., Della Torre, A., Della Sala, F., Manna, L., De Vittorio, M., Calabi, F., Cingolani, R. and Rinaldi, R. (2006). Metal-enhanced fluorescence of colloidal nanocrystals with nanoscale control. *Nat. Nanotech.* **1**: 126-130.
3. Pan, S. L., Wang, Z. J. and Rothberg, L. J. (2006). Enhancement of adsorbed dye monolayer fluorescence by a silver nanoparticle overlayer. *J. Phys. Chem. B* **110**: 17383-17387.
4. Kühn, S., Håkanson, U., Rogobete, L. and Sandoghdar, V. (2006). Enhancement of single-molecule fluorescence using a gold nanoparticle as an optical nanoantenna. *Phys. Rev. Lett.* **97**: 017402.
5. Anger, P., Bharadwaj, P. and Novotny, L. (2006). Enhancement and quenching of single-molecule fluorescence. *Phys. Rev. Lett.* **96**: 113002.
6. Bharadwaj, P., Anger, P. and Novotny, L. (2007). Nanoplasmonic enhancement of single-molecule fluorescence. *Nanotechnology* **18**: 044017.
7. Bharadwaj, P. and Novotny, L. (2007). Spectral dependence of single molecule fluorescence enhancement. *Opt. Express* **15**: 14266-14274.
8. Tam, F., Goodrich, G. P., Johnson, B. R. and Halas, N. J. (2007). Plasmonic Enhancement of Molecular Fluorescence. *Nano Lett.* **7**: 496-501.

9. Chen, Y., Munechika, K. and Ginger, D. S. (2007). Dependence of fluorescence intensity on the spectral overlap between fluorophores and plasmon resonant single silver nanoparticles. *Nano Lett.* **7**: 690-696.
10. Aslan, K., Malyn, S. N. and Geddes, C. D. (2008). Angular-dependent metal-enhanced fluorescence from silver island films. *Chem. Phys. Lett.* **453**: 222-228.
11. Bek, A., Jansen, R., Ringler, M., Mayilo, S., Klar, T. A. and Feldmann, J. (2008). Fluorescence Enhancement in Hot Spots of AFM-Designed Gold Nanoparticle Sandwiches. *Nano Lett.* **8**: 485-490.
12. Chance, R. R., Prock, A. and Silbey, R. J. (1978). Molecular fluorescence and energy transfer near interfaces. *Adv. Chem. Phys.* **37**: 1-65.
13. Barnes, W. L. (1998). Fluorescence near interfaces: the role of photonic mode density. *J. Mod. Opt.* **45**: 661-699.
14. Weber, W. H. and Eagen, C. F. (1979). Energy-Transfer from an Excited Dye Molecule to the Surface-Plasmons of an Adjacent Metal. *Opt. Lett.* **4**: 236-238.
15. Gersten, J. and Nitzan, A. (1981). Spectroscopic Properties of Molecules Interacting with Small Dielectric Particles. *J. Chem. Phys.* **75**: 1139-1152.
16. Dulkeith, E., Morteani, A. C., Niedereichholz, T., Klar, T. A., Feldmann, J., Levi, S. A., van Veggel, F. C. J. M., Reinhoudt, D. N., Möller, M. and Gittins, D. I. (2002). Fluorescence Quenching of Dye Molecules near Gold Nanoparticles: Radiative and Nonradiative Effects. *Phys. Rev. Lett.* **89**: 203002.
17. Dulkeith, E., Ringler, M., Klar, T. A., Feldmann, J., Javier, A. M. and Parak, W. J. (2005). Gold nanoparticles quench fluorescence by phase induced radiative rate suppression. *Nano Lett.* **5**: 585-589.
18. Das, P. C. and Puri, A. (2002). Energy flow and fluorescence near a small metal particle. *Phys. Rev. B* **65**: 155416.
19. Andreussi, O., Corni, S., Mennucci, B. and Tomasi, J. (2004). Radiative and nonradiative decay rates of a molecule close to a metal particle of complex shape. *J. Chem. Phys.* **121**: 10190-10202.
20. Thomas, M., Greffet, J.-J., Carminati, R. and Arias-Gonzalez, J. R. (2004). Single-molecule spontaneous emission close to absorbing nanostructures. *Appl. Phys. Lett.* **85**: 3863-3865.
21. Carminati, R., Greffet, J. J., Henkel, C. and Vigoureux, J. M. (2006). Radiative and non-radiative decay of a single molecule close to a metallic nanoparticle. *Opt. Commun.* **261**: 368-375.
22. Kummerlen, J., Leitner, A., Brunner, H., Aussenegg, F. R. and Wokaun, A. (1993). Enhanced Dye Fluorescence over Silver Island Films - Analysis of the Distance Dependence. *Mol. Phys.* **80**: 1031-1046.
23. Ray, K., Badugu, R. and Lakowicz, J. R. (2006). Distance-dependent metal-enhanced fluorescence from Langmuir-Blodgett monolayers of alkyl-NBD derivatives on silver island films. *Langmuir* **22**: 8374-8378.
24. Ray, K., Badugu, R. and Lakowicz, J. R. (2007). Polyelectrolyte layer-by-layer assembly to control the distance between fluorophores and plasmonic nanostructures. *Chem.Mater.* **19**: 5902-5909.
25. Sokolov, K., Chumanov, G. and Cotton, T. M. (1998). Enhancement of molecular fluorescence near the surface of colloidal metal films. *Anal. Chem.* **70**: 3898-3905.

26. Stoermer, R. L. and Keating, C. D. (2006). Distance-dependent emission from dye-labeled oligonucleotides on striped Au/Ag nanowires: Effect of secondary structure and hybridization efficiency. *J. Am. Chem. Soc.* **128**: 13243-13254.
27. Schneider, G., Decher, G., Nerambourg, N., Praho, R., Werts, M. H. V. and Blanchard-Desce, M. (2006). Distance-dependent fluorescence quenching on gold nanoparticles ensheathed with layer-by-layer assembled polyclectrolytes. *Nano Lett.* **6**: 530-536.
28. Zhang, J., Gryczynski, I., Gryczynski, Z. and Lakowicz, J. R. (2006). Dye-labeled silver nanoshell-bright particle. *J. Phys. Chem. B* **110**: 8986-8991.
29. Hao, E. and Schatz, G. C. (2004). Electromagnetic fields around silver nanoparticles and dimers. *J. Chem. Phys.* **120**: 357-366.
30. FDTD Solutions 5.0.3, Lumerical Solutions, Vancouver, BC
31. Brolo, A. G., Kwok, S. C., Moffitt, M. G., Gordon, R., Riordon, J. and Kavanagh, K. L. (2005). Enhanced fluorescence from arrays of nanoholes in a gold film. *J. Am. Chem. Soc.* **127**: 14936-14941.
32. Shuford, K. L., Ratner, M. A., Gray, S. K. and Schatz, G. C. (2007). Electric field enhancement and light transmission in cylindrical nanoholes. *J. Comput. Theor. Nanosci.* **4**: 239-246.
33. Wenger, J., Gerard, D., Dintinger, J., Mahboub, O., Bonod, N., Popov, E., Ebbesen, T. W. and Rigneault, H. E. (2008). Emission and excitation contributions to enhanced single molecule fluorescence by gold nanometric apertures. *Opt. Express* **16**: 3008-3020.
34. Mock, J. J., Barbic, M., Smith, D. R., Schultz, D. A. and Schultz, S. (2002). Shape effects in plasmon resonance of individual colloidal silver nanoparticles. *J. Chem. Phys.* **116**: 6755-6759.
35. Sherry, L. J., Jin, R. C., Mirkin, C. A., Schatz, G. C. and Van Duyne, R. P. (2006). Localized surface plasmon resonance spectroscopy of single silver triangular nanoprisms. *Nano Lett.* **6**: 2060-2065.
36. Metraux, G. S. and Mirkin, C. A. (2005). Rapid thermal synthesis of silver nanoprisms with chemically tailorable thickness. *Adv. Mater.* **17**: 412-415.
37. Ung, T., Liz-Marzan, L. M. and Mulvaney, P. (1998). Controlled method for silica coating of silver colloids. Influence of coating on the rate of chemical reactions. *Langmuir* **14**: 3740-3748.
38. Jin, R. C., Cao, Y. W., Mirkin, C. A., Kelly, K. L., Schatz, G. C. and Zheng, J. G. (2001). Photoinduced conversion of silver nanospheres to nanoprisms. *Science* **294**: 1901-1903.
39. Xia, Y. N. and Halas, N. J. (2005). Shape-controlled synthesis and surface plasmonic properties of metallic nanostructures. *Mrs Bulletin* **30**: 338-344.
40. Murphy, C. J., San, T. K., Gole, A. M., Orendorff, C. J., Gao, J. X., Gou, L., Hunyadi, S. E. and Li, T. (2005). Anisotropic metal nanoparticles: Synthesis, assembly, and optical applications. *J. Phys. Chem. B* **109**: 13857-13870.
41. Liz-Marzan, L. M. (2006). Tailoring surface plasmons through the morphology and assembly of metal nanoparticles. *Langmuir* **22**: 32-41.
42. Munechika, K., Smith, J. M., Chen, Y. and Ginger, D. S. (2007). Plasmon Line Widths of Single Silver Nanoprisms as a Function of Particle Size and Plasmon Peak Position. *J. Phys. Chem. C* **111**: 18906-18911.

43. Chen, Y., Munechika, K., Plante, I. J., Munro, A. M., Skrabalak, S. E., Xia, Y. and Ginger, D. S. (2008). Excitation Enhancement of CdSe Quantum Dots by Single Metal Nanoparticles. *Submitted*
44. Mock, J. J., Smith, D. R. and Schultz, S. (2003). Local refractive index dependence of plasmon resonance spectra from individual nanoparticles. *Nano Lett.* **3**: 485-491.
45. Sönnichsen, C., Franzl, T., Wilk, T., von Plessen, G. and Feldmann, J. (2002). Plasmon resonances in large noble-metal clusters. *New J. Phys.* **4**: 93.
46. Sönnichsen, C., Franzl, T., Wilk, T., von Plessen, G., Feldmann, J., Wilson, O. and Mulvaney, P. (2002). Drastic Reduction of Plasmon Damping in Gold Nanorods. *Phys. Rev. Lett.* **88**: 077402.
47. Klar, T., Perner, M., Grosse, S., von Plessen, G., Spirkl, W. and Feldmann, J. (1998). Surface-plasmon resonances in single metallic nanoparticles. *Phys. Rev. Lett.* **80**: 4249-4252.
48. Heilweil, E. J. and Hochstrasser, R. M. (1985). Nonlinear Spectroscopy and Picosecond Transient Grating Study of Colloidal Gold. *J. Chem. Phys.* **82**: 4762-4770.
49. Kreibig, U. and Vollmer, M. (1995). Optical Properties of Metal Clusters Springer, New York
50. Scharte, M., Porath, R., Ohms, T., Aeschlimann, M., Krenn, J. R., Ditlbacher, H., Aussenegg, F. R. and Liebsch, A. (2001). Do Mie plasmons have a longer lifetime on resonance than off resonance? *Appl. Phys. B-Lasers Opt.* **73**: 305-310.
51. Jensen, T., Kelly, L., Lazarides, A. and Schatz, G. C. (1999). Electrodynamics of noble metal nanoparticles and nanoparticle clusters. *J. Clus. Sci* **10**: 295-317.
52. Novo, C., Gomez, D., Perez-Juste, J., Zhang, Z. Y., Petrova, H., Reismann, M., Mulvaney, P. and Hartland, G. V. (2006). Contributions from radiation damping and surface scattering to the linewidth of the longitudinal plasmon band of gold nanorods: a single particle study. *Phys. Chem. Chem. Phys.* **8**: 3540-3546.
53. Hu, M., Petrova, H., Sekkinen, A. R., Chen, J. Y., McLellan, J. M., Li, Z. Y., Marquez, M., Li, X. D., Xia, Y. N. and Hartland, G. V. (2006). Optical properties of Au-Ag nanoboxes studied by single nanoparticle spectroscopy. *J. Phys. Chem. B* **110**: 19923-19928.
54. Coronado, E. A. and Schatz, G. C. (2003). Surface plasmon broadening for arbitrary shape nanoparticles: A geometrical probability approach. *J. Chem. Phys.* **119**: 3926-3934.
55. Palik, E. D. (1985). Handbook of optical constants of solids. Academic Press, Orlando
56. Johnson, P. B. and Christy, R. W. (1972). Optical Constants of the Noble Metals. *Phys. Rev. B* **6**: 4370-4379.
57. Aslan, K., Leonenko, Z., Lakowicz, J. R. and Geddes, C. D. (2005). Annealed silver-island films for applications in metal-enhanced fluorescence: interpretation in terms of radiating plasmons. *J. Fluoresc.* **15**: 643-654.
58. Zhang, J. and Lakowicz, J. R. (2007). Metal-enhanced fluorescence of an organic fluorophore using gold particles. *Opt. Express* **15**: 2598-2606.

59. Mirkin, C. A. (2000). A DNA-based methodology for preparing nanocluster circuits, arrays, and diagnostic. *MRS Bull.* **25**: 43-54.
60. Demers, L. M., Ginger, D. S., Park, S.-J., Li, Z., Chung, S.-W. and Mirkin, C. A. (2002). Direct patterning of modified oligonucleotides on metals and insulators by dip-pen nanolithography. *Science* **296**: 1836-1838.
61. McFarland, A. D., Young, M. A., Dieringer, J. A. and Van Duyne, R. P. (2005). Wavelength-Scanned Surface-Enhanced Raman Excitation Spcctroscopy. *J. Phys. Chem. B* **109**: 11279-11285.
62. Munro, A. M., Jen-LaPlante, I., Ng, M. S. and Ginger, D. S. (2007). Quantitative Study of the Effects of Surface Ligand Concentration on CdSe Nanocrystal Photoluminescence. *J. Phys. Chem. C* **111**: 6220-6227.
63. Härtling, T., Reichenbach, P. and Eng, L. M. (2007). Near-field coupling of a single fluorescent molecule and a spherical gold nanoparticle. *Opt. Express* **15**: 12806-12817.
64. Skrabalak, S. E., Au, L., Li, X. D. and Xia, Y. (2007). Facile synthesis of Ag nanocubes and Au nanocages. *Nat. Protoc.* **2**: 2182-2190.
65. Miller, M. M. and Lazarides, A. A. (2006). Sensitivity of metal nanoparticle plasmon resonance band position to the dielectric environment as observed in scattering. *J. Opt. A: Pure Appl. Opt.* **8**: S239-S249.
66. Wang, H. and Halas, N. J. (2008). Mesoscopic Au "Meatball" particles. *Advanced Materials* **20**: 820-825.

5

# Near-IR Metal Enhanced Fluorescence And Controlled Colloidal Aggregation

Jon P. Anderson*, Mark Griffiths, John G. Williams, Daniel L. Grone, Dave L. Steffens, and Lyle M. Middendorf.
LI-COR Biosciences Inc.
4647 Superior Street, Lincoln, NE 68504, USA.

## 5.1 INTRODUCTION

Scientific efforts in systems biology, cancer research, cell signalling, and many other areas are striving to identify less abundant biological targets while continuing to ramp up multiplexing efforts. These research initiatives provide a continual push for reduced detection limits and increased sensitivity for both biological assays and instrumentation. For many molecular biology and clinical laboratories, fluorescence detection is still routinely used as the detection method of choice for a variety of reasons. The small size of most fluorescent dyes produces little interference with the properties of the labelled molecule and allows the dye to infiltrate cellular regions that cannot be labelled by larger molecules [1, 2]. Fluorescent dyes provide a predictable red-wavelength Stokes shift from the excitation to the emission spectra, allowing for very efficient collection of the emission photons away from the excitation beam [1, 3]. Fluorescence can also provide additional information by way of polarization, lifetime, fluorescence resonance energy transfer (FRET), and quenching. DNA and protein microarrays, immunoassays, small animal imaging, and single molecule sequencing are but a few examples of the utility of fluorescence detection [4 - 10]. Though fluorescence detection is a highly informative and well researched method, there is an inherent need for increased sensitivity and reduced detection limits [11]. Improvements in sensitivity and detection will set the groundwork for more highly multiplexed sample sets, as well as helping preserve rare or precious samples for further testing.

A simple and direct way of increasing sensitivity in fluorescence based assays is to utilize fluorophores that emit in the near-infrared (near-IR). Most organic fluorophores emit light in the visible to near-IR spectral region (400 – 900 nm). Fluorescence in the near-IR (700 – 900 nm) has unique advantages over visible fluorescence and is quickly gaining popularity in biological imaging and molecular applications [9]. Near-IR fluorescence offers significantly lower background signals from scatter than those generated by visible wavelength excitation and can generate greater signal-to-noise ratios (Fig. 5.1) [12, 13]. Near-IR wavelengths are not readily absorbed by water and biological compounds, allowing deep penetration through tissues and cells, and reducing the possibility of photodamage to biological samples. Instruments using these longer wavelengths therefore have a distinct advantage, with the near-IR spectral region showing little interference from, or damage to, biological molecules [14, 15].

*Near IR MEF*
Edited by Chris D. Geddes.

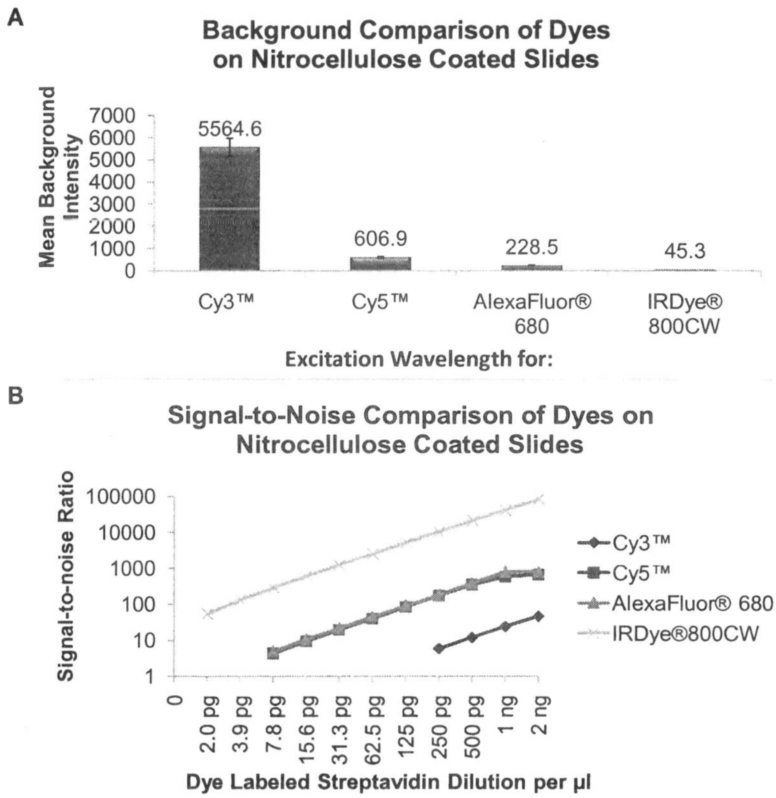


**Figure 5.1:** Comparison of visible and near-IR dyes on nitrocellulose-coated slides. Two-fold serial dilutions of streptavidin conjugates labelled with Cy3, Cy5, Alexa Fluor 680, or IRDye 800CW were spotted onto nitrocellulose-coated glass slides with an automated arrayer. Five replicates were spotted per slide, and three slides were printed for each dye. Arrays were imaged at the appropriate wavelengths, and background and signal-to-noise ratio (SNR) were calculated. A) Membrane background at each wavelength. B) SNR across each dilution series. SNR = (fluorescence intensity – mean background) / standard deviation of background); Data shown for SNR > 3. (Taken from: Osterman, H.L. and Schutz-Geschwender, A. (2007) Seeing beyond the visible with IRDye infrared dyes. LICOR Biosciences., www.licor.com)

The need for greater detection sensitivity in biotechnology has also paved the way for alternative labelling schemes, including quantum dots and metal nanostructures. These probes can overcome some shortfalls that are seen with fluorescence detection, including fast photodegradation of the fluorophore and a small stokes shift. Fluorescent molecules with a small Stokes shift may have the disadvantage of self-quenching when located near another fluorophore by means of homo-FRET, reducing the overall signal. Though these alternative labelling techniques have found their niche in imaging technologies, they may still suffer some additional drawbacks that make fluorescent dyes more appealing in some applications. Quantum dots can suffer blinking characteristics that make them less suitable for single molecule experiments; however, recent data suggests that the blinking can be reduced or eliminated by carefully controlling the chemical environment, including the addition of propyl gallate [16]. Both quantum dots and metallic particles can also be quite large compared to typical fluorophores, but their sizes are becoming more reasonable with sizes <10 nm routinely generated [17 - 20]. Any increase in size can both reduce the diffusion rate of a labelled biomolecule and reduce the rate of binding events in a reaction. Increased size may also limit the ability of labelled molecules to access biologically relevant regions during *in vivo* studies [2]. Though these alternative labelling techniques hold promise for increased sensitivity, the use of fluorescent dyes is still the predominant detection technique used for molecular biology.

The continual wide use of fluorescence detection in biotechnology combined with multiplexing efforts and reduced sample sizes has created a specific need for improved fluorescence sensitivity. In order to improve the sensitivity of fluorescent probes, the signal emissions from target molecules need to increase without increasing the background signal from nonspecific molecules. In other words, the sensitivity and detection limit can be improved by altering the quantum yield or molecular cross section of the fluorophore, reducing the level of the background signal, and increasing the photostability of the fluorophore.

## 5.2 METAL ENHANCED FLUORESCENCE

Metal enhanced fluorescence can be used to alter many of the properties of a fluorophore, ultimately increasing the sensitivity and reducing the detection limit of the system. Metal nanoparticles have been shown to have an influence on both the excitation and emission characteristics of the fluorophore. The molecular cross section of the fluorophore can essentially be increased by positioning the fluorophore near a metal nanostructure, increasing the efficiency of excitation. Furthermore, the excited fluorophore can then induce a surface plasmon on the metal nanostructure, which then emits a photon that can be ultimately detected. Therefore, by moving a fluorophore near a metallic particle, the lifetime properties of the fluorophore can be altered, both improving the sensitivity and lowering the limit of detection (LOD) through an increase of the quantum yield and photostability of the fluorophore [21 - 24]. The inherent characteristics of metal enhanced fluorescence (MEF) make it an appealing supplement for many commonly used fluorescence techniques, and give it the potential to be utilized in a wide range of applications.

Researchers have primarily focused on using either metal island films, or metal colloid coated surfaces as the metal surface of choice for the enhancement of fluorescence, although other metal surfaces exist, including nanorods, nano triangles, fractals, as well as specifically engineered nanofabricated surfaces [25 - 32]. Silver structures have mainly been studied for their ability to increase the relative fluorescence intensity of visible and low quantum yield fluorophores [33 - 35]. The longer wavelength near-IR fluorophores have been studied much less for their ability to be enhanced by metal surfaces, even though these fluorophores may have an inherent advantage over visible fluorophores for MEF [36].

## 5.3 NEAR-INFRARED METAL ENHANCED FLUORESCENCE ADVANTAGES

Near-IR MEF provides an advantage over the enhancement of visible fluorophores by greatly reducing the background scattering component from the measurement. In order to measure the wavelength dependent scattering effect of a silver nanostructure coated surface, we measured the scattering signal generated from a silver island film (SIF) coated glass slide. Silver island films deposited onto a glass coverslip were illuminated with narrow-band light from an Hg arc lamp on an inverted Zeiss Axiovert microscope. Five different standard fluorescence filter sets (e.g. Chroma Technologies) were used to quantify scattered light over a range of wavelengths; in each case, the excitation band is about 10 - 20 nm lower than the cut-on wavelength of the emission filter. Neutral density filters were stacked in each filter set to obtain between 0.11 and 0.74 mW illumination power measured at the objective lens. For each filter set, movies (200 frames each, 40msec exposures) were acquired with a Roper MicroMax 512 CCD camera, and the average pixel intensity was determined using ImageJ software (http://rsb.info.nih.gov/ij/). Intensity values were measured and linearly normalized to what they would be for 1 mW illumination. The results show that the near-IR spectrum generates less background noise than the shorter wavelength visible spectrum (Fig. 5.2). Overall, the background scatter from the SIF coated glass slide is approximately 10-fold lower at the near-IR wavelengths (>700 nm) than at 540 nm. The reduction in background signal is an advantage for near-IR fluorescence and may allow for increased signal-to-noise ratios. The favorable aspect of reduced system noise in the near-IR spectrum can be further combined with enhanced signal emissions by way of MEF to expand the overall utility of the system.

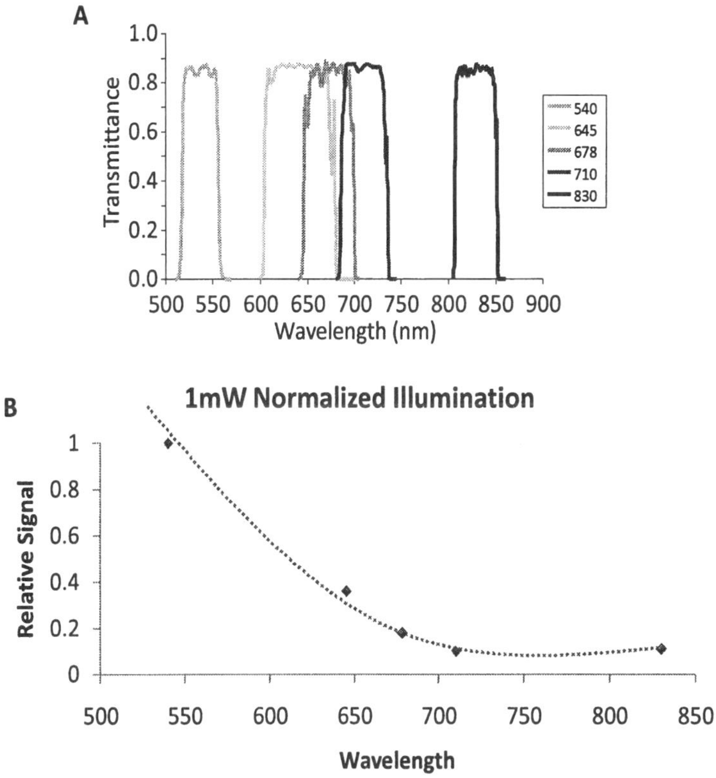


**Figure 5.2:** Background scattered from silver particles is 10-fold lower at near-infrared wavelengths. (A) Emission bandpass transmittance spectra show the tested wavelength ranges. (B) Relative signal intensities from background scatter are shown for the five different tested wavelengths.

## 5.4 NEAR-INFRARED MEF ON SILVER ISLAND FILMS

Experiments on solid surfaces were performed to quantitate the relative enhancement of near-IR fluorophores on silver coated glass surface as compared to the uncoated glass. Subwavelength sized metal island films have been created by numerous methods including chemically deposited dip coating [34], photodeposition [37], fractal-like growth [28, 38], vapor deposition [39], colloidal attachment [40], and nanosphere lithography [32, 41, 42]. Though many methods exist for producing metal coated surfaces, a majority of studies have focused on generating MEF using either metal colloids or metal island films for enhancement on surfaces. These metal nanostructure coated surfaces have been used for over two decades in Raman spectroscopy for signal enhancement [43, 44], and are now more frequently appearing in the field of fluorescence detection [34, 45 - 50]. Metal island films are composed of sub-wavelength sized patches of a highly conducting material located on an inert substrate. Randomly seeded metal islands that ranged in size from 20nm to 500nm have been successfully used to enhance a variety of fluorophores (Fig. 5.3) [34, 45, 49, 51, 52]. Research indicates that the size and the shape of the metal islands may play an important role in the ability to enhance the fluorescent signal [47, 53, 54]. Such metal island films containing a heterogeneous random population of metal islands show distinct regional variations in their ability to enhance fluorescence [47]. Metal composition also plays a major role in the ability to enhance fluorophores. For most visible and near-IR fluorophores, silver nanostructures appear to provide a greater enhancement than gold nanostructures [55]. Aluminum, however, has been recently used to provide efficient enhancement of fluorophores in the ultraviolet-blue spectral region [56].

Research using silver island films (SIF) on glass or quartz surfaces for enhancing visible fluorophores has typically generated a 10 - 12 fold enhancement [27, 40, 57]. The process of creating SIF's by chemical dip coating involves immersing a clean glass slide into a reduced solution of silver nitrate for several minutes [34]. Silver particles are randomly seeded across the glass slide and slowly form into larger metal islands. Throughout our experiments, a known quantity of fluorophore labelled protein or DNA was spotted onto both a silver nanoparticle coated and uncoated portion of a glass slides and then detected using a LI-COR Odyssey near-IR fluorescence imager. Spotting fluorophore onto both types of surfaces allows removal of any affinity biases that may occur in experiments that deposit a layer of fluorophores over the entire surface of a slide. Analyzing individual spots on a scanned image also allows the Odyssey software to account for any increase in background scatter signal caused by the silver surface. Using a SIF coated glass surface with the LI-COR near-IR fluorophores IRDye® 700 (ex max 685 nm; em max 705 nm) and IRDye® 800CW (ex max 774 nm; em max 789 nm), we were able to generate enhancements of 18-fold and 15-fold respectively [36]. These enhancement results are very similar to slightly better than what has been achieved using SIF's on visible fluorophores [22, 34, 49].

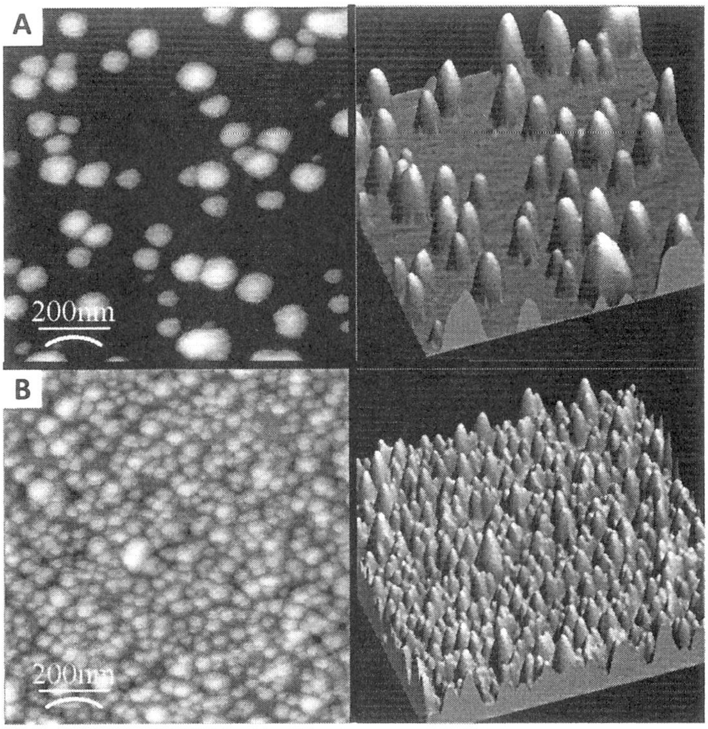


**Figure 5.3:** Atomic Force Microscopy (AFM) images of two Silver Island Film (SIF) coated glass slides (A & B), showing the variation in size and density that can be obtained by altering the dip coating conditions. Slides were produced by LI-COR Biosciences and imaged at the University of Nebraska-Lincoln.

The method of coating a slide with SIF generally produces metal nanostructures that are both randomly distributed and vary in size and shape over the glass surface (Fig. 5.3). Presumably, some fraction of the metal nanostructures on the glass surface will possess the correct dimensions to generate enhancement for the near-IR fluorophores. Producing metal islands of a specific size may be critical for efficient MEF. Although chemical dip coating can successfully create islands over a range of sizes, controlling the reaction and halting the growth of the islands at an exact size and density has proven difficult. The difficulty in controlling the exact nanostructure generation when using SIF's make this method challenging to transfer out of the research lab and into a commercial product, capable of generating consistent and reproducible enhancements.

## 5.5 COLLOID COATED SURFACES

Generating SIF coated surfaces allowed us to control only the overall density of the silver nanostructures on the surface of the glass slide. The size and distribution of the nanoparticles was determined by a somewhat random stochastic process of silver seed generation and metal build up. To better control the metal nanostructure properties, silver colloids were generated to create colloid coated surfaces containing nanostructures of a specific size across the entire slide. The process uses metal nanoparticles (colloids) to effectively coat a glass slide, creating a metal nanostructure film. A clean glass substrate was coated with 3-aminopropyltrimethoxysilane (APS), producing a monolayer of reactive groups that can bind colloidal silver nanoparticles [58, 59]. The silanized glass substrate is then immersed in a colloidal solution, immobilizing the metal nanoparticles on the surface [60].

Using this method to form a nanoparticle coated surface allowed us to better control the size of the nanostructures, but still generated a random distribution of nanoparticles. The density of silver nanostructures could be controlled by both the amount of time the substrate was immersed in the colloidal solution and the concentration of the colloidal solution. Because the size of the metal structures is directly determined by the colloids, sparse to densely packed colloid coated surfaces that maintain a unique metal nanostructure size could be produced by simply varying the incubation time in the colloid solution.

Reports have indicated that colloid coated glass surfaces can enhance the fluorescence of some visible fluorophores more than SIF's, with good results showing a 16-fold enhancement [40]. For our experiments, several methods of producing silver colloids were employed, including sodium citrate reduction [40], the polyol process using ethylene glycol as the reductant [61, 62], small silver seed production followed by repeated rounds of controlled growth [29, 63], silver reduction in the presence of gum arabic [64, 65], as well as the photo induced production of triangular silver particles [30, 66]. Using these methodologies, we were able to generate colloids from 4 nm to greater than 150 nm in diameter. Sizes of the colloids were determined using a Brookhaven 90Plus dynamic light scattering particle size analyzer (Brookhaven Instruments Corporation, Holtsville, NY). The colloids were attached to APS coated glass slides by dip coating the slides for up to 24 hours in the colloid solution. These various sized colloid coated surfaces generated near-IR enhancements of up to 11-fold for IRDye 800CW and only 5-fold for IRDye 700 [36]. Even though we could generate enhancements for the near-IR fluorophores using the colloid coated surfaces, the enhancements were less than that obtained from using SIF's.

## 5.6 NANOPARTICLE INTERACTIONS INCREASE FLUORESCENCE ENHANCEMENTS

Several groups have begun to investigate the interactions that take place between closely spaced nanoparticles. Specifically, relationships between pairs, groups, or arrays of metal colloids have become an area of great interest. A single colloid can generate fluorescence enhancement, however coupled colloids may produce even greater emissions [67, 68]. Much of this work was born from SERS investigations where "hot spots" of enhancement were observes. These hot spots were produced where pairs or groups of metal nanostructures were generated in close proximity to one another. Using discrete dipole approximation (DDA) calculations and finite-difference time-domain (FDTD) methods researchers have calculated that the electric field in the space between a pair of coupled metal nanoparticles is more intense than that of a single particle [68-70]. Coupled metal nanoparticles should produce greater enhancements, since MEF is a property of the near-field interactions of an excited state fluorophore with the induced electric fields generated on a metal nanoparticle [21, 69]. To test this theory, several groups have specifically positioned a fluorescent particle between two adjacent metal nanoparticles, either by tethering the fluorophore between two metal particles or by physically moving the particles into position. Their results show that enhancement is maximized when the fluorophore is positioned directly between the two metal nanoparticles and the distance between the metal particles is minimized [69, 71, 72].

In order to maximize MEF in the near-IR, we proceeded to generate colloidal aggregates on solid surfaces. Our research led to the realization that colloidal aggregates of a specific size produced much greater enhancement of near-IR fluorophores than was observed with the colloid coated surfaces or SIF's. Generation of these specific colloidal aggregates was termed Controlled Colloidal Aggregation (CCA) [73]. Citrate-stabilized silver colloids were used to generate the CCA nanoparticles. The negative citrate ions are bound to the silver colloids and produce an electrostatic interaction that stabilizes the colloidal suspension, the repulsion of like charges preventing aggregation of the colloids. We hypothesize that an addition of a dilute buffer solution provides counter ions that can partially shield the electrostatic interactions between the colloids in suspension. Upon reduction of the electrostatic repulsion between individual colloids, van der Waals forces begin to cluster the colloids, forming aggregates. By controlling the aggregation conditions of the colloids, we were able to produce metal aggregates of similar size that can greatly enhance the fluorescence of near-IR fluorophores.

Surfaces coated with CCA nanostructures were found to be significantly different than SIF's, colloid coated surfaces, or surfaces that had colloids concentrated and dried upon them (Fig. 5.4). The CCA nanostructures form randomly distributed discrete structures on the glass surface that are each composed of multiple colloids. SEM imaging of the CCA nanostructures shows that the individual aggregates are on the order of 500nm to 1.5μm in diameter and are approximately spherical in shape (Fig. 5.5). These aggregates are not unlike aggregates evaluated for surface enhanced Raman scattering [74].

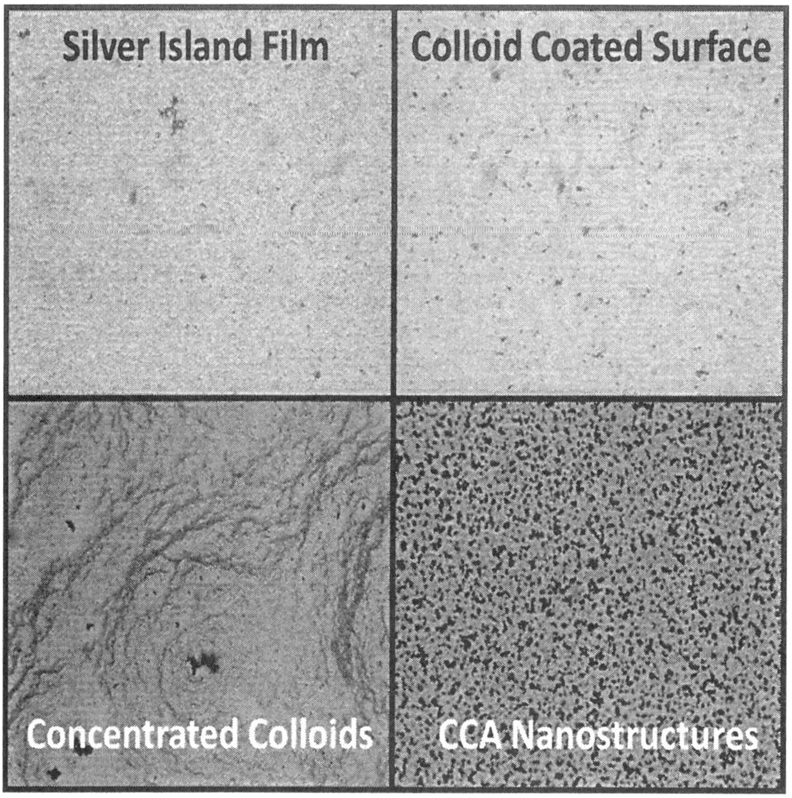


**Figure 5.4:** Light microscope images of glass slides coated with silver island films, colloids, concentrated colloids, and CCA nanostructures. The CCA nanostructures form discrete aggregate structures that are not observed in the other preparations.

To test the effectiveness of our CCA nanoparticles at enhancing near-IR dyes, CCA nanostructures were first produced and adhered onto plasma cleaned plain glass slides. Known concentrations of IRDye 800CW and Alexa Fluor 680 labelled streptavidin were then spotted onto either the CCA coated or uncoated glass. The spots were allowed to dry and the slide was imaged using an Odyssey near-IR imager. Spot integrated intensities were determined using Odyssey software, with any increased background signal generated by the CCA nanostructures subtracted from the calculation. Using these CCA nanostructures, we were able to produce over 200-fold enhancement of IRDye 800CW and over 100-fold enhancement of Alexa Fluor 680 labelled streptavidin (Fig. 5.6). The enhancements obtained using the CCA method on near-IR fluorophores were significantly better than that obtained by using either SIF's or colloid coated surfaces, with enhancements of IRDye 800CW routinely >100-fold over dye spotted on plain glass slides (Table 5.1).

2000X 60,000X

**Figure 5.5:** Scanning electron microscope (SEM) images of silver CCA nanostructures on glass under varying magnifications. The images show that the surface is coated with aggregated colloid particles that have a size of ~500nm – 1.5um.

| | Colloid Coated Surface | Silver Island Film | CCA Nano-structures |
|---|---|---|---|
| Alexa Fluor 680 | 5.2 | 18.2 | 100.8 |
| IRDye 800CW | 11.0 | 15.7 | 184.7 |

**Table 5.1:** Average fold-enhancements observed for either Alexa Fluor 680 or IRDye 800CW spotted on sliver nanostructure coated glass relative to uncoated glass slides. Average enhancements are shown for colloid, silver island film, and CCA nanostructure coated surfaces. The CCA nanostructures provide the greatest enhancements.

To compare the effects of using our CCA enhanced slides on visible fluorophores, we tested the visible fluorophores Cy3 (ex 554 nm; em 568 nm) and Cy5 (ex 649 nm; em 666 nm). Two concentrations of fluorophores were spotted over CCA nanostructures on a plain glass slide and scanned using a GenePix 4100A Scanner (Molecular Devices, Sunnyvale, CA). The results showed that Cy5 was enhanced 9.3 ± 2.35 fold, while Cy3 was enhanced only 2.35 ± 0.31 fold. These results are not surprising, given that the colloidal aggregates were specifically developed to enhance the longer wavelength near-IR fluorophores, and thus may not be effective at enhancing the shorter wavelength visible fluorophores. Furthermore, the increased background signal from the visible fluorophores can additionally contribute to a reduced overall enhancement factor.

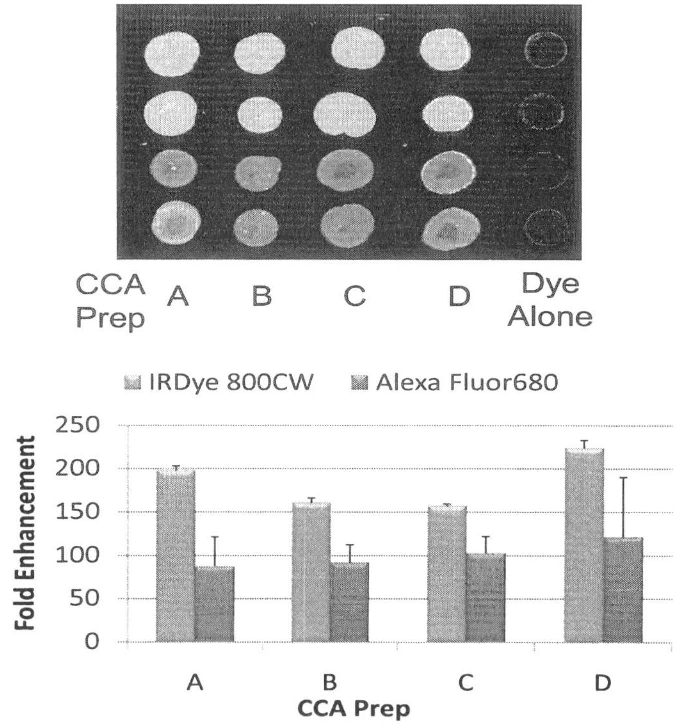


**Figure 5.6:** Odyssey Scanner results showing CCA enhancement of IRDye 800CW (Top two rows) and Alexa Fluor 680 (Bottom two rows) labelled streptavidin on glass slides. Scanner image (top) shows four CCA preparations (A-D) used to enhance the two near-IR fluorophores, as well as the fluorophores spotted on plain glass without CCA nanostructures added (Dye Alone). Bar graph (bottom) shows the relative fold enhancement over the Dye Alone samples for each of the CCA preparations. Error bars are shown that reflect the deviation between two samples for each preparation.

The combination of increased emissions from CCA in the near-IR and the low background signal from near-IR scatter and fluorescence should provide a means for lowering the limit of detection, providing the increased sensitivity that is required for many biological assays.

## 5.7 LIMITS OF DETECTION

The use of MEF for producing increased fluorescence intensity has been successfully demonstrated using a variety of fluorophores [34, 45, 49, 51, 52]. However, along with the dramatic increase in signal, these metal nanostructures may also display an increased amount of background or scatter, as can be seen when imaging SIF's with various wavelengths of light (Fig. 5.2). Any increase in the scatter signal can be detrimental when working with low fluorophore concentrations, hindering detection of a desired signal. Therefore, true utility of the system will be determined by how it can enhance the detection of small amounts of protein, and not just by the scale of relative enhancement seen in higher fluorophore concentrations. We have found that determining the LOD using model systems has proven to be a reliable indicator of the utility of the instrument or technique.

Within the context of this article, the LOD is defined to be the smallest concentration of a sample that can be reliably detected, producing a signal that is three times as large as the standard deviation of the system noise. The ability to improve the LOD can greatly expand the utility of a technique. For example, an improved LOD may allow for the identification of a low abundance sample that would normally be missed in the background of other samples, regardless of the total amount of sample tested.

A 2-fold dilution series of IRDye 800CW labelled streptavidin was spotted on either CCA enhanced or plain glass slides. The slides were imaged using an Odyssey Near-IR imager and integrated fluorescence intensities from each of the spotted dilutions were calculated using Odyssey software with background subtracted. Using CCA enhanced slides, we were able to reliably detect as little as 61 fg (1 attomole) of IRDye 800CW-labelled streptavidin. Using the same, labelled streptavidin on plain glass slides, we were able to reliably detect only 976 fg of the protein. Therefore, we demonstrated a 16-fold decrease in the LOD using the CCA enhanced slides.

## 5.8 DYNAMIC RANGE

Linear dynamic range is also important parameter in fluorescence detection in combination with the increased sensitivity that we have demonstrated. For this methodology to be a useful quantitative tool, it would be advantageous to show that the system is linear over a wide range of fluorophore concentrations. Again, four sets of 2-fold dilution series of IRDye 800CW labelled streptavidin were spotted on either CCA enhanced or plain glass slides. The slides were imaged and the integrated intensities for all spots were calculated. Using this data, we determined that our system remained linear over three orders of magnitude with an $R^2$ value of 0.9886 (Fig. 5.7).

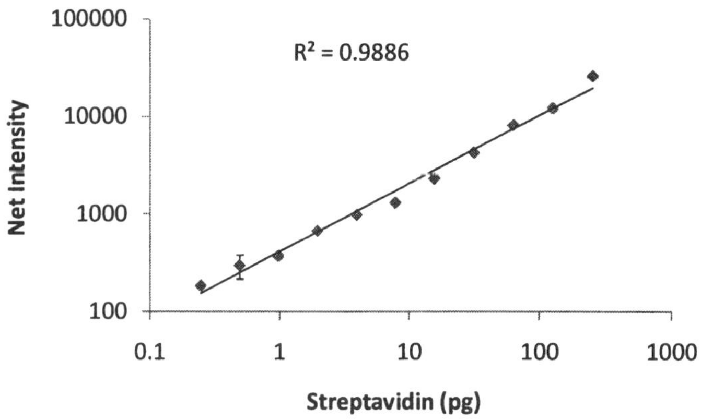


**Figure 5.7:** Linearity of CCA fluorescence enhancement. Two-fold serial dilutions of IRDye 800CW-labelled streptavidin are plotted. Each plotted point represents the average of 4 measurements, with one standard deviation shown by error bars. Note that only the error bars for the sample at 0.5pg is large enough to be visible on the graph.

## 5.9 CONCLUSIONS

Using our near-IR fluorophores combined with MEF, we have shown that the traditional methods of enhancing fluorophores on SIF's and colloid surfaces work as expected, and that our new method of CCA produces significantly better enhancement. CCA based enhancement on our near-IR fluorophores provides both enhancements of >200-fold on glass slides while providing a reduced scattering signal as compared to visible fluorophores. This combination allows us to reduce the LOD on our CCA coated glass slides by 16-fold.

Improvements in sensitivity and LOD can have a profound effect on molecular biology applications. This will equate to reduced sample requirements, allowing for more assays to be done on small sample preparations and allowing precious samples to be better utilized. Improved sensitivity and LOD will also allow the use of antibodies with lower binding affinities and may allow for better quantitation of array data.

As biotechnology research continues to push the limits of our current instrumentation and assays, there arises a need for new technologies and products to come to the forefront and allow the next generation of experiments to continue. Near-IR MEF using CCA is a tool that should complement current fluorescence technology, allowing for greater sensitivity and reduced detection limits.

## 5.10 ACKNOWLEDGEMENTS

This work was supported by the NIH National Center for Research Resources, SBIR Grant numbers RR021785 and RR024266. We would like to thank Dr. Teresa Urlacher for her helpful comments. We also thank Dr. Joseph Lakowicz and The Center for Fluorescence Spectroscopy (University of Maryland, Baltimore) for their assistance.

## 5.11 REFERENCES

1. Cognet, L., et al. (2003) Single metallic nanoparticle imaging for protein detection in cells. *Proc Natl Acad Sci U S A* **100**:11350-5.
2. Seydel, C. (2003) Quantum dots get wet. *Science* **300**:80-1.
3. Moerner, W.E. and Orrit, M. (1999) Illuminating single molecules in condensed matter. *Science* **283**:1670-6.
4. Schena, M., et al. (1995) Quantitative monitoring of gene expression patterns with a complementary DNA microarray. *Science* **270**:467-70.
5. Randolph, J.B. and Waggoner, A.S. (1997) Stability, specificity and fluorescence brightness of multiply-labelled fluorescent DNA probes. *Nucleic Acids Res* **25**:2923-9.
6. Kaiser, R.J., et al. (1989) Specific-primer-directed DNA sequencing using automated fluorescence detection. *Nucleic Acids Res* **17**:6087-102.
7. Smith, L.M., et al. (1986) Fluorescence detection in automated DNA sequence analysis. *Nature* **321**:674-9.
8. Brumbaugh, J.A., et al. (1988) Continuous, on-line DNA sequencing using oligodeoxynucleotide primers with multiple fluorophores. *Proc Natl Acad Sci U S A* **85**:5610-4.
9. Shah, K. and Weissleder, R. (2005) Molecular optical imaging: applications leading to the development of present day therapeutics. *NeuroRx* **2**:215-25.
10. Braslavsky, I., et al. (2003) Sequence information can be obtained from single DNA molecules. *Proc Natl Acad Sci U S A* **100**:3960-4.
11. Lakowicz, J.R. (1999) Principles of fluorescence spectroscopy. 2nd ed, New York: Kluwer Academic/Plenum Publishers.
12. Olive, D.M. (2004) Quantitative methods for the analysis of protein phosphorylation in drug development. *Expert Rev Proteomics* **1**:327-41.
13. Middendorf, L.R., et al. (1998) Near-infrared fluorescence instrumentation for DNA analysis, in *Near-Infrared Dyes for High Technology Applications*, S. Daehne, Editor. 1998, Kluwer Academic Publishers. pp. 21-54.
14. Shealy, D.B., et al. (1995) Synthesis, chromatographic separation, and characterization of near-infrared-labelled DNA oligomers for use in DNA sequencing. *Anal Chem* **67**:247-251.
15. Middendorf, L.R., et al. (1992) Continuous, on-line DNA sequencing using a versatile infrared laser scanner / electrophoresis apparatus. *Electrophoresis* **13**:487-94.

16. Fomenko, V. and Nesbitt, D.J. (2008) Solution control of radiative and nonradiative lifetimes: a novel contribution to quantum dot blinking suppression. *Nano Lett* **8**:287-93.
17. Schultz, S., et al. (2000) Single-target molecule detection with nonbleaching multicolor optical immunolabels. *Proc Natl Acad Sci U S A* **97**:996-1001.
18. Frey, P.A. and Frey, T.G. (1999) Synthesis of undecagold labeling compounds and their applications in electron microscopic analysis of multiprotein complexes. *J Struct Biol* **127**:94-100.
19. Dubertret, B., et al. (2002) In vivo imaging of quantum dots encapsulated in phospholipid micelles. *Science* **298**:1759-62.
20. Chen, X., et al. (2005) Investigation of the crystallization process in 2 nm CdSe quantum dots. *J Am Chem Soc* **127**:4372-5.
21. Lakowicz, J.R. (2001) Radiative decay engineering: biophysical and biomedical applications. *Anal Biochem* **298**:1-24.
22. Aslan, K., et al. (2005) Metal-enhanced fluorescence: an emerging tool in biotechnology. *Curr Opin Biotechnol* **16**:55-62.
23. Aslan, K., Lakowicz, J.R., and Geddes, C.D. (2005) Plasmon light scattering in biology and medicine: new sensing approaches, visions and perspectives. *Curr Opin Chem Biol* **9**:538-44.
24. Lakowicz, J.R., et al. (2002) Enhanced and localized multiphoton excited fluorescence near metallic silver islands: Metallic islands can increase probe photostability. *J Fluoresc* **12**:299-302.
25. Lakowicz, J. and Fu, Y. (2006) Enhanced Fluorescence of Cy5-Labeled DNA Tethered to Silver Island Films: Fluorescence Images and Time-Resolved Studies Using Single-Molecule Spectroscopy *Analytical Chemistry* **78**:6238 - 6245.
26. Fu, Y. and Lakowicz, J.R. (2006) Enhanced fluorescence of Cy5-labelled oligonucleotides near silver island films: a distance effect study using single molecule spectroscopy. *J Phys Chem B* **110**:22557-62.
27. Sabanayagam, C.R. and Lakowicz, J.R. (2007) Increasing the sensitivity of DNA microarrays by metal-enhanced fluorescence using surface-bound silver nanoparticles. *Nucleic Acids Res* **35**:e13.
28. Geddes, C.D., et al. (2003) Silver fractal-like structures for metal-enhanced fluorescence: Enhanced fluorescence intensities and increased probe photostabilities. *J. Fluoresc.* **13**:267-76.
29. Aslan, K., et al. (2005) Fast and slow depositionof silver nanorods on planar surfases: Application to metal-enhanced fluorescence. *J Phys Chem B* **109**:3157-3162.
30. Jin, R., et al. (2001) Photoinduced conversion of silver nanospheres to nanoprisms. *Science* **294**:1901-3.
31. Corrigan, T.D., et al. (2005) Enhanced fluorescence from periodic arrays of silver nanoparticles. *J Fluoresc* **15**:777-784.
32. Jensen, T.R., et al. (2000) Nanosphere lithography: Tunable localized surface plasmon resonance spectra of silver nanoparticles. *J. Phys. Chem.* **104**:10549-56.
33. Gryczynski, I., et al. (2002) The CFS Engineers the Intrinsic Radiative Decay Rate of Low Quantum Yield Fluorophores. *J Fluoresc* **12**:11-13.

34. Lakowicz, J.R., et al. (2002) Radiative decay engineering. 2. Effects of Silver Island films on fluorescence intensity, lifetimes, and resonance energy transfer. *Anal Biochem* **301**:261-77.
35. Malicka, J., et al. (2003) Fluorescence spectral properties of cyanine dye-labelled DNA oligomers on surfaces coated with silver particles. *Anal Biochem* **317**:136-46.
36. Anderson, J.P., Griffiths, M., and Boveia, V.R. (2006) Near-Infrared Fluorescence Enhancement Using Silver Island Films. *Plasmonics* **1**:103-110.
37. Geddes, C.D., Parfenov, A., and Lakowicz, J.R. (2003) Photodeposition of silver can result in metal-enhanced fluorescence. *Appl Spectrosc* **57**:526-31.
38. Shtoyko, T., et al. (2008) Enhanced Fluorescent Immunoassays on Silver Fractal-like Structures. *Anal Chem.*
39. Zhang, J., et al. (2005) Metal-enhanced fluoroimmunoassay on a silver film by vapor deposition. *J Phys Chem B* **109**:7969-7975.
40. Lukomska, J., et al. (2004) Fluorescence enhancements on silver colloid coated surfaces. *J. Fluoresc.* **14**:417-23.
41. Hulteen, J.C. and Van Duyne, P. (1995) Nanosphere lithography: A materials general fabrication process for periodic particle array surfaces. *J. Vac. Sci. Technol.* **13**:1553-58.
42. Haynes, C.L. and Van Duyne, P. (2001) Nanosphere lithography: A versatile nanofabrication tool for studies of size-dependent nanoparticle optics. *J. Phys. Chem.* **105**:5599-611.
43. Albrecht, M.G. and Creighton, J.A. (1977) Anomalously intense Raman spectra of pyridine at a silver electrode. *J. Am. Chem. Soc.* **99**:5215-5217.
44. Fleischmann, M., Hendra, P.J., and McQuillan, A.J. (1974) Raman spectra of pyridine adsorbed at a silver electrode. *Chem. Phys. Letts.* **26**:163-166.
45. Malicka, J., et al. (2003) Metal-enhanced emission from indocyanine green: a new approach to in vivo imaging. *J Biomed Opt* **8**:472-8.
46. Lakowicz, J.R., et al. (2001) Intrinsic fluorescence from DNA can be enhanced by metallic particles. *Biochem Biophys Res Commun* **286**:875-9.
47. Pugh, V.J., et al. (2003) Submicrometer spatial resolution of metal-enhanced fluorescence. *Appl Spectrosc* **57**:1592-8.
48. Malicka, J., Gryczynski, I., and Lakowicz, J.R. (2003) Enhanced emission of highly labelled DNA oligomers near silver metallic surfaces. *Anal Chem* **75**:4408-14.
49. Malicka, J., et al. (2003) Effects of fluorophore-to-silver distance on the emission of cyanine-dye-labelled oligonucleotides. *Anal Biochem* **315**:57-66.
50. Fu, Y. and Lakowicz, J. (2008) Single-molecule studies of enhanced fluorescence on silver island films. *Plasmonics* **2**:1-4.
51. Malicka, J., et al. (2003) Fluorescence spectral properties of cyanine dye labelled DNA near metallic silver particles. *Biopolymers* **72**:96-104.
52. Lakowicz, J.R., Malicka, J., and Gryczynski, I. (2003) Increased intensities of YOYO-1-labelled DNA oligomers near silver particles. *Photochem Photobiol* **77**:604-7.
53. Gersten, J.I. and Nitzan, A. (1985) Photophysics and photochemistry near surfaces and small particles. *Surf. Sci.* **158**:165-189.

54. Zhang, J., et al. (2008) Single-molecule studies on fluorescently labelled silver particles: Effects of particle size. *J phys Chem C* **112**:18-26.
55. Matveeva, E.G., et al. (2005) Plastic versus glass support for an immunoassay on metal-coated surfaces in optically dense samples utilizing directional surface plasmon-coupled emission. *J Fluoresc* **15**:865-71.
56. Ray, K., Chowdhury, M.H., and Lakowicz, J.R. (2007) Aluminum nanostructured films as substrates for enhanced fluorescence in the ultraviolet-blue spectral region. *Anal Chem* **79**:6480-7.
57. Matveeva, E.G., Gryczynski, Z., and Lakowicz, J.R. (2005) Myoglobin immunoassay based on metal particle-enhanced fluorescence. *J Immunol Methods* **302**:26-35.
58. Freeman, G.R., et al. (1995) Self-assembled metal colloid monolayers: An approach to SERS substrates. *Science* **267**:1629-1632.
59. Prikulis, J., et al. (2003) Large-area topography analysis and near-field Raman spectroscopy using bent fibre probes. *J Microsc* **210**:269-73.
60. Shipway, A.N., Katz, E., and Willner, I. (2000) Nanoparticle arrays on surfaces for electronic, optical, and sensor applications. *ChemPhysChem* **1**:18-52.
61. Silvert, P., et al. (1996) Preparation of colloidal silver dispersions by the polyol process. Part 1—Synthesis and characterization. *J. Mater. Chem.* **6**:573-577.
62. Silvert, P., Herrera-Urbina, R., and Tekaia-Elhsissena, K. (1997) Preparation of colloidal silver dispersions by the polyol process. Part 1—Mechanism of particle formation. *J. Mater. Chem.* **7**:293-299.
63. Aslan, K., Lakowicz, J.R., and Geddes, C.D. (2005) Rapid depositionof triangular silver nanoplates on planar surfaces:Application to metal-enhanced fluorescence. *J Phys Chem B* **109**:6247-6251.
64. Graf, C., et al. (2003) A general method to coat colloidal particles with silica. *Langmuir* **19**:6693-6700.
65. Velikov, K.P., Zegers, G.E., and van Blaaderen, A. (2003) Synthesis and Characterization of Large Colloidal Silver Particles. *Langmuir* **19**:1384-1389.
66. Jin, R., et al. (2003) Controlling anisotropic nanoparticle growth through plasmon excitation. *Nature* **425**:487-90.
67. Sweatlock, L.A., Maier, S.A., and Atwater, H.A. (2005) Highly confined electromagnetic fields in arrays of strongly coupled Ag nanoparticles. *Physical Review B* **71**.
68. Hao, E. and Schatz, G.C. (2004) Electromagnetic fields around silver nanoparticles and dimers. *J Chem Phys* **120**:357-66.
69. Zhang, J., et al. (2007) Metal-enhanced single-molecule fluorescence on silver particle monomer and dimer: coupling effect between metal particles. *Nano Lett* **7**:2101-7.
70. Talley, C.E., et al. (2005) Surface-enhanced Raman scattering from individual au nanoparticles and nanoparticle dimer substrates. *Nano Lett* **5**:1569-74.
71. Bek, A., et al. (2008) Fluorescence Enhancement in Hot Spots of AFM-Designed Gold Nanoparticle Sandwiches. *Nano Lett.*
72. Zhang, J., Fu, Y., and Lakowicz, J.R. (2007) Fluorescence images of DNA-bound YOYO between coupled silver particles. *Langmuir* **23**:11734-9.

73. Anderson, J.P. (2006) Colloidal metal aggregates and methods of use, LI-COR, inc. US Patent Application # 60/803,591.
74. Kneipp, K. and Kneipp, H. (2006) Single molecule raman scattering. *Applied Spectroscopy* **60**:322A-334A.

# 6

# Optimisation Of Plasmonic Enhancement Of Fluorescence For Optical Biosensor Applications

Colette McDonagh, Ondrej Stranik, Robert Nooney, Brian D. MacCraith*
Biomedical Diagnostics Institute, Dublin City University, Glasnevin, Dublin 9, Ireland.

## 6.1 INTRODUCTION

Fluorescence is an important tool in many areas of biotechnology and biomedical sciences, including medical diagnostics, DNA sequencing and genomics. In particular, there is increasing interest in fluorescence-based array sensors or biochips, which consist of patterned arrays of biorecognition elements which bind their respective targets in a sample and, through the use of conjugated labels, ultimately yield a fluorescence signal. While fluorescence detection offers high sensitivity, there is generally a low level of fluorescence from the biochip due to the relatively low surface coverage of labelled biomolecules.

The detected fluorescence can be significantly enhanced, however, by exploiting the plasmonic enhancement which can occur when a metal nanoparticle (NP) is placed in the vicinity of a fluorescent label or dye [1-3]. This effect is due to the localised surface plasmon resonance (LSPR) associated with the metal NP, which modifies the intensity of the electromagnetic (EM) field around the dye and which, under certain conditions, increases the emitted fluorescence signal. The effect is dependent on a number of parameters such as metal type, NP size and shape, NP-fluorophore separation and fluorophore quantum efficiency. There are two principal enhancement mechanisms: an increase in the excitation rate of the fluorophore and an increase in the fluorophore quantum efficiency. The first effect occurs because the excitation rate is directly proportional to the square of the electric field amplitude, and the maximum enhancement occurs when the LSPR wavelength, $\lambda_{res}$, coincides with the peak of the fluorophore absorption band [4, 5]. The second effect involves an increase in the quantum efficiency and is maximised when the $\lambda_{res}$ coincides with the peak of the fluorophore emission band [6].

Our work has focussed on two key areas which underpin the eventual exploitation of plasmonic enhancement features in fluorescence-based biosensors:

- Establishment of synthetic and fabrication techniques which enable reproducible implementation of plasmonic enhancement principles in practical biochip systems and,
- Elucidation of the fundamental principles which lead to rational design rules for the key optimisation parameters.

*MEF Biosensor Applications*
Edited by Chris D. Geddes.

In this chapter, we highlight examples of our work that show significant progress in the two areas listed above. In particular, we first describe how it is possible to tune $\lambda_{res}$ across the visible spectrum by altering NP size and shape. We then report on the development of a technique that facilitates uniform deposition of tailored metal NPs on planar polymer substrates in order to achieve reproducible values for $\lambda_{res}$. Building on these synthetic and fabrication capabilities, we provide experimental validation of 2 key optimisation parameters:

(a) the optimal NP radius, and

(b) The optimal NP-fluorophore spatial separation.

Theoretical principles that support the work presented here are outlined in Section 6.2 and this is followed by Sections 6.3 and 6.4 that detail, respectively, fabrication techniques and optimisation strategies.

## 6.2 THEORETICAL BACKGROUND AND MODELLING

The theory describing the interaction of light, fluorescent dyes and metallic nanoparticles is complex and can be solved only for specific systems [7]. The overall interaction can be divided into three stages. The first stage is the interaction of the excitation light with the NP. The second stage is the interaction of the altered EM field in the vicinity of the NP with the dye, and the third interaction is that of the dye fluorescence with the NP. The interaction of an EM wave with a spherical particle has been solved exactly by Mie [8]. This theory predicts the distribution of the EM waves both inside and outside the particle on illumination by a plane wave. The dependence of the extinction coefficients on illumination wavelength and angle can be predicted from these results. If the particle is sufficiently small (smaller than the illuminating wavelength) and if the material of the NP has a negative dielectric constant (as in the case of metals), the theory also predicts that there will be a resonance between the illuminating light and the NP which results in an increase of extinction coefficient for this wavelength range. This constitutes the LSPR effect.

In order to treat this problem mathematically, we consider a system consisting of a gold / silver NP of radius r, placed at the origin of a coordinate system ($e_x$, $e_y$, $e_z$) and with an incident plane wave with x-polarisation and wavelength $\lambda$ propagating in the z-direction. The electric field of the incident wave is expressed as

$$\vec{E}_i(\vec{r},\theta,\phi) = E_0 \exp(irk\cos\phi)\vec{e}_x \tag{1}$$

where $k= 2\pi/\lambda$. The interaction of the field with the NP creates an additional field $E_s$, which is superimposed on the incident field outside the sphere. $E_s$ is expressed as

$$\vec{E}_s(\vec{r},\theta,\phi) = E_0 \sum_{n=1}^{\infty} i^n \frac{2n+1}{n(n+1)}(ia_n\vec{N}_{e1n} - b_n\vec{M}_{o1n}) \tag{2}$$

where $N_{eln}^{(3)}$, $M_{oln}^{(3)}$ are vector spherical harmonics and $a_n$ and $b_n$ are scattering coefficients [9]. An extinction cross section can be derived from knowledge of the exact distribution of the electric field [9]. The expression for the cross section is given in equation 6.3.

$$C_{ext} = \frac{2\pi}{k^2}\sum_{n=1}^{\infty}(2n+1)\,\mathrm{Re}\{a_n + b_n\} \qquad (3)$$

This calculated extinction cross section can then be compared with data measured via absorption spectroscopy.

As an example, extinction, absorption and scattering cross-sections of spherical silver NPs in aqueous solution, are presented in Figure 6.1 for three different values of NP radius. From the diagram, it is seen that the extinction cross section for a 5nm particle is entirely caused by absorption and there is little scattering. With increasing radius, the dipole peak moves to longer wavelengths and the scattering effect starts to be significant. For a 50nm particle, extinction is already dominated by scattering. The dipole plasmon resonance peak ($\lambda_{res}$) occurs at ~500nm and a quadrupole resonance, given by the expansion of coefficient $a_2$, appears at ~400nm.

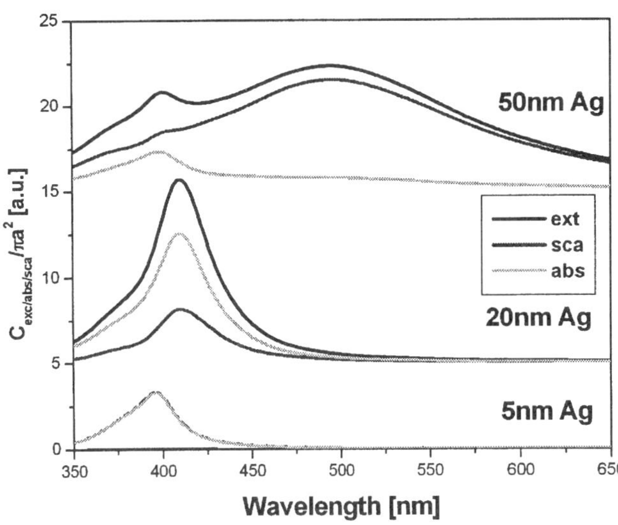


**Figure 6.1:** Theoretical extinction, absorption and scattering spectra of silver NPs with radius 5, 20 and 50nm, respectively.

In order to model the experimental approach which demonstrates the optimal NP radius, we introduce a NP configuration that comprises a metal core surrounded by an outer shell of dielectric material. This thin outer shell provides a buffer layer which prevents fluorophores from residing directly on the NP surface and thereby minimizes fluorescence quenching. In addition, the presence of an outer shell of $SiO_2$ gives rise to a red shift of $\lambda_{res}$. For non-spherical NP shapes, electrodynamic calculations are much more difficult and often only numerical methods can be used. For example, the discrete dipole approximation technique [10, 11] can be used to calculate extinction cross sections for non-spherical structures. In this method, the particle is represented by a 3-D array of point dipoles, where the spacing between the dipoles is smaller than the wavelength of the incident field. The interaction of each dipole with the incident field and the field generated by other dipoles is then calculated. This set of equations leads to the self-consistent solution for the dipole polarizations.

Of the two different types of plasmonic enhancement which were described in section 6.1, the emphasis here is on the excitation enhancement mechanism. There is a linear dependence of the excitation rate of a fluorescent dye on the intensity of the excitation light in the direction of the electric dipole, $e_d$, of the molecule. When the dye molecule is located near the NP, the electric field acting on the dipole changes from $E_i$ to $E_i + E_s$. In this case, the excitation enhancement factor, $f_{enh}$, for one dye molecule is defined as a ratio of intensities:

$$f_{enh}(\vec{r},\theta,\phi) = \frac{|(\vec{E}_i + \vec{E}_s)\vec{e}_d|^2}{|\vec{E}_i\vec{e}_d|^2} \tag{4}$$

Dye molecule orientation at the surface of a NP has a significant influence on the enhancement factor. For an ideal conductor, the electric field is always perpendicular to the surface. As discussed above, in order to demonstrate plasmonic enhancement as a function of NP radius, NPs were coated with a silica shell prior to attachment of a dye in order to minimize dye quenching. The presence of this shell slightly modifies the electric field but it is still almost perpendicular to the surface. It follows from this and from Equation 4 that molecules, which are oriented normally, are considerably more excited than those oriented tangential to the surface. For this model, we assume that there is random dipole orientation at the NP surface. The average intensity over all possible dipole positions due to a EM field E is given by $1/3\ |E|^2$. Hence the averaged enhancement factor can be written as:

$$f_{enh}(\vec{r},\theta,\phi) = \frac{|\vec{E}_i + \vec{E}_s|^2}{E_0{}^2} \tag{5}$$

In the experiments reported later in section 6.4 1, we detected fluorescence from dye molecules distributed uniformly over a NP surface. Therefore, the enhancement was averaged over all possible positions. In the case of a NP with a $SiO_2$ shell, this is given by

$$F^a_{enh} = \frac{1}{4\pi E_0^2}\int_0^{2\pi}\int_0^{\pi}|\vec{E}_i(a+d,\theta,\phi) + \vec{E}_s(a+d,\theta,\phi)|^2 \sin(\theta)d\theta d\phi \tag{6}$$

where *a* and *d* are the NP radius and shell thickness respectively. As a result of the spherical symmetry, the enhancement factors are also valid for illumination of the sphere with unpolarised light.

## 6.3 SYNTHETIC AND FABRICATION TECHNIQUES

### 6.3 1. *Tuning $\lambda_{res}$ Across The Visible Spectrum*

The importance of matching $\lambda_{res}$ of a NP to the spectral properties, especially the absorption or emission peaks, of a fluorophore was discussed in section 6.1. We have investigated a range of techniques for deposition of metal nanostructures on planar surfaces. The technique which is highlighted here is that of Nanosphere Lithography (NSL). NSL is a process which enables fabrication of 2-D arrays of identical metal NPs on a planar substrate. The technique was first used by Deckman [12] and its main advantage is the low-cost production of well defined periodic nanostructures with resolutions which are similar to more expensive techniques such as e-beam lithography[13]. As a result, this technique, or variations on it, has been employed in a range of applications. NSL involves three steps. Firstly, a layer of polystyrene beads (PB) is deposited on a substrate and forms an ordered, close-packed structure due to self-assembly. Secondly, a metal layer of controlled thickness is thermally evaporated over the PBs. The ordered layer of PBs acts as a mask which provides triangular-shaped interstitial spaces between the beads. After deposition, the PBs are removed, leaving an ordered array of metal nanostructures on the surface.

In our work, cleaned glass slides were immersed in suspensions of PBs. A dip-coating system was used to create the self-assembled layers whereby the withdrawal speed of the slides was carefully optimized for ordered, monolayer deposition. PB bead diameters of 200nm, 350nm and 500nm were used. Figure 6.2a shows the ordered hexagonal-packed layer obtained by PBs of diameter 300nm. After deposition of the PB layer, the samples were placed in an evaporation chamber where silver layers of graduated thicknesses were deposited. This was followed by sonication in pure ethanol to remove the PB beads and drying in nitrogen. Figure 6.2b shows an AFM image of the resulting triangular-shaped silver nanostructures. Because of the weak adhesion of the silver to the glass substrates, a 6nm thick layer of SiOx was deposited over the metal structures. This additive dielectric layer, while improving the sample stability, induced a small red-shift of ~20nm in $\lambda_{res}$. In addition, the SiOx layer functioned as a spacer layer between NPs and fluorescent dye in order to avoid the metal-fluorescence quenching effect described earlier.

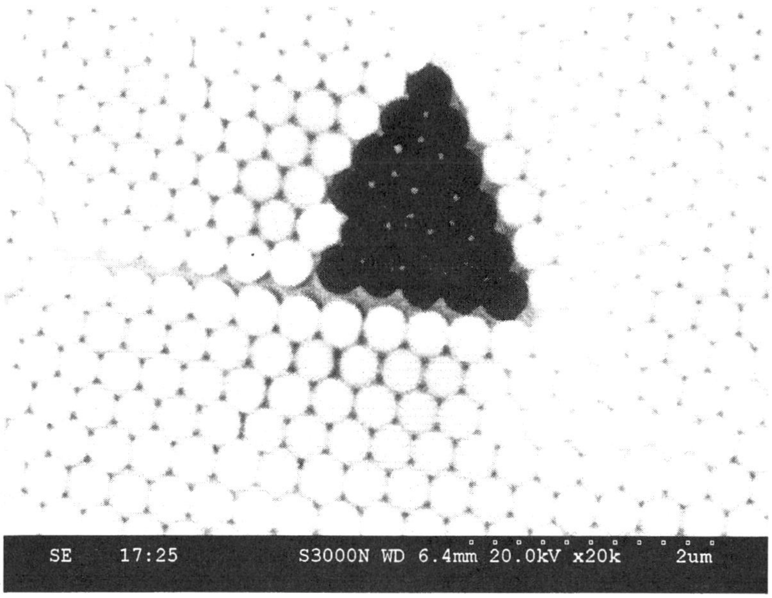


**Figure 6.2a:** SEM of a monolayer of polystyrene beads of diameter 300nm.

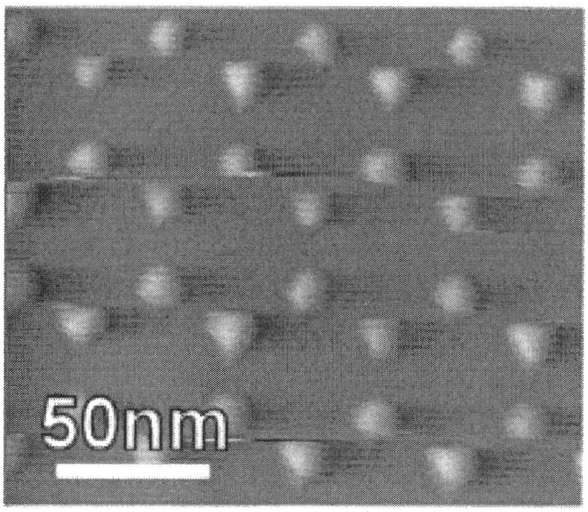


**Figure 6.2b:** AFM image of pyrimidal-shaped NPs formed by the NSL process.

The optical properties of NPs with different in-plane size (resulting from different PB diameters) and different evaporated thickness, were examined. The absorption spectra for a range of NP dimensions are shown in Figure 6.3. It is evident that the peak position of plasmon resonance could be tuned by varying the PB diameter. $\lambda_{res}$ was shifted to the red by increasing the PB diameter. $\lambda_{res}$ could be tuned via the height of the NPs. Because of the high level of accuracy of metal layer thickness achieved by thermal evaporation, it was possible to tailor $\lambda_{res}$ to nanometer precision. Clearly this technique enables tailoring over the entire visible spectrum. However, there are many other strategies such as tailoring the size, shape and composition of the NPs in order to match $\lambda_{res}$ to the spectral properties of the fluorophore. For example, it has been established that $\lambda_{res}$ for triangular NPs, discussed briefly in section 2, can be matched to the spectral properties of the widely used Cy 5 fluorescent label.

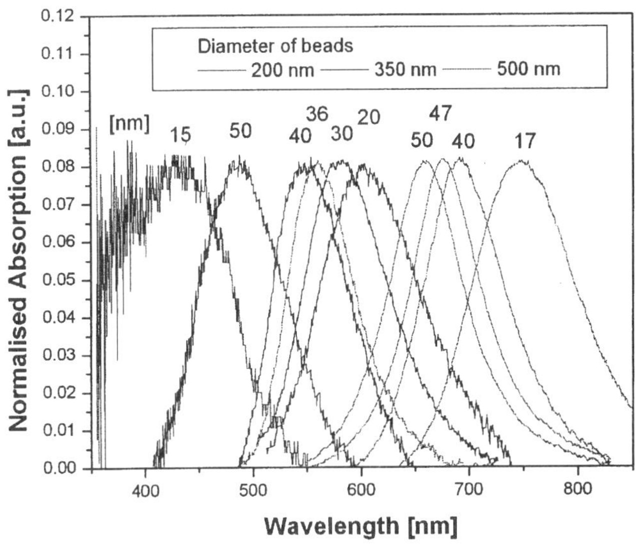


**Figure 6.3:** Absorption spectra of NPs with shapes determined by different diameters of polystyrene spheres and different deposition heights (numbers above the peaks).

## 6.3 2. *Uniform Deposition Of Metal NPs To Achieve Reproducible Values Of* $\lambda_{res}$

In order to achieve plasmonic enhancement of fluorescence it is important first of all to establish a deposition technique which can yield reproducible substrates with the desired $\lambda_{res}$ value. We have carried out a study, where this was achieved using charged polyelectrolyte (PEL) layers which were deposited using a layer-by-layer (LbL) technique and which facilitate uniform and reproducible deposition of metal NPs on a range of surfaces. This is a generic technique which also enables the deposition of highly accurate fluorophore-NP spacer layers as discussed in section 6.4 2.

Spherical NPs made from silver, gold or Ag/Au alloy were prepared by chemical synthesis. Briefly, Au NPs were prepared by the reduction of hydrogen tetracholoraurate with sodium citrate following the method of Turkevich and co-workers [14]. Au/Ag alloy NPs with a range of diameters were prepared using a 2-step process involving co-reduction of silver nitrate and chloroauric acid with sodium citrate followed by seeded growth to increase the NP diameter [14,15]. Pure silver NPs were synthesised by reduction of silver nitrate with sodium citrate in the presence of aniline [15]. These NPs were used to demonstrate plasmonic enhancement of fluorescence (reported in section 6.4). The deposition and enhancement studies were carried out using disposable 96 well microplates.

The PELs used in this work were poly(ethyleneimine) (PEI molecular weight, *Mw* 750,000g), polystyrene sulfonate (PSS *Mw* 250,000g), poly(allyamine) hydrochloride (PAH, *Mw* 70,000g)). After activation by exposure to an oxygen plasma, the polystyrene wells were initially coated with five PEL layers to ensure a uniform coverage of amine groups. The layer sequence used was PEI/PSS/PAH/PSS/PAH. Pure gold, pure silver and gold-silver alloy NPs were deposited into the wells at volumes of 150µl. Each NP type was deposited into 16 wells which were further divided into 4 sets of 4 and coated with different numbers of PEL spacer layers. The thickness of one PAH/PSS bilayer was measured to be 3nm and the NPs were coated with 0, 2, 6 or 12 bilayers. The ruthenium complex bis(2,-2'-bipyridine-(5-isothiocyanato-phenanthroline) ruthenium bis(hexafluorophosphate). Which was initially conjugated to PEI (hereafter referred to as Ru-PEL) was coated onto the spacer layers and bound electrostatically to the top PSS layer. This constituted the NP-fluorophore system with controllable separation which formed the basis of the investigation of the effect of separation distance on plasmonic enhancement of fluorescence (described in section 6.4 2). The quantum efficiency of Ru-PEL was measured to be 0.031 and the excitation and emission maxima were 464 nm and 606 nm, respectively.

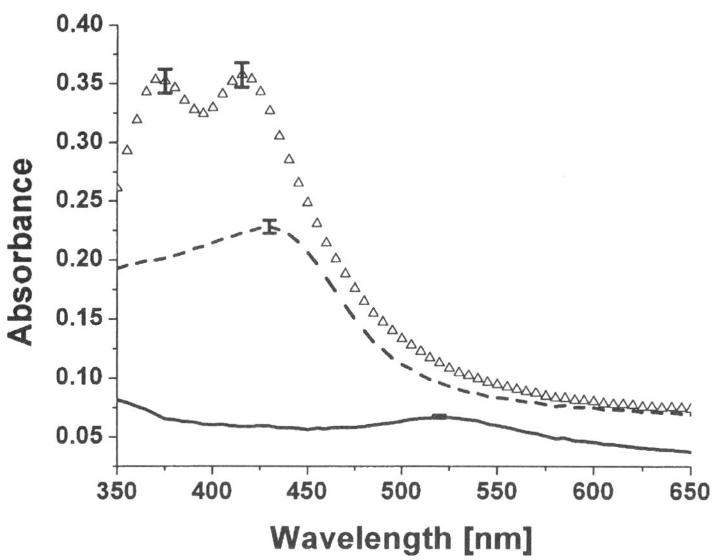

**Figure 6.4:** Optical absorbance of pure silver NPs, (Δ), gold / silver alloy NPs, (dashes) and pure gold NPs (solid line) coated onto polyelectrolyte coated microplate wells analysed under ambient conditions. Forty eight wells were examined in total, sixteen for each NP.

Before the addition of the spacer and dye layers, optical absorption analysis was performed on the 3 different NP types deposited in the microwell plates. Altogether, 48 wells were examined, 16 for each NP type. This data are shown in Figure 6.4 two bands are observed for the silver NPs, the dipole plasmon band occurring at 415 nm [9] and the quadrupole band occurring at 375 nm [16]. For the alloy NPs, a single $\lambda_{res}$ band is observed at 430 nm while for gold NPs, $\lambda_{res}$ is observed at 520nm. The key result from these data is the consistently low value of standard deviation in extinction maxima. For pure silver, alloy and gold NPs, the standard deviation in absorbance, calculated over 16 wells, was 3%, 2.6% and 1.5%, respectively. Figure 6.5 shows an AFM image of silver NPs with extinction properties similar to those shown in Figure 6.4. A surface coverage of $4.2\times10^9$ particles per $cm^2$, corresponding to 12% coverage, was estimated. This is consistent with literature reports for gold NPs [17]. The average NP centre-to-centre distance for the silver NPs was 154nm, and, assuming an average NP size of 60nm, this gives a NP-NP separation of 94nm. We have established that this represents saturation coverage as continued charging of wells with the NP solution did not increase the extinction maximum. It is likely that electrostatic repulsion between the NPs prevents a larger coverage. The data shown in figures 6.4 and 6.5 for NP extinction coefficient and surface coverage, respectively, are an indication of the reproducibility of the PEL layer-by-layer technique and of the subsequent NP deposition process.

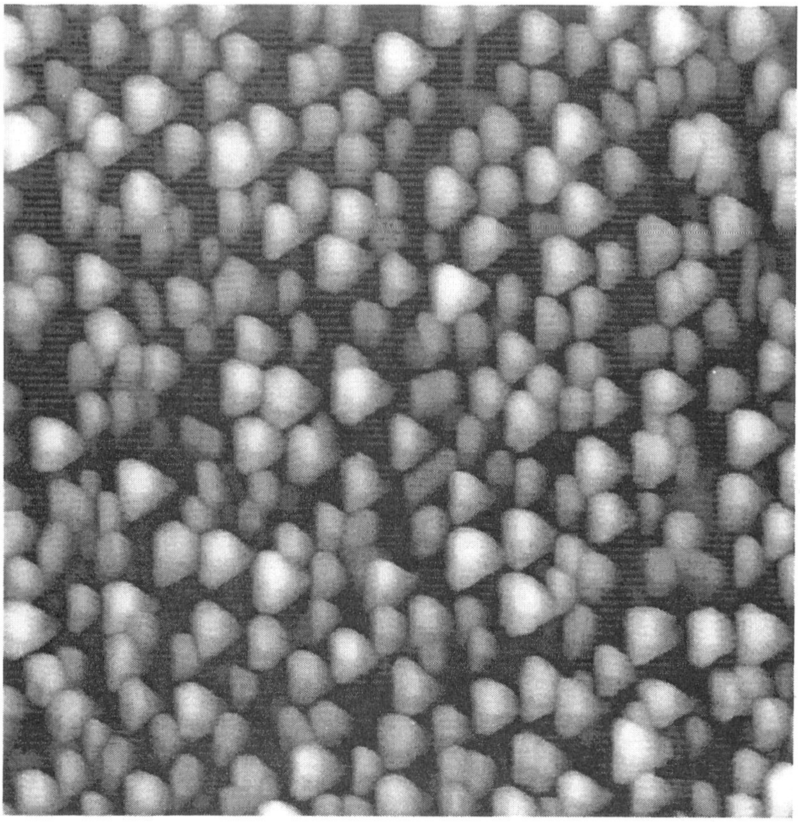

**Figure 6.5:** AFM image of pure silver NPs examined using tapping mode with 12 % total surface coverage.

## 6.4 OPTIMISATION OF DESIGN PARAMETERS

### 6.4 1. *Optimisation Of NP Size*

As discussed in Section 6.1, the scale of plasmonic enhancement depends on many parameters such as NP size, shape and composition. In this section, we report on an investigation of the impact of NP size on the degree of fluorescence enhancement. Spherical metal NPs were coated with a 5nm thick $SiO_2$ shell in order to minimize metal-fluorophore quenching effects and the ruthenium complex, Ru(II)tris(4,7 dephenyl-1,10 phenanthroline dichloride, referred to as $Ru(dpp)_3^{2+}$ was ionically attached to the shell. Gold/silver alloy NPs, fabricated as described in section 3.2, were used. A fixed gold molar ratio of $x_{Au} = 0.2$ was chosen such that $\lambda_{res}$ for a range of NP diameters overlapped the absorption band of the ruthenium complex. The theory was then applied to this specific NP composition and expressions for the enhancement factor were developed, as in section 6.2. The enhanced fluorescence was measured in solution and the dependence on NP diameter was measured and compared to the theoretical predictions.

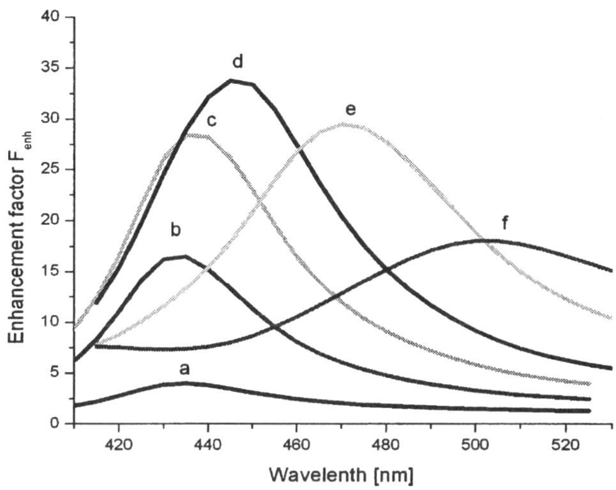


**Figure 6.6:** Theoretical prediction of the dependence of the enhancement on excitation wavelength for different NP sizes. Radius of NPs: a:5nm, b:10nm, c:15nm, d:20nm, e:30nm, f:40nm, (silica shell radius: 5nm).

Figure 6.6 shows the dependence of the enhancement factor, (calculated according to Equation 6), on excitation wavelength for different NP sizes, each surrounded by a 5nm thick silica shell. It can be seen from the data that, for each value of NP radius, an enhancement maximum occurs at a specific wavelength. These curves are not generally the same as the extinction spectra calculated from Equation 3. For the smallest size, the profiles and positions of the extinction and enhancement spectra are similar, but, with increasing radius the maximum enhancement wavelength shifts to longer wavelengths compared to the extinction spectra. The maximum enhancement factor as a function of NP radius, which was calculated from the data in Figure 6.6, is shown in Figure 6.7 and indicates that a maximum fluorescence enhancement of ~32 should occur for a NP radius in the region of 20-25nm.

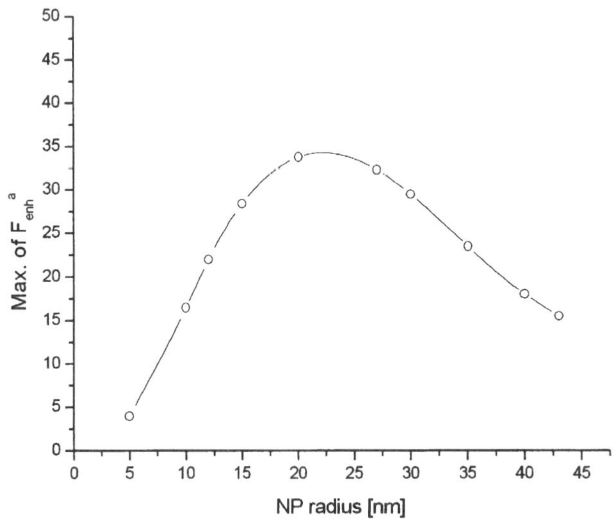


**Figure 6.7:** Dependence of maximum enhancement on NP size. (gold/silver molar ratio: 0.2. Silica shell thickness: 5nm).

As described earlier, NPs of fixed composition and with radii ranging from 13nm to 40nm were synthesized from colloidal solution. The absorption data for four different radii are shown in Figure 6.8. It can be seen that there is a distinct value of $\lambda_{res}$ for each NP size and that $\lambda_{res}$ shifts to longer wavelengths with increasing NP radius. The addition of the 5nm silica shell gives rise to a small red shift as predicted. This shift can be used to confirm the presence of the shell on the NPs. Figure 6.9 shows a TEM image of an alloy NP of radius 27nm surrounded by the silica shell. The NPs are monodispersed with a standard deviation in radius of ~±15%. A comparison of the theoretically computed optical extinction spectra and the experimental extinction data (also shown in Figure 6.8) for a NP with radius of 27nm, is shown in Figure 6.10. There is a broadening of the experimental data due to the small amount of polydispersity present. Computed and experimental values for all NP sizes are shown in Figure 6.11. Clearly there is quantitative agreement between theory and experiment.

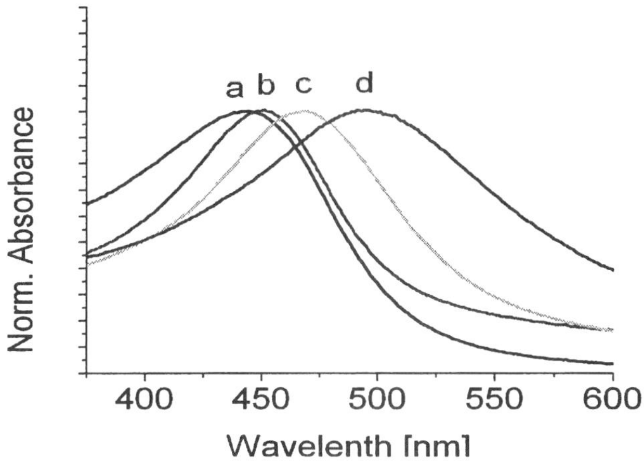


**Figure 6.8:** Extinction spectra for alloy NPs of radius: a: 13nm, b:27nm, c:36nm, d: 40nm.

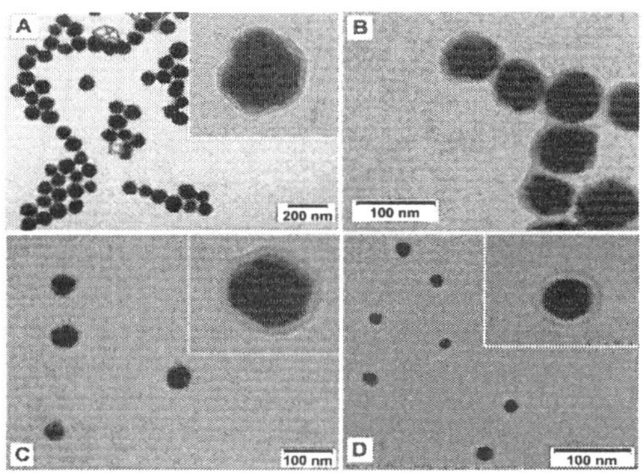


**Figure 6.9:** TEM images of alloy NPs with silica shell.

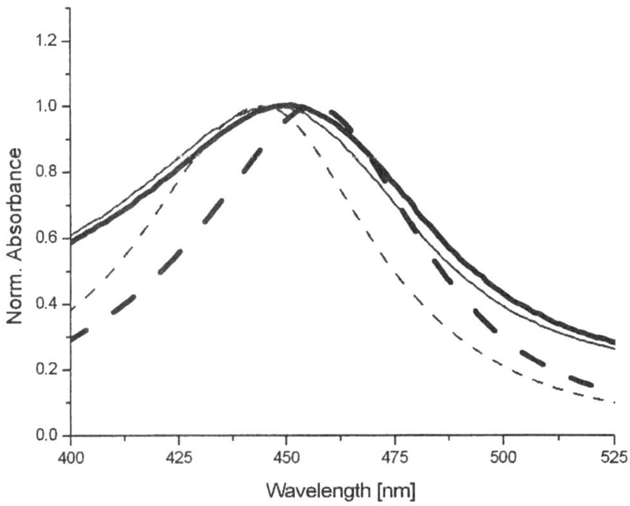


**Figure 6.10:** Theoretical (dashed line) and experimental (solid line) optical extinction spectra of NP of radius 27nm with (thick line) and without (thin line) 5nm silica shell.

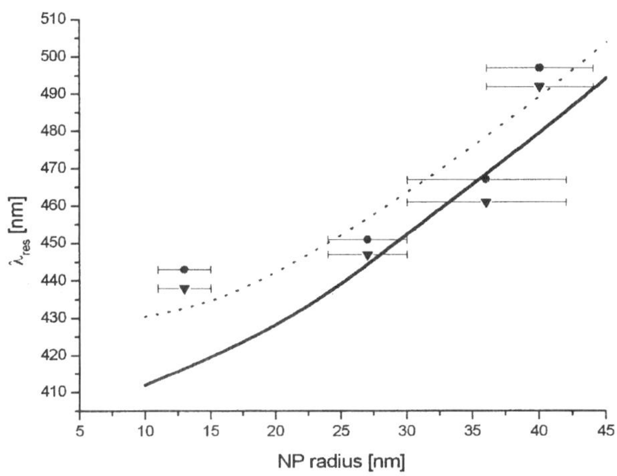


**Figure 6.11:** Theoretical (dotted line with shell and solid line without shell) and experimental (▼ with shell, • without shell) $\lambda_{res}$ values as a function of NP radius.

The fluorescence enhancement was measured by comparing the emission from the dye-NP system to that of the same dye concentration in the absence of the NPs. Prior to this, a protocol was developed for normalization of the fluorescence measurement which involved choosing the optimum dye concentration which would minimize inner filter effects but still be detectable and which was consistcnt with the dye concentration required to form a monolayer on the NP surface.

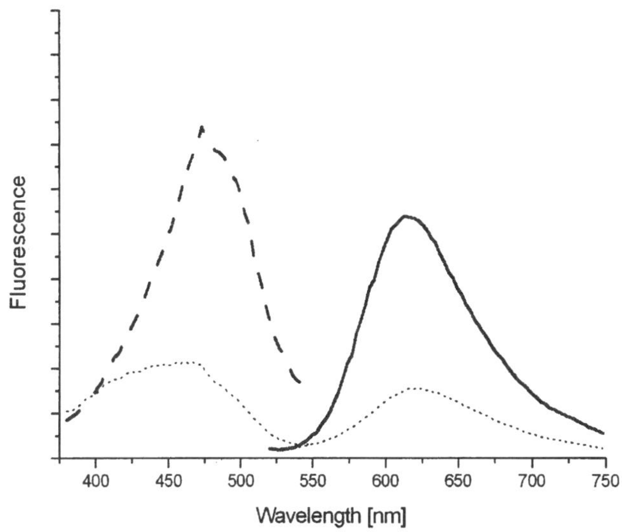


**Figure 6.12:** Emission and excitation spectra of the Ru dye in aqueous solution (lower dashed line) and in solution with 27nm radius NPs (upper dashed and solid lines).

Both the excitation and emission spectra were measured for (i) the dye-NP in solution and (ii) the same dye concentration in solution in the absence of the NPs. The resultant data are shown in Figure 12 for NPs with a radius of 27nm. The fluorescence enhancement, defined as the ratio between the emission intensity for the dye-NP and of the dye alone, measured for all NPs, is shown in Figure 6.13 as a function of NP radius. From these data, it can be seen that a maximum enhancement of ~4 is achieved for NPs of radius 27nm. No enhancement is observed for the smallest NPs while the enhancement decreases gradually for radii > 27nm.

In order to establish that the enhancement was due to the LSPR effect and not to variations in the dye emission when conjugated to the silica shell surface, a separate experiment was performed where the metal NP was replaced by a pure silica NP with the same radius. These NPs were synthesized using a microemulsion technique [18] and the dye was attached as in the case of the metal NPs. The enhancement measurement was repeated and the fluorescence from the dye – silica NP was almost identical to that measured in solution hence confirming the plasmonic nature of the enhancement.

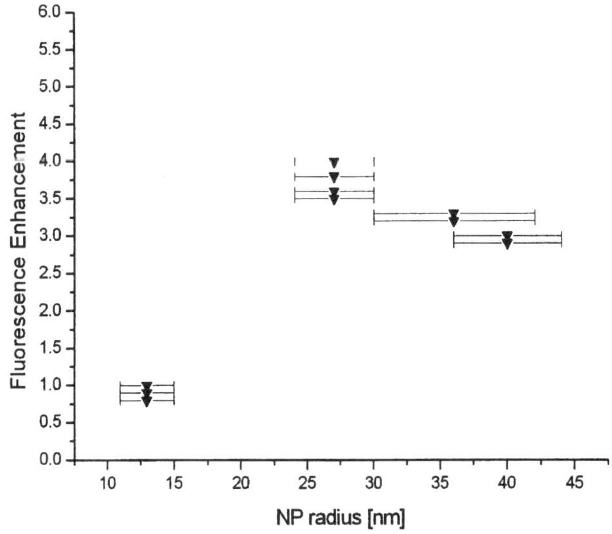


**Figure 6.13:** Experimentally measured dependence of the fluorescence enhancement on NP size.

On comparing the maximum experimental enhancement shown in Figure 6.13 with that predicted in Figure 6.7, it is clear that, while there is qualitative agreement with respect to dependence on NP size, the experimental enhancement maximum is about 9 times less than that predicted. This may be indicative of the induced non-radiative de-excitation rates being still quite significant at the 5nm dye-NP separation, leading to a decrease in quantum efficiency of the dye. The exact source of this discrepancy is not clear.

## 6.4 2. *Distance Dependence Of Enhancement*

Having studied the impact of NP size on plasmonic enhancement as reported in section 6.4 1, we also investigated the impact of another key parameter, namely the NP-fluorophore separation. The enhancement study was carried out on a planar configuration which is compatible with the ultimate application of the enhancement phenomenon in high sensitivity diagnostic chips based on plasmon-enhanced fluorescence. In order to investigate the impact of NP-fluorophore separation on fluorescence enhancement, highly controllable charged polyelectrolyte (PEL) spacer layers were deposited using a layer-by-layer technique as described in section 6.3 2.

| Metal / alloy | Diameter (nm) | $\lambda_{res}$ (nm) |
|---|---|---|
| Silver | 60 +/- 10 | 415 |
| Gold / silver alloy | 45 +/- 7 | 430 |
| Gold | 15 +/- 1 | 520 |
| Ru-PEL | | 412 – 470 |

**Table 6.1**: The diameter and corresponding $\lambda_{res}$ for each metal NP. The $\lambda_{res}$ were calculated from the average absorbance over 16 wells for each NP. Measurements were performed on PEL-coated microplates in ambient conditions. The absorbance peak of the Ru-PEL in aqueous solution is also given.

The physical and optical properties of the NPs used in this investigation are described in Table 6.1. The optical absorption properties of the ruthenium dye complex are also detailed in the table. It can be seen that there is good overlap between $\lambda_{res}$ of the pure silver and alloy NPs and the absorption band of the complex, while the gold NPs lie outside the absorption peak and are used as a negative control. The dependence of the excitation spectra of the dye complex on NP-dye distance is shown in Figure 6.14 for the case of the pure silver NPs. Also included in the figure is the excitation spectrum for the complex coated on the PEL layer in the absence of NPs. From Table 6.1, it can be seen that there is very good overlap between $\lambda_{res}$ of the silver NPs and the dye absorption band which constitutes the optimum plasmonic enhancement condition for the case of excitation enhancement.

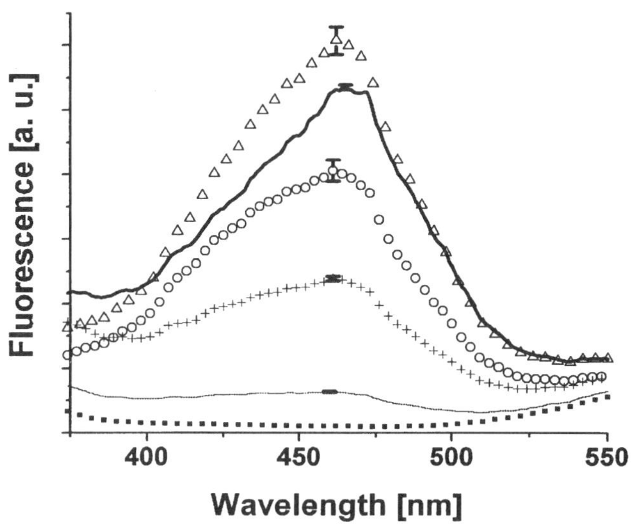


**Figure 6.14:** Excitation spectra of a layer of Ru-PEL dye deposited over silver NPs with intervening polyelectrolyte layers (PEL only (dashed line), pure Ru-

PEL (solid line), 1 bilayer(o), 2 bilayers(Δ), 4 bilayers (thick solid line), 8 bilayers (+).

Figure 6.14 indicates increasing excitation intensity from zero to one bilayer followed by a monotonic decrease for 4 and 8 bilayers. It is interesting to note that there is significant intensity when the dye complex is coated directly on top of the NP layer (0 bilayers). Electromagnetic theory predicts that a dye molecule in close proximity to the NP will experience fluorescence quenching due to radiative energy transfer back to the metal [19]. The presence of significant intensity in this case is likely to be due to a combination of the finite thickness of the dye conjugation layer (1.5nm) and the fact that, as the surface coverage of NPs is only 12%, there will be many dye molecules in the interstitial spaces between the NPs which will not be quenched. The maximum intensity achieved in the presence of 1 bilayer (thickness 3nm) is due to the combination of a reduced quenching effect and enhancement due to the modified electric field and represents the optimum NP-dye separation for this 2-D configuration.

The relative enhancement, $R_{en}$, of fluorescence of Ru-PEL arising from the proximity of metal NPs was calculated using the following equation:

$$R_{en} = \frac{F_{Ru-PEL,NP} - F_{NP}}{F_{Ru-PEL} - F_{PEL}}$$

where $F_{Ru-PEL,NP}$ is the enhancement of Ru-PEL in close proximity to NPs, $F_{NP}$ is the background fluorescence of NPs, $F_{Ru-PEL}$ is the fluorescence of Ru-PEL only and $F_{PEL}$ is the background fluorescence of the polyelectrolyte layers. Using this equation, the maximum enhancement for this system was measured to be 16 for a spacer thickness of 3nm. The fluorescence dropped significantly for larger spacer thicknesses but there was still a factor of ~5 enhancement for a spacer thickness of 16 nm. The relative enhancement as a function of spacer thickness for all 3 NP types is shown in Figure 6.15. Clearly, the largest enhancement of a factor of 16 is obtained with pure silver NPs of diameter 60nm for a spacer thickness of 3 nm. The alloy NPs exhibit a similar behaviour as a function of thickness but the enhancement is significantly less (~3.5) while pure gold NPs exhibit no enhancement. Referring back to section 6.4 1, a similar enhancement factor (~4) was obtained for $Ru(dpp)_3$ using alloy NPs in solution which were of the same composition as used in this section and with a 5nm spacer thickness. Generally, the lower enhancement factor obtained for the alloy NPs compared to pure silver is attributed to the different dielectric properties of gold and silver. The imaginary part of the dielectric constant is higher for gold / silver alloy than for pure silver [20] which results in a broader and less intense plasmon resonance peak for the alloy compared to pure silver [9]. This is clear from the data in figure 6.4.

In section 6.3 2, the reproducibility of the PEL layer-by-layer technique and of the NP deposition was validated by the small standard deviation values obtained for $\lambda_{res}$ absorption values for the NPs. This also applies to the enhancement study discussed above. The enhanced excitation spectra shown in figure 6.4 exhibit standard deviations that are less than 4%, which underlines again the high degree of reproducibility of the PEL, NP and dye deposition techniques.

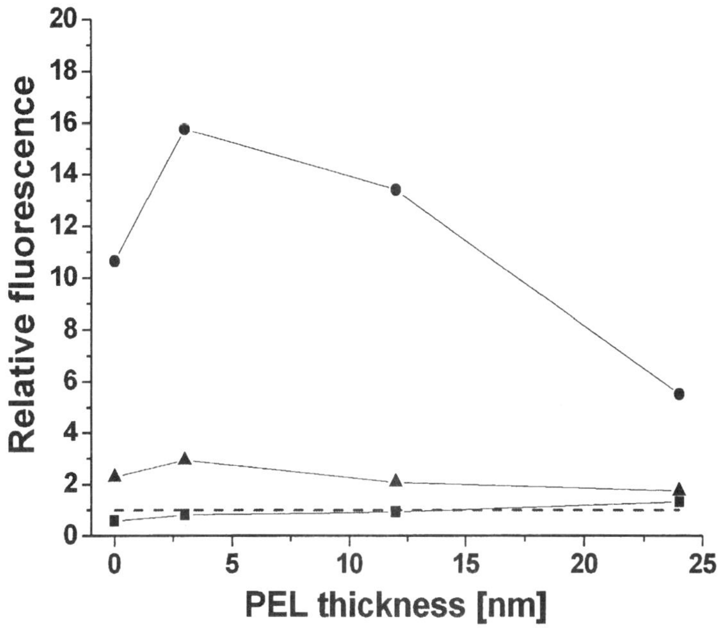


**Figure 6.15:** Relative fluorescence enhancement of Ru-PEL dye deposited over silver NPs (circles), alloy NPs (square) and gold NPs (triangles), as a function of thickness of intermediate PEL spacer layer.

## 6.5 OVERALL CONCLUSIONS AND FUTURE PERSPECTIVES

In the context of developing practical approaches for exploiting plasmonic enhancement of fluorescence on biochip platforms, the key achievements of the work reported here are

- the identification of the optimal NP size for maximum fluorescence enhancement,
- the development of a generic technique for the reproducible deposition of metal NPs on a planar substrate, and
- the elucidation of the optimum dye-NP separation to achieve maximum enhancement in a planar configuration.

These 3 achievements are also useful, however, for pointing out some of the key challenges that remain before one can envisage widespread implementation of plasmonic principles on fluorescence-based biochips.

In the case of (i) and (ii) above, a key issue is whether or not nanoparticles represent the optimal approach to adopt in the low-cost, mass-production of tailored plasmonic surfaces for biochips. An alternative view adopted by many in this field is that the production of silver island films, using low-cost deposition techniques such as vacuum deposition or sputtering, is more appropriate to the needs of the industry. While nanoparticles facilitate both theoretical analysis and precise fabrication control, it may be that they ultimately represent an option that is too complex and too expensive.

A second key issue that remains to be elucidated fully is the metal-fluorophore distance effect and its practical implications. In the case of (iii) above, it is not clear how generic this result is or how dependent it is on the specific experimental configuration used for the study. Nevertheless, most investigations of the metal-fluorophore distance effects conclude that separations in the region of 5 – 10 nm are optimal [21,22]. In more general terms, this raises a broader issue for the implementation of plasmonic principles on fluorescence-based biochips. The key question is how one designs this important feature into generic platforms that may have to accommodate a range of fluorophore-NP distances by virtue of the diversity of biorecognition molecules (antibodies, antibody fragments, DNA etc) and targets employed therein. One can envisage some elegant nano-engineering solutions for specific situations, but these may not be broadly applicable. The more pragmatic approach may well be to provide a broadly efficient plasmonic substrate and to accept enhancement factors which are not optimal but provide significant improvements over a situation where conventional (non-plasmonic) substrates are employed.

Finally, it is clear that plasmonic enhancement of fluorescence is still a developing field. While the factors highlighted in this chapter are important, new approaches and insights are emerging all the time. It may well be that novel substrates such as nano-aperture arrays [23] or hybrid surfaces (employing island films or nanoparticles on continuous thin films) [24] provide a key breakthrough in the pathway to practical implementation. All that can be stated for certain, therefore, is that this field provides a rich vein for fundamental research and that it has significant potential for radically improving current biosensor technology if the issues raised above can be addressed satisfactorily.

## 6.6 REFERENCES

1. Sokolov, K., Chumanov, G., Cotton R.m., (1998). Enhancement of molecular fluorescence near the surface of colloidal metal films. *Anal Chem* **70(18)**: 3898-3905
2. Stich, N., Gandhum, A., Matushin, V., (2001. Nanofilms and nanoclusters:Energy sources driving fluorophores of biochip-bound labels. *J. Nanosci. Nanotechnol* **1(4):** 397-405
3. Lakowicz, J.R., Malicka, J., Gryczynski, I., (2003). Radiative decay engineering: the role of photonic mode density in biotechnology. *Anal. Biochem*. **36(14):** R240-R249.
4. Stranik, O., McEvoy, H. M., McDonagh, C., MacCraith, B.D., (2005). Plasmonic enhancement of fluorescence for sensor applications. *Sensors and Actuators B* ***107*:** 148.
5. Stranik, O., Nooney, R., McDonagh, C., MacCraith,

B.D. (2007). Optimisation of Nanoparticle Size for Plasmonic Enhancement of Fluoresence. *Plasmonics*, **2(1),** 15

6. Lakowicz, J.R., Shen, B., Gryczynski, I., (2001).
7. Intrinsic fluorescence from DNA can be enhanced by metallic particles. *Biochem. Biophys. Res. Commun.* **286:** 875-879
8. Wang, D. S., Kerker, M., (1982). Absorption and luminescence of dye-coated silver and gold nanoparticles. *Phys. Rev. B* **25(4):**2433-2499
9. Mie, G., (1908). Beitrage zur Optik trueber Medien speziell kolloidaler Metalloesungen. *Ann. Phys.* **25**: 377-445
Bohren, C.F., Huffman, D.R., (1998). Absorption and scattering of light by small particles. Wiley, Berlin.
10. Goodman, D. B., Draine, B. T., Flatau, P. J., (1991). Application of fast-Fourier transform techniques to the discrete-dipole approximation. *Optics Letters* **16(15)**: 1198-1200.
11. Shimura, K., Milster, T. D., (2001). Vector diffraction analysis by discrete-dipole approximation. *J. Opt. Soc. Amer. A* **18(11)**: 2895-2900
12. Deckman, H. W., Dunsmuir, J. H., (1982). Natural Lithography. *Appl. Phys. Lett.,* **41(4):** 377-379
13. Rechberger, W., Hohenau, A., Leitner., (2003). Optical properties of two gold nanoparticles. *Opt. Commun.*, **220**: 137-141
14. Turkevich, J., Stevenson, P. C., Hiller, J.(1951) A study of the nucleation and growth processes in the synthesis of colloidal gold. *Discuss. Faraday Soc.* ***58***: 55.
15. Link, S., Wang, Z. L., El-Sayed, (1999). Alloy formation of gold-silver nanoparticles and the dependence of the plasmon absorption on their composition M. A. *J. Phys. Chem. B* ***103***: 3529.
16. Malynych, S.; Chumanov, G. *J. (2003).* Synthesis and Optical Properties of Silver Nanoparticles and Arrays. J. *Am. Chem. Soc.*, ***125***, 2896.
17. Grabar, K. C.; Smith, P. C.; Musick, M. D.; Davis, J. A.; Walter, D. G.; Jackson, M. A.; Guthrie, A. P.; Natan, M. J. *J. Am. Chem. Soc.* **1996**, *118*, 1148.
18. Santra, S., Wang, K., Topec, R., (2001). Development of novel dye doped silica nanoparticles for biomarker applications. *J. Biomed. Opt.* **6(2):** 160
19. Kuhn, S,. Hakkanson, U., Rogobete, L., Sanaghdar, V., (2006), *Phys. Rev. Lett.* Art. No. 017402.
20. Ripken, K. Z.(1972).Die optischen konsstanten von au, ag und ihren legierungen im energiebereich 2.4 bibs 4.4 ev. *Z. Physik*, ***250***, 228.
21. Ray, K., Badugu, R., Lakowicz, J.R., (2007). Sulforhodamine adsorbed Langmuir-Blodgett layers on silver island films: effect of probe distance on the metal-enhanced fluorescence. *J. Phys. Chem. C.* **111**:7091.
22. Fu, Y., Lakowicz, J. R. (2007). Enhanced

Fluorescence of Cy5-labelled DNA tethered to silver island films: fluorescence images and time-resolved studies using single-molecule spectroscopy. *Anal. Chem.*, **78**:6238.

23. Wenger, J., Gerard, D., Dintinger, J., Mahboub, O., Bonod, N., Popov, E., Ebbesen, T.W., Rigneault, H. (2008). Emission and excitation contributions to enhanced single molecule fluorescence by gold nanometric apertures. *Optics Express*, **16(5)**: 3008.

24. Barnctt, A., Matveeva, E. G., Gryczynski, I., Gryczynski, Z., Goldys, E.M.(2007). Coupled plasmon effects for the enhancement of fluorescent immunassays. *Physica B*, **394**:297.

# 7

# Microwave-Accelerated Metal-Enhanced Fluorescence

Kadir Aslan and Chris D. Geddes*

Institute of Fluorescence, laboratory for Advanced Medical Plasmonics and Laboratory for Advanced Fluorescence Spectroscopy, University of Maryland Biotechnology Institute, 701 East Pratt St., Baltimore, 21202, USA.

## 7.1 INTRODUCTION

In the past few years our research laboratory has been working on the applications of a new technique, called microwave-accelerated metal-enhanced fluorescence (MAMEF), as applied to fluorescence-based bioassays. The MAMEF technique couples the benefits of low power microwave heating with metal-enhanced fluorescence (MEF), to address the two major shortcomings of fluorescence-based bioassays currently in use today; i.e., *bioassay sensitivity and rapidity*. In MAMEF, [1] the MEF phenomenon increases the sensitivity of the assays, while the use of low power microwave heating kinetically accelerates assays to completion within only a few seconds.

There are three major components of the MAMEF technique: 1) plasmonic nanoparticles (i.e., silver, gold, copper, nickel, aluminum, zinc, etc.), 2) microwaves and 3) an aqueous assay medium. The plasmonic nanoparticles serve as (i) a platform for the attachment of one of the biorecognition partners (anchor probes) (ii) as an enhancer of the close-proximity fluorescence signatures via surface plasmons (i.e., MEF effect) [2] and (iii) a material not heated by microwaves for the selective heating of the aqueous media with microwave energy.

The crux of the MAMEF technique is the selective heating of water, which occurs as the assay medium is heated by microwaves to a higher temperature than the plasmonic nanoparticles creating a temperature gradient, between the warmer aqueous media and the colder nanoparticles, rapidly driving the biorecognition events to completion. The major criteria for the choice of metal to be employed in the MAMEF technique are that the metals support a plasmon-resonance and demonstrate MEF [3]. The MEF component affords for the improvements in the assay sensitivity and photostability of fluorophores. In MEF, a fluorophore is placed near-to plasmonic nanostructures and two significant observations are made: significantly increased fluorescence emission and a reduced lifetime of the fluorophores [4, 5]. The reduction in fluorescence lifetime results in fluorophores becoming more photostable near-to the plasmonic nanoparticles. More specifically, the excited fluorophores partially transfer energy to the plasmonic nanoparticles, where the energy is in essence amplified, as the emission from the fluorophore-nanoparticle "system" radiates more efficiently than the emission from the fluorophores alone [2, 6, 7].

*Microwave-Accelerated Metal-Enhanced Fluorescence (MAMEF)*
Edited by Chris D. Geddes.

In this review chapter, the reader will find information on the following: a general descriptive section on the MEF phenomenon, a section on the effects of microwave heating on the components of the MAMEF-based assays, and several applications of the MAMEF technique in ultra fast protein detection assays, immunoassays and DNA hybridization assays.

## 7.2 METAL-ENHANCED FLUORESCENCE

Since MAMEF technique utilizes the benefits of the MEF phenomenon, it is important to present a brief description of this phenomenon for the sake of completeness for this review chapter. Early descriptions of MEF can be traced back to the observations made with the interactions of fluorescent species with *planar* metallic surfaces. When placed near a planar metal surface, the spontaneous emission of a fluorescent species follows a radiative and/or a non-radiative decay channel and is mainly dependent on the distance (R) between the emitter and the metal as well as the orientation of the dipole of the fluorescent species with respect to the metal surface [8]. Two effects have been demonstrated as a result of the distance dependence of the emission rate [5]: i) the emission rate oscillates as the distance is increased as the phase of the reflected field changes with distance and ii) the strength of the oscillation decreases since the dipole emitter is a point source. The significance of the dipole orientation can be seen when we consider that the metal surface produces an image dipole on the surface. For a very small distance between the emitter and the metal surface, a dipole that is parallel to the surface is cancelled out by its image and a perpendicular dipole is enhanced. In this regard, the distance dependent spontaneous emission rate can be predicted assuming that the reflecting surface is perfect and the dipole moment of the emitter rotates rapidly within the emission lifetime.

The processes described above were derived for planar metal geometries, and one can find studies in the literature offering a description and applications of spontaneous emission rate near metallic nanoparticles similar to the description for planar metal geometries [9-17]. The major difference between the planar systems and particulate systems is the inclusion of localized modes occurring in particles. The localized modes in particulate systems results in the omission of the oscillations of decay rate observed for the planar systems. The frequency of the localized modes depends on the both the size and the shape of the metallic nanoparticles. For a single emitter placed near metal nanoparticles (using a dipole-dipole model) the non-radiative decay rate is shown to follow an $R^{-6}$ dependence, the radiative decay rate follows an $R^{-3}$ dependence and is also dominated by a dipole polarizability of the metallic nanoparticle [12]. As shown in Figure 7.1A, the energy is partially transferred (via non-radiative coupling) from the excited state of the fluorescent species to surface plasmons of the metallic nanoparticles and is then subsequently radiated by the nanoparticles themselves [2]. The extent of the radiation of the coupled energy by the metallic nanoparticles is also thought to be related to the scattering efficiency of nanoparticles, a component of their complex extinction spectrum [7, 18].

In addition to plasmon coupling, fluorescent species placed near-to metallic nanoparticles experience an increase in their absorption of light (a modification in their absorption cross section), which is thought to be due to increased electric fields around and in between the nanoparticles themselves.[17] The theoretical evidence for enhanced absorption phenomenon is given in Figures 7.1B-D. Figure 7.1B shows a calculated electric field distribution using a finite-difference time-domain (FDTD) numerical approach for a 2-silver nanoparticle array (diameter = 100 nm, gap = 2 nm) illuminated at 400 nm. The maximum electric field intensity reaches up to 1000-fold of the free-space value, in between the silver nanoparticles. It is well known that the interaction of light with metallic nanoparticles is via the free-electrons of the metal and is wavelength-dependent [19, 20]. Figure 7.1C shows the calculated maximum electric field intensity versus wavelength of incident light for 2- and 12-silver nanoparticle arrays. The maximum electric field intensity for 2- and 12-nanoparticle array occurs at 400 nm and 450 nm, respectively. For the 12-nanoparticle silver nanoparticle array (4x3 staggered) maximum electric field intensity is calculated to be 5000-fold the free-space value due to the interaction of multiple silver nanoparticles with the incident light. FDTD calculations also reveal that the electric field enhancement around the surface of silver nanoparticles is at a maximum at short distances (< 1 nm) and completely diminishes at 10 nm from the metal surface a as shown in Figure 1D and Figure 1D-inset.

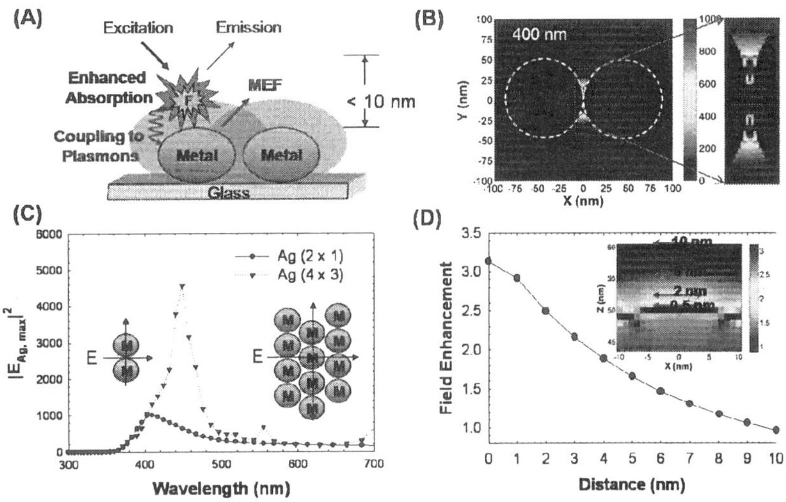


**Figure 7.1:** (A) Schematic representation of the Metal-Enhanced Fluorescence phenomena; (B) FDTD calculations for two silver nanoparticle arrays (d = 100 nm). (C) Wavelength-dependent calculated $|E_z|^2$ maximum intensity for silver nanoparticle arrays (d = 100 nm). Geometries and incident field polarization (*p-polarized*) and propagation direction are shown in the insets. The gap between the nanoparticles was assumed to be 2 nm in the calculations. (D) Calculated field enhancement as a function of distance for a single silver nanoparticle (d = 100 nm). The inset shows these results as an FDTD $|E|^2$ image above the nanoparticle.

Over the last two decades, many surfaces have been developed for MEF based on different metallic nanoparticles, such as those comprised of silver nanoparticles,[4, 21] gold colloids,[22] and even gold,[22] copper, [3] aluminum [23] and zinc particulate films.[24, 25] Several modes of silver deposition have also been developed, such as by wet chemistry,[26] a layer-by-layer deposition technique,[27] deposition by light,[28] electrochemically,[29] on glass,[2] plastic substrates,[30] and even on indium tin oxide [31].

Silver island films (SIFs) has been the substrate of choice in many publications to date [2, 6, 32-36]. In a typical SIFs preparation on a glass support, a solution of silver nitrate (0.5 g in 60 ml of deionized water) in a clean 100-ml glass beaker, equipped with a Teflon-coated stir bar, is prepared and placed on a stirring/hot plate. While stirring at the quickest speed, 200 µL of freshly prepared 5% (w/v) sodium hydroxide solution is added. This results in the formation of dark brown precipitates of silver particles. Approximately 2 ml of ammonium hydroxide is then added, drop by drop, to re-dissolve the precipitates. The clear solution is cooled to 5°C by placing the beaker in an ice bath, followed by soaking the amino silane-coated glass slides in the solution. While keeping the slides at 5°C, a fresh solution of *D*-glucose (0.72 g in 15 ml of water) is added. Subsequently, the temperature of the mixture is then warmed to 30°C. As the color of the mixture turns from yellow-green to yellow-brown, and the color of the slides become green, the slides are removed from the mixture, washed with water, and sonicated for 1 minute at room temperature. SiFs-deposited slides were then rinsed with deionized water several times and dried under a stream of nitrogen gas. A typical photograph for SIFs deposited onto glass microscope slide is given in Figure 7.2A-Top.

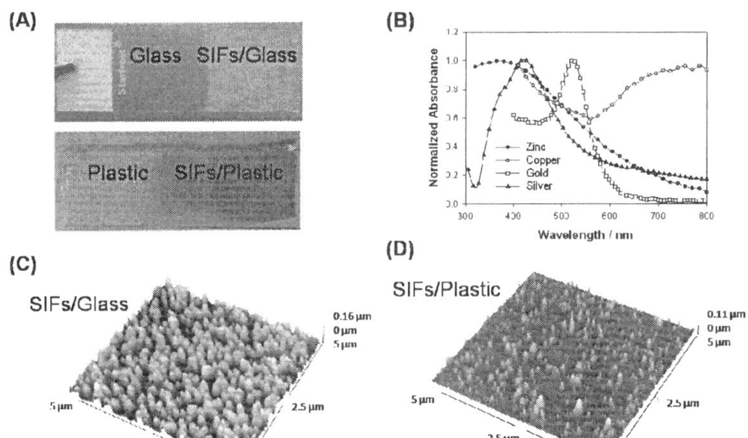


**Figure 7.2**: (A) Photographs of silver island films (SIFs) deposited on to glass and plastic supports. (B) Normalized absorbance of zinc, copper, gold and silver nanostructured particles on a glass support. Atomic force microscope images of SIFs on (C) glass (D) plastic support. Adapted from references 1 (glass) and 30 (plastic).

One can employ the above procedure to any planar surface that presents amine functional groups. In this regard, Figure 7.2A-Bottom shows a photograph of

SIFs deposited onto a plastic surface [30, 37]. The characterization of SIFs on a surface is typically done by employing optical absorption spectroscopy and microscopy techniques such as Atomic Force Microscopy (AFM) and Scanning Electron Microscopy. Figure 7.2B shows the absorbance spectra for metallic surfaces used in MEF applications. Thc surface plasmon resonance (SPR) peaks for silver and zinc films occur in the blue spectral region. The SPR peak for gold and copper occur at longer, red-shifted, wavelengths with respect to silver and zinc films. Figure 7.2C and Figure 7.2D show the AFM images of SIFs on a glass and plastic support, respectively. While the distribution and the size of the silver nanoparticles on the glass is fairly homogeneous, a wide range of sizes of silver nanoparticles are deposited onto the plastic support. This is due to fact that the chemistries of the surface of glass are well established unlike plastic surfaces and therefore the covalent immobilization of silver nanoparticles onto glass less arduous and is subsequently more homogeneous.

## 7.3 LOW POWER MICROWAVE HEATING AND METAL-ENHANCED FLUORESCENCE (MAMEF)

In microwave heating, the electromagnetic energy interacts with the materials at the molecular level, where the electromagnetic energy is transferred and converted to heat through the frustrated motion of the molecules. This results in rapid and uniform heating of materials throughout their volume (also referred as volumetric heating), especially when the size of the materials is smaller than the wavelength of the microwaves. In contrast, the conventional thermal heating of materials proceeds via conduction, convection or radiation of heat from the surfaces of the material. More specifically, heat is transferred from the source to the material due to a temperature gradient via the three heat transfer mechanisms mentioned above and is dependent on several parameters such as the diffusion of heat, conductivity of materials, etc. Conventional heating often requires significantly longer heating times as compared to microwave-based heating.

In addition to the most notable advantage of microwave heating over conventional heating, that is the volumetric heating of the materials; microwave heating can also be utilized for *selective* heating of materials within a mixture or a composite. This arises from the differences in dielectric properties of the materials, in a mixture or a composite, microwaves selectively couple to the material with a higher dielectric loss factor. This selective heating is indeed the crux of the MAMEF technique, where the assay medium is selectively heated to a higher temperature than the metallic nanoparticles: the thermal gradient rapidly driving the biorecognition events to completion. In this review chapter, we offer the applications of MAMEF technique rather than describing the fundamentals of the interactions of microwaves with metals and chemical and biological compounds. The reader is also referred to the review article by Thostenson [38] for the description of dielectric properties and for a summary of electromagnetic theory.

The proof-of-principle of the MAMEF technique, which couples the benefits of MEF with low power microwave heating to kinetically accelerate the bioassays, was first demonstrated with a model protein-fluorophore system,[1] with biotinylated-BSA and fluorophore-labeled streptavidin, as shown in Figure 7.3A. The biotin-streptavidin binding step typically takes up to 30 minutes to complete at room temperature on the assay surface, while low power microwave heating reduces the assays' run time to a staggering 20 seconds. The assay was constructed on a glass support (other materials such as paper, [39] plastics,[37] HTS wells,[40] etc can also be used) coated with silver nanoparticles. In this surface configuration, the use of silver nanoparticles results in 1) enhancement of fluorescence as compared to a blank glass surface (increased emission), 2) creation of a temperature gradient between the bulk and the silver nanoparticles themselves.

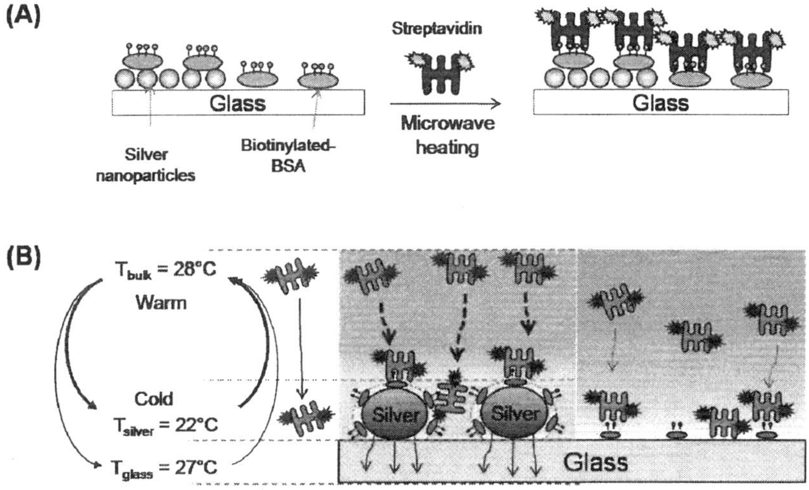


**Figure 7.3:** (A) A Model MAMEF-based Protein Detection Assay. (B) Schematic representation of the effect of microwave heating on the protein detection surface assay.

In the MAMEF technique, while water and glass are selectively heated with microwaves, the silver nanoparticles virtually remain at the same temperature as before the microwave heating is initiated. As shown in Figure 7.3B, the selective heating of these assay components creates a temperature gradient between the water, glass and the silver nanoparticles, which results in the rapid transfer of streptavidin molecules from the warmer bulk to the colder surface due to temperature driven mass transfer. Since the thermal conductivity of silver (429 W / m K) is much larger than that of glass (1.05 W / m K), the transfer of streptavidin happens more efficiently towards silver nanoparticles than to glass. For the duration of the microwave heating (20 seconds), as heat is transferred from the bulk to the surface, and the temperature on the assay surface reaches an equilibrium temperature lower than the bulk, water molecules are continuously circulated between the warmer and the colder regions of the bulk medium.

Since the temperature gradient created between the bulk of the aqueous medium and the metallic nanoparticles during microwave heating is one of the major reasons for the observed faster biorecognition kinetics in MAMEF-based bioassays, it has been informative to determine the temperature of the assay components during the microwave heating process. In this regard, two approaches have been employed

1) an "indirect"[1] and 2) a "direct" method [41]. In the former, the temperature changes during the microwave heating can be determined by monitoring the ratiometric absorbance response of a temperature-sensitive dye (thymol blue). Thymol blue is heated in the microwave cavity and is quickly transferred to the spectrophotometer where the absorption spectrum is measured [1].

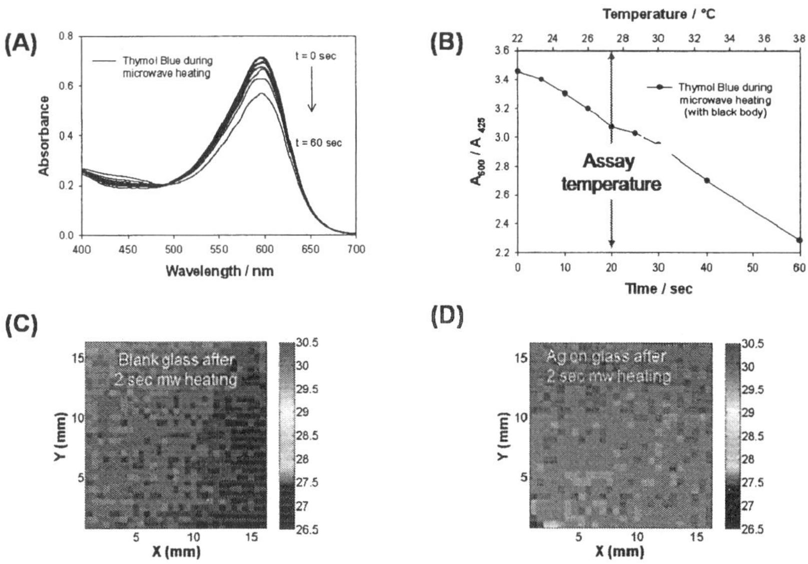


**Figure 7.4:** (A) Absorption spectra as a function of temperature for 30 µl thymol blue measured during microwave heating; (B) the respective absorbance, temperature vs. time ratiometric plot; (C) Real-time temperature distributions of water on a SiFs-deposited sapphire substrate captured using a thermal camera; (D) A thermal image of SiFs during microwave heating. Adapted from references 1 (Figure 2a-b) and 18 (Figure 2c-d). Adapted from references 1 (A, B) and 41 (C,D).

Figure 7.4A and 7.4B show the temperature-dependent absorption spectra of thymol blue and the assay temperature calibration curve for microwave heating up to 60 seconds, respectively. From these calibration plots, a 20 second microwave exposure (140 W, 2.45 GHz), results in a temperature jump of ~ $6^0$C (to ≈ 28 $^0$C) for 30 µl of sample. Hence, with this calibration curve, one can simply change the assay surface temperature by changing the duration of the microwave heating.

In the "direct" method, [41] the determination of temperature changes on the assay surface during microwave heating is undertaken using a thermal imaging camera that captures the infrared (IR) radiation, giving a high speed and high sensitivity determination of the actual temperature. In order to detect the IR radiation (3-6 µm), glass substrates are replaced with sapphire plates that transmit IR radiation, the SiFs simply being deposited on the sapphire plate. Figure 4C and 4D show the results for real-time monitoring of the temperature during microwave heating. From mean temperature versus time plots for blank sapphire (no SiFs) and silvered sapphire sample geometries, a higher thermal gradient is observed for the water on the silvered sapphire substrates (Figure 7.4C). It is important to note that, after the microwave heating is turned off, while the temperature of water remains constant on a blank sapphire substrate, the water starts to cool down more rapidly close-to the silvered sapphire substrate. This implies that the heat is transferred from the warmer water to the colder silver nanoparticles inevitably resulting in the faster diffusion of the biomolecules towards the silver nanoparticles. A thermal image (Figure 7.4D) shows that the temperature distribution is uniform on the silvered sapphire sample geometries, 2 seconds after the onset of microwave heating. *Interestingly*, binding events close-to the cooler plasmonic nanostructures have the potential to protect the surface-bound biomolecules from thermal denaturation.

## 7.3 1. *MAMEF-based protein assays*

The original proof-of-principle of MAMEF-based bioassays were first demonstrated with a model protein detection assay [1]. In this model protein assay, the protein-fluorophore system was coated equally on one-half of a silvered glass substrate, the other half of the glass substrate left intentionally blank to compare the benefits of the MEF phenomenon (Figure 7.4A). Biotinylated-BSA is attached to the surface and is allowed to bind to its binding partner (fluorescein-labeled streptavidin) by incubation at room temperature or through low-power microwave heating. Figure 7.5A shows that the fluorescein emission intensity from the silvered substrate is ≈ 6-fold greater than that from the glass substrate for the assay run at room temperature for 30 minutes. Figure 7.5B shows the combined effect of both low-power microwave heating (20 seconds) and the MEF effect for the identical assay run at room temperature. The microwave-accelerated assay yielded similar fluorescence intensity after just 20 seconds as compared to the assay run at room temperature for 30 minutes. The emission intensity from the silvered substrate is 9-fold greater than that from the glass substrate for the microwave-accelerated assay.

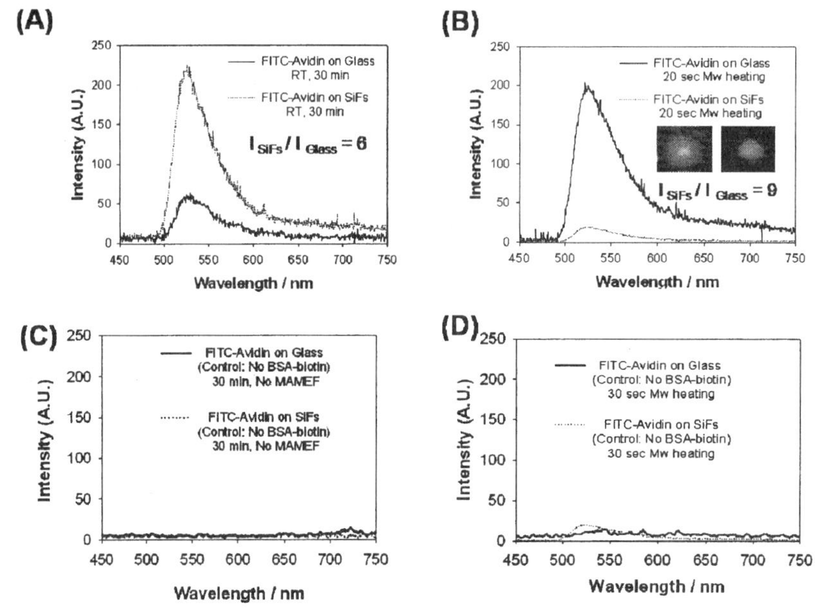

**Figure 7.5:** Emission spectra of FITC for a model protein assay run (A) at room temperature for 30 minutes and (B) with microwave heating for 20 seconds. Control experiments, where one of the protein binding partners, BSA-biotin is omitted from the assay, run (C) at room temperature and (D) with microwave heating for 20 seconds; Room Temp: Room Temperature. Adapted from reference 1.

This is thought to be due to the reduced extent of binding on glass as compared to the silvered side where the thermal gradient is larger. That is, one would need longer microwave heating times on glass substrates to achieve > 95% completion or the equivalence of the assay. Real-color color photographs (Figure 5B-right-inset) taken through an emission filter provides visual evidence for the larger fluorescence emission intensity measured from the silvered substrates.

In addition, two different control experiments were also run to confirm the efficacy of the MAMEF assays. Figure 7.5C shows that no fluorescence emission intensity was detectable from the control assay, where biotinylated-BSA is omitted from the assay, which was run at room temperature for 30 minutes. When the identical control assay was run with microwave heating, once again there was little/no fluorescence emission. These two control experiments nicely demonstrated that the non-specific binding of the fluorescein-labeled streptavidin to the surface was minimal.

The fluorescence lifetime of a fluorophore is indicative of its environment and was proven a useful tool to show the benefits of MEF [5]. It was previously shown that, when placed near-to metallic nanostructures, fluorophores have shorter lifetimes as compared to free-space solution or on the glass substrates. In an MAMEF-based assay, the lifetime information can be useful to assess the extent of completion of the assay: if the assay run at room temperature and with microwave heating goes to > 95% completion (30 minutes and 20 seconds, respectively), then the lifetime of the fluorophores will be very similar. Figure 7.6 shows that the fluorescence intensity decay curves for fluorescein after 30 minutes incubation and 30 seconds microwave heating were almost identical and significantly reduced as compared to the glass control. These results strongly indicate that the assays are virtually identical after 30 minutes incubation at room temperature and after 30 seconds microwave heating.

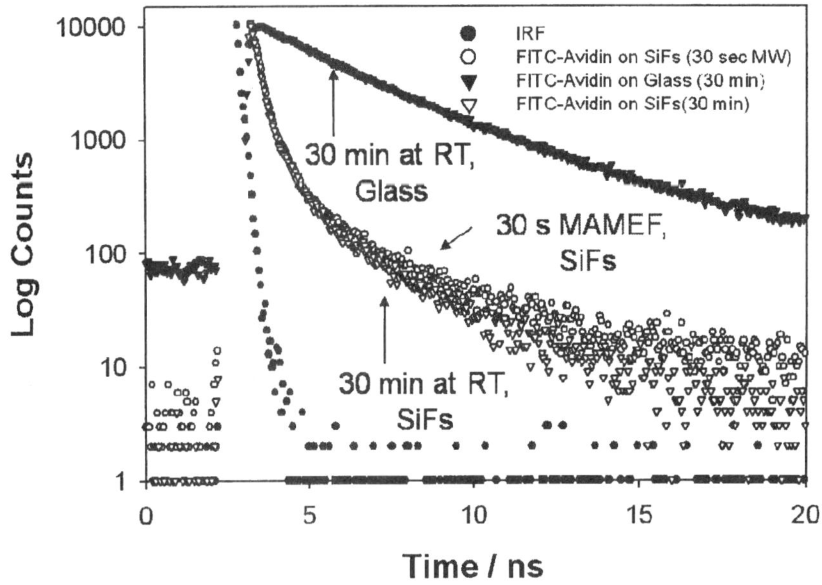


**Figure 7.6:** Intensity decays for FITC-Avidin on both glass and SiFs, both before and after exposure to low power microwave heating. The intensity decays on the SiFs are almost identical. IRF: Instrument Response Function. Adapted from reference 1.

Fluorescence-based detection is also the basis of many assays in High Throughput Screening (HTS) in use today. The sensitivity and rapidity of fluorescence-based HTS applications are underpinned by the quantum yield of the labeling fluorophore and the kinetics of the antigen-antibody recognition step [42-44]. Based on the success of our model MAMEF assay on silvered glass substrates, [1] our group has also shown the application of the MAMEF technology to assays run in an HTS format [40]. In this regard, commercially available poly-lysine coated HTS wells were modified with silver nanoparticles (Figure 7.7A) and a model protein assay run at room temperature and with microwave heating. Figure 7.7B shows the fluorescence emission intensity from both silvered and non-silvered HTS wells after room temperature incubation (for 30 minutes) and after microwave heating (30 seconds). The fluorescence emission intensity is ≈ 5- and 4-fold greater from silvered wells as compared to non-silvered wells, for assays run at room temperature and after microwave heating, respectively. This study [40] shows, in principle, the applicability of the MAMEF technology to high-throughput fluorescence-based applications, such as in drug discovery, biology and general compound library screening.

## 7.3 2. *MAMEF-based immunoassays*

Most commercial protein detection systems in use today are based on the specific recognition of antigens with antibodies in several immunoassay formats [43].

Once again fluorescence is the dominant detection technology in immunoassays due to the availability of fluorophores over a wide range of wavelengths and quantum yields. The immunoassays usually take anywhere from 10 minutes (with an expensive commercial unit) up to a few hours (in HTS format), which involves numerous incubation and washing steps [43].

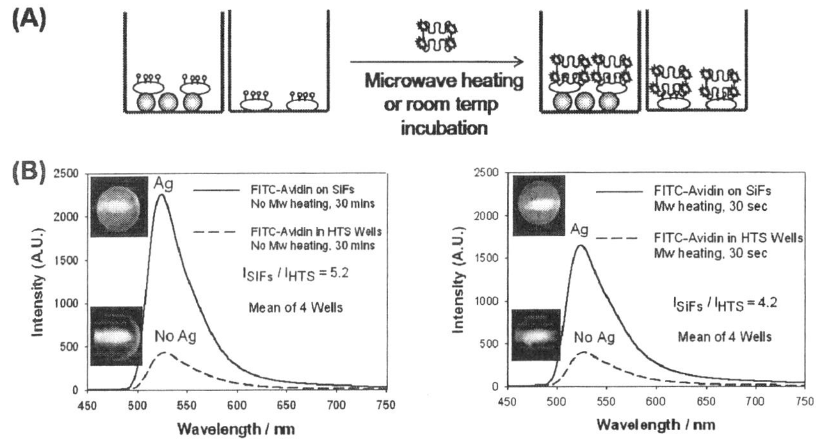


**Figure 7.7:** (A) Model protein-fluorophore system used to demonstrate MAMEF in HTS Wells; (B) Emission spectra of FITC for a model protein assay run at room temperature for 30 minutes and with microwave heating for 20 seconds. Room Temp: Room Temperature. Adapted from reference 40.

Recently, the applicability of the MAMEF technology to a cardiac marker immunoassay to significantly reduce the assay run time and sensitivity was demonstrated [45]. Figure 7.8A shows the experimental details of the myoglobin immunoassay that was constructed on silver-deposited glass microscope slides. In this regard, a capture anti-myoglobin antibody is adsorbed onto silver nanoparticles after an overnight incubation. The subsequent myoglobin and Alexa 647-labeled anti-myoglobin antibody binding steps were carried out either at room temperature or with low-power microwave heating; each step included either 30 minutes incubation at room temperature or a 20 seconds microwave heating and a washing step to remove the unbound myoglobin. Figure 7.8B shows the fluorescence emission spectra of Alexa-647 measured after the final binding step was carried out with low-power microwave heating and at room temperature (in separate experiments) on silver and on glass (a control sample to show the benefits of MEF). After 30-minutes incubation at room temperature, the fluorescence emission intensity of Alexa 647 from the silvered side was ≈ 7.5-fold larger than the intensity from the glass control, showing the benefits of MEF (increased fluorescence emission), Figure 7.8B-left. When the identical immunoassay was run with low-power microwave heating (Figure 7.8B-right), a similar final emission intensity was observed as compared to the assay run at room temperature (Figure 7.8B-left). These results clearly demonstrate the applicability of the MAMEF technology to a myoglobin immunoassay and potentially AMI screening.

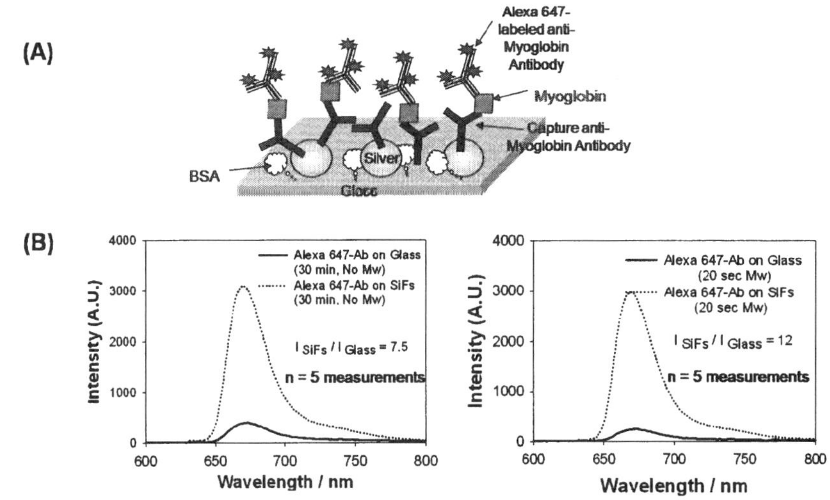


**Figure 7.8:** (A) Myoglobin immunoassay; (B) Fluorescence emission spectra of Alexa-647 for the myoglobin immunoassay on SiFs-deposited glass substrates after low power microwave heating and with no microwave heating. These results were obtained on silvered glass microscope slides. Adapted from reference 45.

## 7.3 3. *MAMEF-based DNA hybridization assays*

Fluorescence-based DNA hybridization assays are routinely used in many diagnostics applications, [46] on gene-chips [47] and fluorescence in-situ hybridization [48]. In all of these applications, usually fluorophores with high-quantum yields are employed to increase the sensitivity of the hybridization assays, raising several issues such as high background emission and photostability of the fluorophores. Ideally, to maximize the efficiency and the sensitivity of the MEF DNA hybridization assays, it would be beneficial to employ low-quantum yield fluorophores which can withstand the long exposure to excitation light [4]. The application of MEF in DNA hybridization assays has been shown to offer improvements in the sensitivity of the DNA hybridization assays as well as in the photostability of the fluorophores [4, 26, 49]. However, one typically has little or no control over the rapidity of the DNA hybridization assays. In this regard, the applicability of the MAMEF technology for rapid and sensitive DNA hybridization assays was recently demonstrated,[50, 51] where two complementary oligonucleotides (one labeled with a fluorophore) was hybridized on silver nanoparticles within 20 seconds, after low power microwave heating with the hybridization believed to be over in < 5 seconds [50, 51].

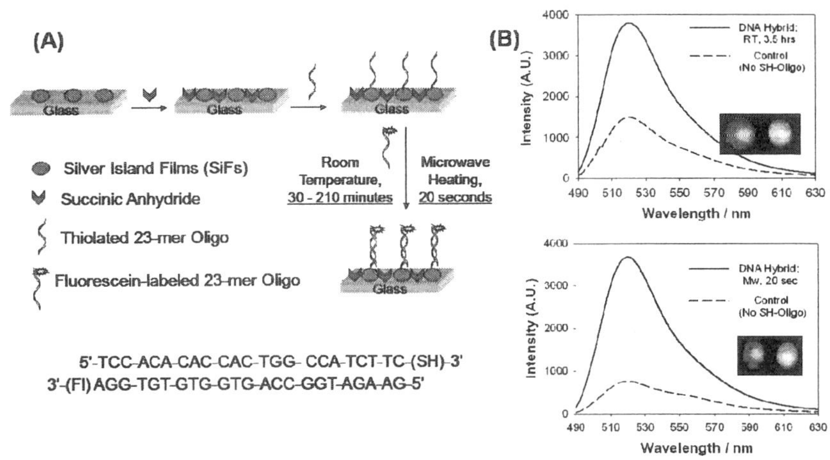


**Figure 7.9:** (A) Model DNA hybridization assay; (B) Emission spectra of fluorescein-oligo (3 □M) after MAMEF-based and room temperature hybridization on SiFs, Insert photographs, both before and after hybridization is completed. Adapted from reference 50.

Figure 7.9A shows the experimental configurations of the DNA hybridization assay constructed on silver-deposited glass microscope slides. In this regard, a 23-mer anchor probe was attached to silver nanoparticles via a sulfhydryl-metal bond after an overnight incubation. The DNA hybridization assay was carried out either by incubation of a fluorescein-labeled complementary oligonucleotide on silvered glass at room temperature for 3.5 hours, or with low-power microwave heating for 20 seconds. Figure 7.9B shows the fluorescence emission spectra of fluorescein after 3.5 hours room temperature incubation and after 20 seconds of microwave heating. Control experiments, where the anchor probe is omitted from the surface, corresponding to the DNA hybridization assays are also shown in Figure 9B. After 3.5 hours incubation at room temperature, fluorescein emission intensity from the hybridization assay is ≈ 2.5-fold larger than the intensity from the corresponding control assay, which is also evident from the real-color photographs taken through an emission filter. Identical fluorescein emission intensity is observed after 20 seconds (a > 600-fold decrease in assay run time) from the MAMEF-based DNA hybridization assay.

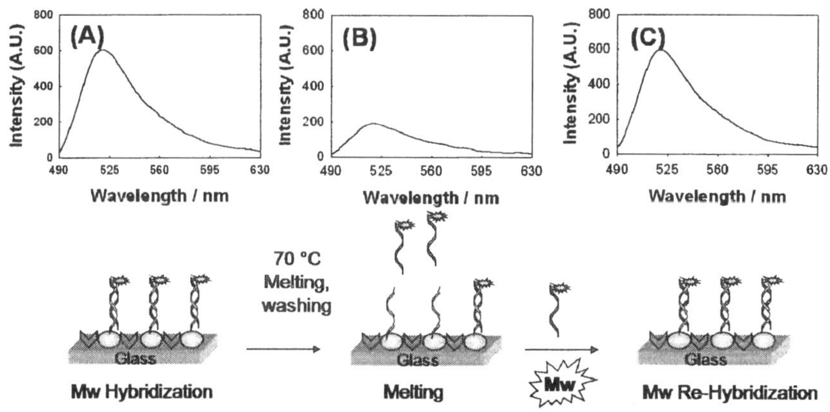


**Figure 7.10:** Emission spectra of fluorescein-DNA (A) after hybridization; (B) after melting at 70°C; and (c) after a further 20 seconds low power microwave heating with an additional 250 nM Fl-DNA. The silvered surface was washed with buffer several times between each measurement. Adapted from reference 50.

One of the most important factors in MAMEF-based DNA hybridization assays is the effect of microwave heating on the oligonucleotides themselves. Our research group has previously shown that proteins do not denature when exposed to low power microwave heating, which was demonstrated using Fluorescence-Resonance Energy Transfer (FRET) studies [1]. In an analogous manner, the effects of low-power microwave heating on the ability of DNA to both melt and re-hybridize with additional complementary target oligonucleotide was also studied [50]. Figure 10 shows the emission spectra of fluorescein-labeled oligonucleotide (A) after 20 seconds of microwave heating (final step of the MAMEF-based DNA hybridization assay), (B) after melting the DNA and removing the fluorescein-labeled oligonucleotide using warm buffer above the melting point of the ds-DNA and (C) after re-hybridization with fresh fluorescein-labeled oligonucleotide using microwave heating. As one can see, after the re-hybridization is complete with microwave heating (Figure 7.9c), a similar fluorescein intensity is observed, indicating that the anchor probe on the silver surface is unaffected during microwave heating. These results imply that silvered surfaces with anchor oligonucleotides are re-usable, an important factor in the preparation of low-cost MAMEF-based DNA hybridization assays.

## 7.3 4. *MAMEF-based Anthrax detection*

In the previous section, the application of the MAMEF technology to DNA hybridization assays using a two-piece oligonucleotide model hybridization assay was demonstrated [50]. However, in a "real-world setting" it would not be practical to label the target oligonucleotide for the detection of target DNA. A common practice is to employ a third oligonucleotide labeled with a fluorophore (fluorescent probe), a specific sequence that hybridizes with the target oligonucleotide at another location close to the anchor probe, i.e., a three-piece DNA hybridization assay. In this regard, the applicability of the MAMEF technology to the detection of target DNA encoding a region of the protective antigen gene of *Bacillus anthracis* (i.e. the disease Anthrax) was recently presented [52].

Figure 7.11A shows the experimental details of the target Anthrax DNA assay constructed on silver-deposited glass microscope slides. In this DNA detection scheme, the anchor probe is attached to silver nanoparticles through sulfhydryl groups. In order to minimize the non-specific binding of DNA to the assay platform, silver nanoparticles and glass are both modified with additional surface-protective chemicals [52]. The subsequent hybridization of the fluorescent probe and target Anthrax DNA contained in the exosporium, a loose-fitting balloon like layer surrounding the spore, was carried out in a single microwave heating step, followed by an orange emission at 585 nm through an emission filter when excited with a green laser (532 nm). The intensity of the emission at 585 nm was directly related to the concentration of the target Anthrax DNA,[52] Figure 7.11B. Control experiments, where the anchor probe was omitted from the assay surface showed that the non-specific binding of target DNA was significantly less than the lowest concentration of exosporium sample detected in the actual assay, c.f. Figure 7.11C [52]. The ability of the MAMEF technology to distinguish between *Bacillus cereus*, a close relative of *B. anthracis* and Anthrax was also demonstrated [52]. Fluorescence emission intensity at 585 nm from this control assay shows constant emission intensity over a wide range of concentrations, indicating that the MAMEF assay platform clearly can distinguish between the two closely genetically related strains.

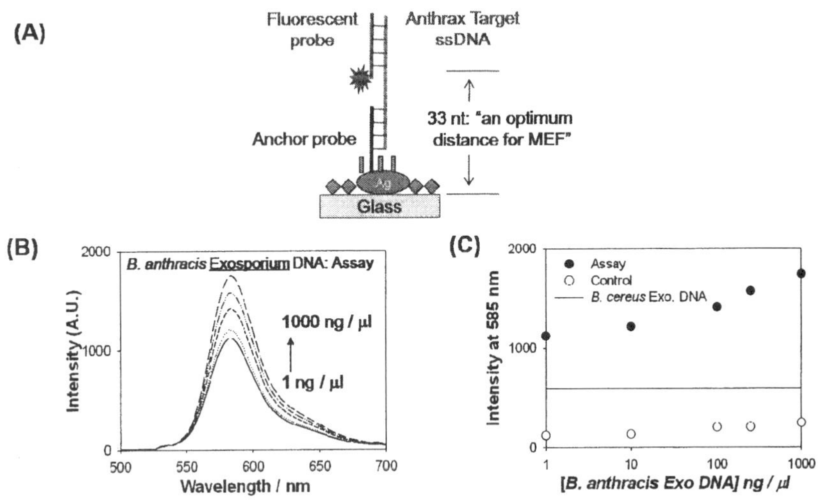


**Figure 7.11:** (A) Experimental design, depicting the organization of the DNA oligomers on SiFs used for the detection of *Bacillus anthracis*. (B) Emission spectra of the TAMRA-Oligo as a function of *B. anthracis* exosporium concentration after 30 s low power microwave heating. (C) Plot of the fluorescence emission intensity at 585 nm for TAMRA-Oligo as a function of target concentration. Data for *B. cereus* (a non-causative strain) is also shown for comparison. Adapted from reference 52.

In addition to the experiments performed for both strains separately as described above, to demonstrate the specificity of the assay for a sample that contains a mixture of DNA from both strains at once, a mixture of (same initial concentrations, 50% v/v) exosporium from *B. anthracis* and *B. cereus* was tested for successful detection of target DNA. Concentration-dependent fluorescence emission data for the assay (Figure 7.12A) shows a significant increase in fluorescence emission as the concentration of exosporium (and fluorescent probe) is increased, while the control assay (Figure 7.12B) shows only a slight increase in intensity. A plot of concentration-dependent emission intensity at 585 shows the range of detection for *B. anthracis* DNA in a mixture of exosporium DNA from two *Bacillus* strains, c.f. Figure 12C.

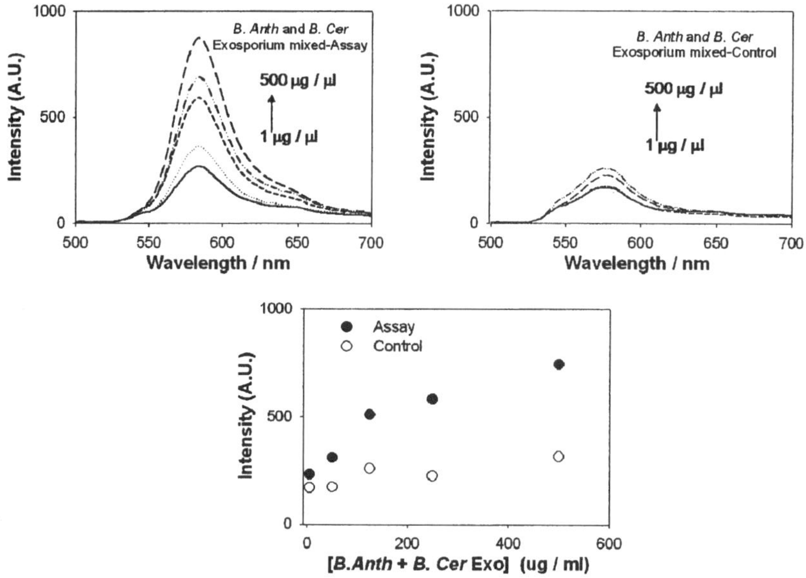


**Figure 7.12:** (A) Emission spectra of the TAMRA-Oligo as a function of concentration of *B. anthracis* and *B. cereus* exosporium after 30 s low power microwave heating. (b) Control experiment where the anchor probe is omitted from the surface (c) Plot of the fluorescence emission intensity at 585 nm for TAMRA-Oligo as a function of target concentration. Adapted from reference 52.

## 7.4 SUMMARY

In this review chapter, the summary of recent work on a unique technique, called microwave-accelerated metal-enhanced fluorescence (MAMEF), which affords for ultra fast and sensitive fluorescence-based detection of biological materials of interest, is presented. MAMEF is a new bioassay platform technique that couples the benefits of Metal-Enhanced Fluorescence (MEF) with low power microwave heating to kinetically accelerate the biorecognition events to completion within a minute. The major components of the MAMEF technique are plasmonic nanoparticles, microwaves and aqueous media. The exposure of these assay components to microwaves results in a temperature gradient between the water and the plasmonic nanoparticles. At 2.45 GHz, microwaves are selectively absorbed by water and thus heat the water, while the plasmonic nanoparticles are transparent to microwaves and remain colder. The temperature gradient created as a result of this selective heating further results in an increase in the transfer of biomolecules towards the colder nanoparticles. Since the binding partners of the biomolecules in the aqueous media are present on the nanoparticles, as the biomolecules move towards them, the biorecognition events occur on the surface of the nanoparticles. In several previous publications, our research group has applied the MAMEF technique to bioassays for the detection of proteins, antigens and DNA hybridization events. In this review chapter, the summary of these papers are presented. The reader is referred to the specific articles for more detailed information.

In summary, the MAMEF technology therefore has several notable advantages including:

- The fluorescence amplification provided by the plasmonic nanostructures has been shown to be applicable to many fluorophores. Hence fluorophores currently employed in assays would still be suitable. However, the use of low quantum yield fluorophores would lead to much larger fluorescence enhancements (i.e. $1 / Q_0$) and could significantly reduce unwanted background emission from fluorophores distal from the metallic surface.
- The MEF phenomenon has been shown to provide for increased emission intensities up to several thousand-fold.[53] This substantially increases detection limits (i.e. lower concentrations detectable), which is a major criterion in assay development today.[44]
- A whole variety of metallic surfaces can be routinely prepared, which do not require the benefits of a nanofabrication lab and sophisticated instrumentation such as electron beam lithography. [2-4, 22, 54, 55]
- The reduced lifetime of fluorophores in close proximity to silver nanostructures provides for a substantially increased fluorophore photostability.[4] In addition, shorter lifetimes allow for higher fluorophore cycling rates, also providing for increased fluorophore and therefore assay detectability.[4]
- The Low power microwaves employed here do not perturb the plasmonic surfaces, do not produce "arcing" which is commonly observed for metallic objects in microwave cavities,[56] or even denature or change protein conformation. Low power microwaves provide for effective rapid heating of the assays, producing identical final fluorescence intensities, fluorophore lifetimes, as well as extents of energy transfer (protein conformation) as compared to room temperature incubation.

## 7.5. ACKNOWLEDGMENTS

The authors acknowledge the Middle Atlantic Regional Center of Excellence for Biodefense and Emerging Infectious Diseases Research (NIH NIAID - U54 AI057168) for financial support.

## 7.6 REFERENCES


1. Aslan, K. and Geddes, C. D. (2005). Microwave-accelerated metal-enhanced fluorescence: Platform technology for ultrafast and ultrabright assays *Analytical chemistry* **77**: 8057-8067.
2. Aslan, K., Leonenko, Z., Lakowicz, J. R., and Geddes, C. D. (2005). Annealed silver-island films for applications in metal-enhanced fluorescence: Interpretation in terms of radiating plasmons *Journal of fluorescence* **15**: 643-654.
3. Zhang, Y., Aslan, K., Previte, M. J. R., and Geddes, C. D., Metal-enhanced fluorescence from copper substrates, vol. 90: AIP, 2007, pp. 173116.
4. Aslan, K., Gryczynski, I., Malicka, J., Matveeva, E., Lakowicz, J. R., and Geddes, C. D. (2005). Metal-enhanced fluorescence: an emerging tool in biotechnology *Current Opinion in Biotechnology* **16**: 55-62.
5. Geddes, C. D. and Lakowicz, J. R. (2002). Metal-enhanced fluorescence *Journal of fluorescence* **12**: 121-129.
6. Aslan, K., Lakowicz, J. R., Szmacinski, H., and Geddes, C. D. (2005). Enhanced ratiometric pH sensing using SNAFL-2 on silver island films: Metal-enhanced fluorescence sensing *Journal of fluorescence* **15**: 37-40.
7. Zhang, Y., Aslan, K., Previte, M. J., and Geddes, C. D. (2007). Metal-Enhanced Fluorescence: Surface Plasmons can Radiate a Fluorophores Structured Emission *Applied Physics Letters* **90**: 053107.
8. Amos, R. M. and Barnes, W. L. (1997). Modification of the spontaneous emission rate of Eu3+ ions close to a thin metal mirror *Physical Review B* **55**: 7249-7254.
9. Gersten, J. and Nitzan, A. (1981). Spectroscopic Properties of Molecules Interacting with Small Dielectric Particles *Journal of Chemical Physics* **75**: 1139-1152.
10. Das, P. C. and Puri, A. (2002). Energy flow and fluorescence near a small metal particle *Physical Review B* **65**.
11. Dulkeith, E., Ringler, M., Klar, T. A., Feldmann, J., Javier, A. M., and Parak, W. J. (2005). Gold nanoparticles quench fluorescence by phase induced radiative rate suppression *Nano Letters* **5**: 585-589.
12. Carminati, R., Greffet, J. J., Henkel, C., and Vigoureux, J. M. (2006). Radiative and non-radiative decay of a single molecule close to a metallic nanoparticle *Optics Communications* **261**: 368-375.
13. Zhang, Y. X., Aslan, K., Previte, M. J. R., and Geddes, C. D. (2006). Metal-enhanced S-2 fluorescence from azulene *Chemical Physics Letters* **432**: 528-532.
14. Chowdhury, M. H., Gray, S. K., Pond, J., Geddes, C. D., Aslan, K., and Lakowicz, J. R. (2007). Computational study of fluorescence scattering by silver nanoparticles *Journal of the Optical Society of America B-Optical Physics* **24**: 2259-2267.
15. Hatling, T., Reichenbach, P., and Eng, L. M. (2007). Near-field coupling of a single fluorescent molecule and a spherical gold nanoparticle *Optics Express* **15**: 12806-12817.
16. Aslan, K. and Perez-Luna, V. H. (2006). Nonradiative Interactions between Biotin Functionalized Gold Nanoparticles and Fluorophore-Labeled Antibiotin *Plasmonics* **1**: 111-119.

17. Hayakawa, T., Selvan, S. T., and Nogami, M. (1999). Field enhancement effect of small Ag particles on the fluorescence from Eu3+-doped SiO2 glass *Applied Physics Letters* **74**: 1513-1515.
18. Zhang, Y., Aslan, K., Previte, M. J., and Geddes, C. D. (2006). Metal-enhanced $S_2$ fluorescence from azulene *Chemical Physics Letters* **432**: 528-532.
19. Yguerabide, J. and Yguerabide, E. E. (1998). Light-scattering submicroscopic particles as highly fluorescent analogs and their use as tracer labels in clinical and biological applications - I. Theory *Analytical biochemistry* **262**: 137-156.
20. Yguerabide, J. and Yguerabide, E. E. (1998). Light-scattering submicroscopic particles as highly fluorescent analogs and their use as tracer labels in clinical and biological applications - II. Experimental characterization *Analytical biochemistry* **262**: 157-176.
21. Chen, Y., Munechika, K., and Ginger, D. S. (2007). Dependence of fluorescence intensity on the spectral overlap between fluorophores and plasmon resonant single silver nanoparticles *Nano letters* **7**: 690-696.
22. Aslan, K., Malyn, S. N., and Geddes, C. D. (2007). Metal-enhanced fluorescence from gold surfaces: angular dependent emission *J Fluoresc* **17**: 7-13.
23. Ray, K., Chowdhury, M. H., and Lakowicz, J. R. (2007). Aluminum nanostructured films as substrates for enhanced fluorescence in the ultraviolet-blue spectral region *Analytical chemistry* **79**: 6480-6487.
24. Dorfman, A., Kumar, N., and Hahm, J. (2006). Nanoscale ZnO-enhanced fluorescence detection of protein interactions *Advanced Materials* **18**: 2685-+.
25. Dorfman, A., Kumar, N., and Hahm, J. I. (2006). Highly sensitive biomolecular fluorescence detection using nanoscale ZnO platforms *Langmuir* **22**: 4890-4895.
26. Aslan, K., Huang, J., Wilson, G. M., and Geddes, C. D. (2006). Metal-enhanced fluorescence-based RNA sensing *Journal of the American Chemical Society* **128**: 4206-4207.
27. dos Santos, D. S. and Aroca, R. F. (2007). Selective surface-enhanced fluorescence and dye aggregation with layer-by-layer film substrates *Analyst* **132**: 450-454.
28. Geddes, C. D., Parfenov, A., and Lakowicz, J. R. (2003). Photodeposition of silver can result in metal-enhanced fluorescence *Applied spectroscopy* **57**: 526-531.
29. Geddes, C. D., Parfenov, A., Roll, D., Fang, J. Y., and Lakowicz, J. R. (2003). Electrochemical and laser deposition of silver for use in metal-enhanced fluorescence *Langmuir* **19**: 6236-6241.
30. Aslan, K., Holley, P., and Geddes, C. D. (2006). Metal-enhanced fluorescence from silver nanoparticle-deposited polycarbonate substrates *Journal of Materials Chemistry* **16**: 2846-2852.
31. Park, H. J., Vak, D., Noh, Y. Y., Lim, B., and Kim, D. Y. (2007). Surface plasmon enhanced photoluminescence of conjugated polymers *Applied Physics Letters* **90**.

32. Aroca, R., Kovacs, G. J., Jennings, C. A., Loutfy, R. O., and Vincett, P. S. (1988). Fluorescence Enhancement from Langmuir-Blodgett Monolayers on Silver Island Films *Langmuir* **4**: 518-521.
33. Garoff, S., Weitz, D. A., Hanson, C. D., Gramila, T. J., and Gersten, J. I. (1981). Deexcitation Channels of Excited Molecules on Silver Island Films *Journal of the Optical Society of America* **71**: 1552-1552.
34. Kummerlen, J., Leitner, A., Brunner, H., Aussenegg, F. R., and Wokaun, A. (1993). Enhanced Dye Fluorescence over Silver Island Films - Analysis of the Distance Dependence *Molecular Physics* **80**: 1031-1046.
35. Hayakawa, T., Selvan, S. T., and Nogami, M. (1999). Enhanced fluorescence from Eu3+ owing to surface plasma oscillation of silver particles in glass *Journal of Non-Crystalline Solids* **259**: 16-22.
36. Shang, L., Chen, H., and Dong, S. (2007). Electrochemical Preparation of Silver Nanostructure on the Planar Surface for Application in Metal-Enhanced Fluorescence *Journal of Physical Chemistry C* **111**: 10780-10784.
37. Aslan, K., Badugu, R., Lakowicz, J. R., and Geddes, C. D. (2005). Metal-enhanced fluorescence from plastic substrates *Journal of fluorescence* **15**: 99-104.
38. Thostenson, E. T. and Chou, T. W. (1999). Microwave processing: fundamentals and applications *Composites Part a-Applied Science and Manufacturing* **30**: 1055-1071.
39. Zhang, Y., Aslan, K., Previte, M. J. R., and Geddes, C. D. (2008). Metal-enhanced fluorescence from paper substrates: Modified spectral properties of dyes for potential high-throughput surface analysis and assays and as an anti-counterfeiting technology *Dyes and Pigments* **77**: 545-549.
40. Aslan, K., Holley, P., and Geddes, C. D. (2006). Microwave-Accelerated Metal-Enhanced Fluorescence (MAMEF) with silver colloids in 96-well plates: Application to ultra fast and sensitive immunoassays, High Throughput Screening and drug discovery *Journal of Immunological Methods* **312**: 137-147.
41. Previte, M. J., Zhang, Y., Aslan, K., and Geddes, C. D. (2007). Real-time Thermal Imaging of Microwave Accelerated Metal-Enhanced Fluorescence (MAMEF) Based Assays on Sapphire Plates *Journal of fluorescence* **17**: 639-642.
42. Bange, A., Halsall, H. B., and Heineman, W. R. (2005). Microfluidic immunosensor systems *Biosensors & bioelectronics* **20**: 2488-2503.
43. Ozinkas, A. J., Principles of Fluorescence Immunoassay, in *Topics in Fluorescence Spectroscopy*, J. R. Lakowicz, Ed. New York: Plenum Press, 1994.
44. Van Dyke, K. and Van Dyke, R., Luminescence Immunoassay and Molecular Applications. Boca Raton, FL.: CRC Press, 1990.
45. Aslan, K. and Geddes, C. D. (2006). Microwave Accelerated and Metal Enhanced Fluorescence Myoglobin Detection on Silvered Surfaces: Potential Application to Myocardial Infarction Diagnosis *Plasmonics* **1**: 53-59.
46. Morrison, L., Fluorescence in nucleic acid hybridization assays, in *Topics in Fluorescence Spectroscopy*, vol. 7, J. R. Lakowicz, Ed. New York: Kluwer Academic Publishers/Plenum Press, 2003, pp. 69-103.

47. Brown, P. O. and Botstein, D. (1999). Exploring the new world of the genome with DNA microarrays *Nature genetics* **21**: 33-37.
48. Difilippantonio, M. and Ried, T., Technicolor genome analysis, in *Topics in Fluorescence Spectroscopy*, vol. 7, J. R. Lakowicz, Ed. New York: Kluwer Academic Publishers/Plenum Press, 2003, pp. 291-316.
49. Malicka, J., Gryczynski, I., and Lakowicz, J. R. (2003). DNA hybridization assays using metal-enhanced fluorescence *Biochemical and biophysical research communications* **306**: 213-218.
50. Aslan, K., Malyn, S. N., Bector, G., and Geddes, C. D. (2007). Microwave-accelerated metal-enhanced fluorescence: an ultra-fast and sensitive DNA sensing platform *Analyst* **132**: 1122-1129.
51. Aslan, K., Malyn, S. N., and Geddes, C. D. (2006). Fast and sensitive DNA hybridization assays using microwave-accelerated metal-enhanced fluorescence *Biochemical and biophysical research communications* **348**: 612-617.
52. Aslan, K., Zhang, Y., Hibbs, S., Baillie, L., Previte, M. J., and Geddes, C. D. (2007). Microwave-accelerated metal-enhanced fluorescence: application to detection of genomic and exosporium anthrax DNA in <30 seconds *Analyst* **132**: 1130-1138.
53. Geddes, C. D., Parfenov, A., Roll, D., Gryczynski, I., Malicka, J., and Lakowicz, J. R. (2003). Silver fractal-like structures for metal-enhanced fluorescence: Enhanced fluorescence intensities and increased probe photostabilities *Journal of fluorescence* **13**: 267-276.
54. Aslan, K., Leonenko, Z., Lakowicz, J. R., and Geddes, C. D. (2005). Fast and slow deposition of silver nanorods on planar surfaces: Application to metal-enhanced fluorescence *Journal of Physical Chemistry B* **109**: 3157-3162.
55. Aslan, K., Malyn, S. N., and Geddes, C. D. (2007). Angular-dependent metal-enhanced fluorescence from silver colloid-deposited films: opportunity for angular-ratiometric surface assays *Analyst* **132**: 1112-1121.
56. Whittaker, A. G. and Mingos, D. M. P. (1993). Microwave-Assisted Solid-State Reactions Involving Metal Powders and Gases *Journal of the Chemical Society-Dalton Transactions*: 2541-2543.

# 8

# Localized Surface Plasmon Coupled Fluorescence Fiber Optic Based Biosensing

Chien Chou[1,3,5], Ja-An Annie Ho[2], Chii-Chang Chen[3], Ming-Yaw Ng[4], Wei-Chih Liu[4], Ying-Feng Chang[5], Chen Fu[2], Si-Han Chen[2], Ting-Yang Kuo[2]

[1]Department of Biomedical Imaging and Radiological Sciences, National Yang Ming University, Taipei, 112, Taiwan

[2]Department of Chemistry, National Tsing Hua University, Hsin-chu, 30013 Taiwan

[3]Department of Optics and Photonics, National Central University, Jhung-li, 320, Taiwan

[4]Department of Physics, National Taiwan Normal University, Taipei, 116, Taiwan

[5]Institute of Biophotonics, National Yang Ming University, Taipei, 112, Taiwan

## 8.1 OPTICAL FIBER

### 8.1 1. *Introduction*

An optical fiber generally refers to either a glass or a plastic fiber. Due to its immunity to electromagnetic interference, optical fibers are widely used in fiber-optic communication. This mode of communication permits transmission over longer distances and at higher data rates than other forms of communications. Fibers supporting multi-propagation paths or transverse modes are called multimode fibers (MMFs), while fibers supporting only a single mode are referred to as single mode fibers (SMFs). Optical fibers can also be used in sensor development and in a variety of other applications (1-4).

An optical fiber is a circular structure composed of three layers inside and out. Its cross section view is shown in Figure 8.1. The innermost layer is called the core, which guides the light and prevents it from escaping. This is enabled by a phenomenon called "total internal reflection" (TIR), which is also designed to guide light along its length. The diameter of the optical fibers' core is generally 5-10 □m for SMFs and >50 □m for MMFs (5). The middle layer is called cladding, which has 1% lower refractive index than the core material. This difference plays a significant role in the TIR phenomenon. The cladding's diameter is usually 125 μm. Lastly, the outer layer is called the coating. This layer provides mechanical protection for the fiber and makes the fiber flexible for handling. It could be epoxy cured by ultraviolet light. Without this coating layer, the fiber will be very fragile and easy to break (6-8).

*Surface Plasmon Coupled Fluorescence Fiber Biosensing.*
Edited by Chris D. Geddes.

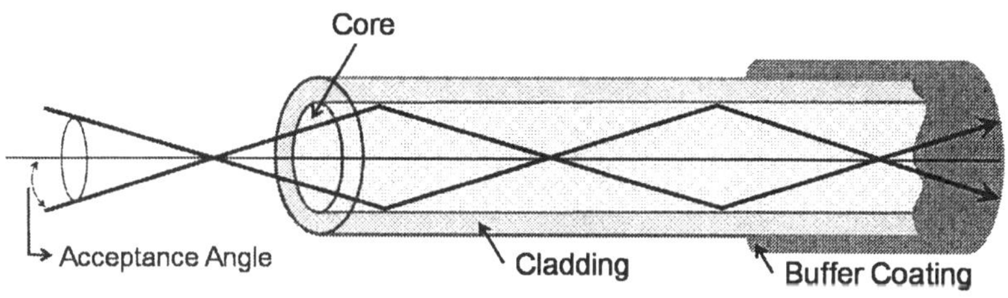


**Figure 8.1:** The cross-section view of optical fiber (drawing is not to scale).

## 8.1 2. *Materials of Optical Fiber*

Based on the formed material, there are two major types of optical fibers: the glass optical fiber and the plastic optical fiber (POF).

### 8.1.2 1. *Glass optical fibers*

Glass optical fibers are made of extremely pure optical glass or silica. Silica has a refractive index of 1.458, which figures to be one of the light-passing materials with the lowest refractive index. Some other materials, such as fluorozirconate, fluoroaluminate, and chalcogenide glasses, can also be used for manufacturing longer-wavelength infrared applications (5). The silicon dioxide based optical fiber is a biocompatible material and can be applied in ultraviolet (UV), visible (VIS), and near infrared (NIR) regions. UV-silica fibers feature high transmission, which is between 180 nm and 300 nm, and possesses resistance to radiation, or laser damage. It can be made with high core-to-clad ratio for improved efficiency. Special metal coating can be manufactured for use in higher temperature regions, which enables it to operate in high vacuum and harsh chemical environments. NIR-silica fibers have similar features as UV-silica fibers, except that they show broader transmission range, which is between 1500 nm and 2600 nm (6). Silicon nitride is often used as a seal for the protection of bare fiber against abrasion and corrosion. The optical absorption of silicon nitride is higher than that of silicon nitrate ($SiO_2$). This provides the advantage of reducing optical crosstalk unique to optical fiber technology in certain configurations of multiple fiber bundles (8).

### 8.1.2. 2. *Plastic optical fiber*

Plastic optical fibers are optical fibers made of plastic. Traditionally, PMMA (acrylic) is the core material (the synthesis of PMMA is shown in Scheme 8.1), and fluorinated polymers serve as the cladding materials. Higher-performance POF based on perfluorinated polymers (mainly polyperfluorobutenylvinylether) has begun to appear in the market since the late 1990s. In large-diameter fibers, 96% of the cross section is the core that allows the transmission of light. Similar to traditional glass fibers, POF transmits light (or digital data) through the core of the fiber. The core size of POF is in some cases 100 times larger than that of the glass fiber. The advantages of POF include the following: (i) serves as a cost-effective alternative to silica fiber and soft glass fibers in the spectral range 400 - 850 nm; (ii) offers high-quality performance in data transmission, illumination, laser, and Light Emitting Diode (LED) therapy; and (iii) features high numeric aperture and a wide range of diameters with high flexibility (9-10).

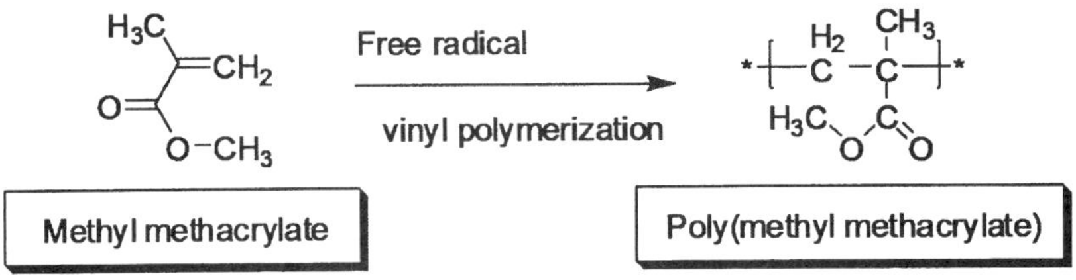


**Scheme 8.1**: PMMA is a vinyl polymer made by free radical vinyl polymerization from the monomer methyl methacrylate.

## 8.1 4. *Applications on Biosensors*

Ever since the first fiber-optic biosensor (FOB) was described in 1979 (11), optical fibers have become an important topic in the field of sensor technology. Their use as a probe or as a sensing element is consistently emerging in clinical, pharmaceutical, industrial, and military applications. A variety of FOBs for the determination of glucose, lactate, penicillin, urea or alcohol has been developed (12). Optical fibers can likewise be used as sensors to measure strain, temperature, pressure, and other parameters. Its small size and the fact that no electrical power is needed at the remote location give FOBs advantage over conventional electrical sensors in certain applications. Trpkovski et al (2003) demonstrated the development of fiber optic sensors which measured co-located temperature and strain, simultaneously and with very high accuracy (13). This biosensor is particularly useful when acquiring information from small complex structures. The main points in favor of the use of optical fibers in developing biosensors are the following: excellent light delivery; interaction length; low cost; and ability not only to excite the target molecules but also to capture the emitted light from the targets (14).

Fiber-optic biosensors are analytical devices in which a fiber optic device serves as a transduction element. The usual aim of fiber-optic biosensors is to produce a signal proportional to the concentration of target analyte to which the biological element reacts. Fiber-optic biosensors are based on the transmission of light along silica glass fiber, or POF to the site of analysis. They can be used in combination with different types of spectroscopic technique, e.g. absorption, fluorescence, phosphorescence, or surface plasmon resonance (SPR) (14).

Optical biosensors based on the use of fiber optics can be classified into two categories: intrinsic sensors where interaction with the analyte occurs within an element of the optical fiber, and extrinsic sensors in which the optical fiber is used to couple light, usually to and from the region where the light beam is influenced by the measurand. Moreover, biosensors become attractive because they can be easily used by non-specialist personnel and they allow accurate determination with either none or little sample treatment. Therefore, fiber-optic biosensors may be especially useful in routine tests, patient home care, surgery, and intensive care, as well as in emergency situations (11-16).

## 8.1 5. *Fiber Optic Evanescent Wave Sensor*

The fiber optic evanescent wave sensor (FO-EWS) belongs to a sensor in which the fiber core interacts with the analyte. This interaction occurs through the attenuated total reflection (ATR) and the evanescent wave excitation in a dielectric medium of smaller refractive index in the vicinity of fiber core. If the surrounding medium is fluorescent, then the fluorescence signal in the reaction region of evanescent wave field is excited and detected. This is illustrated in Figure 8.2.

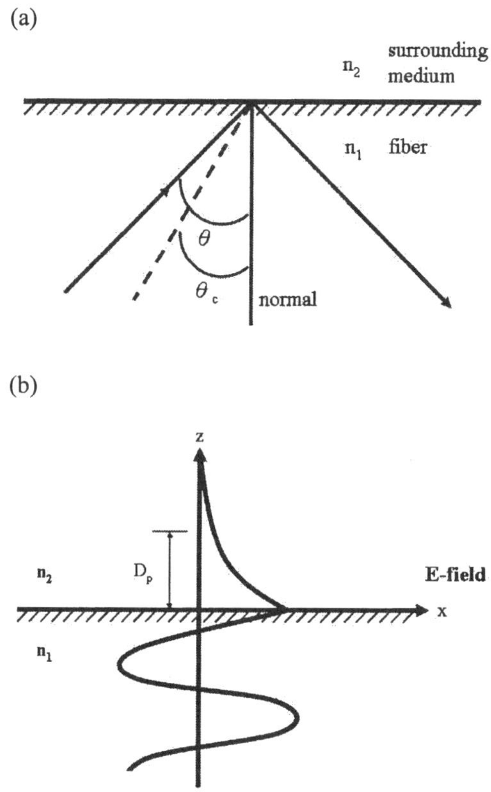


**Figure 8.2:** (a) The incident angle is $\theta > \theta_c$, where $\theta_c$ is critical angle where total reflection on interface starts to produce. (b) The electric field of evanescent wave in the surrounding medium. $D_p$ is the penetration depth of evanescent wave in surrounding medium.

When $n_1 > n_2$ is satisfied, the critical angle of producing total reflection on fiber core interface is expressed by

$$\theta_c = \sin^{-1}(n_2/n_1) \quad (1)$$

where $n_1$ and $n_2$ are refractive indices of fiber core and the surrounding medium, respectively. Once the incident angle $\theta > \theta_c$, the evanescent wave is generated at the fiber core / dielectric medium interface and the electric field of evanescent wave E is presented by

$$E = E_s \exp(-Z/D_p) \quad (2)$$

$E_s$ is the amplitude of electric field at interface (Z=0). $D_p$ means the penetration depth described by

$$D_p = \lambda \Big/ \left\{ 2\pi n_2 \left[ \left( n_{rel}^2 \sin^2\theta \right) - 1 \right]^{1/2} \right\} \quad (3)$$

where $n_{rel} = n_1 / n_2$, $\lambda$ is the wavelength of incident beam in vacuum. In general, $D_p$ is smaller than the wavelength of the light source. The evanescent electric field interacts with bio-molecules in the reaction region surrounding the fiber core surface. Then, the fluorescence is excited by the evanescent wave of appropriate wavelength. Conventionally, the fluorescence signal is detected at the distal end of optical fiber. However, the coaxial propagation with pumping laser beam, which induces a strong background signal, results in reduction of the detection sensitivity of fluorescence signal. In order to avoid the background signal, a narrow band-pass optical filter (Figure 8.3 (a)) or a concave grating (Figure 8.3 (b)) are suggested to separate the pumping beam and the fluorescence signal. Similarly, the fluorescence signal which is opposite propagation to the pumping laser can be detected at front end of the fiber (Figure 8.3 (c)). In this setup, however, the fluorescence signal is attenuated significantly because of the absorption by optical components inserted as can be seen in this arrangement.

(a)

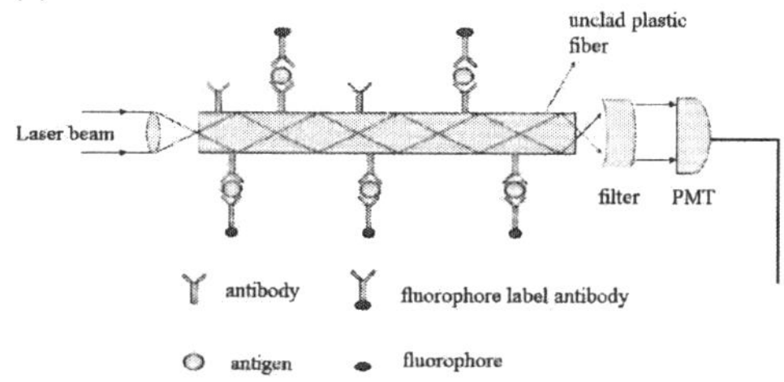
unclad plastic fiber
Laser beam
filter
PMT
antibody
fluorophore label antibody
antigen
fluorophore

(b)

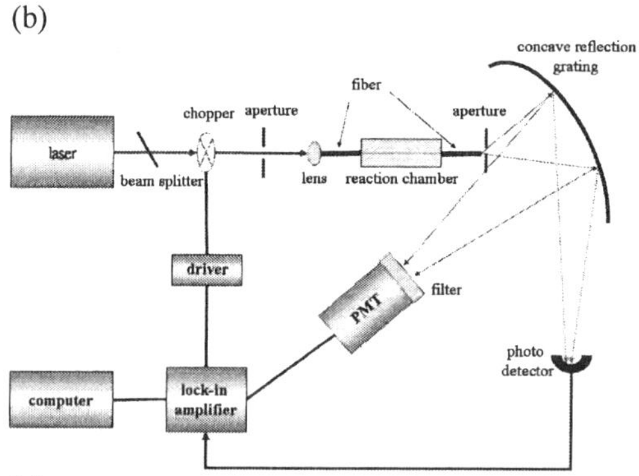
concave reflection grating
fiber
chopper
aperture
aperture
laser
beam splitter
lens
reaction chamber
driver
PMT
filter
photo detector
computer
lock-in amplifier

(c)

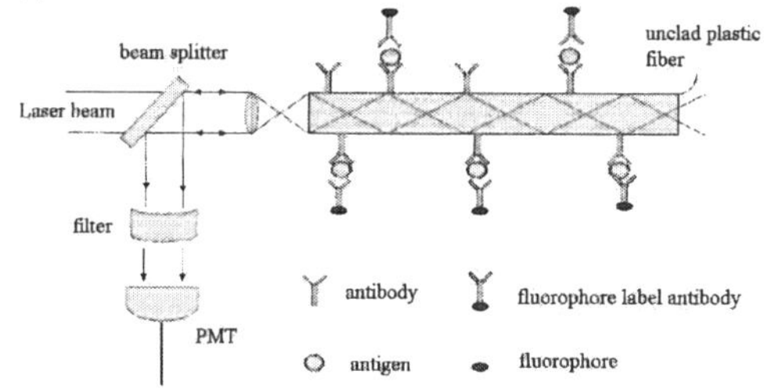
beam splitter
unclad plastic fiber
Laser beam
filter
PMT
antibody
fluorophore label antibody
antigen
fluorophore

(d)

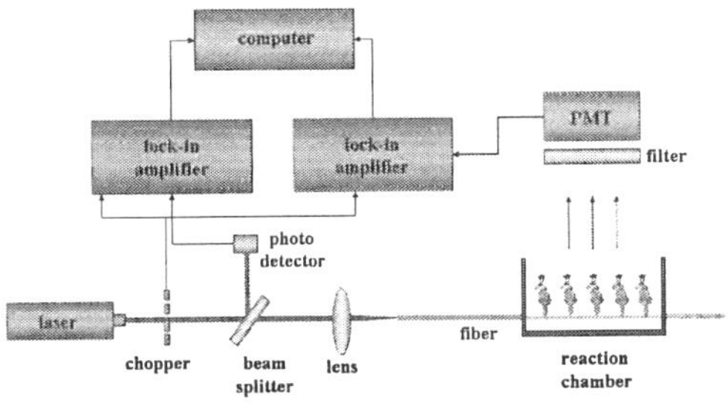


**Figure 8.3:** The optical setup. (a) fluorescence signal detected at distal end of optical fiber; (b) use of grating to separate the pumping beam and the fluorescence signal; (c) fluorescence signal detected at front end of the fiber; (d) fluorescence signal detected beside the optical fiber reaction region

## 8.2 THEORY

### 8.2 1. *Surface Plasmon Wave*

Surface plasmon waves (SPW) have been explored almost half a century ago (23, 24). These are surface-bound electromagnetic waves propagating on the dielectric-metal interface. The existence of SPW strongly depends on the refractive index of the dielectric medium adjacent to the metal, so it is commonly known that the electromagnetic field of SPW is extended into the medium only to a depth of 200 nm or so (25, 26). Therefore, SPW is very suitable for the measurement of small changes in refractive index of dielectric medium in the vicinity of metal for detecting bio-molecules interaction (26-30).

Theoretically, SPW is described as a charge density oscillation that generates highly confined electromagnetic fields on the surface of a metal film (24, 26, 31-35). The criterion for the excitation of SPW is that the incident laser beam must be matched in both frequency and momentum with that of SPW. This can only occur, for example, if P wave (TM wave) is incident from the glass side at a specific angle of which the projection of k vector of the incident photon matches SPW's k vector (26, 36, 37). The dispersion relation for a semi-infinite metal plane surface of complex dielectric constant $\varepsilon_1$ adjacent to a dielectric medium $\varepsilon_2$, can be written as (38, 39)

$$k_x = \frac{\omega}{c}\left(\frac{\varepsilon_1\varepsilon_2}{\varepsilon_1+\varepsilon_2}\right)^{1/2} = k_{sp} \tag{4}$$

Thus, surface plasmons (SPs) are excited on the dielectric medium - metal interface through evanescent wave excitation by attenuated total reflection (ATR) (23, 33, 36-38). One can excite SPs through the Kretschmann–Raether configuration consisting of a prism, a metal film (gold or silver), and the dielectric probed medium (23) as shown in Figure 8.4.

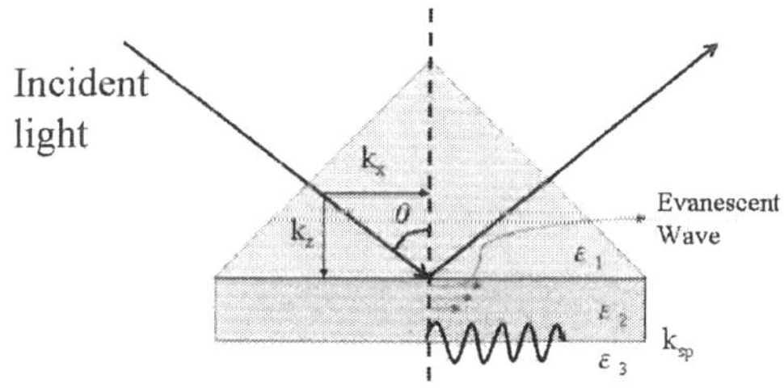


**Figure 8.4:** SPW in the Kretschmann configuration.

On dielectric / metal interface, the in-plane wave vector $k_x$ of the photons is

$$k_x = \frac{\omega}{c}\sqrt{\varepsilon_0}\sin\theta_R \tag{5}$$

$$\theta_R = \sin^{-1}\left(\varepsilon_1\varepsilon_2/(\varepsilon_1+\varepsilon_2)\varepsilon_0\right)^{1/2} \tag{6}$$

The $\theta_R$ is the angle of incidence otherwise known as SPR angle. (43, 44). The position of the SPR angle is strongly sensitive to the changes on refractive index in the dielectric medium close to metal film (27). Theoretically, not only the intensity of the reflected P wave but also its phase is changed at the same time when the incident angle is near the resonance angle (45). A rapid phase change relative to the amplitude of the reflected P polarized laser beam is observed. Conversely, the reflected intensity is minimized at $\theta_R$ as evident in Figure 8.5. Recently, a method for measuring the phase shift instead of the reflected intensity has been demonstrated. This method improves several orders of magnitude on detection sensitivity experimentally (46-50).

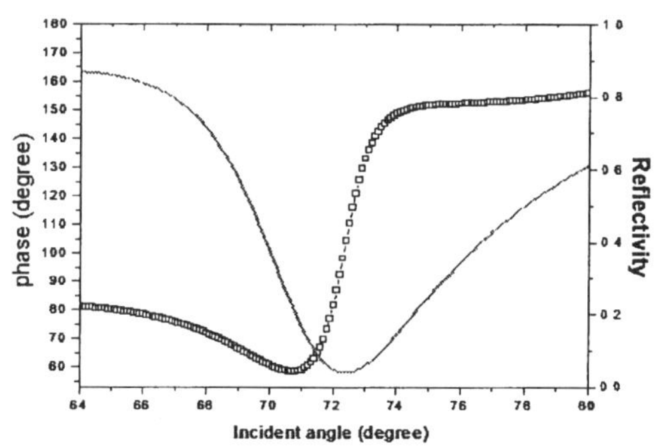

**Figure 8.5:** A rapid phase change relative to amplitude of the reflected intensity near SPR angle. Square: phase curve; line: amplitude curve

A fcaturc of an SPR bioscnsor is its ability to monitor bio-molecular interactions. In doing so, it does not require labeling of the fluorophore, hence, providing a more complete picture of the bio-molecular interactions (51-54). Over the past years, fiber-based SPR biosensors have been demonstrated to be alternative prism-based SPR sensors (56-58). Fiber-optic SPR biosensors have several advantages over coupling prism SPR biosensors such as their low cost, flexible design, portability, and on-site and remote sensing capability (37, 55-57). Equivalently, a decladded optical fiber is coated with thin gold film, which is surrounded by dielectric medium. Then the SPW on gold / dielectric medium interface is excited by an evanescent wave on fiber core / gold film interface. This connection of the evanescent wave with SPs corresponds to the wavelength and fiber parameters such as NA and the number of modes of propagation in optical fiber (37, 58). Recently, research has delved on the application of localized surface plasmon resonance (LSPR) in fiber-optic biosensors (56, 59, 60). This sensor was constructed on the basis of modification of the unclad portion of an optical fiber with hide grating and self-assembled gold nanoparticle (GNP) monolayer as can be seen in Figure 8.6.

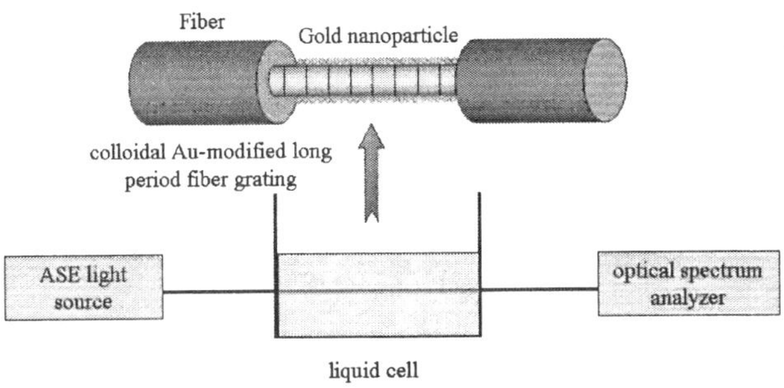


**Figure 8.6:** Schematic representation of the experimental setup used to make measurements with the CMAuLFPG (56).

The transmission spectra of self-assembled gold colloids on the surface of optical fiber change with different refractive indexes of the environment near the colloidal gold surface. The detection limit of the sensor can reach an order of several hundred picomolars (56). Furthermore, Wu et al. integrates the propagation of paired SPWs (61) with localized surface plasmon (LSP) on colloid GNP. This aims to enhance the detection sensitivity of protein A and mouse IgG interaction on CM5 sensor chip at 330 fg/mL (62). Fang et al., on the other hand, proposed the SP fluorescence immunoassay to test free Prostate-specific antigen (f-PSA) in human plasma at femtomolar (63). The development of a surface plasmon fluorescence spectroscopy (SPFS) based sandwich immunoassay is based on the features of the enormous field enhancement achieved through the resonance excitation of SPW. It also took into consideration the combination of this concept with the rather sensitive fluorescence detection technique. In this setup, the fluorescence signal is excited by propagating SPW with the use of a conventional prism-based SPR biosensor. As a result, the detection sensitivity versus fluorescence detection is significantly enhanced.

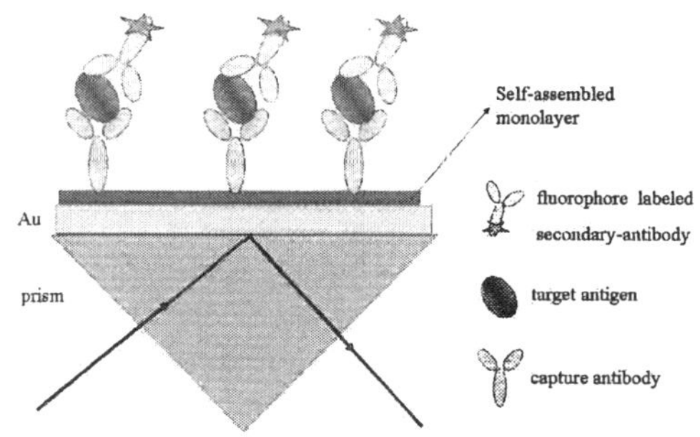


**Figure 8.7:** Schematic drawing of SPFS-based sandwich immunoassay

## 8.2 2. *Localized Surface Plasmon of Metallic Nanoparticles*

According to previous experimental results, the main mechanism of fluorescence signal enhancements is associated strongly with the local-field enhancement in the vicinity of GNPs (64-67). Strongly enhanced local fields are generated around GNPs due to surface plasmon excitation by evanescent waves from surface of the decladded fiber. The enhanced local fields around GNPs likewise become a main implement to enhance the fluorescence signals of fluorophores, which proves binding on the nanoparticles. However, it is not easy to explore the relationship between them directly in practical measurements. This implies the need for a theoretical study based on rigorous scattering theory. Hence, said study was proposed to establish a theoretical background for local-field enhanced fluorescence signals of localized surface plasmon coupled fluorescence fiber (LSPCF) biosensor. The scattering theory of evanescent waves can provide an elementary understanding of local-field-enhanced fluorescence signals. In addition, the magnitude of the enhancement could be estimated according to theoretical calculations. In order to compare with experimental results, local-field enhancements around an individual GNP were averaged to correspond to enhancement of fluorescence signals of experiments.

Solutions for the scattering of evanescent waves by a sphere were studied initially by Chew *et al.* in 1979 (68). Evanescent wave is generated from a dielectric/air plane interface due to total internal reflection and is scattered by a dielectric sphere. Quinten *et al.*, on the other hand, derived the total cross section for extinction and scattering of evanescent waves by a small metal particle (69). In these studies, a particle with smaller size is placed away from the plane interface, and therefore the multiple scattering between the sphere and the plane interface is ignored. Quinten's study further supports that the extinction cross section for the extinction of p-polarized evanescent waves by a metallic nanoparticle is larger than that for *s*-polarized evanescent waves and plane waves. This can be attributed to the ability of the evanescent field to increase the magnitude of contributions of multipoles to the cross section and the local-field enhancement of nanoparticles. In other words, a strongly enhanced local field is excited around the metallic nanoparticle by *p*-polarized evanescent waves, and the local-field enhancement is higher than when using *s*-polarized evanescent-waves or plane-wave illumination. According to the experimental results, the GNPs are far from the dielectric interface of the fiber's core ($d > 2r$) because the size of IgG and anti-IgG is $6.5 \pm 0.9$ nm according to the atomic force microscope (AFM) tapping mode image (70). The sandwich immuno-complex height can be estimated to about 20nm or more. The distance between GNPs and the interface of the optical fiber is larger than the diameter of the nanoparticle, so the influence of the fiber's interface due to induced image charges on the optical properties of metallic nanoparticles is negligible (71). The fluorescence probe solution used in the experiments is a very dilute solution (less than 1 ng/mL), so the near-field interaction between GNPs may not be considered. Consequently, the scattering theory for evanescent waves can be treated as a good first-order approximation to calculate the local-field enhancement of GNPs. Based on theoretical calculations, the enhancement of fluorescence signals can also be estimated.

First, a brief introduction to the localized surface plasmon of spherical metallic nanoparticles is provided. In addition to the surface charge oscillation on the dielectric-metal interface (38), the collective oscillation of bounded charges of metallic nanoparticles can be excited by incident light. A simplified classical model is used to understand the electromagnetic behaviors of the charge oscillation of particles. Consider a particle with a size much smaller than the wavelength of light; the skin effect of the particle can be neglected, and the electric fields inside and outside the particle are treated as uniform. The negative charge density of a particle exhibits oscillation with respect to the lattice ion with positive charges by external harmonic fields. The collective oscillation behavior of the small particle is called particle plasmon excitation. Such surface plasmon excitation associates with the oscillation of bounded charges on the particle, so it is also often called localized surface plasmon excitation. This event strongly depends on the size and geometry of particles.

### 8.2. 3 *Quasi-static approximation*

When a metallic particle is small compared with the illuminating wavelength, the variation of the electric field can be ignored, and the scattering behaviors can be described under the quasi-static approximation. Consider a smaller metallic sphere with radius $a$ and $a << \lambda$ placed in a uniform static electric field $\vec{E} = E_0 \hat{z}$. The field potentials inside and outside the sphere, $\Phi_{in}$ and $\Phi_{out}$, are written by

$$\Phi_{in}(r,\theta) = Ar\cos\theta \qquad r < a \tag{7a}$$

$$\Phi_{out}(r,\theta) = -E_0 r\cos\theta + \frac{B}{r^2}\cos\theta \qquad r > a \qquad (7b)$$

where $r = \sqrt{x^2 + y^2 + z^2}$ , $A$ and $B$ are function coefficients and are associated with the boundary conditions on the surface of the sphere. On the other hand, the potentials $\Phi_{in}$ and $\Phi_{out}$ are solved by Laplace's equation. The potential at the boundary between sphere and surrounding medium must satisfy the boundary conditions as in the following:

$$\Phi_{out}(a,\theta) = \Phi_{in}(a,\theta) \qquad (8a)$$

$$\varepsilon\frac{\partial\Phi_{out}(a,\theta)}{\partial r} = \varepsilon_m\frac{\partial\Phi_{in}(a,\theta)}{\partial r} \qquad (8b)$$

where $\varepsilon$ and $\varepsilon_m$ are the permittivities of the surrounding medium and the sphere, respectively. Here, $\varepsilon_m$ is a function of the frequency $\omega$ of incident light. The coefficients A and B are obtained from Eqs. (8a) and (8b) written as:

$$A = -\left(\frac{3\varepsilon}{\varepsilon_m + 2\varepsilon}\right)E_0 \qquad (9a)$$

$$B = a^3\left(\frac{\varepsilon_m - \varepsilon}{\varepsilon_m + 2\varepsilon}\right)E_0 \qquad (9b)$$

Substituting Eqs. (9a) and (9b) into Eqs. (7a) and (7b), respectively, the potentials inside and outside the sphere are derived.

$$\Phi_{in}(r,\theta) = -\left(\frac{3\varepsilon}{\varepsilon_m + 2\varepsilon}\right)E_0 r\cos\theta \qquad (r < a) \qquad (10a)$$

$$\Phi_{out}(r,\theta) = -E_0 r\cos\theta + \frac{a^3}{r^2}\left(\frac{\varepsilon_m - \varepsilon}{\varepsilon_m + 2\varepsilon}\right)E_0\cos\theta \qquad (r > a)$$

(10b)

If focus is placed on the potential outside the sphere (the scattered field components), the $\Phi_{out}$ can be rewritten as:

$$\Phi_{out}(r,\theta) = -E_0 r\cos\theta + \frac{p}{4\pi\varepsilon r^2}\cos\theta \qquad (11)$$

where $p = \alpha\varepsilon E_0$ and $\alpha = 4\pi a^3\left(\dfrac{\varepsilon_m - \varepsilon}{\varepsilon_m + 2\varepsilon}\right)$. Here, $p$ is called electric dipole moment and $\alpha$ is called polarizability. The results in Eq. (11) show that the potential outside the sphere is the superposition of the potentials of a uniform static electric field and an ideal dipole radiation. In other words, the scattering of an incident plane wave $\vec{E}_{inc} = E_0 e^{i(kx-\omega t)}\hat{z}$ by a metallic sphere can be treated as an ideal dipole radiation, and the scattered field $\vec{E}_{scat}$ in far zone ($r >> \lambda$) of the dipole with time-dependent dipole moment $\vec{p} = \varepsilon\alpha(\omega)E_0 e^{i(kx-\omega t)}\hat{z}$ at the location z = 0 is obtained as follows:

$$\vec{E}_{scat} = i\frac{e^{ik(r-x)}}{kr}\vec{X}E_0 e^{i(kx-\omega t)} \tag{12}$$

where $\vec{X} = \dfrac{ik^3}{4\pi}\alpha(\omega)(\hat{r}\times\hat{r}\times\hat{z})$. The far-field properties of an object are described by the optical total cross section. Extinction $C_{ext}$ and scattering cross sections $C_{scat}$ of the sphere are obtained from Eq. (12) and are given by

$$\begin{aligned} C_{ext} &= k\,\mathrm{Im}\{\alpha\} \\ &= 4\pi a^2 x\,\mathrm{Im}\left(\frac{\varepsilon_m - \varepsilon}{\varepsilon_m + 2\varepsilon}\right) \end{aligned} \tag{13a}$$

$$\begin{aligned} C_{scat} &= \frac{k^4}{6\pi}|\alpha|^2 \\ &= \frac{8}{3}\pi a^2 x^4\left|\frac{\varepsilon_m - \varepsilon}{\varepsilon_m + 2\varepsilon}\right|^2 \end{aligned} \tag{13b}$$

where $x = \dfrac{2\pi a}{\lambda}\sqrt{\varepsilon}$. According to the above-mentioned results, the localized surface plasmon resonance of the metallic sphere appears at $\mathrm{Re}(\varepsilon_m) = -2\varepsilon$. The localized surface plasmon frequency $\omega_{LSP}$ of the sphere can be obtained by applying the Drude model, that is,

$$\omega_{LSP} = \frac{\omega_p}{\sqrt{1+2\varepsilon}} \tag{14}$$

where $\omega_p$ is the plasma frequency of the metallic sphere. It is noted that for particles whose sizes are much smaller than the illuminating wavelength, their optical responses are similar to that of a dipole radiation. However, higher multipole behaviors become more obvious given the increasing size and the low symmetric shape of nanoparticles. This is because the contributions of higher-order mode become significant gradually. Therefore, the metallic particle with larger size or with non-regular shape exhibits complex plasmon spectra. It must be noted that its near-field and far-field optical responses have been extensively studied in theoretical and experimental works, recently.

## 8.2 2. *Local-field enhancement of metallic nanoparticles*

Highly enhanced local fields can be generated in the vicinity of a metallic spherical nanoparticle due to localized surface plasmon excitation. Using the result in Eq. (10b), the electric field components at the surface of the small sphere can be obtained by $\vec{E} = -\nabla\Phi_{out}$ as

$$E_r = \left(1 + \frac{\varepsilon_m - \varepsilon}{\varepsilon_m + 2\varepsilon}\frac{2a^3}{r^3}\right)E_0\cos\theta \tag{15a}$$

$$E_\theta = \left(-1 + \frac{\varepsilon_m - \varepsilon}{\varepsilon_m + 2\varepsilon}\frac{a^3}{r^3}\right)E_0\sin\theta \tag{15b}$$

For a dipole radiation, the strongest field intensity appears at $\theta = 0$. Therefore, the field enhancements $\Lambda(r)$ of the sphere as a function of distance r are obtained from Eq. (15a):

$$\Lambda(r) = \left|1 + \frac{2a^3}{r^3}\left(\frac{\varepsilon_m - \varepsilon}{\varepsilon_m + 2\varepsilon}\right)\right|^2 \tag{16}$$

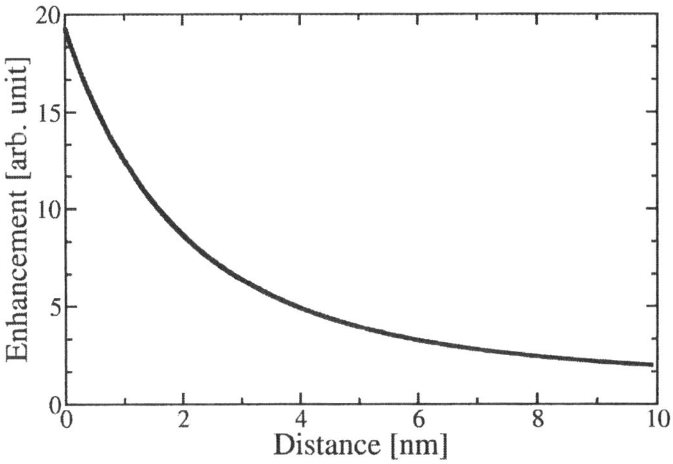


**Figure 8.8:** Electric-field enhancements in the vicinity of a gold nanosphere with radius a = 10 nm as a function of the distance from the surface of the nanosphere at wavelength of 516 nm.

The refractive t and the surrounding medium are 0.608 + 2.12*i* (72) and 1.0, respectively. The strongest field enhancement which is about 19 times of the intensity of the incident light appears on the surface of the nanosphere.

The electric-field enhancements of a GNP with radius of 10 nm as a function of the distance from the surface of the nanoparticle are shown in Figure 8.8. When the GNP is illuminated by the incident field with wavelength $\lambda$ = 516 nm, the strongest field enhancement which is about 19 times of the intensity of the incident field appears on the surface of the nanoparticle. In addition, the field enhancement away from the surface decreases dramatically and becomes close to the intensity of the incident field. It is noted that electric field can be localized and confined effectively in nanoscale region due to localized surface plasmon excitation of metallic nanoparticles. In this chapter, the highly enhanced local fields around the metallic nanoparticles are used to significantly enhance the fluorescence signals in biosensing technologies. The local-field enhancement of metallic nanoparticles is very sensitive to the permittivity of the surrounding medium and exhibits beyond optical diffraction limit capability, so the local-field enhancement of metallic nanoparticles has been proven useful in various applications. These include nanophotonic devices (73-74), high-density data storage techniques (75-77), biosensor (19), and surface-enhanced Raman spectroscopy (SERS) (78).

## 8.2 3. *Scattering of Evanescent Waves by Spherical Metallic Nanoparticles*

The scattering of evanescent waves by an individual GNP is shown in Figure 8.9. As assumed in the experimental setup, the core of a multimode optical fiber is simplified as a semi-infinite dielectric plane medium because the radius of

GNPs is much smaller than the core (1:100000). In addition, the plane medium and the surrounding medium are separated by a plane interface (*y-z* plane). The refractive indices of the core ( $n_{core}$ ) and the surrounding medium (water) of a GNP ( $n_w$ ) are set to be 1.492 and 1.33, respectively. The GNP, as a gold spherical nanoparticle with radius *r*, is placed at a distance *d* above the interface. The refraction index of the GNP is 0.166 + 3.15*i* at λ= 650 nm (72). The evanescent waves $\vec{E}^{s,p}_{inc}$ with *s*- or *p*-polarization from the interface of the fiber core are coupled with the localized surface plasmon of the GNP.

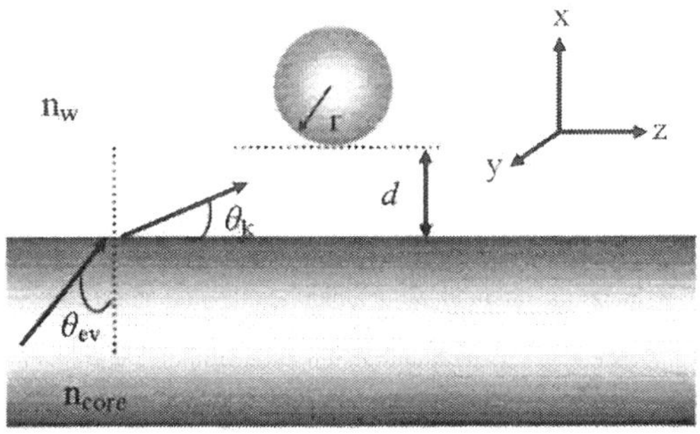


**Figure 8.9:** Scattering of evanescent waves by a gold nanosphere with radius r. The refractive indices of the core of a step-index multimode fiber ( $n_{core}$ ) and of the medium around the gold nanosphere ( $n_w$ ) are 1.492 and 1.33, respectively. The refractive index of the gold nanosphere is 0.166 + 3.15*i* at wavelength of 650 nm and it is placed at a distance d above the surface of the fiber.

This happens when total internal reflection of the plane waves propagating with various propagation constants β in the fiber occurs. $\theta_k$ is the angle between transmitted direction and the interface of the fiber core. The total electric field at the outside of a particle is the sum of the incident field and the scattered field and can be written as:

$$\vec{E}^{s,p}_{total} = \vec{E}^{s,p}_{scat} + \vec{E}^{s,p}_{inc} \tag{17}$$

The illuminated evanescent electric field with *s*- or *p*- polarization can be expanded in spherical coordinates by spherical Bessel functions of the first kind $j_n(\rho)$ and vector spherical harmonics $\vec{X}_{nm}(\theta,\phi)$ (68):

$$\vec{E}^{s,p}_{inc} = \sum_{n=1}^{\infty}\sum_{m=-n}^{n}\frac{i}{n_w k_w}\alpha^{s,p}_{TM}(n,m)\left[\nabla\times j_n(\rho)\vec{X}_{nm}(\theta,\phi)\right]+\alpha^{s,p}_{TE}(n,m)\,j_n(\rho)\vec{X}_{nm}(\theta,\phi) \tag{18}$$

Similar to the expression in Eq. (18), the scattered electric field of a GNP with *s*- or *p*- polarization in spherical coordinates is given by:

$$\vec{E}_{scat}^{s,p}=\sum_{n=1}^{\infty}\sum_{m=-n}^{n}\frac{i}{n_w k_w}a_n\alpha_{TM}^{s,p}(n,m)\left[\nabla\times h_n^{(1)}(\rho)\vec{X}_{nm}(\theta,\phi)\right]+b_n\alpha_{TE}^{s,p}(n,m)h_n^{(1)}(\rho)\vec{X}_{nm}(\theta,\phi) \tag{19}$$

where $a_n$ and $b_n$ are Mie coefficients (79) and functions $\alpha_{TM}^{s,p}$ and $\alpha_{TE}^{s,p}(n,m)$ are expansion coefficients. $\rho=k_w r$ and $k_w$ is the wave number of incident light propagates in water. $h_n^{(1)}(\rho)$ are spherical Hankel functions of the first kind. The vector spherical harmonics $\vec{X}_{nm}=-i\frac{(\vec{r}\times\nabla)Y_{nm}(\theta,\phi)}{\sqrt{n(n+1)}}$ as described by Jackson (80) and the $Y_{nm}(\theta,\phi)$ are scalar spherical harmonics (81). After some arrangements, the component expressions of the Eq. (19) are written as follows:

$$E_r=\sum_{n=1}^{\infty}\sum_{m=-n}^{n}-\frac{a_n\alpha_{TM}^{s,p}(n,m)n(n+1)}{n_w k_w\sqrt{n(n+1)}}\frac{h_n^{(1)}(\rho)Y_{nm}(\theta,\phi)}{r} \tag{20a}$$

$$E_\theta=-\sum_{n=1}^{\infty}\sum_{m=-n}^{n}\frac{1}{\sqrt{n(n+1)}}\left\{\begin{array}{l}\frac{a_n\alpha_{TM}^{s,p}(n,m)n(n+1)}{n_w k_w r}\frac{\partial Y_{nm}(\theta,\phi)}{\partial\theta}\frac{\partial}{\partial r}\left[rh_n^{(1)}(\rho)\right]\\ -ib_n\alpha_{TE}^{s,p}(n,m)\frac{h_n^{(1)}(\rho)}{\sin\theta}\frac{\partial Y_{nm}(\theta,\phi)}{\partial\phi}\end{array}\right\} \tag{20b}$$

$$E_\phi=-\sum_{n=1}^{\infty}\sum_{m=-n}^{n}\frac{1}{\sqrt{n(n+1)}}\left\{\begin{array}{l}\frac{a_n\alpha_{TM}^{s,p}(n,m)n(n+1)}{n_w k_w r\sin\theta}\frac{\partial Y_{nm}(\theta,\phi)}{\partial\phi}\frac{\partial}{\partial r}\left[rh_n^{(1)}(\rho)\right]\\ +ib_n\alpha_{TE}^{s,p}(n,m)\frac{h_n^{(1)}(\rho)\partial Y_{nm}(\theta,\phi)}{\partial\theta}\end{array}\right\} \tag{20c}$$

The expansion coefficients are found by solving the Maxwell's equations with boundary conditions at the surface of the sphere:

$$\alpha_{TM}^{s}(n,m)=2i^{n}n_{w}\sqrt{\frac{\pi(2n+1)(n-m)!}{n(n+1)(n+m)!}}\frac{mP_{nm}(\cos\theta_{k})}{\sin\theta_{k}}E_{0t}^{s} \tag{21a}$$

$$\alpha_{TM}^{p}(n,m)=2i^{n+1}n_{w}\sqrt{\frac{\pi(2n+1)(n-m)!}{n(n+1)(n+m)!}}\frac{dP_{nm}(\cos\theta_{k})}{d\theta_{k}}E_{0t}^{p} \tag{21b}$$

$$\alpha_{TE}^{s}(n,m)=2i^{n+1}\sqrt{\frac{\pi(2n+1)(n-m)!}{n(n+1)(n+m)!}}\frac{dP_{nm}(\cos\theta_{k})}{d\theta_{k}}E_{0t}^{s} \tag{21c}$$

$$\alpha_{TE}^{p}(n,m)=-2i^{n}\sqrt{\frac{\pi(2n+1)(n-m)!}{n(n+1)(n+m)!}}\frac{mP_{nm}(\cos\theta_{k})}{\sin\theta_{k}}E_{0t}^{p} \tag{21d}$$

where $E_{0t}^{s}$ and $E_{0t}^{p}$ are the complex transmitted amplitude of the refracted light with *s*- and *p*- polarization, respectively. $P_{nm}(\cos\theta_{k})$ refers to associated Legendre functions. The electric field around the nanoparticle can be calculated analytically from Eqs. (20a) to (20c). The intensity of the electric field directly affects the intensity of fluorescence signals of LSPCF biosensor.

## 8.2 5. *Fluorescence and Enhanced Local Field of Metallic Nanoparticles*

Consider an incident light propagate in a multimode fiber with various propagation constants $\beta$ ($\sin\theta_{inc}$), and the range of $\beta$ is from 0.0 to NA, where NA is the numerical aperture of the focusing laser beam and is set to be 0.45 in the calculations in this study. The section of the decladded optical fiber is shown in Figure 8.10. The decay length of the evanescent wave is calculated and shown in Figure 8.11, when the incident light propagates in the fiber with different $\beta$ at $\lambda$= 650 nm. The decay length of the evanescent waves from the interface will increase with a larger $\beta$ and the range of the decay lengths will be 153 nm to 205 nm. Figure 8.11 illustrates that the decay lengths are much longer than the distance between the interface of the fiber core and the bound GNPs. Therefore, all bound nanoparticles are illuminated under strong enough evanescent fields.

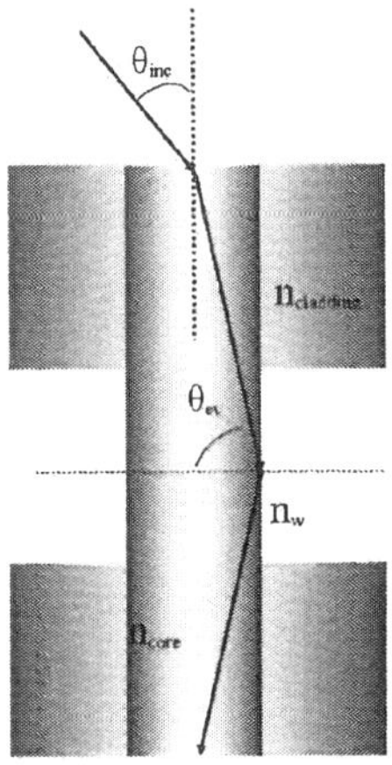


**Figure 8.10:** Section of a decladded fiber. Indices of refraction of the core of the fiber and surrounding medium (water) and 1.492 and 1.33, respectively.

When a GNP is illuminated by evanescent waves, the averaged scattered electric field in the vicinity of the GNP can be obtained by:

$$\left\langle \vec{E}^{p}_{scat} \right\rangle = \frac{1}{NA}\int_{0}^{NA} \vec{E}^{p}_{scat}(\beta)d\beta \tag{22}$$

Only $p$-polarized scattered electric fields $\vec{E}^{p}_{scat}$ are considered in the calculations because the near-field intensity of $s$-polarized $\vec{E}^{s}_{scat}$ of a GNP is much weaker than that of $\vec{E}^{p}_{scat}$. The total averaged electric field outside of the nanoparticle can be given by:

$$\left\langle \vec{E}^{p}_{total} \right\rangle = \frac{1}{NA}\int_{0}^{NA} \left[\vec{E}^{p}_{scat}(\beta) + \vec{E}^{p}_{inc}(\beta)\right]d\beta = \left\langle \vec{E}^{p}_{scat} \right\rangle + \left\langle \vec{E}^{p}_{inc} \right\rangle \tag{23}$$

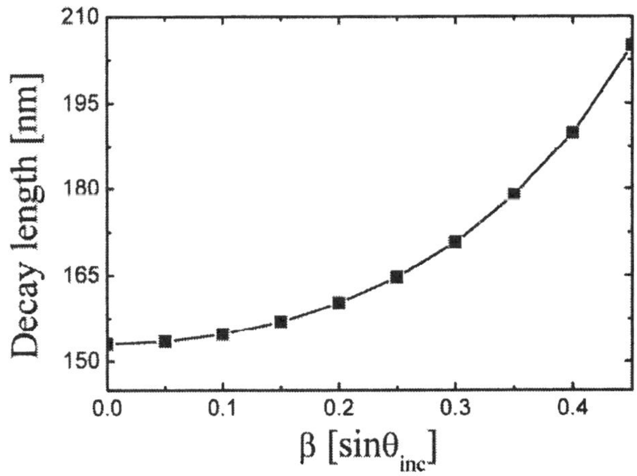


**Figure 8.11:** Decay lengths of an evanescent wave as a function of $\beta$ at t λ= 650 nm. The range of the decay length is changed from 153 to 205 nm.

The fluorescence signals from the bottom half surface of a GNP are screened by the GNP itself, so only the field intensity on top half surface of the GNP is considered in calculations. The field-intensity ratio of the averaged-field intensity of a GNP to the averaged-field intensity of evanescent waves (without GNP) at a distance $r$ can be written as follows:

$$\frac{\langle I_{ev}\rangle}{\langle I_0\rangle}=\frac{\int_{-\pi/2}^{\pi/2}\int_0^{\pi}\left|\langle \vec{E}_{scat}^{p}\rangle+\langle \vec{E}_{inc}^{p}\rangle\right|^2 \sin\theta d\theta d\phi}{\left|\langle \vec{E}_{inc}^{p}\rangle\right|^2} \qquad (24)$$

where the reference $\langle I_0\rangle$ is calculated at a distance $d$ above the interface of the fiber.

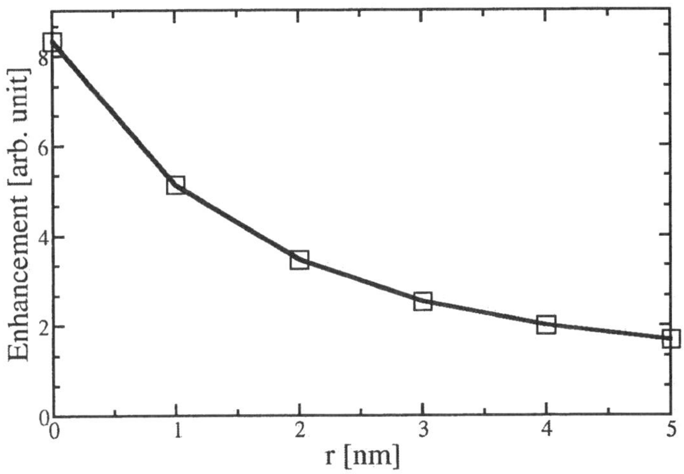


**Figure 8.12:** Averaged-field enhancements on the top half surface of a gold nanosphere as a function of the distance *r* from the surface of the nanosphere.

Although the strongest field intensity appears at the surface of metallic nanoparticles, fluorophores are not placed too close to the surface due to the effect of fluorescence quenching (64, 82-83). Therefore, the influences of the distance *r* from the surface of metallic nanoparticles on averaged-field enhancements should be discussed. Using Eq. (24), the averaged-field enhancements on the top half surface of a gold nanosphere as a function of the distance from the surface are shown in Figure 8.12. The averaged-field intensity is enhanced for about 8 times of the averaged-field intensity of incident evanescent waves on the top half surface of the nanosphere. The intensity decreased dramatically with increasing *r* only in a few nanometers. Compared with the experimental results, the magnification of the near-field intensity of GNP is apparently much weaker than the order of magnitude as there is an increase in fluorescence intensity from the GNP suspensions. This is possibly due to the geometric shape of GNPs, but this is not included in the analytical study. It should be noted that the geometric shape of a realistic GNP is not possibly a perfect sphere; rather, it resembles more an irregular shape. The local-field enhancement of a realistic GNP will be stronger a fewer times than that of the GNP with a perfect spherical shape. This can be attributed to the nanoparticle having complicated geometrical shape that can provide a complex curved surface. The dramatically local-field enhancements can also be generated by the bounded charge oscillation on that surface. However, the first order approximation based on the scattering of evanescent waves of spherical particles can also provide a qualitative understanding for the mechanism of fluorescence signal enhancements of LSPCF fiber-optic biosensor. Using various electromagnetic computational techniques, the averaged-field intensity of realistic nanoparticles with irregular shape can be calculated numerically when its geometric shape is identified. Therefore, more reasonable calculated results can be obtained as compared with the analytical results of spherical particles.

## 8.2 6. *Fluorescence and Absorption of Metallic Nanoparticles*

In the experiments, fluorescence signals cannot only be enhanced by GNPs but can also be absorbed. Therefore, the absorption efficiency of GNPs was considered in this study. As an example, Figure 8.13 shows the absorption efficiency of plane waves by a gold spherical nanoparticle with diameter of 20 nm in water ($n_w = 1.33$) as a function of illuminating wavelength.

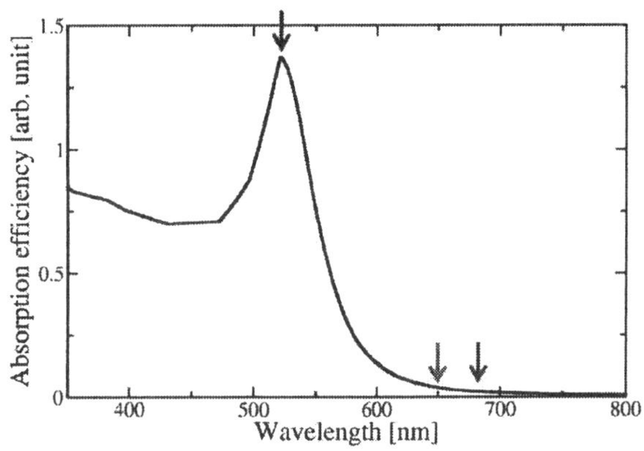


**Figure 8.13:** Absorption efficiency of a gold nanosphere with diameter of 20 nm in the water as a function of the illuminating wavelength.

The plasmon resonant wavelength is about 520 nm (red arrow). The emission wavelength of fluorophores is about 680 nm (blue arrow) when the illuminating wavelength is 650 nm (green arrow).

The cases for evanescent waves are not discussed because the trend of the spectrum for plane waves in Figure 8.13 is similar to the cases for evanescent waves when the particle with $r << \lambda$ (69). The plasmon resonant wavelength of a GNP in water is about 520 nm (the position is denoted by a red arrow in Figure 8.13). Although the exactly effective refractive index of the conjugated forms cannot be found, the best conjecture is in its peak wavelength. This bears close similarity with the case of a GNP in water because the refractive index of protein **A** molecules on the GNPs surface is very similar to that of water. It is noted that the selected frequency of the incident light in the experiments is not the resonant frequency of GNPs for the reason that the GNPs at resonant frequency exhibit strong light absorption (see Figure 8.13). Another consideration is that the fluorescence signals are absorbed strongly by GNPs simultaneously when the emission wavelength of fluorophores is close to the surface plasmon resonant wavelength. To avoid the resonant wavelength of GNPs, the laser light source with wavelength of 650 nm was used in obtaining the fluorescence signals with emission wavelength of 680 nm (the position is denoted by a blue arrow in Figure 8.13). Generally, local-field enhancement around the GNPs at off-resonance (at λ = 650 nm) state is obviously weaker than that at λ= 520 nm. However, the averaged field intensity around a GNP at λ= 650 nm can still be enhanced from 8 to 2 times of the field intensity without nanoparticles when the distance between fluorophores and the surface of nanoparticles is changed within 5 nm (See Figure 8.12).

## 8.2 7. *Localized Surface Plasmon Coupled Fluorescence*

Localized surface plasmon resonance (LSPR) at the metal surface has been exploited to enhance the signal obtained from optical biochips and thereby lower the limits of detection. There are two main enhancement factors: (i) an increase in the excitation of the fluorophore by localizing the optical field on the nanoparticles near the fluorophore; and (ii) an increase in quantum efficiency of the fluorophore. The plasmon resonance wavelength should coincide with the fluorophore absorption band to obtain the maximum emission efficiency. Several parameters concerning the signal detection enhancement are as follows: (84)

1. Shape of the nanoparticles
2. Size of the nanoparticles
3. Metal type of the nanoparticles
4. Polarization of the excitation light
5. Flourophore absorption band
6. Quenching
7. Photobleaching

**Shape:** The radiative emission from molecules confined within metallic nanocavities and on the surface of nanoparticles is of great relevance to biotechnology. In 1986, it has been suggested that fluorescence enhancement and reduced observation volumes could be obtained from small metal apertures (85). Nanocavities of different shapes could induce different surface plasmon (SP) fields. More recently, some studies has been done for different shapes, such as circular (86–90), elliptical (91), coaxial (92), or rectangular (93, 94) metallic nanocavity(95). In 2003, single-molecule detection from a nanocavity was demonstrated (86). However, it might be difficult to position the biospecies in the nanocavities.

Two-dimensional theoretical study has been done for the circular, triangular, and square silver nanowires. Results show that the resonance spectrum strongly depends on the particle shape. The higher the particle symmetry, the simpler the spectrum. For instance, a small cylindrical particle exhibits only one resonance, whereas a square has two while an equilateral triangle at least three distinct resonances. It was determined that the strongest enhancement for nanowires is that with dimensions smaller than 50 nm. Right-angled triangular nanowires can produce field amplitude exceeding 1000 times than that of the incident field at short distances from the surface (96).

The optical properties of spherical, triangular, and ellipsoidal metallic nanoparticles have been reported as illustrated in Figure 8.14. Comparisons with experiment show that the classical electromagnetic theory works well indicating that the complex dielectric environment can be properly characterized and modeled (97, 98).

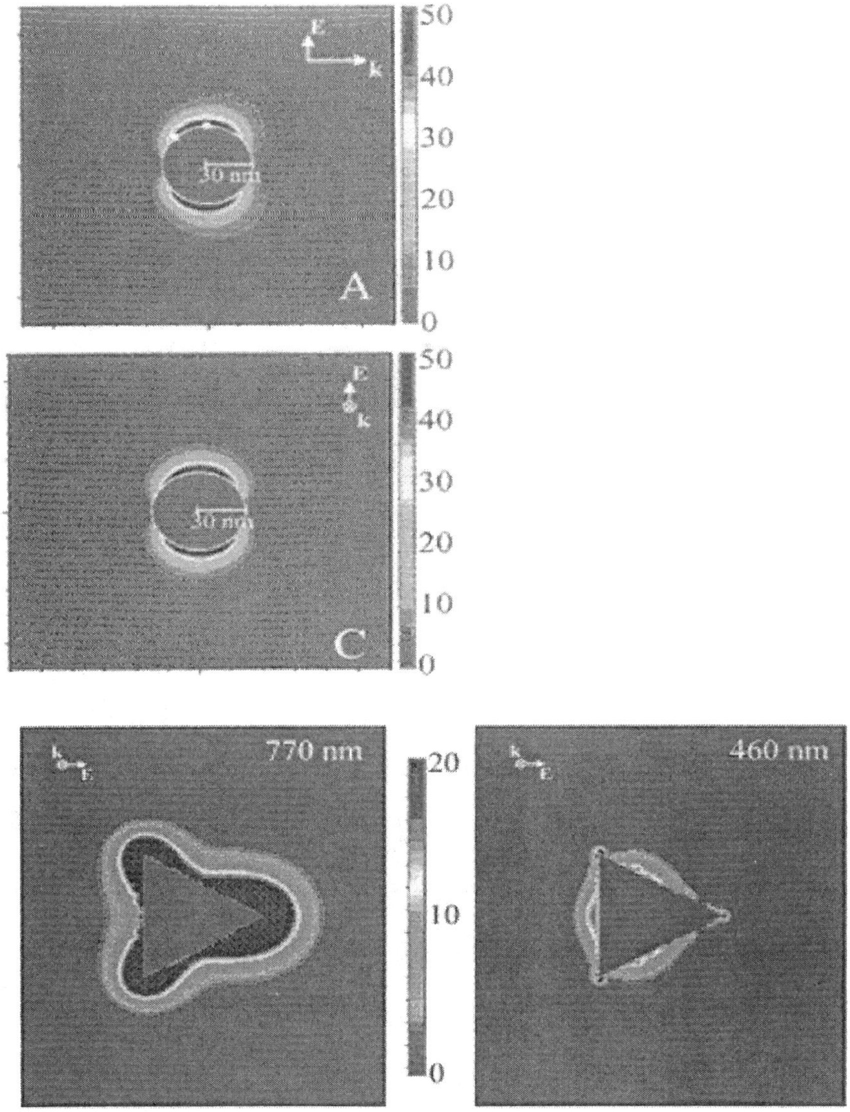


**Figure 8.14:** Optical field distribution of spherical and triangular silver nanoparticle for different polarizations. (97)

It is noted that the optical properties of nanoparticles are generally assumed to be sufficiently dispersed and that they may be treated as isolated. However, in most practical situations, particle interactions are important, and sometimes, they are dominant.

**Size:** Spherical gold/silver alloy nanoparticles, surrounded by a silica spacer shell, to which is attached a fluorescent ruthenium dye, have been studied. The dependence of the fluorescence enhancement on nanoparticles diameter was likewise investigated. By tuning the composition of gold/silver, the plasma frequency is also tuned. On the other hand, tuning the size of the nanoparticles makes it possible to tune the resonant wavelength of the SP. The SPR wavelength shifts to longer wavelength as the size of the nanoparticles increases. It was found that the maximum fluorescent enhancement with a factor of 4 can be obtained as the size of the gold/silver nanoparticles reaches around 27nm. (84)

**Metal type of the nanoparticles:** The plasmon frequency for most metals corresponds to that of an ultraviolet photon. For silver, gold, alkali metals, and a few other materials, the plasmon frequency is low compared to that of a visible or near-ultraviolet range indicating the possibility of exciting plasmon by light.

**Polarization of the excitation light:** In Figure 8.14, it can be observed that the field distribution of the SP on symmetrical spherical nanoparticle for different polarizations is identical. However, if the form of the nanoparticles is not symmetrical, such as a triangle, its field distribution is not identical for different polarizations. In this case, the polarization of the excitation source plays an important role to efficiently excite the SP field.

**Fluorophore absorption band:** To obtain maximum emission efficiency, the wavelength of the excitation light source should coincide both with the SPR wavelength of the nanoparticles and with the fluorophore absorption band of the dye. In addition, the emission wavelength should be far from the absorption band of the nanoparticles.

**Quenching:** At shorter distances, ranging from few nanometers to the physical contact with the metallic structure, a mechanism tends to increase the total decay rate. This effect, which is responsible for fluorescence quenching, is due to the absorption of fluorescence photons in the metallic structure itself (99). Another effect is based on interactions of the fluorophore with free electrons in the metal, wherein the plasmon absorption leads to lower fluorescent emission efficiency (100). Theoretical study asserts that the optimized distance between the excitation source and the fluorophore is around 2-5 nm (99, 101, 102). Nanoparticles coated with a thin shell (e.g. silica, 5nm in thickness) and the dye attached to the dielectric shell could overcome quenching effects (84, 103). The quenching effect can also be found in the quantum dot / GNP system (104). It is noted that as the concentration of fluorophore is high, the self-quenching effect should also be considered. (100)

**Photobleaching:** The decay rate of the fluorophore is around several nanoseconds (ns). By using the SP, the decay time has been measured through time-resolved fluoluminescence. First, pulsed laser excites the SP and the fluorophore. This way, the decay of the intensity of the fluorophore can be measured. The decay time is around 2.33 and 0.61ns for non-nanoparticles and nanoparticles cases, respectively. The lifetime of fluorophores near metal particles can influence their photostability. A shorter lifetime results in a smaller time for photochemistry, while in the excited state, more excitation-emission cycles occur prior to photobleaching (105). The photobleaching of the fluorophore can be reduced more than 30-fold using metallic nanoparticles. The intensity of fluorophore can be enhanced around 50-fold. (106, 107)

## 8.2. 8 *Evanescent Wave in Waveguides*

The evanescent wave can be produced on the waveguides of the integrated optics devices. The materials of the waveguides may be silicon, $SiO_2$, lithium niobate ($LiNbO_3$), gallium arsenide (GaAs), among others. In this section, the design of waveguide and related simulation techniques are discussed. To calculate the evanescent wave on single mode or multimode waveguides, the effective index method (108) serves as a fast and easy way to estimate the field distribution. Nanoparticles on the surface of the waveguide can be simulated using the Finite-Difference Time-Domain (FDTD) method (108). This method can provide the three-dimensional field distribution of the electromagnetic waves. The amplitude and the polarization of the localized field on the metallic particles excited by the evanescent wave can be observed as well. This information is useful in estimating the best distance between the metallic particles and the florescent dye.

In a waveguide, the polarization of the electric field parallel to the sample surface is called the Transverse Electric (TE) mode. The polarization of the magnetic field parallel to the sample surface is called Transverse Magnetic (TM) mode. Figure 8.15 shows the schematic drawing of a waveguide embedded in a substrate and the corresponding polarization definition of the light.

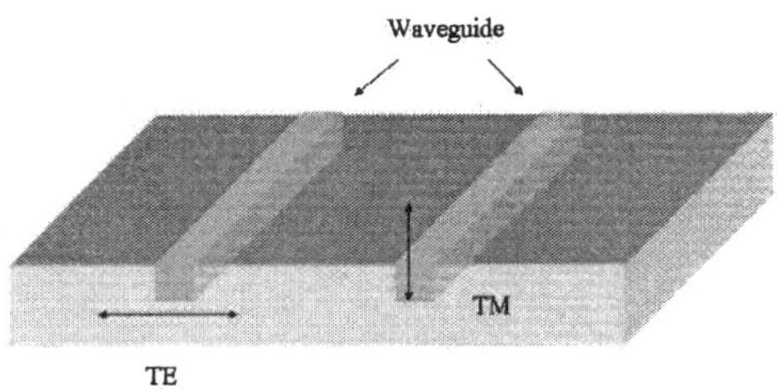


**Figure 8.15:** Polarization definition in waveguide of integrated optics.

The effective index method can provide information on the optical field distribution for different polarizations and the effective refractive index of the light in waveguides. The waveguide structure is limited to consist of three layers as in this example: air, silicon, and $SiO_2$ denoted Layer I, Layer II, and Layer III, respectively. The refractive index of Layer I ($n_1$) and Layer III ($n_3$) should be less than that of Layer II ($n_2$). The light is guided in Layer II where the thickness is $W$.

In Layer II, the optical field distribution is assumed to be a cosine function. In Layers I and III, the optical field distribution is assumed to be exponential decay.

For the TE mode, the electric field distribution in the three layers can be expressed as:

In Layer I

$$E_y(x) = C_1 \exp(\gamma_1 x) \tag{25}$$

Where $\gamma_1 = k_0\sqrt{n_{eff}^2 - n_1^2}$

In Layer II

$$E_y(x) = C_2 \cos(\gamma_2 x + \alpha) \tag{26}$$

where $\gamma_2 = k_0\sqrt{n_2^2 - n_{eff}^2}$

In Layer III

$$E_y(x) = C_3 \exp[-\gamma_3(x - W)] \tag{27}$$

where $\gamma_3 = k_0\sqrt{n_{eff}^2 - n_3^2}$

$k_0$ (=2□/□=□□c) is the propagation number. □□and□□□□are the wavelength of the light in vacuum and the frequency of the light, respectively.

By solving the following equation to find $n_{eff}$ for q=0, 1, 2, and so on, respectively, the corresponding optical field for different modes can be obtained.

$$\tan^{-1}\left(\frac{\gamma_2}{\gamma_1}\right) + \tan^{-1}\left(\frac{\gamma_2}{\gamma_3}\right) + \gamma_2 W - (q+1)\pi = 0 \tag{28}$$

where $q$=0, 1, 2,...

For the TM mode, the magnetic field distribution in the three layers can be expressed as follows:

In Layer I

$$H_y(x) = C_1 \exp(\gamma_1 x) \tag{29}$$

where $\gamma_1 = k_0\sqrt{n_{eff}^2 - n_1^2}$

In Layer II

$$H_y(x) = C_2 \cos(\gamma_2 x + \alpha) \tag{30}$$

where $\gamma_2 = k_0\sqrt{n_2^2 - n_{eff}^2}$

In Layer III

$$H_y(x) = C_3 \exp[-\gamma_3 (x - W)] \tag{31}$$

where $\gamma_3 = k_0 \sqrt{n_{eff}^2 - n_3^2}$

$k_0$ (=2□/□=□□c) is the propagation number. □□and□□□□are the wavelength of the light in vacuum and the frequency of the light, respectively.

By solving the following equation to find $n_{eff}$ for q=0, 1, 2, and so on, respectively, the corresponding optical field for different modes can be obtained.

$$\tan^{-1}\left(\frac{\varepsilon_{r1}}{\varepsilon_{r2}}\frac{\gamma_2}{\gamma_1}\right) + \tan^{-1}\left(\frac{\varepsilon_{r3}}{\varepsilon_{r2}}\frac{\gamma_2}{\gamma_3}\right) + \gamma_2 W - (q+1)\pi = 0 \tag{32}$$

where $q$=0, 1, 2,... , $\varepsilon_{r1} = n_1^2$ , $\varepsilon_{r2} = n_2^2$, and $\varepsilon_{r3} = n_3^2$

Figure 8.16(a) presents the normalized field distribution in a silica waveguide. The top layer is water where the refractive index is $n_1$=1.33. The refractive index of the waveguide is $n_2$=1.49, while its layer's thickness is 2μm. The bottom layer is silica with lower refractive index ($n_3$=1.47). The wavelength of the light is 658nm. The electric field of the evanescent wave is stronger for the TE mode than the TM mode at around 23%. The intensity of the evanescent wave [Figure 8.16(b)] is stronger for the TE mode than the TM mode at around 52%. The depth of the evanescent wave in water is around 500nm. Therefore, if the nanoparticles are located on the surface of the waveguide, the evanescent wave may be coupled on the surface of the nanoparticles and can excite the SPs.

The optical field distribution, *E*, can also be obtained by solving the Helmholtz equation using finite-difference method. (108)

$$\nabla_\perp^2 E + k_0^2(\varepsilon_r - n_{eff}^2)E = 0 \tag{33}$$

where $\nabla_\perp^2 = \partial^2/\partial x^2 + \partial^2/\partial y^2$

Figure 8.17 shows the electric field and the intensity of the light guided in the silica planar waveguide. The thickness of the waveguide layer is 5μm. The wavelength of the light is 632.8nm. Given these, the optical field for different orders can be obtained. It can be observed that the evanescent wave is stronger for higher-order mode. This result reveals that the field coupled by the nanoparticles near the surface of the multimode waveguide may be higher than that on the single mode waveguide.

(a)

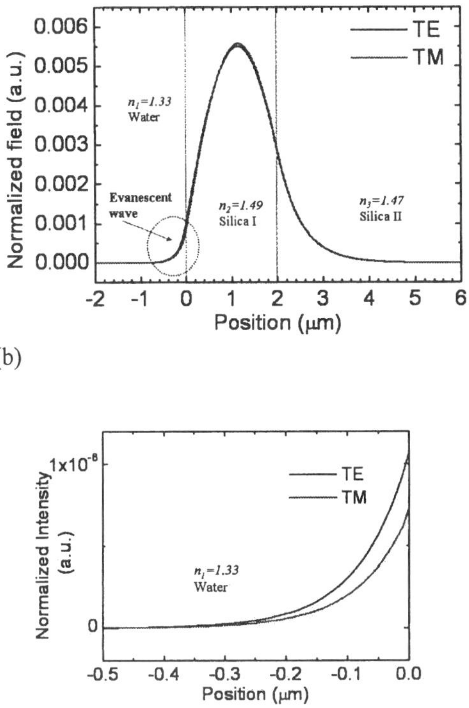


**Figure 8.16:** (a) Normalized field (b) Intensity distribution in a silica waveguide.

The FDTD method can be applied to obtain the optical field distribution and the propagation of the electromagnetic wave. After placing a circular GNP in water, the refractive index of GNP and water is measured at 0.19+3.58i and 1.33, respectively. The diameter of the rod is 40nm. Using the two-dimensional FDTD, the nanoparticle is assumed to be a gold cylindrical rod with infinite length. The magnetic field of the electromagnetic wave parallel to the cylindrical rod can excite the SP. In Figure 8.18(a), the SP on the rod surface can be seen. The SP radiates a dipole mode in water. As the nanorod is positioned near the surface of a silica waveguide (n=1.49), the former can be excited by the evanescent wave of the waveguide. The field distribution coupled between the silica waveguide and the nanoparticle is illustrated in Figure 8.18(b). The distance between the nanorod and the waveguide is 20nm. It can be gleaned that the optical energy is localized on the surface of the rod. The decay time of the energy was measured as shown in Figure 8.19. A quality factor (Q), of 4, was obtained using this equation.

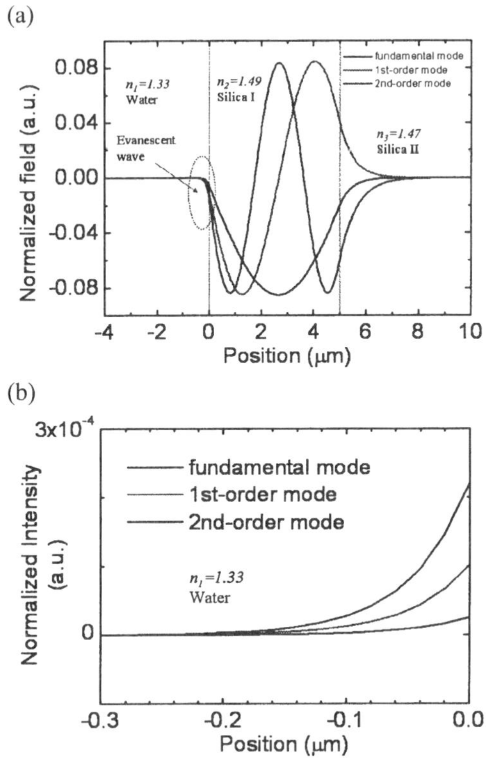


**Figure 8.17:** (a) Optical field of light in waveguide and (b) Intensity distribution of the evanescent wave on the waveguide.

$$U(t) = U(t_0)\exp\left[\frac{-\omega(t-t_0)}{Q}\right] \quad (34)$$

where $U$, $t$, $t_0$, □□, $Q$ are the optical energy, time, initial time, frequency, and quality factor, respectively.

The higher the quality factor, the longer the energy localizes on the surface of the rod. The fluorophore can interact longer with optical energy and can be excited longer, thereby leading to a stronger output signal emission.

(a)

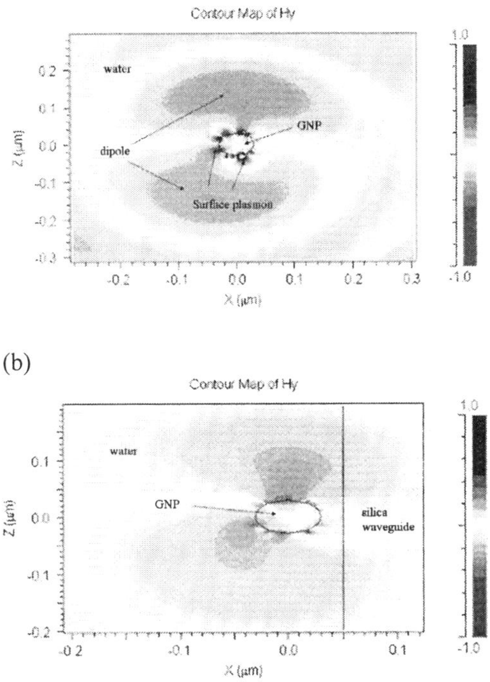


**Figure 8.18:** (a) The magnetic field distribution on the GNP in water. (b) The magnetic field distribution on the GNP near a silica waveguide embedded in water.

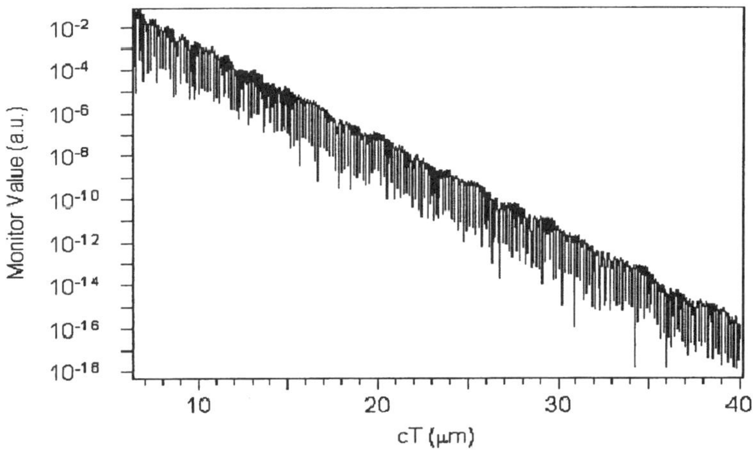


**Figure 8.19:** Decay of the energy on the surface of the nanoparticle where c and T is light velocity in vacuum and escaped time, respectively.

In this study, a two-dimensional FDTD is performed to simulate the optical field on the metallic nanorod. This method also saves calculation time. In calculating the optical field on a nanoparticle, three-dimensional FDTD should be performed. In this case, the SP excited by both TE and TM modes of the waveguide can be studied. However, PC cluster should also be used to have enough memory to launch the simulation.

Q is the main factor to predict the output signal of the LSPCF. Several parameters can tune the value of Q, such as wavelength, diameter of nanoparticle, shape of the nanoparticle, metal type of the nanoparticle, distance between the nanoparticle and the waveguide, polarization, and so on. A detailed and more thorough study using three-dimensional FDTD can prove helpful in choosing the experimental parameters that can obtain higher detection sensitivity.

## 8.3 REQUIRED EXPERIMENTAL TECHNIQUES

### 8.3 1. *Preparation of Optical Fiber Prior to Its Use in Sensor Development De-cladding*

The cladding layer is a protective coating surrounding a glass fiber. It plays an important role in preventing glass core surface of the optical fiber against mechanical and chemical damage. Before the fabrication of fiber-based optical device, such as fiber Bragg gratings, optical fiber couplers, fiber optic-based sensors, connectors, and for splicing and cleaving purposes, it is necessary to remove the protective coating. However, the de-cladding procedure often causes substantial degradation of an optical fiber's strength. Due to the fact that optical fiber strength is one of the key factors that determine the lifetime of fabricated fiber-based devices, it is desired to utilize an appropriate coating removal process that does not seriously affect the original fiber strength (109).

The use of a mechanical blade is a simple and handy means to take off the jacket from the optical fiber without chemical pretreatment. However, this often causes severe degradation of the strength and subjects the optical fiber to a short lifetime because the mechanical contact between the stripper and the optical fiber creates microcracks on the glass core surface. On the other hand, non-contact coating removal involves putting the polymer coating of the optical fiber in the gaseous state through hot air stream. Without any chemical agents utilized in the process, this technique offers a clean alternative that makes the rinsing process unnecessary. In the case of the non-contact laser-assisted jacket removal technique, high power pulsed UV laser sources with wavelengths between 270–300 nm, such as frequency-doubled copper laser, are utilized. This method offers complete removal of the protective polymer coating from the optical fiber without any crack on the surface of the optical fiber in the pristine state. It also has the advantage of preserving the strength and lifetime of the fiber. Still, this method is not a popular alternative in practical use due to several factors: non-portability of the laser stripper, the need for precise optical alignment, and the high-cost required for the process.

Immersion of the optical fibers in the hot sulfuric acid ($H_2SO_4$) bath at temperatures between 180 °C and 220 °C enables the dissolution of the polymer in the coating (109). Some other chemical de-cladding methods are summarized in Table 8.1.

**Table 8.1:** Summary of studies using chemical treatment as the de-cladding method

| **Cladding Materials** | **Core Materials** | **Decladding Chemical Method/Reagent** | **References** |
|---|---|---|---|
| Urethane acrylate polymer | Silica dioxide | Carbon tetrafluoride and oxygen plasma | Company informationa[110] |
| Polyimide | Silica dioxide | 95-98% sulfuric acid/ 150°C | Alfred & Josephine (2006) [111] |
| Fluorinated PMMA | Pure PMMA | Ethyl acetate | Chang et al. (2007)[112] |

## 8.3.1 1. *Cleaning*

Depending on the type of contaminates, various cleaning processes could be employed after the de-cladding process. Standard wash procedure is applied to remove finger grease, adventitious organic deposits, and/or cutting oils (70). This procedure is outlined as follows: (i) Immerse in boiling trichloroethylene (TCE) for 1 min; (ii) Immerse in boiling acetone for 1 min; (iii) Ultrasonicate in isopropyl alcohol

(2-propanol) for 1 min; (iv) Wash in flowing DI water for 1 min, and lastly, (v) Dry in a filtered nitrogen stream.

In order to remove trace organics, the physically adsorbed monolayers of adventitious organics (self-assembled organic layers that are not "covalently bonded" to the surface, such as alkane thiols and silane layers) and some chemically adsorbed monolayers of organics that are not removable by standard wash procedure, necessary actions should be taken. These include the following: immersing optical fiber in a (5:1:1) solution of H2O : NH4OH : H2O2 maintained at 75 80 °C for 10 s, and followed by rinsing under running DI water for 1 min (113).

Immersion of de-cladded optical fiber in a solution of $H_2O$ : HCl : $H_2O_2$ (6:1:1) maintained at 75 – 80 °C for 5 min, followed by washing off the solution under running DI water for 1 min and in DI water for 5 more min proved effective in the removal of heavy metal ions that may have existed as traces adsorbed to the metal or metal oxide surface (114). In addition, UV-Ozone could be used to remove monomolecular layers of chemically bound organics on metallic and oxide surface. On the other hand, oxygen (O2) or argon (Ar) plasma is suitable for taking off most organic films, such as micron-level thick films of various organic materials, including photoresists. Wang's group (2007) (115) and Chou's group (2007) (70) reported that PMMA fiber could be cleaned in an ultrasonic ethanol bath for 5 min.

## 8.3.1 2. *Surface modification of optical fiber*

Surface modification of optical fiber can be categorized into two groups: (i) physical adsorption (non-covalent bonding), and (ii) covalent bonding.

There are several binding forces involved in physical adsorption, namely, hydrophobic interactions, ionic interactions, hydrogen bonds, and van der Waals forces. One of the advantages of physical adsorption is that no reagents and only a minimum of activation steps are required. However, the stability of the adsorbed layer is usually weaker than in the case of covalent bond. Desorption of the ligand resulting from changes in temperature, pH, or ionic strength is also often observed (116).

The formation of covalent bonds between the protein or other biologically important molecules and the support matrix is the most frequently studied immobilization technique (*e.g.*, immobilization in the presence of carbodiimides, cross-linking by glutaraldehyde or cyanogens bromide activation of the support material). It was observed that covalent binding to the activated support led to minimization of the leakage of the immobilized substance. Proteins usually have a number of potential immobilizing sites, which correspond to particular functionalities on the molecules. Lastly, unreacted active groups of solid support must be blocked by reaction with inert moieties provided that the same group was used for ligand immobilization (116).

## 8.3.1 3. *PMMA*

### 8.3.1.3 1. *Physical adsorption*

In 2007, Chou's group (2007) (112) conducted an investigation on the physical adsorption of anti-β-D-glucuronidase (GUD) antibodies onto the unclad region of the PMMA fiber core. This was done by incubating the fiber in the antibody solution (10 μg/mL) at 37 °C for 1 h, followed by the fluorescence detection using photomultiplier tube or PMT.

### 8.3.1.3 2. *Covalent bonding*

PMMA itself does not possess ready-to-use functional groups for covalent binding with biological molecules. The amine-terminated PMMA were often produced by immersing the freshly cleaned PMMA substrate into a 1.0 M ethylenediamine in dimethyl sulfoxide (DMSO) solution for 15 min at room temperature (115); or coated with a thin layer of polyethyleneimine (PEI) or polyallylamine hydrochloride (PAH). This was first treated in 1 N sodium hydroxide (NaOH) solution at 55 °C for 30 min and then immersed in a PEI or PAH solution (0.2%, pH 7) at room temperature for 1 h (117). Tsai and Lin (2005) demonstrated that PEI-derivatized PMMA was used for the determination of alpha-fetoprotein by quartz crystal microbalance (QCM) (118). Furthermore, the amine-terminated PMMA could be generated by reacting with 10% hexamethylene diamine (HMD) (reaction shown in scheme 8.2) or 1,3-diaminopropane (DAP) in 100 mM borate buffer pH 11.5 for 2 h (119) or exposing to *n*-lithioethylenediamine (120) (reaction shown in scheme 8.3).

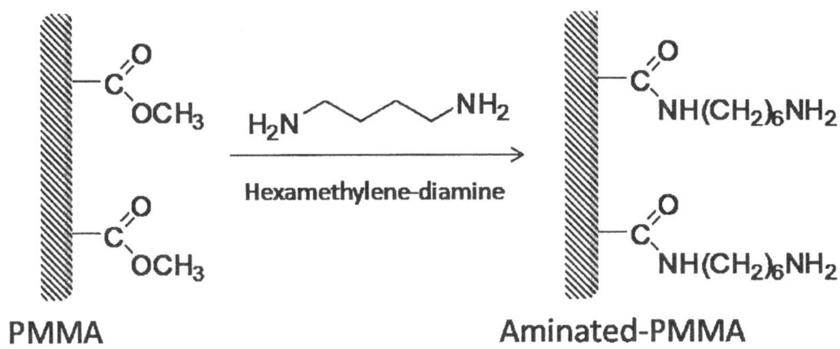


**Scheme 8.2:** The production of amine-terminated PMMA (120)

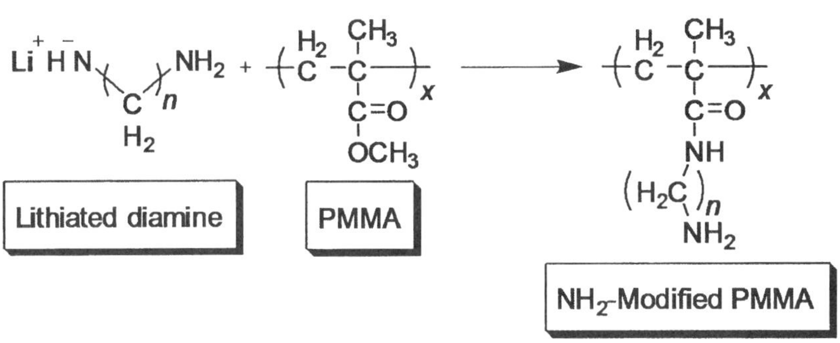

**Scheme 8.3**: Synthesis of amine-modified PMMA, which was initiated by purging diamine with nitrogen for 20 min before the introduction of *n*-butyllithium to form *N*-Lithioethylenediamine (120).

The homobifunctional cross-linker, glutaraldehyde (GA), can be used for conjugating the molecules containing primary amine groups with amine-terminated PMMA. This is where the formation of Schiff bases with possible rearrangement to a stable product or through a Michael-type addition reaction occurred. The reaction takes place at points of double-bind unsaturation created by polymerization of the GA in solution (121-122) (reaction shown in scheme 8.4).

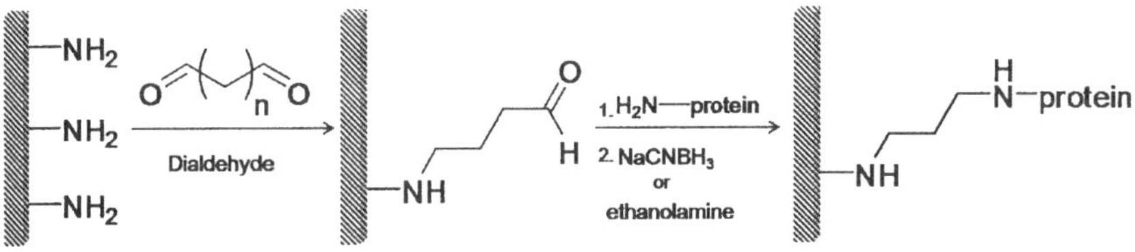


**Scheme 8.4**: Conjugation of amine-containing molecules onto the amine-terminated PMMA

Bulmus et al. (1997) demonstrated the modified PMMA mono-size microbeads for glucose oxidase immobilization, in which PMMA substrate was washed with 10% (w/v) NaOH followed by 50% (v/v) ethanol. Freshly cleaned PMMA was then immersed in a solution of 1 g/L polyvinyl alcohol for 20 min, followed by oxidation reaction with a solution of 1% NaIO4 for 1 h at room temperature. The amino functionality was then added by using a solution of 10% (w/v) hexamethylene diamine in 100 mM borate buffer pH 11.5, for 2 h (123) (shown in scheme 8.5).

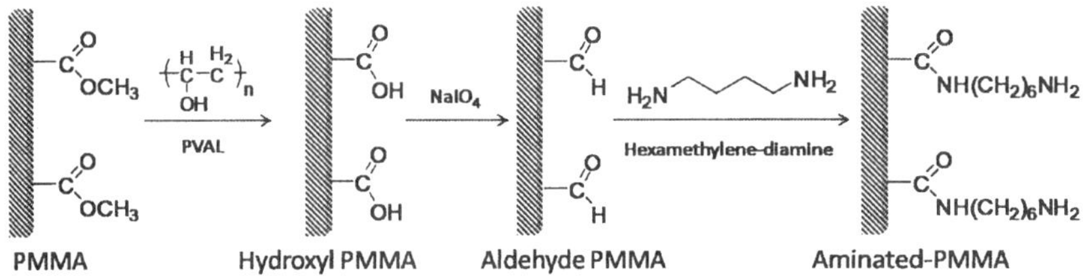


**Scheme 8.5**: Preparation of aminated PMMA via modification of PVAL

On the other hand, the carboxylate-terminated PMMA can be produced by hydrolyzing PMMA in sulfuric acid to form carboxylic acid functional groups on the surface. Proteins or other amine-containing molecules can then be immobilized on the COOH-terminated PMMA surface using carbodiimide as a linker. EDC (1-ethyl-3-[3-dimethylaminopropyl]carbodiimide hydrochloride) is a commonly used carbodiimide linker, which reacted with COOH-terminated PMMA, forming an amine-reactive *O*-acylisourea intermediate. This intermediate may react with an amine-containing molecule, yielding a PMMA conjugate by a stable amide bond (124-126) (reaction shown in scheme 8.6).

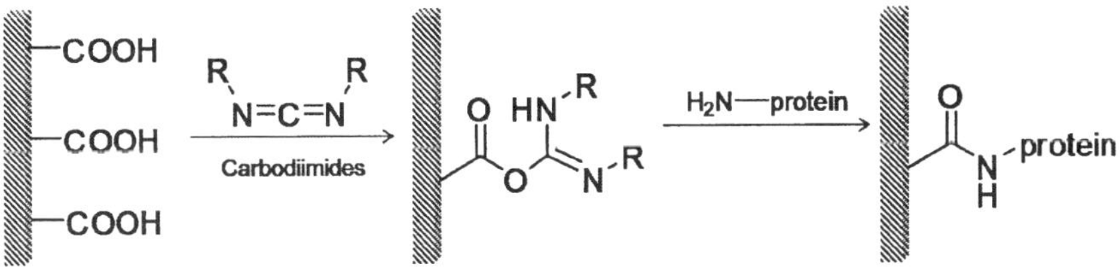


**Scheme 8.6**: Conjugation of amine-containing molecules onto the carboxyl-terminated PMMA

## 8.3.1.3 3. *Silica dioxide*

The silylation process is commonly the first step in modifying silica dioxide surface (127). Silanes may interact with glass fiber surfaces initially through hydrogen bonding, with subsequent condensation and lateral reactions generating siloxane structures. The siloxane film formed on the glass fiber surface may consist of multiple layers. Two factors which may influence the structure of the silane coupling agent inter-phase are (i) the pH of the solution, and (ii) the drying conditions employed. Basic or acidic conditions increase the rate of hydrolysis and condensation of the silane. This will in turn increase the amount of silane adsorbed. The surface potential of the oxide substrate also varies with the pH of the applied solution, affecting the orientation of the adsorbed silane layers. Drying the silane-treated fibers will lead to the formation of siloxane bonds between the coupling agent and the surface. In addition, the number of siloxane bonds with the surface is influenced by the temperature and duration of the drying procedure (128).

Amankwa et al. (1992) (129) successfully derivatized 3-aminopropyl-triethoxysilane (2%) onto the inner wall of fused silica capillary. Following this treatment, the capillary was perfused with a 5.0 mg/mL solution of NHS-LC-biotin (sulfosuccinimidyl 6-(biotinamideo)hexanoate), and it was subsequently treated with avidin in 0.05 M sodium phosphate buffer (pH 7.4). Finally, the avidin-coated capillary was rinsed with distilled water and then treated with a 10 mg/mL solution of biotin-labeled trypsin in sodium phosphate buffer by gravity flow (shown in scheme 8.7).

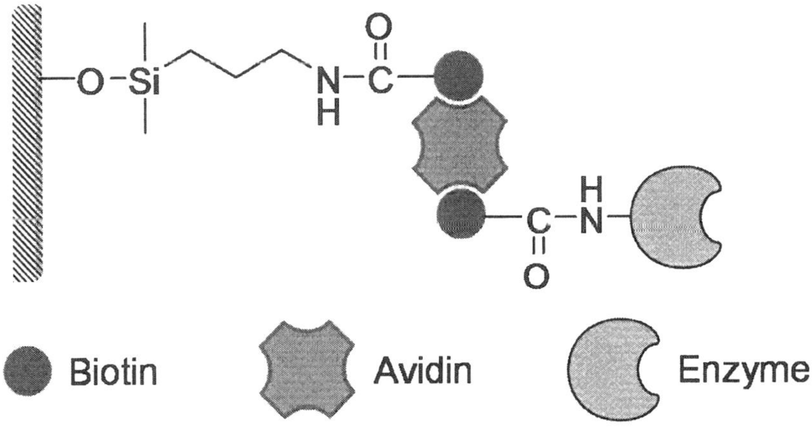


**Scheme 8.7**: Preparation of Trypsin-modified silica surface

### 8.3.1.3 4. *Silica nitride*

Silicon nitride coater is electronically neutral and nonporous, so noncovalent immobilization techniques based on physical adsorption, ionic binding, and entrapment were proven to be ineffective (130). Sheehan et al. (131) deposited a thin film of gold over the top of silicon nitride. Deoxyribonucleic acid (DNA) was modified with thiol linkers and immobilized to gold through thiol–gold bond. A direct surface modification approach based on silanization with aminopropyltriethoxysilane (APTES) and activation with glutaraldehyde is reported by Lee's group in 2005 (132). Karymov et al. (1995) described a covalent attachment of DNA molecules to the surface of silicon nitride without an intermediate polymer layer (133). Redmond's group (2003), on the other hand, demonstrated an efficient activation of silicon nitride substrate for the covalent attachment of amino-terminated probe oligonucleotides using a homobifunctional crosslinker, 1,4 phenylene diisothiocyanate (PDITC) (134).

## 8.3 2. *Signal-amplified Fluorescent Probing Techniques*
## *Metallic nanoparticle enhanced fluorescence*

In the past two decades, studies delving on noble metal nanoparticles, such as silver nanoparticles (AgNPs), have been devoted outstanding efforts. The significance of these lies in the improvement of Raman signals, where the local electric field was enhanced through Surface Enhanced Resonant Raman Scattering (SERRS) (135). Previous study indicated that AgNPs did not only contribute to SERRS, but also played a vital role in metal enhance fluorescence (MEF). Silver nanoparticles were employed by Geddes group (103,136-139) to enhance fluorescent emission for several applications, such as imaging and sensing. Efforts such as this led to a dramatic increase in the use of fluorescent emission since the 1990s (140-141).

Several studies have investigated the effects of MEF with various immobilization strategies or shapes of nanomaterials onto the glass substrates. By employing atomic force microscopy and optical density measurement, Geddes's group (2005) found that irregularly shaped silver nanorod-coated surfaces are much better MEF surfaces as compared with traditional silver island or colloid films (136). A solid substrate-based RNA capture assay was later developed by Geddes's (2006) using AgNPs-based MEF amplification (137). Here, much more sensitive detection of RNA was obtained (less than 25 fmol of RNA, S/N>20), with an approximately 20-fold increase in signal intensity attributed to a plasmon-based luminescent enhancement (142-143).

Furthermore, highly versatile fluorescent core-shell $Ag\text{-}SiO_2$ nanocomposites were developed by Geddes's group for the demonstration of MEF and single nanoparticle sensing platforms (138). The use of plasmonic metal nanostructures, silver island films (SiFs), to enhance the fluorescence emission of five different phycobiliproteins was reported. In this particular case, the phycobiliproteins display up to a 9-fold increase in fluorescence emission intensity, with a maximum 7-fold decrease in lifetime when they are assembled as a monolayer above SiFs (103,139). A novel approach combining the use of metal-enhanced fluorescence with low-power microwave heating was established recently (144-149), in which the sensitivity of surface assays was significantly increased. Due to the excellent sensitivity of microwave-assisted MEF assays, >95 % kinetic completeness of the assay could be observed within a few seconds.

Several other studies (150-153) reported that metal surfaces were able to either enhance or suppress the radiative decay rates of fluorophores. Furthermore, an increase in the extent of resonance energy transfer was also observed. These effects might be due to the interactions of excited-state fluorophores with SPs, which in turn produce constructive effects on the fluorophore. The effects of metallic surfaces include fluorophore quenching at short distances (~0–5 nm), spatial variation of the incident light field (~0–15 nm), and changes in the radiative decay rates (~0–20 nm) (64). The term of metal-enhanced fluorescence could be referred to the appplication of fluorophore and metal interactions in biomedical diagnosis (64).

Variation in fluorescence intensity as a function of the distance between a layer of fluorophores and a number of nanostructured metal surfaces (154,155), suspended colloidal particles (156,157), and adsorbed colloidal particles (158) have been intensively studied.

### 8.3.2 1. *Fluorophores and suspended colloid particles*

The combination of fluorophores and suspended colloid particles could be used in metal-enhanced solution assays. Scheme 8.1 depicts the use of fluorophores and suspended colloid particles. Previous studies on fluorescence intensity enhancement between fluorophores and suspended particles in terms of "metal core of nanoparticles," "fluorophore type," and "spacer used" are summarized in Table 8.2.

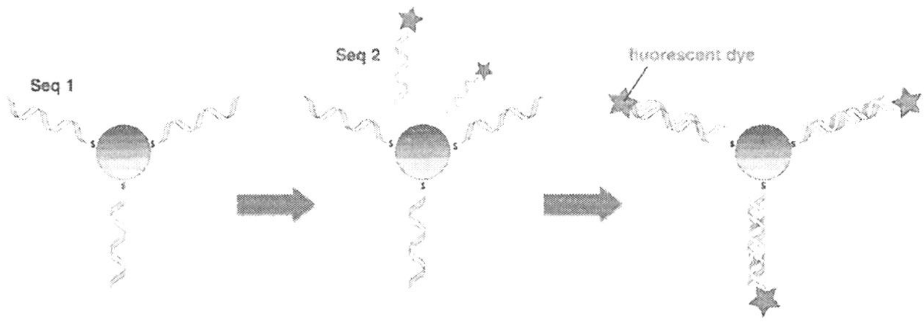


**Scheme 8.8:** The use of fluorophores and suspended colloid particles in the application of sensor development

**Table 8.2:** Studies on fluorescent intensity between fluorophores and suspended colloid particles

| Metal core of nanoparticle | Type of Fluorophore | Spacer | Reference |
|---|---|---|---|
| Au NPs | CdTe Nano Wires | Strepavidin/Biotin | Kotov et al. (2004)[159] |
| Ag NPs | Cy3 | $SiO_2$ shell/biotin-BSA/strepavidin | Geddes et al. (2004)[160] |
| Au NPs or Ag NPs | FAM or CYe | $SiO_2$ shell | Gerritsen et al. (2006)[161] |
| Au/Ag striped nanowires | Cy3, Cy5, 6-FAM, Rhodamine,<br>or TAMRA-labeled oligonucleotides | base pairs | Stoermer & Keating (2006)[162] |
| Spherical gold/silver alloy NPs | Fluorescent ruthenium dye | $SiO_2$ shell | Stranik et al. (2007)[163] |
| Ag NPs | (1) Eu-TDPA<br>(2) Rh800<br>(3) Alexa Fluor 647 | $SiO_2$ shell | Geddes et al. (2007)[103] |
| Ag NPs | Cy5 | dsDNA | Lakowicz et al. (2007)[105] |
| Ag NPs | Alexa Fluor 647 | conA + thiopronin | Lakowicz et al. (2008)[164] |
| Ag NPs | Cy5 | dsDNA | Lakowicz et al. (2008)[165] |

## 8.3.2 2. *Fluorophores and nanostructured metal surfaces or adsorbed colloidal particles*

The use of fluorophores and nano-structured metal surfaces or adsorbed colloidal particles led to the formation of metal-enhanced planar immunoassay, such as sandwich assay or DNA hybridization assay as illustrated in Scheme 8.9.

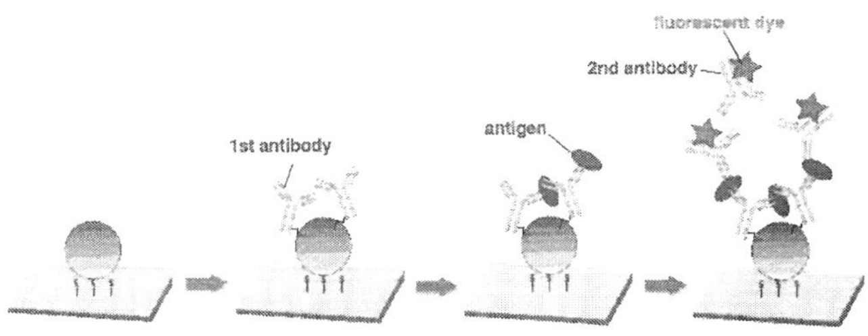


**Scheme 8.9**: The use of fluorophores and nano-structured metal surfaces or adsorbed colloidal particles in the development of sandwich sensor

The combination of metal surface plasmon enhancement of fluorescence and the potential benefits of evanescent wave excitation have been investigated in the field of sensor development. In a study done by Gryczynski et al. (166), dramatic signal enhancements of fluorophores positioned close to surface-bound silver nanostructures were demonstrated. The assay platform utilizes metal particles deposited on glass/quartz surfaces, covered with sub-nanometer layers of a fluorescent biomarker. In such way, the maximum excitation efficiency on the layer of fluorophores just above the metal-island film was obtained. The fluorescence signal is expected to be amplified due to an increase in quantum yield of the fluorophore and the local field effect to the silver island (166). In order to acquire high excitation efficiency in the evanescent wave model, the distance between fluorophore and the surface should be confined to ~ 200 nm (167,168). In designing the format of this assay, surface modification strategies immobilizing metal particles on the evanescent wave substrates, such as gold chip, optical fiber, waveguides, and so on, become very important. A novel fiber-optic biosensor based on a LSPCF system was developed by Chou's group (70). In this method, the fluorophore was excited by LSP on the GNP surface where the evanescent field is applied near the core surface of the optical fiber. Previous studies on fluorescence intensity enhancement between fluorophores and nanostructured metal surfaces or adsorbed colloidal particles in terms of "metal core," "fluorophore type," and "spacer used" are summarized in Table 8.3.

**Table 8.3:** Studies on fluorescent intensity between fluorophores and nanostructured metal surfaces or adsorbed colloidal particles

| Metal Core | Fluorescent molecule / particle | Spacer | Reference |
|---|---|---|---|
| Colloid Ag films | Fluorescin | (1) phospholipid | Cotton et al. (1998)[158] |
| | Fluorescin | (2)BSA-biotin/Avidin multilayers | |
| | NBD | (3) silica scell/IgG/anti-IgG | |
| Au Film | Alexa Fluor 647 | anti-IgG/IgG/thiol | Knoll et al. (2004)[169¶] |
| Ag Islands on glass | Numerous Dyes | anti-IgG/IgG | Lakowicz et al. (2004)[167¶] |
| Au NPs | Cy5 | SAM/anti-Protein C/Protein c/anti-Protein C | Hong & Kang (2006)[170¶] |
| Ag NPs | FITC | avidin/biotinlyated BSA or HSA | Geddes et al. (2006)[168¶] |
| Highly ordered triangular-shaped | (1) CdSe/ZnS quantum dots | PMMA matrix | Pompa et al. (2006)[171] |
| gold nanopatterns | (2) CdSe nanorods | | |
| Particulate gold films | (1) Alexa Fluor 555 | anti-IgG/IgG/$SiO_X$ layer | Zhang & Lakowicz (2007)[172] |
| | (2) Alexa Fluor 680 | | |
| Silver Nanoprisms | Alexa Fluor 532 | dsDNA | Ginger et al. (2007)[173] |
| Au NPs (D = 20 nm) | Cy5 | anti-IgG /IgG/Protein A | Chou et al. (2,007)[70¶] |
| Au NPs | FITC | HSA | Goldys et al. (2008)[174] |
| | DP | BSA | |

Studies involved with evanescent wave excited surface plasmon coupled fluorescence

## 8.3.2 3. *Liposome as signal amplifier in sensor development Overview of liposome*

Liposomes are simple vesicles in which an aqueous media is entirely enclosed by a single phospholipid bilayer (unilamellar liposome) or multiple concentric bilayers (multilamellar liposome) (175). While suspended in an aqueous environment, amphepathic lipid molecules (such as phospholipids, glycolipids) will aggregate in bilayer formation with hydrophilic headgroup residues pointed toward the water- contacting surface and hydrophobic fatty chain aligned within the bilayer. This is done to minimize unfavorable interactions with water. The structure of liposome is illustrated in Figures 8.20 and 8.21.

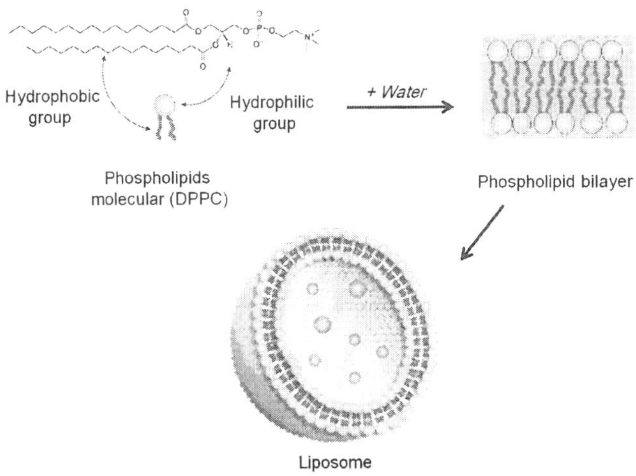


**Figure 8.20:** Liposome structure

Liposomes could be prepared from a combination of saturated or unsaturated lipids. One stable lipid mixture, containing dipalmitoyl phosphatidyl choline (DPPC), dipalmitoyl phosphatidyl glycerol (DPPG), dipalmitoyl phosphatidyl ethanolamine (DPPE), and cholesterol, was developed by O'Connell (1985) (176). In the case of this study, a modified formulation has been intensively used, resulting to several successes in the development of liposome-based immunosensors (177-187). The molecules of DPPC are not soluble in water in an accepted sense. They align themselves closely in planar bilayer sheets in aqueous media in order to minimize unfavorable interactions between the bulk aqueous phase and the long hydrocarbon fatty chain (176). DPPG molecules carrying a negative charge regulate the hydrophilic phosphate headgroup interactions of bilayer membranes by steric hindrance, hydrogen bonding, and electrostatic interaction (21). DPPE molecules bear a primary amine group, which are useful for chemical modification of liposome surface. Cholesterol may also be used as one of the components of liposome. It must be noted that the incorporation of liposome into the phospholipids membrane results in an increase in the phospholipids headgroup separation, liposome diameter, entrapped volume, and rigidity of the bilayer (188).

Since the first discovery and characterization of liposomes in 1965 (Bangham et al.), they have been applied in several technologies, such as drug delivery (189-191), gene delivery (191-193), adjuvant formulation (194-196), vaccine (197-199), skin care (199,200), and immunosensor development (177-187, 201-204).

Based on size and lamellarity, liposomes can be categorized into four groups (175, 205) (as indicated in Figure 8.22): (i) multilanellar vesicles (MLVs); (ii) large unilamellar vesicles (LUVs); (iii) small unilamellar vesicles (SUVs); and (iv) intermediate-size unilamellar vesicles (IUVs), which are also called reverse-phase evaporation vesicles (REV).

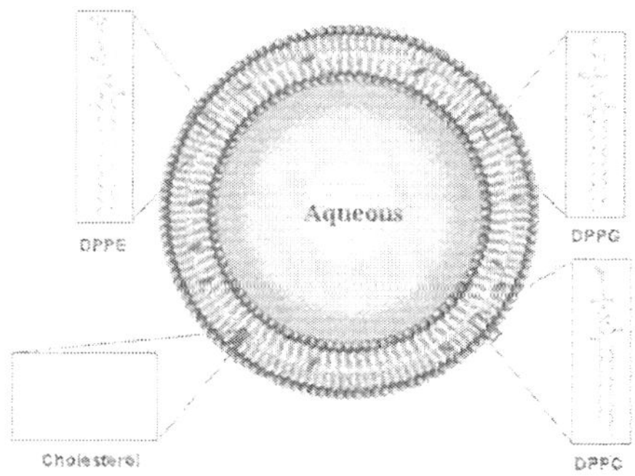


**Figure 8.21:** Chemical composition of liposomes

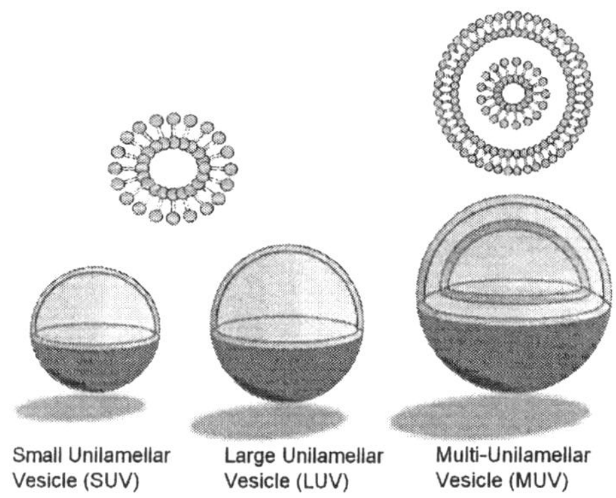


**Figure 8.22:** Classification of liposomes

## 8.3.2 4. *Preparation of liposome*

Based on the modes of lipid dispersion, the methods of liposomes formation can be classified into three categories: mechanical dispersion, solvent dispersion, and detergent solubilization (175). These generally involve the following stages as evident in Figure 8.23.

(i) Dissolution of lipids
(ii) Drying down of lipids from solvent
(iii) Dispersion of lipids in aqueous solution
(iv) Purification and analysis of the liposome

In mechanical dispersion, the lipids are dried down onto a solid support from organic solvents, followed by the dispersion of liposomes by adding the aqueous media through shaking. Other methods using mechanical dispersion techniques include freeze drying, pro-liposome preparation, hand shaking, and the non-shaken method (175).

A common example of solvent dispersion technique is the reverse-phase evaporation method, which is reported to achieve maximum encapsulation volume and efficiency (206). This method involves a complete dissolution of lipids in suitable organic solvents and the subsequent formation of a water-in-oil emulsion by adding aqueous markers to be entrapped in the liposomes. The emulsion is then dried down to a gel-like film in a rotary evaporator under reduced pressure, followed by the conversion of the gel to a homogeneous free-flowing fluid. Traces of organic solvent can be removed by gel filtration or dialysis (204). Reversed-phase evaporation vesicles are usually a heterogeneous mixture of large unilamellar vesicles with diameter ranging from 100 to 1000 nm, so a membrane-assisted extrusion process is helpful in yielding smaller vesicles with better size-homogeneity (207,208). It was reported by Monroe in 1986 (209-210) that sizes of 150-200 nm of REVs are commonly used in liposomal-based immunosensors. Ethanol injection method and ether injection method, proposed by Batzri and Korn in 1973 (211) and Deamer and Bangham in 1976 (212), respectively, also fall under this category.

The third category, detergent solubilization, was described as the most appropriate method for the production of stable homogeneous vesicles (210). Phospholipids are no longer brought into contact with water with the aid of organic solvent, but with that of detergent. The detergent is subsequently removed by dialysis which results in spontaneous formation of liposomes (175). Szoka and Papahadjopoulos assert that liposomes made using this method usually are small sized, unilamellar vesicles (20-50 nm) (213). Triton X-100, bile salt, and cholate are some commonly used detergents (175).

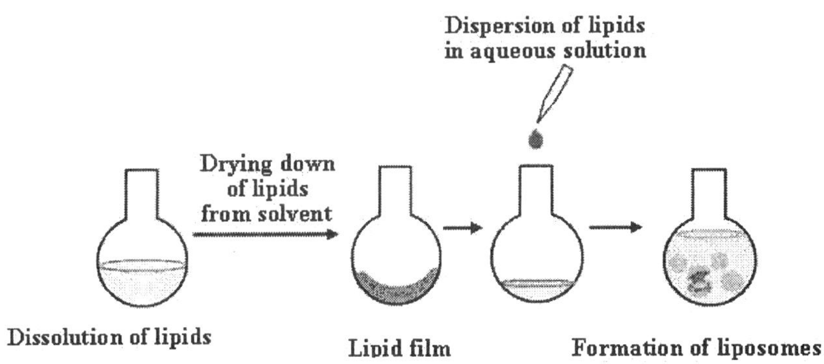


**Figure 8.23:** Common stages for all liposome preparation methods.

## 8.3.2 5. *Liposome-based fluorescence amplification*

Liposomes have been used in several homogeneous assay formats utilizing visible spectrometry for detection. In 1986, Canova-Davis et al. reported a liposomal biosensor for theophylline (214). The release of encapsulated enzymes allows a secondary amplification, where such enzyme is able to facilitate the conversion of numerous substrate molecules into quantifiable products. Liposomes encapsulating sulforhodamine B tagged with Fab' fragments were utilized in a homogeneous sandwich assay for the detection of albumin (215). Several homogeneous assays using fluorescence detection have also been investigated. Carboxyfluorescein-encapsulated, anti-ferritin antibody-tagged liposomes were studied in a sandwich assay for ferritin. The addition of a secondary antibody to a different epitope of the target formed a sandwich complex. The complement triggered the lysis of the lipid bilayer, and the release of carboxyfluorescein was subsequently detected (216).

Liposomes have been used in heterogeneous formats as in the enzyme-linked immunosorbant assay (ELISA). In a convectional ELISA assay for a specific antigen, anti-analyte antibody was first immobilized into microwells (such as wells in 96 microplate), followed by the addition of the antigen. Subsequently a biotinylated antibody specific to another region of the antigen was added. Lastly, an avidin conjugated enzyme or anti-biotin conjugated enzyme and appropriate substrate were introduced to complete the process. In the format of liposome immunoassay, biotinylated dye-encapsulating liposomes or streptavidinylated dye-encapsulating liposomes may replace the role of anti-biotin/avidin conjugated enzyme as in ELISA (217, 218).

Liposomes were reported for use in flow-injection analysis systems as early as 1988 (219). Since then, they have been used for the detection of cholera toxin (177-178), insulin (179, 180), biotin (181), fumonisin B1 (182,183), aflatoxin (184), *E. coli* (185,186), oxygen (187), theophylline (220-222), estrogens (223), alachlor (224-225), and imazethapyr (226,227). In these assays, antibodies against the analyte of interest were immobilized within a capillary column or onto glass beads. The mobile phase and sample were permitted to flow through the immobilized antibody column, followed by the introduction of analyte-tagged liposomes for the competitive assay format. For the sandwich-complex format, on the other hand, they are sensitized with another antibody to the target analyte (so called immunoliposome). In the competitive assay format, the analyte-tagged liposomes competed with the target analyte present in the sample for the available antibody binding sites. The amount of encapsulant released due to lysis of the bound liposomes was inversely proportional to the concentration of target analyte in the tested sample solution. In the sandwich assay, the amount of the bound immunoliposome sandwich complex formed was proportional to the concentration of target in the sample.

Lysis of bound liposomes was achieved by the addition of a detergent solution, such as Triton X-100 (180, 228) or octylglucoside (178,179,181,183), or by complement triggering (216). Furthermore, liposomes encapsulating carboxyfluorescein immobilized in sol-gel films have been reported (229). Due to their mechanical and chemical stability, ease of preparation, and utility for fluorescence assays, sol-gel films have been widely used as solid supports for sensors (229). However, the drawback of this type of assay is the leakage of hydrophilic small molecules such as fluorophores. To minimize dye leakage from sol-gel matrices, fluorophores could be covalently linked to the sol-gel support or conjugated to larger molecules such as dextran. Ruthenium-encapsulating liposomes have been used in an immunoassay relying on electrochemiluminescence (ECL) detection. In this assay, an anti-Legionella antibody was immobilized into the surface of liposomes which were permitted to migrate up a nitrocellulose membrane with immobilized antigen. The nitrocellulose membrane was placed in direct contact with a glass fiber membrane housing electrodes and a dried detergent. Liposomes that did not bind to the nitrocellulose immobilized antigen traveled towards the glass fiber membrane and were lysed by the detergent. The ruthenium released was in proportion to the amount of unbound liposomes which was also proportional to the amount of antigen present in the sample (229). In another chemiluminescence assay, human serum albumin (HSA)-labeled liposomes loaded with Eosin-Y competed with sample HSA for limited anti-HSA antibodies immobilized into glass bead. The supernatant from this assay was subsequently subjected to capillary electrophoresis with chemiluminescence detection yielding a direct proportional signal with increasing HSA in the samples. Compared to fluorophore-labeled HSA, the liposome-based assay was reported to be five times more sensitive (230).

### 8.3.2 6. *Metal enhanced fluorescence with liposomal amplification*

Liposomes carrying numerous marker molecules were believed to boost the detection sensitivity to a higher level. This case was used intensively in Ho's group (177-187, 232) for encapsulating various signal-generating molecules, such as fluorophores, DNA, and photoproteins, with the aim of developing immunodetection systems (177-187, 231). Previous works (177-187, 232), maintain that up to several thousands to millions of fluorescent dye molecules could be retained in the cavity of a liposome, thereby providing greatly enhanced signals. Durst group reported in 2006 that fluorescent quantum dots (QDs) and silica nanoparticles (SNs) were able to be encapsulated in the liposome. Unencapsulated QDs or SNs were separated from the liposomes through size exclusion chromatography (SEC) using sepharose CL-2B column (25 x 1.5cm) (233). It has also been verified by Choquette et al. (234) that the evanescent and scattering components of the propagating laser light in the waveguide caused the amplified fluorescence emission of the specifically bound liposome fraction in competitive immunoassay format. Furthermore, many other groups (235-237) have devoted efforts into the liposomal co-encapsulation of noble metal nanoparticles (such as AgNPs and GNPs) and dye molecules. These cases demonstrated aggregation-preventing effects, which contributed to the protection of lipid bilayers and could be further utilized in the thermal triggered control release. Liposomal nanomaterials (i.e. semiconductor quantum dots (233), silica nanoparticles (238), magnetic nanocrystals (239-241), and polymers (242)) have been extensively researched in recent years with respect to their water solubility, biocompatibility, and potential applications in many fields. However, the optimum NP density loaded in the liposome remains uncertain because not much work was done in this regard. Wijaya and Hamad-Schifferli (241) tried to synthesize high-density NP-loaded vesicles (HNLVs) by introducing hydrophilic $Fe_3O_4$ NPs to the lipid mixture so that they can be spontaneously captured inside the liposomes during vesicle formation with high load. Park et al. proposed an alternative approach for loading metal nanoparticles inside the bilayer region of DPPC liposome (243-244). Successful growth of Au clusters on the bilayer region by means of electroporation was reported by Schelly's group (245). Summary of various studies on encapsulating nanomaterials in liposomes is presented in Table 8.4.

**Table 8.4:** Nanoparticle encapsulated in the liposomes

| Nanoparticle Encapsulated | Reference |
|---|---|
| Quantum dot & silica NP | Chen et al. (2006)[233] |
| Silica NP | Mornet et al.(2005)[238] |
| Ƴ-$Fe_2O_3$ | Martina et al. (2005)[239] |
| Magenetite NP | Giri et al. (2005)[240] |
| $Fe_3O_4$ | Wijaya et al. (2007)[241] |
| Poly (vinyl amine) NP | Kunisawa et al. (2005)[242] |
| Ag NP | Part et al. (2005)[243] |
| Au NP | Part et al. (2006)[244]<br>Wu et al. (2007)[245]<br>Paasonen et al. (2007)[246] |

## 8.4 RESULTS AND DISCUSSION

### 8.4 1. *Immunosensors*

Antibody-based biosensing techniques utilize immunological reactions to measure the presence of a substance. The first format of immunoassay, radio-immunoassay (RIA), was discovered by Yalow and Berson in 1959 (247) for the quantitation of serum insulin. Thereafter, numerous studies have been conducted on the development of sensitive immunoassay (248-249).

Due to the high specificity and sensitivity of immunoassays, there are bioanalytical methods for the measurement of an analyte of interest, with little or without preconcentration or purification of the samples. The principle behind immunoassays is based on an interaction between an antibody and a corresponding antigen, and the detection of the specific interaction using radiolabels (247), enzyme, fluorescent and luminescent compounds (178, 179,181,183), electroactive markers (177,180, 228, 248), or nanomaterials (249-251).

Isotopic labels were the most commonly used probes in the 20$^{th}$ century. The major disadvantages of radio-immunoassays are the short life of the label, the need for special detection equipment, limitations in labeling some antigens and the requirement of a separation step. All these make these assays cumbersome to automate. Hence, non-radioisotope probes are desirable in order to avoid health risks and exacting regulations. Trends in developing probing techniques have focused not only on the development of those probes which are easily adapted to automate systems; but also on the production of amplified signal (249). In recent years, the use of fluorescent probes has increased. Characteristics of these probes include large Stokes shifts, easy labeling and high quantum yield (252-259).

## 8.4 2, *Fluorescence Detection Fiber-optic Biosensor on Bio-molecules Binding Kinetics Measurement*

The fluorescence detection fiber-optic biosensor (FD-FOB) has the capability to measure the kinetics of bio-molecular interactions (260). From time response of the detected fluorescence signal at different concentrations of analyte in solution, the following were obtained at the same time: the association rate constant ($k_a$) and dissociation rate constant ($k_d$) of the interactions between bio-molecules such as antigen and its antibody, the equilibrium dissociation constant $K_D = k_d / k_a$,and the association constant $K_A = k_a / k_d$. In this section, the theory of kinetics constants determination with a sandwich assay is derived and discussed. Conventionally, binding kinetics can be analyzed using the solid phase immobilized antigen and labeled primary antibody in liquid phase or vice versa (261). However, antigens prepared from a complex mixture may lower binding specificity due to high background interference. In contrast, a sandwich assay using two distinct antibodies recognizing different epitopes of the target antigen can largely diminish the chance of intervention from other similar molecules (262). Furthermore, FD-FOB adopted the sandwich assay using fiber immobilized capture antibody, free antigen and fluorophore labeled secondary antibody to form < capture antibody / antigen / fluorophore labeled secondary antibody > complex. The fluorescence of such can only be excited by the evanescent wave through fiber. This allows the study of the binding kinetics of the antigen and secondary antibody using recorded fluorescence upon concentration and time changes. Therefore, the interaction between fluorophore labeled secondary antibody [*A*] and target antigen [*B*] can be described in this equation:

$$[A]+[B]\underset{k_d}{\overset{k_a}{\rightleftarrows}}[AB] \tag{35}$$

where $k_a$ is the association rate constant describing the rate of molecular complex formation, and $k_d$ is the dissociation rate constant describing the stability of the complex, such as in the fraction of complexes that decays per second. In addition, the equilibrium dissociation constant $K_D$ and association constant $K_A$ are defined as (263).

In order to simplify the derivation of rate constants determined by FD-FOB using sandwiched format in the reaction chamber of stagnant (non-flow) setup, the following premises are assumed: (a) the first-order reaction kinetics scheme is applied for antigen / fluorophore labeled antibody interaction at the second reaction that the concentration of fluorophore labeled secondary antibody is assumed to be a constant over the entire course of reaction. Therefore, the reaction proceeds at a rate directly proportional to one of the reactants (261, 263); (b) the binding of target antigen onto the capture antibody immobilized on fiber surface has reached equilibrium and keeps constant during interaction of antigen and fluorophore labeled secondary antibody; (c) the binding sites are limited and the amount of consumed binders is small during antigen / fluorophore labeled secondary antibody interaction to ignore the influence of diffusion (261); and (d) the dissociation and rebinding effect between capture antibody and antigen can be ignored due to the stagnant configuration of the reaction chamber (63, 264).

According to associated rate equation (53),

$$\frac{d[AB]}{dt} = k_a[A][B] - k_d[AB] \tag{36}$$

where $k_a$ is the association rate constant and $k_d$ is the dissociation rate constant. $[A]$ is the ligand, fluorophore labeled secondary antibody, and $[B]$ is the antigen, which is bound with the immobilized capture antibody on the decladding fiber surface. $[AB]$ means the complex of $[A]$ and $[B]$ in the sandwich. In the measurement, the availability of $[B]$ is gradually reduced during the binding process between $[A]$ and $[B]$. Therefore, the number of $[B]$ at time $t$ equals to

$$[A] = [A]_0 - [AB] \tag{37}$$

$$[B] = [B]_0 - [AB] \tag{38}$$

In this setup, $[A]_0$ is the maximum number of $[A]$ at $t = 0$ and $[A]_0 >> [AB]$. Similarly, $[B]_0$ means the maximum number of $[B]$ at $t = 0$ which is identical to the number of antigen bound onto capture antibody immobilized onto the decladded fiber surface. Then Eq. (36) becomes,

$$\frac{d[AB]}{dt} = k_a[A]_0[B]_0 - (k_a[A]_0 + k_d)[AB] \tag{39}$$

As a result, $[A]_0$ and $[B]_0$ are independent of fluorescence signal excitation and $[A]_0$ and $[B]_0$ remain constant during the measurement where $C$ and $C_0$ represent the concentration of $[A]_0$ and $[B]_0$, respectively. The rate equation of the fluorescence signal ($R$) then becomes

$$\frac{dR}{dt} = -(k_aC + k_d)R \tag{40}$$

$$\frac{d}{dR}\left(\frac{dR}{dt}\right) = -(k_aC + k_d) \tag{41}$$

The solution of Eq.(7) can be represented as

$$R(t) = \alpha e^{-(k_aC+k_d)t} + K \tag{42}$$

$$R(t) = R_{bg} + (R_{max} - R_{bg})\left(1 - e^{-(k_aC+k_d)t}\right) \tag{43}$$

$K = R_{max}$ when the fluorescence signal is saturated at $t \rightarrow \infty$, and $\alpha = R_{bg} - R_{max}$ where $R_{bg}$ is the background fluorescence signal at $t = 0$. This result is identical to the output signal of bio-molecules interaction analysis of <antigen / antibody> complex in the real time by using SPR biosensor (265). From Eq. (41), the linear relationship between $\frac{d}{dR}\left(\frac{dR}{dt}\right)$ versus concentration C of fluorophore labeled secondary antibody is measured. The $k_a$ and $k_d$ are obtained simultaneously by the slope and the intersection of the linear response of Eq. (41) apparently.

### 8.4.2 1. *IgG / anti-IgG binding kinetics measurement*

Chou et al. demonstrates the measurement of rate constants between mouse IgG and anti-mouse IgG in sandwich immunoassay (260). Rabbit anti-mouse IgG was immobilized onto the decladded surface of plastic fiber for a start. Before injecting mouse IgG into the reaction chamber to form < anti-IgG / IgG > complex, skim milk was added into the chamber to block mouse IgG or FITC-conjugated anti-mouse IgG adsorbed onto the surface of decladded fiber. Doing so avoids unspecific binding events, allowing only the specific binding between anti-mouse IgG and mouse IgG. By injecting different concentrations of mouse IgG into reaction chamber, < anti-IgG / IgG > complex was formed, which is used for further binding kinetics assay between FITC-conjugated anti-mouse IgG and mouse IgG. Finally, a series of different concentrations of FITC-conjugated rabbit anti-mouse IgG was added to the chamber to form a sandwich composite < anti-IgG / IgG / FITC anti-IgG > complex on the fiber surface. The attached FITC was excited by the evanescent wave in the reaction region near the decladded fiber surface produced by a 488-nm laser beam totally reflected in plastic fiber. The penetration depth of the evanescent wave is about 200 nm at the decladded fiber surface, so only the FITC-conjugated antibody bound with antigen in the solution is excited and detected. The fluorescence signal was detected by a PMT placed beside and facing the fiber wall to improve the fluorescence collection efficiency. A lock-in amplifier (LIA) was incorporated to enhance the signal-to-noise ratio (SNR). Figure 8.24(a) shows the fluorescence intensities of five different mouse IgG concentrations interacting with FITC-labeled anti-IgG. The kinetic theory mentioned above was used to calculate $k_a$, $k_d$ and $K_D$ of <mouse IgG / mouse anti-IgG > interaction. $k_a = 2.49 \times 10^5 M^{-1}s^{-1}$, $k_d = 3 \times 10^{-4} s^{-1}$ and $K_D = 1.2nM$ were measured as shown in Figure 8.24(b).

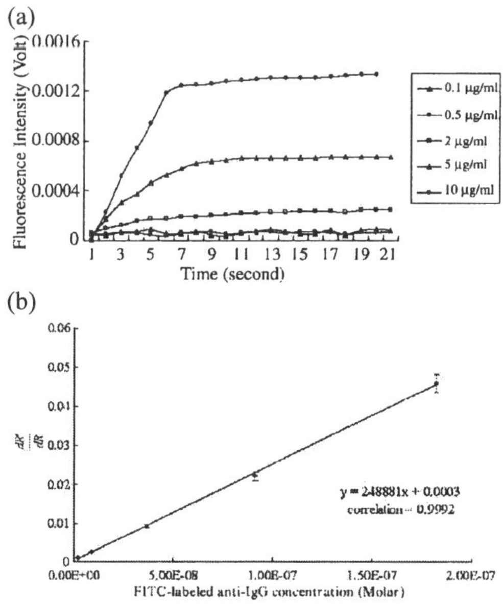


**Figure 8.24:** (a) Fluorescence intensity of five different concentrations of FITC-labeled rabbit anti-mouse IgG interacted with mouse IgG, detected and measured by FOB and (b) Determination of mouse IgG/anti-mouse IgG kinetics.

## 8.4 3. *Localized Surface Plasmon Coupled Fluorescence Fiber-optic Biosensor*

LSPCF fiber-optic biosensor is a novel configuration, which is able to enhance the detection sensitivity significantly and in real time (70). It shows different features compared with SPFS (63, 264) and the fiber-optic sensor on LSPR (52, 55, 56). SPFS is based on the fluorescence detection excited by SPW in the Kretschmann configuration, while LSPR fiber-optic biosensor is independent of fluorescence excitation and detection. However, LSPCF fiber-optic biosensor belongs to LSP excited fluorescence not only with simple geometry but also with highly efficient performance on fluorescence excitation. In addition, the amplification of the fluorescence is produced by the number of fluorophores attached on GNP at the same time. The sensitivity is enhanced significantly by measuring the fluorescence signal of fluorescence probe, which interacted with the target antigen in sandwich format. A poly (methyl methacrylate) (PMMA) MMF of 1mm in diameter (NA=0.467) was used in the said experimental system. As a result, the multiple modes of laser beam propagation in optical fiber produced total reflections that were spread evenly over the fiber core surface. In order to immobilize the captured antibody on the unclad surface, a chemical adsorption method was applied using the covalent binding force (120). Bio-recognition molecules and sandwich immunoassay was constructed in LSPCF fiber-optic biosensor. Here, a sandwiched bio-molecular complex consisting of < capture antibody / antigen / fluorophore labeled secondary antibody conjugated GNP (fluorescence probe) > is built up on the surface of optical fiber (Figure 8.25).

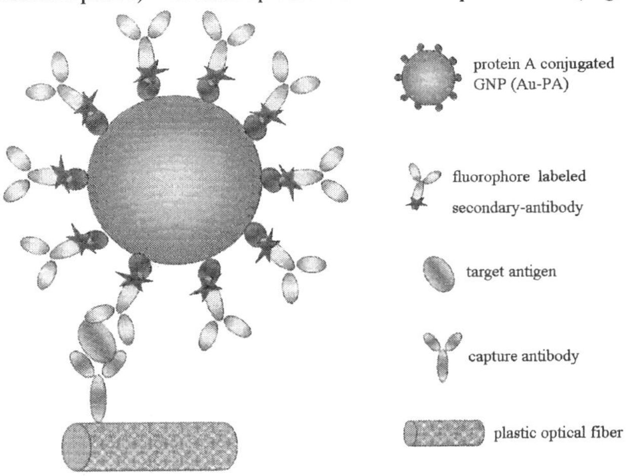


**Figure 8.25:** Scheme of assembled fluorescence probe

Three kinds of fluorescence probe were tested in the experimental system. They were (A) fluorophore labeled secondary antibody without GNP, (B) fluorophore labeled secondary antibody and GNP, which was suspended in PBS solution, and (C) fluorophore labeled secondary antibodies that are connected to protein A conjugated GNP (Au-PA). These are shown in Figure 8.26. Ten protein A molecules are bound to the surface of each GNP (266). Each protein A molecule contains four Fc terminal binding domains, and therefore, nearly 40 fluorophores on GNP are excited by the LSP field simultaneously. This arrangement is able to significantly enhance the intensity of fluorescence signal.

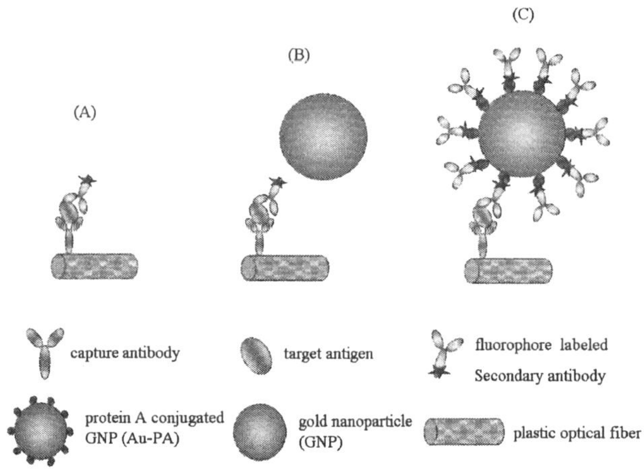


**Figure 8.26:** Kinds of fluorescence probes.

Experiments reveal the fluorescence signals for the binding of mouse IgG (target antigen) at concentrations of 0pg/mL (zero concentration), 1pg/mL, 10pg/mL, 100pg/mL, and 1ng/mL when interacting with the Cy5 labeled anti-mouse IgG (fluorophore labeled secondary antibody), which was bound to the Au-PA ($\psi$=20nm). This analysis was carried out at fixed concentrations of Cy5 labeled anti-mouse IgG and GNP. The sensitivity of the LSPCF biosensor is 1pg/mL (7fM), and a linear relationship between the fluorescent signals versus the logistic scale of mouse-IgG (target antigen) concentration over the range from zero concentration to 1ng/mL can be seen in Figure 8.27.

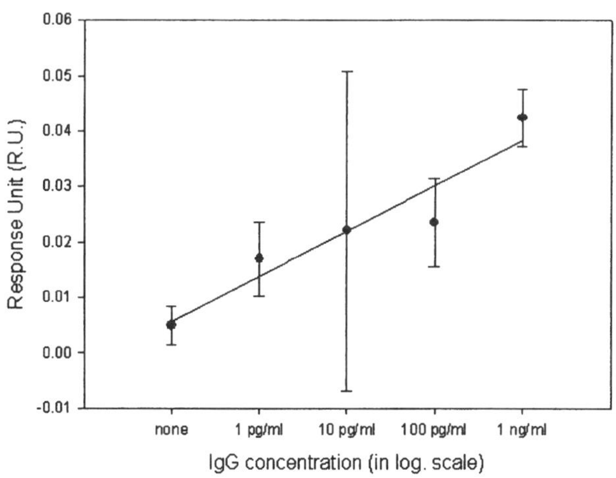


**Figure 8.27:** Linear regression of the mouse IgG concentration dependence. (70)

The LSPCF fiber-optic biosensor which features an integrated FO-EWS with GNPs and which uses a sandwich structure < capture antibody / antigen / fluorophore secondary antibody conjugated GNP (fluorescence probe) > system is demonstrated. Mouse IgG antigen is experimentally detected at a best sensitivity of 1 pg/mL. There are two effects involved in this novel biosensor. First is the LSP excited by the evanescent wave on the GNPs, and this produces a strong local electromagnetic field within 50~60nm around the GNP surface to efficiently enhance the fluorescence signal (60). In the meantime, the non-specific binding can be ignored because a very low concentration of non-specific antigen exists in the region of LSP field. Second, each fluorescence probe contains at least 40 fluorophore, such that the fluorescence signal is amplified significantly through LSP excitation. These two effects synergistically multiply together and produce a significant improvement in the sensitivity of the system compared to a conventional fluorescence probe.

However, fluorescence loss due to the metal-induced quenching (63) might significantly reduce the performance of LSPCF fiber-optic biosensor. According to Lakowicz et al. (99-102), Hong et al. (169) and Borejdo et al. (267), the fluorescence quenching happened within a 2-5nm distance from the surface of GNP in this experiment. Therefore, an appropriate distance between GNP and fluorophore is important in a manner that the GNP can effectively enhance the fluorescence more and can prevent fluorescence quenching at the same time (99-102, 169, 267). This is a significant feature of this LSPCF biosensor when monitoring the binding kinetics of bio-molecular interactions at low concentrations. To circumvent this problem, a protein A which is conjugated on GNP was introduced to displace the interaction platform away from GNP. Thus, the protein A is not only a linker to secondary antibody, but also a spacer to avoid fluorescence quenching.

Recently, LSPCF fiber-optic biosensors are used in clinical diagnosis for alpha-fetoprotein (AFP) identification in human serum. Figure 8.28 shows the relationship between the fluorescent signals and the AFP concentrations of human serum measured by ELISA system (ABBOTT, ARCHITECT i-2000). The preparation and the measurement of human serum were completed at Taipei City Hospital (Taipei, Taiwan). The correlation coefficient $R^2$ is 0.9331.

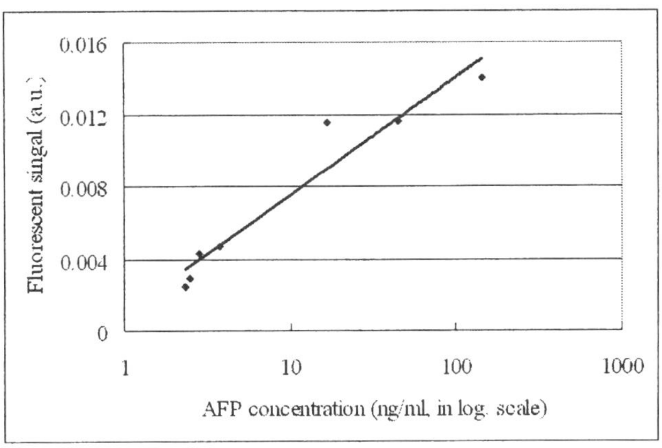


**Figure 8.28:** The linear relationship [y=0.0028 Ln(x) + 0.0010; $R^2$=0.9331]

of different human serum versus fluorescent signal.

The LSP is strongly dependent on nanoparticle size, shape, inter-particle spacing, and the local dielectric environment (96, 97, 268-270). Thus, to optimize the conditions of LSPCF fiber-optic biosensor, finding highly efficient LSPCF excitation becomes vital to the development of a biosensor. This is particularly true when monitoring the binding kinetics of very low abundance protein-protein interaction in real time. Experimentally, the best sensitivity of LSPCF fiber-optic biosensor at 1pg/mL was achieved when detecting mouse IgG interacting with anti-mouse IgG (70). In the meantime, the linear dynamic range of LSPCF fiber-optic biosensor can be from background to 1ng/mL on mouse IgG detection. With proper arrangement, LSPCF fiber-optic biosensor shows great potential on bio-molecule interaction studies at very low concentration.

## 8.4 FUTURE WORK

### 8.4 1. *Micro-array Based on Localized Surface Plasmon Coupled Fluorescence*

In enhancing the throughput of a detection device, micro-array has been a common biotechnique. The multimode planar waveguides can be fabricated in silica ($SiO_2$) materials on silicon substrates. With precise semiconductor fabrication techniques, the thickness of the waveguides can be determined, and the evanescent waves can be well controlled. Figure 8.29 illustrates the structure of the planar waveguides. Silica II and silica I layers are deposited on the silicon wafer by plasma enhanced chemical vapor deposition (PECVD), consecutively. The thickness of the silica I (refractive index=1.49) and silica II (refractive index=1.47) is around 2□m and 3□m, respectively. This ensures the multimode operation in the planar waveguides at the wavelength of 658nm. Evanescent wave propagating on the surface of the waveguide depends on the thickness of the silica I. Higher-order mode can have stronger evanescent wave to excite the nanoparticles. After the fabrication of the micro-array, the biological species can be immobilized on the chip as a microarray as shown in Figure 8.29. After the interaction between biological species, such as hybridization, the detection on the chip surface can be performed by launching the light from the single edge or the four edges of the chip as illustrated in Figure 8.29. Light can be injected from the polished edge into the waveguide using end-fire coupling. All sites in the microarray can be excited. The throughput of the detection can be significantly enhanced. Figure 8.30 (a) presents the output optical field image confirming the fact that the light can be confined in the silica waveguide. $Si_3N_4$ microlenses can also be embedded in the silica waveguides to obtain collimated light (271). As can be seen in Figure 8.30 (b), the propagation of the light can be observed. The collimated light can help obtain a more intense evanescent wave in the devices.

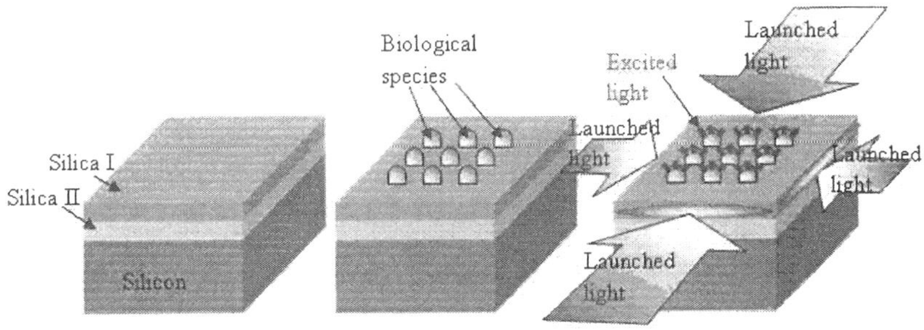


**Figure 8.29:** Schematic drawing of the structure of silica waveguide.

(a)

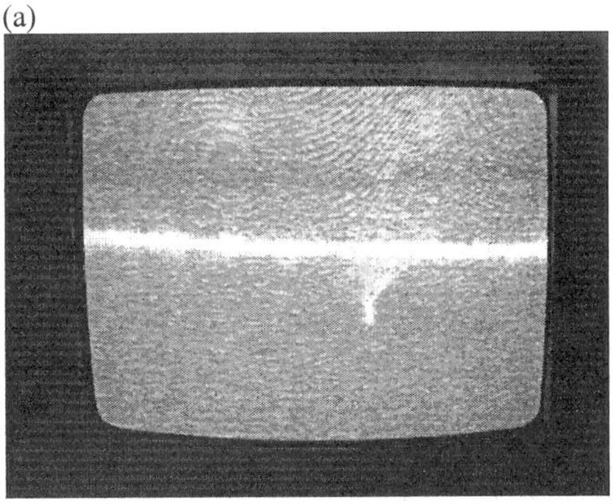

(b)

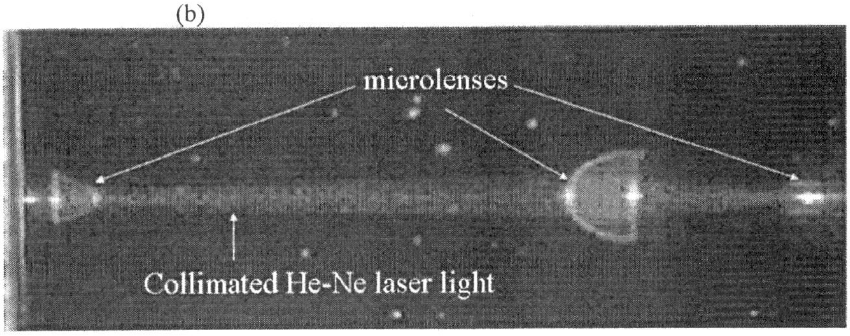


**Figure 8.30:** (a) Output optical field of silica planar waveguide excited by He-Ne laser. (b) Top view of the silica planar waveguide excited by He-Ne laser. Three $Si_3N_4$ microlenses are embedded to obtain the collimated beam and focalized beam.

In the future, it is possible to develop a novel LSPCF-based planar waveguide protein affinity sensor chip where the fluorescence signal is excited by LSP on GNPs. Several GNPs binding with fluorophores are encapsulated in liposomes. The advantages of LSPCF biosensor lie not only in the higher excited efficiency of the fluorescence but also in the specific direction of the fluorescence emission. They are able to enhance the detection sensitivity on fluorescence detection much higher than the conventional TIR method. In addition, the localized enhancement of the electric field of the evanescent wave (EW) near GNPs strengthens SPCF simultaneously. Therefore, combining higher sensitivity of fluorescence detection with the localized properties of EW and SPW enables the real-time monitoring of biomolecules interaction in the near-field region simultaneously. The association rate constant ($K_a$) and dissociation rate constant ($K_d$) of the binding kinetics of biomolecules, as well as the equalization classicization rate constant ($K_D = K_a / K_d$), can be obtained at the same time. The schematic drawing of the micro-array is illustrated in Figure 8.31.

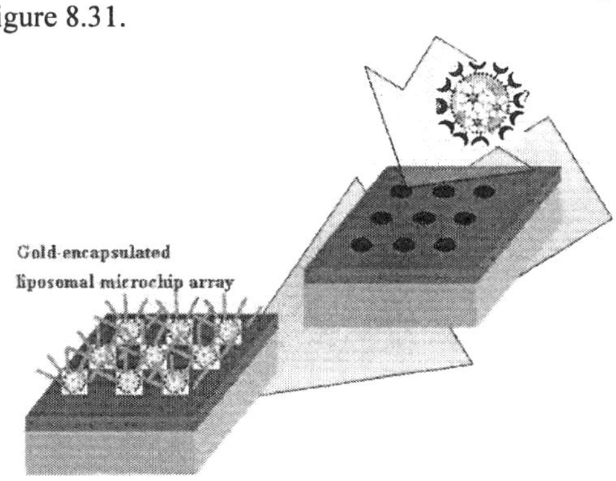


**Figure 8.31:** Schematic drawing of the gold-encapsulated liposomal microarray.

The emission of fluorophore is non-directional, so any conventional optical detection method using lenses or fiber can only receive one part of the light. Directional emission of SP is one of potential solutions to enhance the sensitivity of detection (272). Recently, semiconductor hollow optical waveguides formed by omni-directional reflector (SHOW-ODR) have been realized. (273) (Figure 8.32) Light can be propagated in the hollow core of the SHOW-ODR. The SHOW-ODR can also be used as the micro-fluidic channel. One can combine the micro-fluidic channel (SHOW-ODR) with the gold-encapsulated liposomal detection. A similar idea has been realized using GNPs in microfluidic channel (274). However, in the case of SHOW-ODR, as the light emitted from the fluorophore, only the two ends of the waveguide can receive the signal. This structure can enhance the signal strength.

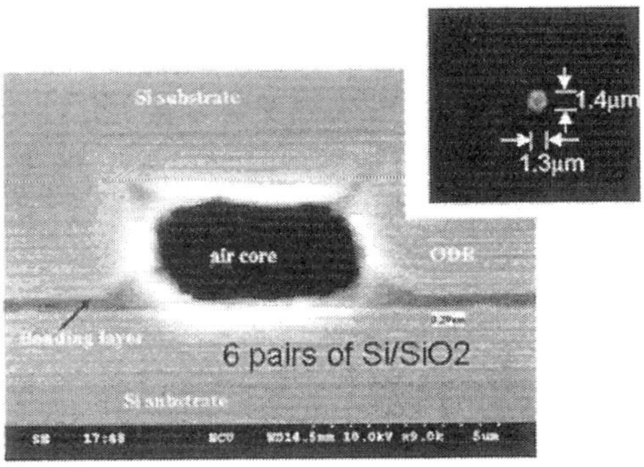


**Figure 8.32:** Cross-section view of SHOW-ODR and the output optical field.

Metal core/dielectric shell nanoparticles have been studied such as Ag/silica and Au/$TiO_2$ nanoparticles. (275, 84) Silver coated PMMA microspheres (276) and magnetic $Fe_3O_4$-encapsulated silica microspheres (277) have likewise been reported. (Figure 8.33) Highly uniform silver coated colloidal microspheres might help to obtain a uniform signal than that obtained from non-spherical and non-uniform GNPs. With the presence of magnetic field, magnetic $Fe_3O_4$-encapsulated silica microspheres binding with bio-species can be attracted toward the surface microarray. This property might accelerate the reaction between bio-species. A similar idea has been forwarded using magnetic beads and gold nanopaticles on immunosensing. (278)

(a)

(b)

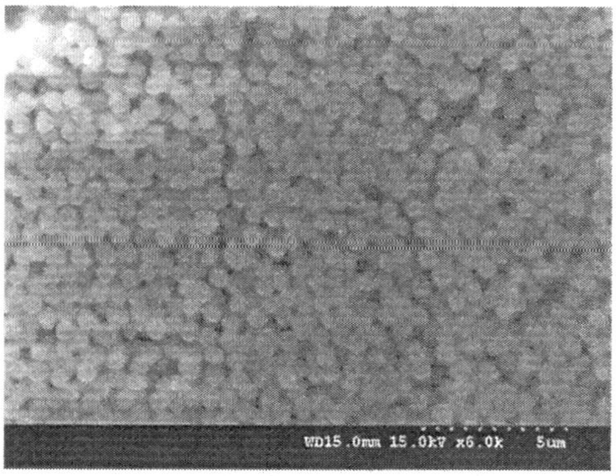

**Figure 8.33:** SEM images of (a) Silver coated PMMA microspheres; (b) $SiO_2/Fe_3O_4$ microspheres.

## 8.5 REFERENCES


1. Bates, R. J. (2001). Optical switching and networking handbook. McGraw-Hill telecommunications. New York, McGraw-Hill: ix, 302 p.
2. Hecht, J. (1999). City of light : the story of fiber optics. New York, Oxford University Press.
3. Gambling, W. A. (2000). The rise and rise of optical fibers. *Ieee Journal of Selected Topics in Quantum Electronics* 6: 1084-1093.
4. Hecht, J. (2002). Understanding fiber optics. Upper Saddle River, NJ, Prentice Hall.
5. http://en.wikipedia.org/wiki/Optical fiber
6. http://electronics.howstuffworks.com/fiber-optic7.htm
7. Murata, H. (1996). Handbook of optical fibers and cables. New York, M. Dekker.
8. Burmeister, R. A., Greene, P. E., Hiskes, R. (1982). US Patent 4,319,803.
9. Kuzyk, M. G. (2007). Polymer fiber optics : materials, physics, and applications. Boca Raton, CRC/Taylor & Francis.
10. Marcou, J. and Club des fibres optiques plastiques. (1997). Plastic optical fibres : practical applications. Chichester ; New York, Paris, John Wiley & Masson.
11. Schultz, J.S., and Sims, G. (1979). Affinity sensors for individual metabolites, Biotechnol. Bioeng. Symp., 9:65.
12. Schaffar, B.P.H. and Wolfbeis, O.S. (1991). Chemically mediated fiberoptic biosensors. In: Biosensor Principles and Application. (Blum, L.J. and Coulet P.R., eds), Marcek Dekker, New York, NY.
13. Trpkovski, S., Wade, S. A., Baxter, G. W. and Collins, S. F. (2003). Dual temperature and strain sensor using a combined fiber Bragg grating and fluorescence intensity ratio technique in Er[sup 3+]-doped fiber. *Review of*

*Scientific Instruments* 74: 2880-2885.

14. Bosch, M. E., Sánchez, A. J. R., Rojas, F. S. and Ojeda, C. B. (2007). Recent Development in Optical Fiber Biosensors. *Sensors* 7: 797-859.
15. Arnold, M.A. (1991). Fluorophore- and chromophore-based fiberoptic biosensors. In: Biosensor Principles and Application. (Blum, L.J. and Coulet P.R., eds), Marcek Dekker, New York, NY.
16. Blum, L.J. and Gautier, S.M. (1991). Chemically mediated fiberoptic biosensors. In: Biosensor Principles and Application. (Blum, L.J. and Coulet P.R., eds), Marcek Dekker, New York, NY.
17. Chang, Y. H., Chang, T. C., Kao, E. F., et al. (1996) Detection of protein A produced by Staphylococcus aureus with a fiber-optic-based biosensor. *Bioscience Biotechnology and Biochemistry* 60: 1571-1574.
18. Loev, W. F., Button, L. J., Slovacek, R. E. (1991) Optical Characteristics of Fiberoptic Evanescent Wave Sensor Theory and Experiment, Biosensor with Fiberoptics, the Humana Press Inc., pp.139
19. P. N. Prasad, "Introduction to Biophotonics", P336 (Wiley Interscience, John Wiley & Sons Inc., 2003)
20. Thompson, R. B., Lakowicz, J. R., et al. (1991) Fluorescence-based fiber-optic seneor. *Topics in Fluorescence Spectroscopy* 2: 345-365.
21. Rowe-Taitt, C. A., Ligler, F. S. (2002) Evanescent wave fiber optic biosensor. Optical Biosensor : Present and Future, Elsevier, Amsterdam, pp. 95-122.
22. D. Robert, "Modern Optics", P 169 (John Wiley & Sons, Inc., 1990)
23. Ritchie, R. H. (1957) Plasma Losses by Fast Electrons in Thin Films. *Physical Review* 106: 874-881.
24. Barnes, W. L., Dereux, A., Ebbesen, T. W. (2003) Surface plasmon subwavelength optics. *Nature* 424 : 824-830.
25. Jung, L. S., Campbell, C. T., Chinowsky, T. M., et al. (1998) Quantitative Interpretation of the Response of Surface Plasmon Resonance Sensors to Adsorbed Films. *Langmuir* 14: 5636-5648.
26. Stark, P. R. H., Halleck, A. E., Larson, D.N. (2005) Short order nanohole arrays in metals for highly sensitive probing of local indices of refraction as the basis for a highly multiplexed biosensor technology. *Methods* 37: 37-47.
27. Kolomenskii, A. A., Gershon, P. D., Schuessler, H. A. (1997) Sensitivity and detection limit of concentration and adsorption measurements by laser-induced surface-plasmon resonance. *Applied Optics* 36: 6539-6547.
28. Sepúlveda, B., Calle, A., Lechuga, L. M., Armelles, G. (2006) Highly sensitive detection of biomolecules with the magneto-optic surface-plasmon-resonance sensor. *Optics Letter* 31: 1085-1087.
29. Liedberg, B., Nylander, C., Lundstrom, I. (1995) Biosensing with surface Plasmon resonance- how it all started. *Biosensors & Bioelectronics* 10: i-ix.
30. Homola, J., Yee, S. S., Gauglitz, G. (1999) Surface plasmon resonance sensors: review. *Sensors and Actuators B* 54: 3-15.
31. Smolyaninov, I. I. and Mazzoni, D. L. (1997) Experimental study of surface-plasmon scattering by individual surface defects. *Physical Review B* 56: 1601-1611.
32. Bouhelier, A., Huser, Th., Tamaru, H., Guntherodt. H.-J., and Pohl, D. W. (2001) Plasmon optics of structured silver films. *Physical Review B* 63:

155404.

33. Weeber, J.-C., Krenn, J. R., Dereux, A., Lamprecht, B., Lacroute, Y., and Goudonnet, J. P. (2001) Near-field observation of surface plasmon polariton propagation on thin metal stripes. *Physical Review B* 64: 045411.
34. Dawson, P., Puygranier, B. A. F., Goudonnet, J-P. (2001) Surface plasmon polariton propagation length: A direct comparison using photon scanning tunneling microscopy and attenuated total reflection. *Physical Review B* 63: 205410.
35. Bozhevolnyi, S. I., Erland, J., Leosson, K., Skovgaard, P. M. W., and Hvam, J. M. (2001) Waveguiding in Surface Plasmon Polariton Band Gap Structures. *Physical Review Letters* 86: 3008-3011.
36. Seidel, J., Grafstrom, S., and Eng, L. (2005) Stimulated Emission of Surface Plasmons at the Interface between a Silver Film and an Optically Pumped Dye Solution. *Physical Review Letters* 94: 177401.
37. Sharma, A. K., Jha, R., and Gupta, B. D. (2007) Fiber-Optic Sensors Based on Surface Plasmon Resonance: A Comprehensive Review. *IEEE Sensors Journal* 7: 1118-1129.
38. Raether, H. (1988) Surface Plasmons on Smooth and Rough Surfaces and on Gratings. Springer-Verlag.
39. Panofsky, W. K. H., Phillips, M. (1962) Classical Electricity and Magnetism. Addison-Wesley, Reading.
40. Hoa, X. D., Kirk, A.G., Tabrizian, M. (2007) Towards integrated and sensitive surface plasmon resonance biosensors: A review of recent progress. *Biosensors & Bioelectronics* 23: 151-160.
41. Kretschmann, E. and Raether, H. (1968) Radiative decay of non-radiative surface plasmons excited by light. *Zeitschrift fur Naturforschung* 23A: 2135–2136.
42. Otto, A. (1968) Excitation of Nonradiative Surface Plasma Waves in Silver by Method of Frustrated Total Reflection. *Zeitschrift fur Physikalische* 216: 398–410.
43. Inagaka, T., Kagami, K. (1981) Photoacoustic observation of nonradiative decay of surface plasmons in silver. *Physical Review B* 24: 3644-3646.
44. Booman, R. A., Olson, G. A., and Sarid, D. (1986) Determination of loss coefficients of long-range surface plasmons. *Applied Optics* 25: 2729-2733.
45. Kochergin, V. E., Beloglazov, A. A., Valeko, M. V., Nikitin, P. I. (1998) Phase properties of a surface-plasmon resonance from the viewpoint of sensor applications. *Quantum Electronics* 28: 444-448.
46. Yu, X. L., Wang, D. G., Yan, Z. B. (2003) Simulation and analysis of surface plasmon resonance biosensor based on phase detection. *Sensors and Actuators B* 91: 285–290.
47. Sheridan, A. K., Harris, R. D., Bartlett, P. N., Wilkinson, J. S. (2004) Phase interrogation of an integrated optical SPR sensor. *Sensors and Actuators B* 97: 114–121.
48. Yu, X. L., Wang, D. G., Wei, X., Ding, X., Liao, W., Zhao, X. S. (2004) A surface plasmon resonance imaging interferometry for protein micro-array detection. *Sensors and Actuators B* 108: 765–771.
49. Su, Y.-D., Chen, S.-J., Yeh, T.-L. (2005) Common-path phase-shift interferometry surface plasmon resonance imaging system. *Optics Letter* 30:

1488-1490.

50. Wu, S. Y., Ho, H. P., Law, W. C., Lin, C. L., Kong, S. K. (2004) Highly sensitive differential phase-sensitive surface plasmon resonance biosensor based on the Mach–Zehnder configuration. *Optics Letter* 29: 2378-2380.
51. Homola, J. (2003) Present and future of surface plasmon resonance biosensors. *Analytical and Bioanalytical Chemistry* 377: 528–539.
52. Shankaran, D. R., Gobi, K. V., Miura, N. (2007) Recent advancements in surface plasmon resonance immunosensors for detection of small molecules of biomedical, food and environmental interest. *Sensors and Actuators B* 121: 158–177.
53. Green, R. J., Frazier, R. A., Shakesheff, K. M., Davies, M. C., Roberts, C. J., Tendler, S. B. (2000) Surface plasmon resonance analysis of dynamic biological interactions with biomaterials. *Biomaterials* 21: 1823-1835.
54. Piliarik, M., Homola, J., Manıkova, Z., Ctyroky, J. (2003) Surface plasmon resonance sensor based on a single-mode polarization-maintaining optical fiber. *Sensors and Actuators B* 90: 236–242.
55. Hernandez, D. M., Villatoro, J., Talavera, D., Moreno, D. L. (2004) Optical-fiber surface-plasmon resonance sensor with multiple resonance peaks. *Applied Optics* 43: 1216-1220.
56. Tang, J. L., Cheng, S. F., Hsu, W. T., Chiang, T. Y., Chau, L. K. (2006) Fiber-optic biochemical sensing with a colloidal gold-modified long period fiber grating. *Sensors and Actuators B* 119: 105–109.
57. Wang, S. F., Chiu, M. H., Chang, R. S. (2006) Numerical simulation of a D-type optical fiber sensor based on the Kretchmann's configuration and heterodyne interferometry. *Sensors and Actuators B* 114: 120–126.
58. Hernandez, D. M., Villatoro, J. (2006) High-resolution refractive index sensing by means of a multiple-peak surface plasmon resonance optical fiber sensor. *Sensors and Actuators B* 115: 227–231.
59. Chau, L. K, Lin, Y. F., Cheng, S. F., Lin, T. J. (2006) Fiber-optic chemical and biochemical probes based on localized surface plasmon resonance. *Sensors and Actuators B* 113: 100–105.
60. Cheng, S. F., Chau, L. K. (2003) Colloidal Gold-Modified Optical Fiber for Chemical and Biochemical Sensing. *Analytical Chemistry* 75: 16-21.
61. Chou, C., Wu, H. T., Huang, Y. C., Chen. Y. L. (2006) Characteristics of a paired surface plasma waves biosensor. *Optics Express* 14: 4307-4315.
62. Wu, H.-T., Chen, Y.-L., Huang, Y.-C., et al. (2007) Colloid-gold nanoparticle enhanced detection sensitivity of paired surface plasma waves biosensor. *Proceedings of SPIE* 6447: 64470S-1－64470S-7.
63. Yu, F., Persson, B., Lofas, S., Knoll, W. (2004) Surface Plasmon Fluorescence Immunoassay of Free Prostate-Specific Antigen in Human Plasma at the Femtomolar Level. *Analytical Chemistry* 76: 6765-6770.
64. Aslan K., Gryczynski I., Malicka J., Matveeva E., Lakowicz J. R., and Geddes C. D. (2005). Metal-enhanced fluorescence: an emerging tool in biotechnology. *Curr. Opin. Biotechnol.* 16: 55-62.
65. Weitz D. A., Garo_ S., Gersten J. I., and Nitzan A. (1983). The enhancement of Raman scattering, resonance Raman scattering and fluorescence from molecules adsorbed on a rough silver surface. *J. Chem. Phys.* 78: 5324-5338.

66. Hayakawa T., Selvan S. T., and Nogami M. (1999). Field enhancement effect of small Ag particles on the fluorescence from Eu3+ -doped SiO2 glass. *Appl. Phys. Lett.* 74: 1513-1515.
67. Lakowicz J. R. (2006). Plasmonics in biology and plasmon-controlled fluorescence. *Plasmonics* 1: 5-33.
68. Chew H., Wang D. S., and Kerker M. (1979). Elastic scattering of evanescent electromagnetic waves. *Appl. Opt.* 18: 2679-2687.
69. Quinten M., Pack A., and Wannemacher R. (1999). Scattering and extinction of evanescent waves by small particles. *Appl. Phys. B* 68: 87-92.
70. Hsieh B.-Y., Chang Y.-F., Ng M.-Y., Liu W.-C., Lin C.-H., Wu H.-T., and Chou C. (2007). Localized surface plasmon coupled fluorescence fiber-optic biosensor with gold nanoparticles. *Anal. Chem.* 79, 3487-3493.
71. Pinchuk A., Hilger A., Plessen G. V., and Kreibig U. (2004). Substrate effect on the optical response of silver nanoparticles. *Nanotechnology* 15: 1890-1986.
72. Palik E. D. (1985). Handbook of optical constants of solids. Academic Press, Inc., New York.
73. Ohtsu M., Kobayashi K., Kawazoe T., Sangu S., and Yatsui T. (2002). Nanophotonics: design, fabrication, and operation of nanometric devices using optical near fields. *IEEE J. Sel. Top. Quantum Electron.* 8: 839-862.
74. Prasad P. N. (2004). Nanophotonics. Wiley, Hoboken, NJ.
75. Liu W.-C., Wen C.-Y., Chen K.-H., Lin W. C., and Tsai D. P. (2001). Near-field images of the $AgO_x$-type super-resolution near-field structure. *Appl. Phys. Lett.* 78: 685-687.
76. Liu W.-C., Ng M.-Y., and Tsai D. P. (2004). Surface Plasmon Effects on the Far-Field Signals of AgOx-Type Super Resolution Near-Field Structure. *Jpn. J. Appl. Phys.* 43: 4713-4717.
77. Ng M.-Y. and Liu W.-C. (2005). Super-resolution and frequency-dependent efficiency of near-field optical disks with silver nanoparticles. *Opt. Express* 13: 9422-9430.
78. Moskovits M. (1985). Surface-enhanced spectroscopy. *Rev. Mod. Phys.* 57: 783-826.
79. Bohren C., and Human D. (1983). Absorption and scattering of light by small particles. Wiley, New York.
80. Jackson J. D. (1999). Classical electrodynamics. Third edition, John Wiley and Sons, Inc., New York.
81. Arfken G. B. and Weber H. J. (2000). Mathematical methods for physicists. Fifth edition, Academic Press.
82. Dulkeith E., Morteani A. C., Niedereichholz T., Klar T. A., Feldmann J., Levi S. A., van Veggel F.C.J.M., Reinhoudt D. N., Moller M., and Gittins D. I. (2002). Fluorescence quenching of dye molecules near gold nanoparticles: radiative and nonradiative effects. *Phys. Rev. Lett.* 89: 203002.
83. Dubertret B., Calame M., and Libchaber A. J. (2001). Single-mismatch detection using gold-quenched fluorescent oligonucleotides. *Nat. Biotechnol.* 19: 365-370.
84. Stanik, O., Nooney, R., McDonagh, C., MacCraith, B. D., (2007). Optimization of nanoparticle size for plasmonic enhancement of

fluorescence. *Plasmonics* 2:15–22.

85. Fischer, U.C. (1986). Submicrometer aperture in a thin metal film as a probe of its microenvironment through enhanced light scattering and fluorescence. *J. Opt. Soc. A. B* 3:1239–1244.
86. Levene, M. J., Korlach, J., Turner, S. W., Foquet, M., Craighead H. G., Webb, W. W. (2003). Zero-mode waveguides for single-molecule analysis at high concentrations. *Science* 299: 682–686
87. Degiron, A., Lezec, H. J., Yamamoto, N,. Ebbesen, T. W. (2004). Optical transmission properties of a single subwavelength aperture in a real metal. *Opt. Commun.* 239:61–66.
88. Yin, L., Vlasko-Vlasov, V.K., Rydh, A., Pearson, J., Welp, U., Chang, S. H., Gray, S. K., Schatz, G. C., Brown, D. B., Kimball, C. W. (2004). Surface plasmons at single nanoholes in Au films. *Appl. Phys. Lett.* 85:467–469.
89. Popov, E., Bonod, N., Neviere, M., Rigneault, H., Lenne, P., Chaumet, P. (2005). Surface plasmon excitation on a single subwavelength hole in a metal sheet. *Appl. Opt.* 44: 2332–2337.
90. Catrysse, P. B., Shin, H., Fan, S. (2005). Propagating modes in subwavelength cylindrical holes. *J. Vac. Sci. Technol. B* 23: 2675–2678.
91. Zakharian, Am, Mansuripur, M., Moloney, J. (2004). Transmission of light through small elliptical apertures. *Opt. Express* 12:2631–2648.
92. Fan, W., Zhang, S., Minhas, B., Malloy, K. J., Brueck, S. R. J. (2005). Enhanced infrared transmission through subwavelength coaxial metallic arrays. *Phys. Rev. Lett.* 94:033902.
93. Gordon, R., Brolo, A. G. (2005). Increased cut-off wavelength for a subwavelength hole in a real metal. *Opt. Express* 13: 1933–1938.
94. Collin, S., Pardo, F., Pelouard, J. L. (2007). Waveguiding in nanoscale metallic apertures. *Opt. Express* 15:4310–4320.
95. Mahdavi, F., Liu, Y., Blair, S., (2007). Modeling Fluorescence Enhancement from Metallic Nanocavities. *Plasmonics* 2:129–141.
96. Kottmann, J. P., Martin, O. J. F. (2001). Plasmon resonances of silver nanowires with a nonregular cross section. *Phys. Rev. B* 64:235402.
97. Kelly, K. L., Coronado, E., Zhao, L. L., Schatz, G. C. (2003). The Optical Properties of Metal Nanoparticles: The Influence of Size, Shape, and Dielectric. *Environment J. Phys. Chem. B*, 107, 668–677.
98. Ambjörnsson, T., Mukhopadhyay, Gautam,. Apell, P., Käll, M. (2006). Resonant coupling between localized plasmons and anisotropic molecular coatings in ellipsoidal metal nanoparticles. *Phys. Rev. B* 73:085412.
99. Baffou, G., Girard, C., Dujardin E., Colas des Francs, G., Martin, O. J. F. (2008). Molecular quenching and relaxation in a plasmonic tunable system. *Phys. Rev. B* 77, 121101.
100. Lakowicz, J. R., Malicka, J., D'Auria, S., Gryczynski, I. (2003). Release of the self-quenching of fluorescence near silver metallic surfaces *Anal. Biochem.* 320:13–20.
101. Anger, P., Bharadwaj, P., Novotny, L. (2006). Enhancement and Quenching of Single-Molecule Fluorescence. *Phys. Rev. Lett.* 96:113002.
102. Bharadwaj, P., Novotny, L. (2007). Spectral dependence of single molecule fluorescence enhancement. *Opt. Express* 15:14266–14274.

103. Aslan, K., Wu, M., Lakowicz, J. R., Geddes, C. D. (2007). Fluorescent Core-Shell Ag@SiO2 Nanocomposites for Metal-Enhanced Fluorescence and Single Nanoparticle Sensing Platforms. *J. Am. Chem. Soc.* 129:1524–1525.

104. Gueroui, Z., Libchaber, A. (2004). Single-Molecule Measurements of Gold-Quenched Quantum Dots. *Phys. Rev. Lett.* 93:166108.

105. Zhang, J., Mustafa, Y. F., Chowdhury, H., Lakowicz, J, R. (2007). Metal-Enhanced Single-Molecule Fluorescence on Silver Particle Monomer and Dimer: Coupling Effect between Metal Particles. *Nano Lett.* 7:2101–2107.

106. Muthu, P., Gryczynski, I., Gryczynski, Z., Talent, J., Akopova I., Jain, K., Borejdo, J. (2007). Decreasing photobleaching by silver island films: application to muscle. *Anal. Biochem.* 366:228–236.

107. Muthu, P., Gryczynski, I., Gryczynski, Z., Talent, J., Akopova I., Jain, K., Borejdo, J. (2008). Decreasing photobleaching by silver nanoparticles on metal surfaces: application to muscle myofibrils. *J. Biomed. Opt.* 13:014023.

108. Kawano, K., Kitoh, T. (2001). Introduction to optical waveguide analysis. Wiley, pp.13–250.

109. Park, H. S., Lee, S. Y., Paek, U. C. and Chung, Y. J. (2005). Noncontact optical fiber coating removal technique with hot air stream. *Journal of Lightwave Technology* 23: 551-557.

110. http://www.marchplasma.com/pdf/cladding%20removale.pdf

111. Andrawis, A. S. and Santiago, J. B. (2006). Fiber optic sensor for simultaneous leak detection of hydrazine and nitrogen dioxide gases. *Optical Society of America.*

112. Chang, Y.-F., Chang, T.-C., Huang, Y.-M., Yu, K.-W., Hsieh, B.-Y., Chung, Y.-C. and Chou, C. (2007). A Fiber Optic Biosensor for the Identification of E. coli in Pure Cultures Isolated from Blood. *Journal of Bionanoscience* 1: 1-5.

*113.* Anthony G. E. (1995-2000). Cleaning and Surface Activation of Microfabricated Interdigitated Microsensor Electrodes (IMEs), Planar Metal Electrodes (PMEs), Independently Addressable Microband Electrodes (IAMEs), and E'Chem "Cell-On-A-Chip". *An ABTECH Application Note*

114. Andrade, A. L., Ferreira, J. M. F. and Domingues, R. Z. (2004). Surface modifications of alumina-silica glass fiber. *J. Biomed. Mater. Res. Part B: Applied Biomaterials* 70B: 378-383.

115. Wang, B., Oleschuk, R. D., Petkovich, P. M. and Horton, J. H. (2007). Chemical force titrations of antigen- and antibody-modified poly(methylmethacrylate). *Colloids Surf. B: Biointerfaces* 55: 107-114.

116. Krˇenková, J. and Foret, F. (2004). Immobilized microfluidic enzymatic reactors. *Electrophoresis* 25: 3550-3563.

117. Bai, Y. L., Koh, C. G., Boreman, M., Juang, Y. J., Tang, I. C., Lee, L. J. and Yang, S. T. (2006). Surface modification for enhancing antibody binding on polymer-based microfluidic device for enzyme-linked immunosorbent assay. *Langmuir* 22: 9458-9467.

118. Tsai, W.-C. and Lin, I. C. (2005). Development of a piezoelectric immunosensor for the detection of alpha-fetoprotein. *Sens. Actuators, B: Chemical* 106: 455-460.

119. Fixe, F., Dufva, M., Telleman, P. and Christensen, C. B. V. (2004). Functionalization of poly(methyl methacrylate) (PMMA) as a substrate for DNA microarrays. *Nucl. Acids Res.* 32: e9.

120. Henry, A. C., Tutt, T. J., Galloway, M., Davidson, Y. Y., McWhorter, C. S., Soper, S. A. and McCarley, R. L. (2000). Surface Modification of Poly(methyl methacrylate) Used in the Fabrication of Microanalytical Devices. *Anal. Chem.* 72: 5331-5337.

121. Avrameas, S. (1969). Coupling of enzyme to proteins with glutaraldehyde. *Immunochem.* 6: 43-52.

122. Bai, Y., Huang, W.-C. and Yang, S.-T. (2007). Enzyme-linked immunosorbent assay of <I>Escherichia coli</I> O157:H7 in surface enhanced poly(methyl methacrylate) microchannels. *Biotechnol. Bioeng.* 98: 328-339.

123. Bulmus, V., Ayhan, H. and Piskin, E. (1997). Modified PMMA monosize microbeads for glucose oxidase immobilization. *Chem. Eng. J.* 65: 71-76.

124. DeSilva, N. S., Ofek, I. and Crouch, E. C. (2003). Interactions of Surfactant Protein D (SP-D) with Fatty Acids. *Am. J. Respir. Cell Mol. Biol.*: 2003-0186OC

125. Grabarek, Z. and Gergely, J. (1990). Zero-length crosslinking procedure with the use of active esters. *Anal. Biochem.* 185: 131-135.

126. Nouaimi, M., Moschel, K. and Bisswanger, H. (2001). Immobilization of trypsin on polyester fleece via different spacers. *Enzyme Microb. Technol.* 29: 567-574.

127. Anderson, A. S., Dattelbaum, A. M., Montano, G. A., Price, D. N., Schmidt, J. G., Martinez, J. S., Grace, W. K., Grace, K. M. and Swanson, B. I. (2008). Functional PEG-modified thin films for biological detection. *Langmuir* 24: 2240-2247.

128. Britcher, L. G., Kehoe, D. C., Matisons, J. G., Smart, R. S. C. and Swincer, A. G. (1993). Silicones on glass surfaces. 2. Coupling agent analogs. *Langmuir* 9: 1609-1613.

129. Amankwa, L. N. and Kuhr', W. G. (1992). Trypsin-Modified Fused-Silica Capillary Microreactor for Peptide Mapping by Capillary Zone Electrophoresis. *Anal. Chem.* 64: 1610-1613.

130. Williams, R. A. and Blanch, H. W. (1994). Covalent immobilization of protein monolayers for biosensor applications. *Biosens. Bioelectron.* 9: 159-167.

131. Sheehan, P. E., Edelstein, R. L., Tamanaha, C. R. and Whitman, L. J. (2003). A simple pen-spotting method for arraying biomolecules on solid substrates. *Biosens. Bioelectron.* 18: 1455-1459.

132. Diao, J. P., Ren, D. C., Engstrom, J. R. and Lee, K. H. (2005). A surface modification strategy on silicon nitride for developing biosensors. *Anal. Biochem.* 343: 322-328.

133. Karymov, M. A., Kruchinin, A. A., Tarantov, Y. A., Balova, I. A., Remisova, L. A. and Vlasov, Y. G. (1995). Fixation of DNA Directly on Optical Wave-Guide Surfaces for Molecular Probe Biosensor Development. *Sens. Actuators, B: Chemical* 29: 324-327.

134. Manning, M., Harvey, S., Galvin, P. and Redmond, G. (2003). A versatile multi-platform biochip surface attachment chemistry. *Mater. Sci. Eng. C*

*Biomimetic Supramol. Syst.* 23: 347-351.

135. Orendorff, C. J., Gole, A., Sau, T. K., and Murphy, C. J. (2005). Surface-Enhanced Raman Spectroscopy of Self-Assembled Monolayers: Sandwich Architecture and Nanoparticle Shape Dependence. *Anal. Chem.* 77: 3261-3266.
136. Aslan, K., Leonenko, Z., Lakowicz, J. R., and Geddes, C. D. (2005). Fast and Slow Deposition of Silver Nanorods on Planar Surfaces: Application to Metal-Enhanced Fluorescence. *J. Phys. Chem. B* 109: 3157-3162.
137. Aslan, K., Huang, J., Wilson, G. M., and Geddes, C. D. (2006). Metal-Enhanced Fluorescence-Based RNA Sensing. *J. Am. Chem. Soc.* 128: 4206-4207.
138. Zhang, Y., Aslan, K., Previte, M. J. R., Malyn, S. N., and Geddes, C. D. (2006). Metal-Enhanced Phosphorescence: Interpretation in Terms of Triplet-Coupled Radiating Plasmons. *J. Phys. Chem. B* 110: 25108-25114.
139. Chowdhury, M. H., Ray, K., Aslan, K., Lakowicz, J. R., and Geddes, C. D. (2007). Metal-Enhanced Fluorescence of Phycobiliproteins from Heterogeneous Plasmonic Nanostructures. *J. Phys. Chem. C* 111: 18856-18863.
140. Bruchez, M., Moronne, M., Gin, P., Weiss, S., and Alivisatos, A. P. (1998). Semiconductor nanocrystals as fluorescent biological labels. *Science* 281: 2013-2016.
141. Chan, W. C. W. and Nie, S. M. (1998). Quantum dot bioconjugates for ultrasensitive nonisotopic detection. *Science* 281: 2016-2018.
142. Chowdhury, M. H., Aslan, K., Malyn, S. N., Lakowicz, J. R., and Geddes, C. D. (2006). Metal-enhanced chemiluminescence. *Journal of Fluorescence* 16: 295-299.
143. Chowdhury, M. H., Aslan, K., Malyn, S. N., Lakowicz, J. R., and Geddes, C. D. (2006). Metal-enhanced chemiluminescence: Radiating plasmons generated from chemically induced electronic excited states. *Appl. Phys. Lett.* 88: 173104.
144. Aslan, K. and Geddes, C. D. (2005). Microwave-Accelerated Metal-Enhanced Fluorescence: Platform Technology for Ultrafast and Ultrabright Assays. *Anal. Chem.* 77: 8057-8067.
145. Aslan, K., Malyn, S. N., and Geddes, C. D. (2006). Multicolor Microwave-Triggered Metal-Enhanced Chemiluminescence. *J. Am. Chem. Soc.* 128: 13372-13373.
146. Previte, M. J. R., Aslan, K., Malyn, S. N., and Geddes, C. D. (2006). Microwave Triggered Metal Enhanced Chemiluminescence: Quantitative Protein Determination. *Anal. Chem.* 78: 8020-8027.
147. Aslan, K. and Geddes, C. D. (2007). Microwave-Accelerated Ultrafast Nanoparticle Aggregation Assays Using Gold Colloids. *Anal. Chem.* 79: 2131-2136.
148. Previte, M. J. R., Aslan, K., and Geddes, C. D. (2007). Spatial and Temporal Control of Microwave Triggered Chemiluminescence: A Protein Detection Platform. *Anal. Chem.* 79: 7042-7052.
149. Previte, M. J. R. and Geddes, C. D. (2007). Microwave-Triggered Surface Plasmon Coupled Chemiluminescence. *J. Am. Chem. Soc.* 129: 9850-9851.
150. Lakowicz, J. R. (2001). Radiative Decay Engineering: Biophysical and

Biomedical Applications. *Anal. Biochem.* 298: 1-24.

151. Lakowicz, J. R., Shen, B., Gryczynski, Z., D'Auria, S., and Gryczynski, I. (2001). Intrinsic Fluorescence from DNA Can Be Enhanced by Metallic Particles. *Biochem. Biophys. Res. Commun.* 286: 875-879.

152. Lakowicz, J. R., Shen, Y., D'Auria, S., Malicka, J., Fang, J., Gryczynski, Z., and Gryczynski, I. (2002). Radiative Decay Engineering: 2. Effects of Silver Island Films on Fluorescence Intensity, Lifetimes, and Resonance Energy Transfer. *Anal. Biochem.* 301: 261-277.

153. Gryczynski, I., Malicka, J., Shen, Y., Gryczynski, Z., and Lakowicz, J. R. (2002). Multiphoton Excitation of Fluorescence near Metallic Particles: Enhanced and Localized Excitation. *J. Phys. Chem. B* 106: 2191-2195.

154. Ku~mmerlen, J., Leitner, A., Brunner, H., Aussenegg, F. R., and Wokaun, A. (1993). Enhanced dye fluorescence over silver island films: analysis of the distance dependence. *Mol. Phys.* 80: 1031 - 1046.

155. Ray, K., Badugu, R., and Lakowicz, J. R. (2006). Distance-Dependent Metal-Enhanced Fluorescence from Langmuir-Blodgett Monolayers of Alkyl-NBD Derivatives on Silver Island Films. *Langmuir* 22: 8374-8378.

156. Schneider, G., Decher, G., Nerambourg, N., Praho, R., Werts, M. H. V., and Blanchard-Desce, M. (2006). Distance-Dependent Fluorescence Quenching on Gold Nanoparticles Ensheathed with Layer-by-Layer Assembled Polyelectrolytes. *Nano Lett.* 6: 530-536.

157. Dulkeith, E., Ringler, M., Klar, T. A., Feldmann, J., MunozJavier, A., and Parak, W. J. (2005). Gold Nanoparticles Quench Fluorescence by Phase Induced Radiative Rate Suppression. *Nano Lett.* 5: 585-589.

158. Sokolov, K., Chumanov, G., and Cotton, T. M. (1998). Enhancement of Molecular Fluorescence near the Surface of Colloidal Metal Films. *Anal. Chem.* 70: 3898-3905.

159. Lee, J., Govorov, A. O., Dulka, J., and Kotov, N. A. (2004). Bioconjugates of CdTe Nanowires and Au Nanoparticles: Plasmon-Exciton Interactions, Luminescence Enhancement, and Collective Effects. *Nano Lett.* 4: 2323-2330.

160. Aslan, K., Lakowicz, J. R., Szmacinski, H., and Geddes, C. D. (2004). Metal-Enhanced Fluorescence Solution-Based Sensing Platform. *Journal of Fluorescence* 14: 677-679.

161. O.G. Tovmachenko, Graf, C., euvel, D. v. d., laaderen, A. v., and Gerritsen, H. (2006). Fluorescence Enhancement by Metal-Core/Silica-Shell Nanoparticles. *Adv. Mater.* 18: 91-95.

162. Stoermer, R. L. and Keating, C. D. (2006). Distance-Dependent Emission from Dye-Labeled Oligonucleotides on Striped Au/Ag Nanowires: Effect of Secondary Structure and Hybridization Efficiency. *J. Am. Chem. Soc.* 128: 13243-13254.

163. Stranik, O., Nooney, R., McDonagh, C., and MacCraith, B. (2007). Optimization of Nanoparticle Size for Plasmonic Enhancement of Fluorescence. *Plasmonics* 2: 15-22.

164. Zhang, J., Fu, Y., Liang, D., Nowaczyk, K., Zhao, R. Y., and Lakowicz, J. R. (2008). Single-Cell Fluorescence Imaging Using Metal Plasmon-Coupled Probe 2: Single-Molecule Counting on Lifetime Image. *Nano Lett.* 8: 1179-1186.

165. Zhang, J., Fu, Y., Chowdhury, M. H., and Lakowicz, J. R. (2008). Single-Molecule Studies on Fluorescently Labeled Silver Particles: Effects of Particle Size. *J. Phys. Chem. C* 112: 18-26.
166. Gryczynski, Z., Malicka, J., Gryczynski, I., Matveeva, E., Geddes, C. D., Aslan, K., and Lakowicz, J. R. (2004). Metal-enhanced fluorescence: a novel approach to ultra-sensitive fluorescence sensing assay platforms. *Biomedical Vibrational Spectroscopy and Biohazard Detection Technologies* 5321: 275-282.
167. Ekgasit, S., Thammacharoen, C., Yu, F., and Knoll, W. (2004). Evanescent Field in Surface Plasmon Resonance and Surface Plasmon Field-Enhanced Fluorescence Spectroscopies. *Anal. Chem.* 76: 2210-2219.
168. Matveeva, E., Gryczynski, Z., Malicka, J., Gryczynski, I., and Lakowicz, J. R. (2004). Metal-enhanced fluorescence immunoassays using total internal reflection and silver island-coated surfaces. *Anal. Biochem.* 334: 303-311.
169. Hong, B. and Kang, K. A. (2006). Biocompatible, nanogold-particle fluorescence enhancer for fluorophore mediated, optical immunosensor. *Biosens. Bioelectron.* 21: 1333-1338.
170. Aslan, K., Holley, P., and Geddes, C. D. (2006). Microwave-Accelerated Metal-Enhanced Fluorescence (MAMEF) with silver colloids in 96-well plates: Application to ultra fast and sensitive immunoassays, High Throughput Screening and drug discovery. *J. Immunol. Methods* 312: 137-147.
171. Pompa, P. P., MartiradonnaL, Torre, A. D., Sala, F. D., MannaL, De Vittorio, M., CalabiF, CingolaniR, and RinaldiR (2006). Metal-enhanced fluorescence of colloidal nanocrystals with nanoscale control. *Nat Nano* 1: 126-130.
172. Zhang, J. and Lakowicz, J. R. (2007). Metal-enhanced fluorescence of an organic fluorophore using gold particles. *Opt. Express* 15: 2598-2606.
173. Chen, Y., Munechika, K., and Ginger, D. S. (2007). Dependence of Fluorescence Intensity on the Spectral Overlap between Fluorophores and Plasmon Resonant Single Silver Nanoparticles. *Nano Lett.* 7: 690-696.
174. Xie, F., Baker, M. S., and Goldys, E. M. (2008). Enhanced Fluorescence Detection on Homogeneous Gold Colloid Self-Assembled Monolayer Substrates. *Chem. Mater.* 20: 1788-1797.
175. New, R. R. C. (1990). Liposomes : a practical approach. Oxford University Press, Oxford, England.
176. Oconnell, J. P., Campbell, R. L., Fleming, B. M., Mercolino, T. J., Johnson, M. D., and Mclaurin, D. A. (1985). A Highly Sensitive Immunoassay System Involving Antibody-Coated Tubes and Liposome-Entrapped Dye. *Clin. Chem.* 31: 1424-1426.
177. Viswanathan, S., Wu, L. C., Huang, M. R., and Ho, J. A. A. (2006). Electrochemical immunosensor for cholera toxin using liposomes and poly(3,4-ethylenedioxythiophene)-coated carbon nanotubes. *Anal. Chem.* 78: 1115-1121.
178. Ho, J. A. A., Wu, L. C., Huang, M. R., Lin, Y. J., Baeumner, A. J., and Durst, R. A. (2007). Application of ganglioside-sensitized liposomes in a flow injection immunoanalytical system for the determination of cholera toxin. *Anal. Chem.* 79: 246-250.

179. Ho, J. A. A., Zeng, S. C., Huang, M. R., and Kuo, H. Y. (2006). Development of liposomal immunosensor for the measurement of insulin with femtomole detection. *Anal. Chim. Acta* 556: 127-132.
180. Viswanathan, S. and Ho, J. A. A. (2007). Dual electrochemical determination of glucose and insulin using enzyme and ferrocene microcapsules. *Biosens. Bioelectron.* 22: 1147-1153.
181. Ho, J. A. and Huang, M. R. (2005). Application of a liposomal bioluminescent label in the development of a flow injection immunoanalytical system. *Anal. Chem.* 77: 3431-3436.
182. Ho, J. A. A. and Durst, R. A. (2000). Preparation of reagents for the determination of fumonisin B1 by flow-injection immunoanalysis. *Anal. Chim. Acta* 414: 51-60.
183. Ho, J. A. A. and Durst, R. A. (2000). Development of a flow-injection liposome immunoanalysis system for fumonisin B1. *Anal. Chim. Acta* 414: 61-69.
184. Ho, J. A. A. and Wauchope, R. D. (2002). A strip liposome immunoassay for aflatoxin B-1. *Anal. Chem.* 74: 1493-1496.
185. Ho, J. A. A. and Hsu, H. W. (2003). Procedures for preparing Escherichia coli O157 : H7 immunoliposome and its application in liposome immunoassay. *Anal. Chem.* 75: 4330-4334.
186. Ho, J. A. A., Hsu, H. W., and Huang, M. R. (2004). Liposome-based microcapillary immunosensor for detection of Escherichia coli O157 : H7. *Anal. Biochem.* 330: 342-349.
187. Lo, L. W., Ho, J.A. A., Chang, Y.H., Chang, C.H, Yang, C.S. (2004). Potential Usage of Liposome-encapulated Phosphor for in vivo Imaging of Tissue Oxygenation. *Biomedical Engineering Applications, Basis and Communications* 16: 224-229.
188. Newman, G. C. and Huang, C.-H. (1975). Structural studies on phosphatidylcholine-cholesterol mixed vesicles. *Biochemistry (Mosc).* 14: 3363-3370.
189. Gregoriadis, G. and Buckland, R. A. (1973). Enzyme-containing Liposomes alleviate a Model for Storage Disease. *Nature* 244: 170-172.
190. Fendler, J. H. and Romero, A. (1977). Liposomes as drug carriers. *Life Sci.* 20: 1109-1120.
191. Schwendener, R. A. (2007). Liposomes in biology and medicine. *Bio-Applications of Nanoparticles* 620: 117-128.
192. Wang, C.-Y. and Huang, L. (1987). pH-Sensitive Immunoliposomes Mediate Target-Cell-Specific Delivery and Controlled Expression of a Foreign Gene in Mouse. *Proceedings of the National Academy of Sciences* 84: 7851-7855.
193. Zelphati, O., Imbach, J. L., Signoret, N., Zon, G., Rayner, B., and Leserman, L. (1994). Antisense Oligonucleotides in Solution or Encapsulated in Immunoliposomes Inhibit Replication of Hiv-1 by Several Different Mechanisms. *Nucleic Acids Res.* 22: 4307-4314.
194. Nicolotti, R. A. and Kinsky, S. C. (1975). Immunogenicity of liposomal model membranes sensitized with mono(p-azobenzenearsonic acid)tyrosylphosphatidylethanolamine derivatives. Antibody formation and delayed hypersensitivity reaction. *Biochemistry (Mosc).* 14: 2331-2337.

195. Karasavvas, N., Beck, Z., Tong, J., Matyas, G. R., Rao, M., McCutchan, F. E., Michael, N. L., and Alving, C. R. (2008). Antibodies induced by liposomal protein exhibit dual binding to protein and lipid epitopes. *Biochem. Biophys. Res. Commun.* 366: 982-987.

196. Allison, A. C. and Gregoriadis, G. (1974). Liposomes as immunological adjuvants. *Nature* 252: 252-252.

197. Gregoriadis, G. and Allison, A. C. (1974). Entrapment of proteins in liposomes prevents allergic reactions in pre-immunised mice. *FEBS Lett.* 45: 71-74.

198. Fenske, D. B. and Cullis, P. R. (2008). Liposomal nanomedicines. *Expert Opinion on Drug Delivery* 5: 25-44.

199. Posner, R. (2002). Liposomes. *J Drugs Dermatol* 1: 161-164.

200. Fiume, Z. (2001). Final Report on the Safety Assessment of Lecithin and Hydrogenated Lecithin. *Int. J. Toxicol.* 20: 21-45.

201. Esch, M. B., Locascio, L. E., Tarlov, M. J., and Durst, R. A. (2001). Detection of viable Cryptosporidium parvum using DNA-modified liposomes in a microfluidic chip. *Anal. Chem.* 73: 2952-2958.

202. Esch, M. B., Baeumner, A. J., and Durst, R. A. (2001). Detection of Cryptosporidium parvum using oligonucleotide-tagged liposomes in a competitive assay format. *Anal. Chem.* 73: 3162-3167.

203. Siebert, S. T. A., Reeves, S. G., and Durst, R. A. (1993). Liposome Immunomigration Field Assay Device for Alachlor Determination. *Anal. Chim. Acta* 282: 297-305.

204. Roberts, M. A. and Durst, R. A. (1995). Investigation of Liposome Based Immunomigration Sensors for the Detection of Polychlorinated-Biphenyls. *Anal. Chem.* 67: 482-491.

205. Gregoriadis, G. (1984). Old Drugs in New Clothing. *Nature* 310: 186-187.

206. Szoka, F. and Papahadjopoulos, D. (1978). Procedure for Preparation of Liposomes with Large Internal Aqueous Space and High Capture by Reverse-Phase Evaporation. *Proceedings of the National Academy of Sciences* 75: 4194-4198.

207. Macdonald, R. C., Macdonald, R. I., Menco, B. P. M., Takeshita, K., Subbarao, N. K., and Hu, L. R. (1991). Small-Volume Extrusion Apparatus for Preparation of Large, Unilamellar Vesicles. *Biochim. Biophys. Acta* 1061: 297-303.

208. Szoka, F., Olson, F., Heath, T., Vail, W., Mayhew, E., and Papahadjopoulos, D. (1980). Preparation of unilamellar liposomes of intermediate size (0.1-0.2 [mu]m) by a combination of reverse phase evaporation and extrusion through polycarbonate membranes. *Biochim. Biophys. Acta* 601: 559-571.

209. Monroe, D. (1986). Liposome immunoassay: a new ultrasensitive analytical method. *Am. Clin. Prod. Rev* 5: 34-41

210. Monroe, D. (1989). Novel Liposome Immunoassays for Detecting Antigens, Antibodies, and Haptens. *J. Liposome Res.* 1: 339 - 377.

211. Batzri, S. and Korn, E. D. (1973). Single bilayer liposomes prepared without sonication. *Biochim. Biophys. Acta* 298: 1015-1019.

212. Deamer, D. and Bangham, A. D. (1976). Large volume liposomes by an ether vaporization method. *Biochim. Biophys. Acta* 443: 629-634.

213. Szoka, F. and Papahadjopoulos, D. (1980). Comparative Properties and Methods of Preparation of Lipid Vesicles (Liposomes). *Annu. Rev. Biophys. Bioeng.* 9: 467-508.
214. Canova-Davis, E., Redemann, C. T., Vollmer, Y. P., and Kung, V. T. (1986). Use of a reversed-phase evaporation vesicle formulation for a homogeneous liposome immunoassay. *Clin. Chem.* 32: 1687-1691.
215. Frost, S. J., Chakraborty, J., and Firth, G. B. (1996). Urinary microalbumin measurement using a homogeneous liposomal immunoassay. *J. Immunol. Methods* 194: 105-111.
216. Ishimori, Y. and Rokugawa, K. (1993). Stable liposomes for assays of human sera. *Clin. Chem.* 39: 1439-1443.
217. Rongen, H. A. H., Vanderhorst, H. M., Hugenholtz, G. W. K., Bult, A., Vanbennekom, W. P., and Vandermeide, P. H. (1994). Development of a Liposome Immunosorbent-Assay for Human Interferon-Gamma. *Anal. Chim. Acta* 287: 191-199.
218. Rongen, H. A. H., Vannierop, T., Vanderhorst, H. M., Rombouts, R. F. M., Vandermeide, P. H., Bult, A., and Vanbennekom, W. P. (1995). Biotinylated and Streptavidinylated Liposomes as Labels in Cytokine Immunoassays. *Anal. Chim. Acta* 306: 333-341.
219. Locascio-Brown, L., Plant, A. L., and Durst, R. A. (1988). Behavior of Liposomes in Flow-Injection Systems. *Anal. Chem.* 60: 792-797.
220. Wu, T. G. and Durst, R. A. (1990). Liposome-based flow injection enzyme immunoassay for theophylline. *Microchimica Acta* 100: 187-195.
221. Locascio-Brown, L., Plant, A. L., Chesler, R., Kroll, M., Ruddel, M., and Durst, R. A. (1993). Liposome-Based Flow-Injection Immunoassay for Determining Theophylline in Serum. *Clin. Chem.* 39: 386-391.
222. Lee, K. S., Kim, T. H., Shin, M. C., Lee, W. Y., and Park, J. K. (1999). Disposable liposome immunosensor for theophylline combining an immunochromatographic membrane and a thick-film electrode. *Anal. Chim. Acta* 380: 17-26.
223. Locascio-Brown, L. and Choquette, S. J. (1993). Measuring Estrogens Using Flow-Injection Immunoanalysis with Liposome Amplification. *Talanta* 40: 1899-1904.
224. Rule, G. S., Palmer, D. A., Reeves, S. G., and Durst, R. A. (1994). Use of Protein A in a Liposome-enhanced Flow-injection Immunoassay. *Analytical Proceedings* 31: 339-340.
225. Reeves, S. G., Rule, G. S., Roberts, M. A., Edwards, A. J., and Durst, R. A. (1994). Flow-Injection Liposome Immunoanalysis (Filia) for Alachlor. *Talanta* 41: 1747-1753.
226. Lee, M. and Durst, R. A. (1996). Determination of Imazethapyr Using Capillary Column Flow Injection Liposome Immunoanalysis. *J. Agric. Food Chem.* 44: 4032-4036.
227. Lee, M., Durst, R. A., and Wong, R. B. (1998). Development of flow-injection liposome immunoanalysis (FILIA) for imazethapyr. *Talanta* 46: 851-859.
228. Baumner, A. J. and Schmid, R. D. (1998). Development of a new immunosensor for pesticide detection: a disposable system with liposome-enhancement and amperometric detection. *Biosens. Bioelectron.* 13: 519-

529.
229. Nguyen, T., McNamara, K. P., and Rosenzweig, Z. (1999). Optochemical sensing by immobilizing fluorophore-encapsulating liposomes in sol-gel thin films. *Anal. Chim. Acta* 400: 45-54.
230. Yoon, C.-H., Cho, J.-H., Oh, H.-I., Kim, M.-J., Lee, C.-W., Choi, J.-W., and Paek, S.-H. (2003). Development of a membrane strip immunosensor utilizing ruthenium as an electro-chemiluminescent signal generator. *Biosens. Bioelectron.* 19: 289-296.
231. Tsukagoshi, K., Akasaka, H., Okumura, Y., Fukaya, R., Otsuka, M., Fujiwara, K., Umehara, H., Maeda, R., and Nakajima, R. (2000). Immunoassay using chemiluminescence detection of dyestuff-containing liposomes as a labeling reagent. *Anal. Sci.* 16: 121-124.
232. Ho, J.-a., Zeng, S.-C., Tseng, W.-H., Lin, Y.-J., and Chen, C.-h.Liposome-based immunostrip for the rapid detection of Salmonella. *Anal. Bioanal. Chem.*
233. Chen, C.-S., Yao, J., and Durst, R. (2006). Liposome encapsulation of fluorescent nanoparticles: Quantum dots and silica nanoparticles. *J. Nanopart. Res.* 8: 1033-1038.
234. Choquette, S. J., Locascio-Brown, L., and Durst, R. A. (1992). Planar waveguide immunosensor with fluorescent liposome amplification. *Anal. Chem.* 64: 55-60.
235. Patil, V., Mayya, K. S., Pradhan, S. D., and Sastry, M. (1997). Evidence for Novel Interdigitated Bilayer Formation of Fatty Acids during Three-Dimensional Self-Assembly on Silver Colloidal Particles. *J. Am. Chem. Soc.* 119: 9281-9282.
236. Zhang, L., Li, P., Li, D., Guo, S., and Wang, E. (2008). Effect of Freeze-Thawing on Lipid Bilayer-Protected Gold Nanoparticles. *Langmuir* 24: 3407-3411.
237. He, P. and Urban, M. W. (2005). Phospholipid-Stabilized Au-Nanoparticles. *Biomacromolecules* 6: 1224-1225.
238. Mornet, S., Lambert, O., Duguet, E., and Brisson, A. (2005). The Formation of Supported Lipid Bilayers on Silica Nanoparticles Revealed by Cryoelectron Microscopy. *Nano Lett.* 5: 281-285.
239. Martina, M. S., Fortin, J. P., Menager, C., Clement, O., Barratt, G., Grabielle-Madelmont, C., Gazeau, F., Cabuil, V., and Lesieur, S. (2005). Generation of Superparamagnetic Liposomes Revealed as Highly Efficient MRI Contrast Agents for in Vivo Imaging. *J. Am. Chem. Soc.* 127: 10676-10685.
240. Giri, J., Guha Thakurta, S., Bellare, J., Kumar Nigam, A., and Bahadur, D. (2005). Preparation and characterization of phospholipid stabilized uniform sized magnetite nanoparticles. *J. Magn. Magn. Mater.* 293: 62-68.
241. Wijaya, A. and Hamad-Schifferli, K. (2007). High-Density Encapsulation of Fe3O4 Nanoparticles in Lipid Vesicles. *Langmuir* 23: 9546-9550.
242. Kunisawa, J., Masuda, T., Katayama, K., Yoshikawa, T., Tsutsumi, Y., Akashi, M., Mayumi, T., and Nakagawa, S. (2005). Fusogenic liposome delivers encapsulated nanoparticles for cytosolic controlled gene release. *J. Controlled Release* 105: 344-353.
243. Park, S.-H., Oh, S.-G., Mun, J.-Y., and Han, S.-S. (2005). Effects of silver

nanoparticles on the fluidity of bilayer in phospholipid liposome. *Colloids Surf. B. Biointerfaces* 44: 117-122.

244. Park, S.-H., Oh, S.-G., Mun, J.-Y., and Han, S.-S. (2006). Loading of gold nanoparticles inside the DPPC bilayers of liposome and their effects on membrane fluidities. *Colloids Surf. B. Biointerfaces* 48: 112-118.

245. Wu, S. X., Zeng, H. X., and Schelly, Z. A. (2005). Growth of Uncapped, Subnanometer Size Gold Clusters Prepared via Electroporation of Vesicles. *J. Phys. Chem. B* 109: 18715-18718.

246. Paasonen, L., Laaksonen, T., Johans, C., Yliperttula, M., Kontturi, K., and Urtti, A. (2007). Gold nanoparticles enable selective light-induced contents release from liposomes. *J. Controlled Release* 122: 86-93.

247. Yalow, R. S. and Berson, S. A. (1959). Assay of Plasma Insulin in Human Subjects by Immunological Methods. *Nature* 184: 1648-1649.

248. Hefle, S. L. (1995). Immunoassay Fundamentals. *Food Technol.* 49: 102-107.

249. Howanitz, J. H. (1992). Overview of Nonisotopic Immunoassay Labels in *Immunochemical Assays and Biosensor Technology for the 1990s*, Yasushi Kasahara Robert M. Nakamura, Garry A. Rechnitz (Ed.), Amer Society for Microbiology, pp. 22-35.

250. Lacharme, F., Vandevyver, C., and Gijs, M. A. M. (2008). Full On-Chip Nanoliter Immunoassay by Geometrical Magnetic Trapping of Nanoparticle Chains. *Anal. Chem.* 80: 2905 -2910.

251. Liu, X., Dai, Q., Austin, L., Coutts, J., Knowles, G., Zou, J. H., Chen, H., and Huo, Q. (2008). A one-step homogeneous immunoassay for cancer biomarker detection using gold nanoparticle probes coupled with dynamic light scattering. *J. Am. Chem. Soc.* 130: 2780-2782.

252. Maier, I., Morgan, M. R. A., Lindner, W., and Pittner, F. (2008). Optical Resonance-Enhanced Absorption-Based Near-Field Immunochip Biosensor for Allergen Detection. *Anal. Chem.* 80: 2694-2703.

253. Soini, E. and Hemmila, I. (1979). Fluoroimmunoassay: present status and key problems. *Clin. Chem.* 25: 353-361.

254. Hemmila, I., Malminen, O., Mikola, H., and Lovgren, T. (1988). Homogeneous Time-Resolved Fluoroimmunoassay of Thyroxine in Serum. *Clin. Chem.* 34: 2320-2322.

255. Briggs, J., Elings, V. B., and Nicoli, D. F. (1981). Homogeneous Fluorescent Immunoassay. *Science* 212: 1266-1267.

256. Bright, F. V. and Mcgown, L. B. (1985). Homogeneous Immunoassay of Phenobarbital by Phase-Resolved Fluorescence Spectroscopy. *Talanta* 32: 15-18.

257. Kronick, M. N. and Little, W. A. (1975). A new immunoassay based on fluorescence excitation by internal reflection spectroscopy. *J. Immunol. Methods* 8: 235-240.

258. Sutherland, R. M., Dahne, C., Place, J. F., and Ringrose, A. S. (1984). Optical-Detection of Antibody Antigen Reactions at a Glass Liquid Interface. *Clin. Chem.* 30: 1533-1538.

259. Okabayashi, Y. and Ikeuchi, I. (1998). Liposome immunoassay by long-lived fluorescence detection. *Analyst* 123: 1329-1332.

260. Chou, C., Hsu, H. Y., et al. (2007) Fiber optic biosensor for the detection of

C-reactive protein and the study of protein binding kinetics. *Journal of Biomedical Optics* 12: 024025.

261. Smith, R. H. et al. (1999) Development of kinetic ligand-binding assays using a fiber optic sensor. *Clinical Chemistry* 45: 1683-1685.
262. Marazuela, D., Moreno-Bondi, M. C. (2002) Fiber-optic biosensors - an overview. *Analytical and Bioanalytical Chemistry* 372: 664-682.
263. Yeung, D., Gill, A., Maule, C. H., Davies, R. J. (1995) Detection and quantification of biomolecular interactions with optical biosensors. *TrAC-Trends in Analytical Chemistry* 14: 49-56.
264. Glaser, R. W. (1993) Antigen-Antibody Binding and Mass Transport by Convection and Diffusion to a Surface: A Two-Dimensional Computer Model of Binding and Dissociation Kinetics. *Analytical Biochemistry* 213: 152-161.
265. Morton, T. A., Myszka, D. G., Chaiken, I. M. (1995) Interpreting Complex Binding Kinetics from Optical Biosensors: A Comparison of Analysis by Linearization, the Integrated Rate Equation, and Numerical Integration. *Analytical Biochemistry* 227: 176-185.
266. Private communication: Ten or more protein A molecules are bound on each gold nanoparticle (GNP, diameter =20nm). This information is confirmed by Dr. S. Bagga, a senior technical scientist of Sigma-Aldrich Technical Services Department in the Sigma-Aldrich Inc. (St. Louis, MO, USA).
267. Borejdo, J.; Gryczynski, Z.; Calander, N.; Muthu, P.; Gryczynski, I. (2006) Application of Surface Plasmon Coupled Emission to Study of Muscle. *Biophysical Journal* 91: 2626-2635.
268. Haes, A.J.; Zou, S.; Schatz, G..C.; Van Duyne, R.P. (2004) A Nanoscale Optical Biosensor The Long Range Distance Dependence of the Localized Surface Plasmon Resonance of Noble Metal Nanoparticles. *Journal of Physical Chemistry B* 108:109-116.
269. Haes, A.J.; Zou, S.; Schatz, G..C.; Van Duyne, R.P. (2004) Nanoscale optical biosensor Short range distance dependence of the localized surface plasmon resonance of noble metal nanoparticles. *Journal of Physical Chemistry B* 108: 6961-6968.
270. Byun, K. M.; Kim, S. J.; Kim, D. H. (2006) Profile effect on the feasibility of extinction-based localized surface plasmon resonance biosensors with metallic nanowires. *Applied Optics* 45: 3382-3389.
271. Chien, D. H., Tsai, C. H., Lo, S. S, Chen, C. C., Chang, J. Y. (2005). Solid immersion lenses in planar waveguides. *IEEE J. Lightwave Technol.* 23:2746–2748.
272. Lakowicz, J. R., Malicka, J., Gryczynski, I., Gryczynski, Z. (2003). Directional surface plasmon-coupled emission: a new method for high sensitivity detection. *Biochem. Biophys. Res. Commun.* 307:435–439.
273. Lo, S. S., Wang, M. S., Chen, C, C. (2004). Semiconductor hollow optical waveguides formed by omni-directional reflectors *Opt. Express* 12:6589–6593.
274. Miao, X., Wilson, B. K., Lin, L. Y. (2008). Localized surface plasmon assisted microfluidic mixing. *Appl. Phys. Lett.* 92:124108.
275. Zheng, Y. B., Huang, T., Amit, J., Desai Y., Wang, S. J., Tan, L. K., Gao, H., Cheng, A., Huan, Hon. (2007). Thermal behavior of localized surface

plasmon resonance of Au/TiO2 core/shell nanoparticle arrays *Appl. Phys. Lett.* 90:183117.

276. Hsiao, F. L., Chan, C. H., Chen, C. C. (2006). Optical properties of metallo-dielectric opals. *Appl. Phys. Lett.* 89: 253123.

277. Huang, C. K. Hou, C. H., Chen, C. C., Tsai, Y. L. Chang, L. M., Wei, H. S., Hsieh, K. H., Chan, C. H. (2008). Magnetic SiO2/Fe3O4 Colloidal Crystal. *Nanotechnol.* 19: 055701.

278. Salvaraju, T., Das, J., Han, S. W., Yang, H. (2008). Ultrasensitive electrochemical immunosensing using magnetic beads and gold nanocatalysts. *Biosensors and Bioelectron.* 23:932–938.

# 9

# Surface Plasmon Enhanced Photochemistry

Stephen K. Gray
Chemistry Sciences and Engineering Division
Argonne National Laboratory, Argonne, Illinois 60439, USA.

## 9.1 INTRODUCTION

Surface plasmons (SPs) are collective electronic excitations near the surfaces of metallic structures. They can usually be described well with classical electromagnetic theory and correspond to electromagnetic fields that are localized and relatively intense near the metallic surfaces [1, 2]. These properties make them potentially useful for a variety of applications in optoelectronics, chemical and biological sensing, and other areas. Metallic nanostructures such as metal nanoparticles and nanostructured thin metal films, particularly those composed of noble metals such as silver or gold, are of special interest because often their SPs can be excited with visible-UV light and are relatively robust.

Figure 1.1 displays an idealized metal nanoparticle / molecule system exposed to an incident light wave, $\mathbf{E}_0$. A dipolar SP excitation can lead to large enhancements in the near-fields around the metal nanoparticle surfaces, particularly at the north and south poles along the axis associated with the incident polarization, which is the *z*-axis in this example (see red-colored regions). Let the typical magnitude of the field due to an SP excitation in such regions be written as

$$\mathrm{E} = g\, \mathrm{E}_0 \tag{1}$$

where $\mathrm{E}_0 = |\mathbf{E}_0|$. There is much variation in estimates of $g$ depending on specific details of a given problem and the level of theory being used. However, realistic amplitude enhancements are
estimated to be on the order of $10\text{-}10^{2.5}$ [3]. Suppose molecules are placed close to a metallic nanostructure. Molecular responses arise from a molecule-field Hamiltonian interaction term,

$$\mathrm{W} = -\,\boldsymbol{\mu} \cdot \mathbf{E}\,(t) \tag{2}$$

where $\mathbf{E}\,(t)$ is the total electric field near the molecule. While quantum mechanically $\boldsymbol{\mu}$ is an operator and molecular transition probabilities are then related to matrix elements of Eq. (2) between initial and final states, a classical model involving an oscillating dipole is instructive. In this picture, the classical dipole is $\boldsymbol{\mu}(t) = \boldsymbol{\mu}_0 + \boldsymbol{\mu}_m(t)$ where the non-static contribution $\boldsymbol{\mu}_m(t)$ is associated with the transition(s). This classical transition dipole is also indicated in Figure 1.1. If a molecule is close to a metallic nanostructure being illuminatcd with light near an SP resonance, an enhanced molecular response is expected due to excitation of this dipole. The most well-known surface plasmon enhanced *spectroscopy* is surface-enhanced Raman spectroscopy, or SERS [4-6], which can exhibit signal enhancements on the order of $g^4$, i.e. $10^4$-$10^8$. This factor can be thought of as arising from a product of enhancement factors on the order of $g^2$ each for absorption and emission. Actually, in addition to the electromagnetic (EM) enhancement being discussed here, there is a chemical enhancement due to molecule-metal electronic interactions modifying the molecular level structure and transition moments. The chemical effect is now believed to be generally smaller than the EM effect, but could still contribute up to an additional order of $10^2$ to the overall enhancement [7]. Surface-enhanced fluorescence [8-11] is another important example of a plasmon-enhanced spectroscopic process.

The topic of this review, however, is surface plasmon enhanced and / or controlled *photochemistry*, which involves SPs influencing how actual chemical change takes place. For an interesting recent discussion and review of photochemistry on metal nanoparticles in general, see also Ref. [12]. The present review begins with a discussion of the theoretical background and then moves on to discuss several case studies involving photodissociation, isomerization and aggregation processes.

## 9.2 THEORETICAL CONSIDERATIONS

Nitzan and Brus first suggested the possibility of surface plasmon enhanced photochemistry in 1981 [13, 14]. They studied a phenomenological model for a molecule interacting with a small spherical metal nanoparticle that can support a SP resonance when irradiated. The model was derived in detail in a subsequent paper by Gersten and Nitzan [15]. I will thus refer to the model as the Nitzan-Brus-Gerstan or NBG model. The calculations based on the NBG model showed that both UV photodissociation and IR multiphoton absorption could be surface plasmon enhanced [13, 14].

While the specifics of the NBG model can be somewhat involved [15], it is in essence an intuitive model of two coupled oscillating dipoles. One dipole is associated with either an electronic or nuclear transition moment of the molecule. The other dipole is chosen to be consistent with the SP excitation of the metal nanoparticle. Let the metal nanoparticle of radius $a$ be at the origin and let the molecule lie on the positive $z$-axis a distance $d$ from the nearest surface of the metal so that $a + d$ is its distance from the origin, as in Figure 9.1. For simplicity I assume both the molecular transition dipole and SP dipole are oriented along the $z$-axis. Let the $z$-components of the molecular and SP dipoles be denoted by $\mu_m$ and $\mu_s$, respectively. The NBG model is then

$$\frac{d^2\mu_m}{dt^2} + \omega_m^2\mu_m + \gamma_m\frac{d\mu_m}{dt} = \alpha_m\omega_m^2E(t) \qquad (3)$$

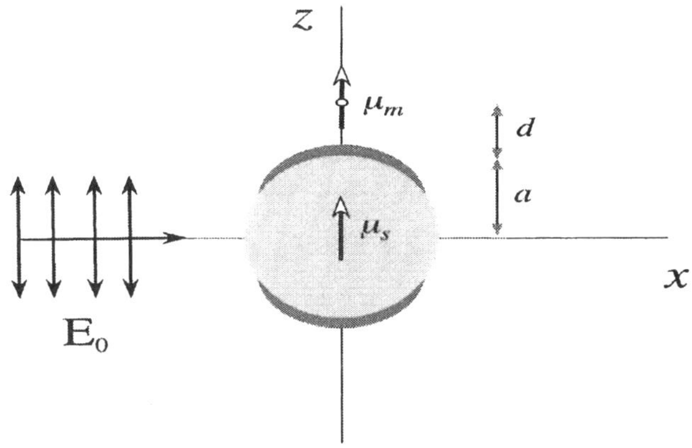


**Figure 9.1:** Schematic diagram of a metal nanoparticle / molecule system with z-polarized incident light. A spherical nanoparticle of radius a (gold-colored) is centered at the origin that, in the small particle limit, is consistent with a oscillating dipole, μs, at the origin. Regions of high near-field intensity are indicated (red). A molecule (small white circle) is assumed to lie along the z-axis a distance d from the metal surface and its induced dipole moment, μm, is also taken to be on the z-axis and centered on the molecule.

$$\frac{d^2\mu_s}{dt^2} + \omega_s^2\mu_s + \gamma_s\frac{d\mu_s}{dt} = \alpha_s\omega_s^2\overline{E}(t) \qquad (4)$$

In Eq. (3), $\omega_m$ is the relevant molecular transition frequency, $\gamma_m$ is a damping rate, $\alpha_m$ is a polarizability, and $E(t)$ is the $z$-component of the total electric field in the vicinity of the molecule. If $E$(t) were simply of the form $E_0\cos(\omega t)$, then Eq. (3) is the well-known phenomenological Lorentzian oscillator model of absorption which leads to an approximate Lorentzian form for the absorption cross section [1]. Similar remarks hold for the SP dipole, $\mu_s(t)$, if $\overline{E}(t) = E_0\cos(\omega t)$, where $\overline{E}(t)$ is the $z$-component of the total electric field near the SP dipole. The parameters $\omega_s$, $\gamma_s$ and $\alpha_s$ in this case are chosen such that the resulting Lorentzian cross section approximates the known exact surface plasmon absorption cross section or its appropriate form in the quasistatic ($a$ << $\lambda$=2 πc/ω) limit. Note that I am using a simplified notation compared to the various notations of Refs. [13-15]. (Relative to Ref. [13], for example, my definitions of surface plasmon dipole

parameters are such that $\gamma_s = 2\,\gamma_1$, $\omega_s^2 = \omega_1^2 + \gamma_1^2$, and $\alpha_s \omega_s^2 = \alpha_1 \left(\omega_1^2 + \gamma_1^2\right)$.)
The two dipoles are coupled through the $E(t)$ and $\overline{E}(t)$ electric field components:

$$E(t) = E_0 \cos(\omega t) + \frac{2\mu_s}{(a+d)^3} \tag{5}$$

$$\overline{E}(t) = E_0 \cos(\omega t) + \frac{2\mu_m}{(a+d)^3} \tag{6}$$

That is, the electric field near the molecule, Eq. (5), is a sum of the incident field and the field that results due to the oscillating SP dipole. Similarly, the electric field near the SP dipole, Eq. (6), is the sum of the incident field and the field that results from the oscillating molecular dipole. (Note: Gaussian units are being assumed for all electromagnetic variables.)

Nitzan and Brus developed an analytical formula for the molecular absorption cross section given the model defined above [14]. Figure 9.2 is taken from Ref. [13] and shows the calculated absorption cross section based on the model associated with the photodissociation of $I_2$. (The $I_2^*$ formed through the absorption process is very short lived.) Photodissociation predicted to be enhanced as the molecule is placed near a silver metal nanoparticle of radius $a$ = 50 nm near the electronic transition resonance position of $\omega_0 \approx$ 22,200 cm$^{-1}$. If $e_{metal}(\omega)$ is the dielectric function for the metal, a small metal nanoparticle plasmon in air will have its dipolar surface plasmon resonance at frequency $\omega_s$ such that [1]

$$e_{metal}(\omega_s) = -2 \tag{7}$$

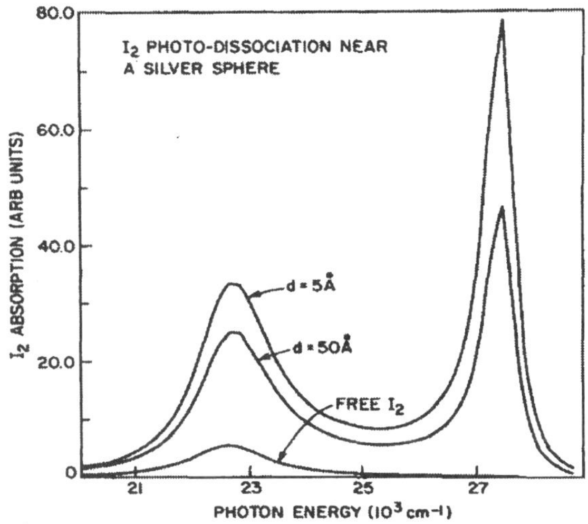


**Figure 9.2:** The photodissociation cross section for $I_2$ is predicted to be considerably enhanced for $I_2$ placed near a small silver metal nanoparticle. Reprinted figure with permission from Ref. [13]. Copyright 1981 by the American Institute of Physics.

Consistent with the Johnson and Christy dielectric constant data for silver [16], $\omega_s$ = 3.5 eV = 28,230 cm$^{-1}$ (or $\lambda_s = 2\pi c/\omega_s \approx$ 354 nm). Figure 9.2 also displays a noticeable absorption enhancement at this SP frequency. Ref. [14] presented more details of these calculations, as well as results for Au, Cu and InSb metal nanoparticles. The authors also considered a similar model for infrared multiphoton dissociation of $SF_6$, also showing that this process can be surface plasmon enhanced. Enhancements, as inferred from comparing relative magnitudes of cross sections and also looking at energy accumulation as a function of time were found to be typically in the 10-10$^3$ range [13, 14]. Such enhancements are not as large as can be achieved with SERS [3-7] owing to the longer time scales involved that allow for energy loss by the molecule to compete with energy pumping.

In addition to Nitzan and Brus's work [13, 14] there have been a number of other important theoretical contributions pertaining to surface enhanced photochemistry. For example, Metiu and Das [17], Gersten and Nitzan [18], and Leung and George [19] developed and explored more sophisticated classical models using Green's function approaches that allow for a more rigorous treatment of the contributions to the total electric field.

## 9.3 CASE STUDIES

### 9.3. b *Photodissociation.*

Shortly after Nitzan and Brus's theoretical prediction [13] of the possibility of surface enhanced photochemistry due to plasmon interactions, Goncher and Harris [20] presented experimental evidence for enhanced photofragmentation of several aromatic molecules (pyridine, pyrazine, and benzaldehyde) on roughened silver surfaces using low intensity UV radiation (364 nm). See also the more detailed, subsequent paper of Goncher, Parsons and Harris [21].

The incident laser intensities used by Goncher and Harris [20] were several orders of magnitude smaller than those required to fragment the molecules in the gas phase, pointing to large enhancements in the absorption cross sections. Furthermore, the roughened surface can be considered to be composed of many small silver nanoparticles and it is significant that the incident wavelength, 364 nm, is close to the small silver particle SP wavelength, $\lambda_s$ = 354 nm, noted in the previous section. However, the photodecomposition processes in question are complex: two or more photons may be absorbed, radical reactions and ionization can occur, and a variety of products can form. Nonetheless other features consistent with plasmon enhanced photochemistry and predictions of the NBG model were found. For example, the multiphoton absorption case of Nitzan and Brus [13] predicted an optimal photochemical rate for a molecule placed at an intermediate distance, $d \approx$ 10-50 Å from the surface of the metal, presumably due to a competition between small $d$ values being advantageous for high evanescent near fields but larger $d$ values being advantageous for reducing energy transfer back from the molecule to the nanoparticle. Consistent with this, the initial photodecomposition rates for pyridine [20, 21] were largest when 10-20 spacer layers were between the roughened metal film and the active molecules ($d \approx$15-20 Å).

Chen and Osgood reported surface enhanced photodissociation of dimethyl cadmium, $Cd(CH_3)_2$ [22, 23], which is believed to involve a more direct, one photon dissociation mechanism. The authors employed a thin dielectric film (carbon) with small deposited metal nanoparticles of either cadmium or gold with sizes in the 10-300 nm range. In the case of small Cd nanoparticles a surface plasmon resonance is expected to exist at wavelengths near the incident UV wavelength, 257 nm, whereas the surface plasmon resonance for small Au nanoparticles occurs at much longer wavelengths ($\approx$ 485 nm). The experiment involved exposing the thin film system to a mixture $Cd(CH_3)_2$ and argon, irradiating it, and studying the resulting thin film / nanoparticle system with electron microscopy. The dissociation of $Cd(CH_3)_2$ creates additional Cd atoms that can add to the existing nanoparticles. The authors observed, in the case of Cd nanoparticles, significant growth of ellipsoidal particles from spherical ones, and that there was a correlation between the incident light polarization and the long axis of the ellipsoids. These observations are consistent with the fact that the highest near fields will occur at the poles of a nanoparticle along the axis of polarization and that enhanced photodissociation will occur in these regions, creating more elongated particles. No such growth was observed in the case of the gold nanoparticles which are not resonant at this wavelength.

It is important to note that the mere presence of SP excitation does not guarantee an enhancement in photochemical processes. Indeed, chemical and physical effects that arise due to molecular interactions with roughened surfaces could conspire to lower photodissociation yields relative to smooth surfaces despite there being SP enhanced local fields. Such a quenching of photodissociation on roughened silver surfaces was observed by Myli, Coon and Grassian in the case of chlorobenzene photodissociation and analyzed with the use of Nitzan and Brus's theory [24]. See also an earlier experimental study of the photochemical degradation of Rhodamine 6G on a silver-island surface [25].

Some more recent surface enhanced photodissociation work has focused on simpler molecules than those discussed so far, allowing for somewhat cleaner conclusions to be made. Kidd, Lennon and Meech [26] studied the photochemical cross sections for OCS, NO and $SO_2$ adsorbed on roughened silver surfaces and on smooth Ag(111) surfaces. Figure 9.3, taken from Ref. [26], shows the CO flux, which is proportionate to the photodissociation cross section, in the OCS case. Remarkably, a significant peak near 350 nm is seen in the roughened silver case that is absent in the smooth silver case. Recall, as noted in the discussion of Goncher and Harris's results [20] above, that the surface plasmon resonance for small silver particles is also expected to be near 350 nm. This interesting result has obvious parallels with the simple theoretical prediction of Nitzan and Brus [13] in relation to $I_2$ photodissociation and shown in Fig. 9.1. Kidd, Lennon and Meech presented a careful discussion of their results, noting that perhaps a somewhat broader plasmon enhanced photodissociation resonance would be expected given the range of effective silver particle sizes that are present in their thin films [26]. They suggested that the SPs could be creating either hot electrons and / or electron-hole pairs and that the interactions of the electrons with the molecules are what ultimately cause photodissociation. Specific details of these electronic processes might limit the spectral range of enhanced photodissociation. However a subsequent study by Meech and coworkers on $Fe(CO)_5$ photodissociation on roughened silver, which involves a more straightforward photodissociation mechanism via an electronically excited state, showed only a slight SP enhancement of less than a factor of two compared to photodissociation on a smooth surface [27].

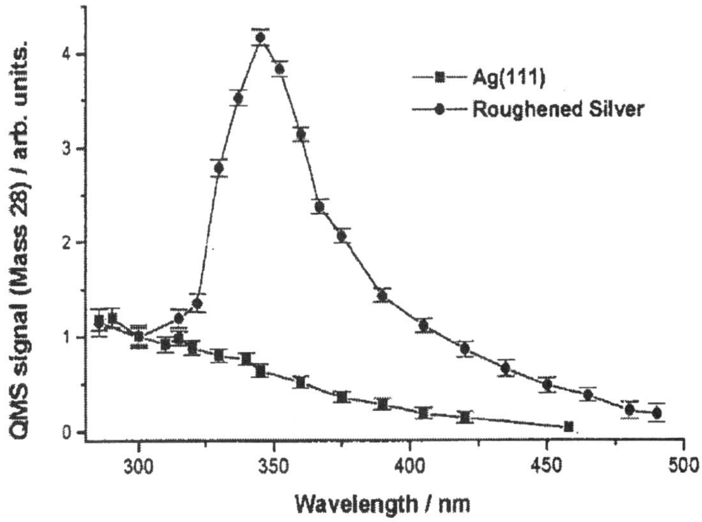


**Figure 9.3:** Initial CO flux, proportionate to the photodissociation cross section, for irradiation of a monolayer of OCS on smooth, Ag(111) and roughened silver surfaces. Reprinted figure with permission from Ref. [26]. Copyright 1981 by the American Institute of Physics.

Kidd, Lennon and Meech [26] also studied the related process of photodesorption, finding that the photodesorption cross sections for NO and $SO_2$ also exhibited peaks near the small particle surface plasmon resonance for silver. For an earlier study that also presents evidence for surface plasmon induced desorption, in this case desorption of atoms from the metal nanoparticles (composed of sodium) themselves, see Ref. [27]. The review article by Watanabe *et al.* [12] also discusses some more recent results on plasmon induced desorption.

## 9.3. b *Isomerization*

The photo-induced trans-cis isomerization reaction of certain azo-dye molecules such as Dispersed Red 1 (DR1),

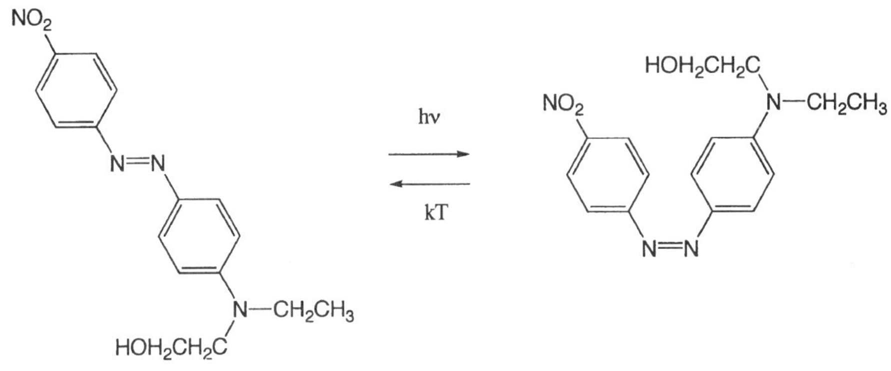


trans cis

it is remarkable because when the molecules are embedded (actually grafted as side chains) in a poly-(methyl methacrylate) or PMMA polymer matrix, the resulting system physically distorts when exposed to light with wavelength in the trans-cis absorption band [29, 30]. The trans-cis isomerization process, which takes place on a picosecond time scale, corresponds to a $\pi$-$\pi^*$ electronic transition followed by internal conversion processes that ultimately lead to the cis state, which is somewhat higher in energy than the trans state. In the case of RD1, however, thermal relaxation back to the trans state occurs on a time scale of seconds. The experiments in question involve exposure times on the order of 20-30 minutes and laser intensities on the order of 50 mW/cm$^2$ so that hundreds of such trans-cis-trans cycles can take place. The resulting material is distorted in a manner corresponding to mass transport away from regions of high light intensity and directed along the axis of polarization. The actual mechanism for the process is not completely understood [29, 30], although a "worm-like" motion along the direction of polarization resulting from the isomerization cycles is one appealing idea [31].

The trans-cis absorption band for the RD1/PMMA system is in the 400-600 nm spectral region. Thus it should be possible to efficiently drive repeated trans-cis-trans cycles in a thin film of RD1/PMMA covering, say, an array of metal nanoparticles with a surface plasmon absorption in this region. This was clearly demonstrated by Hubert *et al.* [32] in experiments that exposed a square array of silver nanoparticles sitting on glass but coated with an 80 nm thick thin film of RD1-doped PMMA and exposed to 532 nm radiation. (Extinction spectra indicated that there is an $\approx$ 80 nm red-shift in the silver nanoparticle SP resonance position to the 500 nm region due to the nanoparticles being interfaced with the PMMA coating, so that both the SP resonance of the silver / film system and the RD1 trans-cis absorption are overlapping.) After exposure to the radiation, atomic force microscopy (AFM) was used to characterize the surface. Figure

9.4 shows a result from Ref. [32]. The white arrow indicates the direction of polarization of the normally incident light from above. The resulting surface topographies show clear indentations or wells near the edges of the nanoparticles along the direction of polarization (Figs. 9.4(a) and 9.4(b)). There is reasonable correlation of these indentations with theoretical estimates of the near field intensity based on finite-difference time-domain (FDTD) calculations [32]. In particular, the negative of the near-field intensity (Fig. 9.4(d)) correlates with the observed topography (Fig. 9.4(b)), consistent with the material displacement away from regions of high intensity. The agreement is not perfect in that there are more wider wells evident in the experimental AFM scan relative to the theoretical result. Nonetheless this procedure represents an interesting new way of probing the nature of near-fields in nanoparticle systems by exploiting the chemistry that the SP excitations induce.

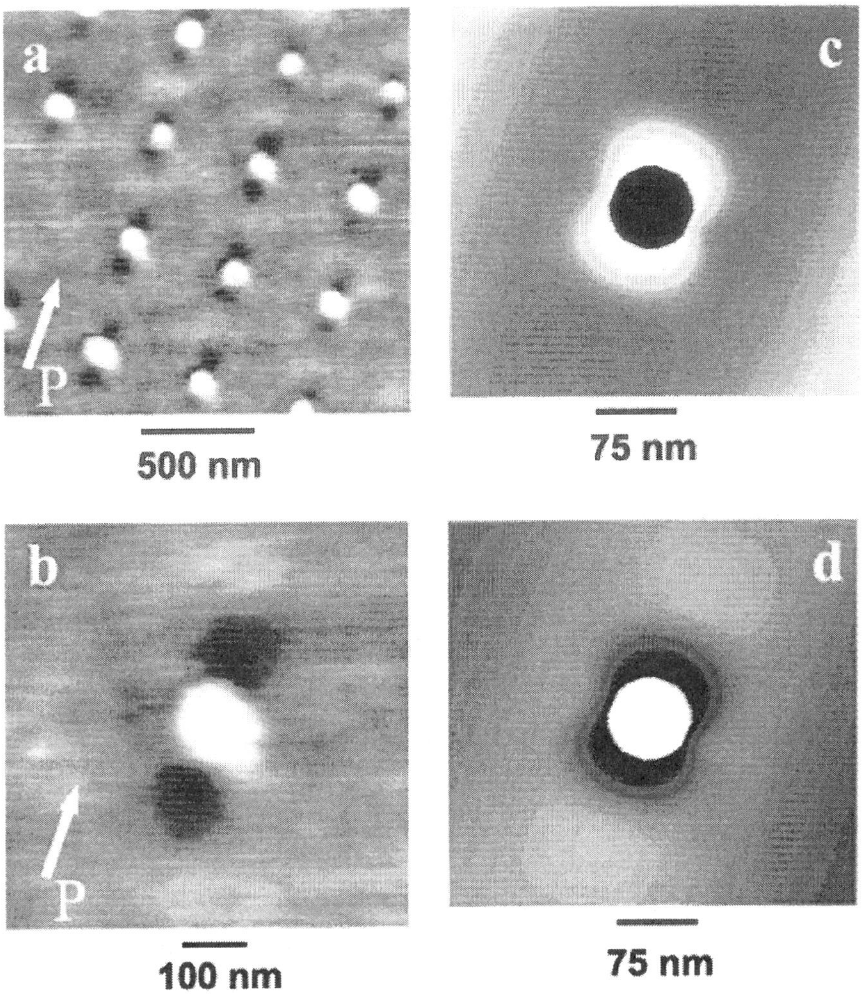


**Figure 9.4:** Photochemical imaging of metallic nanostructures. (a),(b): AFM images showing depletion (dark) areas due to material movement away from local hot spots on the silver nanoparticles (bright) (c): The calculated near-field intensity around a nanoparticle based on the FDTD method. (d) The negative of (c), which can be compared with (b). Reprinted figure with permission from Ref. [32]. Copyright 2005 by the American Chemical Society.

A subsequent study showed how such molecular motion due to plasmon induced isomerization can be used to imprint interference patterns due to counter-propagating surface plasmon polaritons [33], which represents a novel, plasmonics-based means of nanolithography. Most recently, SPs on silver bow-tie structures were imaged with this technique, using different polarizations to map out different electric field components [34].

## 9.3. c *Aggregation Processes*

In addition to dissociation and isomerization, the process of forming new chemical bonds and larger structures can also be enhanced or controlled with surface plasmon excitations. Here I would like to highlight the plasmon-assisted formation of rather larger structures than the simple molecular systems discussed so far: metal nanoparticles themselves and polymeric materials.

Jin *et al.* [35] showed how 5 nm diameter aqueous silver nanosphere systems can be transformed into larger, triangular nanocrystals, or nanoprisms, with size selectivity by exposing the systems to radiation with appropriate wavelength(s). In one experiment, exposure to 550 nm light led to a bimodal distribution of 70 nm and 150 nm edge length silver prisms. Coupled with theoretical calculations of the particle properties it was deduced that the smaller (70 nm) particles have dipolar plasmon resonances that are conducive to them fusing to form a larger particle. Two more additions of the smaller particles are still possible with all relevant particles still exhibiting a dipolar resonance near the incident wavelength. However, by the time a 150 nm edge particle has been formed out of the four smaller ones, the resulting particle no longer has a significant dipolar plasmon resonance at the incident wavelength. The result is a mixture of the 70 nm particles formed out of the 5 nm spherical particles and the 150 nm nanoprisms. Irradiation of the systems with two light sources having different wavelengths allowed greater control over the nanoprism sizes. For example, application of a primary 550 nm light beam with a secondary beam tuned to either 450 nm or 340 nm led to just the 70 nm edge length particles. The 450 and 350 nm wavelengths excite quadrupole resonances that are less conducive to the smaller particles combining, and thus while the primary source is able to photo-induce the smaller 70 nm nanoprisms to form, the secondary source effectively quenches particle fusion. By keeping the secondary wavelength on a quadrupolar resonance and varying the primary source wavelength, a variety of monodisperse nanoprism edge sizes could be achieved.

In a similar spirit, Maillard, Huang and Brus [36] reported how the aspect ratio of silver disks formed in a related process could be controlled by the irradiation wavelength. They correlated their results with the shape dependence of the SP resonance.

Photopolymerization is a process wherein a liquid monomer is converted into a solid polymer after exposure to light of appropriate wavelength and threshold intensity. Recently, by employing incident intensity light below the threshold, it was shown how photopolymerization can be induced around the hot spots (that augment the intensity by the factor $g^2$ of the introduction) leading to solid hills in the vicinity of the hot spots [37].

In addition to being a new route for nanoscale photochemistry, this provides yet another novel means of quantifying the nature of near fields near SPs.

## 9.4 CONCLUDING REMARKS

It has now been over twenty-seven years since the theoretical suggestion of the possibility of SP enhanced photochemistry [13]. Subsequent experimental work presenting evidence for it actually happening followed very quickly [20, 22]. Since this early work, some remarkable results have been achieved, including the results highlighted in the other Case Studies section above.

The general area SP enhanced photochemistry, however, is still not as extensively developed as a surface enhanced spectroscopy like SERS [3-7]. One reason for this situation is that chemical reactivity on rough surfaces or on metal nanoparticles is more complex and thus hard to study experimentally and understand theoretically. However, with the continued rapid advances in nanotechnology, and in our general theoretical understanding of complex processes, one can anticipate that SP enhanced photochemistry will become more widely developed in the coming years. For example, it may be possible to tailor the design of arrays of nanoparticles or other nanoscale features to enhance specific photochemistries in a more clean and reproducible fashion than in the earlier roughened surface experiments. For a recent example of such "plasmonic engineering," Zhang *et al.* [38] reported metal enhanced singlet oxygen generation from silver island films and were able to control its extent by varying system parameters.

## 9.5 ACKNOWLEDGEMENTS

This work was supported by the US Department of Energy, Office of Basic Energy Sciences, Division of Chemical Sciences, Geosciences, and Biosciences under contract DE-AC02-06CH11357. I also am grateful for helpful comments by Stephen R. Meech, Renaud Bachelot, Gary P. Wiederrecht, and L. B. Harding.

## 9.6 REFERENCES

1. Bohren, C. F. and Huffman, D. R. (1983). Absorption and scattering of light by small particles. Wiley, New York.
2. Raether, H. (1988). Surface plasmons. Springer-Verlag, New York.

3. Kelly, K. L., Coronado, E., Zhao, L. L., Schatz, G. C. (2003). The optical properties of metal nanoparticles: the influence of size, shape, and dielectric environment. J. Phys. Chem. B 107: 668-677.
4. Jeanmaire, D. L. and Van Duyne, R. P. (1977). Surface Raman spectroelectrochemistry Part I. Heterocyclic, aromatic, and aliphatic amines adsorbed on the anodized silver electrode. J. Electroanalyt. Chem. 84: 1-20.
5. Moskovits, M. (2005). Surface-enhanced Raman spectroscopy: a brief retrospective. J. Raman Spec. 36: 485-496.
6. Kneipp, K., Kneipp, H., Itzkan, I., Dasari, R. R. and Feld, M. S. (2002). Surface-enhanced Raman scattering and biophysics. J. Phys. Conden. Matter 14: R597- R624.
7. Campion, A., Ivanecky III, J. E., Child, C. M. and Foster, M. (1995). On the mechanism of chemical enhancement in surface-enhanced scattering. J. Am. Chem. Soc. 117: 11807-11808.
8. Ritchie, G. and Burstein, E. (1981). Luminescence of dye molecules adsorbed at a Ag surface. Phys. Rev. B 24: 4843-4846.
9. Glass, A. M., Wokaun, A., Heritage, J. P., Bergman, J. G., Liao, P. F., Olson, D. H. (1981). Enhanced two-photon fluorescence of molecules adsorbed on silver particle films. Phys. Rev. B 24: 4906-4909.
10. Lakowicz, J. R. (2006). Plasmonics in biology and plasmon-controlled fluorescence. Plasmonics 1:5-33.
11. Zhang, Y., Aslan, K., Previte, M. J. R., Geddes, C. D. (2007). Metal-enhanced fluorescence: surface Plasmons can radiate a fluorophore's structured emission. Appl. Phys. Lett. 90: 053107 [3 pages].
12. Watanabe, K., Menzel, D., Nilius, N. and Freund, H.-J. (2006). Photochemistry on metal nanoparticles. Chem. Rev. 106:4301-4320.
13. Nitzan, A. and Brus, L. E. (1981). Can photochemistry be enhanced on rough surfaces? J. Chem. Phys. 74:5321-5322.
14. Nitzan, A. and Brus, L. E. (1981). Theoretical model for enhanced photochemsitry on rough surfaces. J. Chem. Phys. 75: 2205-2214.
15 Gersten, J. and Nitzan, A. (1981). Spectroscopic properties of molecules interacting with small dielectric particles. J. Chem. Phys. 75: 1139-1152.
16. Johnson, P. B. and Christy, R. W. (1972). Optical constants of the noble metals. Phys. Rev. B 6:4370-4379.
17. Metiu, H. and Das, P. (1984). The electromagnetic theory of surface enhanced spectroscopy. Annu. Rev. Phys. Chem. 35: 507-536.
18. Gersten, J. I. and Nitzan, A. (1985). Photophysics and photochemistry near surfaces and small particles. Surf. Science 158: 165-189.
19. Leung, P. T. and George, T. F. (1986). Photodissociation of molecules at structured metallic surfaces. J. Chem. Phys. 85: 4729-4733.
20. Goncher, G. M. and Harris, C. B. (1982). Enhanced photofragmentation on a silver surface. J. Chem. Phys. 77: 3767-3768.

21. Goncher, G. M., Parsons, C. A., and Harris, C. B. 1984. Photochemistry on rough metal surfaces. J. Phys. Chem. 88: 4200-4209.
22. Chen, C. J. and Osgood, R. M. (1983). Direct observation of the local-field-enhanced surface photochemical reactions. Phys. Rev. Lett. 50: 1705-1708.
23. Chen, C. J. and Osgood, R. M. (1983). Surface-catalyzed photochemical reactions of physisorbed molecules. Appl. Phys. A 31: 171-182.
24. Myli, K. B., Coon, S. R. and Grassian, V. H. (1995). Photon-induced reactions of aromatics adsorbed on rough and smooth silver surfaces. J. Phys. Chem. 99:16407-16415.
25. Garoff, S., Weitz, D. A. and Alverez, M. S. (1982). Photochemistry of molecules adsorbed on silver-island films - effects of the spatially inhomogeneous environment. Chem. Phys. Lett. 93: 283-286.
26. Kidd, R. T., Lennon, D. and Meech, S. R. (2000). Surface plasmon enhanced substrate mediated photochemistry on roughened silver. J. Chem. Phys. 113: 8276-8282.
27. Burke, D. J., Vondrak, T. and Meech, S. R. (2002). Photochemistry of Fe(CO)5 adsorbed on single crystal and roughened silver. J. Phys. Chem. B 106: 10205-10214.
28. Hoheisel, W., Jungmann, K., Vollmer, M., Weidenauer, R. and Trager, F. (1988). Desorption stimulated by laser-induced surface-plasmon excitation. Phys. Rev. Lett. 60: 1649-1652.
29. Yager, K. G. and Barrett, C. J. (2006). Novel photo-switching using azobenzene functional materials. J. Photochem. Photobio. A: Chem. 182:250-261.
30. Natansohn, A. and Rochon, P. (2002). Photoinduced motions in azo-containing polymers. Chem. Rev. 102:4139-4175.
31. Lefin, P., Fiorini, C. and Nunzi, J.-M. (1998). Anisotropy of the photo-induced translation diffusion of azobenzene dyes in polymer matrices. Pure Appl. Opt. 7:71-82.
32. Hubert, C., Rumyantseva, A., Lerondel, G., Grand, J., Kostcheev, S., Billot, L., Vial, A., Bachelot, R., Royer, P., Chang, S.-H., Gray, S. K., Wiederrecht, G. P. and Schatz, G.C. (2005). Near-field photochemical imaging of noble metal nanostructures. Nano Letters 5: 615-619.
33. Derouard, D., Hazart, J., Lerondel, G., Bachelot, R., Adam, P. M. and Royer, P. (2007). Polarization-sensitive printing of surface plasmon interferences. Optics Express 15:4238-4246.
34. Hubert, C., Bachelot, R., Plain, J., Kostchev, S., Lerondel, G., Juan, M., Royer, P., Zou, S., Schatz, G. C., Wiederrecht, G. P. and Gray, S. K. (2008). J. Phys. Chem. C. Near-field polarization effects in molecular-motion-induced photochemical imaging. *in press*.
35. Jin, R., Cao, Y. C., Hao, E., Metraux, G. S., Schatz, G. C. and Mirkin, C. A. (2003). Controlling aniostropic nanoparticle growth through plasmon excitation. Nature 425:487-490.

36. Maillard, M., Huang, P.,and Brus, L. (2003). Silver nanodisk growth by surface plasmon enhanced photreduction of adsorbed [Ag+]. Nano Letters 3: 1611-1615.

37. El Ahrach, H. I., Bachelot, R., Vial, A., Lerondel, G., Plain, J., Royer, P. and Spooera, O. (2007). Spectral degeneracy breaking of the plasmon resnance of single metal nanoparticles by nanoscale near-field photopolymerization. Phys. Rev. Lett. 98: 10741 (007) [4 pages].

38. Zhang, Y., Aslan, K., Previte, M. J. R. and Geddes, C. D. (2008). Plasmonic engineering of singlet oxygen generation. Proc. Nat. Acad. Sci. (USA) 105: 1798-1802.

# 10

# Metal-Enhanced Generation of Oxygen Rich Species

Yongxia Zhang, Kadir Aslan and Chris D. Geddes*
Institute of Fluorescence, laboratory for Advanced Medical Plasmonics and Laboratory for Advanced Fluorescence Spectroscopy
University of Maryland Biotechnology Institute, 701 East Pratt St., Baltimore, MD, 21202, USA.

## 10.1 INTRODUCTION TO OXYGEN RICH SPECIES AND THEIR APPLICATIONS

Oxygen is ubiquitous. It comprises nearly 50% Earth's crust and is an essential component in metabolic pathways in organisms. Reactive oxygen species (ROS) including singlet oxygen, oxygen ions, superoxide anion radicals ($O_2^{\cdot -}$), and peroxides are generally very small molecules and are highly reactive. ROSs are natural byproducts of the normal metabolism of oxygen and have important roles in cell signaling.

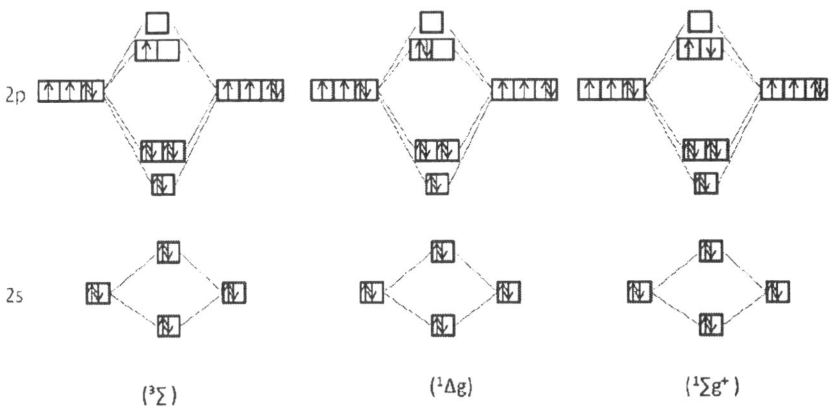


**Figure 10.1:** Orbital diagrams for the three configurations of molecular oxygen, $O_2$. Shown from the left are: The triplet ground state $^3\Sigma$, the singlet oxygen $^1\Delta g$ excited state, and the second singlet oxygen $^3\Sigma g+$ excited state. Adopted from Ref [1].

*Metal-Enhanced Generation of Oxygen Species*
Edited by Chris D. Geddes.

Singlet oxygen is the lowest excited state ($^1\Delta_g$) of Oxygen. Oxygen has two important excited singlet states, $^1\Sigma_g^+$ and $^1\Delta_g$. The electronic energies of $^1\Sigma_g^+$ and $^1\Delta_g$ are 158 and 95 kJ/mol higher respectively, than that of the triplet ground state $^3\Sigma$. For the second excited states $^1\Sigma_g^+$, the electronic configuration is identical to that of the ground state, except that the last two electrons have antiparallel spins (Figure 10.1).(1) Singlet oxygen $^1\Delta_g$ ($^1O_2$) is more important than $^1\Sigma_g^+$, because of its longer lifetime. Singlet molecular oxygen was discovered in 1924 and the physical, chemical and biological properties of it have acquired serious attention by researches since 1963. Since singlet oxygen can readily react with many biological targets and destroy a wide variety of cells, the photosensitized production of singlet oxygen has significance in a range of areas, especially in photodynamic therapy (PDT) where it has been widely used in both oncological, (*e.g.* tumors and dysplasias) and nononcological (*e.g.* age-related macular degeneration, localized infection and non-malignant skin conditions) applications(2-5). Three primary components are involved in PDT: light, a photosensitizing drug and oxygen. The photosensitizer adsorbs light energy, which it then transfers to molecular oxygen to create an activated form of oxygen called singlet oxygen.(2) The singlet oxygen is a cytotoxic agent and reacts rapidly with cellular components to cause damage that ultimately leads to cell death and tumor destruction.(5) PDT treatments are only effective within a specific range of singlet oxygen supply.(6) For example, for solid tumors, too little singlet oxygen cannot effectively treat the tumor cells, but too much singlet oxygen can damage and kill surrounding healthy cells.(7) Currently, the intensity of light is commonly adjusted to control the extent of singlet oxygen generation, but there are some limitations to this method. High fluency rates of the exposure light will lead to oxygen depletion and photosensitizer photo-bleaching.(4) However, low fluency rates of exposure light, lends to a long exposure time and can cause vascular shutdown, a precursory condition to hypoxia in the tissue.(6, 8) One notable approach to control the fluency rate of exposure light is called interstitial PDT, where precise amounts of light is delivered locally to tumors through inserted optical fibers.(9) The interstitial PDT also allows the real-time monitoring of the progression of the treatment via online collection of assessment parameters through the optical fibers.(9) It is important to note that despite the better control over fluency rate, the photobleaching of the photosensitizers remains an issue. In this regard, there is an urgent need to find a way to both optimize and control singlet oxygen generation.

Besides singlet oxygen, the superoxide anion radical is a reactive oxygen species which can further interact with other molecules to generate secondary ROS. They can aid in defense against infectious agents. In addition, superoxide has been widely used in organic chemistry, with several basic modes of action related to superoxide being demonstrated; such as its deprotonation as a base, H-atom abstraction as a radical, nucleophilic attack as an anion, and also as an electron transfer agent.[9,10] Recently, Yong Hae Kim, *et al* have reported that the superoxide anion radical reacts with arenesulfonyl or arenesulfinyl-chlorides to form a arenesulfonylperoxy- or arenesulfinylperoxy radical intermediates.(10) It was reported that these peroxysulfur intermediates showed excellent oxidizing abilities for the regioselective oxidation of olefins, oxidative desulfurizations of thiocarbonyls to carbonyls, cleavage of C=N to C=O and conversion of the benzylic methylene groups to ketones.(10) In biological settings, superoxide is known to interact with not only the bases, but also the deoxyribosyl backbone of DNA causing carcinogenesis.(11) However, the specific mechanism by which oxidative stress contributes to the development of carcinogenesis is for the most part still unknown. In addition, it was previously shown that the determination of superoxide generation *in vitro* can be used to monitor the effect of antibiotics on certain cellular components such as neutrophils.(12) In research settings, the ability to therefore supply abundant superoxide yields, will potentially amplify the uses of superoxide, with abundant supplies being a useful tool to study oxidative stress mechanisms and a deeper understanding of carcinogenesis.

## 10.2 INTRODUCTION TO METAL-ENHANCED FLUORESCENCE (MEF)

### 10.2 1. *Metal Enhanced Fluorescence*

Fluorescence methodologies and technologies are widespread throughout nearly all aspects of biological research. The principles and applications of fluorescence have undergone extensive development and commercialization since its introduction to biochemistry in the early 1950's. To a significant extent, future advances in biology and related disciplines depend on the advances and capabilities of fluorescence measurements. However, there are several well-known limitations to fluorescence technologies, such as the overall detection limits, partly due to the quantum yield of the fluorophore (label), the auto-fluorescence of the sample and the photostability of the fluorophores employed. Although *Plasmonics* emerged in 1964, the potential of surface plasmons lay mostly dormant with regard to fluorescence, until early into the 21$^{st}$ century. Since that time, people have explored and explained a variety of new concepts which combine surface plasmons and close-proximity fluorophores. With *Metal-enhanced fluorescence*, one has the opportunity to modify the intrinsic properties of fluorophores in the near and far-field, a concept described in detail within this book. These opportunities include enhanced quantum yields, photostability and directional emission to name but just a few. Currently, there are several explanations for the near-field interactions of fluorophores with metallic nanoparticles. Fluorophore photophysical properties were originally thought to be modified by a resonance interaction by there close proximity to surface plasmons, which gives rise to a modification of the fluorophore radiative decay rate.(13) This description was fueled by earlier workers who had shown increases in fluorescence emission coupled with a simultaneous drop in radiative lifetime.(14) However, the Geddes laboratories current interpretation of Metal Enhanced Fluorescence is described somewhat differently, by a model whereby non-radiative energy transfer occurs from excited distal fluorophores, to the induced surface plasmon electrons in non-continuous films, in essence a fluorophore induced mirror dipole in the metal.(15, 16) The surface plasmons in turn, radiate the emission of the coupling fluorophores (Figure 10.2-Top). This explanation has been facilitated by the observation of surface plasmon coupled fluorescence (SPCF), whereby fluorophores distal to *a continuous metallic film* can directionally radiate fluorophore emission at a unique angle from the back of the thin film,(17) directly implicating surface plasmons in the mechanism.

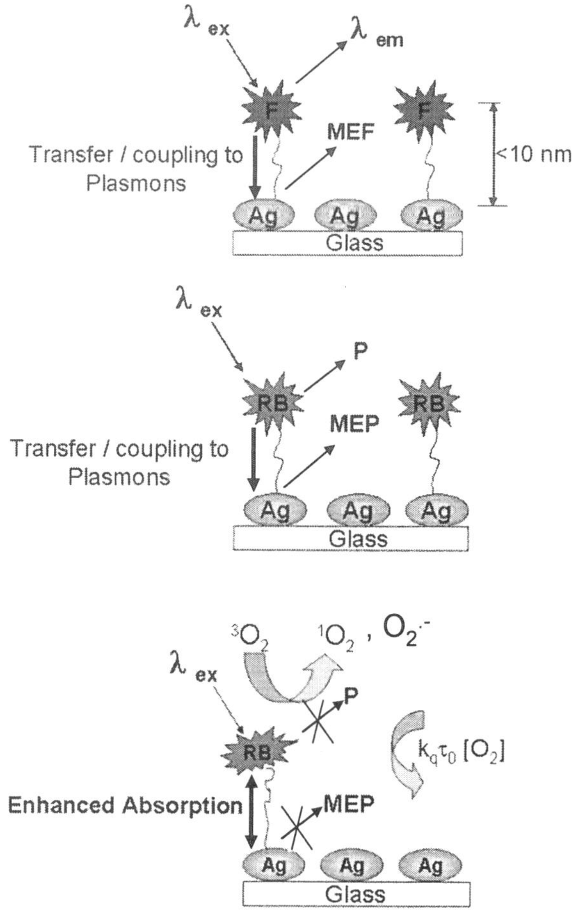


**Figure 10.2:** Graphical representation of Metal-Enhanced Fluorescence (Top), Metal-Enhanced Phosphorescence (Middle), and for the generation of Singlet Oxygen (Bottom). F – Fluorophore, RB – Rose Bengal, P - Phosphorescence and MEP – Metal-Enhanced Phosphorescence, 3O2 – triplet ground state oxygen. 1O2 – singlet oxygen. Adopted from ref [25].

## 10.2 2. *Metal Enhanced Phosphorescence*

In addition to MEF, Metal - Enhanced Phosphorescence (MEP) at low temperature (18, 19) has also been reported, whereby non-radiative energy transfer is thought to occur from excited distal triplet-state luminophores to surface plasmons in non-continuous silver films, which in turn, radiate fluorophore/lumophore phosphorescence emission efficiently (Fig. 10.2-Middle). This observation suggests that photon-induced electronic excited states can both induce and couple to surface plasmons (mirror dipole effect) facilitating both enhanced $S_1$ fluorescence and phosphorescence, $T_1$, emission. Thus, MEP (i.e. enhanced triplet yields) provides researchers with an opportunity to enhanced oxygen rich species such as singlet oxygen and the superoxide anion radical, when a sensitizer is in close proximity to the metal nanoparticles (Figure 10.2-bottom).

## 10.3 PLASMON ENGINEERING OF OXYGEN RICH SPECIES

### 10.3 1. *Metal-Enhanced Singlet Oxygen Generation*

#### 10.3.1 1. *Rose Bengal as a Photosensitizer*

Since Rose Bengal (RB) is a commonly used photosensitizer with a high singlet oxygen yield (0.76), it has been used to demonstrate the properties of metal enhanced singlet oxygen generation.(20) GR which is highly selective for singlet oxygen (21) (Invitrogen) is selected as a singlet oxygen detection probe. The solutions of GR and RB have well-separated fluorescence peaks at 525 nm (Figure 10.3C) and 588 nm (Figure 10.3D). The green sensor (GR) detects singlet oxygen,(21) while Rose Bengal is the photosensitizer that triggers singlet oxygen generation, due to the well-known triplet interaction with ground-state molecular oxygen. Without UV irradiation (sensitization) a green fluorescence emission peak at 525 nm for the GR singlet oxygen sensor on glass (Figure 10.3A) is observed. This result is designated to background solution singlet oxygen and emission of the GR sensor dye itself.(22) Due to the MEF effect,(16) the fluorescence emission peak of GR is enhanced on SiFs (Figure 10.3B), which we correct for, in our calculation of enhanced singlet oxygen yields (MEF Factor, Equation 1). As previously reported, a MEF effect in the RB spectra for the sample on SiFs (Figure 10.3B)(23) is also observed.

After exposure to UV light, the fluorescence emission intensity of GR on SiFs (Figure 10.3B) at 525 nm is ~3.3 times larger than GR emission on glass (Figure 10.3A). This increased intensity suggests that more singlet oxygen was generated from the Rose Bengal system on SiFs. The real color photographs further validate the difference of GR fluorescence emission intensity on glass and SiFs, respectively (Figure 10.3A-B, insets). On glass, real color photographs of GR/RB solutions (Figure 10.3A, insets) before exposure to UV light are visually brighter after exposure to UV light, which reflects an increase in net singlet oxygen yield. On SiFs (Figure 10.3B, insets), this increased brightness of the solution is more pronounced, further suggesting that the presence of the Ag nanoparticles facilitates increased singlet oxygen generation, consistent with previous reports from the Geddes laboratory.(24) (25)

When comparing SiFs to a glass substrate for the production of enhanced singlet oxygen generation, it is important to discuss the similarities in the surface features of these substrate materials. As described in the previously published procedure (26), SiFs are deposited onto the same glass substrate (used for the comparison of singlet oxygen generation) as particles with a diameter in the order of 30 nm and with a surface coverage of ≈ 40%. Thus, the comparison of a *blank* glass substrate with the same glass substrate containing SiFs for singlet oxygen generation is deemed appropriate.

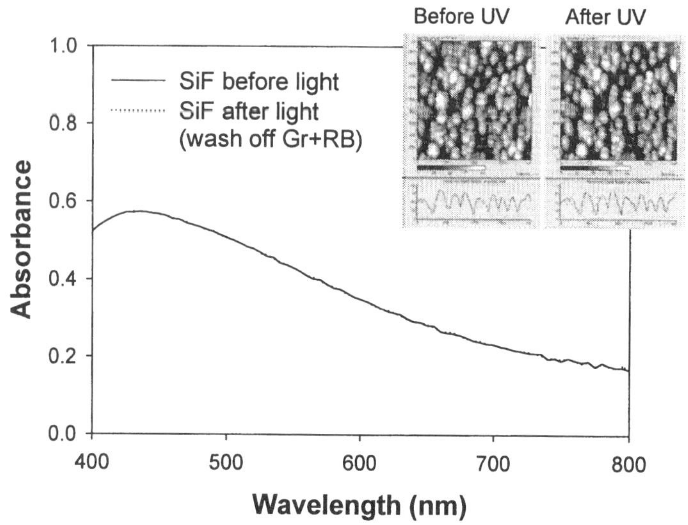


**Figure 10.3:** Fluorescence emission spectra of green sensor (Gr) (A) Rose Bengal (RB) (B) and a mixture of both on (C) glass, (D) SiFs, (E) in cuvette, (F) on SiFs nitrogen purged, (G) on SiFs oxygen purged, (H) on 50 nm thick Ag, before and after light exposure (2 min) at room temperature. Light source is 100 W mercury lamp. λex = 473nm. Adopted from ref [25]

To confirm that the observed signal enhancement in the presence of SiFs is due to an increase in singlet oxygen yield due to silver, the oxygen supply was varied. After extensive purging the RB and GR solution with $N_2$ and subsequent exposure to UV irradiation, the amplitude of the peak at 525 nm did not change in the cuvette (Figure 10.3E) or on SiFs (Figure 10.3F). As expected, these results indicate that no singlet oxygen was generated. However, after purging the mixture of RB and GR solution with oxygen and subsequent exposure to UV irradiation (sensitization), a sharp increase in the GR fluorescence peak intensity from SiFs due to singlet oxygen generation was observed (Figure. 10.3G). A mixture of RB and GR solution on a continuous silver strip was also studied, Figure 10.3-H, since surface plasmons cannot be generated in a *continuous strip of metal* (from the air side) but can of course be in *non-continuous* particulate silver island films. No enhanced singlet oxygen generation was observed on the continuous silver strip giving further evidence for the plasmon enhanced Triplet yield mechanism.

In order to demonstrate that the SiFs are unaffected by $^1O_2$ or UV irradiation, the absorption spectra and AFM images (Figure 10.4 - Insert) of SiFs before UV irradiation and SiFs previously coated with the GR/RB mixture and exposed to UV irradiation (Figure 10.4) were compared. No change in the plasmon absorbance spectra and AFM images for the SiFs samples before and after UV irradiation was observed.

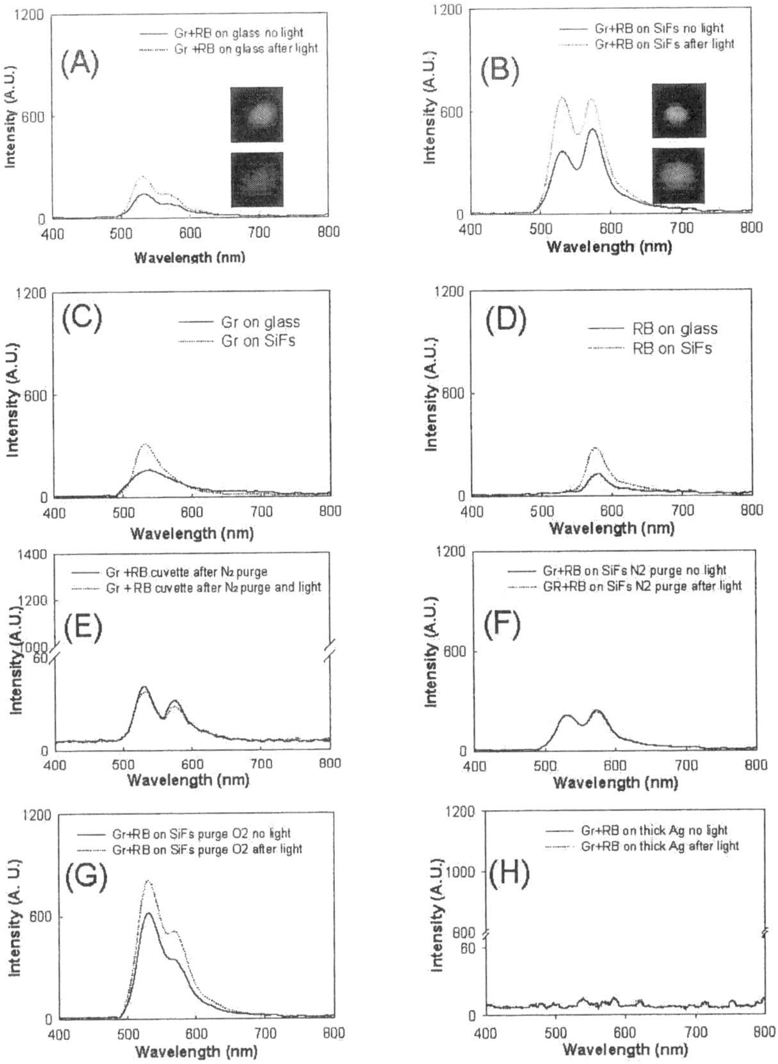


**Figure 10.4:** Absorption spectra and AFM image of SiFs before light & after light showing no effect on the silvered surface by 1O2. SiFs – Silver island Films. Light- UV exposure. Adopted from ref [25].

### 10.3.1 2. *Using different photosensitizers*

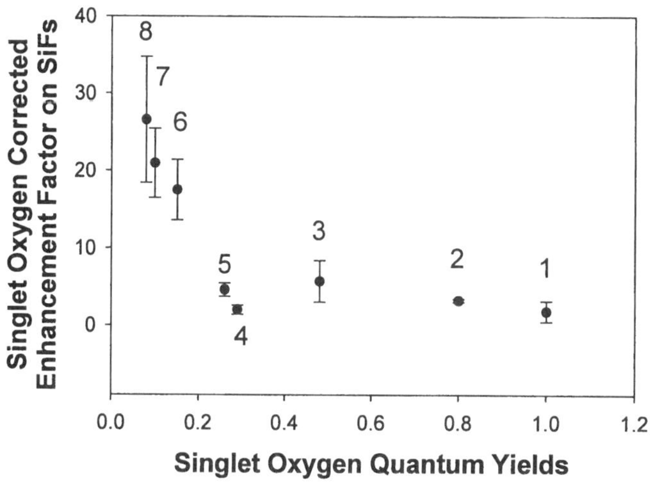


**Figure 10.5:** Singlet Oxygen Corrected Enhancement Factor on SiFs versus Singlet Oxygen Quantum Yield. (1. Acridine 2. Chloroquine 3. Rose Bengal 4. Indomethacin 5. Riboflavin 6. Naproxen 7. Chloropromazine 8. Quinidine). Error bars are based on the standard deviation of 3 spectral measurements. Adopted from ref [25].

Many other photosensitizers with singlet oxygen quantum yields varying from 0.08 to 1.00 (Figure 10.5) have also been tested. In order to calculate the singlet oxygen enhancement of these photosensitizers, the MEF factor of the photosensitizer, GR, (metal enhanced fluorescence from green photosensitizer before exposure to UV light, Figure 10.3A) was first determined. The MEF factor of the photosensitizer, $GR_{MEF,before}$, on silver island films is calculated by:

$$GR_{MEF,before} = \frac{\int_0^\infty GR_{SiF,before} d\lambda}{\int_0^\infty GR_{GL,before} d\lambda} \qquad (1)$$

whereby $\int_0^\infty GR_{SiF,before} d\lambda$ and $\int_0^\infty GR_{GL,before} d\lambda$ are the integrated spectra (Figure 10.3A) for the green sensor dye (GR), before exposure to UV light on silver island films (SiF) and glass substrates (GL), respectively. The calculation of the MEO oxygen yield) of the photosensitizer is as follows

$$^{1}O_{2,MEO} = \frac{\int_0^\infty (GR,RB)_{SiF,after}\, d\lambda - \int_0^\infty RB_{\lambda_n,SiF,after}\, d\lambda}{GR_{MEF,before}\left(\int_0^\infty (GR,RB)_{GL,after}\, d\lambda - \int_0^\infty RB_{GL,after}\, d\lambda\right)} \qquad (2)$$

whereby $\int_0^\infty RB_{SiF,after} d\lambda$ and $\int_0^\infty RB_{GL,after} d\lambda$ and $\int_0^\infty (GR,RB)_{SiF,after} d\lambda$ and $\int_0^\infty (GR,RB)_{GL,after} d\lambda$ are the integrated spectra for Rose Bengal (RB) and the mixture of photosensitizer (GR) with Rose Bengal (RB) after exposure to UV light on silver island films (SiF) and glass substrates (GL), respectively (Figure 10.3B).

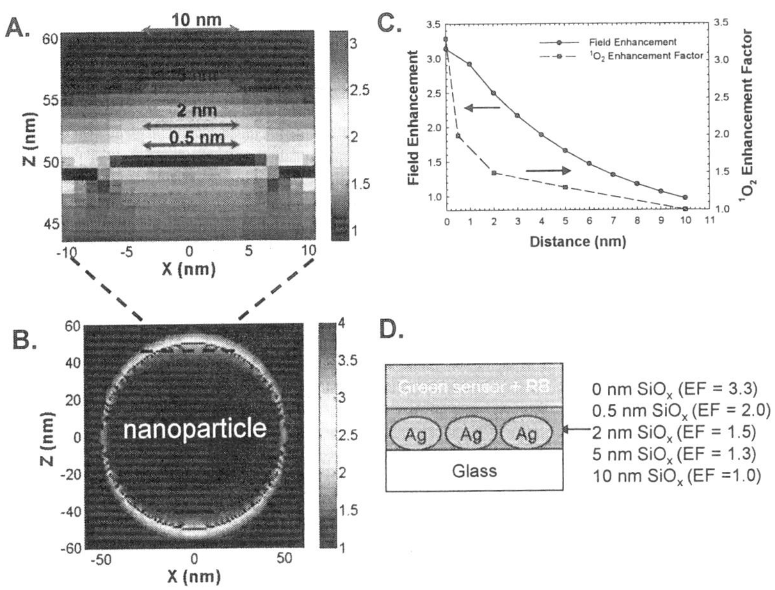


**Figure 10.6:** A) Zoomed image |E|2 field intensity of 10 nm above the surface of the silver sphere. B) |E|2 field intensity (incident plus scatter) distribution in the xz plane around a 100 nm silver sphere due to an incident TFSF wave propagating along the y axis and polarized along the z axis with a wavelength of 365 nm, which corresponds to the max wavelength of the UV source used to excite Rose Bengal and generate singlet oxygen. C) Distance dependence Relationship between for electric field enhancements and singlet Oxygen on 100 nm Ag nanoparticles. D) Distance dependence of Singlet Oxygen Enhancement Factor of Rose Bengal on SiFs. Top layer is mixed solution of Green Sensor and Rose Bengal. SiOx layer was deposited using thermal vapor deposition. RB - Rose Bengal. Ag - Silver island Films. EF – Enhancement Factor. Adopted from ref [25].

The enhancement factor is 26.6 ±8.13 for Quinidine, which has a free-space singlet oxygen quantum yield of 0.08. In contrast, for Acridine which has a high singlet oxygen yield of 1.00, the enhancement factor is only 1.83 ±1.35. Interestingly, the enhanced singlet oxygen yield factor appears to be inversely proportionally to the *free-space* singlet oxygen yield. This finding is consistent with the MEF enhancement factor for fluorophores and (relative intensities in the presence and absence of metal for the fluorophores) increases as the free-space quantum yield ($Q_0$) decreases,(27) i.e. MEF $\propto 1/Q_0$, Where $Q_0$ is the free space quantum yield, in the absence of metal.

### 10.3.1 3. *Distance-Dependence Control of Singlet Oxygen Generation*

In order to control singlet oxygen generation, the distance dependence of ME $^1O_2$ was tested, $SiO_x$ layers of 0.5, 2, 5 and 10 nm thickness were vapor deposited on SiFs (Figure 10.6D). It was observed that the amplitude of the emission spectra of GR and RB solution on SiFs varied with different thicknesses of $SiO_x$. The singlet oxygen enhancement factor of GR and RB solution on SiFs was 2.0-fold for 0.5 nm $SiO_x$ coatings, 1.5-fold for 2.0 nm $SiO_x$, 1.3-fold enhancement for 5 nm $SiO_x$ coatings, and no enhancement was observed on 10 nm $SiO_x$, (Figure 10.6D). These values and distances are consistent with the enhanced absorption effect (enhanced electric field) which partially contributes to the enhanced intensities observed in metal- enhanced fluorescence. (28)

### 10.3.1 4. *Electric Field Enhancement around SIFs*

The enhanced singlet oxygen yields for photosensitizers in proximity to a metallic nanoparticle, is a function of the net system absorption which can be theoretically calculated using FDTD (Finite Difference Time Domain) calculations. The FDTD method has been employed to determine the electric field intensities and distributions at the surface of a 100 nm silver nanoparticle in a Total Field Scattered Field. These results were compared with Mie Theory and previously published reports to verify the accuracy of the model.(29) Total field scattered-field sources are used to divide the computation area or volume into total field (incident plus scattered field) and scattered field only regions.(30) The incident field is defined as a plane wave with a wavevector that is normal to the injection surface and the scattered and total fields are monitored during the simulation such that the total or scattered transmission can be measured. Using Lumerical FDTD Solution software (Canada), the simulation region is set to 700 x 700 x 700 $nm^3$ with a mesh accuracy of 6. To minimize simulation times and maximize the resolution of field enhancement regions around the metal sphere, a mesh override region is set to 1 nm around the 100 nm Ag sphere. The overall simulation time was set to 200 ns and calculated over a frequency range from 300-600 nm, whereby a plasma model is used to represent the properties of the silver nanoparticle in the range from 300 nm to 600 nm (Figure 10.6 A-B). A non-linear relationship was found to exist between the experimentally calculated distance dependent enhancement of singlet oxygen yields for $SiO_x$ films deposited on silver island films, and the simulated electric field enhancements (Figure 10.6C).

### 10.3 2. *Metal Enhanced Superoxide Generation*

#### 10.3.2 1. *Dihydroethidium (DHE) as a Photosensitizor for Superoxide Generation*

Superoxide generation is also effected by a suitable photosensitizer close-to silver nanoparticles(31). Figure 10.7 shows the fluorescence emission spectra of a mixture of DHE and acridine solutions on glass and SiFs, before and after UV light exposure. On glass no fluorescence was detected both before and after light exposure, where exposure (from 10 cm away for 2 minutes) with a 100 W Mercury lamp was used with the acridine photosensitizer for the generation of superoxide. This suggests too little superoxide was generated to be detected in the glass sandwich using the DHE probe and the optical system. However, on SiFs before light, one broad peak at 595 nm was observed, which is attributed to the amplified fluorescence peak of DHE.

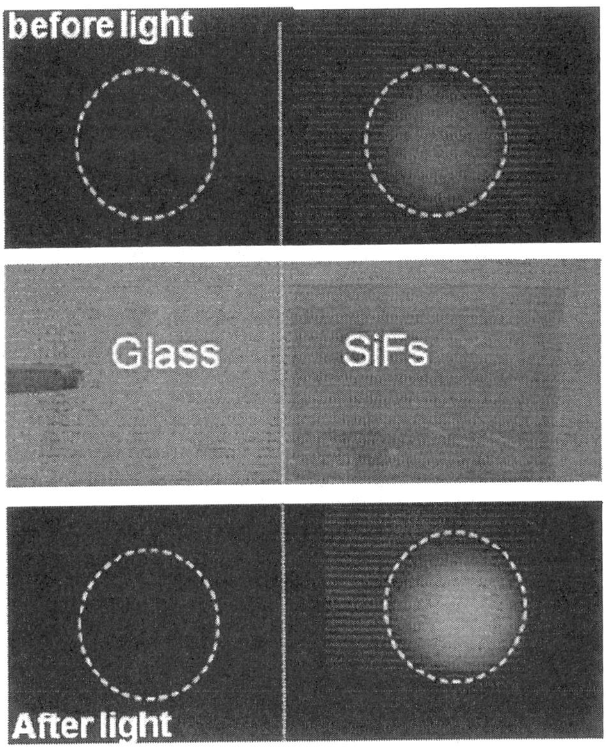


**Figure 10.7:** Schematic representation of the sample geometry (Insert) and fluorescence emission spectra of a mixture of the DHE probe and Acridine on glass,

and on SiFs, before and after light exposure (2 min) at room temperature. Light source was a 100 W mercury lamp. λex = 473 nm. DHE- dihydroethidium. SiFs – Silver Island Films. Adopted from ref [31].

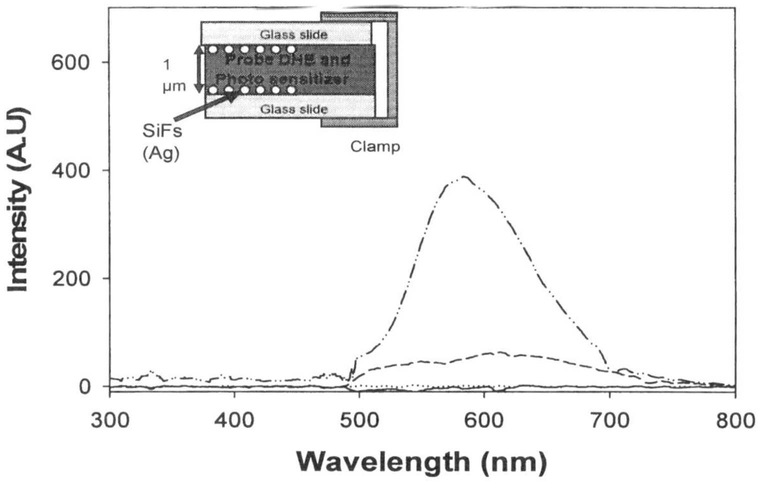


**Figure 10.8:** Real color photographs of dihydroethidium (DHE) and Acridine emission from glass and SiFs, before and after 2 mins light exposure (sensitization). Light exposure source was a 100 W mercury lamp. λex = 473 nm. **SiFs** - Silver Island Films. Adopted from ref [31].

This peak becomes apparent on the SiFs but is not visible on the glass control sample, due to the MEF effect, which the Geddes group has shown can significantly enhance the emission intensity for nearly every fluorophore tested to date.(32) After UV light exposure, a significant increase in the fluorescence emission of the DHE probe at 595 nm was evident from SiFs, which strongly indicates enhanced superoxide generation as compared to the glass control sample, which contains no silver nanostructures (*Note:* We have corrected for the MEF effect on the DHE probe in the absence of acridine). These enhancements can also be evidenced visually from the Figure 10.8 photographs. On glass, the DHE fluorescence emission was not observed before and after light, top left and bottom left respectively. However on SiFs, the DHE fluorescence emission was much more intense after light exposure in the presence of the Acridine photosensitizer, indicating that more superoxide anion radical is generated on SiFs than on the glass slide, c.f. top right panel and bottom right panel of Figure 10.8. It is interesting to note that the photographs were taken through an emission filter and the intensities observed are not due to backscattering of the excitation light by silver. The middle panel shows a photograph of the silver island films, coated on only half of the glass slide.

### 10.3.2 2. *Distance Dependence of Metal Enhanced Superoxide Generation*

The metal-enhanced fluorescence phenomenon is distance dependent(16, 33, 34) with a maximum enhancement factor (for emission intensities) for fluorophores positioned somewhere between 5-10 nm from the surface.(16, 34) Subsequently, it has been investigated whether the generation of superoxide would similarly be influenced by the distance of both the sensitizer and DHE probe from the surface. Using thermal vapor deposition, $SiO_2$ coatings were deposited on the surface of the SiFs, effectively distancing the probes from the silvered surface when in a sandwich geometry, Figure 10.9 top (correcting for the DHE MEF effect).

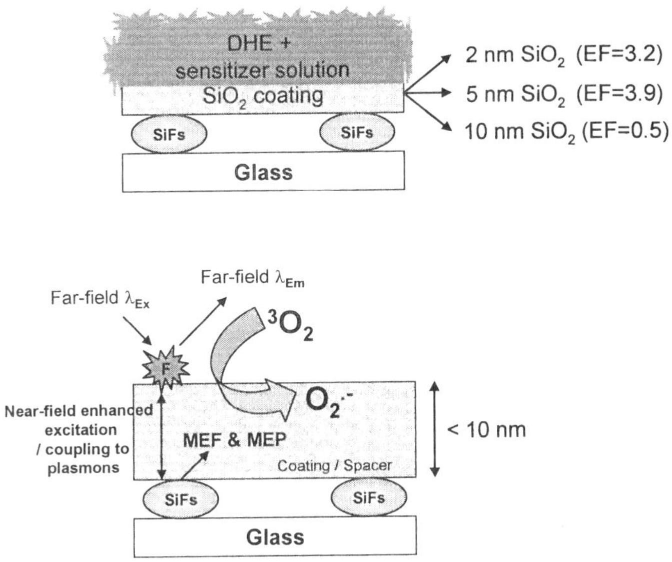


**Figure 10.9:** Sample architecture for the distance dependence of metal-enhanced Superoxide generation (Top), and graphical representation of the interpretation of metal-enhanced superoxide generation with an enhanced and distance dependent excitation rate (Bottom). F – Fluorophore, MEF – Metal-Enhanced Fluorescence, MEP – Metal-enhanced Phosphorescence, SiFs – Silver Island Films. EF- Enhancement factor = I Silver /I Glass. Adopted from ref [31].

Similarly to MEF findings,(16, 34) close proximity to silver results in only modest enhancements of superoxide as compared to the glass control sample also supporting $SiO_2$ layers, (enhancement factor, EF = 3.2 for 2 nm $SiO_2$ coatings). For 10 nm $SiO_2$ coatings the enhancement factor was the smallest, ≈ 0.5, as compared to 5 nm $SiO_2$ which yielded an ≈ 4-fold enhancement in superoxide generation as compared to the glass slide control sample, Figure 10.9. While at first this finding appears completely consistent with MEF findings and indeed the current interpretation of MEF, it should be noted that it is thought that an increase in the *net system absorption* facilitates metal-enhanced superoxide generation, where enhanced

$S_1$ and $T_1$ emission are competitive with superoxide generation, which is known to be due to a physical process of an excited state collision with $^3O_2$ (ground state triplet oxygen). Subsequently, it appears that the enhanced absorption component of the sensitizer near-to silver is also distance dependent, with a maximum value near-to 5 nm. While the emission of fluorophores near-to silver is well known to be distance dependent (efficiency of plasmon coupling(35)), this observation strongly suggests that the enhanced excitation component of fluorophores near-to silver is also distance dependent. Subsequently, both ME $^1O_2$ and ME superoxide generation are distance dependent in an analogous manner.

## 10.4 CONCLUSIONS

Enhanced fluorescence, or MEF, is a result of both a *net system absorption* and plasmon coupling and subsequently efficient emission, but to date, it has not been possible to quantify the relative contributions of enhanced emission and net increase in the system absorption to the MEF phenomena.(23) Due to the increase in the population of the singlet excited state or net system absorption, the very presence of MEP has also suggests an increase in the population of the triplet state.(23) The presence of Metal-Enhanced Fluorescence, Phosphorescence, Metal-Enhanced singlet oxygen and superoxide anion radical generation in the same system is an effect of the enhanced absorption and emission effects of the fluorophores near-to silver, although these processes are effectively competitive and ultimately provide a route for deactivation of electronic excited states.

The observations of creating surface architectures to optimize singlet oxygen generation and enhancements in the generation of superoxide for fluorophores / sensitizers in close-proximity to silver nanoparticles and other noble metals are helpful in understanding the interactions between plasmons and lumophores / fluorophores, The distance dependent manner, similar to reports for Metal-Enhanced Fluorescence, manifests itself in an increased triplet and therefore singlet oxygen and Superoxide anion radical yield. This is a most helpful observation and indeed technology, and this approach may well be of significance for enhancing triplet-state reactive oxygen based assays, especially for those used in PDT.

## 10.5 ACKNOWLEDGEMENTS

The authors like to thanks the IoF, UMBI and the NIH NINDS R21 NS055187 for their support.

## 10.6 REFERENCES

1. Ameta, S., Punjabi, P., Chobisa, C., Mangal, N., and Bhardwaj, R. (1990). Singlet Molecular Oxygen *Asian Journal of Chemistry Recies* **1**: 106-124.

2. Demidova, T. N. and Hamblin, M. R. (2004). Photodynamic therapy targeted to pathogens *International Journal of Immunopathology and Pharmacology* **17**: 245-254.
3. Brown, S. B., Brown, E. A., and Walker, I. (2004). The present and future role of photodynamic therapy in cancer treatment *Lancet Oncology* **5**: 497-508.
4. Kendall, C. A. and Morton, C. A. (2003). Photodynamic therapy for the treatment of skin disease *Technology in Cancer Research & Treatment* **2**: 283-288.
5. Dougherty, T. J. (2002). An update on photodynamic therapy applications *Journal of Clinical Laser Medicine & Surgery* **20**: 3-7.
6. Jarvi, M. T., Niedre, M. J., Patterson, M. S., and Wilson, B. C. (2006). Singlet oxygen luminescence dosimetry (SOLD) for photodynamic therapy: Current status, challenges and future prospects *Photochemistry and Photobiology* **82**: 1198-1210.
7. Moan, J., Peng, Q., Sorensen, R., Iani, V., and Nesland, J. M. (1998). The biophysical foundations of photodynamic therapy *Endoscopy* **30**: 387-391.
8. Bauer, J., Chen, K. H., Hiltbunner, A., Wehrli, E., Eugster, M., Schnell, D., and Kessler, F. (2000). The major protein import receptor of plastids is essential for chloroplast biogenesis *Nature* **403**: 203-207.
9. Thompson, M. S., Johansson, A., Johansson, T., Andersson-Engels, S., Svanberg, S., Bendsoe, N., and Svanberg, K. (2005). Clinical system for interstitial photodynamic therapy with combined on-line dosimetry measurements *Applied optics* **44**: 4023-4031.
10. Kim, Y. H., Lim, S. C., and Kim, K. S. (1993). Activation of Superoxide - Application of Peroxysulphur Intermediates to Organic-Synthesis *Pure and Applied Chemistry* **65**: 661-666.
11. Valko, M., Leibfritz, D., Moncol, J., Cronin, M. T., Mazur, M., and Telser, J. (2007). Free radicals and antioxidants in normal physiological functions and human disease *Int J Biochem Cell Biol* **39**: 44-84.
12. Sato, K., Sato, N., Shimizu, H., Tsutiya, T., Takahashi, H., Kakizaki, S., Takayama, H., Takagi, H., and Mori, M. (1999). Faropenem enhances superoxide anion production by human neutrophils in vitro *Journal of Antimicrobial Chemotherapy* **44**: 337-341.
13. Gersten, J. and Nitzan, A. (1980). Electromagnetic theory of enhanced Raman scattering by molecules adsorbed on rough surfaces *The Journal of Chemical Physics* **73**: 3023-3037.
14. Weitz, D. A., Garoff, S., Hanson, C. D., Gramila, T. J., and Gersten, J. I. (1982). Fluorescent Lifetimes of Molecules on Silver-Island Films *Optics Letters* **7**: 89-91.
15. Das, P. C. and Puri, A. (2002). Energy flow and fluorescence near a small metal particle *Physical Review B* **65**.
16. Geddes, C. D. and Lakowicz, J. R. (2002). Metal-enhanced fluorescence *Journal of Fluorescence* **12**: 121-129.
17. Liebermann, T. and Knoll, W. (2000). Surface-plasmon field-enhanced fluorescence spectroscopy *Colloids and Surfaces A-Physicochemical and Engineering Aspects* **171**: 115-130.
18. Zhang, Y., Aslan, K., Malyn, S. N., and Geddes, C. D. (2006). Metal-enhanced phosphorescence (MEP) *Chemical Physics Letters* **427**: 432-437.

19. Zhang, Y., Aslan, K., Previte, M. J. R., Malyn, S. N., and Geddes, C. D. (2006). Metal-enhanced phosphorescence: Interpretation in terms of triplet-coupled radiating plasmons *Journal of Physical Chemistry B* **110**: 25108-25114.
20. Redmond, R. W. and Gamlin, J. N. (1999). A compilation of singlet oxygen yields from biologically relevant molecules *Photochemistry and Photobiology* **70**: 391-475.
21. Flors, C., Fryer, M. J., Waring, J., Reeder, B., Bechtold, U., Mullineaux, P. M., Nonell, S., Wilson, M. T., and Baker, N. R. (2006). Imaging the production of singlet oxygen in vivo using a new fluorescent sensor, Singlet Oxygen Sensor Green (R) *Journal of Experimental Botany* **57**: 1725-1734.
22. Stiel, H., Teuchner, K., Paul, A., Leupold, D., and Kochevar, I. E. (1996). Quantitative comparison of excited state properties and intensity-dependent photosensitization by rose bengal *Journal of Photochemistry and Photobiology B: Biology* **33**: 245-254.
23. Zhang, Y. X., Aslan, K., Previte, M. J. R., Malyn, S. N., and Geddes, C. D. (2006). Metal-enhanced phosphorescence: Interpretation in terms of triplet-coupled radiating plasmons *Journal of Physical Chemistry B* **110**: 25108-25114.
24. Zhang, Y. X., Aslan, K., Previte, M. J. R., and Geddes, C. D. (2007). Metal-enhanced singlet oxygen generation: A consequence of plasmon enhanced triplet yields *Journal of Fluorescence* **17**: 345-349.
25. Zhang, Y., Aslan, K., Previte, J. R. P., and Geddes, D. C. (2008). Plasmonic engineering of singlet oxygen generaton *PNAS* **105**: 1798-1802.
26. Aslan, K., Leonenko, Z., Lakowicz, J. R., and Geddes, C. D. (2005). Annealed silver-island films for applications in metal-enhanced fluorescence: Interpretation in terms of radiating plasmons *Journal of Fluorescence* **15**: 643-654.
27. Lakowicz, J. R. (1999) *Principles of Fluorescence Spectroscopy* 2nd ed. Kluwer Academic, New York.
28. Aslan K, Previte M J R, Zhang YX, and CD, G. (2007). Metal-enhanced fluorescence (MEF): Progress towards a unified plasmon-fluorophore theory *BIOPHYSICAL JOURNAL* 371A-371A
29. Challener, W. A., Sendur, I. K., and Peng, C. (2003). Scattered field formulation of finite difference time domain for a focused light beam in dense media with lossy materials *Optics Express* **11**: 3160-3170.
30. Anantha, V. and Taflove, A. (2002). Efficient modeling of infinite scatterers using a generalized total-field/scattered-field FDTD boundary partially embedded within PML *Ieee Transactions on Antennas and Propagation* **50**: 1337-1349.
31. Zhang, Y., Aslan, K., Previte, M. J. R., and Geddes, C. D. (2007). Metal-Enhanced Superoxide Generation: A Consequence of Plasmon Enhanced Triplet Yields *Applied Physics Letters* **91**: 0234114.
32. Aslan, K., Gryczynski, I., Malicka, J., Matveeva, E., Lakowicz, J. R., and Geddes, C. D. (2005). Metal-enhanced fluorescence: an emerging tool in biotechnology *Current Opinion in Biotechnology* **16**: 55-62.
33. Sokolov, K., Chumanov, G., and Cotton, T. M. (1998). Enhancement of molecular fluorescence near the surface of colloidal metal films *Analytical Chemistry* **70**: 3898-3905.

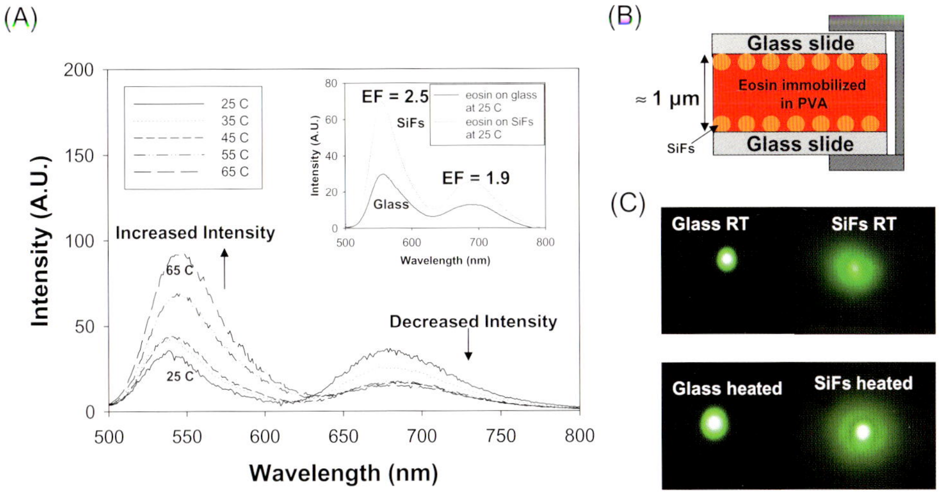


**Figure 1.9:** (A) E-type Fluorescence and phosphorescence emission spectra, $\lambda_{ex} = 473$ nm, of Eosin in a cuvette at different temperatures. **Insert** Eosin immobilized in PVA sandwiched between two silvered and unsilvered slides at 25C EF – Enhancement Factor. RT – Room Temperature. (B) experimental sample geometry. (C) Real-color photographs of Eosin emission from glass and SIFs, before and after 2 mins heating. $\lambda_{ex} = 473$ nm. SIFs – Silver Island Films. The real- color photographs were taken through an emission filter (488 nm razor edge).

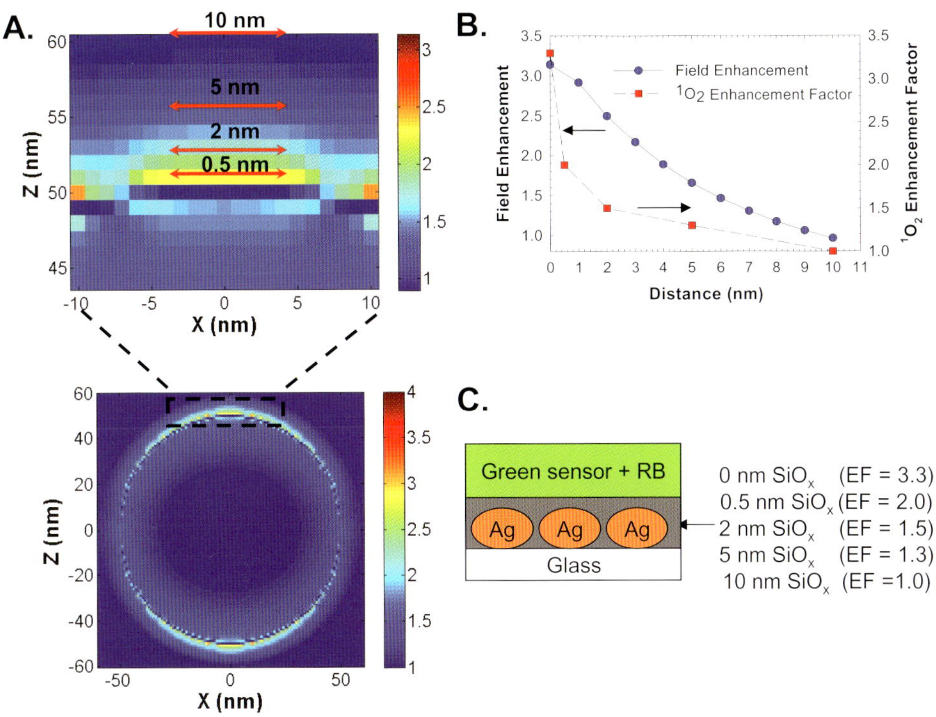


**Figure 1.15:** (A) FDTD calculations for field enhancements around a silver sphere. Zoomed image of 10 nm above the surface of the silver sphere maximum field intensity at z = 10 nm to correlate increased field enhancements in proximity to sphere surface with increased singlet oxygen generation. (B) Distance dependence relationship between for electric field enhancements and singlet oxygen on 100 nm Ag nanoparticles. (C) Distance dependence of singlet oxygen Enhancement Factor of Rose Bengal on SiFs. Top layer is mixed solution of Green Sensor and Rose Bengal. SiOx layer was deposited using thermal vapor deposition.

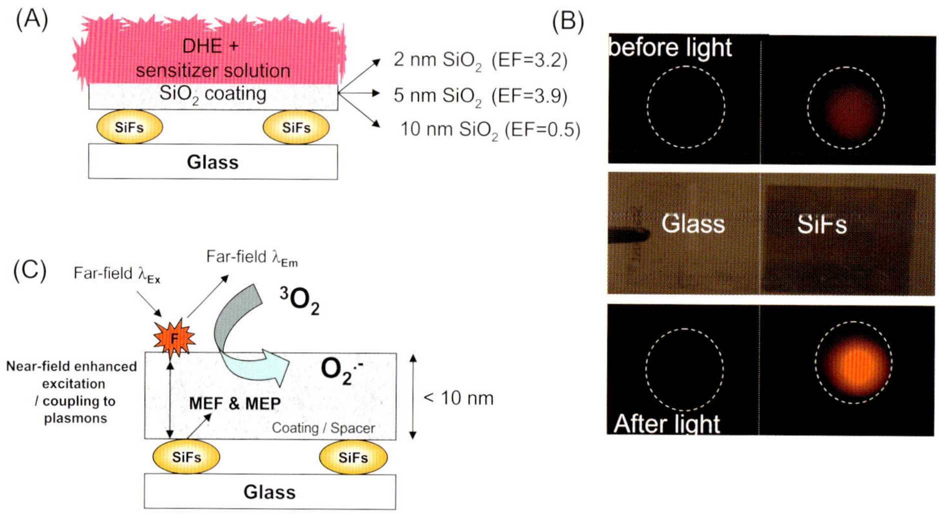


**Figure 1.16:** (A) Sample architecture for the distance dependence of metal-enhanced Superoxide generation. (B) Real color photographs of dihydroethidium (DHE) and Acridine emission from glass and SiFs, before and after 2 mins light exposure (sensitization). (C) Graphical representation of the interpretation of metal-enhanced superoxide generation with an enhanced and distance dependent excitation rate. Light exposure source was a 100 W mercury lamp. $\lambda_{ex}$ = 473 nm.

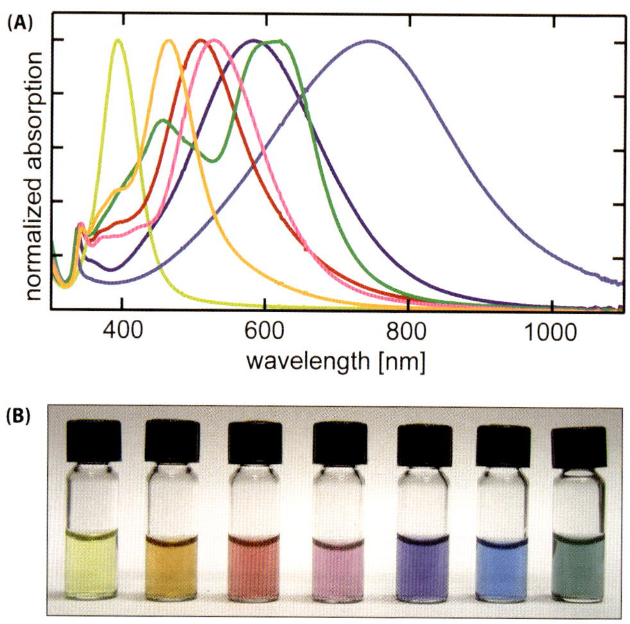


**Figure 4.2:** (A) Series of normalized extinction spectra of chemically synthesized colloidal Ag nanoparticle solutions, showing the tuning of the plasmon resonance across the visible region. (B) Images of colloidal nanoparticle solutions. The differences in color are due to variations in the size and shape of the nanoparticles within each solution.

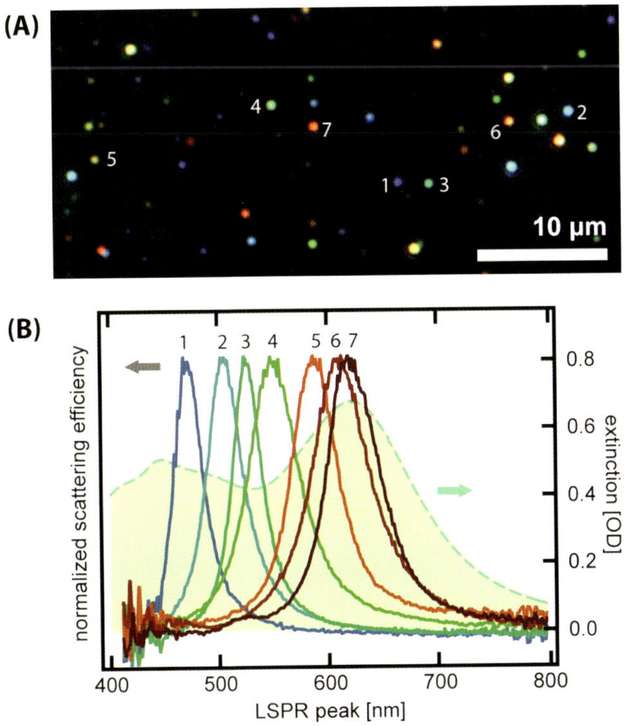


**Figure 4.5:** (A) Darkfield optical micrograph of a typical distribution of single Ag nanoparticles immobilized on a glass cover slip. (See text for full caption.)

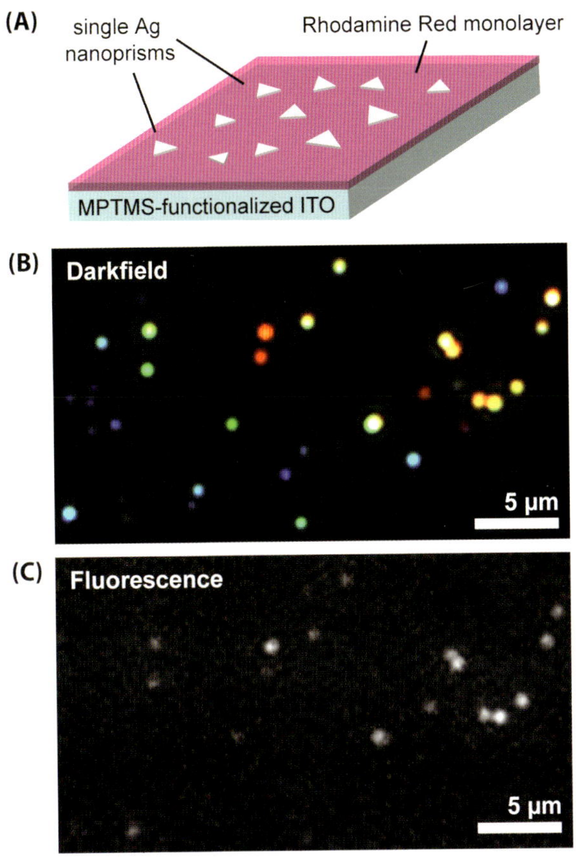


**Figure 4.7:** (A) Schematic illustration of fluorescence enhancement experiment; Ag nanoprisms are adsorbed on top of monolayer of Rhodamine red on glass slide. (See text for full caption.)

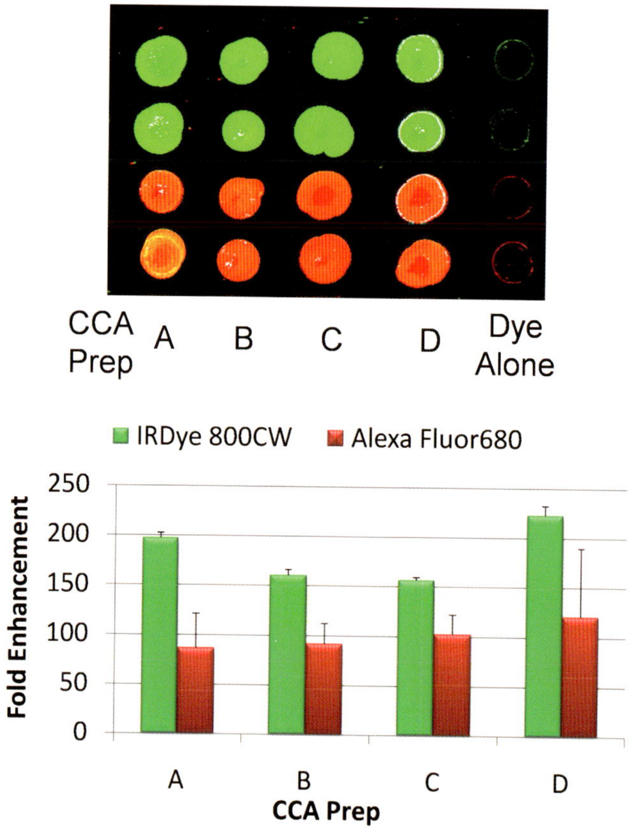


**Figure 5.6:** Odyssey Scanner results showing CCA enhancement of IRDye 800CW (Top two rows) and Alexa Fluor 680 (Bottom two rows) labelled streptavidin on glass slides. Scanner image (top) shows four CCA preparations (A-D) used to enhance the two near-IR fluorophores, as well as the fluorophores spotted on plain glass without CCA nanostructures added (Dye Alone). Bar graph (bottom) shows the relative fold enhancement over the Dye Alone samples for each of the CCA preparations. Error bars are shown that reflect the deviation between two samples for each preparation.

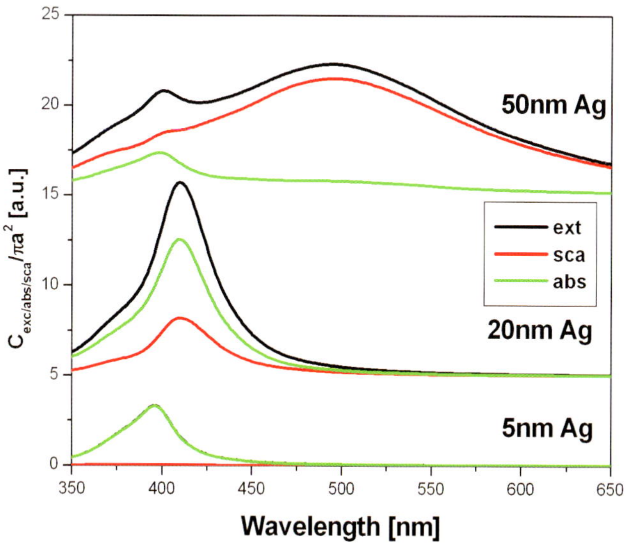


**Figure 6.1:** Theoretical extinction, absorption and scattering spectra of silver NPs with radius 5, 20 and 50 nm, respectively.

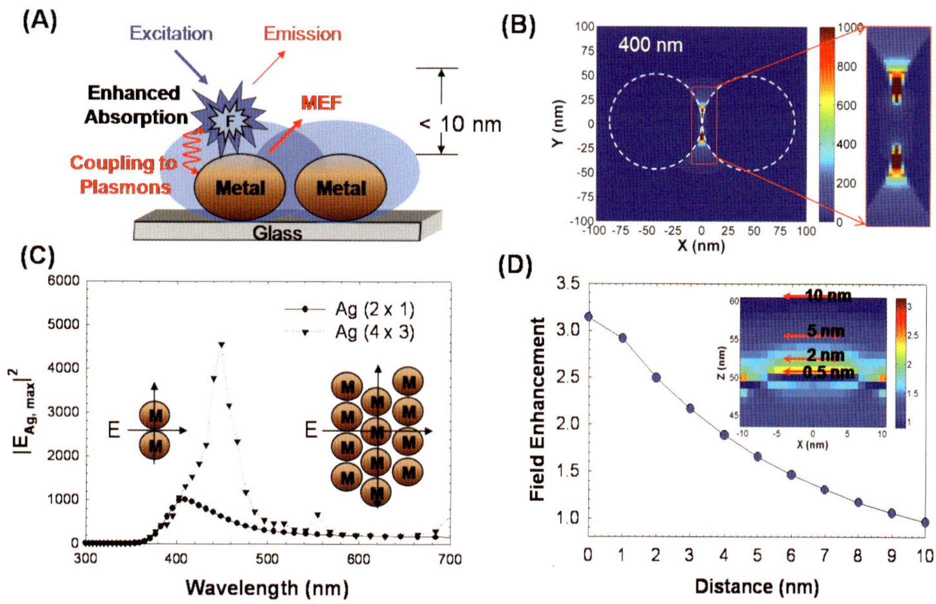


**Figure 7.1:** (A) Schematic representation of the Metal-Enhanced Fluorescence phenomena; (B) FDTD calculations for two silver nanoparticle arrays (d = 100 nm). (C) Wavelength-dependent calculated $|E_z|^2$ maximum intensity for silver nanoparticle arrays (d = 100 nm). Geometries and incident field polarization (*p-polarized*) and propagation direction are shown in the insets. The gap between the nanoparticles was assumed to be 2 nm in the calculations. (D) Calculated field enhancement as a function of distance for a single silver nanoparticle (d = 100 nm). The inset shows these results as an FDTD $|E|^2$ image above the nanoparticle.

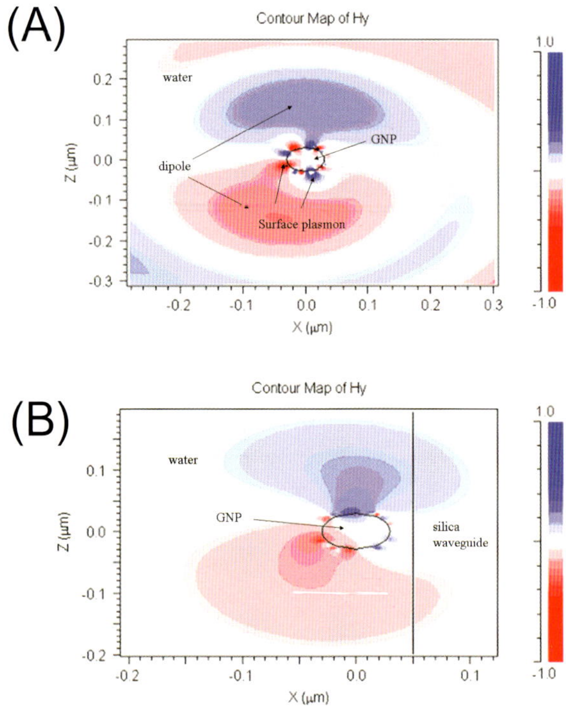


**Figure 8.18:** (a) The magnetic field distribution on the GNP in water. (b) The magnetic field distribution on the GNP near a silica waveguide embedded in water.

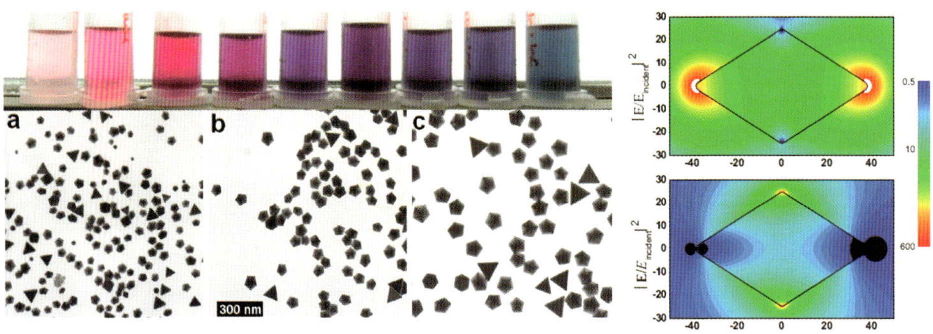


**Figure 11.34:** Top left: photograph of samples withdrawn from the reacting solution at various times during a synthesis of gold nanodecahedra. Bottom left: TEM images of decahedral gold nanoparticles prepared using different amounts of gold-seed solution (a: 1.4 mL, b: 0.7 mL, c: 0.3 mL). The scale is the same in all TEM images. Right: plots of the calculated near-field enhancement ($|E/E\text{incident}|^2$) for bicones (computation model for decahedron) with 40 nm radius and 25 nm height. Light is coming from below in the upper plot and from the left in the lower one, with the electric field contained in the plane of the plots. For the calculations, the wavelengths of maximum extinction cross-section (620 and 522 nm, respectively) were used. The vertical and horizontal axes are in nanometres. Reproduced with permission from reference [92]. © (2006) Wiley-VCH Verlag GmbH & Co. KGaA.

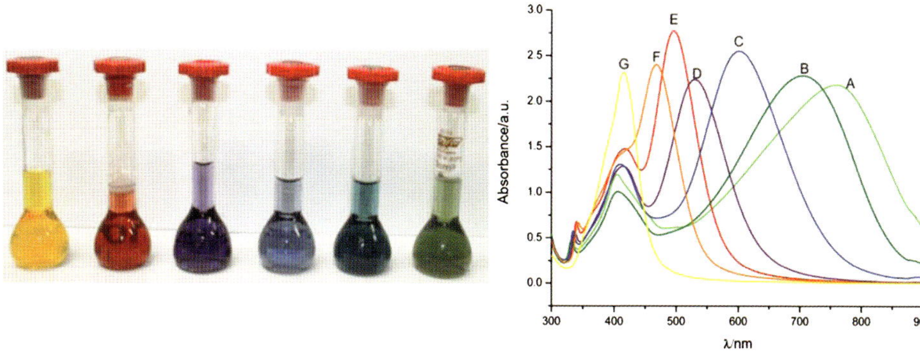


**Figure 11.47:** UV–vis spectra and photograph of different coloured nanoprism samples prepared by varying only the concentration of citrate in the growth step. Citrate concentrations used are: A) 0.7 mM; B) 0.35 mM; C) 0.175 mM; D) 0.07 mM; E) 0.024 mM; F) 0.012 mM; G) 0.004 mM. Reference [119] – Reproduced by permission of The Royal Society of Chemistry.

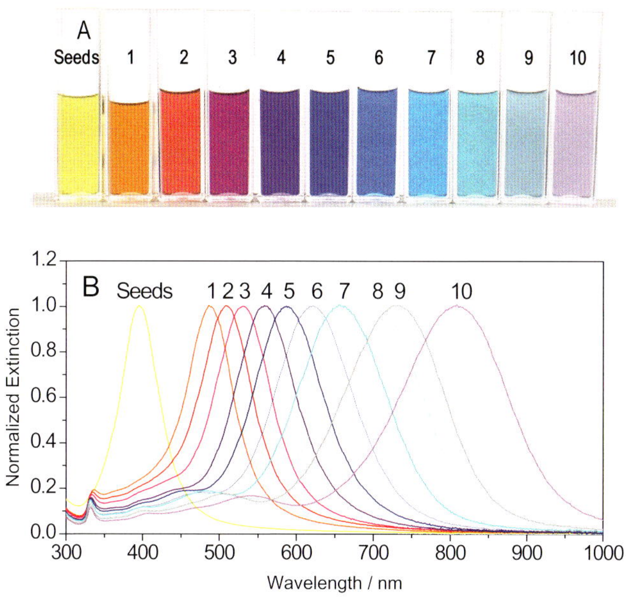


**Figure 11.48:** A) Photograph of series of samples illustrating range of colours obtained. B) Normalized spectra of a series of *as prepared* samples obtained using different volumes of seed solution: 1) 650, 2) 500, 3) 400, 4) 260, 5) 200, 6) 120, 7) 90, 8) 60, 9) 40, 10) 20 μl. Reproduced with permission from reference [128]. © (2008) Wiley-VCH Verlag GmbH & Co. KGaA.

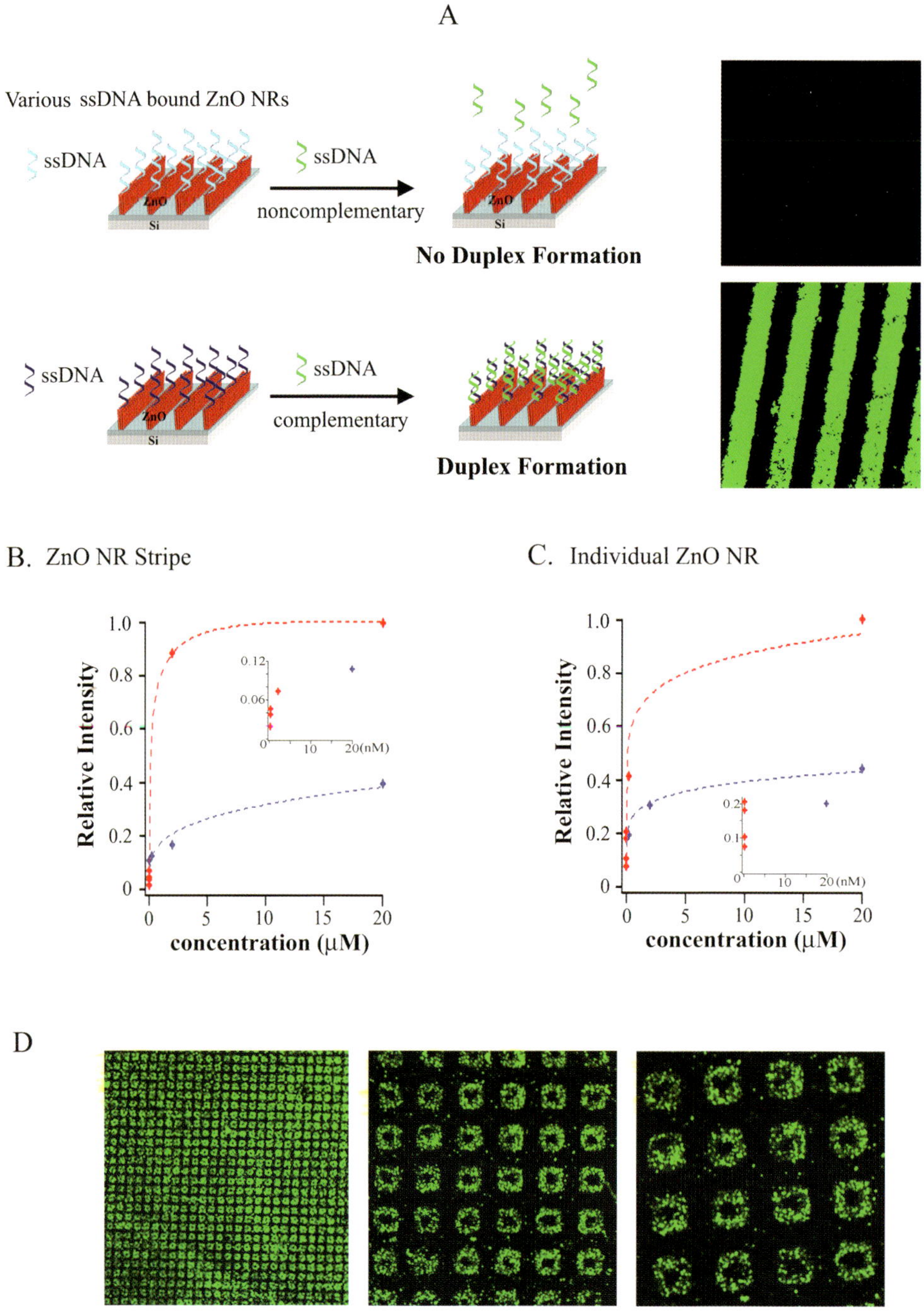


**Figure 12.5:** DNA hybridization reactions performed on ZnO NR arrays. (A) Strong fluorescence emission is observed from a sample containing fully complementary ssDNA strands whereas no signal is detected from noncomplementary strands. (B and C) Concentration dependent assays displaying the detection sensitivity of ZnO NR platforms. Data shown in red and blue correspond to assays empolying a covalent and non-covalent linking scheme of DNA strands on ZnO NRs, respectively. (D) Fluorescence emission due to duplex DNA formation on open-squared ZnO NR arrays. The easy integration potential of ZnO NR arrays into high density platforms is demonstrated. Copyright American Chemical Society, Inc. Reproduced with permission.

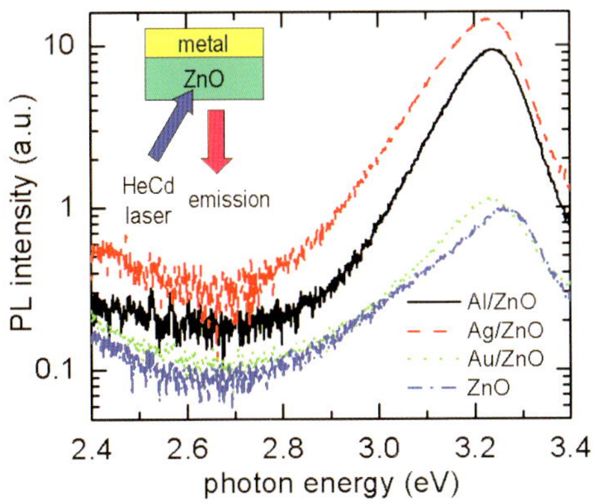


**Figure 13.3:** The experimental band-edge emission spectra of Al, Ag and Au capped ZnO. The backward photoluminescence geometry is shown in the inset. The emission of bare ZnO is also displayed as reference.

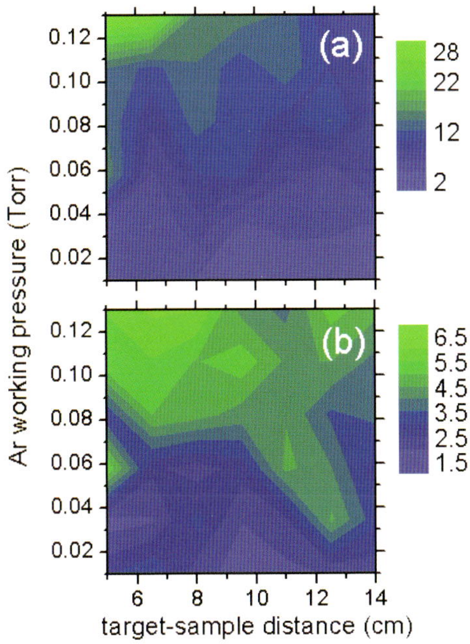


**Figure 13.8:** The contour plots of a) band-edge emission enhancement ratio and b) deep-level emission suppression ratio with Ar working pressure and target-sample distance.

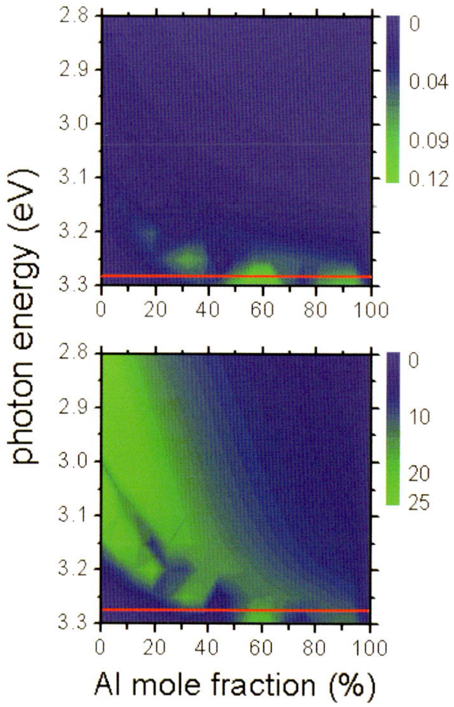


**Figure 13.11:** The dependence of a) plasmonic DOS and b) Purcell factor of air / $Al_xAg_{1-x}$ / ZnO on photon energy and Al mole fraction. The solid lines are ZnO emission. The thickness of the alloys is 30 nm [37].

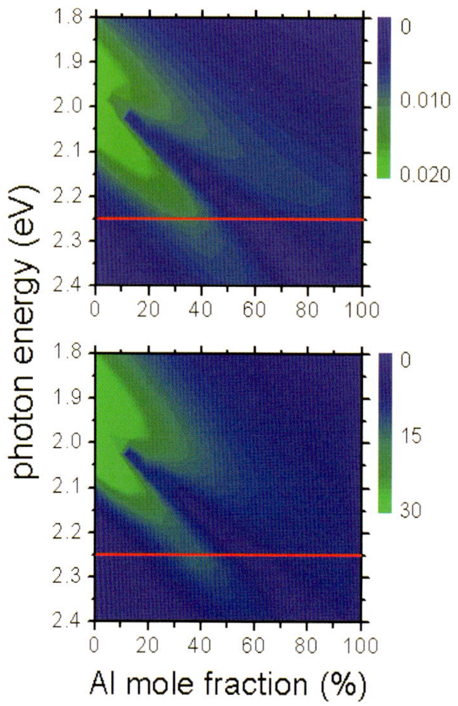


**Figure 13.12:** The dependence of a) plasmonic DOS and d) Purcell factor of air / AlxAu1−x / ZnTe on photon energy and Al mole fraction. The solid lines are ZnTe emission. The thickness of the alloys is 30 nm.

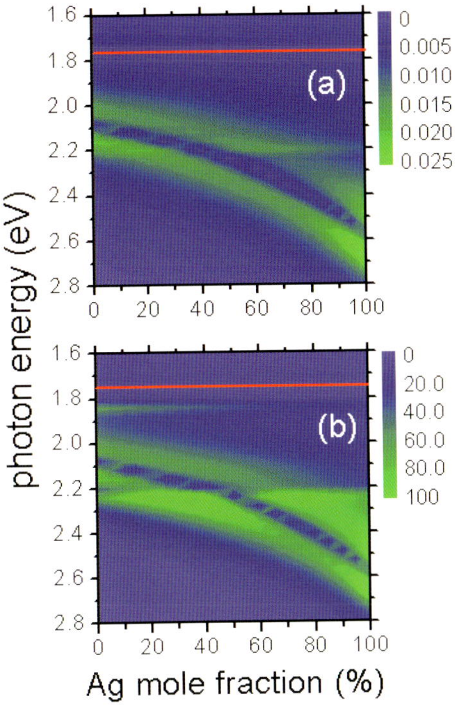


**Figure 13.13:** The dependence of a) plasmonic DOS and b) Purcell factor of air / $Ag_xAu_{1+x}$ / CdSe on photon energy and Ag mole fraction. The solid lines are CdSe emission. The thickness of the alloys is 30 nm.

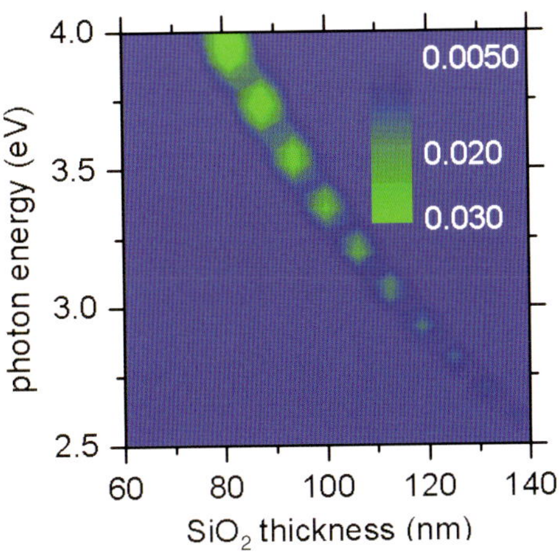


**Figure 13.19:** The counterplot of the dependence of transmittance of Al / $SiO_2$ / Al on photon energy and $SiO_2$ thickness.

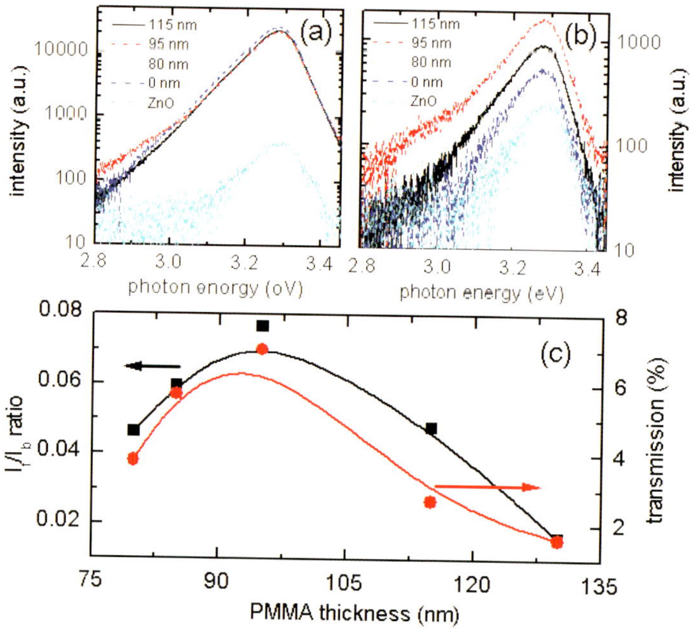


**Figure 13.21:** The a) backward and b) forward PL spectra air / Al / PMMA / Al / ZnO / substrate for different PMMA thickness. The spectra of bare ZnO are also shown. c) The plots of forward to backward emission intensity ratio (If / Ib) (square) and transmission (circle) of air / Al / PMMA / Al / ZnO / substrate with PMMA thickness [45].

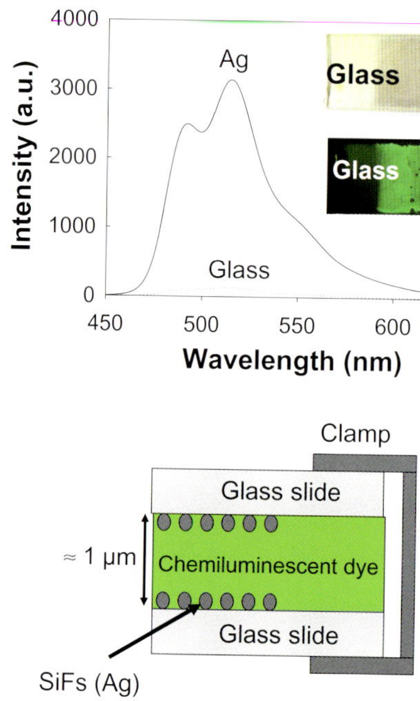


**Figure 15.2:** Chemiluminescence emission intensity from both the glass and the silvered surface (Ag) (Top). Insert – photographs of the silvered and glass surfaces, with (insert – Bottom) and without (insert – Top) chemiluminescence material in the experimental sandwich. The enhancement factor was >20, i.e. intensity on Ag / intensity on glass. Experimental sample sandwich (Bottom). Reproduced from *Applied Physics Letters* 88: 173104. (2006).

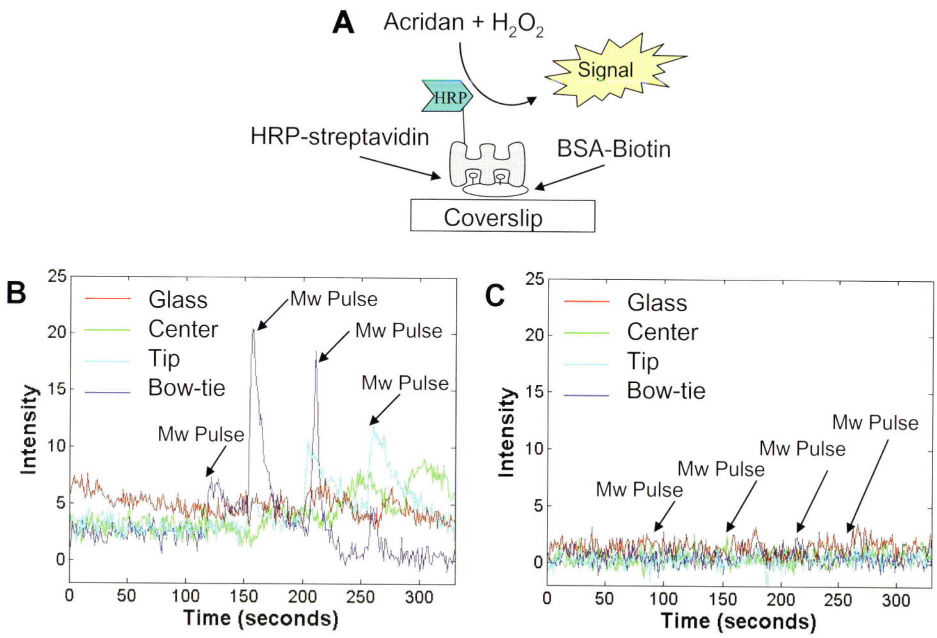


**Figure 15.15:** A) Model BSA-biotin, HRP-streptavidin chemiluminescent assay scheme. (See text for full caption.)

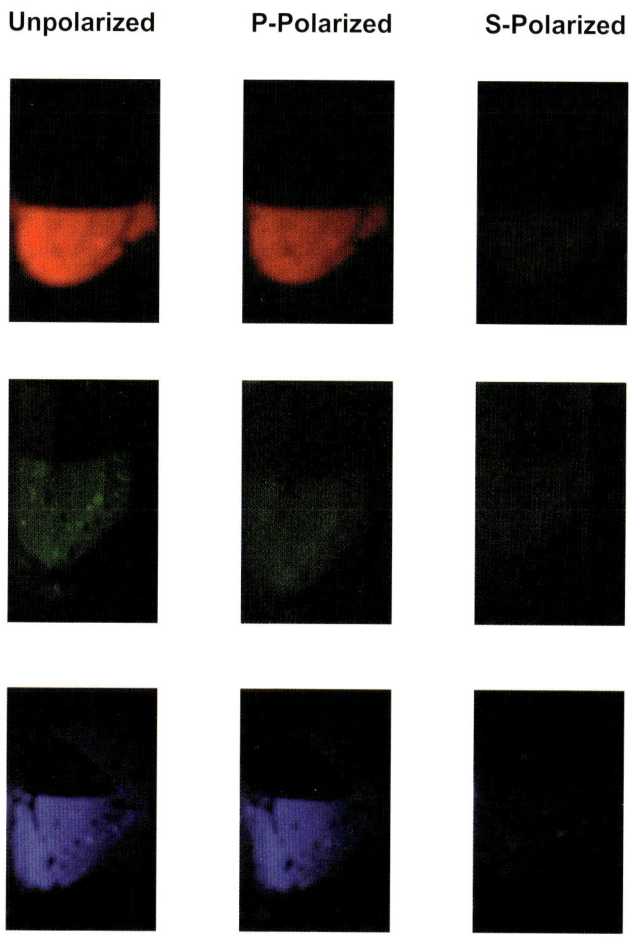


**Figure 15.19:** Photographs of the coupled Chemilumiescence emission at various polarizations for Gold, Silver and Aluminum films, top to bottom respectively, taken at their respective SPCC peak angles. Reproduced from *Journal of Physical Chemistry B* 110: 22644–22651, 2006.

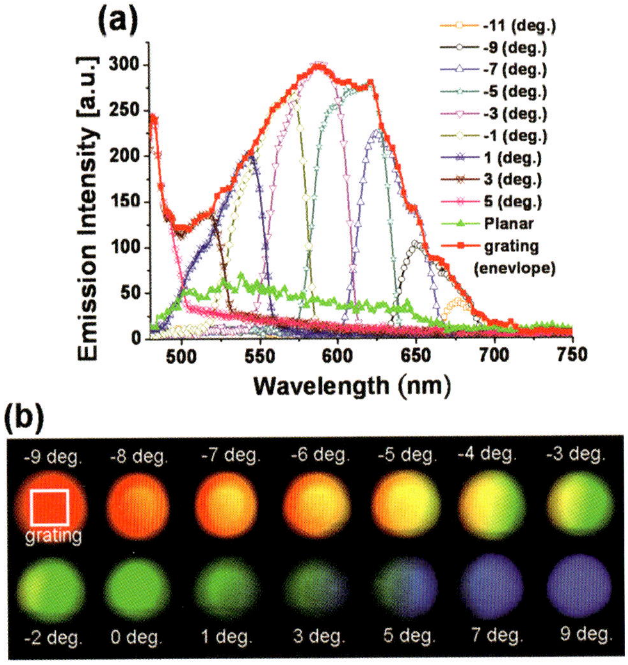


**Figure 16.10:** (a) The PL emission obtained from a grating of 4-layer sample, grating size (line 400 nm, pitch 800 nm, area size 1.2 × 1.2 mm). (b) The changes in SPR were measured from the colour of different angular spectra is shown here. (See text for full caption.)

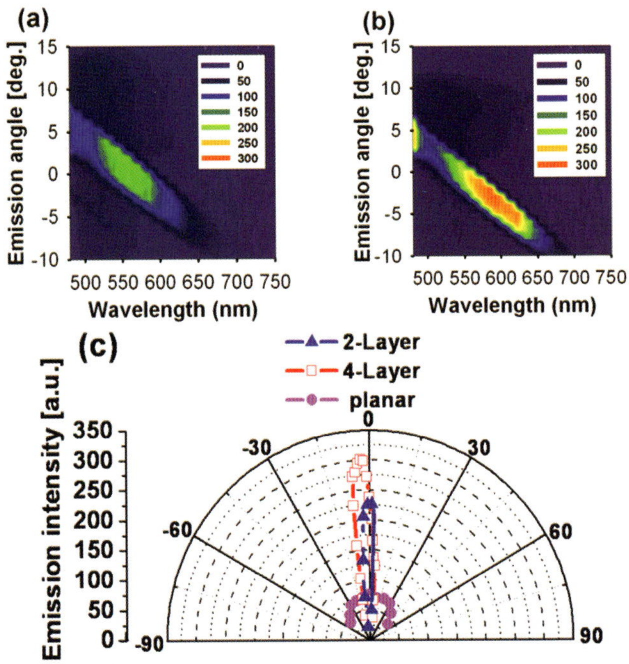


**Figure 16.11:** The PL emission obtained from a grating sample having 2-layer and 4-layer structure (grating size: line 400 nm, pitch 800 nm, area size 1.2 × 1.2 $mm^2$). (See text for full caption.)

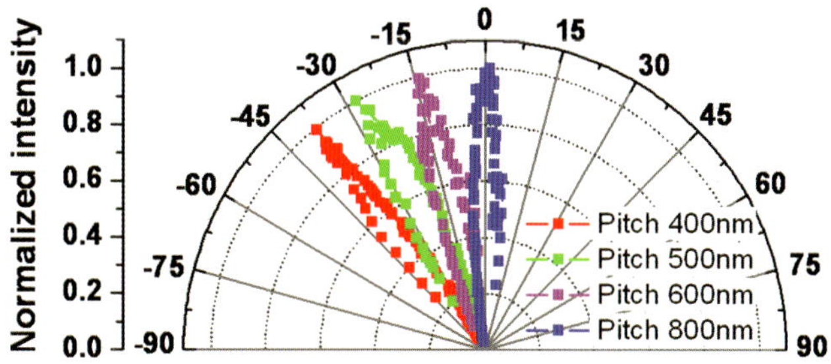


**Figure 16.16:** The experimental PL emission obtained from a sample with grating structure shows SPGCE diagram in directional emission spectra.

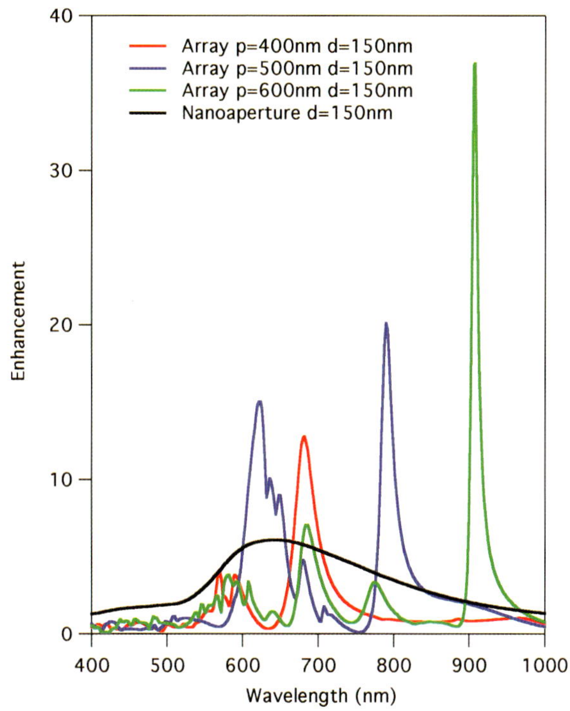


**Figure 17.19:** Calculated excitation intensity enhancement as a function of wavelength at the bottom 10 nm slice (V1) of a single 150 nm diameter nanoaperture in 100 nm thick gold compared to an array of nanoapertures of spacings 400 nm, 500 nm and 600 nm. Excitation occurs from the substrate side and the upper region is air.

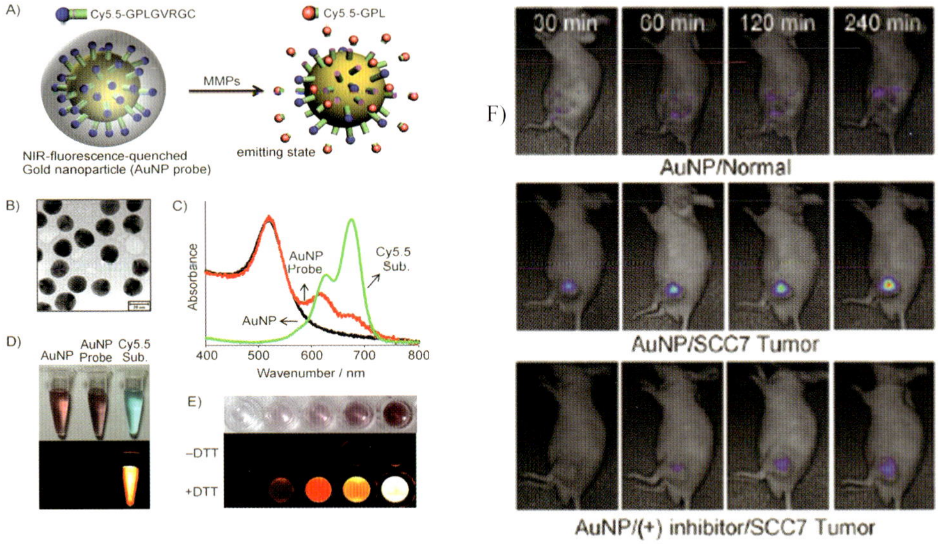


**Figure 20.8:** A) The MMP-sensitive AuNP probe. (See text for full caption.)

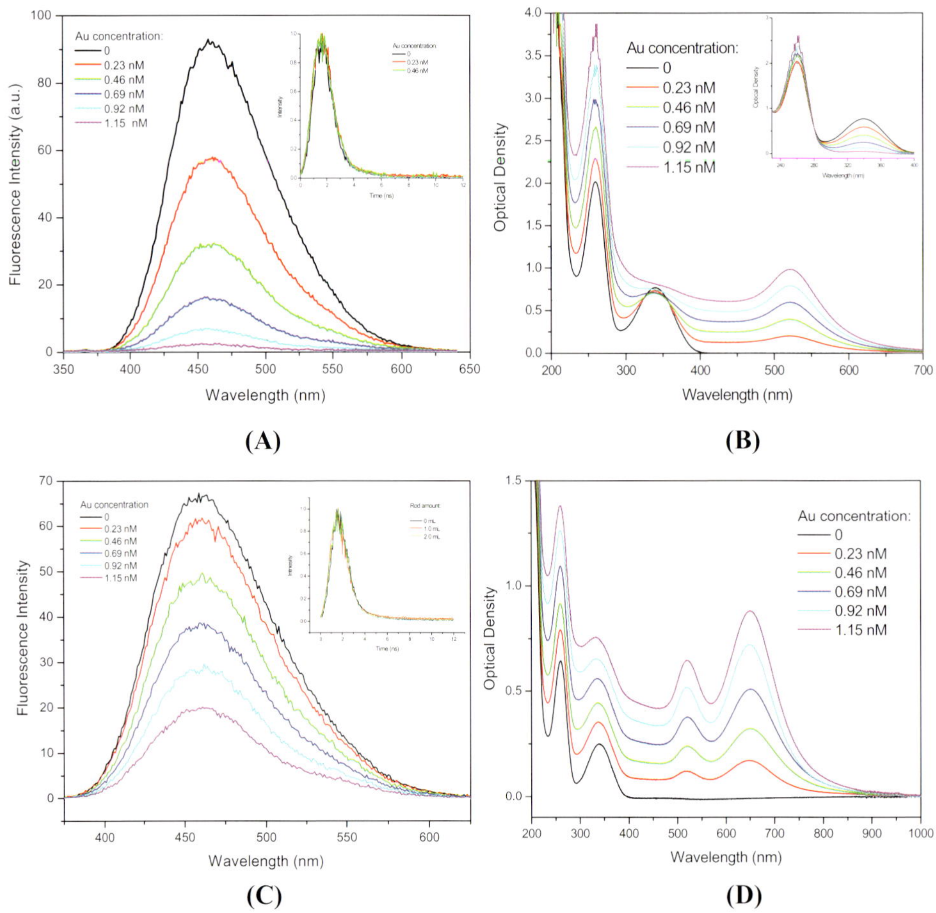


**Figure 20.12:** (A) Fluorescence spectra of NADH in the presence of different concentrations of gold nanospheres (OD = 1.0). (See text for full caption.)

34. Lakowicz, J. R. (2001). Radiative decay engineering: Biophysical and biomedical applications *Analytical Biochemistry* **298**: 1-24.
35. Aslan, K., Lakowicz, J. R., and Geddes, C. D. (2005). Plasmon light scattering in biology and medicine: new sensing approaches, visions and perspectives *Current Opinion in Chemical Biology* **9**: 538-544.

# 11

# Synthesis Of Anisotropic Noble Metal Nanoparticles

Damian Aherne,* Deirdre M. Ledwith, John M. Kelly
School of Chemistry, Trinity College Dublin,
Dublin 2, Ireland.

## 11.1 INTRODUCTION

Metal nanoparticles and nanostructures are at the core of research devoted to the investigation of the phenomenon of Metal Enhanced Fluorescence (MEF) and will likely be key components in any MEF technique or device that is developed. This is chiefly a result of the existence of the surface plasmon resonance (SPR) in metal nanoparticles [1-5]. The SPR results from the collective oscillation of the conduction electrons in the metal in resonance with certain frequencies of light. In any given sample of metal nanoparticles, the SPR manifests itself as an extinction spectrum of light passing through the sample. The spectral position of the SPR is highly dependent on nanoparticle size, shape and aspect ratio, as well as the refractive index of the metal and the surrounding medium. The SPR underpins MEF and can lead to enhancement of fluorescence in a twofold manner:

- At the SPR of a metal nanoparticle, the electric field intensity near the surface of the nanoparticle is enhanced strongly relative to the applied field, which causes increased excitation of the fluorophore that is in proximity to the metal nanoparticle surface [6];

- Coupling of the excited state of a fluorophore to the SPR can result in an increase of the fluorescence quantum yield [6-12].

To maximize fluorophore excitation and increase the fluorescence quantum yield, the spectral properties of the metal nanoparticles need to be optimized. While spherical colloidal nanoparticles of noble metals have been well known for many years, it is only recently that there has been an explosion of reports on the preparation and properties of anisotropically-shaped materials. As will be discussed in the following sections, a wide range of morphologies can be produced, including triangular nanoplates (nanoprisms), cubes, octahedra, nanowires, nanorods and bi-pyramids. The last few years have also seen major developments in our understanding of the growth processes involved, so that now it is possible to prepare many types of shaped particles in a controlled fashion.

At the same time, the possibility of exploiting the unusual properties of these nanomaterials has been recognised, with potential applications ranging from biosensors, non-linear optical devices [13], surface-enhanced Raman spectroscopy [14,15] to antibacterial agents [16-18]. In this review we discuss the synthesis of anisotropic gold and silver nanoparticles with many examples, illustrating a wide of range of shapes and sizes that are possible and highlighting the latest and most successful synthetic

*The Role Of Plasmonic Engineering In Surface-Enhanced Fluorescence.*
Edited by Chris D. Geddes.

approaches for anisotropic growth. We also endeavour to provide an understanding of the underlying mechanisms behind anisotropic growth in solution.

### *Abbreviations Used Throughout This Work*

| | | |
|---|---|---|
| BDAC | = | Benzyldimethylammonium chloride |
| BSPP | = | Bis(*p*-sulfonatophenyl)phenyl phosphine |
| CN | = | Coordination Number |
| CTAB | = | Cetyltrimethylammonium bromide |
| CTBAB | = | Cetyltributylammonium bromide |
| CTPAB | = | Cetyltripropylammonium bromide |
| CTAT | = | Cetyltrimethylammonium tosylate |
| DDA | = | Discrete Dipole Approximation |
| DMF | = | Dimethylformamide |
| fcc | = | face centred cubic |
| FFT | = | Fast Fourier Transform |
| hcp | = | hexagonally close packed |
| HRTEM | = | High Resolution TEM |
| ICG | = | Indocyanine green |
| LB | = | Langmuir-Blodgett |
| LbL | = | Layer-by-Layer |
| MEF | = | Metal Enhanced Fluorescence |
| NIR | = | Near Infra-Red |
| NMR | = | Nuclear Magnetic Resonance |
| ODA | = | Octadecyl amine |
| PVP | = | Poly(vinyl pyrrolidone) |
| QD | = | Quantum Dot |
| QY | = | Quantum Yield |
| SA | = | Stearic Acid |
| SAED | = | Selected Area Electron Diffraction |
| SEM | = | Scanning Electron Microscopy |
| SIF | = | Silver Island Film |
| SPR | = | Surface Plasmon Resonance |
| SRB | = | Sulforhodamine B |
| TEM | = | Transmission Electron Microscopy |
| XRD | = | X-Ray Diffraction |

## 11.2 BACKGROUND EXPERIMENTAL WORK ON METAL ENHANCED FLUORESCENCE (MEF)

At very short metal nanoparticle-fluorophore distances (~ 1 to 3 nm), a large decrease in fluorescence, known as quenching, is expected [8,19,20]. At greater distances however, the fluorescence can undergo enhancement or continue to experience a degree of quenching. The examples outlined below will illustrate that whether enhancement or quenching is observed depends on nanoparticle size and shape, the distance between the fluorophore and the metal nanoparticle surface, and on the overlap between the SPR and the excitation and/or emission transitions in the fluorophore.

Recently, there has been a flurry of experimental work investigating the phenomenon of MEF with a range of metal nanoparticle shapes and sizes. In this Section, we provide a brief overview of some recent experiments that have shown quenching and enhancement effects, and demonstrate a correlation between observed effects on fluorescence and the morphology of the nanoparticles employed.

## 11.2 1. *Fluorescence Quenching By Metal Nanoparticles*

Many experiments have been carried out where the distance between the fluorophore and the metal nanoparticle surface is varied yet only quenching is observed. In these cases the nanoparticles are usually small spherical metal nanoparticles. The following examples demonstrate distance-dependent quenching in a couple of nanoparticle-fluorophore systems.

In 2006, Schneider *et al.* [21] prepared core-shell gold nanoparticle-polymer nanocolloids fabricated by electrostatic layer-by-layer (LbL) assembly. The outer polymer layer was fluorescently labelled with fluorescein isothiocyanate. The LbL approach produced coatings of well-defined thickness and thus it was possible to carefully control the distance between the nanoparticle surface and the outermost fluorescent polymer layer by varying the number of non-fluorescent layers between the gold nanoparticle core and the outer fluorescent layer. Photophysical investigations revealed strongly distance-dependent fluorescence quenching, see Figure 11.1.

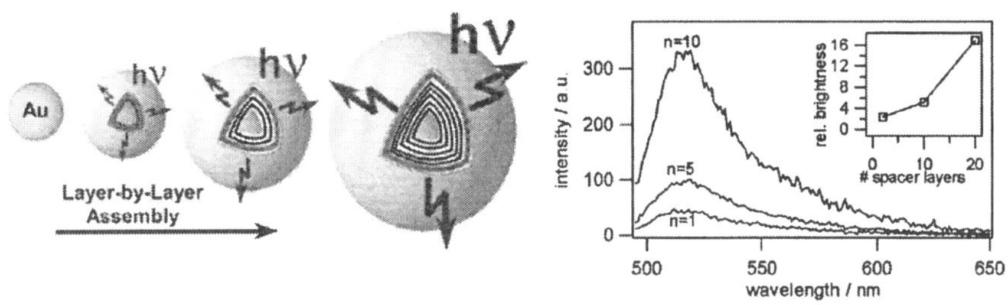


**Figure 11.1:** Left: Layer-by-Layer assembly for the construction of core-shell nanoparticles containing fluorescent outer layers. Right: Fully corrected emission spectra of core-shell nanoparticles. The decrease in emission with decreasing distance of the fluorescent layer from the surface is direct evidence of quenching. Reprinted with permission from reference [21]. © (2003) American Chemical Society.

More recently, Seelig *et al.* [22] employed combined interferometric detection of single gold nanoparticles, single molecule microscopy, and fluorescence lifetime measurement to study the modification of the fluorescence decay rate of an emitter close to a metal nanoparticle surface. In their experiment, gold particles with a diameter of 15 nm were attached to single dye molecules via double-stranded DNA of different lengths. The smaller the distance between the fluorophore and the nanoparticle surface the shorter the lifetime and the lower the measured quantum yield, see Figure 11.2.

One may ask why these experiments only showed quenching and not enhancement. The first thing to take note is the fact that the metal nanoparticles here are spherical, and therefore the SPR does not produce a very large enhancement of the local field. Also, the nanoparticles are small, which, as will be explained in Section 11.3, indicates that most of the extinction is due to absorption, which means that most of the decay of the SPR is by non-radiative pathways. As will be made clear in Section 11.3, this increases the likelihood that energy will be dissipated away instead of being emitted. This results in quenching of the fluorescence and a reduction in the emission lifetime.

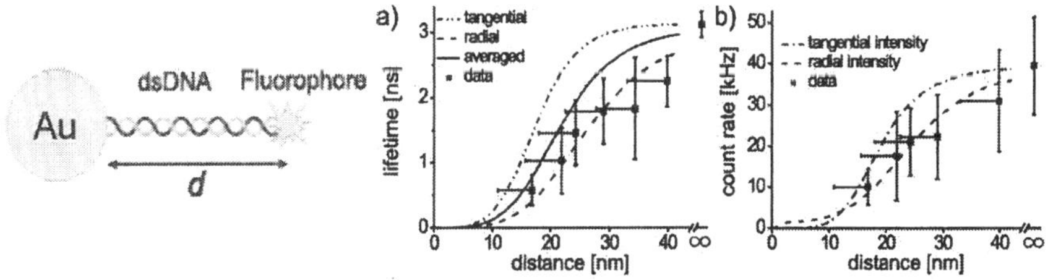


**Figure 11.2:** Left: Schematic illustrating use of double stranded DNA to mediate distance between fluorophore and nanoparticle surface. Middle: Symbols show the average values of experimentally measured fluorescence lifetime. The last measurement was done in the absence of gold nanoparticles as a calibration. The dashed and dashed-dotted curves display the calculated fluorescence lifetime for the molecular dipole oriented radially or tangentially with respect to the gold nanoparticle. Right: Fluorescence signal corresponding to the measurements presented the middle panel. Reprinted with permission from reference [22]. © (2007) American Chemical Society.

## 11.2 2. *Fluorescence Enhancement By Metal Nanoparticles*

The Novotny group have performed confocal measurements of the fluorescence of single molecules at defined distances from 80 nm silver and gold nanoparticles [19, 23]. The sample is illuminated with a laser and the confocal measurements taken with an optical fibre tip with a gold nanoparticle placed at the tip, see Figure 11.3. The fluorophore molecules are well-dispersed on a cover-slip below the tip and so the nanoparticle-fluorophore distance is controlled by vertical movement of the optical fibre.

As expected, quenching dominates over the enhancement at very short distances for both silver and gold nanoparticles, leading to a drop in the overall fluorescence. The maximum fluorescence enhancement can be readily estimated by measuring the fluorescence as a function of the distance between the tip and the sample. Typical results are shown in the right hand side panel of Figure 11.3. It was found that the maximum enhancement for an 80 nm silver nanoparticle is approximately ten-fold. Slightly weaker enhancement (nine-fold) was found for an 80 nm gold nanoparticle.

Geddes and co-workers have developed core-shell nanoparticles with various silica shell thicknesses around 130 nm silver nanoparticle cores. These feature a variety of fluorophores attached to the surface of the silica shell and have demonstrated their applicability for MEF [24]. To show the benefit of using a silver core in the fluorescent core-shell nanoparticles, rather than doping the fluorophores directly onto silica nanoparticles without a silver core, they prepared control sample probes without the silver core. The control fluorescent probes, (hollow fluorescent nanobubbles) are prepared by dissolving the silver core away (etching) with cyanide from the fluorescent Ag@$SiO_2$ nanocomposites, see Figure 11.4. Since the fluorophores are hydrophobic and retained in the hydrophobic pockets of the silica shell or covalently linked to the silica shell, the etching of the silver core with cyanide did not cause the removal of fluorophores from the shell (thickness >10 nm). Thus, it is possible to compare the fluorescence emission and lifetime of the fluorescent core-shell Ag@$SiO_2$ nanocomposites and of the fluorescent nanobubbles in a quantitative manner. It was shown that the fluorescent nanoparticles with core-shell architecture yielded up to twenty-fold enhancement of the fluorescence signal.

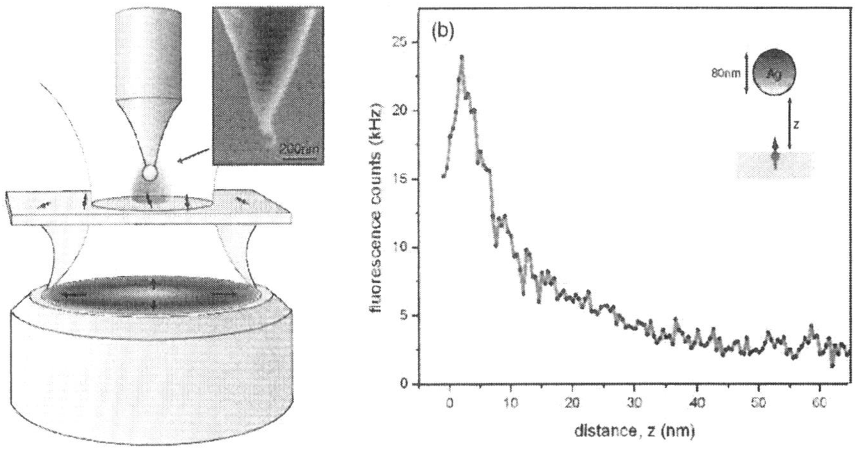


**Figure 11.3:** Left: Schematic diagram of experiment by Novotny group for controlling distance between fluorophores and metal nanoparticle surface. A radially polarized laser beam is focused on the surface of a glass cover slip with well-separated single dye molecules (Nile Blue). The fluorescence is monitored as a function of the proximity of a single gold or silver nanoparticle attached to the end of a glass tip. The inset shows an SEM image of a nanoparticle tip. Right: Single molecule fluorescence rate as a function of nanoparticle–fluorophore distance for an 80 nm silver particle. Similar results were obtained for an 80 nm gold nanoparticle. Reprinted with permission from reference [23]. © (2007) Institute of Physics.

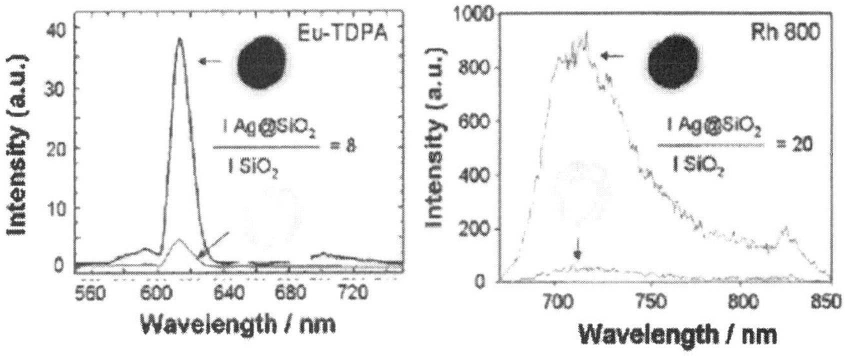

**Figure 11.4:** Fluorescence emission intensity of Eu-TDPA-doped Ag@$SiO_2$ and Rhodamine 800-doped Ag@$SiO_2$ and from the corresponding fluorescent nanobubbles (control samples), Eu-TDPA-doped $SiO_2$ and Rhodamine 800-doped $SiO_2$. The diameter of the Ag nanoparticle core is 130 ± 10 nm and the thickness of the shell is 11 ± 1 nm (optimized) for all the samples. Reprinted with permission from reference [24]. © (2007) American Chemical Society.

The above two examples that demonstrate enhancement utilized spherical nanoparticles. Significantly, these were much bigger than any of the spherical nanoparticles used in the previous two examples that showed quenching. It is most likely that at least as far as spherical metal nanoparticles are concerned, this size difference is a key factor in determining whether fluorescence enhancement or quenching is observed. As will be explained in Section 11.3, for larger nanoparticles, scattering is the major contribution to the extinction spectrum. This is important as it means that radiation damping is the major decay pathway for the SPR and thus there is a higher possibility of an increase in quantum yield.

Tam *et al.* [6] have investigated the role of SPR energy in the enhancement of poorly emitting indocyanine green (ICG) molecules independent of scattering cross-section. This is achieved through the use of nanoshells. In many ways nanoshells are the ideal plasmonic nanoparticles for these experiments since their SPRs are easily tuned across a large visible and infrared wavelength region by varying their thickness. The SPR energy is controlled by the ratio of the inner and outer radius of the metallic shell layer and can be varied independently of the nanoparticle scattering cross-section, which is controlled by absolute nanoparticle size.

In Figure 11.5 it can be seen that the weakest enhancement is, unsurprisingly, for the case of the small gold nanoparticles (plot 1). Plots 2 and 3 are for nanoshells that have a higher scattering cross-section than the small gold nanoparticles but have the same scattering cross-section as each other. The energy of the thinner nanoshell of these (plot 3) is closer to the energy of the fluorophore emission and for this reason shows a slightly greater fluorescence enhancement than for the thicker nanoshell (plot 2). Plots 4 and 5 are for a pair of much larger nanoshells with much larger scattering cross-sections and so decay of the SPR will proceed mostly via radiation damping (see Section 11.3). As before, these have the same scattering cross-section as each other and therefore the

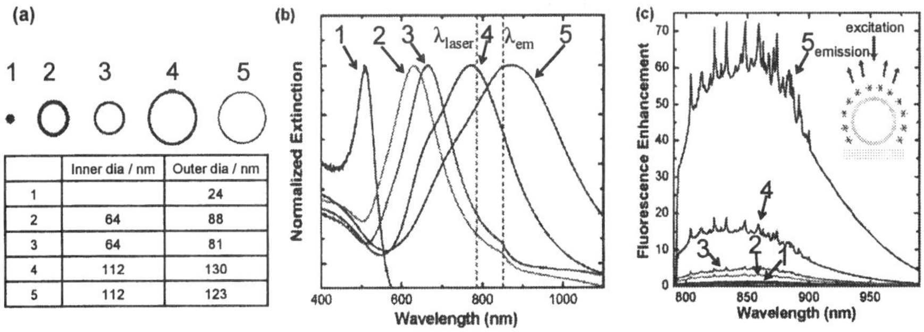


| | Inner dia / nm | Outer dia / nm |
|---|---|---|
| 1 | | 24 |
| 2 | 64 | 88 |
| 3 | 64 | 81 |
| 4 | 112 | 130 |
| 5 | 112 | 123 |

**Figure 11.5:** a) Schematic of samples used as fluorescence enhancement substrates (one Au colloid, and four nanoshells of various inner and outer radii), arranged from short to long plasmon resonance wavelength, corresponding to the spectra in (b). b)

Normalized extinction measurements from nanoparticle substrates corresponding to (a) in air prior to sample preparation. The laser excitation is at 785 nm, and the emission wavelength of ICG is 850 nm. c) Corresponding fluorescence emission from ICG conjugated to the nanoshell substrates adjusted for surface area available for fluorophore conjugation and normalized to the fluorescence from a control sample with no nanoparticles. Inset schematic illustrates experimental geometry. Reprinted with permission from reference [6]. © (2007) American Chemical Society.

surface plasmons in both will radiate with the same efficiency. Since the thicknesses of the nanoshells are different, the SPR energies are also different with the SPR of the thinner nanoshell tuned very close to the emission of the fluorophore. As can be seen in Figure 11.5, there is a much greater enhancement of the fluorescence for the nanoshell whose SPR is tuned to the emission of the fluorophore (plot 5). In addition, it should be noted that the SPR of the thicker nanoshell (plot 4) is tuned to the laser excitation so therewould be a much larger enhancement of the local field at this wavelength compared to the thinner nanoshell. Clearly, in this case, tuning the SPR to the emission, and thereby maximizing the coupling of the fluorophore excited state to the SPR, rather than tuning the SPR to the excitation wavelength, is more effective at enhancing the fluorescence. Overall, this is empirical evidence that for a low quantum yield fluorophore, fluorescence enhancement is optimized by increasing the nanoparticle scattering efficiency while tuning the SPR to the emission wavelength of the fluorophore.

The enhancement properties of highly-shaped nanoparticles such as nanoprisms have also been investigated. Chen *et al.* [25] have investigated the fluorescence from fluorophore-labelled oligonucleotides coupled to immobilized silver nanoprisms. They utilized single nanoparticle dark field scattering and fluorescence microscopy to correlate the fluorescence intensity of the fluorophores with the SPR of the individual nanoprisms to which they are attached. For each of the three high quantum yield fluorophores investigated, they observed a strong correlation between the fluorescence intensity of the fluorophore and the degree of spectral overlap with the SPR. On average they observed the brightest fluorescence from fluorophores attached to nanoprisms that have a SPR peak 40 to 120 meV higher in energy than the emission peak of the fluorophore. This is clear from Figure 11.6 below.

So, for all three dyes, most fluorescence intensity is observed when the dye emission peak is red-shifted from the SPR peak. In addition, it would seem that the optimal SPR location is between the absorption and emission maxima of the dyes, since for two of the three dyes studied, the maximum brightness occurs when the SPR peak is in between the dye absorption and emission maxima. This could be explained if both the dye excitation and emission rates are being enhanced. This is not unexpected as enhancement of fluorescence by increasing the excitation of nearby fluorophores would be the main enhancement mechanism for fluorophores that have a high intrinsic quantum yield.

## 11.2 3. *Fluorescence Enhancement By Aggregates Of Metal Nanoparticles*

As will be seen in Section 11.3, multi-particle arrangements of nanoparticles are expected to produce very high field enhancements. Aggregates are also expected to exhibit very high field enhancements but are more difficult to characterize since the optical properties of the aggregate can deviate immensely and unpredictably from the optical properties of the constituent nanoparticles.

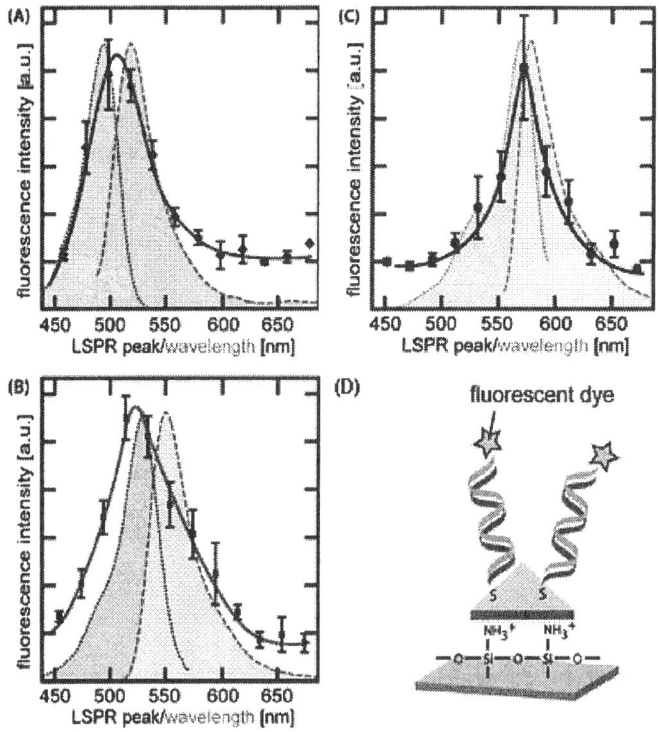


**Figure 11.6:** Summary of single particle fluorescence vs. SPR peak position measurements by Chen *et al.* with three different fluorescent dyes. The SPR peak positions are binned in 20 nm intervals along the x-axis. The average fluorescence intensity observed from particles within each bin is then plotted as a function of the SPR position for silver nanoprisms functionalized with (A) Alexa Fluor 488, (B) Alexa Fluor 532, and (C) Rhodamine Red dyes. The excitation spectra (dotted lines) and emission spectra (dashed lines) are plotted for reference for each dye. The solid line is a guide to the eye. Y-error bars represent the standard deviation of the mean fluorescence intensity observed from particles with SPR peaks within each 20 nm bin. D) Schematic illustrating use of DNA oligonucleotides to conjugate fluorophores a finite distance from the nanoprism surface. Reprinted with permission from reference [25] © (2007) American Chemical Society.

Both Geddes and the Lakowicz group's have investigated the metal-enhanced fluorescence of fluorophores on silver island films (SIFs) [11,26,27] and variously aggregated silver nanoparticles in solution [28,29]. One example of enhancement on SIFs is discussed below [26]. In this work the distance-dependent MEF of a monolayer of sulforhodamine B (SRB) on SIFs was studied. A SRB monolayer was electrostatically incorporated into the Langmuir-Blodgett (LB) layers of octadecylamine (ODA) deposited on glass and SIF substrates (see Figure 11.7). The distances between SRB molecules and the SIFs or glass surfaces were controlled by depositing a variable number of inert stearic acid (SA) spacer layers. SRB molecules were incorporated into positively charged LB layers of ODA by immersing the ODA-coated substrates into an aqueous solution of SRB.

Dye-incorporated ODA layers with 10 nm separation distance from the SIF surface showed maximum enhancement with about a 7-fold increase in intensity as compared to that from the glass surface. Additionally, SRB molecules on SIF surfaces showed reduced lifetimes. It was also observed that the shortest lifetime from the SRB monolayer was with a distance of 5 nm from the SIF surface and the lifetime increased consistently with increasing the distances between the fluorophore and the SIF surface.

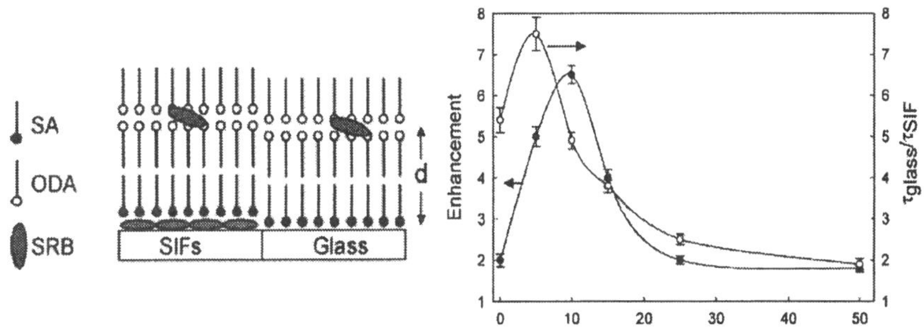


**Figure 11.7:** Left: Schematic representation of LB layers construction and subsequent electrostatically adsorbed SRB dye molecules at octadecylamine layers. Octadecylamine layers were spaced from the substrate by inert stearic acid layers. *d* is the distance between fluorophore and SIF surface that can be varied with the number of inert stearic acid layers at a resolution of ~ 2.5 nm. Right: Fluorescence enhancement factor (filled circles) and corresponding ratio of lifetimes (open circles) of SRB on glass and SIF surfaces versus distance from the SIF surface. Reprinted with permission from reference [26]. © (2007) American Chemical Society

Similar results have been obtained with monolayers of small gold nanoparticles. As we saw above, only quenching was observed in the case of individual small spherical gold nanoparticles. However the work by Komarala *et al.* [30] has shown that a three monolayer thick layer of gold nanoparticles can enhance the fluorescence from CdTe quantum dots (QDs) that are separated from the gold nanoparticle layer by a polyelectrolyte layer, see Figure 11.8. The polyelectrolyte layer is comprised of bilayers of positively charged poly(diallyldimethylammonium chloride) and negatively charged poly(sodium styrenesulfonate) with thicknesses of approximately 1.4 and 11.7 nm for one and nine polyelectrolyte bilayers respectively.

QDs that had an emission spectrum red-shifted (at 667 nm) compared to the SPR of gold nanoparticles (although the SPR of the gold nanoparticle layer is not shown) showed the maximum enhancement, which was reached with six polyelectrolyte bilayers.

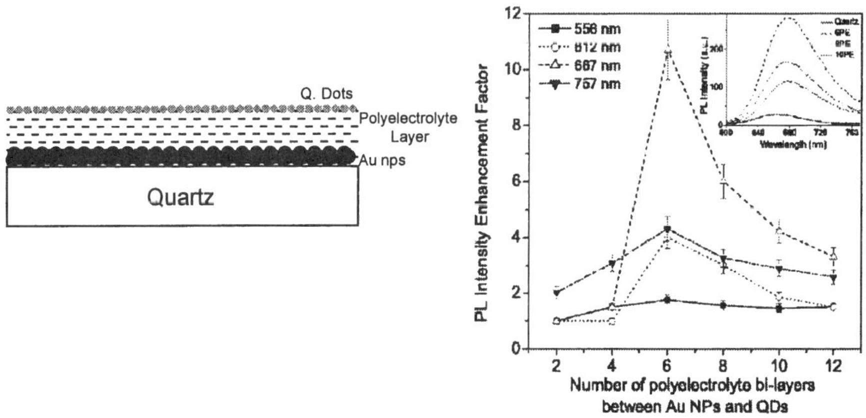


**Figure 11.8:** Left: Schematic illustration of gold nanoparticle, polyelectrolyte layer and QD layer construction for experiments. Right: Plots of fluorescence of QDs with various emission maxima as a function of number of polyelectrolyte bilayers between QDs and gold nanoparticle layer. The emission spectra are shown in the inset. Reprinted with permission from reference [30]. © (2006) American Institute of Physics.

## 11.3 THEORY

In 1908, Mie presented a solution to Maxwell's equations that describes the extinction spectra (extinction = scattering + absorption) of spherical particles of arbitrary size. Mie's solution remains of great interest to this day, although there are no analytical solutions for metal nanoparticles of interesting shapes such as nanorods, nanocubes and nanoprisms. Numerical approaches such as the discrete dipole approximation (DDA) have been employed to investigate the effect of nanoparticle size and shape on the extinction spectra, and the calculation of the intensity and spatial distribution of the local electromagnetic field [1-5,31].

### *Tuning Of The SPR*

The spectral position of the SPR is highly dependent on nanoparticle size, shape and aspect ratio, and also depends on the refractive index of the metal and the surrounding medium. For example in Figure 11.9 below [1], the extinction spectra of silver nanoprisms with different degrees of truncation of the tips are shown. With sharper tips and longer edge lengths, the main SPR (in-plane dipole) undergoes a large red-shift. The strong dependence of the spectra on shape is illustrated in Figure 11.10 [3]. DDA calculations have been performed for a range of shapes including nanocubes and nanoprisms and the absorption and scattering contributions to the extinction spectra have been obtained.

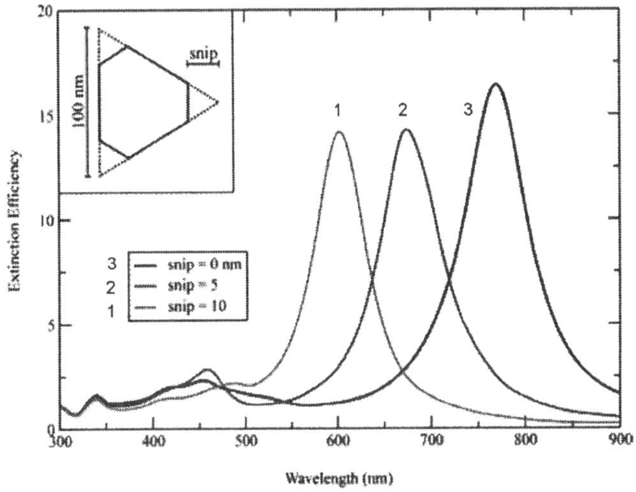


**Figure 11.9:** Orientation-averaged extinction efficiency for silver nanoprisms based on DDA calculations for a 100 nm edge dimension with snips of 0, 10, and 20 nm. The inset shows the shape of a snipped prism. The prism thickness is 16 nm. Reprinted with permission from reference [1]. © (2003) American Chemical Society.

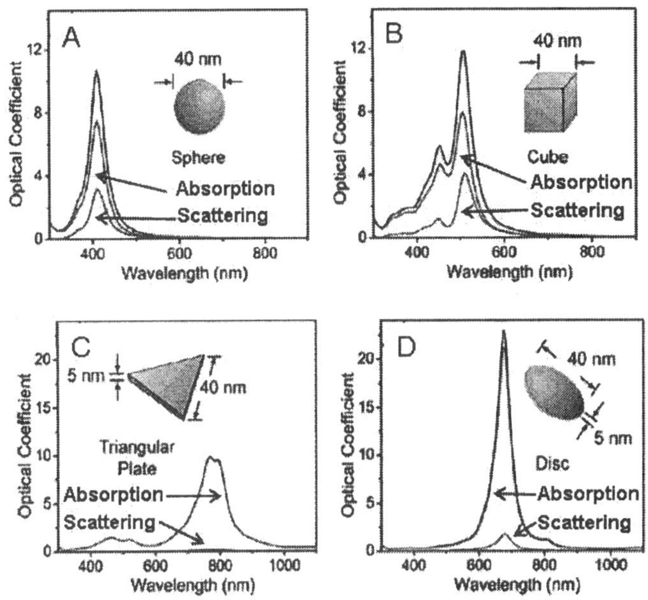


**Figure 11.10:** DDA simulations of extinction (black), absorption, and scattering spectra of silver nanostructures, illustrating the effect of a nanostructure's shape on its SPR. (A) isotropic sphere, (B) isotropic cubes, (C) nanoprisms and (D)

discs. Reprinted with permission from reference [3]. © (2006) American Chemical Society.

## *Enhancement of the local field*

An increase in excitation of the fluorophore depends on the spectral overlap between the SPR and the excitation spectrum of the molecule; and on the enhancement of the local field which, as can be seen below, depends on the position of the fluorophore and its distance from the metal surface. The distribution of the local (enhanced) fields for a nanoprism and nanorod are illustrated in Figure 11.11 [5]. The largest field intensities occur at the tips of the nanoprism and at the ends of the nanorods. The field intensities are calculated to be approximately 4000 times the applied field. These field enhancements are much larger than can be obtained with spheres. Even larger field enhancements can be obtained at the interface of nanoparticles in very close proximity to one another, as shown in Figure 11.12 [5].

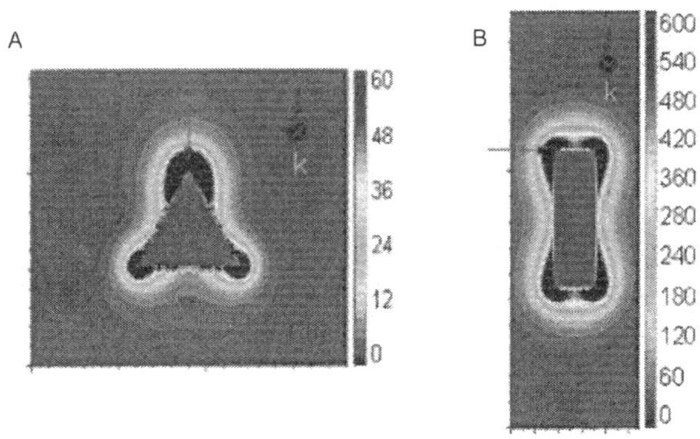


**Figure 11.11:** E-field enhancement contours external to differently shaped silver nanoparticles for the in-plane dipole SPR from DDA calculations, for (A) a nanoprism (edge length = 60 nm, thickness = 12 nm.) and (B) a nanorod (aspect ratio = 2.8:1, effective radius = 15 nm). The arrows indicate the regions of maximum field intensity. Reprinted with permission from reference [5]. © (2004) American Institute of Physics.

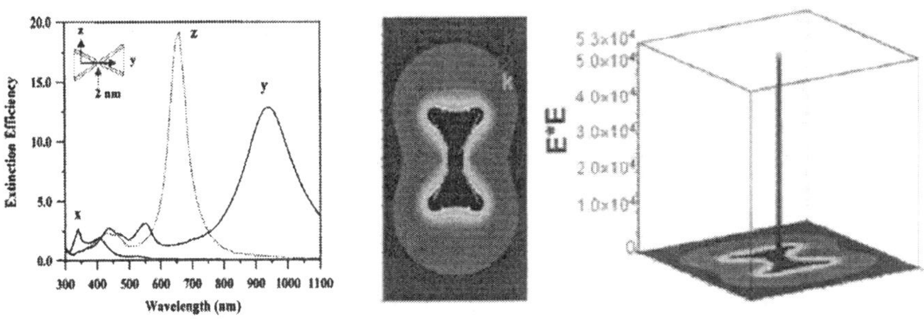


**Figure 11.12:** (Left) Calculated extinction spectra for a tip-tip dimer arrangement (2 nm spacing) of silver nanoprisms along three primary symmetry axes. Nanoprisms have an edge length of 60 nm and 2 nm snip. (Centre) Calculated spatial

distribution of local electric field at 932 nm. (Right) 3D plot showing a significantly greater enhancement at the interface between the two nanoprisms. Reprinted with permission from reference [5]. © (2004) American Institute of Physics.

We have seen from the above that the degree of enhancement clearly depends on shape and the close proximity of other metal nanoparticles. The degree of enhancement can be quantified by a field enhancement factor $|f|$ [32], which is directly proportional to the dephasing time, $T_2$ of the SPR ($|f| \alpha T_2$) [33]. The overall extent of damping can be measured from the linewidths, $\Gamma_{hom}$, of the SPRs of individual nanoparticles. Since $T_2 = 2\hbar/\Gamma_{hom}$, where $\Gamma_{hom}$ is the homogeneous line width [33-35], we can say $|f| \alpha 1/\Gamma_{hom}$. Thus, a narrower homogeneous linewidth implies less damping and a larger local field enhancement. The field enhancement can also be expressed by another important quantity, the quality factor of the resonance, $Q = E_{res}/\Gamma_{hom}$, where $E_{res}$ is the energy of the resonance, and it clearly follows that $Q \alpha T_2$ also [34]. The quality factor is the enhancement of the oscillation amplitude of a driven oscillating system with respect to the driving amplitude, *i.e.*, the local-field enhancement in the case of particle plasmons [34]. There are two decay pathways for the SPR: radiation damping which involves transformation of the plasmons into photons (scattering) and non-radiative decay into electron-hole excitations (absorption) [12,25,26,28], *i.e.* $\Gamma_{hom} = \Gamma_{rad} + \Gamma_{non\text{-}rad}$. This is illustrated schematically in Figure 11.13 [34].

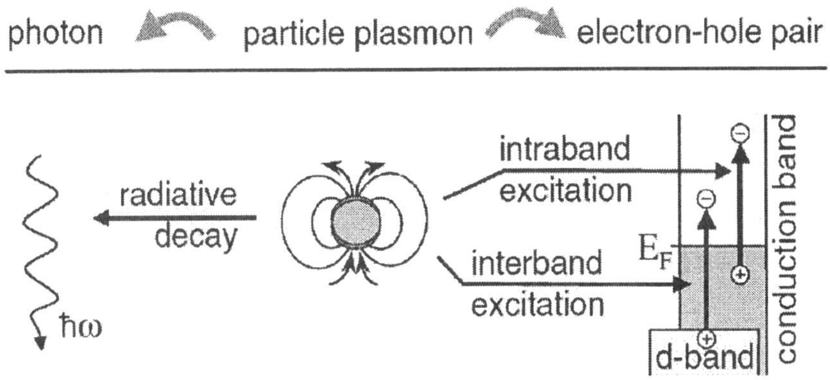


**Figure 11.13:** Schematic representation of radiative (left) and non-radiative (right) decay of particle plasmons in noble-metal nanoparticles. The non-radiative decay occurs via excitation of electron-hole pairs either within the conduction band (intraband excitation) or between the *d* band and the conduction band (interband excitation). Reprinted with permission from reference [34]. © (2002) American Physical Society.

To optimize local enhancement of the electric field we need to minimize all damping as much as possible and the suitability of certain nanoparticle morphologies for MEF by increased excitation of fluorophores can be estimated from measurements of the homogeneous line width of individual nanoparticles. For example, a series of experiments comparing nanospheres and nanorods (see Figure 11.14) has shown that nanorods typically display dramatically reduced plasmon damping compared to spheres, *i.e.* narrower line widths [34], and therefore produce a stronger field enhancement.

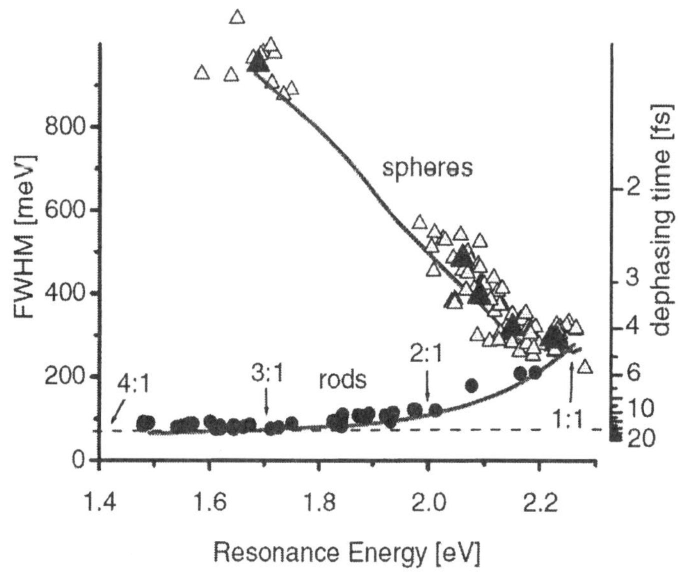


**Figure 11.14:** (a) Measured linewidth Γ of plasmon resonances in single nanorods (dots) and nanospheres (open triangles) *vs.* resonance energy $E_{res}$. The right scale gives the dephasing times $T_2$ calculated from Γ. Black triangles: averages for spherical particles of the same nominal size (150, 100, 80, 60, 40, and 20 nm from left to right). Lines: theory. Some selected aspect ratios *a*/*b* are indicated. Reprinted with permission from reference [34]. © (2002) American Physical Society.

At low plasmon resonance energies this difference is a result of the nanorods exhibiting much lower radiation damping. This is a consequence of the nanorods having a much lower volume than the corresponding nanospheres with the same plasmon resonance energy, since the radiative dephasing rate (radiation damping) is proportional to nanoparticle volume [12,24,26,28], *i.e.* $\Gamma_{rad} \alpha V$. Since different nanoparticle shapes result in different nanoparticle volumes for a given plasmon resonance energy, it is clear that the degree of plasmon damping is highly influenced by nanoparticle shape and this is another route for nanoparticle shape to influence the degree of enhancement of the local field. Non-radiative decay depends on frequency-dependent dielectric properties of the metal [36,37,38].

## *Coupling Of Fluorophore Excited State To SPR.*

There are essentially two models that describe the interaction between an excited fluorophore and the SPR of the metal to account for quenching and enhancement of the fluorescence. They both depend on coupling of the fluorophore excited state to the SPR and this is dependent of the spectral overlap of the emission of the fluorophore and the SPR, and the distance between the fluorophore and the metal nanoparticle surface.

The first model was originally postulated by Gersten and Nitzan [7], and has been reviewed more recently by Lakowicz [9,10]. This model accounts for quenching by considering the metal as providing additional non-radiative pathways for decay of the excited state of the fluorophore; depopulation of the excited state through resonant energy transfer to the metal. This results in an increase the non-radiative decay rate. Enhancement, on the other hand, is accounted for by considering the interaction between the excited state of the fluorophore and the SPR of the metal nanoparticle to increase the intrinsic decay rate of the fluorophore, and thereby increase the quantum yield of the fluorescence.

A more recent model is what is known as the radiative plasmon model. It is a relatively simple but useful approach for consideration of the effects of the metal nanoparticle on fluorescence [8,11,12]. Within this model, the coupling of the fluorophore to the SPR can result in fast transfer of energy from the excited fluorophore to the SPR. Depending on the degree of coupling between the SPR and the excited state of the fluorophore, this energy transfer can be much faster than the radiative life time of the molecule and therefore compete even more effectively with non-radiative recombination pathways of the molecule. Although this results in a faster depopulation of the excited state of the fluorophores, unlike the earlier model this does not necessarily increase the non-radiative decay rate of the fluorophore and result in quenching. This is because, as we saw earlier, the plasmon itself can decay through emission of a photon into the far field (radiation damping). If this process is very efficient with respect to non-radiative plasmon decay pathways, as is the case with highly scattering nanoparticles, then there may be an overall increase in the quantum yield and a reduction in the measured radiative life time (increase in measured radiative decay rate). In fact we can say "...the scattering component of the extinction is a measure of the extent to which the plasmons can radiate into the far field..."[8] So, although all damping lowers the enhancement of the local field, radiation damping (scattering) plays an important role in boosting quantum yield, especially for poorly emitting fluorophores.

If the metal nanoparticle is a poor scatterer then most of the decay of the plasmon is by non-radiative pathways (absorption). If the fluorophore is strongly coupled to the plasmon, this can result in a decrease in the overall quantum yield, *i.e.* quenching of the fluorescence, as in this case all the nanoparticle has done is provide additional fast non-radiative pathways for deexcitation of the fluorophore. This is important as even if there is a very large enhancement of the local field, quenching may still be observed if enough energy from the excited molecules is dissipated by non-radiative decay of the plasmon. Essentially, in the radiative plasmon model, it is the ability or inability of the plasmon to radiate that makes the difference between fluorescence enhancement and quenching.

The contribution of scattering to the overall extinction of a sample scales as $V^2$ while the contribution of absorption scales linearly with $V$. For this reason, the larger a nanoparticle is, the greater the fraction of incident photons that will be scattered instead of being absorbed. Therefore, for a given SPR energy, larger volume nanoparticles are more suitable for enhancing the quantum yield, although the enhancement of the local field will be diminished. This means that there may be a trade-off between enhancement of the field (by minimizing all plasmon damping) and enhancing the quantum yield (by maximizing radiation damping). For example, for a molecule with a very high quantum yield it might be very difficult to avoid quenching and to do so without compromising enhancement of the local field.

## 11.4 SYNTHESIS OF HIGHLY-SHAPED GOLD AND SILVER NANOPARTICLES

Due to the desire for highly anisotropic metal nanoparticles for field enhancement applications, there has been extensive research investigating the different factors that influence particle size and shape. Indeed, the desire to independently control the position of the SPR and the contribution of scattering to the extinction spectrum highlights the importance and necessity of obtaining a high degree of control over synthesis. Until recently, some explanations for the existence of anisotropic growth in an isotropic medium were based upon the assembly of surfactant molecules into a template whose shape then defines the growth of the crystal [39-41], particularly for nanorods and nanowires. It is now thought that in many cases, there is preferential absorption of organic molecules, such as polymers and surfactants, to {100} and {110} crystal faces [42,43]. In this selective binding model, the result is a much faster rate of addition of metal atoms to the more exposed {111} faces, thus resulting in preferred growth directions. Indeed, recent computational work has successfully predicted one-dimensional growth based on the face-selective binding of surfactants [44].

Nevertheless, it is clear that any anisotropic growth that results from the preferential binding of organic species to certain crystal faces relies on the crystal structure of the seed nanoparticles. Whether the seeds are single crystalline or whether they possess any twin planes or other defects, will determine the type and orientation of the crystal faces that are exposed to the growth medium in the first place. This is all the more apparent when we consider that in most syntheses a range of particle shapes are observed and yet the same shaped particle can be the major product of very different syntheses.

Furthermore, anisotropic structures such as nanoprisms present a particular challenge to the face-selective binding model in that gold and silver nanoprisms typically have large flat {111} faces, with two-dimensional growth from the edges. Many syntheses for nanoprisms take place in the presence of stabilizers such as poly(vinyl pyrrolidone) (PVP) or surfactants, yet growth is restricted in the <111> direction. This would suggest that in certain circumstances, it is quite possible that although the organic stabilizers that are often present in the syntheses of nanoprisms provide a general stabilization of the growing nanoprism, they may play little or no shape-directing role.

In this article we will categorize anisotropic nanoparticle shapes into two categories:

- Those whose growth mechanism can be explained by the selective binding growth model and
- Those whose growth mechanisms cannot be explained so readily by the selective binding growth model, despite the presence of the surfactants or polymers that are normally utilized to direct growth by selective binding.

In each category we will briefly review methods for the synthesis of a range of highly-shaped nanoparticles with a view to providing an understanding of the critical factors that drive the anisotropic growth in each case, rather than providing an exhaustive review of available synthetic procedures. We note that we make this categorization in spite of the degree of uncertainty in the literature about the exact nature of the selective binding process, as mentioned in the recent excellent review of anisotropic metal nanoparticles by Tao *et al.* [45]. We will focus only on gold and silver. For details of any of the methods discussed, the reader is referred to the relevant literature.

Gold and silver both have a face-centred cubic (fcc) crystal structure, which is illustrated in Figure 11.15. Gold and silver are by far the most common noble metals and the most studied when it comes to metal nanoparticle synthesis. The first reported nanoparticle preparation was for gold by Faraday, who theorized that the samples he prepared consisted of metals "*in a state of extreme division*" [46]. Today, methods for the synthesis of nanoparticles of all shapes and sizes from gold and silver are in great abundance.

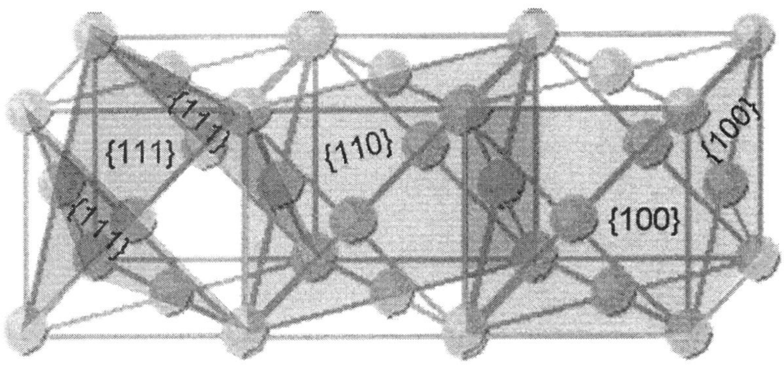


**Figure 11.15:** Face-centred cubic (fcc) lattice showing {100}, {110} and {111} planes.

## 11.4 1. *Nanoparticle Shapes That Can Be Explained By Selective Binding Model For Anisotropic Growth*

### 11.4 1.1 *Physical Basis Of Selective Binding*

The surface energies associated with the different crystal faces of the fcc crystal structure are expected to differ due to the different degrees of "openness" in each face. The {111} face is expected to be the most stable due to the close-packed nature of the atoms in the layer (the coordination number (CN) is 9). The {100} face, which consists of one side of the fcc unit cell, has a more open arrangement of atoms (CN = 8). The {110} face is even more open (CN = 7) and so this face is expected to be the most unstable. The more unstable a crystal face, the greater the stabilization there is to be gained through absorption of chemical species.

Calculations comparing the {111} and {100} faces of gold have shown the {100} face to be less stable [47]. The energy of the {111} face was calculated to be 0.523 $J.m^{-2}$ while the energy of the {100} face was calculated to be 0.606 $J.m^{-2}$. Direct experimental evidence of the particular instability of the {110} face has been uncovered by high-resolution TEM analysis of single crystal gold nanorods. Exposed {110} faces were observed to undergo reconstruction of the surface atoms that resulted in small areas of {111} faces being exposed [48], as shown in Figure 11.16. This reconstruction lowers the total energy of the surface despite slightly increasing the overall surface area.

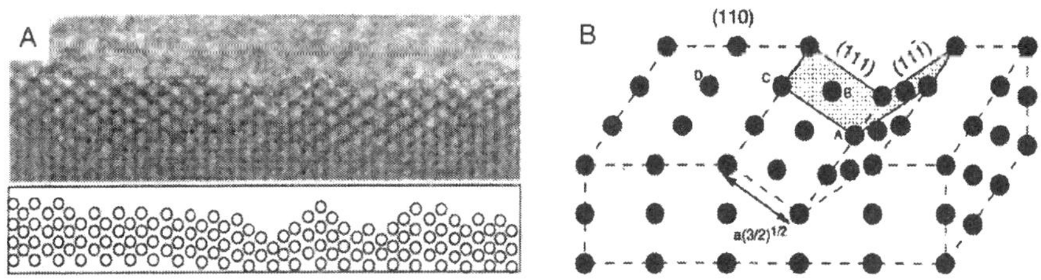


**Figure 11.16:** A) HRTEM image recorded along <110> of a gold nanorod and the corresponding positions of the projected atom rows, showing the rearrangement of the surface atoms. B) The reconstructed {110} surface with missing rows. Reprinted with permission from reference [48]. © (2000) American Chemical Society.

Thus it is entirely plausible that surface atoms in {100} and {110} crystal faces can be stabilized to a greater extent by coordinating organic species compared to atoms in {111} crystal faces. In other words, the energy of interaction would be higher and thus the binding would be stronger, leaving the {111} faces more exposed to the reactive growth medium. The interaction energy between a coordinating organic species and a {111} face would probably not be hugely different than the interaction energy between the organic species and a less stable {100} or {110} face, at least compared to *k*T. Therefore, for selective binding to occur it is necessary to amplify this slight difference in interaction energy through cooperation between the interacting organic species. Generally this is achieved through linking the organic species together in a continuous chain, *i.e.* a polymer, or utilizing coordinating species, such as surfactants, that interact strongly with each other to form layered assemblies. This will become clear from the anisotropic structures discussed below and the discussion on computational studies of one-dimensional growth in Section 11.4 1.2 4.

By selectively binding to less stable crystal faces, the more stable crystal faces are more available for growth. Depending of the underlying crystal structure of the seed, growth can then proceed anisotropically into nanowires, nanorods, nanocubes or bipyramids. Thus the selective binding of organic species is a thermodynamically driven process that leads to kinetic control over the direction of growth.

## 11.4 1.2 *Nanowires And Nanorods*

### 11.4 1.2 1. *Introduction*

Nanorods and nanowires are characterized by their long, linear shape. The aspect ratio defines the difference between a nanorod and a nanowire: It is considered that nanorods have aspect ratios < 20, while nanowires have aspect ratios > 20 [40]. In the context of MEF, nanorods are of greater interest. There are a number of reasons for this. Firstly, the enhancement of the field occurs largely at the end of the nanorods, and for a given amount of material in a sample there are more nanorods than nanowires, and therefore more ends in the sample. Secondly, the optical properties of metal nanorods can be tuned by varying the length (and thickness) of a nanorod. Thirdly, nanowires are easily entangled due to their length and therefore more difficult to manipulate for whatever application or device that is envisaged. Despite the clear preference for nanorods over nanowires, methods for the synthesis of nanowires must be covered since these methods and the underlying growth mechanisms are directly applicable to the synthesis of nanorods. An earlier review of nanowire and nanorod synthesis and characterization can be found in the literature [49].

A number of nanowire and nanorod morphologies are available but the most common are single crystalline and those with a pentagonal symmetry that results from a five-fold twinning in the seed nanoparticles. The structure of the single crystal [50-52] and pentagonal [53-55] morphologies has been established by extensive TEM analysis.

## 11.4 1.2 2. *Polyol Methods*

### *Pentagonal Silver Nanowires*

The Xia group has extensive experience in the synthesis of silver nanowires. Their polyol process involves the formation of nanoparticles in refluxing ethylene glycol at 160° C, which serves as both solvent and reducing agent. Early work involved seeding the reaction with Ag [56] or Pt [42, 56] nanoparticles or later involved a "self-seeding" process [57]. In this process, $AgNO_3$ (silver nitrate) and poly(vinyl pyrrolidone) (PVP) solutions (in ethylene glycol) are then added to the refluxing solution. The silver nanowires produced by the polyol process (see Figure 11.17) have a five-fold (pentagonal) symmetry with {100} faces exposed along the side, and {111} faces, where growth occurs, at the ends [55]. The five-fold symmetry of the nanowires derives from the five-fold symmetry that exists in decahedral seeds that are the result of a five-fold twinning pattern. These decahedral seeds have {100} sides and {111} ends. The linear growth of the nanowires is promoted by PVP preferentially binding to the {100} sides, restricting growth to the {111} ends, thus resulting in nanowire formation.

In addition, decahedral nanoparticles are strained. This is because the preferred angle between the twin planes is 70.5°, yet 72° is available [58]. Thus, a total of 7.5° needs to be filled by strain of the crystal lattice. As a result, the twin boundaries represent high energy sites that may promote growth at the uncoated {111} end faces.

### *Bicrystalline Silver Nanowires*

It is clear that controlling the crystallinity of the seeds is crucial, with the five-fold symmetry of the decahedral seeds clearly promoting the growth of nanowires with pentagonal symmetry. So, by controlling the crystallinity of the seeds, the polyol process can be adapted to produce other silver nanostructures. Addition of a corrosive anion such as chloride selectively etches twinned seeds and enables the production of pure single crystal seeds that can then grow to form nanocubes, see Section 1.4 1.3 If a less corrosive anion such as bromide is substituted for chloride, then there is still adequate etching to eliminate the multiply twinned seeds, but seeds with a single twin remain. The singly twinned seeds grow to form right bipyramids, see Section 11.4 1.4. By decreasing the rate of atomic addition to these same single twinned seeds it was found that bicrystalline silver nanowires (nanobeams) were formed, see Figure 11.18 [59].

## *Single Crystal Silver Nanorods*

Higher concentrations of bromide result in increased etching and, as mentioned for chloride, the single crystal morphology is preferred. However, unlike with chloride, the nanoparticles formed are not cubes but are elongated into nanorod-like structures, termed nanobars [60]. This is discussed in more detail in Section 11.4 1.3 2, which deals with the synthesis of nanocubes by the polyol process.

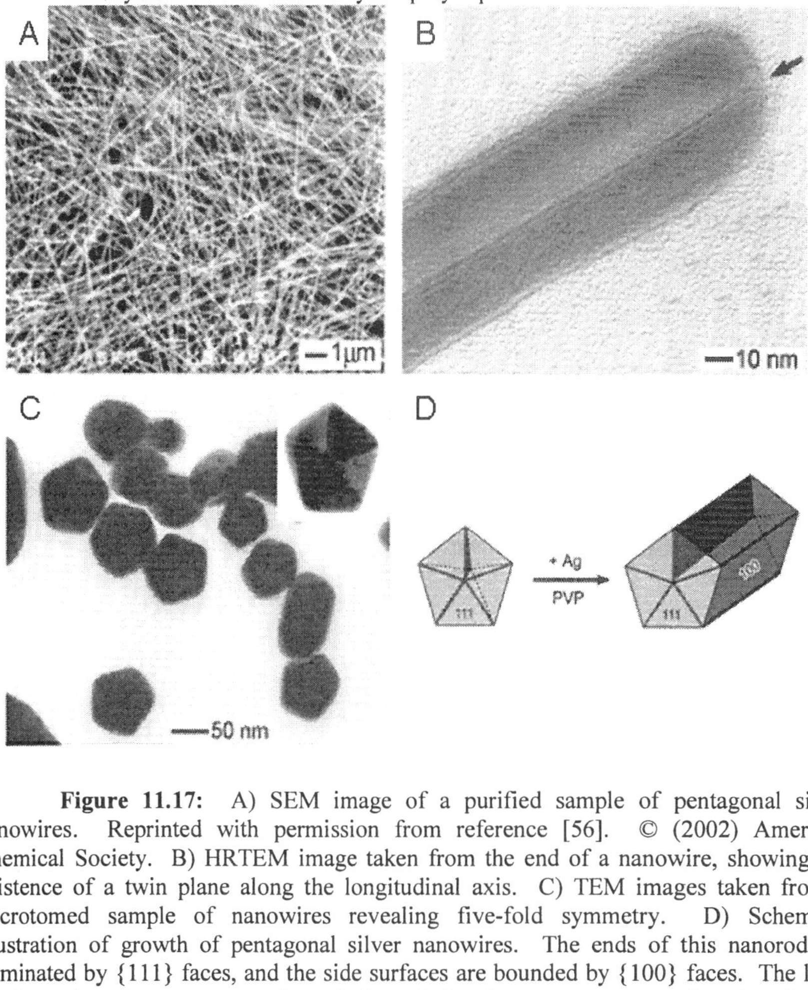


**Figure 11.17:** A) SEM image of a purified sample of pentagonal silver nanowires. Reprinted with permission from reference [56]. © (2002) American Chemical Society. B) HRTEM image taken from the end of a nanowire, showing the existence of a twin plane along the longitudinal axis. C) TEM images taken from a microtomed sample of nanowires revealing five-fold symmetry. D) Schematic illustration of growth of pentagonal silver nanowires. The ends of this nanorod are terminated by {111} faces, and the side surfaces are bounded by {100} faces. The lines on the end surfaces represent the twin boundaries that can serve as active sites for the addition of silver atoms. Reprinted with permission from references [55]. © (2003) American Chemical Society.

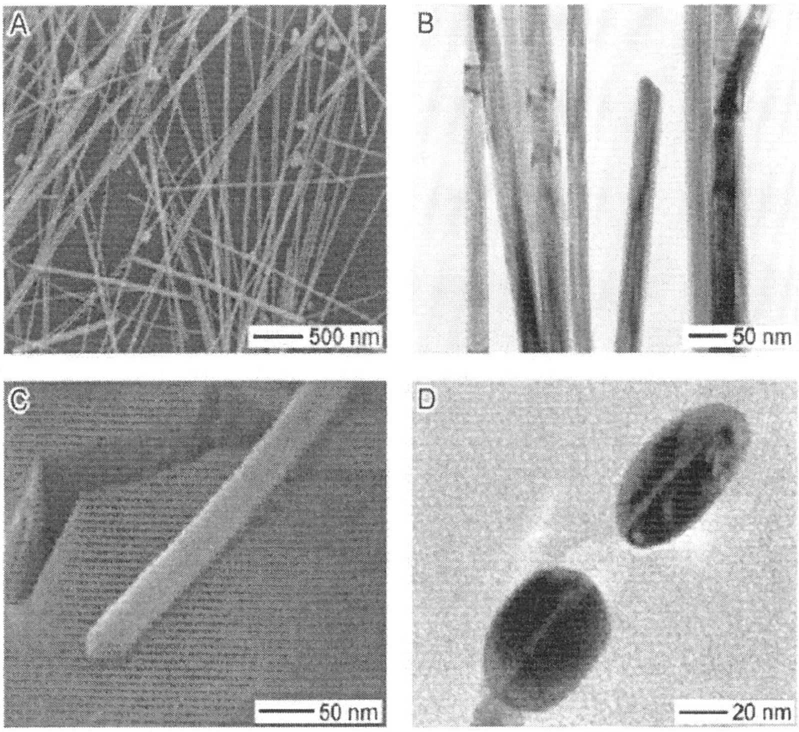


**Figure 11.18:** (A) SEM and (B) TEM images of bicrystalline silver nanowires (nanobeams). (C) SEM image of a nanobeam tilted at 65° relative to the electron beam, where its rounded profile is visible (the scale bar only applies to the horizontal axis). (D) TEM image of a microtomed sample of silver nanobeams showing their cross-sectional profile. This image suggests that the nanobeam is bisected by a twin plane parallel to the base. Reprinted with permission from reference [59]. © (2006) American Chemical Society.

## 11.4 1.2 3. *Aqueous Surfactant Methods*

### *Pentagonal Gold Nanowires And Nanorods*

Gold nanorods and nanowires can also be produced by an approach that involves the use of an aqueous surfactant such as cetyltrimethylammonium bromide (CTAB) to direct the linear growth of the nanostructures from citrate-stabilized gold seeds. Initially it was thought that the anisotropic growth was directed by the assembly of the surfactant molecules into a template whose shape then defined the growth of the crystal [39-41]. However, it has become clear that the CTAB surfactant has a preference for the {100} face of gold, similar to the behaviour of PVP with silver in the polyol process, and so directs the growth of decahedral seeds into gold nanorods and nanowires with pentagonal symmetry [54,61-64]. Generally, the length of the nanorods, and thus the aspect ratio, can be tuned by adjusting the ratio of seeds to gold salt. The aspect ratio can also be tuned by varying the length of the long alkyl chain of the surfactant molecule [62].

Surfactants with longer alkyl chains form a more cohesive monolayer on the preferred crystal faces, thus enhancing the selectivity for these faces. Thus surfactants with longer alkyl chains result in nanorods with higher aspect ratios. This is clearly illustrated in Figure 11.19, where the aspect ratio of the nanorods scales with alkyl chain length [62]. For alkyl chains of 10 carbons (or less) the attractive interactions between the surfactant molecules are much too weak. There is not enough cooperation between the surfactant molecules to amplify the small difference in the interaction energy between the surfactants and the different crystal faces, and as a result no nanorods are formed.

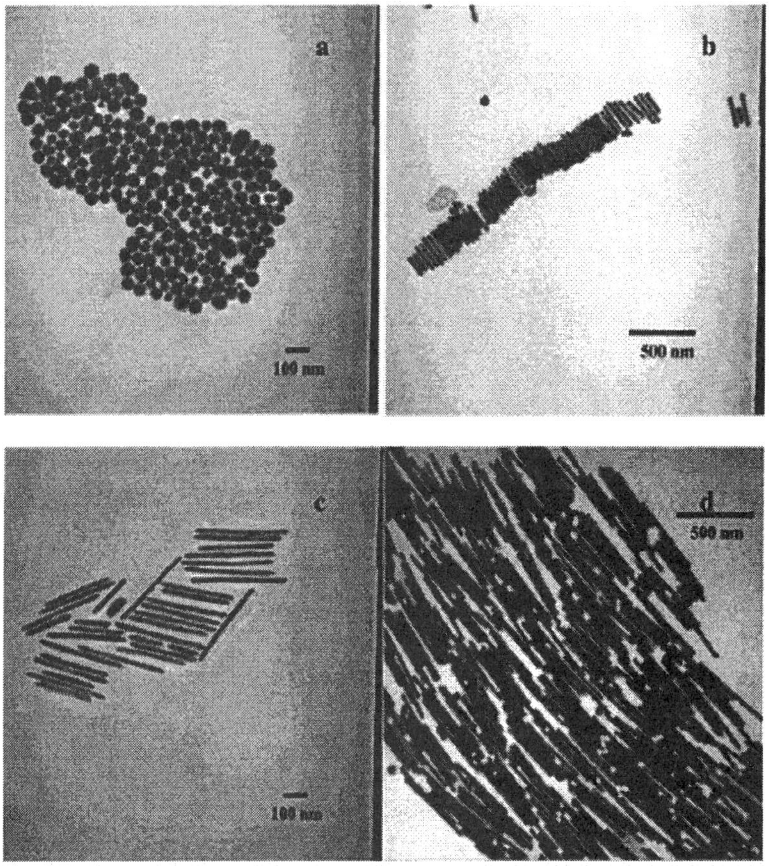


**Figure 11.19:** TEM micrographs of gold nanorods prepared in the presence of (a) $C_{10}$TAB, (b) $C_{12}$TAB, (c) $C_{14}$TAB, and (d) $C_{16}$TAB after purification to remove the spheres. Scale bars are 500nm (b and d) and 100 nm (a and c). Reprinted with permission from reference [62]. © (2003) American Chemical Society.

## *Silver Nanorods*

There are relatively few examples in the literature of using aqueous surfactants for the synthesis of silver nanorods. In these cases the aqueous surfactant approach has been applied successfully, via the use of CTAB, to generate silver nanorods of a range of aspect ratios [65,66]. In addition, pentagonal silver nanorods have been obtained by using a cetyltrimethylammonium tosylate (CTAT) surfactant [67]. By switching the surfactant counterion from bromide to tosylate the surfactant-metal surface interaction is still optimised to yield the preferred binding to the {100} side faces and promote one-dimensional growth.

## *Single Crystal Gold Nanorods And Nanowires*

Experiments have shown that if the citrate in the gold seed preparation step (pentagonal gold nanowire and nanorod synthesis) is replaced by CTAB then predominantly single crystalline seeds are formed. These can then be grown into single crystalline nanorods (and nanowires), see Figure 11.21, although this usually requires the presence of $Ag^+$ ions in the growth mixture [51,52,68-71]. The role of the $Ag^+$ ion is uncertain but it might enhance the cohesion of the monolayer of surfactant on the higher energy {100} and {110} side faces of the single crystalline nanorods, thus promoting growth at the predominantly {111} end faces, see Figure 11.20. The single crystalline nature of the nanorods is clearly visible in Figure 11.22.

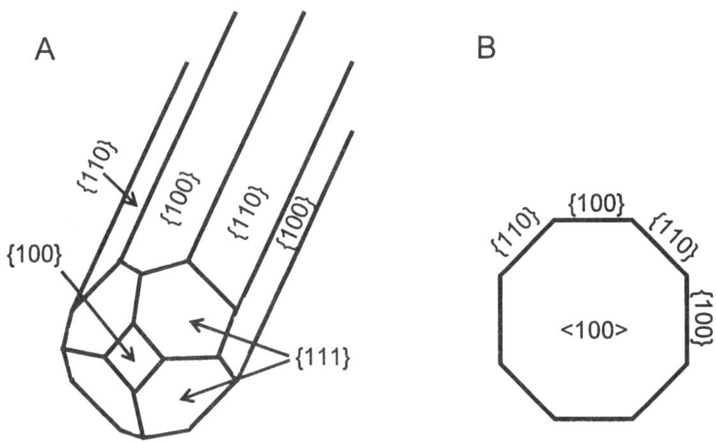


**Figure 11.20:** A) Single crystal nanorod showing predominantly {111} end faces. B) Cross-sectional view of single crystalline nanorod looking along <100>. Side faces are alternating less stable {100} and {110} crystal faces.

The aspect ratio can be varied between 1.5 and 4.5 by varying the $Ag^+$ content, with higher aspect ratios up to 10 being achieved when a binary surfactant mixture was used [68]. Recently, single crystalline gold nanorods and nanowires with aspect ratios up to 70 have been obtained with cetyltripropylammonium bromide (CTPAB) and cetyltributylammonium bromide (CTBAB) solutions in the presence of $AgNO_3$ [51].

The role of surfactants in promoting one-dimensional growth from single crystal seeds has been analysed computationally and this is discussed in Section 11.4 1.2 4.

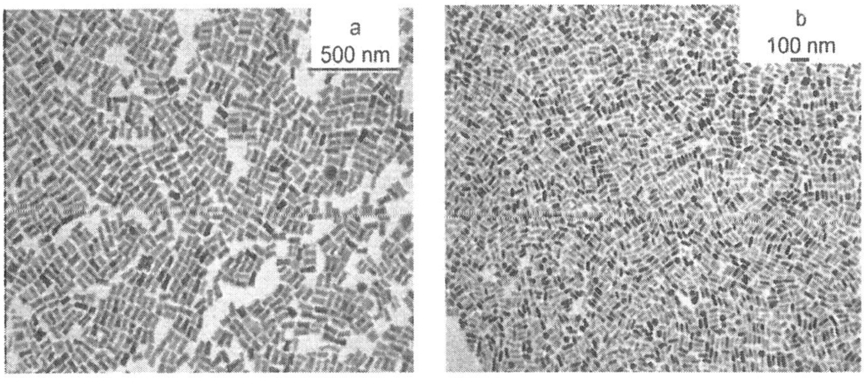


**Figure 11.21:** TEM images of single crystal gold nanorods synthesized with different amounts of gold salt for a constant seed concentration. Reprinted with permission from reference [69]. © (2004) American Chemical Society.

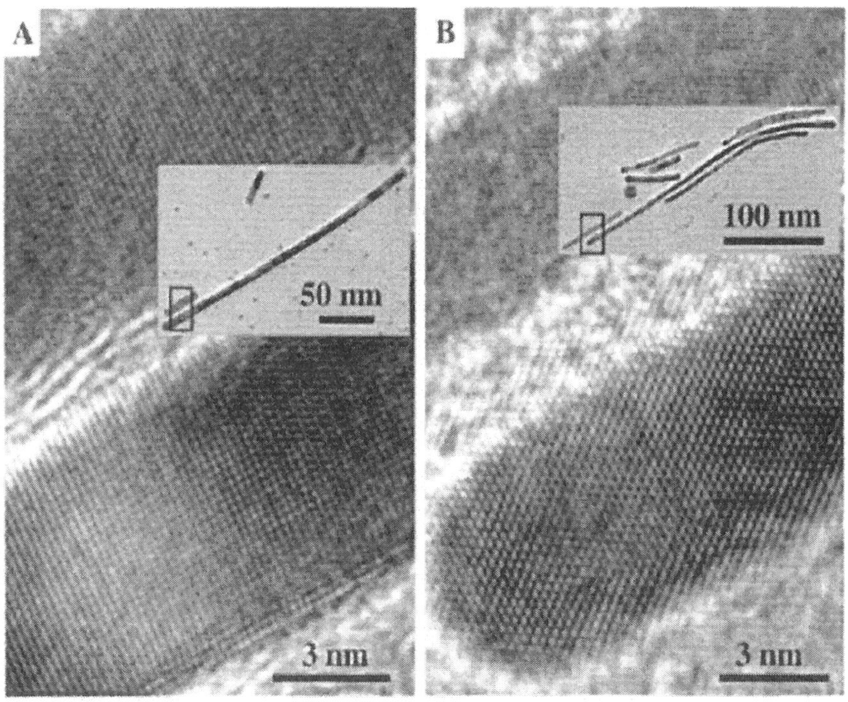


**Figure 11.22:** A) HRTEM image of two <100>-oriented single crystal gold nanorods whose aspect ratios are 4.7 and 33. Inset is the low-magnification TEM image of the nanorods. B) HRTEM image of one <110>-oriented single crystal gold nanorod whose aspect ratio is 43. Inset is the low-magnification TEM image of the nanorod.

Reproduced with permission from reference [51]. © (2007) Wiley-VCH Verlag GmbH & Co. KGaA.

### 11.4 1.2 3.1 *Reproducibility And Role Of Impurities In Aqueous Surfactant Preparations*

Nanorod preparation procedures that employ the aqueous surfactant approach have persistently had issues with reproducibility of shape and size, but only recently has this been acknowledged in the literature [66,67]. Jiang *et al.* have found that there is always some significant variability in shape and size in simply carrying out a reaction repeatedly [72].

More significantly, Durr *et al.* have found that the CTAB source is a critical factor in the success of a nanorod synthesis [73]. It was found that it was necessary to use lower purity CTAB, while preparations with higher purity CTAB did not work at all. A more thorough investigation of this phenomenon has been carried out by Smith and Korgel [74]. They have found that the highest purity CTAB supplied sometimes succeeded in producing nanorods and that there is an impurity involved that varies in amount and/or nature from batch to batch of a given CTAB source of the same nominal purity. Despite analysis of the CTABs by size exclusion chromatography (SEC), x-ray diffraction (XRD), NMR and mass spectrometry, it has not been possible to identify any particular impurity that would explain the variability in the preparations. Additionally, they tried adding different "impurities" including NaBr, KBr, cetyldimethylamine and surfactants with differing head groups (benzyldimethylammonium chloride (BDAC) and cetyltrimethylammonium chloride (CTAC)) in small amounts to the reactant solutions that did not yield nanorods, but could not induce nanorod formation. Moreover, they have found that using the "right" CTAB is most important in the growth step but that it is also important in the seed production step. It is possible that due to varying impurities, the different batches of CTAB have slightly different binding strengths to the seed particles, which in turn affects the selectivity of the CTAB for the less stable crystal faces and therefore the size and shape of the final grown nanostructure.

### *Added Halide Ions In Aqueous Surfactant Preparations*

There has previously been some amount of investigation into the role played by the halide counterion of the positively charged ammonium surfactant. It has been found that syntheses can be directed to produce nanoparticles of different shapes by the addition of various halide ions [75].

### 11.4 1.2 4. *Computational Studies Of One-Dimensional Growth*

Further support for the model of one-dimensional growth being driven by preferential binding of organic species to higher energy crystal faces is provided by recent computational work.

Grochola *et al.* have undertaken a molecular dynamics study to successfully reproduce the growth of gold nanorod morphologies from starting "spherical" seeds in the presence of model surfactants [44]. The surfactant model was developed through extensive systematic attempts aimed at inducing anisotropic nanoparticle growth in strictly isotropic computational growth environments. A two-component surfactant mixture was considered. One surfactant (SuA) is "monatomic" and the other (SuB) is "diatomic" consisting of a headgroup atom and a "tail".

The study identified two key properties of the surfactants which were most important for successful anisotropic growth. Firstly, selective adsorption to open crystal faces ({100} and {110}), which could only be induced by inclusion of a metal-like electron density function for the model headgroup atom of SuB. Secondly, attractive like-like surfactant interactions and non-attractive interactions for unlike surfactants, which were found to be very important as this drove the segregation of surfactants.

Essentially, a difference in the interaction energy of the headgroup of SuB on different crystal faces is a prerequisite for preferential adsorption, however this interaction energy is not differentiated sufficiently for individual SuB surfactants to produce significant degree of selectivity in SuB adsorption. Rather, it was found that the segregation of the surfactants that was induced by attractive like-like surfactant interactions led to a semi-cohesive, grouped selective adsorption, which amplified the small degree of selectivity in the adsorption of SuB on {100} and {110} crystal faces. This resulted in an effective and stable selective adsorption of SuB along the length of the nanorods throughout the entire growth process.

Interestingly, the model not only reproduced the growth of nearly all known nanorod morphologies when starting from an initial decahedral seed (see Figure 11.23) or single crystal fcc seed (see Figure 11.24), but also reproduced the experimentally observed failure of nanorod growth when starting from spherical nanoparticles with an icosahedral morphology.

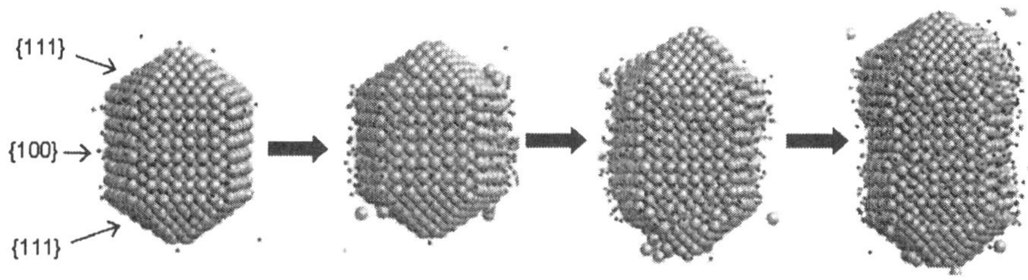


**Figure 11.23:** A sequence of snapshots (at irregular intervals) of fivefold nanorod growth in the <100> direction starting from an ideal spherical decahedral (*m-D*$_h$) seed. In all figures, only gold atoms are shown in full (grey), SuB headgroup atoms are shown as (black) point spheres, while SuA atoms are hidden from view. Reprinted with permission from reference [44]. © (2007) American Institute of Physics.

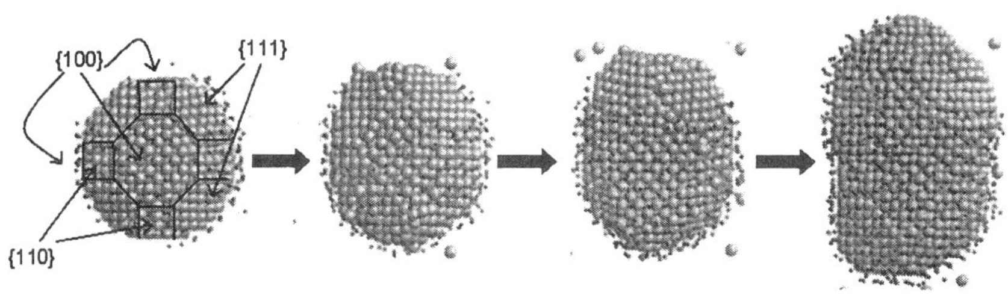


**Figure 11.24:** A sequence of snapshots (at irregular intervals) of fcc nanorod growth in the <100> direction starting from a cuboctahedral (TO$_h$) spherical seed. In

these snapshots, only gold atoms are shown in full (grey), SuB headgroup atoms are shown as (black) point spheres, while SuA atoms are hidden from view. Reprinted with permission from reference [44]. 

The importance of the like-like attractive interactions between surfactants in this model reflects the observed effect of controlling the aspect ratio of gold nanorods by adjusting the length of the alkyl chain of the surfactant [62], see Section 11.4 1.2 3 Also the required metal-like electron density of the SuB headgroup is suggestive of the important but little understood role played by silver ions in the growth process of single crystal gold nanorods [45,62,63,65]. This computational study is also highly relevant to the polyol process for silver nanowire synthesis. In that process there is no surfactant but there is the polymer PVP which is understood to bind preferentially to {100} and {110} crystal faces. The selectivity of individual functional groups on the polymer for these crystal faces may well be quite low but the cooperative effect mentioned above would greatly enhance the selectivity. Such cooperation is certainly present in the case of the polymer since each polymer molecule consists of a large number of the relevant functional groups connected together in a chain, and as such there is a low entropic barrier to the wrapping of the polymer around a nanostructure.

### 11.4 1.3 *Nanocubes*

#### 11.4 1.3 1. *Physical Aspects Of Nanocube Growth*

A key factor in all of the preparations for gold and silver nanocubes is the production and stabilization of single crystalline seeds during initial nucleation steps. These single crystalline seeds are cuboctahedral and possess higher energy {100} crystal faces that are then preferentially stabilized by the surfactant or polymer that is present, promoting growth on the {111} faces, as illustrated in Figure 11.25 below. This is certainly the case for single crystalline seeds of silver. However, as we saw earlier, in the case of gold, growth on the {111} faces of single crystals can produce nanorods, and even computational work has predicted one-dimensional nanostructures. Nevertheless, gold nanocubes are still possible. Ultimately, for gold, it is an empirical fine tuning of the interactions between the surfactant and the single crystal seed nanoparticles, by varying reaction conditions, that determines whether we observe growth of gold single crystals into nanorods or nanocubes.

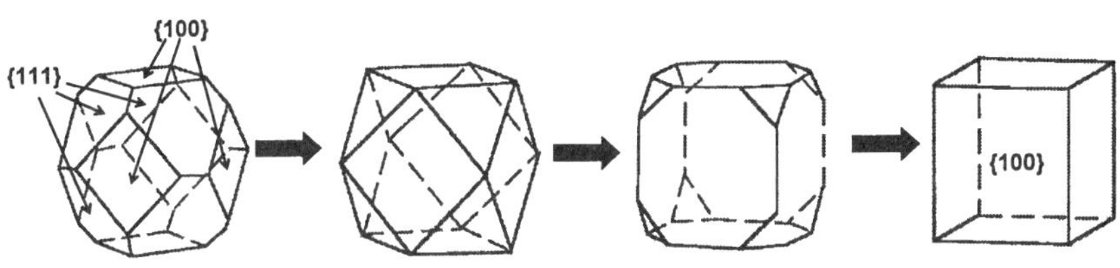


**Figure 11.25:** Preferential growth on {111} faces results in "spherical" single crystal cuboctahedron growing into a cube bounded by {100} faces as the {111} faces grow out of existence.

#### 11.4 1.3 2. *Polyol Methods*

## *Silver Nanocubes*

Early reports of silver nanocube synthesis by the polyol route involved empirically adjusting the reaction conditions, with PVP being used to stabilize {100} crystal faces [76]. Later, as mentioned above in the discussion on the polyol synthesis for silver nanorods and nanowires, corrosive etching with chloride was employed to remove all twinned particles and leave only single crystal (cuboctahedral) seeds. Due to the preferential binding of PVP to {100} faces, faster growth occurs on the {111} faces resulting in nanocubes [77]. The growth process is illustrated schematically in Figure 11.25. If growth is incomplete then truncated cubes (cuboctahedra) will result. This approach has been modified to greatly improve the yield and rate of production of silver nanocubes [78-81]. A recent improvement involves the use of sodium sulphide that acts both as an etchant of multiply twinned seed nanoparticles and as a growth accelerator [80,81]. A typical sample of silver nanocubes is shown in Figure 11.26.

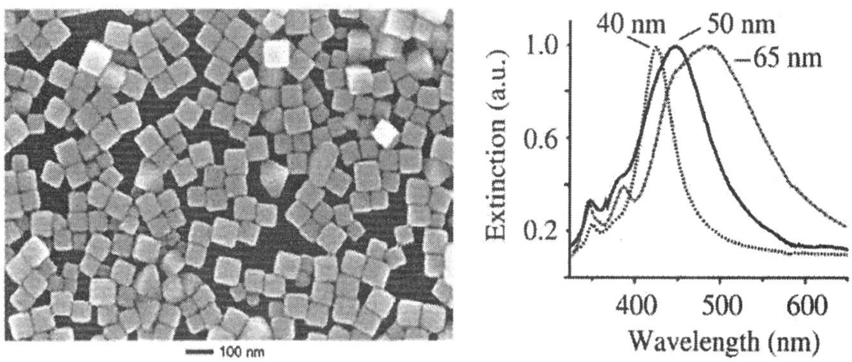


**Figure 11.26:** Left: SEM image of a sample of silver nanocubes produced by the polyol method with a trace amount of $Na_2S$. Reprinted by permission from Macmillan Publishers Ltd: Nature Protocols [81], © (2007). Right: UV–Vis spectra taken from silver nanocubes of different sizes. Reprinted from reference [80] © (2006) with permission from Elsevier.

## *Silver Nanobars And Nanorice*

Instead of using chloride it is possible to favour a single crystalline morphology for silver seed nanoparticles by using a high enough concentration of bromide (twice that needed for bipyramid formation). However, by using bromide, there is a significant preference for growth in one direction resulting in elongated nanocubes or "nanobars" as shown in Figure 11.27 [60]. After synthesis, the sharp corners can be rounded down to yield grain-like structures or "nanorice". It is uncertain why one-dimensional growth occurs in this case. As has been discussed above, single crystals of gold can grow into nanorods, yet the mode of growth here seems to be different. The nanobars have square cross-section and unlike the case of single crystal gold nanorods, do not seem to have {111} faces at the end which would explain one-dimensional growth.

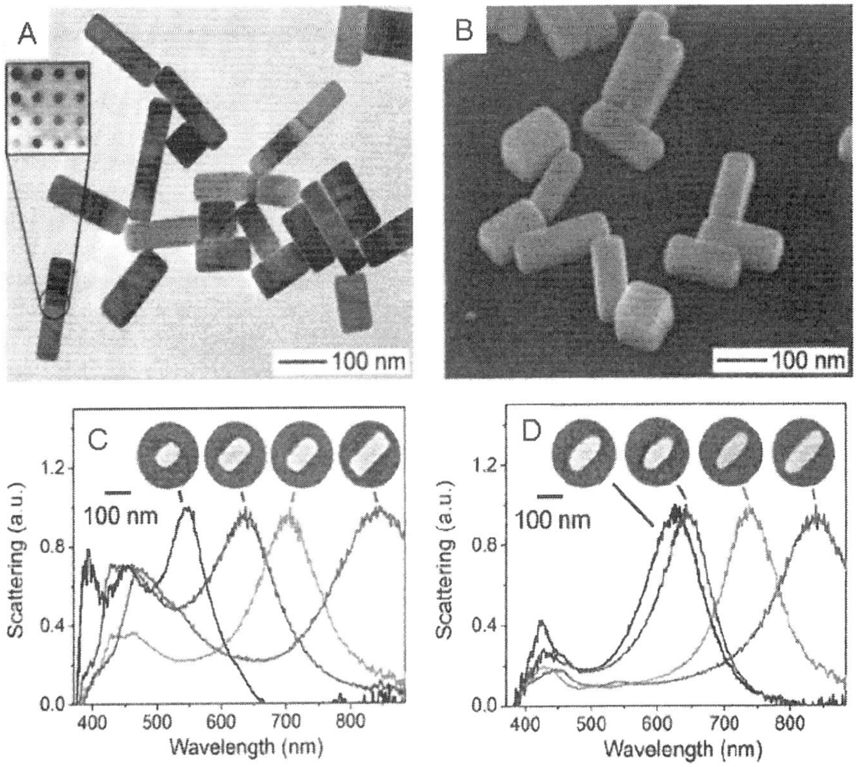


**Figure 11.27:** A). TEM image of silver nanobars. The inset is a convergent beam electron diffraction pattern, indicating the nanobars are single crystals bounded by (100) faces. B) SEM image of nanobars tilted by 45°. C) SEM images of individual nanobars with their corresponding normalized scattering spectra. The longitudinal plasmon peak of the nanobars red shifts with increasing aspect ratio. D) SEM images of individual nanorice with their corresponding normalized scattering spectra. Reprinted with permission from reference [60]. © (2007) American Chemical Society.

## *Gold Nanocubes*

The polyol process has been adapted for gold but this often produces a range of shapes including nanocubes [82]. As in the case of silver, injection of ethylene glycol solutions of PVP and $HAuCl_4$ into refluxing ethylene glycol resulted in multiply twinned particles being formed. Depending on reaction parameters, the products of the reactions were mostly tetrahedra or icosahedra with some decahedra; nanostructures with predominantly {111} faces. Unlike with silver, chloride will not provide the role of oxidative etchant to leave only single crystal seeds. However, the addition of some $AgNO_3$ into the reaction mixture beforehand resulted in gold nanocubes (with {100} faces) being formed. It would seem that the combination of $Ag^+$ and PVP provides a sufficient stabilization of single crystal nuclei when they are initially formed, preventing them from aggregating into multiply twinned seeds. The PVP, in the presence of $Ag^+$, preferentially binds to the {100} faces of these to produce nanocubes.

### 11.4 1.3 3. *Aqueous Surfactant Methods*

#### *Gold Nanocubes*

In Section 11.4 1.2 3 it was seen how the aqueous surfactant approach was modified to produce single crystalline gold nanorods instead of nanorods with pentagonal symmetry. Further modification of this synthesis has enabled the production of nanocubes, *i.e.* nanorods with aspect ratio = 1, and other shapes, from the single crystalline seeds [83,84], see Figure 11.28. In this case the surfactant is directing the growth of the single crystal gold seeds in the <111> direction in the same way that PVP does to single crystal seeds in the polyol process.

**Figure 11.28:** TEM images of gold nanoparticles of different shapes, all prepared with CTAB and in the presence of silver ion. Scale bars are 100 nm for A and B. Reprinted with permission from reference [83]. © (2004) American Chemical Society.

#### *Silver Nanocubes*

The synthesis of silver nanocubes by the reduction of $Ag^+$ with glucose in the presence of CTAB has also been reported, see Figure 11.29 [85]. In this synthesis there is no separate seed preparation step. The $Ag^+$ combines with the $Br^-$ of the CTAB to produce AgBr which maintains a low concentration of $Ag^+$ in solution. During the reaction, single crystal silver seed nanoparticles are formed. The CTAB stabilizes the less stable {100} faces of these, as PVP does in the polyol process, and so nanocubes are the preferred final morphology.

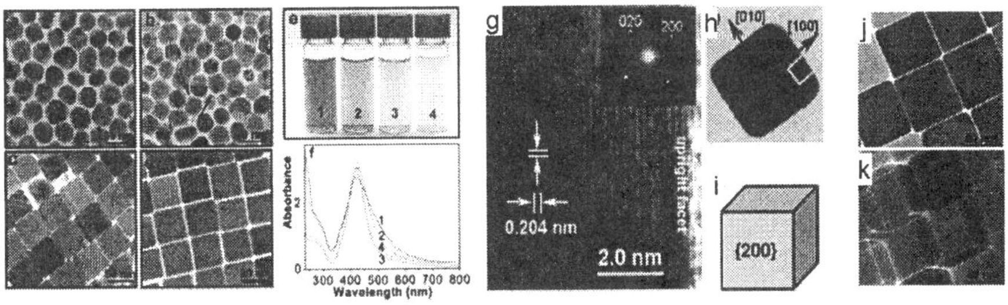


**Figure 11.29:** (a to d) TEM images of Ag nanocubes synthesized with various molar ratios of CTAB/$[Ag(NH_3)_2]^+$: (a) 1, (b) 1.5, (c) 2.5, (d) 3. (e, f) photos and extinction spectra of the aqueous solutions of the Ag nanocubes, marked with 1, 2, 3, and 4 corresponding to the Ag nanocubes shown in panels a, b, c, and d, respectively. (g) A typical HRTEM image of a selected area of an individual Ag nanocube with the SAED pattern shown in the inset. (h) The nanocube used for the HRTEM and SAED studies with the area marked where the HRTEM image was recorded. (i) Schematic illustration of the facets of an individual cube ({200} is equivalent to {100}). (j, k) TEM images of Ag nanocubes before and after rotation of the TEM grid by 30°. Reprinted with permission from reference [85]. © (2004) American Chemical Society.

## 11.4 1.3 4. *Electrochemical + Surfactant Methods*

### *Gold Nanocubes*

High quality gold nanocubes can also be prepared by electrochemical methods in the presence of the appropriate surfactant and solvent [86,87]. As illustrated in Figure 11.30 below [87], gold for nanoparticle growth is supplied through oxidation of a gold anode. Reduction takes place at an inert platinum cathode in the presence of the surfactant.

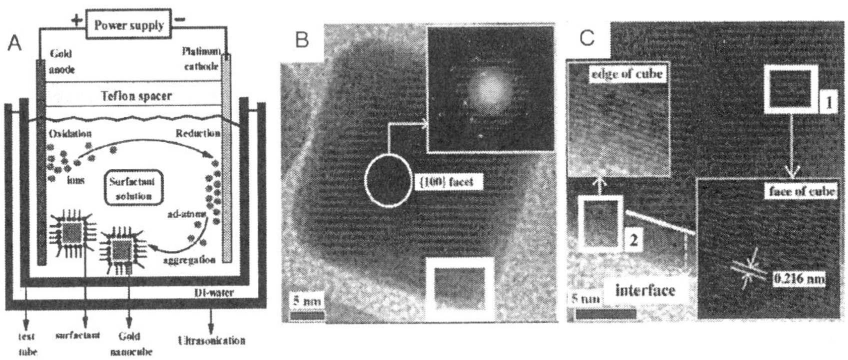


**Figure 11.30:** A) Schematic diagram of the electrochemical apparatus for the synthesis of gold nanocubes. B) Typical TEM image of single gold nanocube: the upper inset shows the SAED pattern. TEM image at high magnification, of the point marked with a white square mark in (B). C). High-resolution TEM images of points marked with

white squares in the inset in (C). Reprinted from reference [87] © (2006) with permission from Elsevier.

### 11.4 1.4 *Bipyramids*

Singly twinned silver nanoparticles can be obtained by oxidative etching of polycrystalline seeds with bromide in the polyol process. As mentioned earlier, if the reaction conditions are optimized, these nanoparticles can then undergo one-dimensional growth to produce silver nanowires possessing a single twin plane (nanobeams). Otherwise the initial bipyramid seeds can be grown into much larger bipyramid-shaped nanoparticles, as shown in Figure 11.31 [88]. The growth mechanism is consistent with that discussed so far for nanowires, nanorods and nanocubes. PVP selectively binds to less stable {100} faces promoting deposition of silver on the {111} faces. Essentially each half of the bipyramid is a truncated corner of a nanocube.

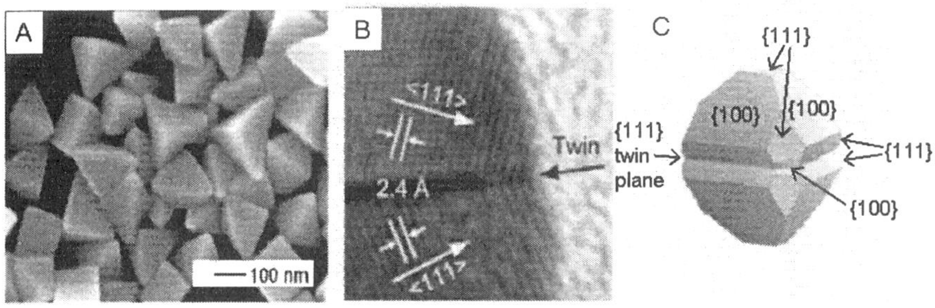


**Figure 11.31:** A) SEM of bipyramids approximately 150 nm in edge length. B) HRTEM of a twinned seed shows the lattice fringes reflecting across the {111} twin plane. C) Model of bipyramidal seed showing {111} truncation of corners, reentrant {111} surfaces at the twin boundary and reentrant {100} surfaces at the twin boundary corners. Reprinted with permission from reference [88]. © (2006) American Chemical Society.

## 11.4 2. *Nanoparticle Shapes Not Readily Explained By Selective Binding Model For Anisotropic Growth*

### 11.4 2.1 *Aspects of Growth Not Readily Explained by Selective Binding Model*

In this category, the shapes that cannot be explained by the selective binding model tend to maximize the surface area of the most stable {111} crystal face. In most cases this simply involves a reverse of the growth anisotropy that is observed with the selective binding model. In the case of nanoprisms (and nanoplates), flat {111} faces are favoured but there is clearly another factor at work giving rise to the typical flat morphology of these nanostructures.

It is unlikely that it is possible to simply reverse the selective binding of organic species to less stable crystal faces because the stability of each face is an intrinsic property of the metal and its crystal structure. However, instead of preferred binding of organic species to the less stable {100} faces, it may be possible, under appropriate reaction conditions, to utilize the lower stability of these faces to drive growth itself.

That is, in the absence of stabilizing species or under reaction conditions in which the selectivity of stabilizer-surface interactions is largely diminished, there may be a preferred (faster) binding of metal atoms to the less stable crystal faces, and thus preferred growth on these less stable crystal faces. These would grow out of existence with the more stable {111} faces left behind. The end result would be anisotropic growth that is consistent with a reverse of the selective binding model, with the organic stabilizers playing no immediate shape-directing role and simply acting to maintain a stable dispersion of nanoparticles.

Since we expect different crystal faces to have different growth rates leading to anisotropic growth, clearly the underlying crystal structure of the seeds is extremely important. Indeed, depending on reaction conditions, surfactants and polymers could quite possibly play a significant role in determining the crystal structure of the seed by limiting the aggregation of initially formed nuclei. For this reason, even without preferential binding, surfactants and polymers can still influence the ultimate nanoparticle morphology. The role of organic species in influencing the structure of the seed during the seed preparation step is highlighted by the case of the aqueous surfactant preparation of gold nanorods that was driven by selective binding, see Section 11.4 1.2 3. When CTAB was present in the production of the seeds, the grown nanorods were single crystalline as opposed to pentagonal when no CTAB was used in the seed production step.

### 11.4 2.2 *Nanooctahedra*

#### *Gold Nanooctahedra*

To produce an octahedron, one needs to reverse the direction of preferential growth that was employed to generate gold and silver nanocubes, see Figure 11.25 and Figure 11.32.

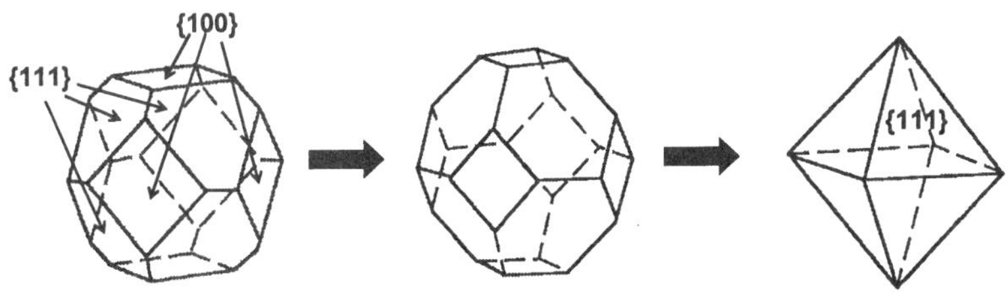


**Figure 11.32:** Preferential growth on {100} faces results in "spherical" single crystal cuboctahedron growing into an octahedron bounded by {111} faces as the {100} faces grow out of existence.

A recent paper has successfully reversed the nanocube mode of growth to produce nanooctahedra of gold [89]. In that process, a less than stoichiometric amount of sodium borohydride ($NaBH_4$) was added to a PEG 600 solution of PVP prior to the addition of gold (III) chloride ($AuCl_3$) aqueous solution. The solution was preheated at 75° C for more than 24 hrs, then further heated at 125° C for different reaction times from 6 hrs to 48 hrs, resulting in the formation of gold nanooctahedra as can be seen in Figure 11.33 [89]. At the higher temperature, the PEG 600 acted as a reducing agent.

It was found that PVP was essential for stable nanoparticle growth and indeed it is claimed in reference [89] that PVP preferentially binds to the {111} faces, thus directing growth in the {100} direction. It should be noted that this implies a reversal of the behaviour of PVP, which normally preferentially bind to {100} faces. It is also

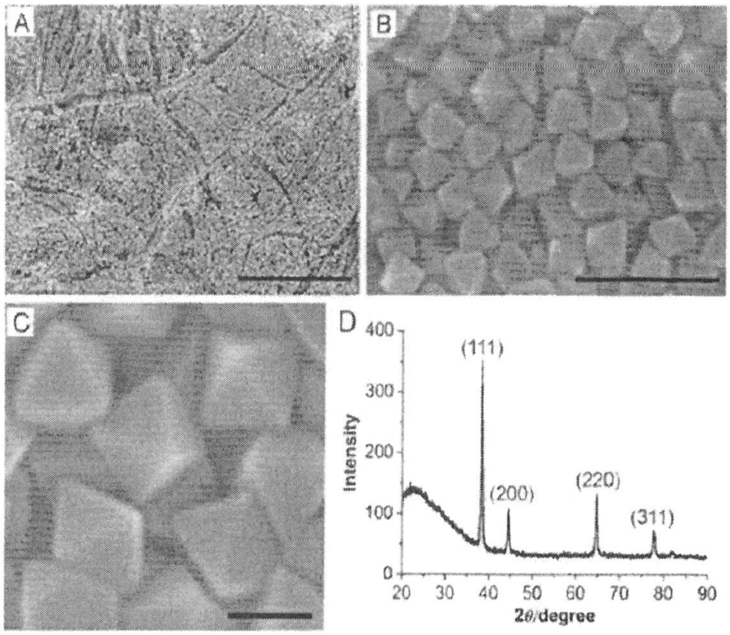


**Figure 11.33:** A) Low-magnification and B, C) high-magnification SEM images of the gold nanooctahedra. D) X-ray diffraction pattern. Scale bars for (A), (B), and (C) are 2 μm, 200 nm, and 50 nm, respectively. Reproduced with permission from reference [89]. © (2007) Wiley-VCH Verlag GmbH & Co. KGaA.

suggested that there may be a slightly higher supersaturation [90] at the corners than the surfaces of the initially formed seeds that enhances the preferential deposition of atoms at the vertexes of the as-formed nanocrystals.

A key difference between this reaction and the polyol procedure that yielded gold nanocubes (see Section 11.4 1.3 2) is the absence, in this case, of $Ag^+$. In the polyol procedure it was found that the presence of $Ag^+$ resulted in single crystal seeds, with the formation of nanocubes requiring the selective binding of PVP to the {100} faces so that preferential growth would take place on the {111} faces. In the gold nanooctahedra synthesis above, single crystal seeds are also generated but preferential growth occurs on the {100} faces instead, resulting in {111}-bounded nanooctahedra. It seems that $Ag^+$ may be necessary for a selective binding interaction between PVP and different crystal faces of gold; when $Ag^+$ is absent the interaction between PVP and gold may be too weak for there to be any significant selectivity in binding thus allowing faster growth on the higher energy {100} faces.

### *Silver Nanooctahedra*

A report that describes a modified polyol process that is employed to produce silver nanooctahedra in the presence of PVP [91], will be discussed in Section 11.4 2.4 below.

## 11.4 2.3 *Gold Nanodecahedra*

As argued above, in the absence of selective binding, the more unstable crystal faces can grow faster. In the case of decahedral seeds (see the left hand side of Figure 11.23), this means that growth would occur faster on the {100} side faces, instead of on the {111} faces as happens when the selective binding growth model applies and gives rise to one-dimensional nanostructures as discussed earlier. Growth on the {100} faces results in these crystal faces growing out of existence leaving only the ten {111} faces. Growth then proceeds on all the {111} faces at the same rate, leading to larger and larger decahedra (with no sides).

Nanodecahedra are very often observed as "impurity" nanoparticles in the synthesis of nanorods, nanowires and nanoparticles of other shapes. But to synthesize these as a majority product in a controlled manner is more of a challenge. Recently, Sánchez-Iglesias *et al.* have reported the formation of gold nanodecahedra of a range of sizes, see Figure 11.34 [92]. This procedure starts with the formation of gold seeds in DMF, which are then grown by the addition of an aliquot of seeds to a DMF solution of $Au^{3+}$ and PVP followed by ultrasonication. The main effect of the ultrasound is a noticeable temperature increase (up to 100 °C). During growth DMF acts as the reducing agent.

Although the nanodecahedra synthesis takes place in the presence of PVP, the PVP is clearly not behaving in a normal shape-directing manner as there is no one-dimensional growth that would be expected from a decahedral seed. Clearly any selectivity for lower energy crystal faces in the interaction between PVP and the nanoparticles has been overcome. In part this could be because of the high temperature used or the highly coordinating character of the DMF solvent. Indeed, structures that could be explained by a selective binding model are exceedingly rare in DMF

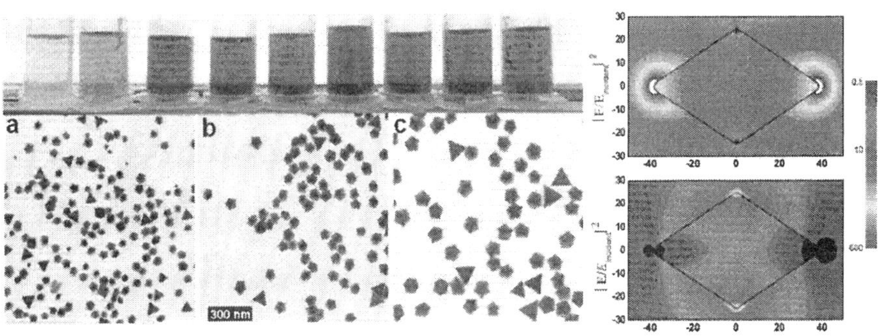

**Figure 11.34:** Top left: photograph of samples withdrawn from the reacting solution at various times during a synthesis of gold nanodecahedra. Bottom left: TEM images of decahedral gold nanoparticles prepared using different amounts of gold-seed solution (a: 1.4 mL, b: 0.7 mL, c: 0.3 mL). The scale is the same in all TEM images. Right: plots of the calculated near-field enhancement ($|E/E\text{incident}|^2$) for bicones

(computation model for decahedron) with 40 nm radius and 25 nm height. Light is coming from below in the upper plot and from the left in the lower one, with the electric field contained in the plane of the plots. For the calculations, the wavelengths of maximum extinction cross-section (620 and 522 nm, respectively) were used. The vertical and horizontal axes are in nanometres. Reproduced with permission from reference [92]. © (2006) Wiley-VCH Verlag GmbH & Co. KGaA.

preparations despite the ubiquitous presence of PVP. This is not surprising given the similarity in amide functionality between DMF and PVP. For example, the only DMF nanowire preparation that we are aware of, generates silver nanowires through the aggregation of small silver nanoparticles [93]. In addition, we are aware of only one report where nanocubes were observed in a DMF preparation [94].

Another, and more recent report, also demonstrates a synthesis of gold nanodecahedra in ethylene glycol in the presence of PVP, see Figure 11.35 [95]. Lowering the PVP concentration results in nanoicosahedra being formed and lowering the PVP concentration even further results in nanoprisms being the preferred morphology.

The optical properties of gold nanodecahedra have been studied in detail by Pastoriza-Santos *et al.* [96].

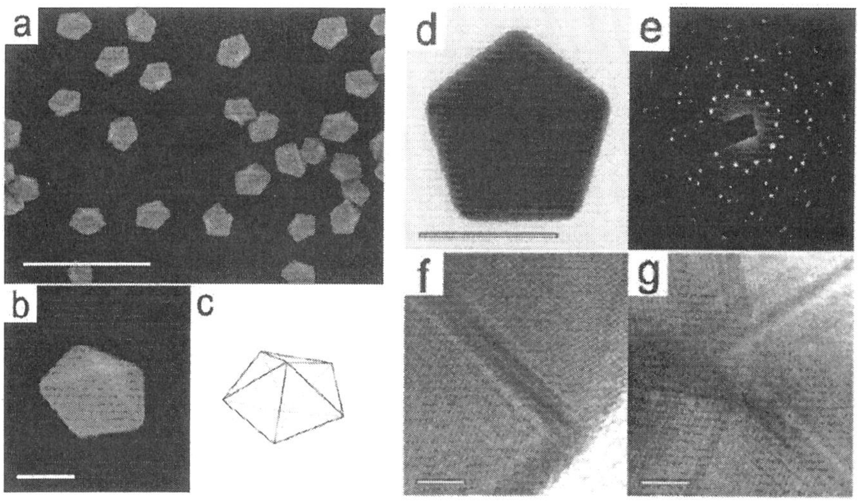


**Figure 11.35:** (a) SEM image of 88-nm gold decahedra. (b) SEM, (c) ideal model, and (d) TEM images of a decahedron. (e) Electron diffraction pattern of a decahedron along <110> zone axis. HRTEM images of (f) edge and (g) centre areas in a decahedron. The bars represent 500 nm (a), 100 nm (b, d), and 2 nm (f, g). Reprinted with permission from reference [95]. © (2008) American Chemical Society.

## 11.4 2.4 *Combining Selective Binding And Non-Selective Binding Growth Modes.*

PVP is rarely successful at producing anisotropic gold nanoparticle structures that can be explained by the selective binding growth model, unless $Ag^+$ is also present. Indeed, for the gold nanooctahedra and nanodecahedra above, PVP was present but did not play the normal shape directing role. On the other hand, PVP is very effective at selectively binding to less stable crystal faces of silver and thus promoting growth on the more stable {111} crystal faces, as has been discussed earlier.

There is one very notable, and somewhat unexplained, exception to this where a polyol process was developed that could produce silver nanooctahedra, where growth occurs on the less stable {100} faces [91]. There are two key differences between this approach and the polyol procedure for silver nanocubes. Firstly $CuCl_2$ is used as the source of chloride instead of NaCl, and secondly 1,5 pentanediol is used as the solvent and reducing agent, instead of ethylene glycol.

This preparation is unusual as one would expect PVP to selectively bind to less stable crystal faces of single crystals of silver, such as the {100} face, to generate nanocubes. Indeed, in this reaction silver nanocubes are the initial product as expected, but by continuing the reaction, after nanocubes have already formed, further growth leads to cuboctahedra, truncated octahedra and then finally to octahedra, see Figure 11.36. It is quite uncertain why the growth anisotropy switches in this manner.

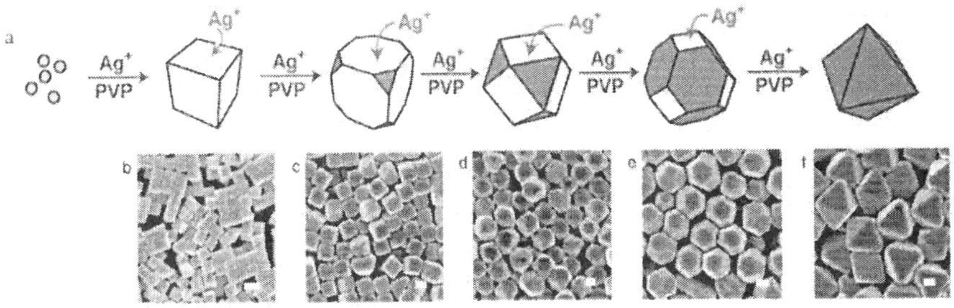


**Figure 11.36:** By extending the polyol reaction for a given time period, various polyhedral shapes capped with {100} and {111} faces can be obtained in high yield. a) A schematic of the nucleation and growth process, in which silver continuously deposits onto the {100} faces to eventually result in a completely {111}-bound octahedron. b to f) SEM images of cubes, truncated cubes, cuboctahedra, truncated octahedra, and octahedra, respectively (scale bar: 100 nm). Reproduced with permission from reference [91]. © (2006) Wiley-VCH Verlag GmbH & Co. KGaA.

Other results that show a similar switching in growth anisotropy are more readily explained. Recent experiments have found that certain reaction conditions that include PVP and produce gold nanoparticles that cannot be explained by the selective binding model, *i.e.* are {111}-bound, can then be used to grow silver shells, over the gold nanoparticle cores, into shapes that can be explained by the selective binding growth model [97].

This is illustrated in the three figures that follow below. It is worth pointing out that it is claimed that in the case of the gold nanoparticle cores, the PVP has a reversed selectivity, i.e. it binds preferentially to the {111} instead of the {100} crystal faces of gold. Alternatively, the observed change in growth anisotropy is consistent with the idea that in cases where PVP (or any coordinating organic species) shows too low a degree of selectivity between crystal faces, growth will occur faster on the less stable crystal faces as discussed in Section 11.4 2.1. Indeed, this is how we explained the growth of the gold nanooctahedra and nanodecahedra in Sections 11.4 2.2 and 11.4 2.3. Upon deposition of silver onto the gold nanoparticles, the expected selective binding to less stable crystal faces is re-established and growth continues on the more stable crystal faces. In the context of the examples below, this means that gold decahedra can grow and then act as templates for pentagonal silver nanorods; gold octahedra can grow and then act as templates for silver nanocubes; gold nanoprisms can grow and then act as templates for silver bipyramids.

In Figure 11.37 the templating of pentagonal silver nanorods from decahedral gold seeds is clearly illustrated. The reaction conditions are optimal for growth of the gold decahedra but on addition of $AgNO_3$ one-dimensional growth of silver on the gold decahedral cores takes place due to the binding of the PVP to the {100} side faces of silver.

A similar pattern emerges in the case of single crystals. Under conditions where the selective binding model does not apply and growth is dominated by the different growth kinetics of different crystal faces, single fcc crystals prefer to grow into octahedra bounded by the most stable {111} faces. As can be seen in Figure 11.38 the single crystal gold nanoparticles that are produced here have this octahedral structure. After addition of $AgNO_3$ and subsequent deposition of silver on the octahedral gold cores, the selective binding model now applies as PVP preferentially binds to the {100} faces of silver, resulting in faster growth on the {111} faces which then grow out of existence to yield the familiar nanocube shape bounded by {100} faces.

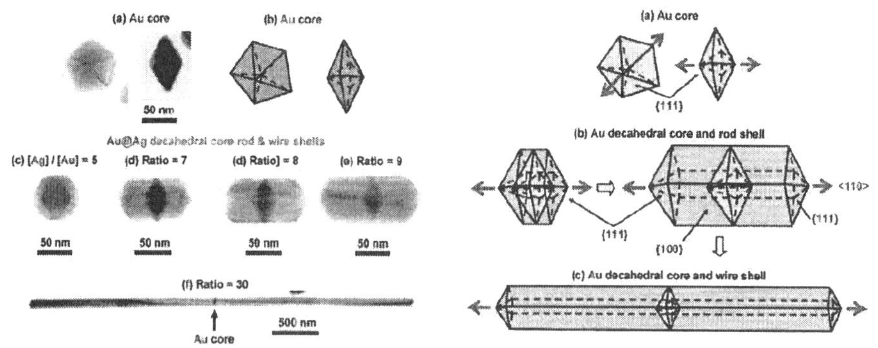


**Figure 11.37:** Left: (a) TEM image of and (b) schematic diagrams of the Au decahedral core. (c to f) TEM images of Au@Ag nanorods and nanowires prepared by addition of various amounts of $AgNO_3$ to Au cores with microwave heating for 2 min. Right: Schematic for growth of silver nanowires from Au decahedral core. Reprinted with permission from reference [97]. © (2006) American Chemical Society.

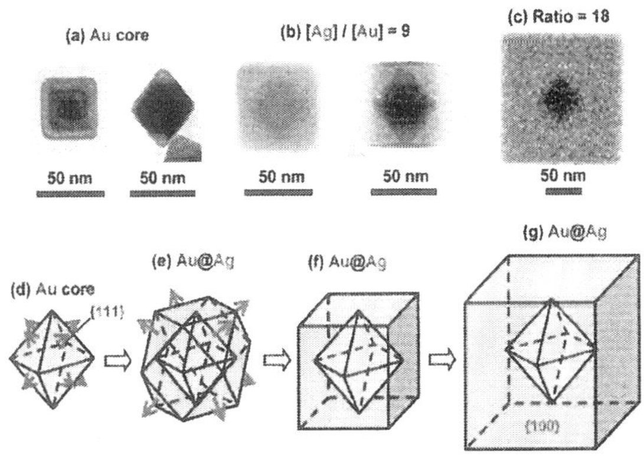


**Figure 11.38:** TEM images of (a) the octahedral Au core and (b, c) Au@Ag nanocrystals prepared by addition of different amounts of $AgNO_3$ to Au cores with microwave heating for 2 min. (d to g) Schematic for growth of cubic Au@Ag nanocrystals from the octahedral Au core. Reprinted with permission from reference [97]. © (2006) American Chemical Society.

Nanoprisms provide an interesting template for anisotropic growth within the selective binding growth model. The nanoprisms have two large flat {111} faces, which upon addition of $AgNO_3$ in the presence of PVP, become preferred growth surfaces. As can be seen in Figure 11.39 this results in growth of a silver pyramid on each flat triangular face producing a bipyramid bounded by the {100} faces stabilized by the PVP.

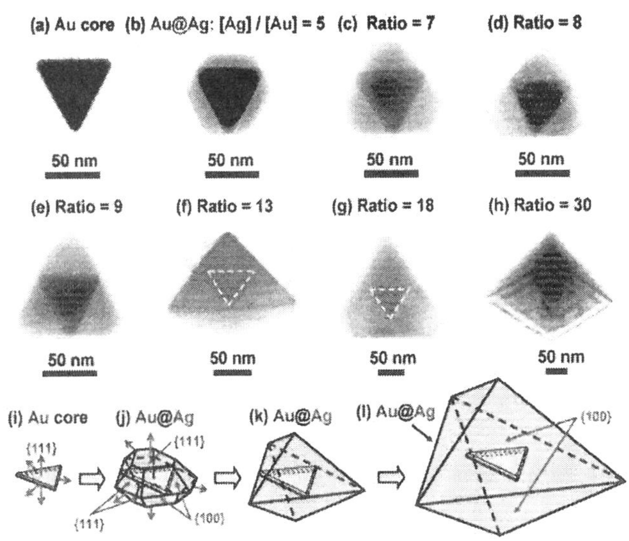


**Figure 11.39.** TEM images of (a) the triangular twin Au core and (b to h) Au@Ag nanocrystals prepared by addition of various amounts of $AgNO_3$ to Au cores with microwave heating for 2 min. Dotted lines represent triangular Au core nanoplates, which can be observed using photographs with better contrast. (i to l) Schematic for

growth of triangular-bipyramidal Au@Ag nanocrystals from triangular Au nanoprism cores. Reprinted with permission from reference [97] © (2006) American Chemical Society.

## 11.4 2.5 *Dendritic Structures*

Multipodal nanoparticles have been produced by the aqueous surfactant method [83,98], but these methods are difficult to reproduce accurately. Kumar *et al.* have recently demonstrated the reliable production of gold nanostars [99]. These nanostars (see Figure 11.40) are formed by a rapid injection of an ethanolic solution of 15 nm gold seeds into a solution of PVP and $HAuCl_4$ in DMF. It was found that the sharpness of the gold nanostars scaled with the amount of PVP used.

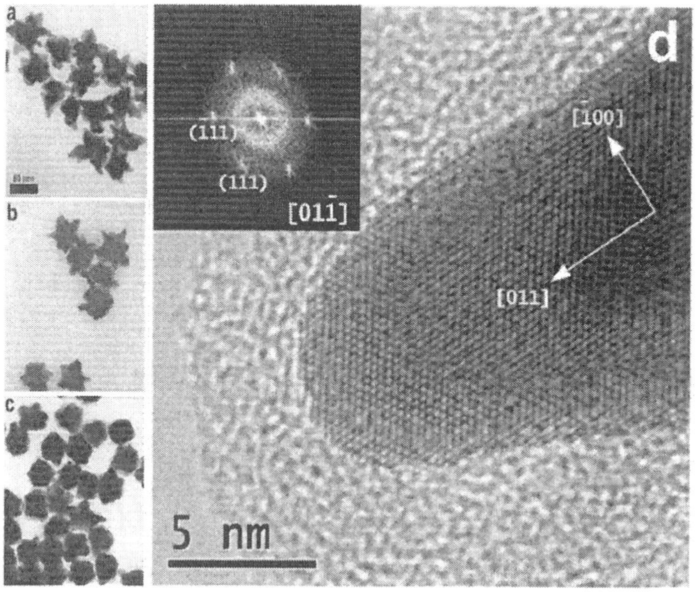


**Figure 11.40:** TEM images of gold nanostars synthesized through reduction of $HAuCl_4$ in a PVP/DMF mixture, in the presence of preformed gold seeds, using different PVP (Mw = 10 000) concentrations: a) 10.0, b) 5.0, and c) 2.5 mM. d) HRTEM image of one single tip in the nanostar where the <110> growth direction of the tip can be clearly identified (the inset is the corresponding FFT pattern, demonstrating that the image in (d) was obtained in the <110> zone axis). Reprinted with permission from reference [99]. © (2008) Institute of Physics.

The growth mechanism is not well understood but analysis of the limbs of the nanostars by HRTEM shows that growth occurs in the <110> direction with each limb bounded by at least one {100} face. This mode of growth is consistent with growth on the less stable face.

Tang *et al.* have also recently reported the synthesis of gold nanostars and more highly branched nanostructures, see Figure 11.41 [100]. In their method, $HAuCl_4$ and a reducing agent, ammonium formate are added to an aqueous solution of PVP and then heated to 120° C for over 5 hours. It was found that the PVP concentration had little effect on the outcome of the reaction but that the gold nanostructures were more highly branched as more reducing agent was used. A growth mechanism is difficult to discern but Tang *et al.* suggest that a diffusion limited aggregation model [101] cannot fully explain the results.

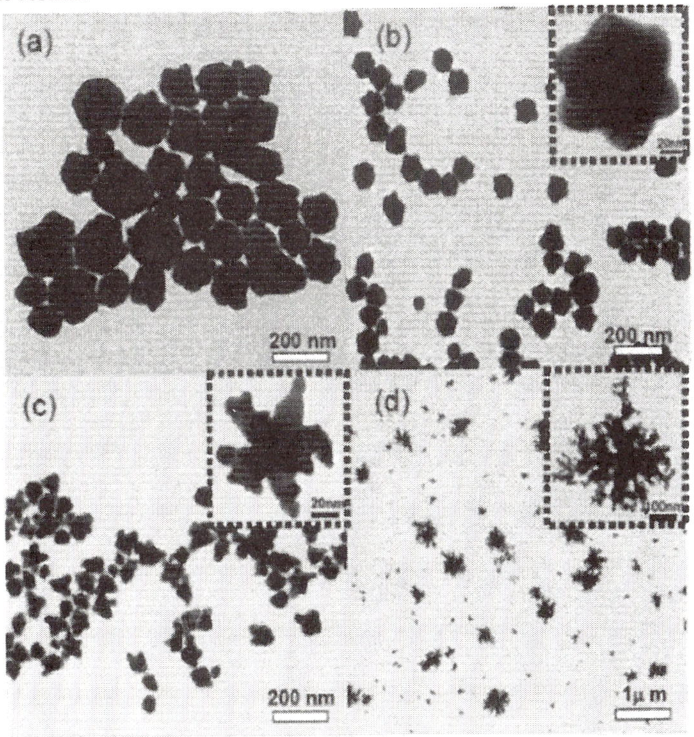


**Figure 11.41:** TEM images of dendritic gold nanostructures produced at various ammonium formate concentrations: a) 0.06, b) 0.1, c) 0.3, and d) 0.5 M. The insets are higher magnification TEM images of individual gold nanostructures prepared under the corresponding reaction conditions. Reprinted with permission from reference [100]. © (2008) American Chemical Society.

### 11.4 2.6 *Nanoprisms*

#### 11.4 2.6 1. *Physical Aspects*

Nanoprisms have received considerable attention due to the ability to tune the main (in-plane dipole) SPR by controlling the edge length and thickness (aspect ratio) [2]. In particular, the main SPR of silver can be tuned across the entire visible spectrum from ~ 400 nm to near infra-red (NIR) wavelengths [14].

Nanoprisms typically have an equilateral triangular shape (Figure 11.42A) [102] and are usually relatively thin compared to their edge length (Figure 11.42C) [102]. Extensive characterization has confirmed that the relatively large flat faces are {111} fcc planes, although in the literature there is no consistent identity of the crystal faces at the edges. Electron diffraction of flat-lying nanoprisms typically shows spots corresponding to diffraction from the {220} and {422} planes of a fcc crystal. There is also usually a series of spots normally assigned to formally forbidden $^1/_3\{422\}$ reflections [58,102-105], see Figure 11.42B [102]. X-ray diffraction (XRD) spectra typically show a dominant sharp peak for {111} indicative of relatively large {111} planes in the sample, see Figure 11.42D [14].

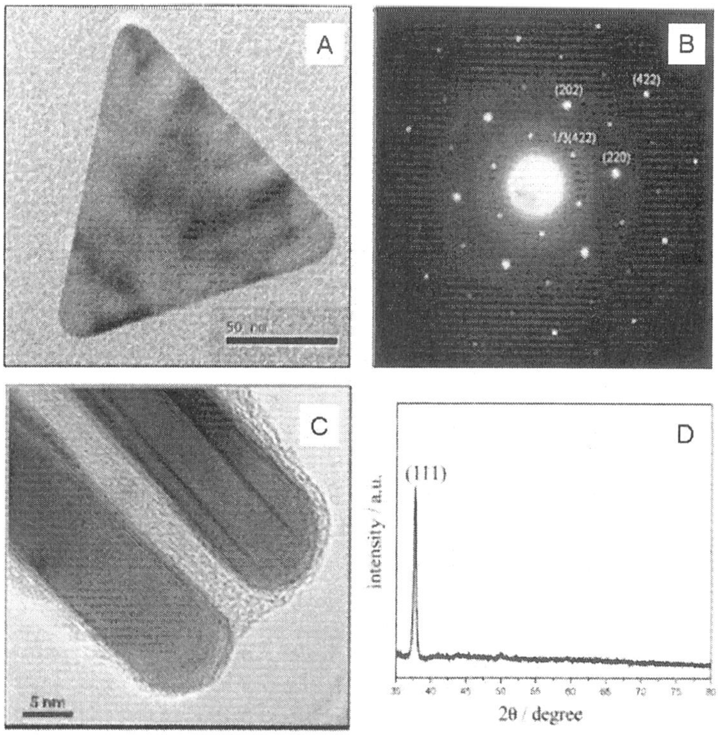


**Figure 11.42:** A) TEM image of <111> oriented silver nanoprism. B) Electron diffraction pattern taken from an individual silver nanoprism showing the assigned reflection indices. C) Edge-on view of silver nanoprisms illustrating flat structure. Reproduced with permission from reference [102]. © (2006) Wiley-VCH Verlag GmbH & Co. KGaA. D) Typical XRD pattern for a sample of nanoprisms. Reprinted with permission from reference [14]. © (2006) American Chemical Society.

Further confirmation of the <111> orientation and fcc structure of flat-lying nanoprisms is given by HRTEM studies. In the HRTEM image of edge-oriented nanoprisms in Figure 11.42C, we can see the lattice fringes for {111} planes co-planar with the flat faces of the nanoprisms. In flat-lying nanoprisms, lattice fringes with a spacing of 2.50 Å are commonly observed (Figure 11.43A) and are often assigned to the formally forbidden $^1/_3\{422\}$ reflections mentioned above. Lattice spacings corresponding to reflections from {220} planes (Figure 11.43B) and {311} planes (Figure 11.43C) can also be observed.

As mentioned earlier, nanoprism growth cannot really be explained by the selective binding growth model, although it is still sometimes postulated that the large {111} faces in nanoprisms are stabilized to a greater extent than the other crystal faces by whatever surfactants or polymers that are present. Nevertheless, it still has not been adequately explained why growth is restricted in the <111> direction perpendicular to the flat {111} face of the nanoprisms. Since the growth of nanoprisms cannot be explained properly by the selective binding growth model it is possible that no growth-directing role is being played by the various organic species in the synthesis. However, the

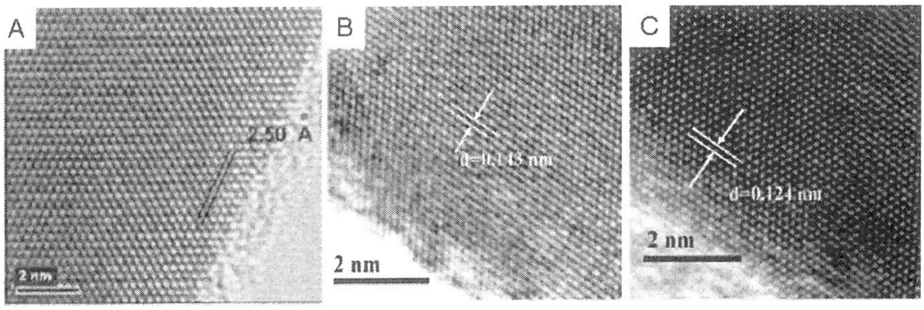


**Figure 11.43:** HRTEM images of silver nanoprisms showing A) 2.50 Å lattice fringes assigned to formally forbidden $^{1}/_{3}${422} reflections of fcc silver. Reproduced with permission from reference [102]. © (2006) Wiley-VCH Verlag GmbH & Co. KGaA. B) 1.43 Å lattice fringes indexed as {220} of fcc silver, C) 1.24 Å lattice fringes indexed as {311} of fcc silver. Reprinted with permission from reference [106]. © (2007) American Chemical Society.

nanoprism morphology cannot be explained by simply reversing the anisotropy of the selective binding growth model either. For these reasons, it has become increasingly plausible to consider that there is an underlying defect structure in the seeds that is driving two-dimensional growth leading to the nanoprism shape.

Indeed, recent reports have implicated defects as a direct factor influencing crystal growth. Specifically, defects such as twinning that arise during the early stages of particle formation give rise to preferred growth directions where the defects are exposed to the growth medium. In the case of nanoprisms, parallel stacking faults in the <111> direction have been observed with these making contact with the growth medium at the edges, precisely where growth occurs [103].

The silver halide growth model has been resurrected as a way of explaining particle growth in many synthesis methods [58,104]. In this model, twin planes form reentrant grooves (A-type faces in Figure 11.44), which are favourable sites for the

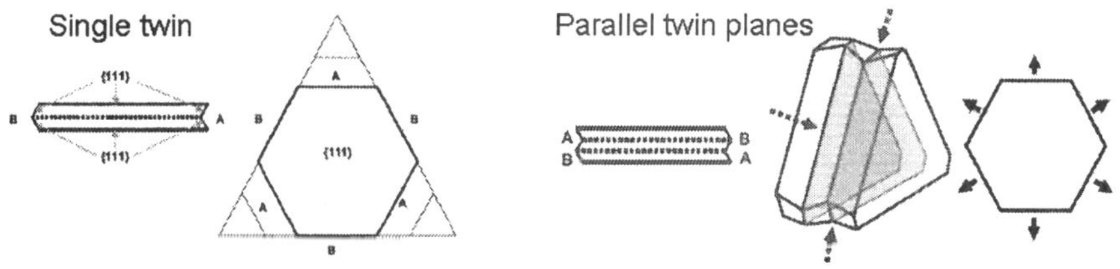


**Figure 11.44:** Silver halide model for nanoplate growth. Left: silver halide model for the case of a single twin plane. Alternating sides contain A-type and B-type faces. The re-entrant grooves of the A-type faces cause rapid growth that is arrested when the face grows itself out, leaving a triangular prism with slow-growing B-type faces. Right: silver halide model for the case of a particle with two parallel twin planes. The second twin plane causes all six sides to contain A-type faces (dashed arrows) with re-entrant grooves. This leads to each A-type face regenerating those adjacent to it, allowing for rapid growth in two dimensions (solid arrows). Reproduced with permission from reference [58]. © (2005) Wiley-VCH Verlag GmbH & Co. KGaA.

attachment of adatoms. A single twin plane is expected to direct growth in two dimensions but limit the final size of the nanoprism, while the presence of two parallel twin planes would allow the fast growing edges to regenerate one another, allowing shapes such as hexagonal nanoplates to form.

Recently, Rocha and Zanchet have studied the defects in silver nanoprisms in some detail and have shown that the internal structure can be very complex with many twins and stacking faults [107]. These defects are parallel to each other and the flat {111} face of the nanoprism, subdividing it into lamellae which are stacked in a <111> direction, and are also present in the silver seeds. In that paper, it was demonstrated how the planar defects in the <111> direction could give rise to local hexagonally close-packed (hcp) regions. These could in turn explain the 2.50 Å lattice fringes that are observed in <111> orientated nanoprisms, which have hitherto been attributed to formally forbidden 1/3{422} reflections as mentioned above.

Although the growth mechanism for nanoprisms is not yet fully understood, there is no question that the flat shape of silver nanoprisms results from highly selective two-dimensional growth from the edges. A lamellar defect structure, as mentioned above, would certainly produce a circular band on the surface of a nanoparticle seed where the defects are exposed to the growth medium. If these defect sites caused accelerated growth then a two-dimensional lateral growth pattern could emerge. It is thus reasonable to suspect that a lamellar defect structure in nanoparticle seeds is a key factor in the growth of nanoprisms. This will be addressed again in Section 11.4 2.6 2.1.

### 11.4 2.6 2. *Synthetic Approaches*

The syntheses that exist for the production of silver nanoprisms can be generally placed into either of two categories: photochemical (plasmon-driven synthesis) [102,108-111] and thermal [13,14,106,112-119]. Photochemical syntheses have until recently produced the highest quality samples to date but this approach typically involves days for the preparation of a sample. Thermal approaches are much quicker but often produce samples with a range of shapes and sizes.

## 11.4 2.6 2.1 *Thermal Methods*

Most, although not all, thermal methods employ some sort of coordinating organic species such as polymers or surfactants.

### *DMF Reduction*

Some early methods involved using the solvent as the reducing agent, similar to the polyol approach discussed earlier. Typically, solvent reduction experiments are quite slow and can take hours to complete. Dimethylformamide (DMF), in the presence of PVP, has been successfully utilized to produce silver nanoprisms [112]. Such preparations usually result in a mixture of shapes being obtained. More recently [106], it has been found that adjustment of the concentrations of PVP and $AgNO_3$ and the mole ratio between the two, can direct the synthesis to yield samples that are enriched in one particular shape, including one sample that was up to 95 % nanoprisms. The mechanism for the shape dependence on these parameters is not clear. Nor, is it clear that the preference of PVP for higher energy crystal faces should change dramatically by simply varying the concentrations of PVP and $AgNO_3$ or the mole ratio between them. Also, given the contrary growth mechanisms of nanoprisms on the one hand and nanocubes and nanorods on the other, it seems likely that modifying the reaction parameters has the effect of influencing the defect structure of the initially formed seeds and this in turn determines the final reaction output.

### *PVP Reduction*

Some researchers have exploited the reducing ability of PVP itself, or rather its –OH end groups as the reducing agent in the synthesis of silver nanoparticles. There has been some success in employing this approach to produce nanoprisms. The Xia group has demonstrated this. The initial flat particles are very rounded and sharpen up over time as the nanoparticles grow larger, after about 7 hours, see Figure 11.45 [113,120]. The extinction spectra tend to be quite broad, indicating a wide size distribution, and TEM analysis shows that there are plenty of "spherical" nanoparticles present. This approach has also been extended to gold and palladium [120], and has also been explored by other researchers [115].

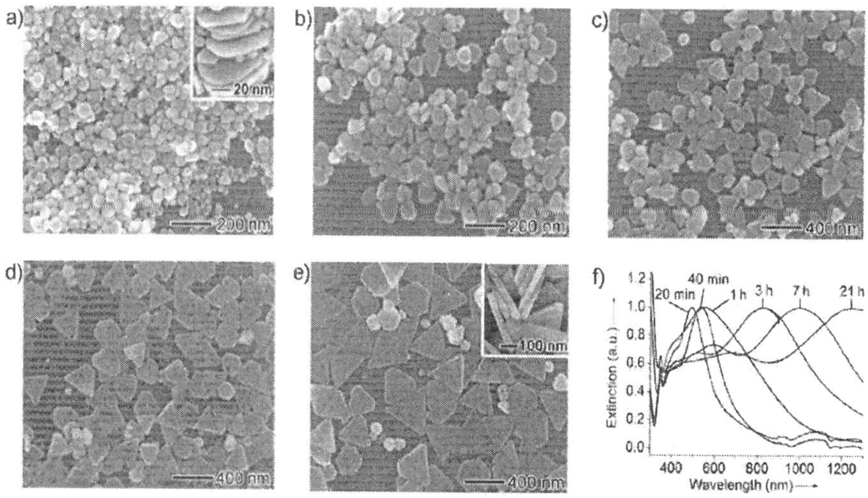


**Figure 11.45:** SEM images of nanoparticles sampled at different stages of a synthesis of silver nanoprisms by reduction by PVP: a) $t$ = 40 min, b) $t$ = 1 h, c) $t$ = 3 h; d) $t$ = 7 h and e) $t$ = 21 h. f) Extinction spectra of these products dispersed in water. The insets of (a) and (e) are SEM images taken from tilted samples. The PVP had an average molecular weight of 29 000 g $mol^{-1}$ and its molar ratio (in terms of the repeating unit) to $AgNO_3$ was 30. Reproduced with permission from reference [113]. © (2006) Wiley-VCH Verlag GmbH & Co. KGaA.

## *Aqueous Surfactant Preps*

Many nanoprism syntheses employ the use of surfactants such as CTAB. The exact role of surfactants is unlikely to be the same as in syntheses of nanorods and nanocubes where preferential adsorption to {100} and sometimes {110} crystal faces dictates growth. It was shown earlier that in the case of gold and silver, this leads to growth in the <111> direction, yielding nanocubes, nanorods and nanowires, yet nanoprisms have large flat {111} faces where growth has been inhibited. In any case, early preparations by Chen *et al.* resulted in highly truncated nanoprisms or nanodisks [118,121] of silver and it has been possible to obtain some size control by varying reaction parameters [122]. Nevertheless, samples typically have a wide size distribution and have plenty of spherical nanoparticles present.

The Mirkin group have obtained samples of gold nanoprisms by modifying the reaction parameters of the nanorod preparation procedure developed by the Murphy group [123]. There are plenty of spherical nanoparticles present but the nanoprisms are large and it is possible to see the in-plane quadrupole SPR. Even larger gold nanoprisms can be obtained by successive additions of $HAuCl_4$ and ascorbic acid (reducing agent) as can be seen in Figure 11.46 [124].

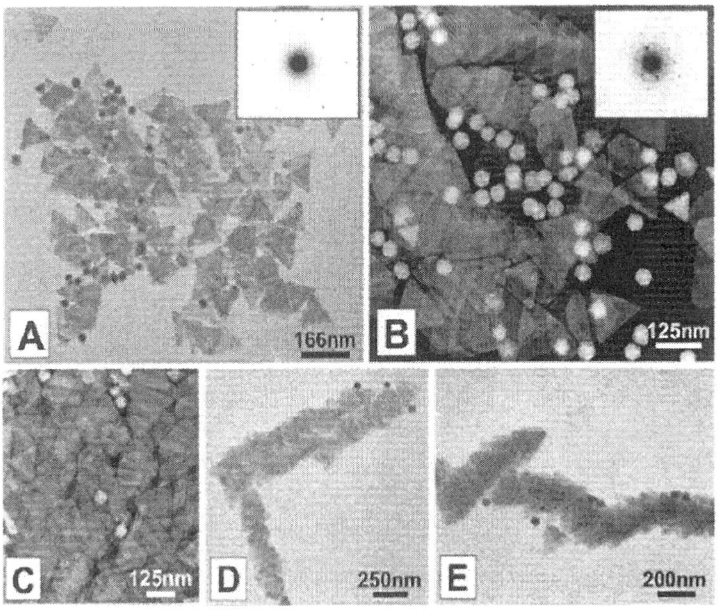


**Figure 11.46:** TEM images of Au nanoprisms with increasing edge lengths. A) TEM image of Au nanoprisms made without additional growth (control sample). The inset shows a diffraction pattern of the nanoprisms. B) TEM image of nanoprisms after two growth additions (x2). The inset shows diffraction pattern of x2 prisms. C) TEM image of x4 nanoprisms, D) x6 nanoprisms, and E) x8 nanoprisms. Reproduced with permission from reference [124]. © (2006) Wiley-VCH Verlag GmbH & Co. KGaA.

## *Other Various Aqueous Nanoprism Preparations*

The Mirkin group has developed an interesting hydrogen peroxide-catalyzed silver nanoprism synthesis that in addition to providing control over nanoprism edge length provides a way of controlling nanoprism thickness [116]. Higher amounts of reducing agent ($NaBH_4$) result in thinner nanoprisms. PVP is present is this procedure although the authors state that it most likely does not play a shape-directing role. In the absence of PVP, ill-defined aggregates are formed, indicating that PVP is necessary to generate a stable colloid, although nanoprisms of comparable shape and size were obtained when bis(*p*-sulfonatophenyl)phenyl phosphine dipotassium dihydrate (BSPP) was used instead of PVP.

More recently there has been a variety of improved aqueous procedures producing nanoprisms with less morphological impurities and reasonable control over nanoprism size. Zou and Dong's procedure produces silver nanoprisms without the need for any coordinating organic species besides citrate [14,125]. Ledwith *et al.*'s procedure, with PVP present, shows a variation in position of the in-plane dipole SPR with amount of added citrate, see Figure 11.47 [119]. Jiang *et al.*, using sodium bis(2-ethylhexyl) sulfosuccinate (NaAOT) as a capping molecule, have also been able to produce reasonably monodisperse samples of nanoprisms that subsequently etch to disks [126,127].

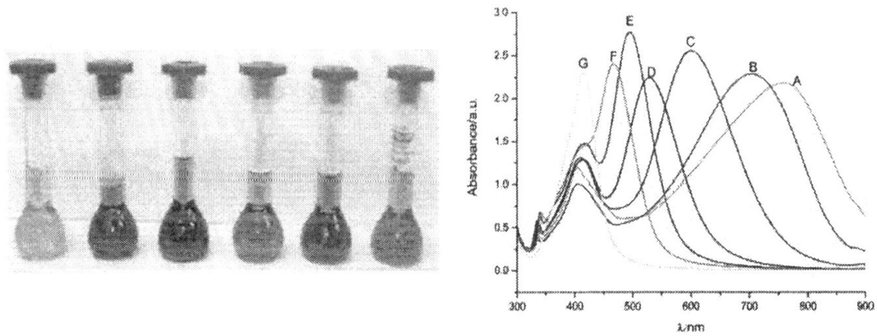


**Figure 11.47:** UV–vis spectra and photograph of different coloured nanoprism samples prepared by varying only the concentration of citrate in the growth step. Citrate concentrations used are: A) 0.7 mM; B) 0.35 mM; C) 0.175 mM; D) 0.07 mM; E) 0.024 mM; F) 0.012 mM; G) 0.004 mM. Reference [119] – Reproduced by permission of The Royal Society of Chemistry.

Aherne *et al.* have recently developed a rapid and readily reproducible method for the production of silver nanoprisms in high yield [128]. The method involves the silver seed-catalyzed reduction of $Ag^+$ by ascorbic acid, and contrasts with a previously reported procedure from the group [119] in that the concentration of spherical nanoparticles produced is minimal, and PVP has been eliminated from the synthesis. The nanoprisms are well-defined triangular plates and the spectral position of the SPR can be tuned through the visible to the near infra red (NIR), see Figure 11.48, by controlling the edge length of the nanoprisms, without any significant variation in thickness, *i.e.* by varying their aspect ratio. This can be achieved through adjustment of the number of seeds in the growth mixture. A typical example of the nanoprisms produced with this method is shown in Figure 11.49

To characterize the nanoprisms produced by this method and explore the relationship between nanoparticle dimensions and the position of the main SPR, TEM analysis of statistically significant numbers of nanoprisms from four samples was carried out. TEM grids of samples were prepared such that many of the particles were arranged in a stacked formation with their flat faces parallel to the electron beam. TEM images of nanoprisms of 4 different sizes (Samples 1 to 4) are shown in Figure 11.50. It is clear that the triangular shape of the nanoprisms is established early on in the growth process and that growth proceeds through enlargement of these nanoprisms, while nanoprism thickness remains relatively constant between samples.

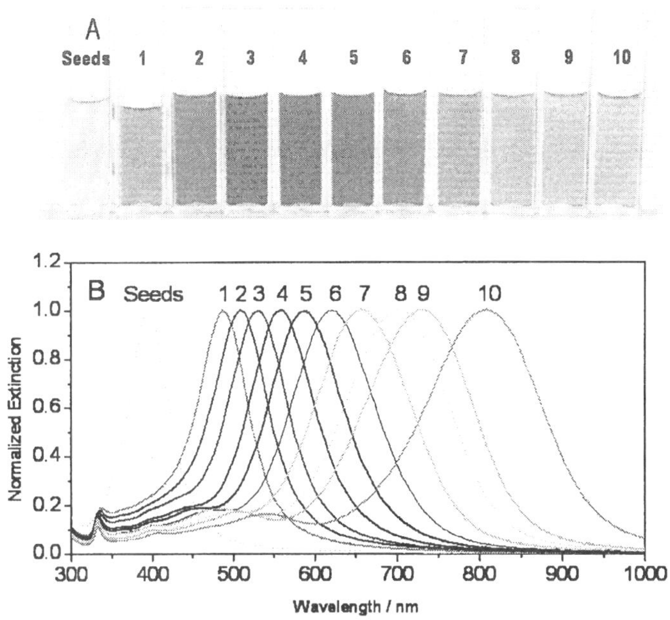


**Figure 11.48:** A) Photograph of series of samples illustrating range of colours obtained. B) Normalized spectra of a series of *as prepared* samples obtained using different volumes of seed solution: 1) 650, 2) 500, 3) 400, 4) 260, 5) 200, 6) 120, 7) 90, 8) 60, 9) 40, 10) 20 μl. Reproduced with permission from reference [128]. © (2008) Wiley-VCH Verlag GmbH & Co. KGaA.

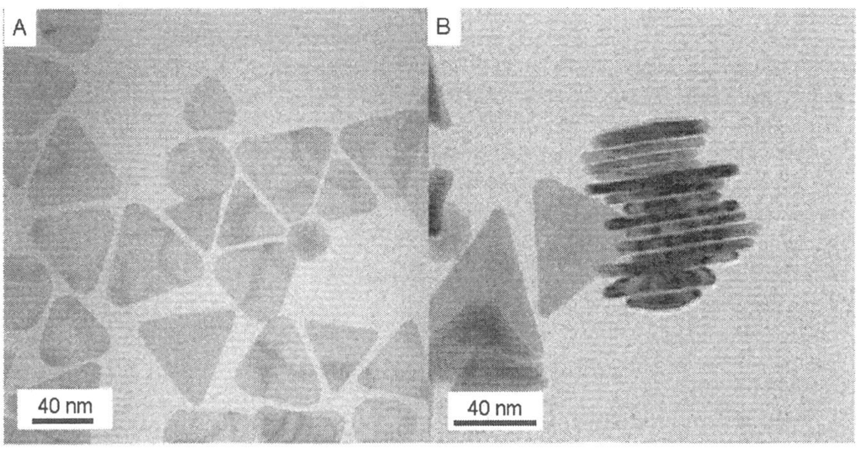


**Figure 11.49:** A) TEM image of flat-lying silver nanoprisms from a typical sample produced by method described in reference [128]. B) TEM image of silver nanoprisms from another sample, made by the same procedure, that are stacked together

and are oriented such that they are standing vertically on their edges. Reproduced with permission from reference [128]. © (2008) Wiley-VCH Verlag GmbH & Co. KGaA.

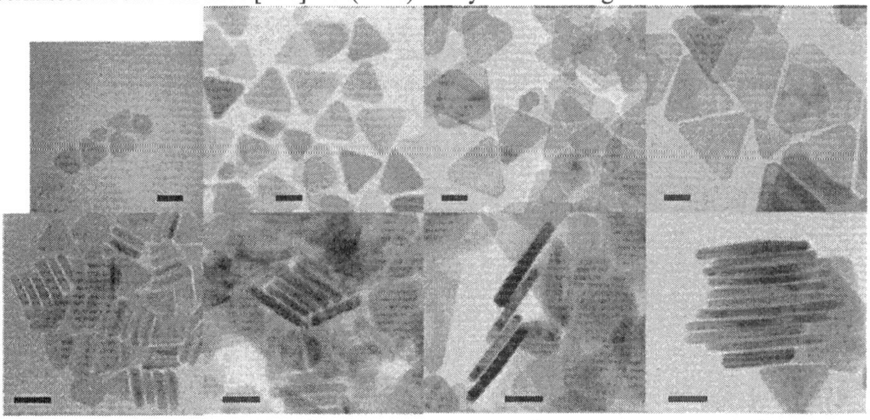


**Figure 11.50:** TEM images of flat-lying and stacked silver nanoprisms for Samples 1 to 4. There is a clear trend of increasing edge length of nanoprisms. Scale bar is 20 nm. Reproduced with permission from reference [128]. © (2008) Wiley-VCH Verlag GmbH & Co. KGaA.

As mentioned earlier, the defect structure of silver nanoprisms is a current area of investigation as it most likely is the source of the anisotropic growth that leads to the nanoprism shape. TEM analysis of silver nanoprisms in this Section provides direct evidence of a defect-induced arrangement of silver atoms that results in a hcp structure in the vicinity of the defects and also shows multiple defects combining to yield a continuous hcp lamellar region of about 1.5 nm in thickness. As shown in detail below, this hexagonal arrangement of atoms propagates perpendicular to the flat {111} face of the nanoprism with a spacing of 2.50 Å and thereby explains the commonly observed 2.50 Å lattice fringes in flat-lying silver nanoprisms.

To investigate this possible hcp arrangement of atoms, HRTEM studies of vertically oriented silver nanoprisms were conducted. For a defect in the <111> direction to be observed in the TEM, it is necessary that the nanoprism is oriented such that a {110} plane is in the plane of the image. In this orientation, two {111} planes and a {100} plane are aligned vertically with respect to the electron beam. The defects can then be detected as discontinuities in either the {100} or {111} planes that propagate away from the flat face of the nanoprism. This is illustrated schematically in Figure 11.51.

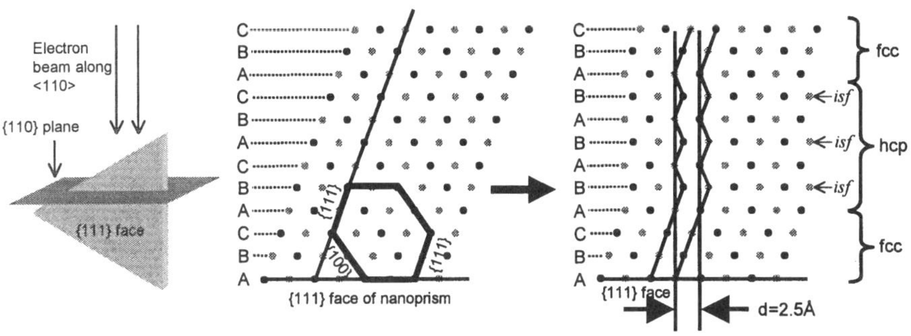


**Figure 11.51:** Schematic illustrating how intrinsic stacking faults along <111>, i.e. faults in the successive stacking of the ABC layers ({111} planes) of an fcc crystal, give rise to a hcp region. The black dots represent atoms in the {110} plane while the grey dots represent atoms immediately below. Reproduced with permission from reference [128]. © (2008) Wiley-VCH Verlag GmbH & Co. KGaA.

For the correct orientation to occur, the nanoprisms need to be vertically orientated and secondly need to have one edge parallel to the electron beam (see the left hand side of Figure 11.51). This means that few nanoprisms will have the {110} plane correctly aligned since most nanoprisms are probably resting on one of their edges on the TEM grid. However, some nanoprisms do have the right orientation and a layered defect structure is visible in two of the stacked silver nanoprisms in Figure 11.52A. Closer inspection of the nanoprism on the right reveals that it is indeed being observed along <110> as the internal defect structure of the crystal is visible (Figure 11.52B). An analysis of the defects is shown in Figure 11.52C. The flat {111} face of the nanoprism is clearly indicated and lattice fringes corresponding to {111} planes can be seen propagating away from the face of the nanoprism, parallel to the {111}-labelled side of the hexagon. The spacing between these fringes was measured to be 2.35 ± 0.05 Å, the correct spacing for {111} planes. Further away from the face of the nanoprism, these {111} planes show discontinuities due to repeated stacking faults between the {111} planes parallel to the face of the nanoprism. There is now an arrangement of atoms that propagates perpendicular to the flat face of the nanoprism, indicated by the two white lines. Significantly, this perpendicular arrangement of atoms has a periodicity of 2.50 ± 0.05 Å, corresponding to the lattice spacing that is observed when a flat-lying nanoprism is observed along <111>.

In fact, there are so many defects in the nanoprism here that a significant continuous portion of the crystal has a hcp arrangement; a lamellar region about 1.5 nm thick. This is highlighted by the superposition of a zigzag pattern on the TEM image in the top of Figure 11.52C. Assigning each apex on this pattern to an atom in alternate A and B layers (atomic planes) of the hcp lattice, the average measured distance between an A and B layer in this region is 2.35 Å, which is the spacing between {111} planes in an fcc lattice, which are stacked in an ABCABC... configuration. Since the spacing between alternate layers in an ABABAB... configuration is the same as that in an ABCABC... configuration, each A and B point on the zigzag pattern therefore corresponds to atoms in alternate A and B layers of a hcp lattice.

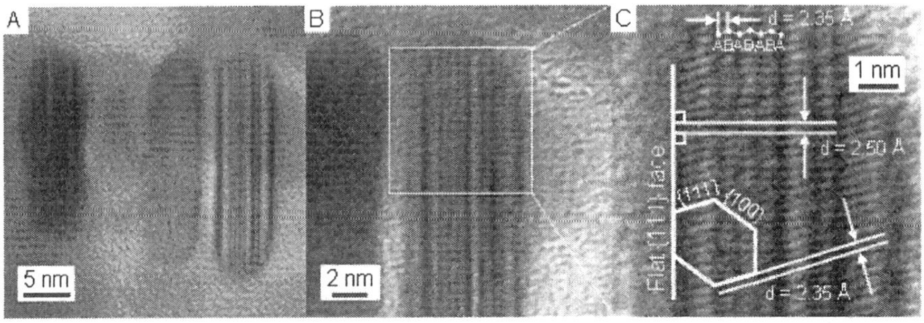


**Figure 11.52:** A) TEM image of a stack of vertically oriented silver nanoprisms. B) High resolution image of the nanoprism on the right hand side of (A) showing defect structure. This nanoprism is oriented such that the {110} plane is in the plane of the image, *i.e.* the electron beam is along <110>. C) Analysis of internal structure of nanoprism in (B). A series of intrinsic stacking faults has resulted in a hexagonally close packed pattern emerging and gives rise to an arrangement of atoms that is aligned perpendicular to the surface with a spacing of 2.50 Å. The correct spacing of 2.35 Å has been obtained for {111} planes and also for the alternate ABAB... layers of the hcp region. Reproduced with permission from reference [128]. © (2008) Wiley-VCH Verlag GmbH & Co. KGaA.

The reconstruction of the silver lattice is illustrated schematically in Figure 11.51. By introducing a series of intrinsic stacking faults (*isf*) it is easy to see how these defects give rise to an ABABAB... stacking arrangement of the atomic planes in a region of the nanoprism. The perpendicular arrangement of atoms with respect to the flat {111} face of the nanoprism is indicated and has a 2.50 Å spacing.

Due to the lamellar defect structure of the nanoprisms, it is precisely at the edges where a hcp crystal structure is exposed to the growth solution and it is clear that the flat morphology of the nanoprisms arises from much faster growth here than on any other crystal face that is present. Moreover, since the hcp structure is not the natural crystal structure for silver, it must therefore be less stable than the fcc structure, making it likely that the edges where the hcp structure is exposed are less stable than the {111} or {100} faces. This higher degree of instability may be the basis of the faster two-dimensional growth at the edges.

The hcp and fcc crystal structures have a hexagonal symmetry so it remains to be explained why triangles, and not hexagonal nanoplates, are the preferred outcome of two-dimensional growth. To explain this let's consider a thin, <110> oriented, fcc single crystal as shown in the schematic in Figure 11.53A below. It is not proposed that a single fcc crystal would take up such an anisotropic structure, but it is clear that it is possible to cut a flat crystal such that opposite sides could have alternating {111}/{100} pairs of faces. The fcc crystal has six-fold symmetry around the <111> axis so a hexagonal platelet could have the alternating faces as outlined in Figure 11.53B below, although the relative sizes of each face at an edge would not necessarily be as fixed as suggested here and a structure as illustrated in Figure 11.53C would be perfectly possible. Next consider a more realistic version of a hexagonal nanoplate that could be the result of initial two-dimensional growth from the seed, see Figure 11.53D. This possesses the hcp region sandwiched between two fcc regions, corresponding to what our TEM data suggest. The schematic is drawn such that the regions on either side of the central hcp region are

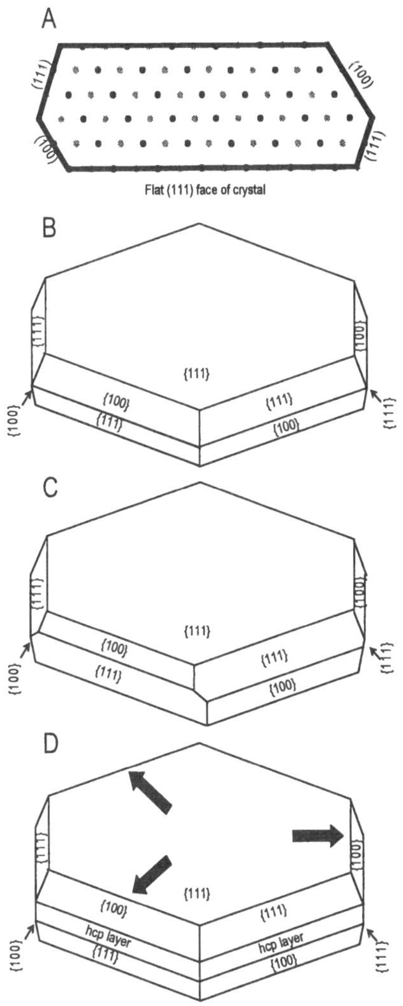


**Figure 11.53:** A) Schematic illustrating a <110> oriented segment of fcc crystal. The edges of a crystal cut in this manner have alternating pairs of {100} and {111} faces. B) Schematic of a nanoplate constructed from a single fcc crystal (no twin planes or defects). A singe crystal would not normally take up this structure but the schematic illustrates that a nanoplate cut from a fcc crystal could have edges consisting of alternating pairs of {100} and {111} faces. C) Without a defect-induced hcp layer in the middle, the faces are free to rearrange and each edge could well be equivalent. D) Schematic of a nanoplate with a defect-induced hcp layer sandwiched between two fcc layers of unequal thicknesses. The thickness of each fcc layer defines the size of the {100} and {111} faces at each edge. The hcp layer drives the lateral growth. Within the two-dimensional growth plane, growth is preferred on the edges with the larger {100}

faces. The block arrows indicated the proposed directions of preferred growth that lead to the familiar triangular shape of nanoprisms. Reproduced with permission from reference [128]. 

asymmetric, *i.e.* one is thicker than the other. The thickness of each fcc layer would then define the size of each of the respective crystal faces on each edge and the faces would not rearrange as in Figure 11.53C. This would mean that not all of the edges of the nanoplate are identical; three of them have a larger {100} face than the {111} face while the other three have a larger {111} face than the {100} face. The three edges with the larger, more stable {111} faces will grow more slowly than the other three with the larger, less stable {100} faces. This is consistent with the lack of growth of the nanoprism on the flat {111} face of the nanoprisms and is consistent with the idea that in the absence of selective binding of organic species, the less stable crystal faces should grow faster.

Thus three of the edges grow faster than the other three, as indicated in Figure 11.53D, leading to the formation of a triangular nanoprism early on during growth. Thus, the asymmetry in thickness between the fcc layers on either side of the hcp layer defines triangular as opposed to hexagonal growth. After a triangular shape is formed, growth continues on the less-preferred edges with the smaller {100} faces, and as it does so, it opens up the preferred growth edges at the apices of the nanoprism for continued growth. Since these preferred edges always grow faster, the nanoprism maintains its triangular shape, with both types of edges growing in a concerted fashion. In this manner smaller triangular nanoprisms grow continuously into larger triangular nanoprisms. In cases where there is no asymmetry in thickness between the fcc layers on either side of the hcp layer, hexagonal nanoplates are expected.

The data and the model presented here differ markedly from what is expected with the silver halide growth model (see Section 11.4 2.6 1). Firstly, this data, and that of others [107], shows that several stacking faults can be present in a nanoprism. Indeed the data presented here shows that the stacking faults can combine to yield continuous hcp regions. Secondly, it is clear that nanoprisms do not stop growing once the triangular shape has been established. As can be seen in Figure 11.50, the triangular shape is established early on in the synthesis and larger nanoprisms can be formed, long after any re-entrant grooves (silver halide model) at the edge would have grown out of existence.

### 11.4 2.6 2.2 *Photochemical Methods*

The photochemical route to nanoprism formation was first reported by Jin *et al.* [108]. In that work, a sample of silver nanospheres was exposed to a white light source for an extended period of time; over 70 hours the sample was observed to change from yellow to blue. A much deeper understanding of this process was gained through experiments that involved excitation of the samples by narrow-band light sources [109]. As can be seen in Figure 11.54, excitation with a single wavelength produced samples with a bimodal distribution of type 1 and type 2 nanoprisms, with type 2 nanoprisms having edge lengths about twice those of type 1 nanoprisms. The observed bimodal growth process occurs through an edge-selective nanoprism fusion mechanism with four type 1 nanoprisms coming together in a step-wise fashion to form a type 2 nanoprism.

It was found that excitation of the samples during the growth process with a second narrow-band light source corresponding to the $\lambda_{max}$ of either the in-plane or out-of-plane quadrupole SPR halted the nanoprism fusion so that a unimodal growth occurred. As can be seen in Figure 11.55 [109], this resulted in the formation of a single distribution of type 1 nanoprisms. Indeed, it is possible to generate a solution of nanoprisms of desired average size with the edge lengths scaling with the wavelength of the primary light source.

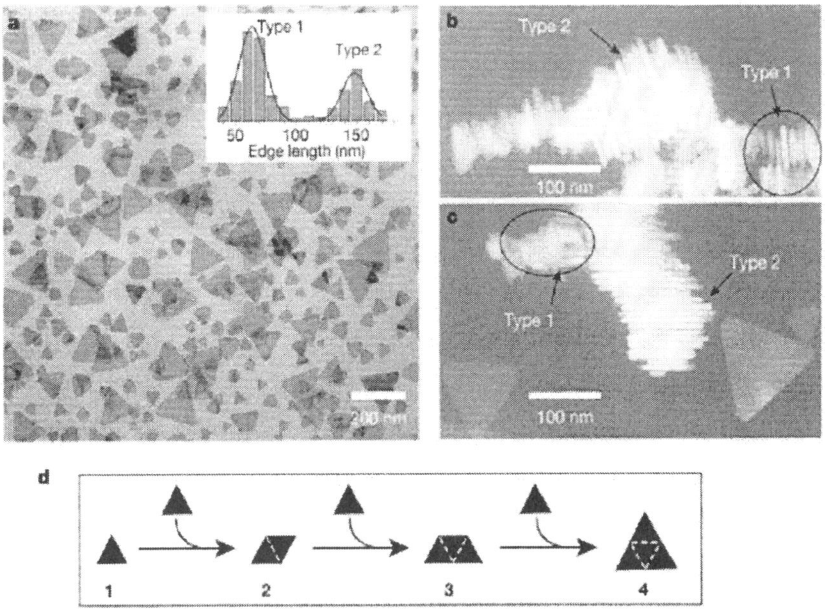


**Figure 11.54:** The bimodal growth of Ag nanoprisms. a) TEM image of a sample of Ag nanoprisms formed using single-beam excitation (550 ± 20 nm); inset, histograms used to characterize the size distribution as bimodal. b, c) TEM images of nanoprism stacks showing that nanoprisms have nearly identical thicknesses (9.8 ± 1.0 nm). d) Schematic diagram of the proposed light-induced fusion growth of Ag nanoprisms. Reprinted by permission from Macmillan Publishers Ltd: Nature [109] © (2003).

Follow-up work has confirmed the dependence of nanoprism size on excitation wavelength. The latest theories for the mechanism of anisotropic growth in photochemically grown nanoprisms will not be dealt with here in detail but a discussion of recent results and ideas can be found in the literature [129,130]. One theory is that it is likely that a lamellar defect structure in the original seed sample is crucial as in thermal methods. Irradiation with light in the presence of oxygen generates $Ag^+$ which is later reduced to $Ag^0$, and is then deposited preferentially at the defect sites at the edge in a similar manner to the mechanism as described above for thermal methods. Also, the SPR may play an important role in reduction and deposition of material at the edge.

Recent experiments have exploited the photochemical approach to produce silver nanoprisms with SPRs at communication wavelengths in the near infra red (NIR) [102]. Of particular interest is recent work by Xue and Mirkin that shows that when the pH is raised to a high enough value, unimodal growth is observed analogous to the dual beam approach mentioned above [110]. The high pH results in the initial nanoprisms being sufficiently negatively charged that the nanoprism fusion process is inhibited due to coulombic repulsion between the nanoprisms. This renders a second light source for unimodal growth unnecessary.

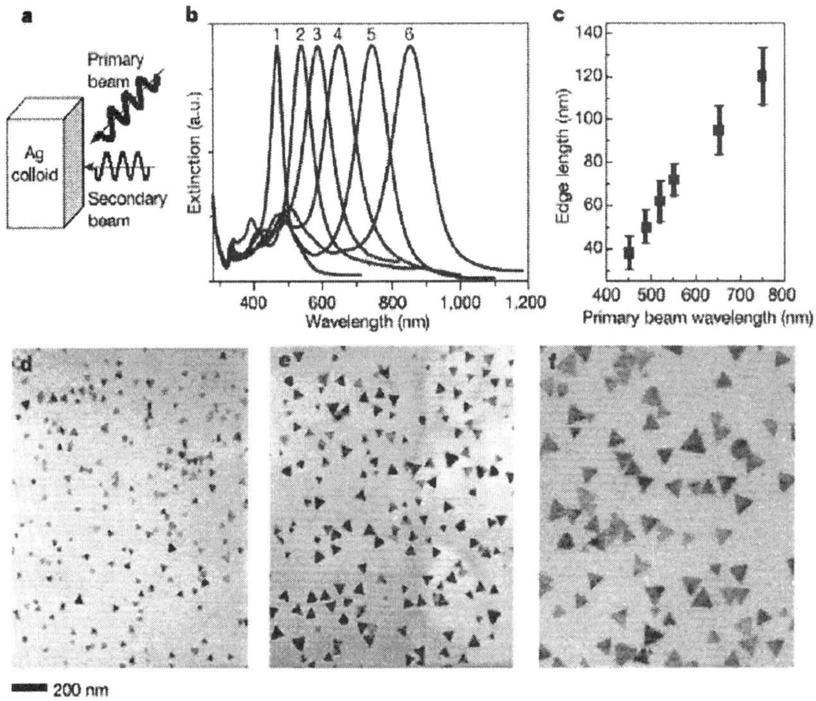


**Figure 11.55:** The unimodal growth of Ag nanoprisms. a) Schematic diagram of dual-beam excitation. b) The optical spectra (normalized) for six different-sized nanoprisms (1 to 6, edge length: 38 ± 7 nm, 50 ± 7 nm, 62 ± 9 nm, 72 ± 8 nm, 95 ± 11 nm and 120 ± 14 nm) prepared by varying the primary excitation wavelength (central wavelength at 450, 490, 520, 550, 650 and 750 nm, respectively; width, 40 nm) coupled with a secondary wavelength (340 nm; width, 10 nm). c) The edge lengths as a function of the primary excitation wavelength. d, e, f) TEM images of Ag nanoprisms with average edge lengths of 38 ± 7 nm (d), 72 ± 8 nm (e) and 120 ± 14 nm (f). Scale bar applies to panels d to f. Reprinted by permission from Macmillan Publishers Ltd: Nature [109] © (2003).

## *Plasmon-Driven Deposition of Silver On Gold Nanoprisms*

Xue *et al.* have explored the SPR excitation-mediated deposition of silver on gold nanoprisms that exhibit SPRs in the NIR, see Figure 11.56 [131]. With NIR excitation, core-shell nanoprisms with triangular gold cores and triangular silver shells are formed. In a control experiment, when the solution was irradiated with 550 nm light, only silver nanoprisms without gold cores are formed. These results show that it is the activation of the SPR of the core nanoprism that leads to preferential silver deposition on the gold core and subsequent triangular silver shell growth.

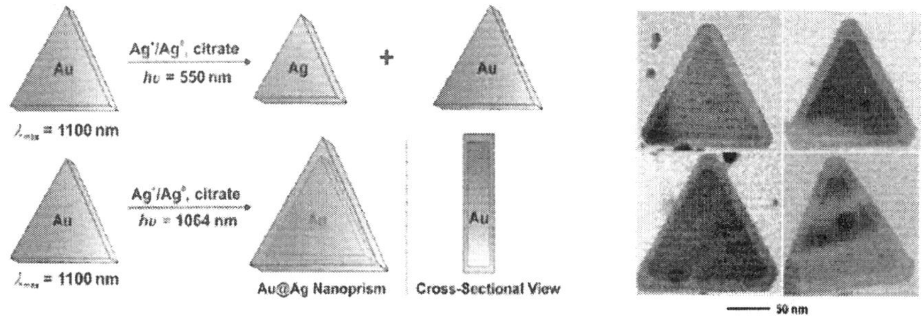


**Figure 11.56:** Left, Schematic illustration of the growth pathways for the Au@Ag core–shell nanostructures. Right, representative TEM images of Au@Ag core–shell nanoprisms with a gold prism core. The scale bar is the same for all images. Reproduced with permission from reference [131]. © (2007) Wiley-VCH Verlag GmbH & Co. KGaA.

## 11.4 3 *Templated Nanostructures*

Many experiments have been carried out to investigate the use of silver nanoparticles as a sacrificial template for the deposition of an outer shell of material. The silver is oxidized away during the deposition of the outer shell in a process known as galvanic replacement. This is a process whereby a metal is deposited as a result of the reduction of relevant ions by another metal with a lower reduction potential. Silver is a choice template material for this as it easily forms a whole range of anisotropic shapes.

The outer shell material is usually gold but can be another noble metal such at palladium or platinum. In the case of gold and silver, $AuCl_4^-$ ions oxidize the silver and are reduced to $Au^0$ during the gold deposition (the standard reduction potential of the $AuCl_4^-/Au$ pair is 0.99 V versus standard hydrogen electrode whereas that of the $Ag^+/Ag$ pair is 0.80 V versus standard hydrogen electrode). This results in a Au/Ag alloy shell with a hollow interior. Dealloying of the shell to gold by further oxidation of silver takes place upon addition of increased quantities of $AuCl_4^-$.

Galvanic replacement reactions with small silver nanoparticles have been studied in detail by Lu *et al.* [132]. They found that a complete gold shell did not form on the surface of each individual silver nanoparticle template. Instead, the replacement reaction resulted in the formation of alloy nanorings and nanocages from multiply-twinned silver nanoparticles of decahedral or icosahedral shape.

### 11.4 3.1 *Nanoboxes And Nanocages*

There has been considerable success in employing silver nanocubes as sacrificial templates for gold deposition. Coating silver nanocubes with a layer of gold yields hollow, porous, Au/Ag alloy nanoboxes [76,81,133,134]. Additionally, it has proven possible to control the size of the pores. A typical example is shown in Figure 11.57 below [134].

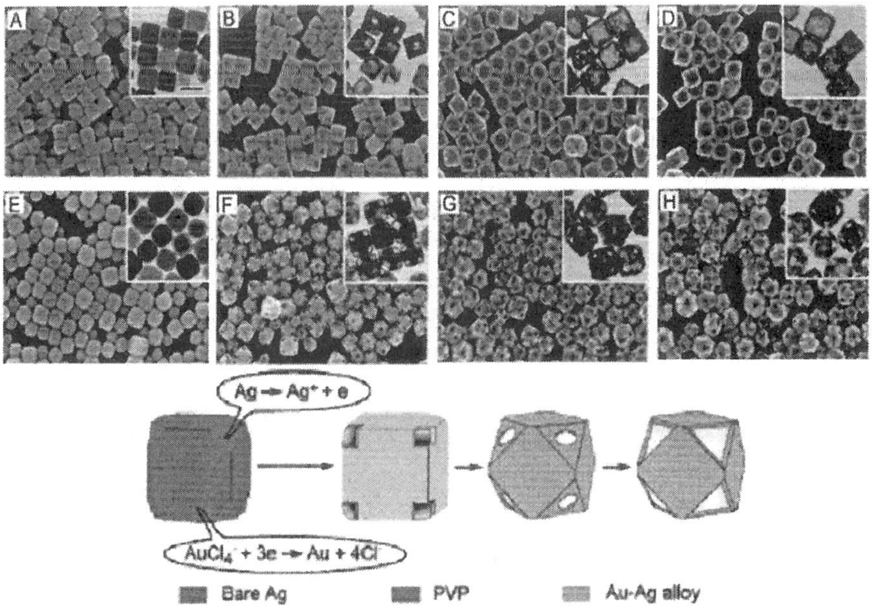


**Figure 11.57:** SEM images of four different stages of galvanic replacement reaction with Ag nanocubes serving as the sacrificial template. (A to D) Ag nanocubes with sharp corners titrated with 0.1 mM $HAuCl_4$, 0, 0.6, 1.6, and 3.0 mL, respectively. (E to H) Ag nanocubes with truncated corners reacted with the same volumes of 0.1 mM $HAuCl_4$ as for the sharp cubes. (Inset) TEM image of each respective sample. Reprinted with permission from reference [134]. © (2006) American Chemical Society.

## 11.4 3.2 *Nanorings*

If silver nanoprisms are used as the scaffold template, then upon addition of $HAuCl_4$ gold nanorings result from the ensuing galvanic replacement [126,135]. A ring forms due to initial deposition of gold at the edge. This indicates that the silver atoms here being more easily oxidized that those on the flat {111} faces. This is consistent with the {111} face being the most stable. The ring grows in thickness as more $HAuCl_4$ is added and silver is oxidized away from the centre. A sample of nanorings is shown in Figure 11.58 below [135].

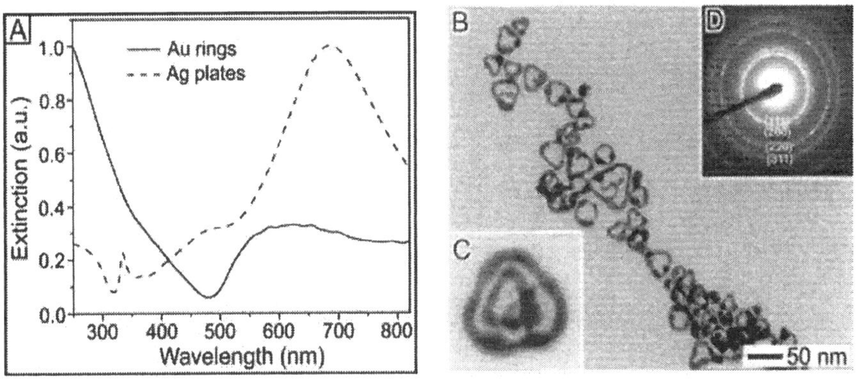


**Figure 11.58:** A) Extinction spectra of silver nanoprisms before and after they had reacted with an aqueous solution of $HAuCl_4$. B) TEM image of the gold product, clearly indicating the formation of ring-like nanostructures. C) TEM image of two coaxial rings, further confirming the existence of ring-like morphology. D) A SAED pattern obtained from a random assembly of gold nanorings. Reproduced with permission from reference [135]. © (2003) Wiley-VCH Verlag GmbH & Co. KGaA.

## 11.5 CONCLUSIONS

The survey of methods for the synthesis of highly shaped gold and silver nanoparticles presented here shows that an enormous variety of shapes are possible and that experimental approaches for their synthesis are now well developed. The most important factor in determining nanoparticle shape is the defect structure of the seed nanoparticles. In addition, coordinating organic species are generally required to maintain a stable dispersion of nanoparticles during growth. In many cases these organic species may selectively bind to more unstable crystal faces to direct crystal growth onto more stable crystal faces to generate structures such as nanorods, nanowires, nanocubes and bipyramids. In the absence of such selective binding, the anisotropy of nanoparticle growth can in many cases be essentially reversed with preferred growth taking place on the less stable crystal faces giving rise to shapes such as decahedra and octahedra. The flat triangular shape of nanoprisms arises from a lamellar defect structure that is present in the seed nanoparticles.

The many experiments carried out so far have shown that MEF is a phenomenon that shows promise as the basis of a sensing technology. Furthermore, we have gained important experimental insight into many of the parameters such as nanoparticle shape and size, and distance between fluorophore and nanoparticle surface that determine the degree of fluorescence enhancement or quenching. It has been shown that to maximize fluorescence enhancement, we generally need larger nanoparticles that scatter more rather than smaller nanoparticles. It is also desirable that we utilize metal nanoparticles that are very highly shaped, i.e. nanorods or nanoprisms. As far as the spectral position of the SPR is concerned, tuning the SPR of the metal nanoparticles to the emission of the fluorophore is necessary to maximize the fluorescence enhancement.

Nevertheless, research into this phenomenon is still very much in its early stages and much more work needs to be done to optimize all the relevant parameters for MEF to display the level of sensitivity required for it to be a viable basis for a biosensing technique. For MEF the challenge lies less in the development of further syntheses of anisotropic metal nanoparticles and more in the development of methodologies that will permit a much higher degree of control over important parameters such as the distance between the fluorophore and the nanoparticle surface; the position of fluorophores with respect to shaped features of the nanoparticles, e.g. nanoprism tips; control over the assembly of nanoparticles into higher order structures. Future work in this regard will need to focus on the continued development and application of advanced coating technologies for nanoparticles and perhaps take advantage of robust supramolecular recognition motifs such as those provided by oligonucleotides of DNA.

## 11.6 REFERENCES


1. Kelly, K. L., Coronado, E., Zhao, L. L. and Schatz, G. C. (2003). The optical properties of metal nanoparticles: The influence of size, shape, and dielectric environment. *J. Phys. Chem. B* 107:668-677.
2. Shuford, K. L., Ratner, M. A. and Schatz, G. C. (2005). Multipolar excitation in triangular nanoprisms. *J. Chem. Phys.* 123:114713.
3. Wiley, B. J., Im, S. H., Li, Z.-Y., McLellan, J., Siekkinen, A. and Xia, Y. (2006). Maneuvering the surface plasmon resonance of silver nanostructures through shape-controlled synthesis. *J. Phys. Chem. B* 110:15666-15675.
4. Liz-Marzán, L. M. (2006). Tailoring surface plasmons through the morphology and assembly of metal nanoparticles. *Langmuir* 22:32-41.
5. Hao, E. and Schatz, G. C. (2004). Electromagnetic fields around silver nanoparticles and dimers. *J. Chem. Phys.* 120:357-366.
6. Tam, F., Goodrich, G. P., Johnson, B. R. and Halas, N. J. (2007). Plasmonic enhancement of molecular fluorescence. *Nano Lett.* 7:496-501.
7. Gersten, J. and Nitzan, A. (1981). Spectroscopic properties of molecules interacting with small dielectric particles. *J. Chem. Phys.* 75:1139-1152.
8. Lakowicz, J. R. (2005). Radiative decay engineering 5: Metal-enhanced fluorescence and plasmon emission. *Anal. Biochem.* 337:171-194.
9. Lakowicz, J. R. (2001). Radiative decay engineering: biophysical and biomedical applications. *Anal. Biochem.* 298:1-24.
10. Geddes, C. D. and Lakowicz, J. R. (2002). Editorial: Metal-enhanced fluorescence. *J. Fluorescence* 12:121-129.
11. Zhang, Y., Aslan, K., Previte, M. J. R. and Geddes, C. D. (2007). Metal-enhanced fluorescence: Surface plasmons can radiate a fluorophore's structured emission. *Appl. Phys. Lett.* 90:053107.
12. Aslan, K., Leonenko, Z., Lakowicz, J. R. and Geddes, C. D. (2005). Annealed silver-island films for applications in metal-enhanced fluorescence: Interpretation in terms of radiating plasmons. *J. Fluorescence* 15:643-654.
13. Whelan, A. M., Brennan, M. E., Blau, W. J. and Kelly, J. M. (2004). Enhanced third-order optical nonlinearity of silver nanoparticles with a tunable surface plasmon resonance. *J. Nanosci. Nanotechnol.* 4:66-68.

14. Zou, X. Q. and Dong, S. J. (2006). Surface-enhanced Raman scattering studies on aggregated silver nanoplates in aqueous solution. *J. Phys. Chem. B* 110:21545-21550.
15. Bell, S. E. J. and Sirimuthu, N. M. S. (2006). Surface-enhanced Raman spectroscopy (SERS) for sub-micromolar detection of DNA/RNA mononucleotides. *J. Am. Chem. Soc.* 128:15580-15581.
16. Morones, J. R., Elechiguerra, J. L., Camacho, A., Holt, K., Kouri, J. B., Ramirez, J. T. and Yacaman, M. J. (2005). The bactericidal effect of silver nanoparticles. *Nanotechnology* 16:2346-2353.
17. Panáček, A., Kvítek, L., Prucek, R., Kolář, M., Večeřová, R., Pizúrová, N., Sharma, V. K., Nevěčná, T. and Zbořil, R. (2006). Silver colloid nanoparticles: Synthesis, characterization, and their antibacterial activity. *J. Phys. Chem. B* 110:16248-16253.
18. Pal, S., Tak, Y. K. and Song, J. M. (2007). Does the antibacterial activity of silver nanoparticles depend on the shape of the nanoparticle? A study of the gram-negative bacterium Escherichia coli. *Appl. Environ. Microbiol.* 73:1712-1720.
19. Anger, P., Bharadwaj, P. and Novotny, L. (2006). Enhancement and quenching of single-molecule fluorescence. *Phys. Rev. Lett.* 96:113002.
20. Dulkeith, E., Morteani, A. C., Niedereichholz, T., Klar, T. A., Feldmann, J., Levi, S. A., van Veggel, F., Reinhoudt, D. N., Moller, M. and Gittins, D. I. (2002). Fluorescence quenching of dye molecules near gold nanoparticles: Radiative and nonradiative effects. *Phys. Rev. Lett.* 89:203002.
21. Schneider, G., Decher, G., Nerambourg, N., Praho, R., Werts, M. H. V. and Blanchard-Desce, M. (2006). Distance-dependent fluorescence quenching on gold nanoparticles ensheathed with layer-by-layer assembled polyelectrolytes. *Nano Lett.* 6:530-536.
22. Seelig, J., Leslie, K., Renn, A., Kuhn, S., Jacobsen, V., van de Corput, M., Wyman, C. and Sandoghdar, V. (2007). Nanoparticle-induced fluorescence lifetime modification as nanoscopic ruler: Demonstration at the single molecule level. *Nano Lett.* 7:685-689.
23. Bharadwaj, P., Anger, P. and Novotny, L. (2007). Nanoplasmonic enhancement of single-molecule fluorescence. *Nanotechnology* 18:044017.
24. Aslan, K., Wu, M., Lakowicz, J. R. and Geddes, C. D. (2007). Fluorescent core-shell Ag@$SiO_2$ nanocomposites for metal-enhanced fluorescence and single nanoparticle sensing platforms. *J. Am. Chem. Soc.* 129:1524-1525.
25. Chen, Y., Munechika, K. and Ginger, D. S. (2007). Dependence of fluorescence intensity on the spectral overlap between fluorophores and plasmon resonant single silver nanoparticles. *Nano Lett.* 7:690-696.
26. Ray, K., Badugu, R. and Lakowicz, J. R. (2007). Sulforhodamine adsorbed Langmuir-Blodgett layers on silver island films: Effect of probe distance on the metal-enhanced fluorescence. *J. Phys. Chem. C* 111:7091-7097.
27. Wu, M., Lakowicz, J. R. and Geddes, C. D. (2005). Enhanced lanthanide luminescence using silver nanostructures: opportunities for a new class of probes with exceptional spectral characteristics. *J. Fluorescence* 15:53-59.
28. Zhang, J., Malicka, J., Gryczynski, I. and Lakowicz, J. R. (2004). Oligonucleotide-displaced organic monolayer-protected silver nanoparticles and enhanced luminescence of their salted aggregates. *Anal. Biochem.* 330:81-86.

29. Zhang, J., Malicka, J., Gryczynski, I. and Lakowicz, J. R. (2005). Surface-enhanced fluorescence of fluorescein-labeled oligonucleotides capped on silver nanoparticles. *J. Phys. Chem. B* 109:7643-7648.
30. Komarala, V. K., Rakovich, Y. P., Bradley, A. L., Byrne, S. J., Gun'ko, Y. K., Gaponik, N. and Eychmuller, A. (2006). Off-resonance surface plasmon enhanced spontaneous emission from CdTe quantum dots. *Appl. Phys. Lett.* 89:253118.
31. Prescott, S. W. and Mulvaney, P. (2006). Gold nanorod extinction spectra. *J. Appl. Phys.* 99:123504.
32. Wokaun, A., Gordon, J. P. and Liao, P. F. (1982). Radiation damping in surface-enhanced Raman-scattering. *Phys. Rev. Lett.* 48:957-960.
33. Klar, T., Perner, M., Grosse, S., von Plessen, G., Spirkl, W. and Feldmann, J. (1998). Surface-plasmon resonances in single metallic nanoparticles. *Phys. Rev. Lett.* 80:4249-4252.
34. Sonnichsen, C., Franzl, T., Wilk, T., von Plessen, G., Feldmann, J., Wilson, O. and Mulvaney, P. (2002). Drastic reduction of plasmon damping in gold nanorods. *Phys. Rev. Lett.* 88:077402.
35. Link, S. and El-Sayed, M. A. (2003). Optical properties and ultrafast dynamics of metallic nanocrystals. *Annu. Rev. Phys. Chem.* 54:331-366.
36. Melikyan, A. and Minassian, H. (2004). On surface plasmon damping in metallic nanoparticles. *Appl. Phys. B: Laser Opt.* 78:453-455.
37. Munechika, K., Smith, J. M., Chen, Y. and Ginger, D. S. (2007). Plasmon line widths of single silver nanoprisms as a function of particle size and plasmon peak position. *J. Phys. Chem. C* 111:18906-18911.
38. Coronado, E. A. and Schatz, G. C. (2003). Surface plasmon broadening for arbitrary shape nanoparticles: A geometrical probability approach. *J. Chem. Phys.* 119:3926-3934.
39. Jana, N. R., Gearheart, L. and Murphy, C. J. (2001). Wet chemical synthesis of high aspect ratio cylindrical gold nanorods. *J. Phys. Chem. B* 105:4065-4067.
40. Jana, N. R., Gearheart, L. and Murphy, C. J. (2001). Seed-mediated growth approach for shape-controlled synthesis of spheroidal and rod-like gold nanoparticles using a surfactant template. *Adv. Mater.* 13:1389-1393.
41. Murphy, C. J. and Jana, N. R. (2002). Controlling the aspect ratio of inorganic nanorods and nanowires. *Adv. Mater.* 14:80-82.
42. Sun, Y., Gates, B., Mayers, B. and Xia, Y. (2002). Crystalline silver nanowires by soft solution processing. *Nano Lett.* 2:165-168.
43. Murphy, C. J., Gole, A. M., Hunyadi, S. E. and Orendorff, C. J. (2006). One-dimensional colloidal gold and silver nanostructures. *Inorg. Chem.* 45:7544-7554.
44. Grochola, G., Snook, I. K. and Russo, S. P. (2007). Computational modeling of nanorod growth. *J. Chem. Phys.* 127:194707.
45. Tao, A. R., Habas, S. and Yang, P. (2008). Shape control of colloidal metal nanocrystals. *Small* 4:310-325.
46. Faraday, M. (1857). Experimental relations of gold (and other metals) to light. *Philos. Trans. R. Soc. London* 147:145-181.
47. Barnard, A. S., Lin, X. M. and Curtiss, L. A. (2005). Equilibrium morphology of face-centered cubic gold nanoparticles >3 nm and the shape changes induced by temperature. *J. Phys. Chem. B* 109:24465-24472.

48. Wang, Z. L., Gao, R. P., Nikoobakht, B. and El-Sayed, M. A. (2000). Surface reconstruction of the unstable {110} surface in gold nanorods. *J. Phys. Chem. B* 104:5417-5420.
49. Perez-Juste, J., Pastoriza-Santos, I., Liz-Marzan, L. M. and Mulvaney, P. (2005). Gold nanorods: Synthesis, characterization and applications. *Coordin. Chem. Rev.* 249:1870-1901.
50. Wang, Z. L., Mohamed, M. B., Link, S. and El-Sayed, M. A. (1999). Crystallographic facets and shapes of gold nanorods of different aspect ratios. *Surf. Sci.* 440:L809-L814.
51. Kou, X., Zhang, S., Tsung, C.-K., Yang, Z., Yeung, M. H., Stucky, G. D., Sun, L., Wang, J. and Yan, C. (2007). One-step synthesis of large-aspect-ratio single-crystalline gold nanorods by using CTPAB and CTBAB surfactants. *Chem. Eur. J.* 13:2929-2936.
52. Liu, M. Z. and Guyot-Sionnest, P. (2005). Mechanism of silver(I)-assisted growth of gold nanorods and bipyramids. *J. Phys. Chem. B* 109:22192-22200.
53. Nepijko, S. A., Ievlev, D. N., Schulze, W., Urban, J. and Ertl, G. (2000). Growth of rodlike silver nanoparticles by vapor deposition of small clusters. *ChemPhysChem* 1:140-142.
54. Johnson, C. J., Dujardin, E., Davis, S. A., Murphy, C. J. and Mann, S. (2002). Growth and form of gold nanorods prepared by seed-mediated, surfactant-directed synthesis. *J. Mater. Chem.* 12:1765-1770.
55. Sun, Y. G., Mayers, B., Herricks, T. and Xia, Y. (2003). Polyol synthesis of uniform silver nanowires: A plausible growth mechanism and the supporting evidence. *Nano Lett.* 3:955-960.
56. Sun, Y., Yin, Y., Mayers, B. T., Herricks, T. and Xia, Y. (2002). Uniform silver nanowires synthesis by reducing $AgNO_3$ with ethylene glycol in the presence of seeds and poly(vinyl pyrrolidone). *Chem. Mater.* 14:4736-4745.
57. Sun, Y. and Xia, Y. (2002). Large-scale synthesis of uniform silver nanowires through a soft, self-seeding, polyol process. *Adv. Mater.* 14:833-837.
58. Lofton, C. and Sigmund, W. (2005). Mechanisms controlling crystal habits of gold and silver colloids. *Adv. Funct. Mater.* 15:1197-1208.
59. Wiley, B. J., Wang, Z., Wei, J., Yin, Y., Cobden, D. H. and Xia, Y. (2006). Synthesis and electrical characterization of silver nanobeams. *Nano Lett.* 6:2273-2278.
60. Wiley, B. J., Chen, Y., McLellan, J. M., Xiong, Y., Li, Z.-Y., Ginger, D. and Xia, Y. (2007). Synthesis and optical properties of silver nanobars and nanorice. *Nano Lett.* 7:1032-1036.
61. Busbee, B. D., Obare, S. O. and Murphy, C. J. (2003). An improved synthesis of high-aspect-ratio gold nanorods. *Adv. Mater.* 15:414-416.
62. Gao, J. X., Bender, C. M. and Murphy, C. J. (2003). Dependence of the gold nanorod aspect ratio on the nature of the directing surfactant in aqueous solution. *Langmuir* 19:9065-9070.
63. Murphy, C. J., San, T. K., Gole, A. M., Orendorff, C. J., Gao, J. X., Gou, L., Hunyadi, S. E. and Li, T. (2005). Anisotropic metal nanoparticles: Synthesis, assembly, and optical applications. *J. Phys. Chem. B* 109:13857-13870.
64. Gole, A. and Murphy, C. J. (2004). Seed-mediated synthesis of gold nanorods: Role of the size and nature of the seed. *Chem. Mater.* 16:3633-3640.

65. Jana, N. R., Gearheart, L. and Murphy, C. J. (2001). Wet chemical synthesis of silver nanorods and nanowires of controllable aspect ratio. *Chem. Commun.*:617-618.
66. Orendorff, C. J., Gearheart, L., Jana, N. R. and Murphy, C. J. (2006). Aspect ratio dependence on surface enhanced Raman scattering using silver and gold nanorod substrates. *Phys. Chem. Chem. Phys.* 8:165-170.
67. Ni, C., Hassan, P. A. and Kaler, E. W. (2005). Structural characteristics and growth of pentagonal silver nanorods prepared by a surfactant method. *Langmuir* 21:3334-3337.
68. Nikoobakht, B. and El-Sayed, M. A. (2003). Preparation and growth mechanism of gold nanorods (NRs) using seed-mediated growth method. *Chem. Mater.* 15:1957-1962.
69. Sau, T. K. and Murphy, C. J. (2004). Seeded high yield synthesis of short Au nanorods in aqueous solution. *Langmuir* 20:6414-6420.
70. Gou, L. and Murphy, C. J. (2005). Fine-tuning the shape of gold nanorods. *Chem. Mater.* 17:3668-3672.
71. Wu, H. Y., Huang, W. L. and Huang, M. H. (2007). Direct high-yield synthesis of high aspect ratio gold nanorods. *Cryst. Growth Des.* 7:831-835.
72. Jiang, X. C., Brioude, A. and Pileni, M. P. (2006). Gold nanorods: Limitations on their synthesis and optical properties. *Colloid Surface A* 277:201-206.
73. Durr, N. J., Larson, T., Smith, D. K., Korgel, B. A., Sokolov, K. and Ben-Yakar, A. (2007). Two-photon luminescence imaging of cancer cells using molecularly targeted gold nanorods. *Nano Lett.* 7:941-945.
74. Smith, D. K. and Korgel, B. A. (2008). The importance of the CTAB surfactant on the colloidal seed-mediated synthesis of gold nanorods. *Langmuir* 24:644-649.
75. Ha, T. H., Koo, H.-J. and Chung, B. H. (2007). Shape-controlled syntheses of gold nanoprisms and nanorods influenced by specific adsorption of halide ions. *J. Phys. Chem. C* 111:1123-1130.
76. Sun, Y. and Xia, Y. (2002). Shape-controlled synthesis of gold and silver nanoparticles. *Science* 298:2176-2179.
77. Wiley, B., Herricks, T., Sun, Y. and Xia, Y. (2004). Polyol synthesis of silver nanoparticles: Use of chloride and oxygen to promote the formation of single-crystal, truncated cubes and tetrahedrons. *Nano Lett.* 4:1733-1739.
78. Wiley, B., Sun, Y. and Xia, Y. (2005). Polyol synthesis of silver nanostructures: Control of product morphology with Fe(II) or Fe(III) species. *Langmuir* 21:8077-8080.
79. Im, S. H., Lee, Y. T., Wiley, B. and Xia, Y. (2005). Large-scale synthesis of silver nanocubes: The role of HCl in promoting cube perfection and monodispersity. *Angew. Chem., Int. Ed.* 44:2154-2157.
80. Siekkinen, A. R., McLellan, J. M., Chen, J. Y. and Xia, Y. (2006). Rapid synthesis of small silver nanocubes by mediating polyol reduction with a trace amount of sodium sulfide or sodium hydrosulfide. *Chem. Phys. Lett.* 432:491-496.
81. Skrabalak, S. E., Au, L., Li, X. and Xia, Y. (2007). Facile synthesis of Ag nanocubes and Au nanocages. *Nat. Protocols* 2:2182-2190.
82. Kim, F., Connor, S., Song, H., Kuykendall, T. and Yang, P. (2004). Platonic gold nanocrystals. *Angew. Chem., Int. Ed.* 43:3673-3677.

83. Sau, T. K. and Murphy, C. J. (2004). Room temperature, high-yield synthesis of multiple shapes of gold nanoparticles in aqueous solution. *J. Am. Chem. Soc.* 126:8648-8649.
84. Hernandez, J., Solla-Gullon, J., Herrero, E., Aldaz, A. and Feliu, J. M. (2007). Electrochemistry of shape-controlled catalysts: Oxygen reduction reaction on cubic gold nanoparticles. *J. Phys. Chem. C* 111:14078-14083.
85. Yu, D. and Yam, V. W. W. (2004). Controlled synthesis of monodisperse silver nanocubes in water. *J. Am. Chem. Soc.* 126:13200-13201.
86. Huang, C.-J., Chiu, P.-H., Wang, Y.-H., Chen, W. R. and Meen, T. H. (2006). Synthesis of the gold nanocubes by electrochemical technique. *J. Electrochem. Soc.* 153:D129-D133.
87. Huang, C.-J., Wang, Y.-H., Chiu, P.-H., Shih, M.-C. and Meen, T.-H. (2006). Electrochemical synthesis of gold nanocubes. *Mater. Lett.* 60:1896-1900.
88. Wiley, B. J., Xiong, Y., Li, Z.-Y., Yin, Y. and Xia, Y. (2006). Right bipyramids of silver: A new shape derived from single twinned seeds. *Nano Lett.* 6:765-768.
89. Li, C., Shuford, K. L., Park, Q.-H., Cai, W., Li, Y., Lee, E. J. and Cho, S. O. (2007). High-yield synthesis of single-crystalline gold nano-octahedra. *Angew. Chem., Int. Ed.* 46:3264-3268.
90. Herricks, T., Chen, J. and Xia, Y. (2004). Polyol synthesis of platinum nanoparticles: Control of morphology with sodium nitrate. *Nano Lett.* 4:2367-2371.
91. Tao, A., Sinsermsuksakul, P. and Yang, P. (2006). Polyhedral silver nanocrystals with distinct scattering signatures. *Angew. Chem., Int. Ed.* 45:4597-4601.
92. Sánchez-Iglesias, A., Pastoriza-Santos, I., Pérez-Juste, J., Rodríguez-González, B., García de Abajo, F. J. and Liz-Marzán, L. M. (2006). Synthesis and optical properties of gold nanodecahedra with size control. *Adv. Mater.* 18:2529-2534.
93. Giersig, M., Pastoriza-Santos, I. and Liz-Marzán, L. M. (2004). Evidence of an aggregative mechanism during the formation of silver nanowires in N,N-dimethylformamide. *J. Mater. Chem.* 14:607-610.
94. Chen, Y., Gu, X., Nie, C.-G., Jiang, Z.-Y., Xie, Z.-X. and Lin, C.-J. (2005). Shape controlled growth of gold nanoparticles by a solution synthesis. *Chem. Commun.*:4181-4183.
95. Seo, D., Yoo, C. I., Chung, I. S., Park, S. M., Ryu, S. and Song, H. (2008). Shape adjustment between multiply twinned and single-crystalline polyhedral gold nanocrystals: Decahedra, icosahedra, and truncated tetrahedra. *J. Phys. Chem. C* 112:2469-2475.
96. Pastoriza-Santos, I., Sánchez-Iglesias, A., García de Abajo, F. J. and Liz-Marzán, L. M. (2007). Environmental optical sensitivity of gold nanodecahedra. *Adv. Funct. Mater.* 17:1443-1450.
97. Tsuji, M., Miyamae, N., Lim, S., Kimura, K., Zhang, X., Hikino, S. and Nishio, M. (2006). Crystal structures and growth mechanisms of Au@Ag core-shell nanoparticles prepared by the microwave-polyol method. *Cryst. Growth Des.* 6:1801-1807.
98. Chen, S., Wang, Z. L., Ballato, J., Foulger, S. H. and Carroll, D. L. (2003). Monopod, bipod, tripod, and tetrapod gold nanocrystals. *J. Am. Chem. Soc.* 125:16186-16187.

99. Kumar, P. S., Pastoriza-Santos, I., Rodríguez-González, B., García de Abajo, F. J. and Liz Marzán, L. M. (2008). High-yield synthesis and optical response of gold nanostars. *Nanotechnology* 19:015606.
100. Tang, X. L., Jiang, P., Ge, G. L., Tsuji, M., Xie, S. S. and Guo, Y. J. (2008). Poly(N-vinyl-2-pyrrolidone) (PVP)-capped dendritic gold nanoparticles by a one-step hydrothermal route and their high SERS effect. *Langmuir* 24:1763-1768.
101. Witten, T. A. and Sander, L. M. (1981). Diffusion-limited aggregation, a kinetic critical phenomenon. *Phys. Rev. Lett.* 47:1400-1403.
102. Bastys, V., Pastoriza-Santos, I., Rodríguez-Gonzáles, B., Vaisnoras, R. and Liz-Marzán, L. (2006). Formation of silver nanoprisms with surface plasmons at communication wavelengths. *Adv. Funct. Mater.* 16:766-773.
103. Germain, V., Li, J., Ingert, D., Wang, Z. L. and Pileni, M. P. (2003). Stacking faults in formation of silver nanodisks. *J. Phys. Chem. B* 107:8717-8720.
104. Elechiguerra, J. L., Reyes-Gasga, J. and Yacaman, M. J. (2006). The role of twinning in shape evolution of anisotropic noble metal nanostructures. *J. Mater. Chem.* 16:3906-3919.
105. Rodríguez-González, B., Pastoriza-Santos, I. and Liz-Marzán, L. M. (2006). Bending contours in silver nanoprisms. *J. Phys. Chem. B* 110:11796-11799.
106. Yang, Y., Matsubara, S., Xiong, L., Hayakawa, T. and Nogami, M. (2007). Solvothermal synthesis of multiple shapes of silver nanoparticles and their SERS properties. *J. Phys. Chem. C* 111:9095-9104.
107. Rocha, T. C. R. and Zanchet, D. (2007). Structural defects and their role in the growth of Ag triangular nanoplates. *J. Phys. Chem. C* 111:6989-6993.
108. Jin, R. C., Cao, Y. W., Mirkin, C. A., Kelly, K. L., Schatz, G. C. and Zheng, J. G. (2001). Photoinduced conversion of silver nanospheres to nanoprisms. *Science* 294:1901-1903.
109. Jin, R., Cao, C., Hao, E., Métraux, G. S., Schatz, G. C. and Mirkin, C. A. (2003). Controlling anisotropic nanoparticle growth through plasmon excitation. *Nature* 425:487-490.
110. Xue, C. and Mirkin, C. A. (2007). pH-switchable silver nanoprism growth pathways. *Angew. Chem., Int. Ed.* 46:2036-2038.
111. Callegari, A., Tonti, D. and Chergui, M. (2003). Photochemically grown silver nanoparticles with wavelength-controlled size and shape. *Nano Lett.* 3:1565-1568.
112. Pastoriza-Santos, I. and Liz-Marzán, L. M. (2002). Synthesis of silver nanoprisms in DMF. *Nano Lett.* 2:903-905.
113. Washio, I., Xiong, Y., Yin, Y. and Xia, Y. (2006). Reduction by the end groups of poly(vinyl pyrrolidone): A new and versatile route to the kinetically controlled synthesis of Ag triangular nanoplates. *Adv. Mater.* 18:1745-1749.
114. Xiong, Y., Washio, I., Chen, J., Sadilek, M. and Xia, Y. (2007). Trimeric clusters of silver in aqueous $AgNO_3$ solutions and their role as nuclei in forming triangular nanoplates of silver. *Angew. Chem., Int. Ed.* 46:4917-4921.
115. Hoppe, C. E., Lazzari, M., Pardinas-Blanco, I. and Lopez-Quintela, M. A. (2006). One-step synthesis of gold and silver hydrosols using poly(N-vinyl-2-pyrrolidone) as a reducing agent. *Langmuir* 22:7027-7034.
116. Metraux, G. S. and Mirkin, C. A. (2005). Rapid thermal synthesis of silver nanoprisms with chemically tailorable thickness. *Adv. Mater.* 17:412-415.

117. Sun, Y., Mayers, B. and Xia, Y. (2003). Transformation of silver nanospheres into nanobelts and triangular nanoplates through a thermal process. *Nano Lett.* 3:675-679.
118. Chen, S. H., Fan, Z. Y. and Carroll, D. L. (2002). Silver nanodisks: Synthesis, characterization, and self-assembly. *J. Phys. Chem. B* 106:10777-10781.
119. Ledwith, D. M., Whelan, A. M. and Kelly, J. M. (2007). A rapid, straight-forward method for controlling the morphology of stable silver nanoparticles. *J. Mater. Chem.* 17:2459-2464.
120. Xiong, Y., Washio, I., Chen, J., Cai, H., Li, Z.-Y. and Xia, Y. (2006). Poly(vinyl pyrrolidone): A dual functional reductant and stabilizer for the facile synthesis of noble metal nanoplates in aqueous solutions. *Langmuir* 22:8563-8570.
121. Chen, S. H. and Carroll, D. L. (2002). Synthesis and characterization of truncated triangular silver nanoplates. *Nano Lett.* 2:1003-1007.
122. Chen, S. and Carroll, D. L. (2004). Silver nanoplates: size control in two dimensions and formation mechanisms. *J. Phys. Chem. B* 108:5500-5506.
123. Millstone, J. E., Park, S., Shuford, K. L., Qin, L. D., Schatz, G. C. and Mirkin, C. A. (2005). Observation of a quadrupole plasmon mode for a colloidal solution of gold nanoprisms. *J. Am. Chem. Soc.* 127:5312-5313.
124. Millstone, J. E., Metraux, G. S. and Mirkin, C. A. (2006). Controlling the edge length of gold nanoprisms via a seed-mediated approach. *Adv. Funct. Mater.* 16:1209-1214.
125. Zou, X., Ying, E., Chen, H. and Dong, S. (2007). An approach for synthesizing nanometer- to micrometer-sized silver nanoplates. *Colloid Surface A* 303:226-234.
126. Jiang, X. C., Zeng, Q. H. and Yu, A. B. (2006). A self-seeding coreduction method for shape control of silver nanoplates. *Nanotechnology* 17:4929-4935.
127. Jiang, X. C., Zeng, Q. H. and Yu, A. B. (2007). Thiol-frozen shape evolution of triangular silver nanoplates. *Langmuir* 23:2218-2223.
128. Aherne, D., Ledwith, D. M., Gara, M. and Kelly, J. M. (2008). Optical properties and growth aspects of silver nanoprisms produced by a highly reproducible and rapid synthesis at room temperature. *Adv. Funct. Mater.* 18:2005-2016.
129. Rocha, T. C. R. and Zanchet, D. (2007). Growth aspects of photochemically synthesized silver triangular nanoplates. *J. Nanosci. Nanotechnol.* 7:618-625.
130. Rocha, T. C. R., Winnischofer, H., Westphal, E. and Zanchet, D. (2007). Formation kinetics of silver triangular nanoplates. *J. Phys. Chem. C* 111:2885-2891.
131. Xue, C., Millstone, J. E., Li, S. Y. and Mirkin, C. A. (2007). Plasmon-driven synthesis of triangular core-shell nanoprisms from gold seeds. *Angew. Chem., Int. Ed.* 46:8436-8439.
132. Lu, X., Tuan, H. Y., Chen, J., Li, Z. Y., Korgel, B. A. and Xia, Y. (2007). Mechanistic studies on the galvanic replacement reaction between multiply twinned particles of Ag and $HAuCl_4$ in an organic medium. *J. Am. Chem. Soc.* 129:1733-1742.
133. Chen, J., Wiley, B., Li, Z.-Y., Campbell, D., Saeki, F., Cang, H., Au, L., Lee, J., Li, X. and Xia, Y. (2005). Gold nanocages: Engineering their structure for biomedical applications. *Adv. Mater.* 17:2255-2261.

134. Chen, J., McLellan, J. M., Siekkinen, A., Xiong, Y., Li, Z.-Y. and Xia, Y. (2006). Facile synthesis of gold-silver nanocages with controllable pores on the surface. *J. Am. Chem. Soc.* 128:14776-14777.

135. Sun, Y. and Xia, Y. (2003). Triangular nanoplates of silver: Synthesis, characterization, and use as sacrificial templates for generating triangular nanorings of gold. *Adv. Mater.* 15:695-699.

# 12

# Enhanced Fluorescence Detection Enabled By Zinc Oxide Nanomaterials

Jong-in Hahm
Department of Chemical Engineering, The Pennsylvania State University, 160 Fenske Laboratory, University Park, PA 16802, USA.

## 12.1 INTRODUCTION

Fluorescence is one of the most widely used detection mechanisms in the fields of biology, biophysics, biochemistry, gene profiling, proteomics, drug discovery, disease diagnostics and environmental analysis. Factors that make fluorescence methods the technique of choice are reasonable sensitivity to target components in complex biomolecular assemblies, versatility in accommodating a range of sample types for investigation, and modest instrumentation requirement for signal detection. One of the major challenges, still facing many fluorescence techniques in biomolecular detection, is attaining improved detection sensitivity to target biomolecules while simultaneously reducing background signal [1-4]. Overcoming this challenge will lead to a breakthrough in biology and medicine by advancing key areas such as population-level genetic screening, system-wide study of proteins, and early disease diagnosis. Therefore, novel methods enabling rapid, high-throughput, ultrasensitive, and specific fluorescence detection are in great demand for these burgeoning areas.

## 12.2 ROLE OF NANOMATERIALS IN FLUORESCENCE DETECTION

In order to improve the capability and resolution of fluorescence detection, numerous research efforts have been made on three main aspects:

- Molecular design of better fluorophores,
- Development of improved detection apparatus, and
- Engineering of advanced substrates.

New organic, inorganic, and hybrid labels were developed to prevent photobleaching of fluorescing dyes while allowing measurements of multiple fluorophores with a single excitation source at very low concentration levels [5-10]. The use of metallized substrates has been also explored to increase quantum yield and photostability of fluorophore labels [11-13]. At the same time, the field of

MEF by Zinc Oxide Nanomaterials
Edited by Chris D. Geddes.

fluorescence has undergone a number of improvements in instrumentation, data analysis, and data interpretation in recent years. For these efforts, advanced fluorescence optics and more reliable miniaturized detection devices were developed in order to increase detection sensitivity, in some cases down to the single molecule level [14-19]. Owing to the much higher computing power now available, improvements in software development have led to fast, simple, and more accurate data analysis and interpretation.

Recently, progresses made in the area of nanoscience also joined the efforts to push the field of fluorescence detection further. The most recognizable contribution of nanoscience to the field of biomolecular fluorescence detection so far has been made mainly in the areas of developing new fluorescent probes. A well-known example of this contribution is the design of semiconductor nanocrystals known as quantum dots, whose emission spectrum at a specific wavelength can be tuned by simply changing the physical size of the nanomaterials. Continuing research efforts in these fields have led to improved fluorophores that are less subject to photo bleaching, while displaying high quantum yield [5, 6, 8-10, 20-22]. Despite the benefits of quantum dot-based labels, potential hurdles in their applications for facile biodetection are still being assessed and means for circumventing previously identified challenges of biotoxicity and environmental concerns are being developed currently [23].

In addition to its contribution described above, nanoscience can provide an alternative step forward to promoting biomolecular fluorescence detection. Recent advances in nanoscience permits innovative assembly and fabrication of nanomaterials for use as advanced biosensor substrates in fluorescence detection. Therefore, nanoscience may offer a much simpler and convenient route in promoting biomolecular fluorescence detection, even when using as-grown nanomaterials without any downstream modifications following their synthesis. In order to design such substrates comprised of nanomaterials, four key characteristics of the candidate nanoscale materials should be carefully considered;

i) No spectral overlap,

ii) Ease of fabrication,

iii) Stability and

iv) Surface chemistry.

The presence of candidate nanomaterials should not hinder the absorption and emission processes of fluorophores. Ideal candidate nanomaterials should not display autofluorescence. This requirement is important for reducing background noise in fluorescence detection and permitting signal collection only from the target biomolecules under study. In addition, simple and straightforward synthesis and assembly routes should be established to yield the successful growth and fabrication of these nanomaterials. This criterion is important for increasing the high-throughput capability of biomolecular fluorescence detection, especially when screening a large number of samples. In addition, the candidate nanomaterials should be biocompatible and chemically inert in detection environments involving most biologically relevant assays. Lastly, the surfaces of these nanomaterials should exhibit the chemical reactivity needed for covalent derivatization enabling the linking of specific biomolecules onto the nanomaterials and maximizing the specificity of biomolecular interaction. It will be even more desirable if the candidate materials, while having all the criteria discussed above, exhibit an appropriate optical property to foster the fluorescence signal from fluorophore-linked biomolecules. This property will be critical in promoting biomolecular detection at very low concentrations.

From the view point of nanoscience, many research efforts have been put forward to achieve simplicity in the steps leading from the material's synthesis to final applications of nanomaterials. Due to the small size inherent to nanomaterials, manipulation and assembly processes following their synthesis are much more complicated and time consuming than their bulk counterparts. Therefore, numerous studies focused on novel synthetic methods which allow spatial and orientational control of nanomaterials during their growth process. These efforts will be discussed more in Section 12.4. These approaches enable the use of nanomaterials directly upon their synthesis, without going through complex and costly fabrication steps to assemble nanomaterials into application-ready devices and platforms.

## 12.3 PROPERTIES AND APPLICATIONS OF ZINC OXIDE

Very recently, as-grown nanomaterials have been demonstrated as potentially suitable platforms for much improved fluorescence detection involving a variety of biological systems. High-quality zinc oxide (ZnO) nanomaterials are used in these applications. This chapter highlights the newly demonstrated, promising roles of metal oxide nanomaterials in fluorescence detection. The following sections will outline the desired properties of ZnO nanomaterials, in-place synthesis methods of ZnO nanomaterials, and their as-synthesized applications in biomolecular fluorescence detection. Possible fluorescence enhancement mechanisms pertaining to ZnO nanomaterial platforms will be also discussed.

Existing applications of ZnO are very diverse, ranging from rubber production, to food additives, to pigmentary components, to cosmetic ingredients, to medical products. ZnO used in rubber goods such as car tires promotes effective dissipation of heat during their manufacturing processes as well as under road-driving conditions. ZnO is added in some food products as a source of zinc which is a nutritive substance. ZnO is also used as a pigment in paints and as a coating material in papers. Lotions and creams, that contain ZnO as an active ingredient, serve as protective elements against ultraviolet (UV) rays. In addition, the antibacterial and antifungal property of ZnO is put to use as topical ointments.

More related to electronic and photonic applications, ZnO materials have received considerable attention in the past particularly due to their desirable optical properties. These properties include a wide, direct band gap of 3.37 eV and a large exciton binding energy of 60 meV at room temperature. The wide band gap of ZnO may permit higher blocking voltages, switching frequencies, operating temperature, efficiency, and reliability of devices. The direct bandgap of ZnO enables more efficient absorption and emission of light. The large exciton binding energy of ZnO permits lower temperature operation of devices, unlike most other semiconductor materials. When compared to other wide band gap materials such as SiC, GaN, and diamond, ZnO shows more promising potential in device and sensor applications. For example, ZnO has higher chemical and thermal resistance, higher electromechanical coupling efficiency, and lower growth temperature. For these reasons, ZnO thin films and micro/nano structures have been previously demonstrated as candidate materials for use in a broad range of optical and optoelectric applications. ZnO applications in photonic and electronic areas include short-wavelength light-emitters [24, 25], field-emitters [26, 27], luminescence [28], and UV lasers [29]. In addition, sensitization of ZnO by organic dye molecules has been extensively studied for use as highly efficient solar cells [30-35]. Sub-wavelength waveguiding property of ZnO nanomaterials has also been demonstrated recently [36-39].

In addition to these attractive optical properties, nanometer scale ZnO is stable in typical biomolecular detection environments and ZnO nanomaterials can be easily synthesized through many established processing routes. ZnO is biocompatible and biosafe, as evidenced by their current applications in food, cosmetic, and medical products. Since ZnO is nontoxic, as-grown ZnO can be used for biomedical applications without adding a protective coating layer. Although these characteristics of ZnO suggest that they may be an ideal candidate material for aiding optical detection of many important bioconstituents, applications of ZnO nanomaterials in biomolecular detection have not been realized until very recently.

## 12.4 SYNTHESIS AND CHARACTERISATION OF ZINC OXIDE NANOMATERIALS FOR BIODETECTION PLATFORMS

For their rich potential in various applications described in the previous section, the synthesis and assembly of various ZnO micro and nanostructures have been extensively explored using both gas-phase and solution-based approaches. The most commonly used gas-phase growth approaches for synthesizing ZnO structures at the nanometer and micrometer scale include physical vapor deposition (40, 41), pulsed laser deposition (42), chemical vapor deposition (43), metal-organic chemical vapor deposition (44), vapor-liquid-solid epitaxial mechanisms (24, 28, 29, 45), and epitaxial electrodeposition (46). In solution-based synthesis approaches, growth methods such as hydrothermal decomposition processes (47, 48) and homogeneous precipitation of ZnO in aqueous solutions (49-51) were pursued.

Depending on specific growth conditions, ZnO exhibits a variety of nano- and micro-structures resembling dots, rods, wires, belts, bows, tubes, bridges, helixes / springs, propellers, combs, seamless rings, and polyhedral cages [52, 53]. Such variations in the morphologies of ZnO are governed by factors such as substrates, carrier gases, local gas concentrations, growth temperature, and growth time. Catalysts also play an important role in their synthesis and a wide variety of catalysts such as Au, Ag, Co, Sn, Ni, NiO, Ge, Al, Zn, and Pt have been previously used for gas-phase growth to produce the rich family of ZnO structures [17, 28, 40, 53-60]. Also recently, biogenic catalysts such as magnetic bacteria have been successfully employed as active catalysts in producing ZnO nanomaterials [61]. Growth substrates for ZnO can be selected from a wide range of materials as well. They include glass, sapphire, diamond, silicon, silicon oxide, and gallium nitride [17, 28, 40, 53-60].

Although such a variety of synthetic methods can be used to produce ZnO nanomaterials, the following section will provide an overview of synthetic procedures to produce ZnO nanomaterials that are further demonstrated for fluorescence detection of biomolecules [61-65]. Specifically, the following section will focus on a gas-phase synthetic route exploiting microcontact-printed catalysts and describe an *in situ* method for producing ZnO nanorod (ZnO NR) platforms in an array format. The physical and optical properties of as-synthesized ZnO NRs will be also discussed.

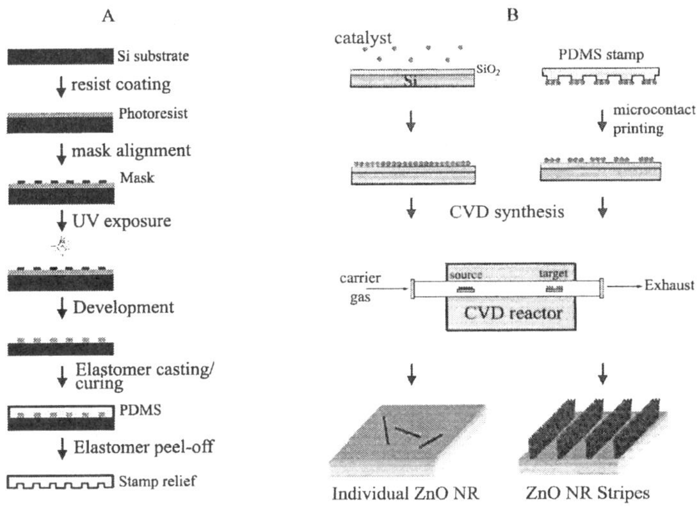


**Figure 12.1:** Schematic illustration showing the overall experimental design of synthesizing ZnO NR arrays that are subsequently used for biomolecular fluorescence detection. (A) A series of steps to produce elastomeric PDMS stamps to print catalyst particles on predetermined locations on Si substrates. (B) Individual and patterned arrays of ZnO NRs assembled on the catalytic sites upon materials' growth in a chemical vapor deposition reactor.

Microcontact printing techniques can facilitate the assembly of nanomaterials where these materials can be produced in a high-density array format at prearranged locations. Microcontact printing is often used to assemble biological, inorganic, and organic materials at predetermined sites on a large area of a substrate, especially involving materials whose sizes are too small to be conveniently handled for easy assembly [66-69]. An elastomeric polymer, polydimethylsiloxane (PDMS), is typically used as a stamping tool in the microcontact printing process, Figure 12.1A. PDMS stamps are constructed by casting and curing the elastomeric polymer against a photoresist-micropatterned silicon master which is prefabricated using standard photolithography procedures. Catalysts effective for growing ZnO nanomaterials can be delivered to predetermined locations of substrates using the microcontact printing method. The typical growth of ZnO NRs is carried out in a chemical vapor deposition reactor where ZnO NRs are grown over the predelivered catalytic areas. Thermally reduced, source materials in the upper stream of the reactor travel down with the help of a carrier gas to the target region, where ZnO NRs are subsequently produced and assembled into predetermined arrays *in situ*.

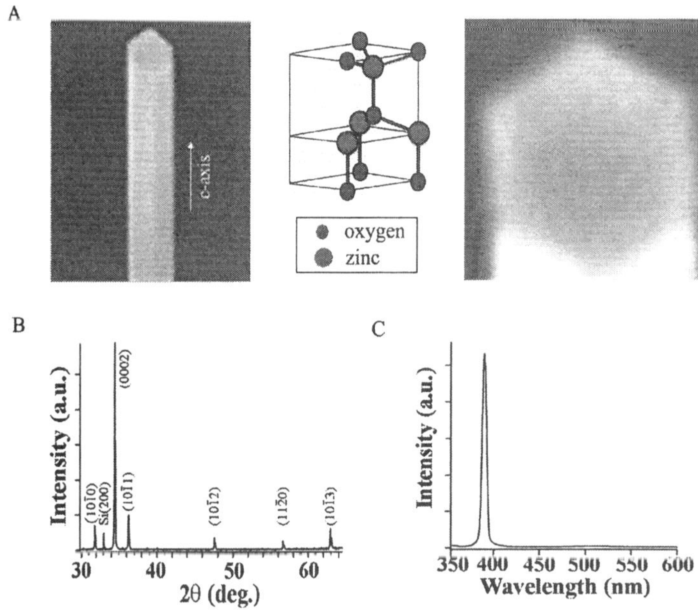


**Figure 12.2:** Characterization of as-grown ZnO NRs. (A) 420 x 1000 nm and 200 x 200 nm scanning electron microscopy images showing the side and end facets of a ZnO NR along the preferential growth axis (c-axis). The width and length of the ZnO NR is 180 nm and 5 μm, respectively. (B) X-ray diffraction data displaying the high crystalline quality of as-synthesized ZnO NRs. (C) Room-temperature photoluminescence spectrum showing the extremely strong and narrow band-edge emission of ZnO NRs at 390 nm.

The quality of as-grown nanomaterials plays a crucial role in their applications. It is imperative that ZnO nanomaterials are produced into a uniform size and shape and that their crystalline structures are not only atomic defect-free but also free of other chemical constituents.

Characterization techniques such as X-ray diffraction (XRD) and photoluminescence (PL) allow further investigation of as-grown ZnO nanomaterials. The XRD data, showing the crystalline structures of the ZnO nanostructures grown on silicon substrates, are presented in Figure 12.2B. The XRD pattern of ZnO NRs indicates that these ZnO materials exhibit wurtzite structures. The pronounced peak at $2\theta = 34.5°$ which corresponds to (0002) facet of wurtzite ZnO specifies <0001> as the preferential growth direction. In the wurtzite arrangement of ZnO crystalline structures, each cation is surrounded by four anions at the corner of a tetrahedron and vice versa, as shown in the cartoon in Figure 12.2A. The [0001] axis points from the face of the oxygen plane to the zinc plane and, therefore, ZnO exhibits crystallographic polarity; Zn polarity and O polarity. As shown in the XRD data in Figure 12.2B, other secondary planes and directions do exist in the wurtzite ZnO crystal structure besides the primary polar plane (0001) and associated direction <0001>.

Optical properties of as-grown ZnO materials can be assessed by conducting photoluminescence (PL) of these ZnO nanostructures. Room-temperature PL spectrum of as-synthesized ZnO nanomaterials in Figure 12.2C shows extremely strong and sharp UV emission at 390 nm. ZnO nanomaterials with atomic defects are known to emit in the visible range. This phenomenon results from radiative recombination of a photo-generated hole with an electron in an oxygen vacancy site and causes undesired deep-level or trap-state emission [70-72]. However, ZnO nanomaterials synthesized by the above method lead to only the UV emission, corresponding to near band-edge emission of the semiconductor ZnO. These atomic defect-free ZnO nanomaterials do not show any absorption and emission in the visible or near-infrared range. The PL spectrum of the ZnO nanomaterials in Figure 12.2C indicates that these as-grown materials are free of any atomic defect sites. These data show that the outlined synthetic procedures can produce ZnO nanostructures of high crystalline and optical quality that can be further applied for biological detection platforms.

The width and diameter of individual NRs are a few hundreds of nanometers and several micrometers, respectively. Various ZnO NR array structures shown in Figure 12.3 are produced by controlling the size and shape of the features engraved on the PDMS stamps. Typical repeat distance of striped and squared patterns is a few to tens of micrometers. The exact dimensions of these ZnO NR platforms can be controlled during their growth by changing the growth and catalyst delivery conditions. As displayed in Figure 12.3, the microcontact printing method is highly efficient for the growth of ZnO nanostructures at predetermined locations over large areas of substrates. Shapes and sizes of the resulting ZnO NR array patterns can be arranged into any user-controlled geometry and dimension, thereby making the synthetic method very versatile.

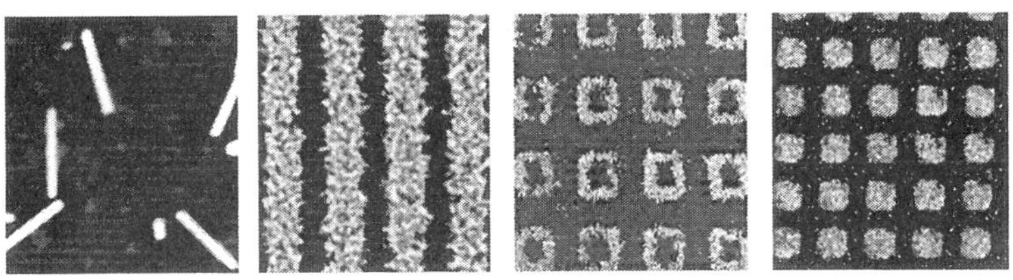

**Figure 12.3:** Scanning electron micrographs of various ZnO NR platforms used for biomolecular fluorescence detection, ranging from individual ZnO NRs as well as striped, open-squared, and filled-squared patterns of ZnO NR arrays. The width of each pattern and the repeat distance between patterns are the same as 20 μm, 10 μm, and 5 μm for the striped, open-squared, and filled-squared ZnO NR arrays, respectively.

## 12.5 FLUORESCENCE ENHANCEMENT EFFECT

The high quality ZnO nanomaterials synthesized using the aforementioned growth routes can meet many important criteria for serving as ideal candidate substrates in biomolecular fluorescence detection, as discussed in Section 12.2. Based on these potentials, the suitability of the ZnO nanomaterials as optical signal enhancing platforms is initially assessed. For this initial evaluation, fluorescence signal is measured from simple biological assays and compared between various platforms [63].

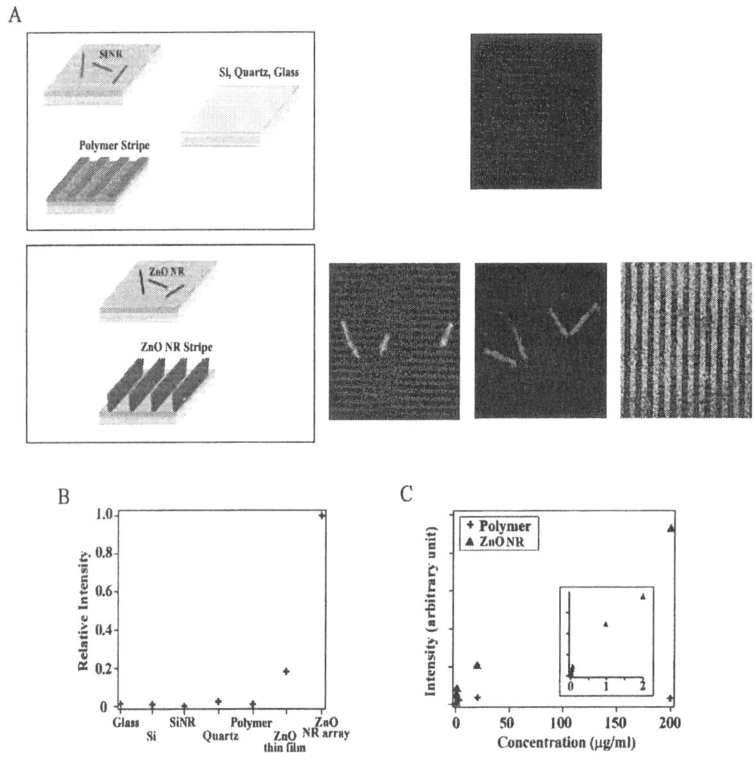


**Figure 12.4:** Fluorescence emission from biomolecules on ZnO NRs versus various control substrates after performing the identical biotreatment processes. (A) No fluorescence signal is detected on control substrates including glass, quartz, silicon oxide, silicon nanorods (SiNRs), and polymeric surfaces. On the other hand, strong fluorescence signal is observed from individual and striped ZnO NR platforms regardless of the spectroscopic properties fluorophores. (B) Normalized fluorescence intensity observed from biomolecules on various substrates. (C) Fluorescence

intensity measured from concentration-dependent assays, showing the detection sensitivity of a ZnO NR platform compared to that of a polymeric platform. Copyright American Chemical Society, Inc. Reproduced with permission.

The test platforms consist of ZnO nanorods, silicon nanorods, and other conventional materials that are often employed as solid supports in biological detection. Commonly used, biodetection substrates include glass, quartz, silicon, and polymers. Polymeric biodetection supports are typically constructed out of polystyrene (PS), polymethylmethacrylate (PMMA), polyethylene, and polycarbonate. In the comparison study, polymeric supports are produced into stripe arrays using nanoimprint lithography so that the array dimension of polymeric substrates is comparable to that found in ZnO NR stripe platforms. The use of nanoimprint lithography allows detection of fluorescence signal from these polymeric supports that mimic the surface geometries of patterned arrays of ZnO NR platforms. Homogeneous thin films of PS and PMMA are also tested as control substrates. Evaluation of additional one-dimensional (1D) nanomaterials other than ZnO NRs permits the investigation of possible contribution from the increased surface area of 1D nanomaterial supports when compared to conventionally used, two-dimensional (2D) supports. Increased surface to volume ratio is inherent to nanoscale materials, due to their reduced dimensionality when compared to their 2D counterparts. Silicon nanorods (SiNRs) are chosen as the additional nanomaterial platform. During SiNR growth, the physical dimensions of SiNRs are controlled to be comparable to those of ZnO NRs. In addition, other conventional 2D supports such as glass and silicon with a native oxide layer are also evaluated. Figure 12.4 displays data comparing fluorescence signal obtained from various platforms when these platforms are exposed to the identical biological reaction processes. Commercially available confocal fluorescence microscope can be effectively used in the signal detection. During the fluorescence measurements, the excitation and detection wavelengths are determined by the specific absorption and emission properties of the fluorophores.

No measurable fluorescence signal is detected from model protein molecules deposited on various control supports when the protein concentration is kept low. However intense fluorescence signal is observed from the detection platforms when these substrates are replaced by various types of ZnO NR platforms while keeping all other assay conditions the same. Many other intriguing observations have been made from these control experiments. In an assay scheme involving nonspecific adsorption of biomolecules, much stronger fluorescence signal is observed from biomolecules on the areas of ZnO NRs whereas negligible signal is detected from the neighboring Si regions. In another assay involving two or three orders of magnitude higher concentrations of biomolecules on the control supports, no fluorescence is readily detected from the control platforms whereas strong fluorescence signal is obtained from ZnO NRs although the concentration of biomolecules is much lower on ZnO NR platforms. Results from direct comparison of fluorescence signal from biomolecules on the two different 1D nanomaterial systems, comparably-sized ZnO NRs versus SiNRs, show extremely strong fluorescence emission from those on ZnO NR platforms. On the contrary, negligible emission is detected from biomolecules on SiNR platforms. In quantitative assays designed to compare normalized fluorescence signals of biomolecules between various platforms, fluorescence signal obtained from ZnO NR supports is at least two or three orders of magnitude higher than that from control supports. When 2D ZnO thin films are used as supports in the fluorescence detection, higher fluorescence signal is detected when compared to all other control supports. However, the signal increase is only about 20% of that detected using 1D ZnO NR platforms. For an assay designed to assess the detection sensitivity on various platforms, proteins show much higher signal on ZnO NR supports than PMMA supports. Therefore the detection sensitivity of the protein on ZnO NRs in this adsorption case is much higher than on PMMA.

These initial results demonstrate that, compared to conventional platforms, application of ZnO NR platforms in fluorescence detection effectively leads to two or three order of magnitude higher emission signal from biomolecules. Therefore, the use of ZnO NR platforms in biomolecular fluorescence assays may facilitate highly sensitive biodetection. Yet, challenging tasks still lie ahead. For these platforms to be useful in more demanding biological assays, they need to be further tested with different types of biomolecules, buffers, and biological fluids. The compatibility and robustness of ZnO NR platforms need to be determined. Throughput and multiplexing capability of the newly identified ZnO NR supports in fluorescence detection should also be carefully assessed, if a widespread utility in biodetection is to be demonstrated. The following sections introduce recent efforts in demonstrating the robustness, compatibility, high sensitivity, multiplexing and high throughput capability of ZnO NR platforms in biomolecular fluorescence detection.

## 12.6 SIMPLE BIOLOGICAL REACTIONS

The enhanced fluorescence detection capability of ZnO NR platforms is further assessed in relatively simple biological assays involving pure DNA and proteins. Unlike the previously described tests involving single layer adsorption of biomolecules, biological assays discussed from here on pertain to interactions between multilayered biomolecules. The following sections describe the results of fluorescence enhancement effect observed in these relatively simple but multilayered bioassays on ZnO NRs.

### 12.6.-1. *DNA Hybridization Reaction*

DNA sequence analysis is widely applied to the areas of mapping genes, determining genetic variations, and detecting genetic diseases. Novel techniques which can perform rapid and accurate genetic sequence analyses on a large scale are specially warranted as the need for fast, inexpensive, ultrasensitive, and high throughput DNA detection escalates in the areas of medicine and public health. Fluorescence detection is the dominant mechanism and extensively utilized in state-of-the-art, miniaturized DNA sensors such as DNA arrays and gene chips [73-79, 80]. The emerging need for high throughput genetic detection will continue to push the limit of fluorescence detection sensitivity. These sequencing assays require the use of lower DNA concentrations as well as smaller amounts of fluorophores in order to cope better with the increasing demands for effectively screening human genes or biological agents at large scale. At the same time, these DNA detection platforms need to eliminate high costs associated with large numbers of samples and biomedical reaction steps. Therefore, advanced techniques are presently warranted in order to enhance the fluorescence detection sensitivity of DNA beyond the limits that current technologies offer.

In the previous section, innovative assembly and fabrication of ZnO nanomaterials for use as advanced bioassay substrates prove to be greatly beneficial in increasing the detection sensitivity of biomolecular fluorescence assays. To further demonstrate the applicability of ZnO nanomaterials in the highly sensitive detection of DNA duplex formation, the overall experimental design shown in Figure 12.5A is used. In this scheme, two different target oligonucleotides are linked to ZnO NRs noncovalently or covalently. Test sample solution containing a probe oligonucleotide, that is preconjugated with a fluorophore and fully complementary to one of the target strands, is subsequently introduced. Formation of duplex DNA resulting from successful hybridization reactions is then monitored by observing fluorescence signal.

As shown in Figure 12.5, the combined use of ZnO NR platforms and a covalent linking scheme allows ultrasensitive genetic sequence detection at DNA concentration levels down to a few femtomolar range [62]. The detection limit of ZnO NR platforms coupled with DNA through a non-covalent linking scheme is in the tens of nanomolar range. The lowest detection limit in these cases is defined by the DNA concentration for which the observed fluorescence signal exceeds the baseline noise by a factor of three. With the use of ZnO NR platforms, such sensitivity range is accessible even when using a conventional confocal microscope and a conventional fluorophore.

The results further demonstrate that ZnO nanomaterials exhibit an optical property useful in fostering the fluorescence signal from fluorophore-linked DNA molecules and promoting DNA duplex detection at ultratrace concentrations. ZnO NR platforms have an excellent integration potential into high density arrays directly upon their synthesis. Although all addressable spots in the ZnO NR array displayed in Figure 12.5 are treated equally using the same biosample, conventional automatic sample handling apparatus can permit addressing individual spots independently in the array. A large number of samples can be independently delivered in a manner where individual spots in the ZnO NR array contain different samples for high throughput measurements. When such automated sample delivery and fluorescence detection are realized, these ZnO NR platforms can be even more beneficial in accomplishing rapid, high throughput, highly sensitive detection of genetic variations.

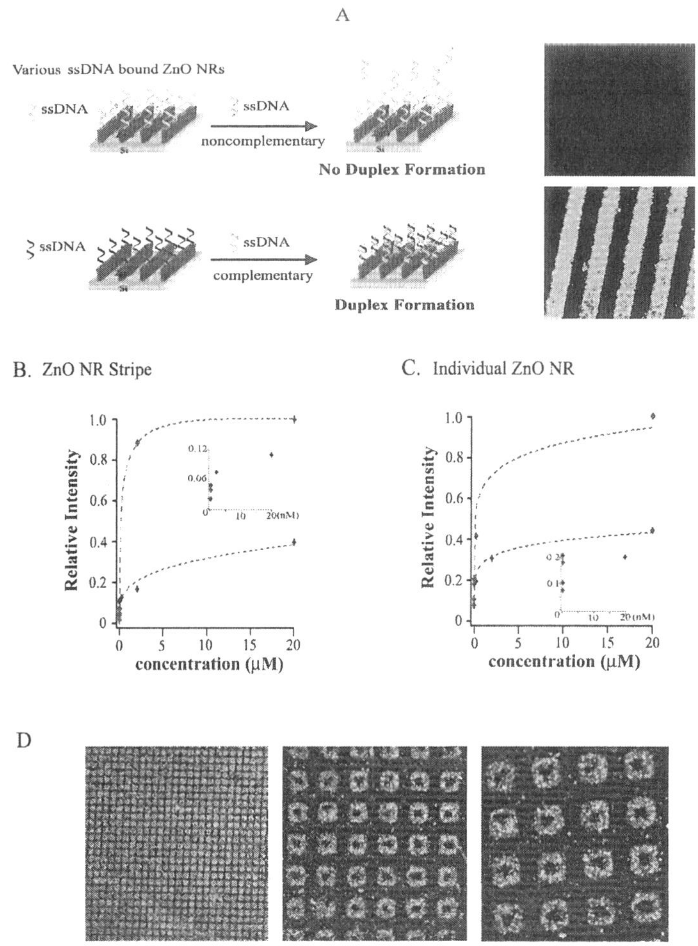


**Figure 12.5:** DNA hybridization reactions performed on ZnO NR arrays. (A) Strong fluorescence emission is observed from a sample containing fully complementary ssDNA strands whereas no signal is detected from noncomplementary strands. (B and C) Concentration dependent assays displaying the detection sensitivity of ZnO NR platforms. Data shown in red and blue correspond to assays empolying a covalent and non-covalent linking scheme of DNA strands on ZnO NRs, respectively. (D) Fluorescence emission due to duplex DNA formation on open-squared ZnO NR arrays. The easy integration potential of ZnO NR arrays into high density platforms is

demonstrated. Copyright American Chemical Society, Inc. Reproduced with permission.

## 12.6.2. *Protein – Protein Reaction*

Proteins are the key components of the cellular machinery responsible for the processing of detailed biological functions decoded from genetic information. The rapid pace in discovery of new gene products by large-scale genomics demands significant improvements in current technology pertaining to quantitative and functional proteomics. Specifically, the design of alternative strategies for analyzing protein functions via novel high-throughput approaches is highly warranted. Biomolecular fluorescence is the most widely used detection mechanism in both laboratory-scale and high-throughput proteomics research, as evidenced by its use in essential techniques such as fluorescent gel staining and protein microarrays. Rapid, low-cost, high-throughput, ultrasensitive, and specific protein detection is much needed in the areas of basic protein research as well as in large-scale clinical testing and screening of protein markers.

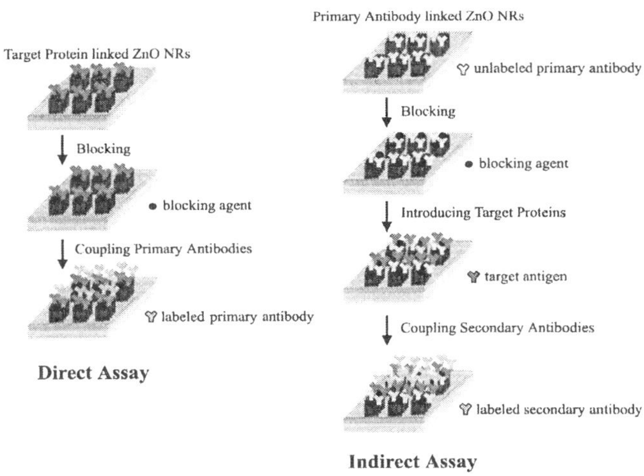


**Figure 12.6:** Schematics showing two common types of protein immunoassays. (Left) Direct assay exploiting interactions between target proteins and labeled primary antibodies. (Right) Indirect, sandwich assay relying on interactions, firstly between unlabeled primary antibodies and target proteins, and secondly between target proteins and labeled secondary antibodies.

As shown in Figure 12.6, two types of assays are often used in analyzing possible protein interactions and evaluating protein expression levels; direct and indirect (or sometimes called sandwich) assays. Direct assays typically involve a target protein that are prelinked onto an assay platform such as a well-plate array or a cover slip. Subsequently, detection of the target protein is carried out with a primary antibody that is labeled with a fluorochrome. On the other hand, more target-specific sandwich assays typically start with prelinking of an unlabeled primary antibody to an assay platform. The platform is then treated with a target protein and the target protein is detected by a fluorochrome-labeled secondary antibody in the next step. Enzymes can be used instead of or in conjunction with fluorophores in these two types of assays. Enzyme-linked immunosorbent assay (ELISA) or related assay techniques are widely adapted in numerous biological fluorescence detection settings. This approach is used commonly to obtain a higher signal to noise ratio in the detection scheme.

The capability of ZnO nanomaterials for reliable, multipurpose, and multiplexed fluorescence detection of interacting protein molecules is tested with a variety of model proteins [64]. As a proof-of-concept, different pairs of proteins are sequentially introduced to NR platforms and screened for fluorescence. The approach involving ZnO nanostructures in the enhanced fluorescence detection is then extended to identify the presence or absence of multiple protein / protein interactions on the same substrate. In some cases, microfluidic chambers made out of PDMS are used in order to carry out multiple protein interaction assays on the same ZnO NR supports.

Two possible contributing factors to fluorescence detection, other than the signal enhancing effect of ZnO NR platforms are evaluated in a series of experiments. Specifically, a control experiment is designed to rule out potential errors that may arise from substrate to substrate variations. The variations in the size and density of ZnO NRs may have been introduced during the batch growth process of ZnO NR platforms. Another control experiment is carried out to ascertain that the observed fluorescence signal from the model systems of interacting protein pairs is solely due to the formation of the protein/protein complexes. Additional control experiments are designed to ensure that the signal is not attributable to nonspecific adhesion of fluorophore-labelled proteins to the underlying ZnO nanostructures. Figures 12.7 and 12.8 display the data from these series of control experiments. On the ZnO NR platforms, interacting protein pairs lead to strong fluorescence emission whereas no signal is detected from non-interacting protein pairs. For the simultaneous screening of multiple interacting protein pairs using a microfluidic chamber on the same ZnO NR platform, only the chambers containing interacting proteins lead to fluorescence.

The data demonstrate that engineered nanoscale ZnO can serve as ideal substrates for identifying and screening protein-protein interactions. Another advantage of these ZnO NR platforms is that the nanomaterials significantly enhance detection capability of biomolecular fluorescence regardless of the emission properties of fluorophores [64]. Therefore, versatile and highly sensitive optical detection of many commonly used fluorophores spanning all visible wavelengths can be easily realized. This characteristic of ZnO NRs enables their applications in multiplexed detection when screening protein / protein interactions and determining protein marker levels. High detection sensitivity of ZnO NR platforms is achieved without the need for extended target and signal amplification steps that other assay methods require.

In addition to its exquisite sensitivity, other key advantages of ZnO NR platforms include ease of array fabrication, mechanical and chemical robustness, no autofluoroescence, and direct correlation of observed signal to protein concentration. Unlike other commonly used biosupport materials, this unique property of ZnO NRs exhibiting no spectral overlap with fluorophores can be conveniently used in fluorescence data analysis. Fluorescence signal in the ZnO NR-assisted assays shown in Figures 12.7 and 12.8 is obtained directly from target proteins interacting with other proteins in the assays, which differs from indirect signal monitoring of enzyme-substrate reactions in ELISA-based assays. These combined advantages suggest that ZnO NR platforms can be efficiently used for rapid identification of interacting protein pairs in an array format, especially for screening large libraries of protein molecules and biochemical studies of multiple protein activities.

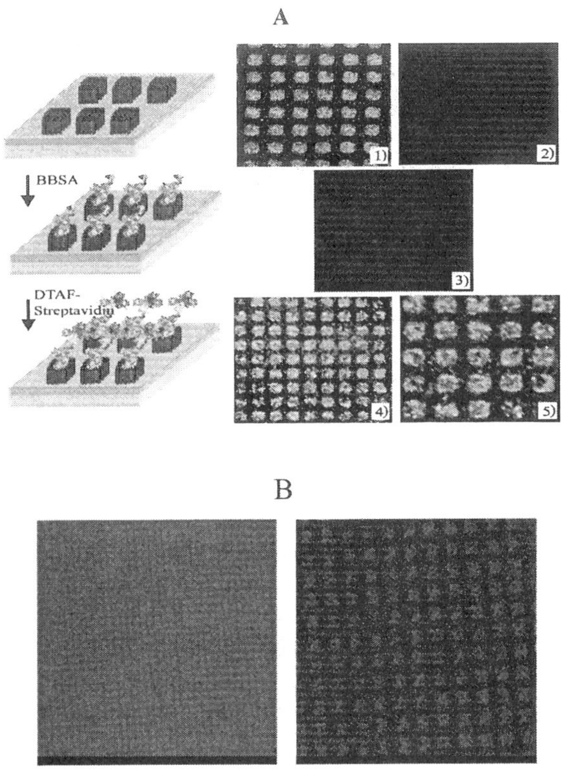


**Figure 12.7:** Protein-protein interaction assays performed on filled-squared ZnO NR arrays. (A) Interaction between biotinylated bovine serum albumin (BBSA) and dichlorotriazinylaminofluorescein (DTAF) conjugated streptavidin leads to strong fluorescence emission shown in panels 4 and 5. As-grown ZnO NR arrays do not show any autofluorescence. SEM micrograph of as-synthesized NR arrays is displayed in panel 1 and the corresponding fluorescence emission of as-grown NR arrays is shown in panel 2. (B) Strong fluorescence signal is visible when an alternative fluorophore, exhibiting different absorption and emission profiles than

DTAF, is used in a similar protein assay. A. Dorfman, N. Kumar, J. Hahm, Nanoscale ZnO-enhanced fluorescence detection of protein interactions, Adv. Mater. 2006, 18, 2685. Copyright Wiley-VCH Verlag GmbH & Co. KGaA. Reproduced with permission.

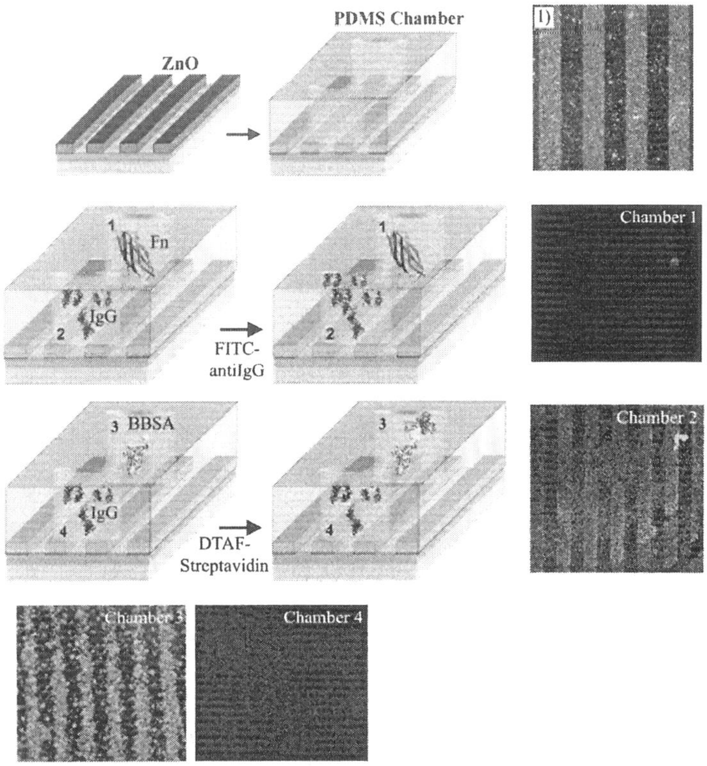


**Figure 12.8:** Protein protein interaction reactions carried out in parallel on the same, striped ZnO NR platform using a microfluidic chamber. Panel 1 displays SEM micrograph of the striped ZnO NR arrays. Fluorescence signal is observed only from the chambers containing interacting pairs of proteins, chambers 2 and 3. No fluorescence is detected from ZnO NRs in chambers 1 and 4. The model protein pairs used in the control experiments include fibronectin (Fn), immunoglobulin G (IgG), biotinylated bovine serum albumin (BBSA), DTAF-streptavidin.
A. Dorfman, N. Kumar, J. Hahm, Nanoscale ZnO-enhanced fluorescence detection of protein interactions, Adv. Mater. 2006, 18, 2685. Copyright Wiley-VCH Verlag GmbH & Co. KGaA. Reproduced with permission.

## 12.7 ORIGIN OF FLUORESCENCE ENHANCEMENT

The exact mechanism leading to enhanced fluorescence signal on ZnO NRs is not clearly understood yet and requires further study. This section provides discussion on some plausible mechanisms for the aforementioned ZnO NR-enabled fluorescence enhancement effect.

One pathway leading to enhanced fluorescence results from reducing resonance energy transfer between fluorophores themselves. Fluorophores that may show good quantum yield and high extinction coefficients can have non-radiative trap sites in their excited states. These fluorophores can go through undesired self-quenching processes, as observed frequently in molecules such as fluorescein and rhodamine. In spectrally sensitized solar cells containing a wide band-gap metal oxide and an organic fluorophore, electron transfer processes are known to occur between the two. In such settings, fluorophore molecules are in direct contact with metal oxides. Upon photo-excitation of fluorophores, electrons from the excited states of fluorophores may transfer to the conduction band of metal oxides [81, 82]. A similar electron transfer process may occur in the biomoelcular detection system of ZnO NRs described in the previous sections. This electron transfer process from the excited levels of fluorophores to the conduction band of ZnO NRs may prevent self-quenching that may have been otherwise widely present in fluorophore-only systems. However, considering the relatively high quantum efficiency (>90%) of the fluorophores used in ZnO NR assays, this effect alone cannot explain the degree of the enhancement discussed earlier.

It is possible that surface enhancement effects, similar to the observations made earlier in metal-fluorophore systems [11, 83-85] may occur. Metal surfaces are known to have effects on fluorophores such as increasing or decreasing rates of radiative decay or resonance energy transfer. A similar effect may take place in ZnO nanomaterial platforms. However, decay lengths of fluorescence enhancement observed in the semiconducting ZnO NRs are not commensurate with the length scale seen on metals such as Au or Ag. For effective metal enhanced fluorescence, fluorophores should be placed approximately between 5-20 nm away from the metal surface. However, fluorescence enhancement effect on ZnO NRs is observed even when fluorophores are located well beyond 20 nm away from the NR surface. At the same time, no quenching effect is seen when they are placed directly onto ZnO NR surfaces. In addition, there is no overlap between the absorption and emission spectra between ZnO NRs and fluorophores used in the bioassays. In surface enhanced fluorescence of the metal-fluorophore systems, metals and fluorophores show overlapping regions in their absorption and emission spectra, resulting in local changes in the electromagnetic field around metal-fluorophore. When surface plasmon of metallic thin films is used for controlling the local electromagnetic environment of fluorophores, corrugated or roughened surface is known to yield better enhancement effect. However, the surface morphology of ZnO NR is atomically flat and smooth. In fact, the hexagonal crystalline facets of ZnO NRs are high enough quality to function effectively as Fabry-Perot laser cavity. When considering these aspects necessary for the enhancement effect seen in metal-fluorophore systems, factors other than surface plasmon (a collective oscillation of electrons at the interface between conducting and insulating surfaces) or polariton (hybrid interaction between a transverse electromagnetic field and a resonant oscillation of an active material) may play a more important role in ZnO NR-enabled fluorescence detection system.

Another pathway leading to enhanced fluorescence may originate from the evanescence wave-enhancing nature of wide band-gap metal oxides. In an earlier work, a metal oxide-incorporated waveguide has been predicted for enhancing evanescent wave fields up to 1500 times higher than a waveguide with no metal oxide [86]. In addition, metal oxide nanostructures have also shown as exceptional UV and visible light-guiding mediums [37-39]. In these studies, metal oxide NRs are capable of guiding visible light in and out of fluorophores as well as along NRs.[39] This phenomenon results in direct fluorescence from the fluorophores as well as guided fluorescence on NRs. The wave-guiding property of ZnO NRs and their ability to enhance the intensity of evanescent field may explain the remarkable fluorescence sensitivity observed in ZnO NR-aided biodetection. Other experimental observations that corroborate this explanation include the dimensions of ZnO NRs used in biological assays and the relatively large decay length of fluorescence enhancement in biomolecular system. The predicted decay distance of the evanescent wave field on the metal oxide NRs for visible wavelengths is approximately up to 100 nm [39]. The average diameter of ZnO NRs that are used in biological assays is commensurate to the predicted dimensions of ZnO NRs to guide visible light effectively. The absorption and emission characteristics of common fluorophores operate in this visible wavelength range. Therefore, it is highly likely that ZnO NRs serve as efficient waveguides considerably enhancing the absorption and emission processes of fluorophores which, in turn, enables the extremely high detection sensitivity in the previously discussed biodetection scheme.

## 12.8 COMPLEX BIOLOGICAL REACTIONS

The use of ZnO NR platforms in enhanced fluorescence detection of simple biological systems is discussed in Section 12.6. The model biological systems in the previous section involve purified DNA and proteins where biomolecular interaction is limited only between the same type of biomolecules. For these ZnO NR platforms to be widely applicable to biomedical and clinical applications, their performance needs to be further assessed in more biologically complex and clinically relevant systems. The platform should be robust and stable in multi-step assays including different types of biomolecules. The platform should yield high sensitivity for target biomolecules even for assays performed in biological fluids such as blood or urine. This following section discusses both previously demonstrated and future applications of ZnO NR platforms in such complex biological reactions. Telomerase and cytokines are described as example biomarker systems for complex bioassays.

### 12.8. 1. *Telomerase Assay*

Telomere, a chromosomal structure consisting of tandem GT-rich repeats of $(TTAGGG)_n$, protects the termini of linear chromosomes from degradation [87]. However, the natural division processes in normal somatic cells result in reduction of the original telomeric repeats where each cell division progressively shortens the length of chromosomes by losing about 50~200 nucleotides of telomeric sequence. When the telomeric sequence reaches a certain short length, cells stop dividing and enter a state known as replicative senescence or die [88]. In contrast, immortal cancer cells have a mechanism to keep their telomeric length. Telomerase is a ribonucleoprotein complex that catalyzes the addition of telomeric repeats to the 3' end of chromosomal DNA and, thus, is responsible for the elongation of telomere ends in cancer cells [87]. As telomerase prevents the loss of telomeric sequences after each cell division, it leads to the uncontrollable and indefinite growth often associated with cancer cells. Therefore, activity of telomerase is linked to the expansive proliferation of abnormal cancer cells [88].

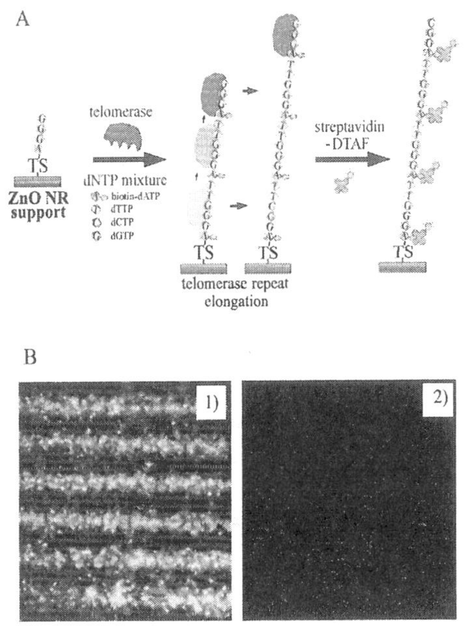


**Figure 12.9:** (A) Schematic illustration of overall assay design for ZnO NR-based telomeric repeat elongation assays. TS is an oligonucleotide whose sequence is recognized by telomerase. (B) Fluorescence panel obtained from positive (1) versus negative (2) samples after performing telomeric repeat elongation assays on ZnO NR stripe platforms.

Due to its involvement in carcinogenesis, activity of telomerase can serve as a promising biomarker in cancer diagnosis and therapy. Accurate and rapid assays for detecting telomerase activity are highly warranted in order to promote its potential uses as prognostic markers in cancer diagnosis and anti-telomerase drugs in chemotherapy. Both solution-based as well as surface-based telomerase assays have been developed in the past [89-93]. Telomerase activity is usually detected in cellular protein extracts by the telomeric repeat amplification protocol (TRAP) assay which is a solution-based technique. Though still in its infancy, several surface-based techniques have recently been used for measuring telomerase activity. Detection platforms of these surface-based techniques include surface plasmon resonance [92], nanowire field effect transistors [93], and total internal reflection [94].

Nanoscale ZnO biosensor-based platforms are recently used for a fast and straightforward assay for determining telomerase activity via telomeric repeat elongation (TRE) assays [65]. TRE assays are useful since they can eliminate polymerase chain reaction (PCR)-related artifacts as well as post-PCR procedures such as separating PCR products by gel electrophoresis and evaluating them by phosphorimager or densitometry. As a model biological system, telomerase-expressing HeLa cells are chosen to carry out ZnO-NR based TRE assays. HeLa, derived from an adenocarcinoma of the cervix 1952, is the first human epithelial cancer cell line established in continuous culture [95, 96]. ZnO NR-based TRE assay permits more accurate assessment of the applicability of ZnO NR platforms in more biologically meaningful as well as clinically relevant settings. The experimental design of this model TRE analysis, shown in Figure 12.9, relies on the physical integrity and biological functionality of many different bioconstituents on ZnO NR platforms. Typical biological components in TRE assays include cell lysates, proteins, oligonucleotides, and deoxyribonucleotide triphosphate (dNTPs). The results in Figure 12.9 indicate that ZnO NR platforms can allow native functions of all biocomponents employed in the TRE assays which, in turn, effectively enhances fluorescence signal for the sensitive detection of active telomerases in subject cells.

### 12.8. 2. *Cytokine Assay*

Cytokines are small proteins and comprised of interleukins, chemokines, tumor necrosis factors, growth factors, interferons, and colony-stimulating factors. They serve an important role as intercellular mediators in producing immune responses of cells. In addition to immune reaction, cytokines are involved in a variety of functions including reproduction, growth and development, blood clotting, normal homeostatic regulation, and response to injury and repair. Therefore, changes in the production process of cytokines are implicated in the pathogenesis of many complex diseases and elevated levels of cytokines may serve as protein markers of either disease severity or diagnosis.

Early detection of such disease markers can provide higher diagnostic power and improve disease prognosis. Early diagnosis of diseases is crucial in increasing the effectiveness of treatment and reducing mortality. However, accurate diagnosis is extremely hard to make at the early stages of disease progress due to the lack of physical symptoms. Therefore, in many clinical and laboratory settings, efforts are being made to develop alternative screening methods, rather than checking for physical symptoms associated with diseases. Detection of protein biomarkers is very useful in compiling diseased-related versus healthy-related footprints for disease diagnosis. Protein biomarkers such as cytokines are increasingly used for screening, monitoring and diagnosing diseases as well as for guiding therapy [97]. Current detection sensitivity for a certain type of cytokine, human interleukin-18 (IL-18), ranges between tens of picogram/milliliter and hundreds of nanogram/milliliter [98-101]. Conventional antibody-based assays provide decent levels of sensitivity and specificity for target proteins. However, these assays may lack sufficient sensitivity to accurately quantify low abundance proteins. Yet, quantifying IL-18 levels in very low concentrations can be extremely important in early detection of various diseases as well as in elucidation of the relationships between various low-levels of IL-18 and disease progress. In these areas of detection involving low abundance proteins, the applications of ZnO NR platforms can be very useful as ZnO NR-based assay platforms demonstrate enhanced fluorescence detection of proteins at extremely low concentrations. Figure 12.10 shows preliminary results from human IL-18 assays performed on ZnO NR platforms. The detection sensitivity of the NR platform is well below the sensitivity range that conventional technologies can currently offer.

The demonstrated ZnO NR-enabled fluorescence sensitivity is even more remarkable, when considering the fact that such fluorescence enhancement of ZnO NR platforms is achieved without the use of chemical and biological amplification processes, major improvements of detection apparatus and analysis software, or application of specially designed fluorophores. Although telomerase and cytokine assays are discussed in this section as example cases, the versatility and general applicability of ZnO NR-based assays can be extended to other disease-marker or biomarker systems.

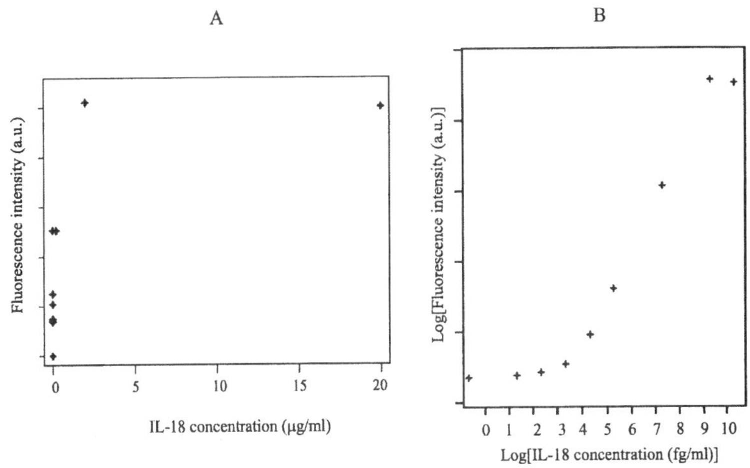


**Figure 12.10:** Preliminary results showing IL-18 assay performed on ZnO NR platforms. (A) In the assay, target concentrations of IL-18 are systematically varied from μg/ml to fg/ml and fluorescence intensity is measured. (B) The log-log plot of the data shown in (A) is displayed in order to show clearly the detection sensitivity and the linear response range.

## 12.9 SUMMARY AND OUTLOOK

In this chapter, recently developed ZnO NR arrays and their application in enhanced fluorescence detection are reviewed. Topics covered include

- an effective synthetic route of ZnO NRs promoting a seamless integration of as-grown ZnO NRs into high-density biodetection arrays,

- the physical, chemical, and optical properties of the as-synthesized ZnO NRs,
- the fluorescence enhancing capability of ZnO NR platforms in comparison to conventional biosupports,
- applications of ZnO NR platforms in simple DNA hybridization reactions and protein-protein interaction studies, and
- their applications in more complex biological assays.

Additional topics covered in this chapter are possible mechanisms leading to the fluorescence enhancement effect observed on ZnO NRs and key advantages of ZnO NR platforms in biomolecular detection.

Although ZnO NR platforms demonstrate promise for rapid, low-cost, multiplexed, high-throughput, and highly sensitive biomedical detection in both laboratory and clinical settings, further study is required from the standpoints of nanomaterials design and broadly applicable bioassays. From the purpose of engineering novel and useful nanomaterials, elucidating the exact mechanism leading to fluorescence enhancement is critical. Research efforts on this front will guide the smart design of other or similar nanomaterials that may be used as signal enhancing platforms in improved biomolecular fluorescence detection. From the viewpoint of advanced clinical application of ZnO NRs, their performance still needs to be carefully evaluated in a variety of complex assays involving multiple bioconstituents and multi-step detection in the same assay. The high detection sensitivity of ZnO NRs needs to be further tested using biological fluids, as clinical testing often involves profiling biomarker levels in sample forms of blood, urine, and plasma. As fluorescence is a widely used detection technique, other ZnO NR-based assays, that are beyond what has been demonstrated so far, may be developed in the future. In addition to the *in vitro* applications of ZnO NRs discussed in this chapter, potential capability of ZnO NRs in *in vivo* detection and real-time monitoring of biomolecules awaits exploration.

## 12.10 ACKNOWLEDGEMENTS

The author acknowledges the Materials Research Institute and the Huck Institutes of the Life Sciences at the Pennsylvania State University. The author acknowledges use of facilities at the Pennsylvania State University site of National Science Foundation National Nanotechnology Infrastructure Network. The author thanks W. Brian Reeves, M.D. for help with cytokine assays. The author acknowledges grants from Grace Woodward and Johnson & Johnson for collaborative research in engineering and medicine. The author also acknowledges support from the Pennsylvania State Cancer Institute at the Hershey Medical School. Lastly, the author thanks grants from the division of Chemistry and the division of Chemical, Bioengineering, Environmental, and Transport Systems at the National Science Foundation.

## 12.11 REFERENCES


1. Li, Y., Cu, Y. T. H., and Luo, D. (2005). Multiplexed detection of pathogen DNA with DNA-based fluorescence nanobarcodes. *Nat. Biotechnol.* **23:**885-889.
2. Emili, A. Q., and Cagney, G. (2000). Large-scale functional analysis using peptide or protein arrays. *Nat. Biotechnol.* **18:**393-397.
3. Sauer, S., Lange, B. M. H., Gobom, J., Nyarsik, L., Seitz, H., and Lehrach, H. (2005). Miniaturization in functional genomics and proteomics. *Nat. Rev. Genet.* **6:**465-478.
4. MacBeath, G. (2002). Protein microarrays and proteomics. *Nat. Genet. Suppl.* **32:**526-532.
5. Taton, T. A., Mirkin, C. A., and Letsinger, R. L. (2000). Scanometric DNA array detection with nanoparticle probes. *Science* **289:**1757-1760.
6. McFarland, A. D., and Van Duyne, R. P. (2003). Single silver nanoparticles as real-time optical sensors with zeptomole sensitivity. *Nano Lett.* **3:**1057-1062.
7. Zimmer, M. (2002). Green fluorescent protein (GFP): Applications, structure, and related photophysical behavior. *Chem. Rev.* **102:**759-781.
8. Bailey, R. E., Smith, A. M., and Nie, S. (2004). Quantum dots in biology and medicine. *Physica E* **25:**1-12.
9. Sönnichsen, C., Reinhard, B. M., Liphardt, J., and Alivisatos, A. P. (2005). A molecular ruler based on plasmon coupling of single gold and silver nanoparticles. *Nat. Biotech.* **23:**741-745.
10. Dubertret, B., Calame, M., and Libchaber, A. J. (2001). Single-mismatch detection using gold-quenched fluorescent oligonucleotides. *Nat. Biotech.* **19:**365-370.
11. Geddes, C. D., and Lakowciz, J. R. (2002). Metal-enhanced fluorescence. *J. Fluoresc.* **12:**121-129.
12. Badugu, R., Lakowicz, J. R., and Geddes, C. D. (2005). Enhanced fluorescence cyanide detection at physiologically lethal levels: Reduced ICT-based signal transduction. *J. Am. Chem. Soc.* **127:**3635-3641.
13. Aslan, K., Lakowicz, J. R., Szmacinski, H., and Geddes, C. D. (2004). Metal-enhanced fluorescence solution-based sensing platform. *J. Fluoresc.* **14:**677-679.
14. Eigen, M., and Rigler, R. (1994). Sorting single molecules: Application to diagnostics and evolutionary biotechnology. *PNAS* **91:**5740-5747.
15. Barnes, M. D., Ng, K. C., Whitten, W. B., and Ramsey, J. M. (1993). Detection of single rhodamine-6G molecules in levitated microdroplets. *Anal. Chem.* **65:**2360-2365.
16. Shera, E. B., Seitzinger, N. K., Davis, L. M., Keller, R. A., and Soper, S. A. (1990). Detection of single fluorescent molecules. *Chem. Phys. Lett.* **174:**553-557.
17. Wang, T., Peng, Y., Zhang, C., Wong, P. K., and Ho, C. (2005). Single-molecule tracing on a fluidic microchip for quantitative detection of low-abundance nucleic acids. *J. Am. Chem. Soc.* **127.**

18. Effenhauser, C. S., Bruin, G. J. M., Paulus, A., and Ehrat, M. (1997). Integrated capillary electrophoresis on flexible silicone microdevices: Analysis of DNA restriction fragments and detection of single DNA molecules on microchips. *Anal. Chem.* **69:**3451-3457.

19. Haab, B. B., and Mathies, R. A. (1999). Single-molecule detection of DNA separations in microfabricated capillary electrophoresis chips employing focused molecular streams. *Anal. Chem.* **71:**5137-5145.

20. Bruchez, M., Moronne, M., Gin, P., Weiss, S., and Alivisatos, A. P. (1998). Semiconductor nanocrystals as fluorescent biological labels. *Science* **281:**2013-2016.

21. Larson, D. R., Zipfel, W. R., Williams, R. M., Clark, S. W., Bruchez, M. P., Wise, F. W., and Webb, W. W. (2003). Water-soluble quantum dots for multiphoton fluorescence imaging in vivo. *Science* **300:**1434-1436.

22. Michalet, X., Pinaud, F. F., Bentolila, L. A., Tsay, J. M., Doose, S., Li, J. J., Sundaresan, G., Wu, A. M., Gambhir, S. S., and Weiss, S. (2005). Quantum dots for live cells, in vivo imaging, and diagnostics. *Science* **307:**538-544.

23. Hardman, R. (2006). A toxicologic review of quantum dots: Toxicity depends on physicochemical and environmental factors. *Environ. Health Perspect.* **114:**165-172.

24. Liu, C. H., Zapien, J. A., Yao, Y., Meng, X. M., Lee, C. S., Fan, S. S., Lifshitz, Y., and Lee, S. T. (2003). High-density, ordered ultraviolet light-emitting ZnO nanowire arrays. *Adv. Mater.* **15:**838-841.

25. Saito, N., Haneda, H., Sekiguchi, T., Ohashi, N., Sakaguchi, I., and Koumoto, K. (2002). Low-temperature fabrication of light-emitting zinc oxide micropatterns using self-assembled monolayers. *Adv. Mater.* **14:**418-421.

26. Jo, S. H., Banerjee, D., and Ren, Z. F. (2004). Field emission of zinc oxide nanowires grown on carbon cloth. *Appl. Phys. Lett.* **85:**1407-1409.

27. Bai, X. D., Wang, E. G., Gao, P. X., and Wang, Z. L. (2003). Measuring the work function at a nanobelt tip and at a nanoparticle surface. *Nano Lett.* **3:**1147-1150.

28. Yang, P. D., Yan, H. Q., Mao, S., Russo, R., Johnson, J., Saykally, R., Morris, N., Pham, J., He, R. R., and Choi, H. J. (2002). Controlled growth of ZnO nanowires and their optical properties. *Adv. Funct. Mater.* **12:**323-331.

29. Huang, M. H., Mao, S., Feick, H., Yan, H. Q., Wu, Y. Y., Kind, H., Weber, E., Russo, R., and Yang, P. D. (2001). Room-temperature ultraviolet nanowire nanolasers. *Science* **292:**1897-1899.

30. Lagowski, J., Gatos, H. C., and Balestra, C. L. (1978). Charge transfer in ZnO surfaces in the presence of photosensitizing dyes. *J. Appl. Phys.* **49:**2821-2826.

31. Bauer, W. (1977). Experimental studies on the spectral sensitization of photoconductivity on zinc oxide crystals. *J. Phys. Chem. Solids* **38:**463-467.

32. Yang, M., Wang, D., Lin, Y., Li, Z., and Zhang, Q. (2004). A study of photoinduced charge transfer at the interface of ZnO/Azo. *Materials Chem. Phys.* **88:**333-338.

33. Oekermann, T., Yoshida, T., Schlettwein, D., Sugiura, T., and Minoura, H. (2001). Photoelectrochemical properties of ZnO/tetrasulfophthalocyanine hybrid thin films prepared by electrochemical self-assembly. *Phys. Chem. Chem. Phys.* **3:**3387-3392.

34. Rensmo, H., Keis, K., Lindstrom, H., Sodergren, S., Solbrand, A., Hagfeldt, A., Lindquist, S.-E., Wang, L. N., and Muhammed, M. (1997). High light-to-energy conversion efficiencies for solar cells based on nanostructured ZnO electrodes. *J. Phys. Chem. B* **101:**2598-2601.
35. Gordillo, G. (2002). New materials used as optical window in thin film solar cells. *Surf. Rev. Lett.* **9:**1675-1680.
36. Johnson, J. C., Yan, H., Schaller, R. D., Haber, L. H., Saykally, R. J., and Yang, P. (2001). Single Nanowire Lasers. *J. Phys. Chem. B* **105:**11387-11390.
37. Johnson, J. C., Yan, H., Yang, P., and Saykally, R. J. (2003). Optical Cavity Effects in ZnO Nanowire Lasers and Waveguides. *J. Phys. Chem. B* **107:**8816-8828.
38. Law, M., Sirbuly, D. J., Johnson, J. C., Goldberger, J., Saykally, R. J., and Yang, P. (2004). Nanoribbon Waveguides for Subwavelength Photonics Integration. *Science* **305:**1269-1273.
39. Sirbuly, D. J., Law, M., Pauzauskie, P., Yan, H., Maslov, A. V., Knutsen, K., Ning, C.-Z., Saykally, R. J., and Yang, P. (2005). Optical routing and sensing with nanowire assemblies. *PNAS* **102:**7800-7805.
40. Lyu, S. C., Zhang, Y., Ruh, H., Lee, H. J., Shim, H. W., Suh, E. K., and Lee, C. J. (2002). Low temperature growth and photoluminescence of well-aligned zinc oxide nanowires. *Chem. Phys. Lett.* **363:**134-138.
41. Pan, Z. W., Dai, Z. R., and Wang, Z. L. (2001). Nanobelts of semiconducting oxides. *Science* **291:**1947-1949.
42. Choi, J. H., Tabata, H., and Kawai, T. (2001). Initial preferred growth in zinc oxide thin films on Si and amorphous substrates by a pulsed laser deposition. *J. Cryst. Growth* **226:**493-500.
43. Wu, J. J., and Liu, S. C. (2002). Low-temperature growth of well-aligned ZnO nanorods by chemical vapor deposition. *Adv. Mater.* **14:**215-218.
44. Park, W. I., Kim, D. H., Jung, S. W., and Yi, G. C. (2002). Metalorganic vapor-phase epitaxial growth of vertically well-aligned ZnO nanorods. *Appl. Phys. Lett.* **80:**4232-4234.
45. Ng, H. T., Li, J., Smith, M. K., Nguyen, P., Cassell, A., Han, J., and Meyyappan, M. (2003). Growth of epitaxial nanowires at the junctions of nanowalls. *Science* **300:**1249.
46. Liu, R., Vertegel, A. A., Bohannan, E. W., Sorenson, T. A., and Switzer, J. A. (2001). Epitaxial electrodeposition of zinc oxide nanopillars on single-crystal gold. *Chem. Mater.* **13:**508-512.
47. Vayssieres, L., Keis, K., Hagfeldt, A., and Lindquist, S. E. (2001). Three-dimensional array of highly oriented crystalline ZnO microtubes. *Chem. Mater.* **13:**4395-4398.
48. Vayssieres, L. (2003). Growth of arrayed nanorods and nanowires of ZnO from aqueous solutions. *Adv. Mater.* **15:**464-466.
49. Vayssieres, L., Keis, K., Lindquist, S. E., and Hagfeldt, A. (2001). Purpose-built anisotropic metal oxide material: 3D highly oriented microrod array of ZnO. *J. Phys. Chem. B* **105:**3550-3552.
50. Tian, Z. R., Voigt, J. A., Liu, J., Mckenzie, B., and Mcdermott, M. J. (2002). Biomimetic Arrays of Oriented Helical ZnO Nanorods and Columns. *JACS* **124:**12954-12955.

51. Li, W. J., Shi, E. W., Zhong, W. Z., and Yin, Z. W. (1999). Growth mechanism and growth habit of oxide crystals. *J. Cryst. Growth* **203:**186-196.
52. Wang, Z. L. (2004). Nanostructures of zinc oxide. *Materials Today* **June:**26-33.
53. Wang, Z. L. (2004). Zinc oxide nanostructures: growth, properties and applications. *J. Phys. Condensed Matter* **16:**R829-R858.
54. Lee, C. J., Lee, T. J., Lyu, S. C., Zhang, Y., Ruh, H., and Lee, H. J. (2002). Field emission from well-aligned zinc oxide nanowires grown at low temperature. *Appl. Phys. Lett.* **81:**3648-3650.
55. Ding, Y., and Wang, Z. L. (2004). Structure analysis of nanowires and nanobelts by transmission electron microscopy. *J. Phys. Chem. B* **108:**12280-12291.
56. Gao, P. X., and Wang, Z. L. (2004). Substrate atomic-termination-induced anisotropic growth of ZnO nanowires/nanorods by the VLS process. *J. Phys. Chem. B* **108:**7534-7537.
57. Koh, Y. W., Lin, M., Tan, C. K., Foo, Y. L., and Loh, K. P. (2004). Self-assembly and selected area growth of zinc oxide nanorods on any surface promoted by an aluminum precoat. *J. Phys. Chem. B* **108:**11419-11425.
58. Pan, Z. W., Dai, S., Rouleau, C. M., and Lowndes, D. H. (2005). Germanium-catalyzed growth of zinc oxide nanowires: a semiconductor catalyst for nanowire synthesis. *Angew. Chem.* **117:**278-282.
59. Rybczynski, J., Banerjee, D., Kosiorek, A., Giersig, M., and Ren, Z. F. (2004). Formation of super arrays of periodic nanoparticles and aligned ZnO nanorods-simulation and experiments. *Nano Lett.* **4:**2037-2040.
60. Wang, X., Li, Q., Liu, Z., Zhang, J., and Liu, Z. (2004). Low-temperature growth and properties of ZnO nanowires. *Appl. Phys. Lett.* **84:**4941-4943.
61. Kumar, N., Dorfman, A., and Hahm, J. (2005). Fabrication of optically enhanced ZnO nanorods and microrods using novel biocatalysts. *J. Nanosci. Nanotech.* **5:**1915-1918.
62. Kumar, N., Dorfman, A., and Hahm, J. (2006). Ultrasensitive DNA sequence detection of *Bacillus anthracis* using nanoscale ZnO sensor arrays. *Nanotechnology* **17:**2875-2881.
63. Dorfman, A., Kumar, N., and Hahm, J. (2006). Highly sensitive biomolecular fluorescence detection using nanoscale ZnO platforms. *Langmuir* **22:**4890-4895.
64. Dorfman, A., Kumar, N., and Hahm, J. (2006). Nanoscale ZnO-enhanced fluorescence detection of protein interactions. *Adv. Mater.* **18:**2685-2690.
65. Dorfman, A., Parajuli, O., Kumar, N., and Hahm, J. (2008). Novel telomeric repeat elongation assay performed on ZnO nanorod array supports. *J. Nanosci. Nanotech.* **8:**410-415.
66. Weibel, D. B., DiLuzio, W. R., and Whitesides, G. M. (2007). Microfabrication meets microbiology. *Nature Rev. Microbiol.* **5:**209-218.
67. Jackman, R. J., Wilbur, J., and Whitesides, G. M. (1995). Fabrication of submicron features on curved substrates by microcontact printing. *Science* **269:**664-666.
68. Kumar, A., and Whitesides, G. M. (1993). Features of gold having micrometer to centimeter dimensions can be formed through a combination

of stamping with an elastomeric stamp and an alkanethiol ink followed by chemical etching. *Appl. Phys. Lett.* **63**:2002-2004.

69. Xia, Y., and Whitesides, G. M. (1997). Extending microcontact printing as a microlithographic technique. *Langmuir* **13**:2059-2067.

70. Özgür, Ü., Alivov, Y. I., Liu, C., Teke, A., Reshchikov, M. A., Doðan, S., Avrutin, V., Cho, S.-J., and Morkoç, H. (2005). A comprehensive review of ZnO materials and devices. *J. Appl. Phys. 041301 2005*7 **98**:041301.

71. Vanheusden, K., Seager, C. H., Warren, W. L., Tallant, D. R., and Voigt, J. A. (1996). Correlation between photoluminescence and oxygen vacancies in ZnO phosphors. *Appl. Phys. Lett.* **68**:403-405.

72. Vanheusden, K., Warren, W. L., Seager, C. H., Tallant, D. R., Voigt, J. A., and Gnade, B. E. (1996). Mechanisms behind green photoluminescence in ZnO phosphor powders. *J. Appl. Phys.* **79**:7983-7990.

73. Hacia, J. G. (1999). Resequencing and mutational analysis using oligonucleotide microarrays. *Nat. Genet.* **21**:42-47.

74. Schorkabce, N. J., Fallina, D., and Lanchburyd, J. S. (2000). Single nucleotide polymorphisms and the future of genetic epidemiology. *Clinical Genetics* **58**:250.

75. Ramsay, G. (1998). DNA chips: State-of-the-art. *Nature Biotech.* **16**:40-44.

76. DeRisi, J., Penland, L., Brown, P. O., Bittner, M. L., Meltzer, P. S., Ray, M., Chen, Y., Su, Y. A., and Trent, J. M. (1996). Use of a cDNA microarray to analyse gene expression patterns in human cancer. *Nat. Genet.* **14**:457-460.

77. Bordoni, R., Consolandi, C., Castiglioni, B., Busti, E., Bernardi, L. R., Battaglia, C., and Bellis, G. D. (2002). Investigation of the multiple anchors approach in oligonucleotide microarray preparation using linear and stem–loop structured probes. *Nucleic Acids Research* **30**:34.

78. Loftus, S. K., Chen, Y., Gooden, G., Ryan, J. F., Birznieks, G., Hilliard, M., Baxevanis, A. D., Bittner, M., Meltzer, P., Trent, J., and Pavan, W. (1999). Informatic selection of a neural crest-melanocyte cDNA set for microarray analysis. *PNAS USA* **96**:9277-9280.

79. Marshall, A., and Hodgson, J. (1998). DNA chips: An array of possibilities. *Nature Biotechnology* **16**:27-31.

80. White, K. P., Rifkin, S. A., Hurban, P., and Hogness, D. S. (1999). Microarray analysis of Drosophila development during metamorphosis. *Science* **286**:2179-2184.

81. Jiang, C. Y., Sun, X. W., Lo, G. Q., Kwong, D. L., and Wang, J. X. (2007). Improved dye-sensitized solar cells with a ZnO-nanoflower photoanode. *Appl. Phys. Lett.* **90**:263501.

82. Pasquier, A. D., Chen, H., and Lu, Y. (2006). Dye sensitized solar cells using well-aligned zinc oxide nanotip arrays. *Appl. Phys. Lett.* **89**:253513.

83. Lakowicz, J. R. (2001). Radiative Decay Engineering: Biophysical and Biomedical Applications. *Anal. Biochem.* **298**:1-24.

84. Lakowicz, J. R., Shen, Y., D'Auria, S., Malicka, J., Fang, J., Gryczynski, Z., and Gryczynski, I. (2002). Radiative Decay Engineering: 2. Effects of Silver Island Films on Fluorescence Intensity, Lifetimes, and Resonance Energy Transfer. *Anal. Biochem.* **301**:261-277.

85. Lakowicz, J. R., Malicka, J., D'Auria, S., and Gryczynski, I. (2003). Release of the self-quenching of fluorescence near silver metallic surfaces. *Anal. Biochem.* **320:**13.

86. Kaiser, R., Levy, Y., Vansteenkiste, N., Aspect, A., Seifert, W., Leipold, D., and Mlynek, J. (1994). Resonant enhancement of evanescent waves with a thin dielectric waveguide. *Optics Comm.* **104:**234-240.

87. Greider, C. W. (1996). Telomere length regulation. *Annu. Rev. Biochem.* **65:**337-365.

88. Hanahan, D. (2000). Cancer: Benefits of bad telomeres. *Nature* **406:**573-574.

89. Lackey, D. B. (1998). A homogeneous chemiluminescent assay for telomerase. *Anal. Biochem.* **263:**57-61.

90. Kim, N. W., Piatyszek, A. M., Prowse, K. R., Harley, C. B., West, M. D., Ho, P. L. C., Coviello, G. M., Wright, W. E., Weinrich, S. L., and Shay, J. W. (1994). Specific association of human telomerase activity with immortal cells and cancer. *Science* **266:**2011-2015.

91. Sun, D., Hurley, H. L., and Hoff, D. D. (1998). Telomerase assay using biotinylated-primer extension and magnetic separation of the products. *Biotechniques* **25:**1046-1051.

92. Maesawa, C., Inaba, T., Sato, H., Iijima, S., Ishida, K., Terashima, M., Sato, R., Suzuki, M., Yashima, A., Ogasawara, S., Oikawa, H., Sato, N., Saito, K., and Masuda, T. (2003). A rapid biosensor chip assay for measuring of telomerase activity using surface plasmon resonance. *Nucleic Acids Res.* **31:**e4.

93. Zheng, G., Patolsky, F., Cui, Y., Wang, W. U., and Lieber, C. M. (2005). Multiplexed electrical detection of cancer markers with nanowire sensor arrays. *Nature Biotech.* **23:**1294-1301.

94. Schmidt, P. M., Lehmann, C., Matthes, E., and Bier, F. F. (2002). Detection of activity of telomerase in tumor cells using fiber optical biosensors. *Biosens. Bioelectron.* **17:**1081-1087.

95. Gey, G. O., Coffman, W. D., and Kubicek, M. T. (1952). Tissue culture studies of the proliferative capacity of cervical carcinoma and normal epithelium. *Cancer Res.* **12:**264-265.

96. Goodwin, E. C., and DiMaio, D. (2001). Induced senescence in HeLa cervical carcinoma cells containing elevated telomerase activity and extended telomeres. *Cell Growth Differ.* **12:**525-534.

97. Altar, C. A. (2008). The Biomarkers Consortium: on the critical path of drug discovery. *Clin. Pharmacol. Ther.* **83:**361-364.

98. Furuya, D., Yagihashi, A., Yajima, T., Kobayashi, D., Oritab, K., Kurimotoc, M., and Watanabea, N. (2000). An immuno-polymerase chain reaction assay for human interleukin-18. *J. Immunol. Methods* **238:**173-180.

99. Klein, S. A., Ottman, O. G., Dobmeyer, T. S., Pape, M., Weidman, E., Hoelzer, D., and Kalina, U. (1999). Quantification of human interleukin 18 mRNA expression by competitive reverse transcriptase polymerase chain reaction. *Cytokine* **11:**451.

100. Konishi, K., Tanabe, F., Taniguchi, M., Yamauchi, M., Tanimoto, T., Ikeda, M., Orita, K., and Kurimoto, M. (1997). A simple and sensitive bioassay for the detection of human interleukin-18/interferon-gamma-inducing factor using human myelomonocytic KG-1 cells. *J. Immunol. Methods* **209:**187.

101. Taniguchi, M., Nagaoka, K., Kunitake, T., Kayano, T., Yamauchi, H., Nakamura, S., Ikeda, M., Orita, K., and Kurimoto, M. (1997). Characterization of anti-human interleukin-18 (IL-18) /interferon-gamma-inducing factor monoclonal antibodies and their application in the measurement of IL-18 by ELISA. *J. Immunol. Methods* **206:**107.

# 13

# ZnO Platforms For Enhanced Directional Fluorescence Applications

H.C. Ong*, D.Y. Lei, J. Li, J.B. Xu[b]
Department of Physics, The Chinese University of Hong Kong, Shatin, Hong Kong, People's Republic of China.
[b]Department of Electronic Engineering

## 13.1 INTRODUCTION

Since its introduction in last century, light-emitting diode (LED) has made significant contributions in solid-sate lighting and display technology [1,2]. However, the external quantum efficiency ($\eta_{external}$) of currently used LEDs is ~ 50 lm / W and is still far below that of fluorescent tube of 100 lm / W. Knowing $\eta_{external} = \eta_{int}\times\eta_{extraction}$, both the internal quantum efficiency and light extraction efficiency have to be increased in order to improve $\eta_{external}$ [2]. While $\eta_{int}$ is primarily dependent on the crystal quality of semiconductor, $\eta_{extraction}$ is determined by many factors such as the refractive index of the materials, device geometry, etc. In particular, the high refractive index, n, of the semiconductors prevents makes light from leaving the semiconductors as the output scales approximately as $1/2n^2$ [3].

In fact, a large amount of radiative energy is trapped as guided modes within the semiconductors via total internal reflection and eventually is lost as heat due to re-absorption [3]. In addition to that, placing metallic electrodes adjacent to the semiconductors also introduces electromagnetic modes at the interface that increase additional energy loss [3]. As a result, these undesirable power dissipations limit $\eta_{external}$ of LEDs. Numerous efforts have been devoted to overcome these drawbacks. For example, the application of optical microcavity [4], photonic crystal [5], microlens [6], and micro-structured surfaces [7] has proven a certain extent to increase $\eta_{extraction}$. However, these methods are usually time-consuming and expensive and are not scalable to industrial standard. Therefore, the development of other simple approaches could revolutionize this area. Recently, surface plasmon polaritons (SPPs) arising from metal / semiconductor interface have been applied to increase both $\eta_{external}$ provided that the light emitter is sufficiently close to metal [8,9]. In this contribution, we summarize the statue of using SPPs to increase the external quantum efficiency and forward emission of ZnO.

## 13.2 BACKGROUND

The basic principle of SPP mediated emission involves two steps, SPP coupling and scattering. When radiating dipole is placed in close proximity to the

*Zino Oxide for Enhanced Directional Fluorescence*
Edited by Chris D. Geddes.

metal surface, it is possible that the recombination energy from the dipole is coupled into the SPP modes instead of being released as free space photons. The SPP coupling rate, $\Gamma_p(\omega)$, can be determined by using Fermi's Golden rule given as: [10]

$$\Gamma_p(\omega) = \frac{2\pi}{\hbar}\langle \mathrm{d} \cdot \mathrm{E(a)} \rangle^2 \rho(\hbar\omega) \tag{1}$$

where d is the electron-hole pair dipole moment, $\hbar\omega$ is the photon energy, a is the location of the dipole relative to the metal / dielectric interface, E(z) is the local electric field profile and $\rho(\hbar\omega)$ is the plasmonic density-of-states (DOS). If the local field is strong and the plasmonic DOS is large, $\Gamma_p(\omega)$ will be strongly enhanced. In addition, the spontaneous recombination rate, $\Gamma_s$, can be increased significantly since $\Gamma_s = \Gamma_p + \Gamma_o + \Gamma_{nr}$, where $\Gamma_o$ and $\Gamma_{nr}$ are radiative and nonradiative recombination rates, respectively. In fact, theoretical results have predicted that the spontaneous recombination rate can be enhanced by more than two orders of magnitude due to SPP coupling [10]. Once the SPPs are generated and propagating at the interface, they scatter and are eventually recovered as photons provided the interface is sufficiently rough [11]. As a result, the external quantum efficiency of the dipole will then be increased.

Using ZnO as an example, the plasmonic DOS can be expressed as [10]:

$$\rho(\hbar\omega) = \frac{2\pi k dk}{(2\pi)^2 d(\hbar\omega)} L^2 = \frac{L^2}{4\pi}\frac{d(k^2)}{d(\hbar\omega)}, \tag{2}$$

where $L^2$ is the in-plane quantization area and k is the wave-vector parallel to the interface. Therefore, the DOS is proportional to the derivative of the dispersion relation, i.e. $\omega$ vs. k, which can be solved by using a three-layer model (air / metal / ZnO) given as [10]

$$\left(\frac{\gamma_1}{\varepsilon_1} + \frac{\gamma_2}{\varepsilon_2}\right)\left(\frac{\gamma_3}{\varepsilon_3} + \frac{\gamma_2}{\varepsilon_2}\right) - \left(\frac{\gamma_1}{\varepsilon_1} - \frac{\gamma_2}{\varepsilon_2}\right)\left(\frac{\gamma_3}{\varepsilon_3} - \frac{\gamma_2}{\varepsilon_2}\right)e^{-2\gamma_2 t} = 0, \tag{3}$$

where $\gamma_i = k^2 - \varepsilon_i\omega^2/c^2$ with $\varepsilon_i$ at i = 1, 2 and 3 stand for the dielectric functions of air, metal [12] and ZnO [13], t is the thickness of metal and $\omega$ is the angular frequency. Here, both air and ZnO are considered semi-infinitely thick. Fig. 13.1 shows the plasmonic DOS of 30 nm thick Al, Ag and Au capped on ZnO. Considering ZnO has the band-edge emission at 3.25eV, it is clear from Fig. 13.1 that Al and Ag have large DOS while Au has almost none. As a result, the emission energy of ZnO can be coupled to SPPs of Ag and Al but not to those of Au.

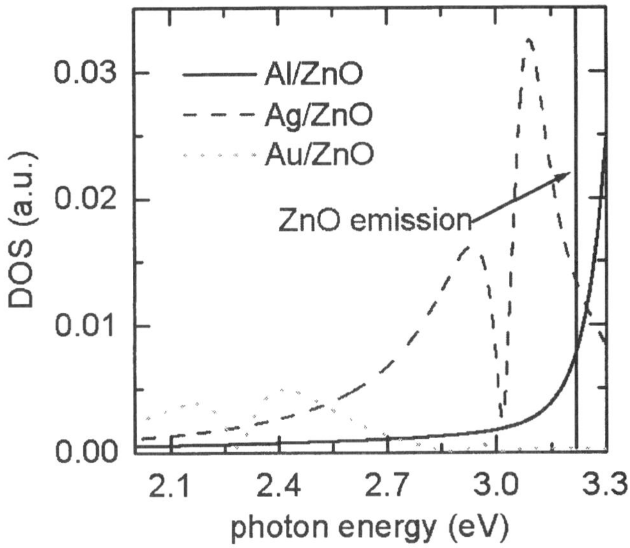


**Figure 13.1:** Calculated plasmonic DOS of air / metal / ZnO system. The thickness of Al, Ag and Au is 30 nm.

The Purcell factor, which is defined as the ratio of the spontaneous recombination rate of ZnO with SPP coupling to that of without coupling, can be expressed as [10]:

$$F_P(\omega,T) = \frac{\Gamma_P(\omega,T)+\Gamma_o(\omega,T)+\Gamma_{nr}(\omega,T)}{\Gamma_o(\omega,T)+\Gamma_{nr}(\omega,T)} = 1+\frac{\Gamma_P(\omega,T)}{\Gamma_o(\omega,T)+\Gamma_{nr}(\omega,T)} \quad (4)$$

All the rates are frequency- (ω), and temperature- (T) dependent. Eq. (4) can also be rewritten as [14]:

$$F_P(\omega,T) = 1+\frac{\Gamma_P(\omega,T)\Gamma_o(\omega,T)}{(\Gamma_o(\omega,T)+\Gamma_{nr}(\omega,T))\Gamma_o(\omega,T)} = 1+\frac{\Gamma_P(\omega,T)}{\Gamma_o(\omega,T)}\eta(\omega,T) \quad (5)$$

where η is the quantum efficiency of semiconductor without SPP coupling. Therefore, Eq. (5) enables us to calculate the emission enhancement of ZnO via SPP mediation assuming the scattering efficiency is unity. Eq. (5) is slightly different from the previous one because $\Gamma_{nr}$ is no longer assumed to be small [10]. Small $\Gamma_{nr}$ is only valid for a high quality crystal in which defect density is low but certainly is not true in most of the cases. $\Gamma_o$ can be expressed as [10]:

$$\Gamma_o(\omega,T) = \frac{4n(\omega,T)d^2\omega^3}{3\hbar c^3} \quad (6)$$

where n is the refractive index of semiconductor and c is the speed of light. Therefore, $\Gamma_p / \Gamma_o$ can be expressed as [10]:

$$\frac{\Gamma_p(\omega,T)}{\Gamma_o(\omega,T)} = \frac{\pi c^3 E_o(a,T)^2}{2n(\omega,T)\omega^2 \int_{-\infty}^{\infty} \frac{\partial(\omega\varepsilon(\omega,T))}{\partial\omega} E_o(z,T)^2 dz} \frac{dk(\omega,T)^2}{d\omega} \qquad (7)$$

where ε is the dielectric functions of the corresponding materials. Fig. 13.2 shows the calculated $F_p$ for Ag, Al, and Au capped ZnO at room temperature; $\eta(\omega,T)$ can be determined by measuring the temperature-dependent photoluminescence (PL) of bare ZnO and we assume the quantum efficiency at 10 K is 1, which only gives an approximation here. Large Purcell factor can be obtained from Al- and Ag-capped ZnO whereas negligible Purcell enhancement ($F_P = 1$) is seen from Au-capped ZnO. In consequence, the DOS and $F_p$ calculations show it is possible to realize SPP mediated emission from ZnO by using Al and Ag as the capping layer.

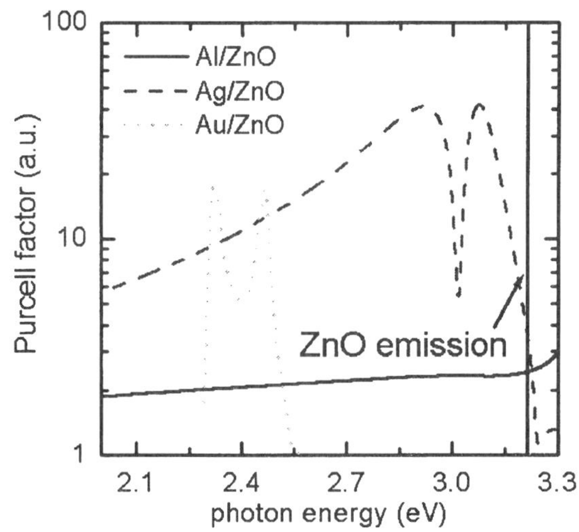


**Figure 13.2:** Calculated Purcell factor of air / metal / ZnO system. The thickness of Al, Ag and Au is 30 nm.

## 13.3 EXPERIMENTAL

ZnO films were grown on fused silica substrate by using a radio-frequency (RF) magnetron sputtering system [15]. Metallic Zn (99.99%) target was used for deposition. A working pressure of 8 mTorr was employed in the depositions by mixing an equal ratio of Ar and $O_2$. The growth temperature was kept at 550°C and the RF power was fixed at 90W. Both XRD and TEM showed that the films were highly c-axis textured without any in-plane preferred orientation. In addition, AFM showed the root mean square surface roughness of the films was less than 15 nm [16]. For metallization, only half of the ZnO films were covered by metal film so that direct comparison could be made between the bare and capped regions. Ag, Au, and Al with thickness varied from 10 to 250 nm, were deposited on the films by sputtering and thermal evaporation. The metal film thickness was then measured by a profilometer. No thermal treatment was involved during or after metallization. For optical measurements, the samples were placed on a home-built computer controlled goniometer so that the incident and detection angles can be varied independently. For photoluminescence (PL), the backside of the films was excited by a 60 mW HeCd laser in normal direction and the signal was collected through the transparent substrate at the detection of 30$^o$ and dispersed by a 0.25m spectrometer and finally captured by a CCD detector. For temperature-dependent PL and reflectivity measurements, the samples were loaded into an Oxford closed cycle He cryostat and the temperature was varied from 10 to 300 K. Since the absorption coefficient of ZnO reaches 17$\mu m^{-1}$ at 325 nm [13], ZnO films with thickness of less than 70 nm was employed so that the excitation laser beam could irradiate almost the entire film. EDX mapping carried out in a LEO 1450VP scanning electron microscope showed that no detectable impurities were found in both films and metals.

## 13.4 SURFACE PLASMON POLARITON MEDIATED EMISSION FROM ZnO

To support our theoretical prediction, Fig. 13.3 shows the PL spectra taken from bare and metal-capped ZnO films. The strong emission peak observed at 3.24 eV is identified as the band-edge emission of ZnO. Expectedly, 15- and 9-fold of integrated emission enhancement are observed from Ag and Al capped ZnO while negligible enhancement is seen from Au-capped sample [15]. These experimental results support the emission enhancements are due to SPP mediation.

Other dipolar interactions also exist that interfere SPP mediated emission enhancement. Fig. 13.4(a) displays the plot of band-edge emission enhancement ratio of Al / $AlO_x$ / ZnO as a function of $AlO_x$ thickness [17]. The enhancement ratio is defined as the integrated intensity of coated-ZnO to that of bare ZnO. The thickness of ZnO and Al are kept at 67 nm and 55 nm, respectively. One can see the ratio increases substantially right after the introduction of $AlO_x$ spacer and decreases slowly afterwards. The decrease of ratio at thick $AlO_x$ is understandable if the dependence of SPP generated local field on distance is considered. As the SPP field penetration depth (Z) in ZnO is given as $Z = \lambda/2\pi[(\varepsilon_{AlO_x} - \varepsilon_{Al})/\varepsilon_{AlOx}^2]^{0.5}$, where $\varepsilon_{AlO_x}$ and $\varepsilon_{Al}$ are the real part of the dielectric constants of $AlO_x$ [18] and Al [12], respectively, and $\lambda$ is the emission wavelength, Z is calculated to be ~92 nm, which indicates the field decays significantly above this distance [19]. Therefore, the SPP coupling is reduced to almost negligible if ZnO is placed too far away from the metal.

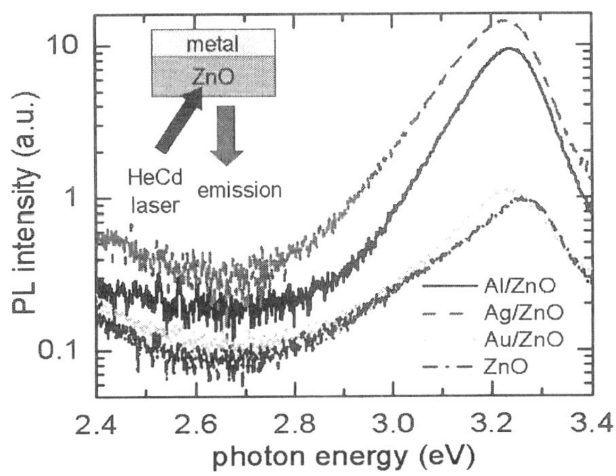


**Figure 13.3:** The experimental band-edge emission spectra of Al, Ag and Au capped ZnO. The backward photoluminescence geometry is shown in the inset. The emission of bare ZnO is also displayed as reference.

## 13.4 1. *Effects Of Spacer*

However, to account for the initial rise of emission, the dependence of SPP coupling rate on $AlO_x$ thickness is estimated. By using three layer model given in Eq. (3) to represent Al / $AlO_x$ / ZnO, we have calculated the SPP dispersion curves for different $AlO_x$ thickness. The dispersion relations and DOS are illustrated in Fig. 13.4(b)&(c). The results show that the DOS decreases with increasing $AlO_x$ and reduces to negligible at very thick $AlO_x$. By assuming the SPP local field (E(t)) decays exponentially within $AlO_x$ / ZnO [19], we approximate the SPP coupling rate, $\Gamma_P$, at the band-edge (3.24 eV) of ZnO simply as: $\Gamma_P \propto |E(t)|^2 \times \text{DOS} = \exp(-2t/Z) \times \rho(3.24\text{eV})$ [10]. After multiplying $\Gamma_P$ with an arbitrary constant to match the tail of $\Gamma_{SP}$ with that of the enhancement ratio, the results are illustrated in Fig. 13.4(a) showing $\Gamma_P$ decreases monotonically with $AlO_x$, which agrees well with our experimental data when $AlO_x$ is thick. However, the large discrepancy between the calculated $\Gamma_P$ and the experimental data shown in Fig. 13.4(a) indicates an additional quenching channel that is extremely distance dependent is present in our system. We attribute this quenching effect to the nonradiative Förster energy transfer between ZnO and Al and it can be eliminated simply by introducing a thin layer of $AlO_x$ spacer [20]. As shown in Fig. 13.4(a), the quenching efficiency can be obtained by subtracting the enhancement ratio from $\Gamma_{SP}$. The normalized quenching efficiency is shown in Fig. 13.4(d) and one can see the process has a very short range of order and basically decays rapidly after several nanometers. This quenching process correlates with the Förster energy transfer given as: $1/(1 + (R/R_o)^6)$ [20], where R is the $AlO_x$ thickness and $R_o$ is the Förster distance. The best fit is shown in Fig. 13.4(d) and $R_o$ is determined to be 3.14 nm, which is on the same order of magnitude as other reported values [21].

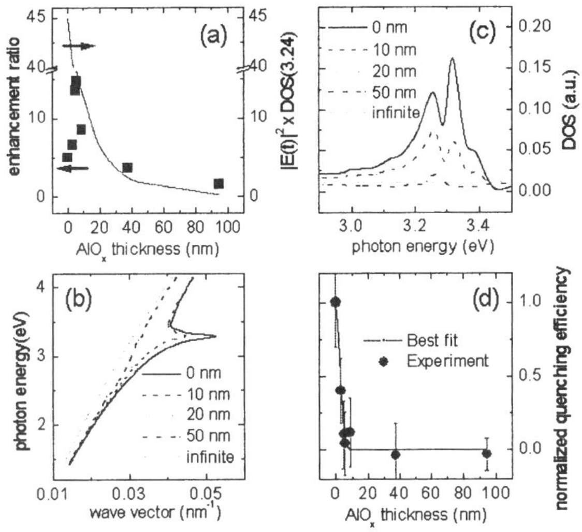


**Figure 13.4:** a) The dependence of integrated band-edge emission enhancement ratio of Al / $AlO_x$ / ZnO on $AlO_x$ thickness (square dot). The dependence of $\Gamma_p$ on $AlO_x$ thickness (solid line). b) The dispersion relation of Al / $AlO_x$ / ZnO at different $AlO_x$ thickness. c) The plasmonic DOS of Al / $AlO_x$ / ZnO at different $AlO_x$ thickness. d) The dependence of normalized quenching efficiency on Al / $AlO_x$ / ZnO (solid circle). The best fit of experimental data using Förster energy transfer [17].

## 13.4 2. *Effect Of Temperature*

More evidence of SPP mediation can also be provided from the temperature-dependent PL measurements [14]. The PL spectra of bare and Al / AlOx / ZnO films taken at different temperatures are shown in Fig. 13.5(a)&(b) and their integrated intensities and enhancement ratios are plotted in Fig. 13.5(c)&(d). We see the ratio increases linearly with decreasing temperature suggesting either the coupling between ZnO and SPPs or the radiative scattering of SPPs at Al / AlOx interface becomes stronger at lower temperature.

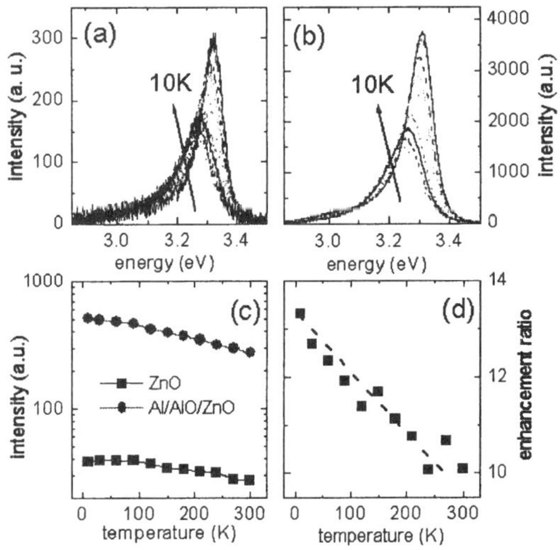


**Figure 13.5:** Temperature-dependent PL spectra of (a) bare ZnO and (b) metal-capped ZnO obtained from 10 to 300 K. (c) The variation of integrated band edge emission intensity of bare (square) and capped (circle) ZnO as a function of temperature. (d) The dependence of emission enhancement on temperature [14].

The temperature-dependent Purcell factor has been determined to explain the PL data. By taking into account the dependence of electron-electron and phonon-electron scatterings on temperature, Drude model is used for calculating the temperature-dependent dielectric functions of Al and the results are shown in Fig. 13.6(a) indicating the functions are rather insensitive to temperature [22]. On the other hand, for ZnO, reflectivity measurements have been performed and the reflectance spectra are shown in Fig. 13.6(b) for different temperatures. To extract the dielectric functions, Sellmeir dispersion model with $n^2 = a + \frac{b\lambda^2}{\lambda^2 - c}$ and k = 0 [13], where *a*, *b* and *c* are the fitting parameters, have been used below the gap where absorption is negligible and Lorentzian model with [23]

$$\varepsilon(k,\omega) = \varepsilon_b + \sum_{j=1}^{3} \frac{A_j}{(\hbar\omega_{0j})^2 - (\hbar\omega)^2 - i\Gamma_j \hbar\omega}, \tag{8}$$

where $\varepsilon_b$ is the background dielectric constant, $\hbar\omega_{0j}$ is the excitonic energy, $A_j$ is oscillator strength and $\Gamma_j$ is the broadening parameter corresponding to the interaction of excitons with phonons and intrinsic defects, has been employed for above the gap [23]. Fig. 13.6(b) shows the best fits of the reflectance spectra by using these two models. One can see the experimental data is well described by two models across the entire energy range. The resulting complex dielectric functions of ZnO are shown in Figs 13.6(c)&(d) and the excitonic features are clearly observed at low temperature due to the increase of exciton oscillator strength [14].

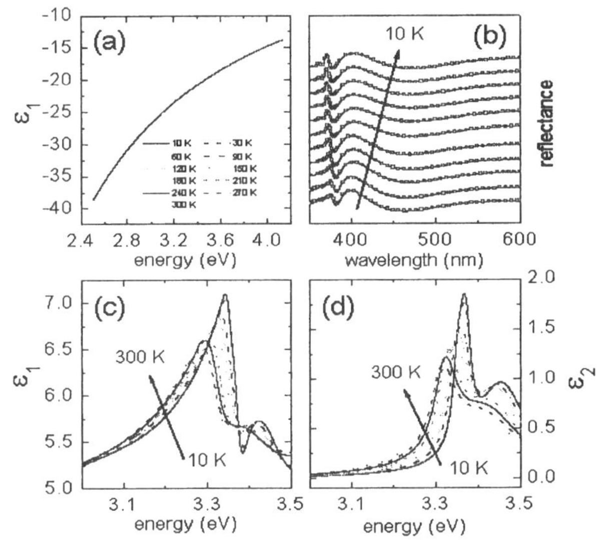


**Figure 13.6:** a) The real part dielectric function of Al at various temperatures. b) The reflectance of bare ZnO together with best fits at various temperatures. The curves are vertical shifted for visualization. The real c) and imaginary d) parts of dielectric function of bare ZnO at various temperatures [14].

The plasmonic DOS of Al / AlOx / ZnO for different temperatures are then determined by solving the dispersion relation and the results are shown in Fig. 13.7(a). The dielectric constant of AlOx is assumed to be temperature independent. By comparing with Fig. 13.6(c), we see the DOS increases with decreasing temperature due to the increase of oscillator strength. The Purcell factors taken at emission peak position and at a = 5 nm are shown in Fig. 13.7(b). Clearly, Fp increases with decreasing temperature, which suggests the SPP coupling should become stronger at low temperature due to the increase of the DOS. The PL enhancement ratios at the same peak position extracted from the experimental data (the values are divided by 2 to take into account of the double excitation caused by the Al reflection) are also plotted in Fig. 13.7(b) showing reasonable agreement with the calculated Fp. As a result, the consistence between theory and experiment indicates the emission enhancement in fact arising from SPP mediation [14].

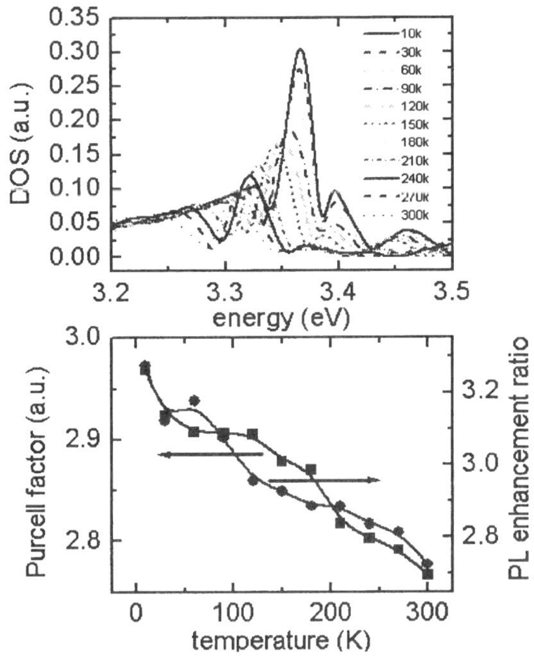


**Figure 13.7:** a) The temperature-dependent plasmonic DOS. b) Purcell factor taken at emission peak position at various temperatures. The experimental enhancement ratios taken at the same energy position, divided by 2 for the consideration of double excitation, are also illustrated [14].

## 13.4 3. *Effects Of Morphology*

Although SPP coupling is known to depend on the dielectric constants of semiconductor and metal and the surface corrugation [19], its dependence on the microstructure of materials is rarely known. Chen et al have studied the effects of Ag microstructure on the emission enhancement of Ag / InGaN quantum well and find Ag nanocrystals substantially decrease the radiative decay rate of SPP leading to weak SPP mediated emission [24].

The SPP resonance of Au / ZnO occurs at 2.4 eV and thus no SPP mediation is expected for ZnO [15]. However, Lin et al recently have proposed an alternative channel that the deep-level emissions at ~ 2-2.5 eV can be coupled to the SPP modes via Au nanoparticles [25]. The coupled energy then excites electrons within Au to higher energy states. The energetic electron can be transferred to the conduction band of ZnO and decay as band-edge emission provided the upper energy level of Au is higher than the conduction band located at 0.8 eV [26]. As a result, from their model, stronger deep-level suppression will eventually lead to higher band-edge emission enhancement because of more electron excitation. In addition, their results also imply microstructure may play a major role in controlling SPP mediated emission.

Various Au films are prepared on ZnO under different processing conditions for illustrating the effects of microstructure on SPP mediated emission. Figures 8(a)&(b) plot out the dependences of band-edge emission enhancement and deep-level emission suppression ratios on Ar pressure and target-sample distance for 42 samples. All the thickness of Au cap layer was kept at ~ 40-50 nm for ease of comparison. A phenomenological relationship can be seen from the figures that stronger deep-level suppression always leads to higher band-edge emission enhancement. In other words, the band-edge emission is enhanced at the expense of deep-level emissions. In addition, we see that higher Ar pressure and shorter distance are required to generate stronger effect. As the nucleation and growth process of Au is highly dependent on deposition condition, different microstructures could result from changing working pressure and target-sample distance [27].

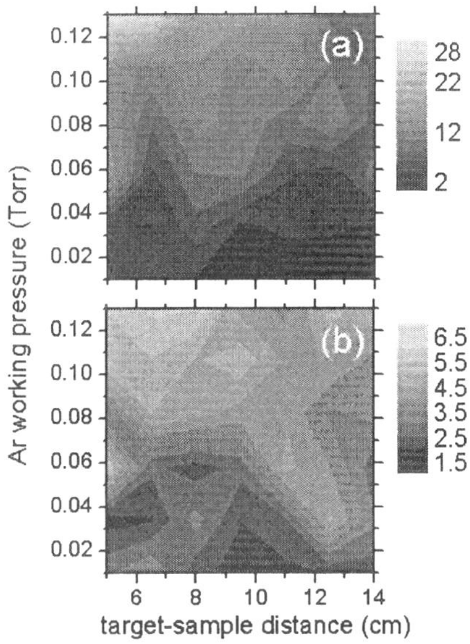


**Figure 13.8:** The contour plots of a) band-edge emission enhancement ratio and b) deep-level emission suppression ratio with Ar working pressure and target-sample distance.

We have examined the microstructure of Au films by TEM and two examples are being shown here. Fig. 13.9 illustrate the plane-view and cross-section images of Au / ZnO films grown at low pressure, (a)&(b), and high pressure, (c)&(d). The ZnO film is absent in the images since it peels off during sample preparation. Although two films show polycrystalline structure, two distinct microstructures are revealed. The structure of low pressure Au is very continuous with all the grains coalesce fairly well together to produce fully dense film while the structure of high pressure Au is relatively porous and contains a large number of nanometers-sized grain separated by large air gaps, which appear as bright or "white" spot in the image throughout the entire film. This porous structure arises from incomplete grain coalescence. The average size of the grains is estimated to be ~ 50 nm. The high brightness of the spots indicates the holes actually go through the film. As illustrated in Figs. 13.9(b) & (d), the channels are clearly seen from high pressure Au image indicating some grains are well apart from each other. In fact, Terauchi et al have found higher Ar pressure and lower power are more favorable to the growth of nanometers-sized Au particles whereas lower Ar pressure and higher power produce continuous films [27]. In addition, Eisenmenger-Sittner et al have proposed higher working pressure and shorter target-sample distance promote non-istropic flux of the arrival species and therefore leads to stronger self-shadowing effect [28]. Porous instead of continuous films are therefore produced.

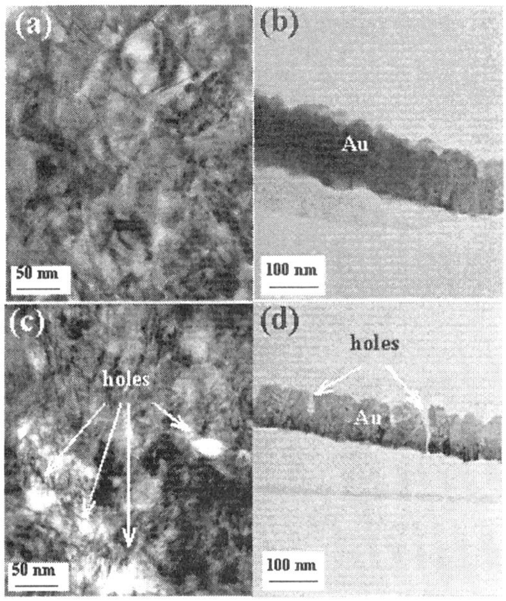


**Figure 13.9:** Plane-view TEM images of dense, continuous a) and porous, nanocrystalline c) Au films. Cross-section TEM images of dense, continuous b) and porous, nanocrystalline d) Au films.

As a result, different microstructures possibly lead to different degrees of SPP mediated emission. We have attributed the need of nanocrystalline Au films for generating strong band-edge emission enhancement to the reduction of SPP radiative decay rate due to the presence of Au nanocrystals. Chen et al have used time-resolved PL to study the SPP coupling in Ag / InGaN and find the lifetime of SPP decay process is highly dependent on Ag microstructure [24]. In particular, they show that while continuous Ag supports propagating SPPs that have short SPP radiative decay lifetime, localized SPPs supported by nanocrystalline Ag have much longer decay lifetime. In our case, we believe similar situation occurs in our nanocrystalline Au. The long radiative decay process is not able to compete with the short electron transfer process so that the energetic electrons will be transferred to the conduction band of ZnO and therefore produces strong band-edge emission.

## 13.5 OPTIMISING THE ENERGY MATCHING BETWEEN SPP AND ZnO

### 13.5 1. *Tuning SPP Resonance By Metal Alloy*

In order to maximize the SPP coupling rate, the emission energy should coincide with the SPP resonance energy so that energy can be coupled to the largest possible DOS. Deviated from this criterion would reduce the resulting emission enhancement. Therefore, one can see from Fig. 13.1 that on-resonance SPP coupling can never occur for pure metal (Al and Ag) on ZnO. Following from Eq. (3), it is understood that the SPP resonance energy is largely determined by the dielectric functions of the metal cap layer as well as the underlying semiconductor and therefore not much control can be accomplished from these fixed quantities [19]. Many approaches have been proposed in attempt to increase the tunability of the resonance energy but they are found to be costly and time-consuming [29-31]. Therefore, to develop a simple but effective approach in tuning the SPP resonance energy for any semiconductor has become a critical step in moving towards the direction of device fabrication.

We have used metal alloy to tune the SPP resonance energy and the idea is based on the results from Link et al. [32] and Gaudry et al. [33] that the SPP absorption band of Ag-Au alloys is found to blueshift linearly with Ag concentration, which suggests the resonance energy is strongly dependent on the relative composition of Ag and Au. To determine the plasmonic DOS of alloy / semiconductor, the dielectric functions of the alloys must first be determined. They are calculated under the framework of the composition-weighted average model, which has been widely used for calculating the optical properties of alloyed nanoparticles [33]. We believe this model is sufficient to estimate the effective dielectric functions of the alloys so that the principle can be illustrated. Composition weighted average model is based on the assumption that the mixture $A_xB_{1-x}$ can be pictured as the stacking of small homogeneous A and B domains, whose optical properties are equal or close to those of the corresponding bulk materials [32-33].

Under this assumption, the effective dielectric functions of metal alloy, $\varepsilon_{alloy}$, are given as:

$$\varepsilon_{alloy}(x,\omega) = x'\varepsilon_A(\omega) + (1-x')\varepsilon_B(\omega),$$

where x is the mole fraction of metal A, x' is the relative volume composition, defined as $x' = x[r_s(A)/r_s(B)]^3$ with $r_s$ is the Wingner-Seitz radius per electron in the bulk material, ω is the frequency, and $\varepsilon_A(\omega)$ and $\varepsilon_B(\omega)$ are the dielectric constant of bulk A and B metals, respectively. For Al, Ag and Au, their radii are $r_s(Al) = 2.07$ a.u., $r_s(Ag) = 3.02$ a.u., and $r_s(Au) = 3.01$ a.u., respectively [Ref. 34]. All the dielectric functions of metals are obtained from Ref. 12. Fig. 13.10 shows the dispersion relations for $Al_{0.5}Ag_{0.5}$ and $Ag_{0.8}Au_{0.2}$ alloys and pure Al, Ag and Au on

$SiO_2$ based on the two-layer model given as $\hbar k = \frac{\hbar\omega}{c}\sqrt{\frac{\varepsilon_{alloy}\varepsilon_{SiO_2}}{\varepsilon_{alloy}+\varepsilon_{SiO_2}}}$ [19]. From the figure, one can see the resonance energies are located between their corresponding metals implying they indeed can be tuned by controlling the relative composition of metal alloys.

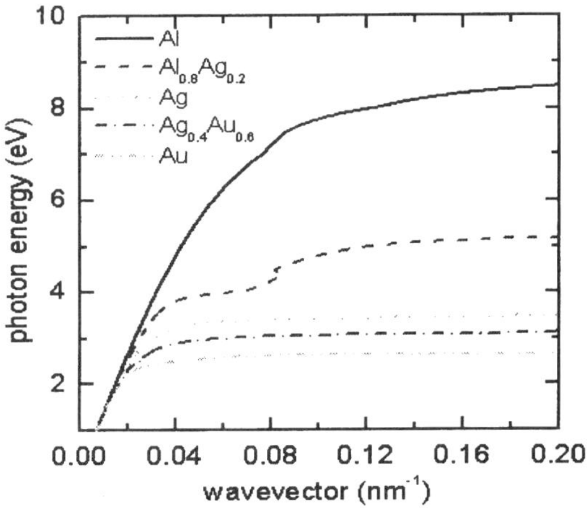


**Figure 13.10:** The dispersion relations of curves two layer $SiO_2$ / binary alloy structures. For simplicity, only the real parts of dielectric functions of different metal alloys are considered [37].

## 13.5 2. *Plasmonic DOS And $F_P$ Of Metal Alloy / Semiconductor*

To prove the feasibility of using alloys on any semiconductor, we have considered ZnO, ZnTe and CdSe, whose emission energies cover from UV to visible range. ZnO has a band-gap at ~3.3 eV [13] while ZnTe and CdSe have a gap at ~ 2.26 eV and 1.74 eV, respectively [35]. Therefore, they represent UV, green and red emitting materials. For the calculation of $F_P$, the location of the emitting dipole relative to the alloy / semiconductor interface, a, is taken as 5 nm for simplicity. In addition, $\Gamma_{nr}$ is assumed to be negligible. The dielectric functions of ZnO are obtained from Ref. 13 and those of ZnTe and CdSe are obtained from Ref. 36. The thickness of the alloys is kept at 30 nm while the thickness of semiconductor is assumed to be semi-infinite. Fig. 13.11 shows the contour plot of the dependence of (a) DOS and (b) $F_P$ of air / $Al_xAg_{1-x}$ / ZnO on photon energy and Al mole fraction [37]. $Al_xAg_{1-x}$ is chosen because of its good tunability in the UV range. It is clear from the Fig. 13.11(a) that at the ZnO emission energy of ~3.28 eV, large DOS can be obtained from ~50 to 65% and ~85 to 95% of Al mole fraction, which indicates the emission energy from ZnO can be coupled to the SPP modes. Correspondingly, the highest $F_P$ of ~ 18-9 can be obtained from ~ 55-65% Al mole fraction for ZnO as shown in Fig. 13.11(b).

Therefore, at this Al mole fraction large emission enhancement is expected if the scattering efficiency of SPPs is unity. We also see high DOS does not always warrant high $F_P$ as the SPP coupling is proportional to not only the plasmonic DOS but also the field strength at that point. On the other hand, for air / $Al_xAu_{1-x}$ / ZnTe, Fig. 13.12(a) & (b) show similar contour plots and one can see large DOS can be obtained from ~30 to 50% of Al mole fraction for ZnTe at emission energy of ~2.26 eV and the optimal $F_p$ of ~ 15 can be achieved. Fig. 13.13 shows the contour plot of (a) DOS and (b) $F_P$ of air / $Ag_xAu_{1-x}$ / CdSe and no optimal DOS and $F_P$ can be found for CdSe although moderately high $F_P$ is still obtained across the entire Ag mole fraction. The lack of optimal DOS and $F_P$ is due to the fact that both the SPP resonance energies of Ag- and Au-CdSe lie above the emission of CdSe and therefore no tuning can be achieved.

To verify our proposition that alloys can actually induce resonant coupling of SPP, we have deposited ~30 nm thick $Al_xAg_{1-x}$ alloys on ZnO films by co-sputtering of Al and Ag. The composition of the alloys is controlled by adjusting the power of Al and Ag cathodes and is determined by EDX. Fig. 13.14 shows the emission enhancement ratio of $Al_xAg_{1-x}$ / ZnO at 3.28 eV for different Al mole fractions. It is seen that the highest emission enhancement can be obtained from ~ 70-90% Al mole fraction. To compare the experimental results with theory, we have plotted out the dependence of $F_p$ at a = 1 nm on Al mole fraction in Fig. 13.14 for comparison. We see a reasonably good agreement between theory and experiment [37].

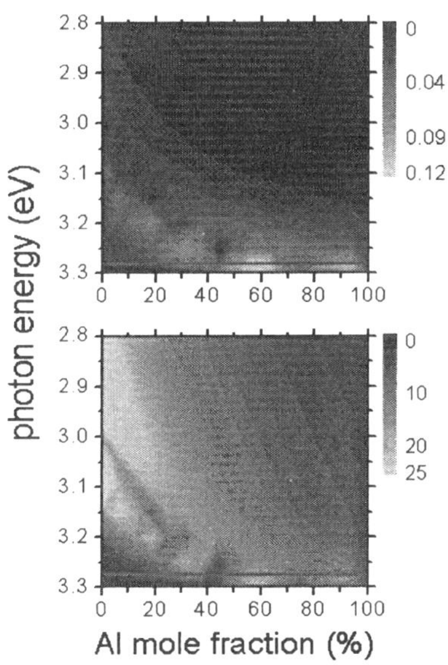


**Figure 13.11:** The dependence of a) plasmonic DOS and b) Purcell factor of air / $Al_xAg_{1-x}$ / ZnO on photon energy and Al mole fraction. The solid lines are ZnO emission. The thickness of the alloys is 30nm [37].

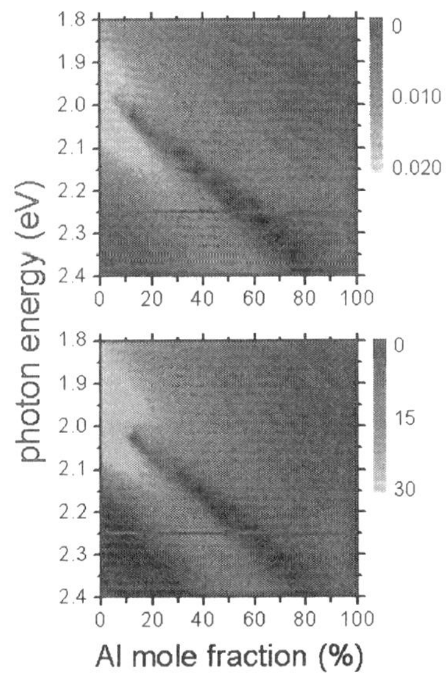


**Figure 13.12:** The dependence of a) plasmonic DOS and d) Purcell factor of air / AlxAu1-x / ZnTe on photon energy and Al mole fraction. The solid lines are ZnTe emission. The thickness of the alloys is 30nm.

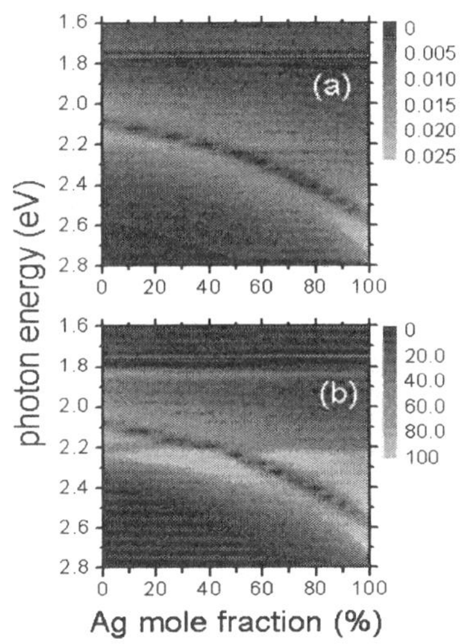


**Figure 13.13:** The dependence of a) plasmonic DOS and b) Purcell factor of air / $Ag_xAu_{1-x}$ / CdSe on photon energy and Ag mole fraction. The solid lines are CdSe emission. The thickness of the alloys is 30nm.

### 13.5 3. *Experimental Verification*

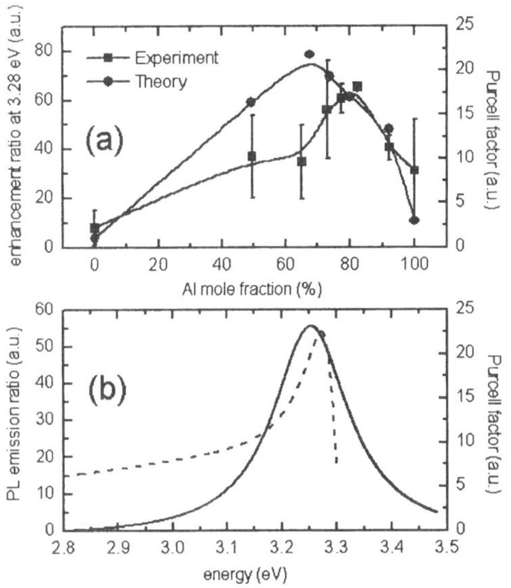


**Figure 13.14:** a) The variation of theoretical Purcell factor (solid circle) and experimental PL enhancement ratio (solid square) with Al mole fraction in air / AlxAg1-x / ZnO structure. The lines are for visualization. b) The PL emission ratio (solid) and Fp (dash) at a = 1 nm spectra for Al0.77Ag0.23 / ZnO [37].

## 13.6 USE OF SPP TO ENHANCE FORWARD EMISSION OF ZnO

For the fabrication of top-emitting LEDs, light emission is output through the metal electrode. To increase this forward emission via metal layer, energy has to be transferred effectively from the semiconductor to the air without much energy loss. One possible way is to employ metal-insulator-metal (MIM) structure that invokes the radiative SPPs so that the optical transmission of the metal system can be increased. The forward emission can then be enhanced.

### 13.6 1. *Radiative SPPs in MIM*

Although it is impractical, a simple semi-infinite MIM with air as the insulating middle layer can serve as an illustrative example. Fig. 13.15 displays the calculated dispersion relation of different SPP modes for the Ag / air / Ag structure [38]. For simplicity, the lossless Drude model $\varepsilon(\omega) = \varepsilon_\infty(1-\omega_p^2/\omega^2)$, where $\varepsilon_\infty$ is the static background dielectric (= 9.6 for Ag) and $\omega_p$ is the plasma frequency (= 3.76

eV), is used. The dielectric function of air is set as 1. The thickness of the insulator (air) is $0.5\lambda_p$, where $\lambda_p$ is the plasma wavelength (= $2\pi c/\omega_p$,). From the figure, it is noted that one of the braches is below the light line. In other words, these SPP modes are radiative in nature and direct light excitation is possible without the need of prism coupler or grating.

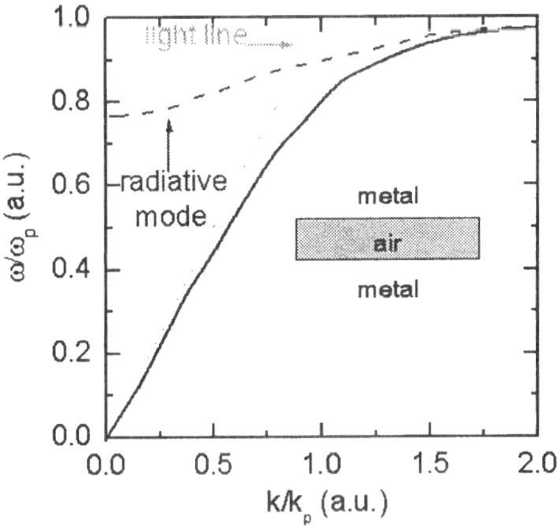


**Figure 13.15:** Dispersion relation of MIM structure with the insulator region is air and is $0.5\lambda p$ thick and two metal layers are semi-infinitely thick. The frequency and the wave vector are normalized with respect to $\omega p$ and kp where $kp=\omega p/c$; c is the speed of light in air and $\omega p$ is the plasma frequency. Inset shows the MIM structure. $\varepsilon\infty$ and $\omega p$ are chosen as 9.6 and 3.76 for Ag, respectively.

We can understand the origin of these radiative SPP modes by picturing the MIM structure exactly just like the metal-clad microcavity. In microcavity, light bounds back and forth within the cavity to form the guided-wave modes propagating along the interface [39,40]. In particular, the long propagation length and large field decay length allow the SPP waves at two interfaces to establish plasmonic standing wave within the insulating layer and thus supports radiative SPP guided modes [40]. As the field is maximized within the insulating layer where optical absorption is negligible, energy loss is minimized in the entire system. As a result, an increase of transmission is expected from this system

However, for experimentally observing the radiative SPP modes, the thickness of the metal layers must be sufficiently thin so that light can be coupled to the system for excitation. For this five-layer system, the dispersion relation is solved by using the following determinant [41]:

$$Det\begin{vmatrix} 1 & -1 & -1 & 0 & 0 & 0 & 0 & 0 \\ \frac{\varepsilon_{air}}{k_{air}} & \frac{\varepsilon_m}{k_m} & -\frac{\varepsilon_m}{k_m} & 0 & 0 & 0 & 0 & 0 \\ 0 & e^{jk_m d_m} & e^{-jk_m d_m} & -1 & -1 & 0 & 0 & 0 \\ 0 & \frac{\varepsilon_m e^{jk_m d_m}}{k_m} & \frac{\varepsilon_m e^{-jk_m d_m}}{k_m} & \frac{\varepsilon_i}{k_i} & -\frac{\varepsilon_i}{k_i} & 0 & 0 & 0 \\ 0 & 0 & 0 & e^{jk_i d_i} & e^{-jk_i d_i} & -1 & -1 & 0 \\ 0 & 0 & 0 & \frac{\varepsilon_i e^{jk_i d_i}}{k_i} & \frac{\varepsilon_i e^{-jk_i d_i}}{k_i} & \frac{\varepsilon_m}{k_m} & -\frac{\varepsilon_m}{k_m} & 0 \\ 0 & 0 & 0 & 0 & 0 & e^{jk_m d_m} & e^{-jk_m d_m} & -1 \\ 0 & 0 & 0 & 0 & 0 & \frac{\varepsilon_m e^{jk_m d_m}}{k_m} & \frac{\varepsilon_m e^{-jk_m d_m}}{k_m} & \frac{\varepsilon_{ZnO}}{k_{ZnO}} \end{vmatrix} = 0 \quad (9)$$

where dm and di are the thickness of metal and insulator (air in this case), εns are the corresponding dielectric constants and kns are the z-components of the wave vectors given as $k_n^2 = \varepsilon_n\omega^2/c^2 - k_x^2$. Here, x- and z-axes are parallel to and perpendicular to the interface. Since the thickness of each layer is smaller than their corresponding penetration depth, all the SPP modes, including radiative and nonradiative, are coupled together. Fig. 13.16 shows the dispersion relation of the finite air / MIM / air structure by using the same parameters given above except the thickness of two metal layers is kept at $0.25\lambda_p$. Similar SPP radiative modes are observed.

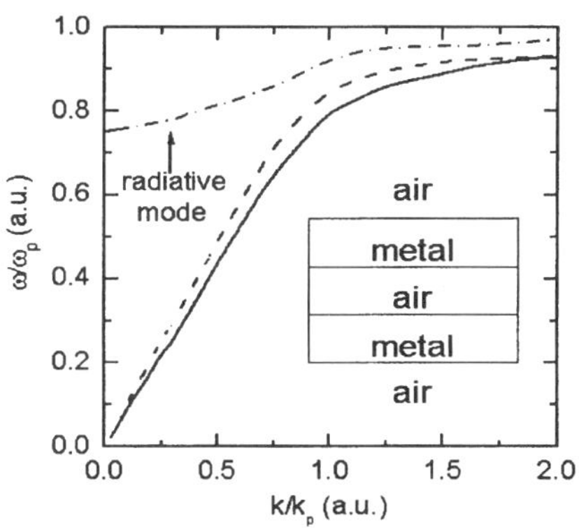


**Figure 13.16:** Dispersion relation of MIM structure with the insulator region is air and is $0.5\lambda_p$ thick and two metal layers are both $0.25\lambda_p$ thick. Inset shows the MIM structure.

The radiative SPP resonance energy is highly dependent on the thickness of the insulator [42]. We have computed the dispersion curves for the air gaps of $0.5\lambda_p$, $0.75\lambda_p$ and $\lambda_p$ and part of the results are shown in Fig. 13.17. As one can see, the SPP resonance energy decreases with increasing the thickness of the air gap implying the radiative resonance energy can be tuned simply by changing the insulator thickness.

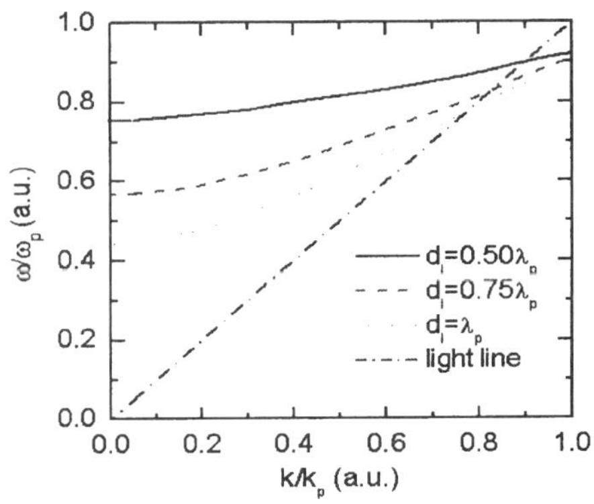


**Figure 13.17:** The dispersion relations of radiative mode of MIM structure at different air thickness.

## 13.6 2. *Enhanced Transmission In MIM*

To confirm the MIM structure can excite radiative SPPs, we have simulated the transmission spectra of Al / SiO2 / Al structure on fused silica substrate and the results for p-polarization are shown in Fig. 13.18. The total thickness of the Al layers is taken to be 50 nm. Al is chosen because it forms an Ohmic contact with ZnO so that it is advantageous for device fabrication [43].

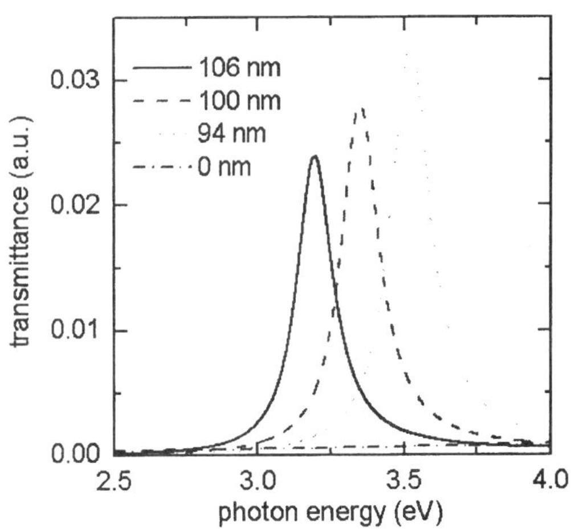


**Figure 13.18:** The simulated transmission spectra of Al / $SiO_2$ / Al with the Al thickness of 25 nm and four different $SiO_2$ thicknesses at 94, 100, 106 and 0 nm.

Consistent with the dispersion relation, strong transmission peak is observed due to the excitation of radiative SPP mode. However, no such peak is observed from the single Al layer. The transmission peak is found to be strongly dependent on the thickness of $SiO_2$ and it red-shifts with increasing the thickness, as shown in Fig. 19 for the contour plot of the dependence of transmission of Al / $SiO_2$ / Al on photon energy and the thickness of $SiO_2$. One can see that the transmission peak can be tuned from 4 eV to 2.7 eV simply by increasing the thickness from 80 nm to 120 nm. More importantly, this thickness dependence allows one to choose an optimal thickness of $SiO_2$ to maximize the transmission at any given energy. Therefore, for the fabrication of top-emitting LEDs, it is possible to identify the emission energy of the active semiconductor and then choose the proper MIM thickness for maximizing the forward emission.

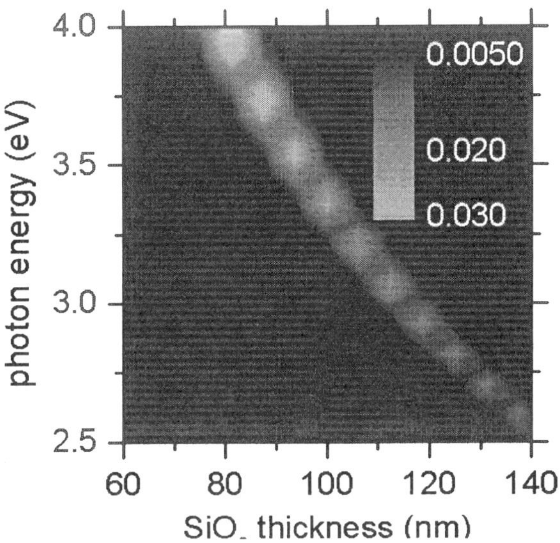


**Figure 13.19:** The counterplot of the dependence of transmittance of Al / $SiO_2$ / Al on photon energy and $SiO_2$ thickness.

## 13.6 3. *Enhanced forward emission from ZnO*

Experimentally, we have fabricated Al / PMMA / Al at different PMMA thickness on ZnO [44]. Fig. 13.20 shows the un-polarized normal incident transmission spectra of MIM / ZnO films at different PMMA thickness. In general, only single transmission peak is observed from all the spectra and the peak position is found to shift to higher energy when decreasing the PMMA thickness. Both observations agree with theory. However, close inspection reveals that the peak shape appears to be broader than expected and the peak position is also found to be lower than the theoretical values. In addition, the transmission of all samples is also found to be ~10 times higher than the simulations. Although the reason remains unknown, we attribute these discrepancies to primarily the imperfection of the MIM structure such as thickness inhomogenity, interfacial roughness, etc, and uncertainty in the dielectric constants of ZnO, Al and PMMA [45]. The low transmission above 3.25 eV, which causes the asymmetric profile, is due to the intrinsic absorption of the underlying ZnO.

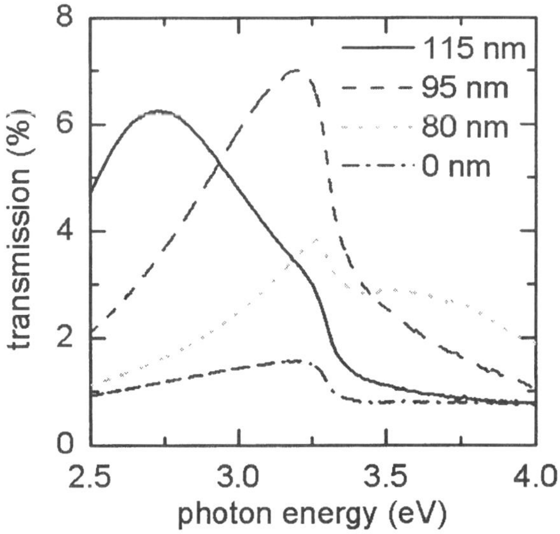


**Figure13.20:** The experimental transmission spectra of air / Al / PMMA / Al / ZnO / substrate for different PMMA thickness [45].

Knowing that the band gap of ZnO is at ~ 3.25 eV, our results indicate the MIM structure with 95 nm thick PMMA should have the transmission peak aligned well with the emission from ZnO so that the forward emission can be enhanced. To show that MIM indeed can be used to increase the forward emission, we have measured both the backward and forward photoluminescence spectra of the corresponding MIM / ZnO films [45]. Figs. 13.21(a) & (b) show the backward and forward emission spectra of MIM / ZnO films and the spectra of uncapped ZnO are also shown for reference. Surface plasmon mediated emission enhancement is evident from the spectra as shown in Fig. 13.21(a) for which the integrated backward band-edge emission intensity ($I_b$) of all MIM / ZnO is on average increased by 50 times compared to that of uncapped ZnO [15]. The insertion of PMMA does not alter the enhancement much suggesting the coupling and scattering of SPs take place primarily at the Al / ZnO interface. The coupling between ZnO and Al / PMMA SPs is too weak to have any observable effect. For the forward emission, on the other hand, significant difference is found between different MIM / ZnO, as shown in Fig. 13.21(b). Without any PMMA insertion, the intensity ratio, which is defined as the ratio of integrated forward emission intensity to backward emission intensity, ($I_f$/ $I_b$), is found to be 0.0216 due to the intrinsic optical absorption of Al.

However, the ratio increases after the insertion of PMMA. For example, the intensity ratio increases to 0.046 after the introduction of 80 nm thick PMMA. In addition, this ratio continuously increases to 0.076 for 95 nm thick PMMA and then gradually decreases to 0.016 for 130 nm despite the total thickness of Al is kept constant at 60 nm. This finding is expected due to the transmission enhancement resulted from the excitation of radiative SPs. Fig. 13.21(c) compares the intensity ratio of MIM / ZnO with the corresponding transmission values at ~3.26 eV. Similar trend is obtained between the intensity ratio and transmission demonstrating the importance of aligning the position of transmission peak with the ZnO emission. Misalignment would cause reduction in emission enhancement. As a result, the forward emission from 95 nm thick PMMA MIM / ZnO is found to be 7 and 3.5 times stronger than that from the bare ZnO and Al / ZnO.

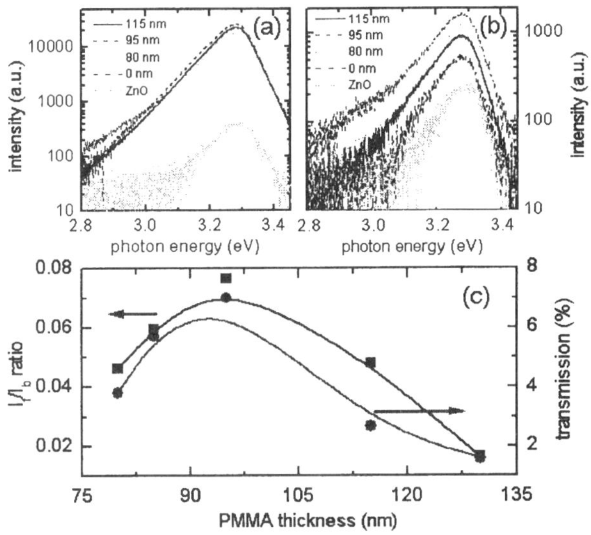


**Figure 13.21:** The a) backward and b) forward PL spectra air / Al / PMMA / Al / ZnO / substrate for different PMMA thickness. The spectra of bare ZnO are also shown. c) The plots of forward to backward emission intensity ratio (If / Ib) (square) and transmission (circle) of air / Al / PMMA / Al / ZnO / substrate with PMMA thickness [45].

## 13.7 CONCLUSIONS

By using metal as the capping layer, SPPs have been demonstrated as an effective means for increasing the external quantum efficiency of ZnO. The use of thin spacer and metal alloy can eliminate the unwanted Förster energy transfer and support on-resonance SPP coupling. In addition, nanocrystalline Au can on one hand suppress the deep-level emission while on the other hand increase the band-edge emission of ZnO. Finally, the radiative SPP arising from MIM can be used to increase the forward emission of ZnO.

## 13.8 ACKNOLEDGMENT

The authors thank J. An, C.W. Lai and W.H. Ni for carrying out part of the experimental works. This research was supported by the Chinese University of Hong Kong through the RGC Competitive Earmarked Research Grants (402004 and 402606) and Direct Grant (2060263).

## 13.9 REFERENCES

1. Schubert E.F. (2003) Light-Emitting Diodes, Cambridge University Press.
2. Zukauskas A., Shur M.S., and Gaska R. (2002) Introduction to Solid-State Lighting, Wiley-Interscience Press.
3. Barnes W.L. (1999) Electromagnetic crystals for surface plasmon polaritons and the extraction of light from emissive devices, J. Light. Tech. 17: 2170-2182.
4. Huang G.S., Lu T.C., Kuo H.C., Wang S.C., and Chen H.G. (2007) Fabrication of microcavity light-emitting diodes using highly reflective AlN-GaN and Ta2O5-SiO2 distributed Bragg mirrors, IEEE Photonics Technol. Lett. 19: 999-1001.
5. Kim J.Y., Kwon M.K., Lee K.S., Park S.J., Kim S.H., Lee K.D. (2007) Enhanced light extraction from GaN-based green light-emitting diode with photonic crystal Appl. Phys. Lett. 91: 181109.
6. Lee M.K., Ho C.L., Fan C.H. (2008) High light extraction efficiency of gallium nitride light emitting diode with silicon oxide hemispherical microlens Appl. Phys. Lett. 92: 061103.
7. Cheng Y.H., Wu J.L., Cheng C.H., Syao K.C., Lee M.C.M. (2007) Enhanced light outcoupling in a thin film by texturing meshed surfaces Appl. Phys. Lett. 90: 091102.
8. Yeh D.M., Huang C.F., Chen C.Y., Lu Y.C., Yang C.C. (2007) Surface plasmon coupling effect in an InGaN / GaN single-quantum-well light-emitting diode Appl. Phys. Lett. 91: 171103.
9. Barnes W.L. (2004) Light-emitting devices - Turning the tables on surface plasmons Nat. Mater. 3: 588-589.
10. Gontijo I, Boroditsky M, Yablonovitch E, Keller S, Mishra U.K., DenBaars S.P. (1999) Coupling of InGaN quantum-well photoluminescence to silver surface plasmons Phy. Rev. B 60: 11564-11567.
11. Okamoto K, Niki I, Shvartser A, Narukawa Y, Mukai T, Scherer A (2004) Surface-plasmon-enhanced light emitters based on InGaN quantum wells Nat. Mater. 3: 601-605.
12. Palik E.D. (1985) Handbook of optical constants of solid, Vol 1, Academic Press, London.
13. Washington P.L., Ong H.C., Dai J.Y., Chang R.P.H. (1998) Determination of the optical constants of zinc oxide thin films by spectroscopic ellipsometry Appl. Phys. Lett. 72: 3261-3263.

14. Li. J. and Ong H.C. (2008) Temperature dependence of surface plasmon mediated emission from metal-capped ZnO films Appl. Phys. Lett. 92: 121107.
15. Lai C.W., An J, Ong H.C. (2005) Surface-plasmon-mediated emission from metal-capped ZnO thin films Appl. Phys. Lett. 86: 251105.
16. Wang R.S., Gu Q.L., Ling C.C., and Ong H.C. (2008) Studies of oxide / ZnO near-interfacial defects by photoluminescence and deep level transient spectroscopy Appl. Phys. Lett. 92: 042105.
17. Ni W.H., An J, Lai C.W., Ong, H.C., Xu J.B. (2006) Emission enhancement from metallodielectric-capped ZnO films J. Appl. Phys. 100: 026103.
18. By ellipsometry measurement (SOPRA ES4G).
19. Raether H, (1988) Surface Plasmons, Springer, Berlin.
20. Lakowicz J.R. (1999) Principles of Fluorescence Spectroscopy, Kluwer Academic / Plenum Publisher, New York.
21. Gueroui Z. and Libchaber A., (2004) Single-molecule measurements of gold-quenched quantum dots Phys. Rev. Lett. 93: 166108.
22. Chiang H.P., Chen C.W., Wu J.J., Li H.L., Lin T.Y., Sánchez E.J. and Leung P.T. (2007) Effects of temperature on the surface plasmon resonance at a metal-semiconductor interface, Thin Solid Films, 515, 6957-6961.
23. L. Siozade, S. Colard, M. Mihailovic, J. Leymarie, A. Vasson, N. Grandjean, M. Leroux, J. Massies, (2000) Temperature dependence of optical properties of h-GaN films studied by reflectivity and ellipsometry Jpn. J. Appl. Phys. 39, 20-25.
24. Chen C.Y., Yeh D.M., Lu Y.C., Yang C.C. Dependence of resonant coupling between surface plasmons and an InGaN quantum well on metallic structure Appl. Phys. Lett. 89: 203113.
25. Lin H.Y., Cheng C.L., Chou Y.Y., Huang L.L., Chen Y.F., Tsen K.T. (2006) Enhancement of band gap emission stimulated by defect loss, Opt. Exp. 14: 2372-2379.
26. Kamat P.V. and Shanghavi B. (1997) Interparticle electron transfer in metal / semiconductor composites. Picosecond dynamics of CdS-capped gold nanoclusters J. Phys. Chem. B, 101: 7675-7679.
27. Terauchi S., Koshizaki M. and Umehara H., (1995) Fabrication of Au nanoparticles by radiofrequency magnetron sputtering, Nanostruct. Mater. 5, 71-78.
28. Eisenmenger-Sittner C, Bergauer A. and Bangert H., (1994) Measurement of the angular distribution of sputtered neutrals in a planar magnetron geometry, J. Vac. Sci. Technol. A 12, 536-541
29. Chen J., Shen N.H., Cheng C., Fan X.Y., Ding J.J. and Wang H.T., (2006) Tunable resonance in surface-plasmon-polariton enhanced spontaneous emission using a denser dielectric cladding, Appl. Phys. Lett. 89: 051916; Paiella R., (2005) Tunable surface plasmons in coupled metallo-dielectric multiple layers for light-emission efficiency enhancement, Appl. Phys. Lett. 87: 111104 (2005).
30. Jensen T.R., Malinsky M.D., Haynes C.L. and Van Duyne R.P., (2000) Nanosphere lithography: Tunable localized surface plasmon resonance spectra of silver nanoparticles J. Phys. Chem. B 104, 10549-10556.

31. Abe S. and Kajikawa K., (2006) Linear and nonlinear optical properties of gold nanospheres immobilized on a metallic surface Phy, Rev. B 74: 035416.
32. Link S, Wang Z.L. and El-Sayed M.A., (1999) Alloy formation of gold-silver nanoparticles and the dependence of the plasmon absorption on their composition, J. Phys. Chem. B 103: 3529-3533.
33. Gaudry M., Lerme J., Cottancin E., Pellarin M., Vialle J.L., Broyer M., Prevel B., Treilleux M. and Melinon P., (2001) Optical properties of (AuxAg1-x)(n) clusters embedded in alumina: Evolution with size and stoichiometry Phy, Rev. B 64: 085407.
34. Kittel C., (1983) Introduction to Solid State Physics, Wiley, New York.
35. Schroder D.K., (1990) Semiconductor Material and Device Characterization, Wiley, New York.
36. Palik E.D. (1991) Handbook of optical constants of solid, Vol 2, Academic Press, London.
37. Lei D.Y., Li J. and Ong H.C. (2007) Tunable surface plasmon mediated emission from semiconductors by using metal alloys Appl. Phys. Lett. 91, 021112.
38. Economou E.N. (1969) Surface plasmons in thin films, Phys. Rev. 182, 539-549.
39. Basko D.M., Bassani F., La Rocca G.C., and Agranovich V.M., (2000) Electronic energy transfer in a microcavity, Phy, Rev. B 62, 15962-15977.
40. Barnes W.L., Dereux A., and Ebbesen T.W., (2003) Surface plasmon subwavelength optics Nature 424, 824-830.
41. Gilmore M.A., Johnson B.L., (2003) Forbidden guided-wave plasmon polaritons in coupled thin films J. Appl. Phys. 93, 4497-4504.
42. Y. Wang (2003), Wavelength selection with coupled surface plasmon waves, Appl. Phys. Lett. 82, 4385-4387.
43. Ozgur U., Alivov Y.I., Liu C., Teke A., Reshchikov M.A., Dogan S., Avrutin V., Cho S.J., Morkoc H. (2005) A comprehensive review of ZnO materials and devices J. Appl. Phys. 98: 041301.
44. Lei D.Y. and Ong H.C. (2007) Enhanced forward emission from ZnO via surface plasmons Appl. Phys. Lett. 91, 211107.
45. Feng J., Okamoto T., Simonen J., and Kawata S., (2007) Color-tunable electroluminescence from white organic light-emitting devices through coupled surface plasmons, Appl. Phys. Lett. 90, 081106.

14

# E-Beam Lithography And Spontaneous Galvanic Displacement Reactions For Spatially Controlled MEF Applications

Luigi Martiradonna, S. Shiv Shankar, and Pier Paolo Pompa*
National Nanotechnology Laboratory of CNR-INFM, IIT Research Unit, ISUFI, University of Salento, Via Arnesano, 73100 Lecce, Italy.

## 14.1 INTRODUCTION

The employ of metallic nanostructures to modify the photophysical properties of fluorophores is a quite recent issue which is gaining enormous interest in the scientific community. The possibility to engineer the spectral properties of fluorophores by exploiting proximity effects with metallic surfaces (also called "surface-enhancement" effects) has been demonstrating its potential for a wide range of applications, from material science to opto-electronics, biophysics, molecular electronics and biomedical investigations [1-7]. In the last few years, numerous works discussing the theoretical and experimental implications of the coupling of surface plasmons (SPs) with both organic and inorganic fluorophores have been reported, trying to clarify the physical mechanisms underlying metal enhanced fluorescence (MEF) processes.

As widely discussed in literature [1, 8], the kind, the shape and the dimensions of the metallic species, as well as the spectral overlap of the SPs bands with the excitation and / or emission bands of the fluorophore, are key parameters for the enhancement of fluorophores emission.

While the proximity to flat metallic layers is typically proved to be detrimental for the radiative emission of fluorophores, mainly causing quenching of the luminescence [9, 10], nanoscale roughness of noble metal films may indeed be able to enhance the emission intensity significantly [6, 11-15]. Analogously, increased luminescence can be also observed when more controlled nanoclusters, either synthesized by colloidal wet-chemistry [16-19] or defined by lithographic techniques [20], are put in proximity to organic or inorganic emitters. Systematic theoretical and experimental studies of the plasmonic properties of colloidal silver and gold nanoparticles, in particular, demonstrated that precise spectral tuning of the SPs can be obtained by properly varying their geometrical features [19, 21-24]. The control of the size and shape of metallic nanoparticles can, therefore, allow to optimize the spectral overlap between the excitation and emission spectra of the fluorophores with the surface plasmons bands [8].

The approaches to metal enhanced fluorescence proposed in literature are manifold. Depending on the strategy adopted for the fabrication of metallic-dielectric interfaces where plasmons can be localized, it is possible to divide such approaches in two main categories: surface roughness-based and nanoclusters-based techniques.

*Spatially Controlled MEF Applications*
Edited by Chris D. Geddes.

In the first category, we can include, for example, the chemical or physical deposition (by means of e-beam evaporation, thermal evaporation, sputtering, electrochemistry, dealloying and so on) of metallic layers [11, 14, 25-30]. The accurate optimization of the deposition conditions allows the control of both the average thickness and the average roughness of the fabricated samples (although the maximum roughness that can be achieved by simple thermal deposition methods does not typically lead to high fluorescence enhancements).

Interesting results have been obtained by depositing uniform metallic layers which are subsequently locally eroded by etching processes [13]. Concerning the second category, the immobilization of silver and gold colloidal particles onto suitable substrates is an alternative and widely diffused strategy to realize nanoclusters for MEF applications [17, 18, 31-33]. In this latter case, the nanoparticles can be produced with different shapes and dimensions, providing plasmonic bands tunable from the near-UV to the red spectral region. Notably, nano-objects with various shapes (e.g., spherical, spheroidal or rod-like) have been deposited onto solid substrates through several deposition strategies [3, 17, 31, 32, 34]; very recently, Langmuir-Blodgett technique has been proved to be a very effective method to realize dense and homogeneous monolayers [35] of metallic particles. Moreover, lithography-based approaches have been also proposed in order to define highly-controlled metallic nanostructures onto solid substrates [20].

The approaches belonging to the first category clearly allow the fabrication of very broad area MEF substrates with fast and well-consolidated techniques borrowed from the fifty-year old experienced fabrication of integrated circuits. On the other hand, no precise local control of the roughness of the metal film can be obtained, and therefore the overall enhanced fluorescence is determined by the surface optical properties averaged on the whole detection area.

The use of colloidal metallic nanoclusters deposited onto solid substrates can provide a higher degree of control over the SPs spectral properties: state-of-the-art results in the chemical synthesis showed, in fact, the possibility to grow nano-objects with high uniformity and low size dispersion [36-38]. This can allow to fabricate extended substrates, in which the local morphology, and thus the resulting MEF effect, can be controlled with good precision [19, 39]. As a counterpart, still some randomness is unavoidable in this approach, since it is not simple to define the position of the nanoclusters on the substrate with micrometer precision to realize, for instance, ordered arrays of metallic particles.

A completely opposite perspective underlies metal-enhanced fluorescence with lithography-defined metallic nanoclusters. In this case, in fact, high-resolution lithographic techniques, such as electron beam lithography (EBL), are employed to define, with nanometer resolution, the geometry and the position of each individual nano-object. The reproducibility of the single feature on the whole substrate is clearly far above the uniformity allowed by self-assembly based approaches; moreover, the positioning of the different nano-objects on the substrate is fully customable, thus allowing selective MEF effect from nanometer to millimeter sized patterns. Such a high control of the metallic shape and, therefore, over the spectral properties of the surface plasmons is, anyway, counterbalanced by the inherent complexity of the EBL process and by the limited suitability for large-area applications: each feature composing the overall pattern is, in fact, drawn serially, so that the realization of extended MEF substrates, i.e. larger than several millimeters squared, with nanometer resolution, is a particularly demanding procedure, in terms of both costs and time consumption.

As described above, each MEF strategy shows peculiar advantages and significant drawbacks; the choice of the proper approach strongly depends on the application to be pursued. In the following paragraphs, we will show the results obtained by two different approaches for the study of metal enhanced fluorescence, based, respectively, on E-beam lithography and galvanic displacement reactions. In the first case, we propose a strategy based on the fabrication, by electron beam lithography (EBL), of ordered arrays of gold nanostructures. These nano-objects are coupled to colloidal nanocrystals (NCs), semiconductor emitters whose absorption and emission spectra can be tailored from the near-UV to the infrared spectral range by controlling their size and shape during the chemical synthesis. We demonstrate that this approach can be exploited to obtain a spatially controlled enhancement of the fluorescence signal down to the nanoscale. In the second case, we show the realization of large-area rough metallic patterns by means of an interesting approach, namely spontaneous galvanic displacement reactions. This strategy is based on the controlled displacement of a metallic layer, previously deposited on a substrate, with a second metal through a chemical reaction performed in wet solution. Such reaction can be optimized to give a final metallic film with a surface roughness suitable for MEF applications.

Noteworthy, this approach is very effective and low-cost and allows the fabrication of extended substrates (up to several centimeters squared), while maintaining an appreciable uniformity over the entire sample. Moreover, if combined with lithographic techniques, such strategy may lead to the realization of micro- and nano-patterned MEF substrates.

## 14.2 E-BEAM LITHOGRAPHY APPROACH TO METAL ENHANCED FLUORESCENCE

The EBL-based MEF approach has been originally developed by us in order to tailor the optical properties of colloidal semiconductor nanocrystals (NCs), although it can be generally applicable to a wide class of emitters. Nanocrystals have recently gained recognition as highly efficient and stable emitters for photonics applications as well as versatile bio-probes for advanced molecular and cellular imaging techniques [40-45]. As compared to organic fluorophores, NCs show a higher photobleaching threshold, good chemical stability, a broad absorption spectrum, along with the peculiar possibility of accurately tuning their spectral properties by controlling their size and shape [46-49]. Although their quantum efficiency is high in solution, it is severely reduced when the emitters are deposited in solid films either directly or dispersed in organic matrices. In order to fully exploit these fluorophores for the realization of advanced solid-state optical devices and diagnostic tools such as ultra-bright light sources, single-photon sources, microarrays, or single-molecule sensors, a further enhancement of their emission intensity through coupling with surface plasmons is desirable.

Our approach for the emission enhancement of NCs, by means of MEF mechanisms, was based on the coupling of highly ordered metallic nanopatterns, defined by EBL, with NCs dispersed in a polymeric blend. A sketch of the adopted configuration is shown in Figure 14.1.

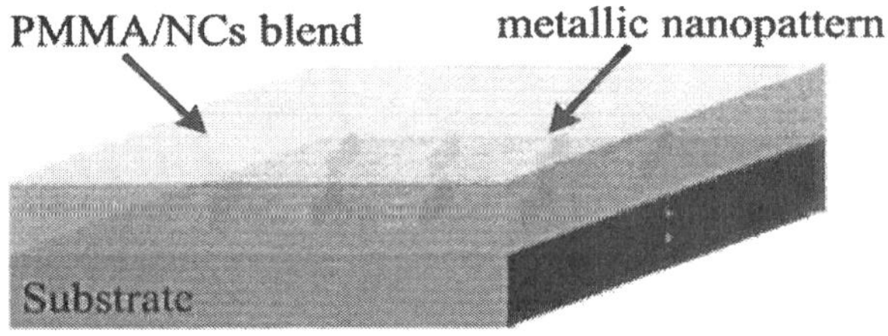


**Figure 14.1:** Sketch of our EBL-based MEF approach: a metallic nanopattern is fabricated by e-beam lithography and a layer of PMMA embedding colloidal NCs is spin-coated on top of it.

The semiconductor nanoclusters were synthesized with traditional wet-chemistry based methods and dispersed into a polymeric matrix (polymethylmethacrylate, PMMA), which is nearly transparent in the spectral range of NCs absorption and emission bands. After the realization of the metallic pattern by EBL, the polymeric blend containing the nanoemitters is deposited onto the whole substrate by spin-coating. This simple approach provides several advantages. As already mentioned, due to the high resolution of EBL processes, the shape of the single metallic nanostructures can be precisely engineered with nanometer resolution. Therefore, the plasmonic resonances can be tuned in a wide spectral range, thus improving the spectral overlap of the SPs with the excitation and / or emission spectra of the quantum emitters. Moreover, the use of a polymeric matrix permits to change and optimize the average distance between the NCs and the metallic surface, also avoiding the direct deposition of the emitters onto the metal surface, which would lead to emission quenching. Furthermore, spin-coating allows us to obtain a uniform dispersion of the nanocrystals onto the substrates even for large areas, while maintaining a good control of the concentration of the active material in the blend.

In order to investigate the potential of this approach, we performed several experiments by fabricating periodic patterns of gold clusters with cylindrical or triangular prism shape by EBL. Several thicknesses, lateral dimensions and pattern periodicities were tested in order to find the best conditions for an efficient MEF coupling with the emitters [50, 51]. In Figure 14.2, two representative examples of the morphology of the fabricated patterns, assessed by scanning electron microscopy, are shown.

Two different nanomaterials, namely colloidal core / shell Quantum Dots (QDs) and Quantum Rods (QRs) were synthesized as described in [51]. In the case of CdSe / ZnS QDs, the synthesis yielded samples emitting at $l_{max}$ = 580 nm with a spectral width of the fluorescence emission of 40 nm. CdSe quantum rods showed an emission peak centered at $l_{max}$ = 567 nm with similar linewidth. The NCs were subsequently dispersed in PMMA and deposited onto the substrate by spin-coating. In order to study the influence of the average fluorophore-metallic surface distance on the MEF effect, several thicknesses of the active layer were investigated, finding an optimum value of 35 nm, as measured from the surface of the metallic nanostructures.

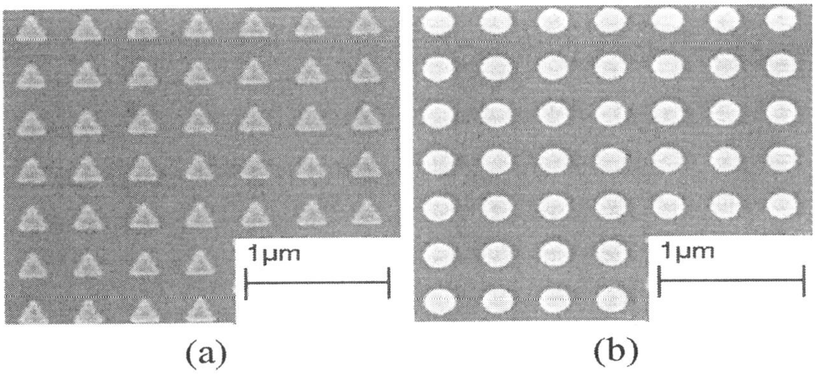


**Figure 14.2:** Two typical examples of highly ordered arrays of metallic nanoclusters fabricated by electron beam lithography and lift-off. a) Triangular prisms with lateral dimensions of 200 nm; b) Cylinders with a diameter of 200 nm. In both cases the period is equal to 400 nm.

A direct experimental measurement of the surface plasmons bands of the gold nanoclusters was not straightforward, since the metallic patterns were quite small and stuck onto an opaque substrate ($SiO_2$). Therefore, we performed theoretical calculations in order to infer the spectral features of the SPs used in our MEF experiments. The absorption spectra of gold triangular prisms and cylinders with thicknesses of 35 nm (according to the dimensions of the fabricated patterns) were calculated by using the Discrete dipole approximation (DDA). Further details on the method are given in references [52-57].

In Figure 14.3 (lower graph) the calculated spectra for triangular prisms with lateral size of 100 nm and 200 nm (small triangle line and large triangle line, respectively) and for cylinders with diameters of 100 nm and 200 nm (small circle line and large circle line, respectively) are reported. These graphs can be directly compared, in the top figure, with the emission and absorption spectra of the QDs. According to this theoretical analysis, the best overlap of the SPs band with the emission and absorption spectra of the exploited quantum dots was found with the 100 nm-wide, 35-nm thick triangular prisms.

Further simulations showed also that the optimal thickness of the gold nanostructures was in the 30-35 nm range, while lower thicknesses (< 20 nm) were characterized by a significant red-shift of the SPs resonance band, thus lowering the spectral overlap with the colloidal nanocrystals.

In order to verify the enhancement of the NCs emission induced by MEF coupling with the underlying gold nanostructures, spatially resolved photoluminescence measurements were performed on the fabricated samples by confocal microscopy. The inset in Fig. 14.4 (a) shows the fluorescence map of the QDs / PMMA blend film deposited onto a gold nanopattern (triangular prisms with 200 nm lateral dimensions). The fluorescence emission of the QDs was strongly enhanced in the central square (due to the MEF interaction with the underlying gold nanostructures) as compared to the surrounding zones in which the NCs are directly deposited onto the $SiO_2$ substrate. The same nanopattern was then cleaned and covered with the second nanomaterial, i.e., CdSe nanorods; also in this case (Fig. 14.4(b), inset), higher luminescence was collected from the nanorods deposited onto the metallic pattern with respect to the nanorods on $SiO_2$. A quantitative analysis of the MEF effects for the two different nanocrystals showed a fluorescence increase as high as $F_{QDs}$ ~30, after normalization over the spatial duty cycle of the metallic pattern (Fig. 14.4(a)), whereas a weaker increase of $F_{QRs}$ ~19 was detected for QRs (Fig. 14.4(b)). In both cases, spectral analyses also revealed that the SPs-coupled NCs preserve the same emission line-shape of the original sample, with no detectable MEF-induced emission shifts. According to the MEF theory [1], these results suggest that the enhancement mechanism is mostly due to the increased intensity of the excitation field localized onto the metallic surface, while a minor role of the increase in the radiative decay rate of the two fluorophores may be envisaged (even though lifetime measurements with high spatial resolution would further support such interpretation).

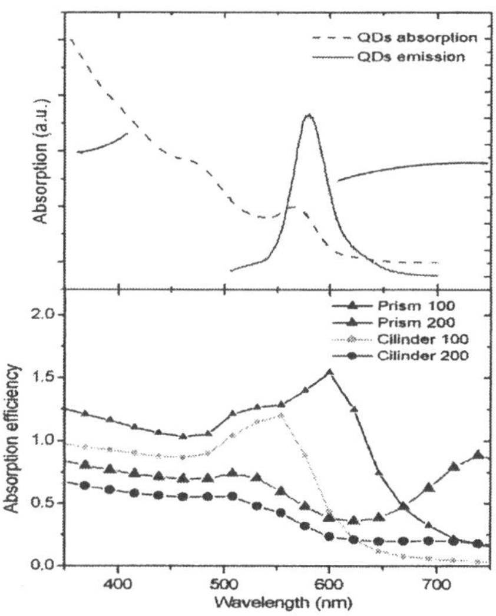


**Figure 14.3:** Comparison of the calculated absorption efficiencies of the different gold nanostructures (bottom) with the absorption and emission spectra of the colloidal QDs used in the MEF experiments (top).

The systematic measurement of the emission enhancement as a function of the nanoclusters shape and dimensions (the enhancement factors experimentally determined in the case of QDs with the different gold patterns is reported in Table 14.I) confirmed in general the results predicted by the theoretical calculations, with minor discrepancies, likely due to slight variations in the geometry of the fabricated nanostructures, and / or to the effects of the sharp shapes on the localization of the excitation electromagnetic field [50]. Moreover, we observed that the increase of the NCs fluorescence intensity was almost independent of the interspacing between the different metallic clusters (at least for interspacing distances higher than 100 nm).

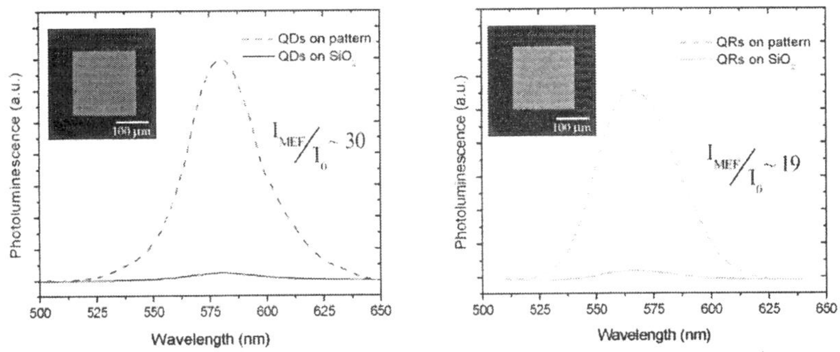


(a) (b)

**Figure 14.4:** (a) and (b): Fluorescence spectra of the QDs and QRs, respectively, collected both on $SiO_2$ (solid lines) and on the metallic nanopattern (dashed lines). As shown, c.a. 30-fold (for QDs) and 19-fold (for QRs) fluorescence increases were observed. Insets: fluorescence maps of (a) QDs / PMMA and (b) QRs / PMMA blend films deposited on the same gold nanopattern (200x200 $\mu^2$), collected by a confocal microscope. The bright regions correspond to the area in which the luminescence of the NCs is strongly enhanced, due to the coupling with the underlying gold nanoarray.

| *Metallic nanostructure* | *Lateral size (nm)* | *MEF factor* |
|---|---|---|
| Triangular prisms | 100 | 33 |
| | 200 | 30 |
| Cylinders | 100 | 26 |
| | 200 | 21 |

**Table 14.I:** Experimental enhancement factors of QDs fluorescence emission observed with the different gold nanopatterns: triangular prisms and cylinders with 100- and 200-nm lateral dimensions (35 nm height). The reported enhancement factors were normalized over the correspondent spatial duty cycles.

Besides the precise tuning of the geometry of the metallic nanoprism in order to improve the MEF coupling mechanism, the use of E-beam lithography in our approach allows to localize the enhancement process with high spatial control. We have demonstrated the potential of this technique by fabricating different pattern geometries with micron- and sub-micron dimensions. This is shown in Figure 14.5, where the fluorescence maps (bottom line) of the NC / PMMA blend films deposited onto various stripe-like metallic nanopatterns (top line, SEM images of the fabricated patterns) are displayed.

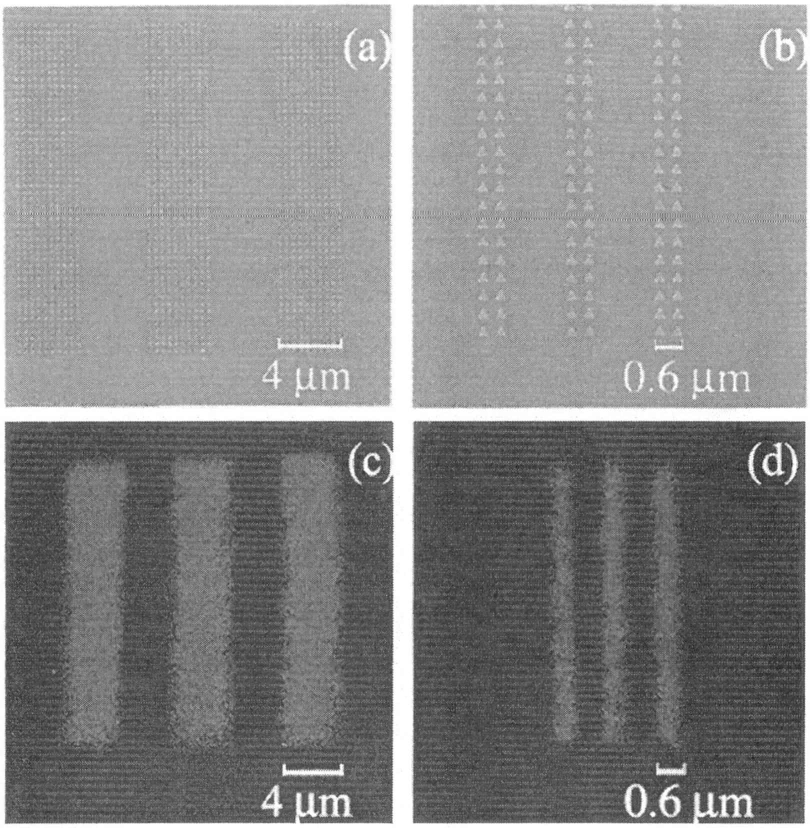


**Figure 14.5:** (a-b) SEM images of stripe patterns with 4 μ and 0.6 μ width, respectively, realized by aligning 200-nm-wide triangular nanoprisms by e-beam lithography; (c-d) Confocal microscopy images of the NC / PMMA blend film deposited onto the same arrays.

Moreover, the same method was followed to reproduce accurately, on the micrometer scale, the acronym of our laboratory (National Nanotechnology Laboratory, NNL). A strong enhancement of NC fluorescence, along with a precise localization of the phenomenon (due to the underlying pattern), was observed (Fig. 14.6).

**Figure 14.6:** Confocal microscopy image collected on colloidal quantum dots deposited on a metallic nanopattern reproducing the acronym of our laboratory.

Although the proposed approach is based on a quite demanding and time-consuming technology, so that the fabrication of broad-area devices is hindered, the EBL-based MEF strategy offers several interesting properties:

- Both the shape and size of the metallic nanopatterns can be accurately designed, at the nanoscale;
- Intense multicolor (or even white) fluorescence emission can be obtained by using NCs of different sizes;
- Single metallic nanostructures can be properly engineered / optimized owing to the nanometer resolution of the EBL. This allows accurate tailoring of SPs over a wide range of wavelengths, and may enable very strong enhancements while preserving a very precise spatial control.

Lastly, it has been demonstrated that NCs dispersed into polymeric blends sensitive to the electron beam radiation (such as PMMA itself) can be localized with nanometer resolution by exploiting conventional EBL processes [58]. Nanoscale engineering of metal enhanced fluorescence with a complete control over the shape and dimensions of the metallic particle and of the nanoemitter, together with the complete control over the relative vertical and lateral position of the two interacting objects is, therefore, an intriguing target. This may open up attractive perspectives in various optical-based technologies, such as highly sensitive optical sensors and photonic devices operating at nanoscale and / or requiring very low densities of fluorescent emitters.

## 14.3 SPONTANEOUS GALVANIC DISPLACEMENT REACTIONS FOR MEF SUBSTRATES

An alternative and very promising approach to realize substrates suitable for MEF applications is based on spontaneous galvanic displacement reactions (SGDR). Similarly to other deposition techniques previously described, SGDR permits the fabrication of extended metallic substrates; additionally, it allows to control and optimize the surface roughness of the metal film in order to obtain efficient MEF processes, while maintaining an appreciable uniformity of the morphology over the entire sample.

SGDR is based on the reaction of a metallic substrate with an electrolytic solution of metal ions having sufficiently high reduction potential. In this process the metal ions from the electrolytic solution get reduced to the metallic state while simultaneously displacing the metal from the substrate by oxidizing it preferably to a soluble form [59]. Such displacement usually leads to the formation of rough / porous metal films [60, 61]. This process is commercially used to coat desired metals with a thin layer of a more air-stable metal [62]. In electronics industry it is also used to make metal interfaces, interconnects, patterns on semiconducting substrates [59, 63, 64]. However, evolution of porous / nanostructured surfaces is less desirable in such applications and usually modifications are employed to achieve a more continuous film [65]. J.M. Buriak and coworkers have used SGDR process to deposit thin metal films on semiconducting surfaces such as Silicon and GaAs [66]. Furthermore, they demonstrated that this method is also suitable to pattern gold and silver on silicon by using the selectivity of a self-assembling block copolymer over layer [67].

Recently, nanostructured metal surfaces fabricated by using SGDR process have been applied for surface enhanced Raman spectroscopy (SERS) [68]. In our laboratory we are exploiting such process for metal enhanced fluorescence applications. Using SGDR method it is possible to fabricate rough metal surfaces on a number of substrates and fairly control the nanostructure at the surface. This can be achieved by controlling the deposition conditions of the first metal film, and thus its morphological features, as well as the parameters of the SGDR reaction. Notably, we have demonstrated that this procedure can be conveniently applied to fabricate desired micro- or nano-scale patterns of metals by using an intermediate lithographic process, such as optical or electron beam lithography.

In this case, it is also possible to realize patterns of different metals onto the same substrate, preserving the capability to separately control the local surface roughness of each region. Furthermore, this method can be applied to a wide variety of substrates (e.g., silicon, silica, glass, quartz, plastic, etc.), by taking into account eventual differences in the metals adhesion to the substrate and by properly adjusting the SGDR reaction conditions.

In a typical experiment, for instance to prepare a nanostructured metal film on a silica or glass substrate, the surface is preliminary treated by chemical or physical methods to promote the adhesion of the following materials. Subsequently, a metallic layer with controlled thickness is deposited on the substrate by either thermal evaporation or sputter coating (usually, the former technique is preferred since the final roughness is better assessed). By varying the deposition rate and the amount of the deposited material, it is possible to realize an island film or a continuous film with various degrees of roughness (although in a limited range). Immersion of these substrates into a solution containing ions or complexes of a second metal, with a reduction potential higher than that of the pre-deposited metal, leads to the initiation of the SGDR process, i.e., the first layer is replaced by the second metal present in the solution. The extent of replacement on the surface can be controlled by varying the amount and the concentration of the second metal species in the solution, as well as the reaction time (although the presence of trace amount of the first metal in the replaced layer is possible). The reaction proceeds until all the metal ions present in the solution are consumed or till the surface is completely covered by the second metal. Remarkably, by using this method, the replacement of the metal is quite uniform throughout the sample.

We exploited SGDR technique to displace thin silver films (typically in the 15-50 nm thick range) deposited onto glass to obtain rough gold substrates suitable for MEF experiments. The growth process of the nanostructured gold by displacement of silver was monitored in real-time by holographic microscopy and was found to occur within few minutes at ambient conditions (Fig. 14.7). Interestingly, we observed that the growth of the second nanostructured metal occurs by an initial nucleation stage, followed by further growth around the nucleation sites, until a complete replacement of the initial flat layer of silver is achieved.

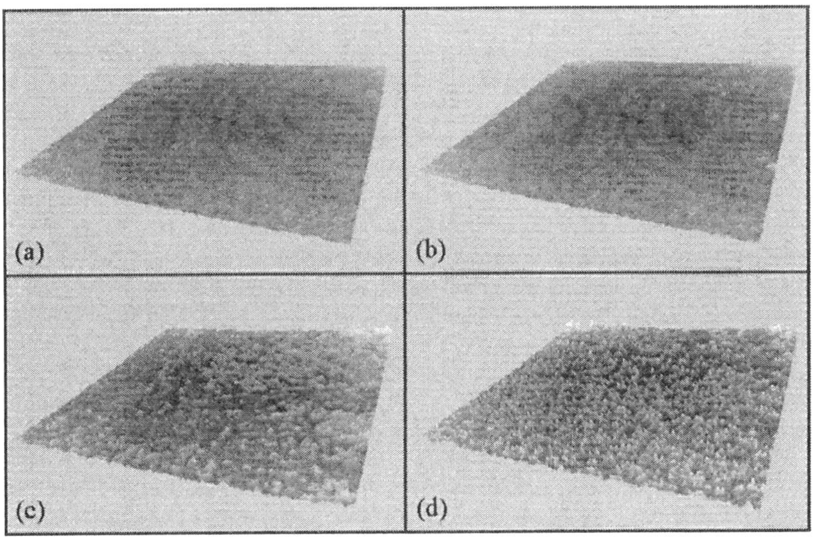


**Figure 14.7:** Real-time investigation of a SGDR reaction by holographic microscopy: a 15 nm thick silver film deposited onto a glass cover-slip was displaced by gold. The images show the evolution of the SGDR reaction, starting from a flat Ag substrate (a), to the initial formation of some gold nuclei (b), followed by further growth of gold around the nucleation sites (c), finally leading to a complete replacement of silver by a rough gold film (d). The entire process (from a to d) is typically completed in a few minutes. The region analyzed in the figure is ca. 50x50 microns.

As shown in Fig. 14.7(d), the SGDR-based growth results in the formation of a nanostructured gold film, characterized by a significant and quite uniform roughness.

Since the displacement reaction initiates at the surface of the metal, it is also possible to use a resist mask to restrict the replacement of the metal in desired confined regions. In general, the SGDR process can be carried out using either aqueous or organic solutions, therefore no specific limitations on the choice of both substrate and mask materials are imposed by the chemical environment of the reaction.

The schematic sequence of the approach used to obtain localized SGDR processes is reported in Fig. 14.8.

- A resist mask is realized onto a silver film through standard lithographic techniques, such as optical or E-beam lithography (Fig. 14.8a);

- then, the sample is immersed in a suitable gold ions solution, so that the displacement reaction occurs only in the unmasked regions (Fig. 14.8b);

- lastly, the resist mask is removed, leading to the final sample configuration, in which nanostructured gold stripes alternate with flat silver regions (Fig. 14.8c).

The entire procedure was assessed, step by step, by holographic microscopy (Fig. 14.8, right column): the initial flat silver film, localized between the resist stripes, is replaced by gold clusters, exhibiting a significant and uniform roughness. The remarkable contrast in the surface morphology of the two materials is particularly evident in the last image, after the resist removal.

Noteworthy, this approach allows the realization of arbitrary features, thus opening the possibility to pattern two different metals onto the same substrate with desired geometries. Moreover, as also discussed in the previous section, the use of lithographic techniques permits a high resolution control of the masking layer and, therefore, a precise spatial localization of MEF effects on the micro- and nanoscale. This latter possibility is demonstrated, for instance, in Fig. 14.9, where we exploited E-beam lithography to realize a pattern composed of a few micrometer wide alternating stripes of gold and silver. In this case, the conditions of thermal evaporation of the silver layer were set to obtain an island film, instead of a continuous flat layer, as shown in the darker regions of the high magnification scanning electron microscope image of Fig. 14.9. It is evident that the brighter contrast gold zone, where the SGDR process is efficiently confined, thanks to the lithographic mask, indeed exhibits a more pronounced nanostructured / porous morphology. The nanoscale structures observed in the gold region may lead to a local strong enhancement of the electromagnetic field, in turn eliciting significant increase of the emission intensity of fluorophores distributed over such rough metallic surfaces.

Fluorescence enhancement studies were carried out on gold substrates prepared using the SGDR method. Fluorescein-5-isothiocyanate (FITC) was used as a model fluorophore for this purpose. These MEF experiments were carried out by using a nanostructured gold film made by SGDR process over a 50 nm thick silver layer thermally deposited onto a silica substrate. Silver was deposited only in a selected central region of the substrate (and then completely replaced by gold) in order to compare the increase of the fluorescence intensity of FITC localized onto the nanostructured gold with respect to the FITC emission onto $SiO_2$ (since the FITC emission onto the flat silver film is significantly quenched). Atomic Force Microscopy (AFM) analyses assessed a low surface roughness of the thermally evaporated silver layer (mean surface roughness r~2 nm), while the SGDR-fabricated gold film exhibited a remarkable roughness, with a typical peak-to-valley ratio of 70-80 nm and r~18 nm (not shown here). A monolayer of FITC was then uniformly deposited onto the entire sample (both on gold and silica) by chemical functionalization of the two materials with short spacer molecules (~1 nm) exposing amine groups. We observed by confocal microscopy a significant enhancement (higher than 10-fold) of the fluorescence emission of FITC conjugated over the nanostructured gold substrate with respect to the reference emission onto the $SiO_2$ substrate (Fig. 14.10).

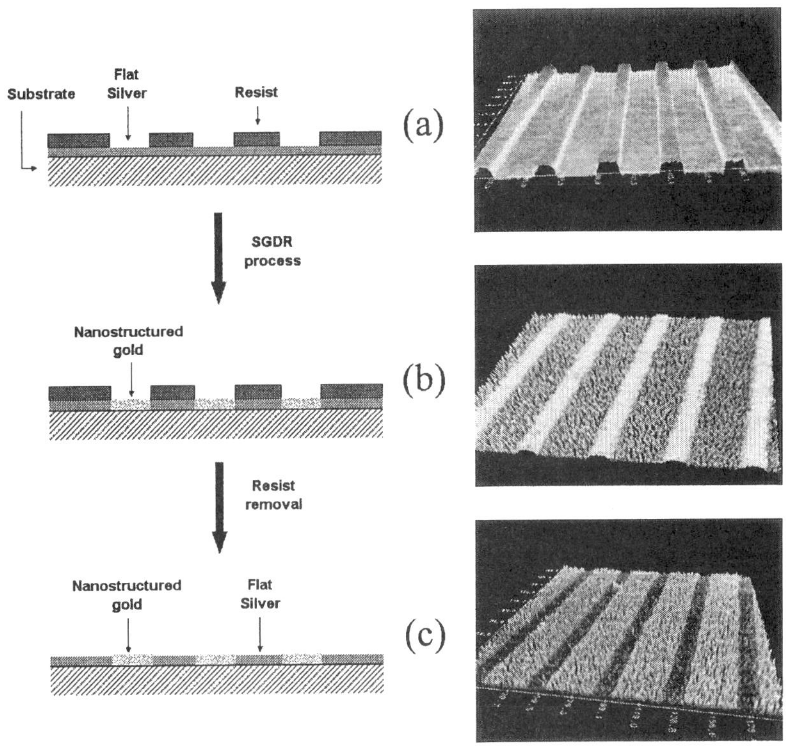


**Figure 14.8:** Localized SGDR process. (Left) schematic and (right) holographic microscopy images of the different steps of the procedure: (a) deposition of a resist mask layer on a thermally evaporated silver film; (b) local displacement of silver by gold ions resulting in the formation of nanostructured gold stripes; (c) removal of the resist, showing the almost unperturbed flat morphology of the masked Ag regions. The pattern shown in the right column was realized by optical lithography (the resist pattern was composed by 15 μ wide stripes with a period of 40 μ).

The enhancement effect was found to be quite uniform over the entire gold substrate, as expected from the appreciable uniformity of the surface morphology obtainable by SGDR processes. The spectral line-shape of the MEF-increased FITC emission was very similar to that of the reference fluorophore, with no detectable modifications / shifts caused by the interaction with gold plasmons. We also observed that, upon prolonged exposure to high intensity radiation, the FITC localized on the gold substrate underwent a significantly faster photobleaching as compared to the fluorescein on silica (under otherwise identical conditions). These data indicate that the enhancement mechanism is mostly due to an increase in the excitation rate of FITC molecules, elicited by a large increase in the local density of the excitation field close to the gold nanostructures [1]. It is likely that the close proximity of the fluorophore to the metallic surface (~1 nm) used in our experiments may also induce some quenching effect of the FITC emission; a systematic investigation by exploiting longer spacer molecules is, therefore, of great interest and may allow to find an optimal FITC-gold distance to maximize the MEF enhancement factor [6, 15].

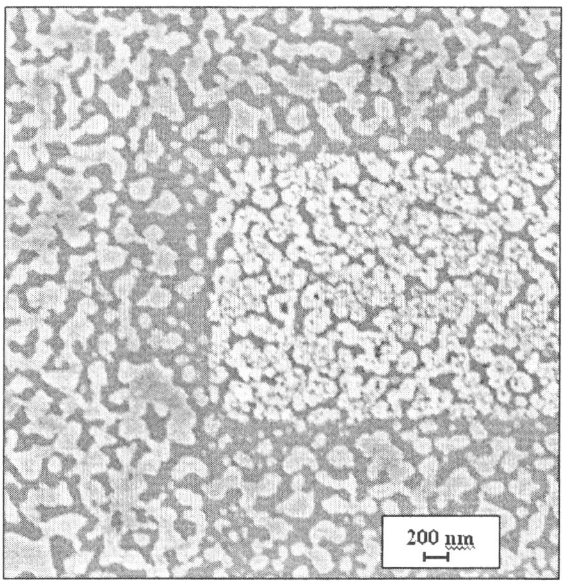


**Figure 14.9:** Scanning electron microscope image of a silver islands film on a silicon substrate, after selective masking with an EBL-patterned resist layer, subsequent immersion in a gold ions aqueous solution and final resist removal. The brighter region corresponds to a ~2 μm wide Au stripe, while the surrounding area corresponds to the initially evaporated Ag layer.

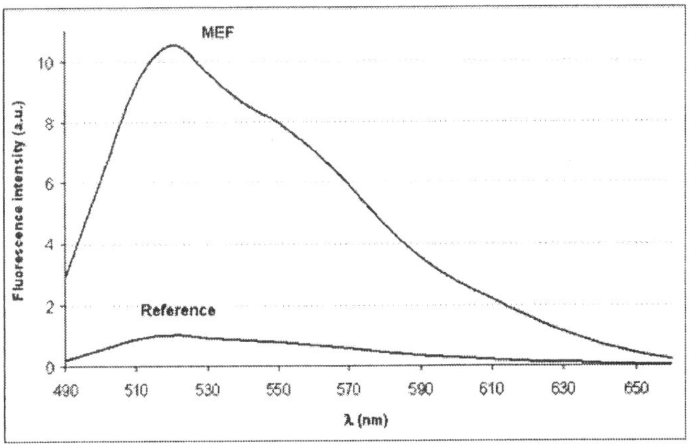


**Figure 14.10:** Fluorescence spectra of FITC collected on the nanostructured gold (MEF, upper line) and on $SiO_2$ substrate (reference, lower line) by confocal microscopy. Both the curves were scaled so that the peak intensity of the reference emission was normalized to 1. The excitation wavelength was 458 nm.

The spontaneous Galvanic displacement reaction has, therefore, proved to be an effective and convenient way to realize MEF-suitable metallic substrates. The high uniformity of the roughness over extended areas, along with the possibility to precisely localize different metals (with different roughness) and thus the MEF effect with nanoscale resolution makes this process a very attractive technique for both research and commercial applications. Moreover, since the SGDR process just depends on the relative redox potential of the pre-deposited and the solvated metal ions present in the solution, it can be extended to a wide range of different metals and / or repeatedly performed on the same sample thus allowing the fabrication of patterned substrates made by two or more than two different metals.

## 14.4 REFERENCES

1. Lakowicz, J. R. (2001). Radiative decay engineering: biophysical and biomedical applications. *Anal. Biochem.* **298**: 1-24.
2. Feldstein, M. J., Golden, J. P., Rowe, C. A., MacCraith, B. D., and Ligler, F. S. (1999). Array biosensor: Optical and fluidics systems. *J. Biomed. Microdevices* **1:** 139-153.
3. Schalkhammer, T., Aussenegg, F.R., Leitner, A., Brunner, H., Hawa, G., Lobmaier, C., and Pittner, F. (1997). Detection of fluorophore-labelled antibodies by surface-enhanced fluorescence on metal nanoislands. *SPIE* **2976**: 129-136.
4. Hobson, P. A., Wedge, S., Wasey, J. A. E., Sage, I., and Barnes, W. L. (2002). Surface plasmon mediated emission from organic light-emitting diodes. *Adv. Mater.* **14**, 19: 1393-1396.

5. Malicka, J., Gryczynski, I., and Lakowicz, J. R. (2003). DNA hybridization assays using metal-enhanced fluorescence. *Biochem. Biophys. Res. Comm.* **306**: 213-218.
6. Zhang, J., Matveeva, E., Gryczynski, I., Leonenko, Z., and Lakowicz, J. R. (2005). Metal-Enhanced Fluoroimmunoassay on a Silver Film by Vapor Deposition. *J. Phys. Chem. B* **109**, 7969-7975.
7. Mackowski, S., Wolrmke, S., Maier, A. J., Brotosudarmo, T. H. P., Harutyunyan, H., Hartschuh, A., Govorov, A. O., Scheer, H., and Braluchle, C. (2008). Metal-Enhanced Fluorescence of Chlorophylls in Single Light-Harvesting Complexes. *Nano Lett.* **8**, 2: 558-564.
8. Chen, Y., Munechika, K., and Ginger, D. S. (2007). Dependence of fluorescence intensity on the spectral overlap between fluorophores and plasmon resonant single silver nanoparticles. *Nano Lett.* **7**, 3: 690-696.
9. Drexhage, K. H. (1970). Influence of a dielectric interface on fluorescence decay time, *J. Luminesc.* **1,2**: 693-701.
10. Barnes, W. L. (1998). Fluorescence near interfaces: the role of photonic mode density. *J. Modern Optics* **45**, 4: 661-669.
11. Weitz, D. A., Garoff, S., Gersten, J. I., Nitzan, A. (1983). The enhancement of Raman scattering, resonance Raman scattering, and fluorescence from molecules adsorbed on a rough silver surface. *J. Chem. Phys.* 78, 9: 5324-5338.
12. Amos, R. M., and Barnes, W. L. (1999). Modification of spontaneous emission lifetime in the presence of corrugated metallic surfaces. *Phys. Rev. B* **59**, 11: 7708-7714.
13. Shimizu, K.T., Woo, W.K., Fihier, B.R., Eisler, H.J., and Bawendi, M.G. (2002). Surface-enhanced emission from single semiconductor nanocrystals. *Phys. Rev. Lett.* **89**, 11: 117401.
14. Neal, T.D., Okamoto, K., and Scherer, A. (2005). Surface plasmon enhanced emission from dye doped polymer layers. *Optics Express* **13**, 14: 5522-5527.
15. Ray, K., Badugu, R., and Lakowicz, J. R. (2007). Polyelectrolyte layer-by-layer assembly to control the distance between fluorophores and plasmonic nanostructures. *Chem. Mater.* **19**, 24: 5902-5909.
16. Weitz, D. A., Garoff, S., Hanson, C. D., and Gramila, T. J. (1982). Fluorescent lifetimes of molecules on silver-islands films. *Optics Lett.* **7**,2: 89-91.
17. Hayakawa, T., Selvan, S. T., and Nogami, M.(1999). Field enhancement effect of small Ag particles on the fluorescence from $Eu^{3+}$-doped $SiO_2$ glass. *Appl. Phys. Lett.* **74**, 11: 1513-1515.
18. Tovmachenko, O. G., Graf, C., van den Heuvel, D., van Blaaderen, A., and Gerritsen, H. C. (2005). Fluorescence enhancement by metal-core / silica-shell nanoparticles. *Adv. Mater.* **18**, 1: 91-95.
19. Nakamura, Y., and Hayashi, S. (2005). Enhancement of dye fluorescence by gold nanoparticles: analysis of particle size dependence. *Jap. J. Appl. Phys.* **44**, 9A: 6833-6837.
20. Song, J. -H., Atay, T., Shi, S., Urabe, H., and Nurmikko, A. V. (2005). Large enhancement of fluorescence efficiency from CdSe / ZnS Quantum Dots induced by resonant coupling to spatially controlled surface plasmons. *Nano Lett.* **5**, 8: 1557-1561.

21. Barber, P.W., Chang, R.K., and Massoudi, H. (1983). Surface-enhanced electric intensities on large silver spheroids. *Phys. Rev. Lett.* **50**, 13: 997-1000.
22. Kreibig, U., and Vollmer, M. (1995). *Optical properties of Metal Clusters*, Springer Series in Materials Science (Ed.), Springer Verlab, Berlin.
23. Link, S., and El-Sayed, M.A. (1999). Spectral Properties and Relaxation Dynamics of Surface Plasmon Electronic Oscillations in Gold and Silver Nanodots and Nanorods. *J. Phys. Chem. B.* **103**: 8410-8426.
24. Link, S., and El-Sayed, M.A (2000). Shape and size dependence of radiative, nonradiative and photothermal properties of gold nanocrystals. *Int. Rev. Phys. Chem.* **19**: 409-453.
25. Wokaun, A., Lutz, H. -P., King, A. P., Wild, U. P., and Ernst, R. R. (1983). Energy transfer in surface enhanced luminescence. *J. Chem. Phys.* **79**, 1: 509-514.
26. Pinchuk, A., Hilger, A., von Plesse, G., and Kreibig, U. (2004). Substrate effect on the optical response of silver nanoparticles. *Nanotechnology*, **15**: 1890-1896.
27. Gryczynski, I., Malicka, J., Jiang, W., Fischer, H., Chan, W.C.W., Gryczynski, Z., Grudzinski, W., and Lakowicz, J.R. (2005). Surface-plasmon-coupled emission of quantum dots. *J. Phys. Chem. B* **109:** 1088-1093.
28. Okamoto, K., Vyawahare, S., and Scherer, A. (2006). Surface-plasmon enhanced bright emission from CdSe quantum-dot nanocrystals. *J. Opt Soc. Am. B* **23**, 8: 1674-1678.
29. Erlebacher, J., Aziz, M. J., Karma, A., Dimitrov, N., and Sieradzki, K. (2001). Evolution of nanoporosity in dealloying. *Nature* **410**: 450-453.
30. Biteen, J. S., Pacifici, D., Lewis, N. S., and Atwater, H. A. (2005). Enhanced Radiative Emission Rate and Quantum Efficiency in Coupled Silicon Nanocrystal-Nanostructured Gold Emitters. *Nano Lett.* **5**, 9: 1768-1773.
31. Geddes, C.D., Cao, H., Gryczynski ,I., Gryczynski, Z., Fang, J., and Lakowicz, J.R. (2003). Metal-enhanced fluorescence due to silver colloids on a planar surface: potential applications of indocyanine green to in vivo imaging. *J. Phys. Chem. A* **107**: 3443-3449.
32. Aslan, K., Lakowicz, J.R., and Geddes, C.D. (2005). Rapid deposition of triangular silver nanoplates on planar surfaces: application to metal-enhanced fluorescence. *J. Phys. Chem. B* **109**¨6247-6251.
33. Aslan, K., Lakowicz, J.R., and Geddes, C.D. (2005). Metal-enhanced fluorescence using anisotropic silver nanostructures: critical progress to date. *Anal. Bioanal. Chem.* **382**: 926-933.
34. Xie, F., Baker, M. S., and Goldys, E. M. (2008). Enhanced Fluorescence Detection on Homogeneous Gold Colloid Self-Assembled Monolayer Substrates. *Chem. Mater.* **20**, 5: 1788–1797.
35. Tao, A., Sinsermsuksakul, P., and Yang. P (2007). Tunable plasmonic lattices of silver nanocrystals. *Nature Nanotech.* **2**: 435-440.
36. Smith, D. K., and Korgel, B. A. (2008). The Importance of the CTAB Surfactant on the Colloidal Seed-Mediated Synthesis of Gold Nanorods. *Langmuir* **24**, 644-649.

37. Pyrpassopoulos, S., Niarchos, D., Nounesis, G., Boukos, N., Zafiropoulou, I., and Tzitzios, V. (2007). Synthesis and self-organization of Au nanoparticles. *Nanotechnology* **18**: 485604.
38. Xiang, Y., Wu, X., Liu, D., Feng, L., Zhang, K., Chu, W., Zhou, W., and Xie, S. (2008). Tuning the Morphology of Gold Nanocrystals by Switching the Growth of {110} Facets from Restriction to Preference. *J. Phys. Chem.* C **112**: 3203-3208.
39. Xie, F., Baker, M. S., and Goldys, E. M. (2006). Homogeneous Silver-Coated Nanoparticle Substrates for Enhanced Fluorescence Detection. *J. Phys. Chem. B* **110**: 23085-23091.
40. Martiradonna, L., Carbone, L., De Giorgi, M., Manna, L., Gigli, G., Cingolani, R., and De Vittorio, M. (2006). High Q-factor colloidal nanocrystal-based vertical microcavity by hot embossing technology. *Appl. Phys. Lett.* **88**: 181108.
41. Martiradonna, L., Carbone, L., Tandaechanurat, A., Kitamura, M., Iwamoto, S., Manna, L., De Vittorio, M., Cingolani, R., and Arakawa, Y. (2008). Two-Dimensional photonic crystal resist membrane nanocavity embedding colloidal dot-in-a-rod nanocrystals. *Nano Lett*, **8**, 1: 260-264.
42. Gur, I., Fromer, N. A., Geier, M. L., and Alivisatos, A. P. (2005). Air-stable all-inorganic nanocrystal solar cells processed from solution, *Science* **310**: 462-465.
43. Medintz, I. L., Clapp, A. R., Mattoussi, H., Goldman, E. R., Fisher, B., and Mauro, J. M. (2003). Self-assembled nanoscale biosensors based on quantum dot FRET donors, *Nature Materials* **2**: 630-638.
44. Medintz, I. L., Uyeda, H. T., Goldman, E. R., and Mattoussi, H., (2005). Quantum dot bioconjugates for imaging, labelling and sensing, *Nature Materials* **4**: 435-446.
45. So, M. K., Xu, C., Loening, A. M., Gambhir, S. S., and Rao, J., (2006). Self-illuminating quantum dot conjugates for in vivo imaging, *Nature Biotechnol.* **24**: 339-343.
46. Bruchez, M., Moronne, M., Gin, P., Weiss, S., and Alivisatos, A. P., (1998). Semiconductor nanocrystals as fluorescent biological labels, *Science* **281**: 2013-2016.
47. Chan W. C. W., and Nie, S., (1998). Quantum dot bioconjugates for ultrasensitive nonisotopic detection, *Science* **281**: 2016-2018.
48. Mattoussi, H., Mauro, J. M., Goldman, E. R., Anderson, G. P., Sundar, V. C., Mikulec, F. V., and Bawendi, M. G. (2000). Self-assembly of CdSe-ZnS Quantum Dot bioconjugates using an engineered recombinant protein. *J. Am. Chem. Soc.* **122**: 12142-12150.
49. Klimov, V. I. (2004) *Semiconductor and metal nanocrystals*, Marcel Dekker, Inc., New York.
50. Pompa, P. P., Martiradonna, L., Della Torre, A., Della Sala, F., Manna, L., De Vittorio, M., Calabi, F., Cingolani, R., and Rinaldi, R. (2006). Metal enhanced fluorescence of colloidal nanocrystals with nanoscale control. *Nature Nanotechnology* **1**: 126-130.
51. Pompa, P. P., Martiradonna, L., Della Torre, A., del Mercato, L. L., Manna, L., De Vittorio, M., Calabi, F., Cingolani, R., and Rinaldi, R. (2007). Fluorescence enhancement in colloidal semiconductor nanocrystals by metallic nanopatterns. *Sens. Actuators B: Chem.* **126**, 1: 187–192.

52. Purcell, E. M., and Pennypacker, C. R. (1973). Scattering and absorption of light by nonspherical dielectric grains. *Astrophys. J.* **186:** 705-714.
53. Draine, B. T., and Flatau, P. J. (1994). Discrete dipole approximation for scattering calculations. *J. Opt. Soc. Am. A* **11:** 1491-99.
54. Boren, C. F., and Huffman, D. R. (1983). *Absorption and Scattering of Light by Small Particles.* Wiley-Interscience, New York.
55. Yang, W. -H., Schatz, G. C., and Van Duyne, R. P. (1995). Discrete dipole approximation for calculating extinction and Raman intensities for small particles with arbitrary shapes. *J. Chem. Phys.* **103:** 869-875.
56. Sosa, I. O., Noguez, C., and Barrera, R. G. (2003). Optical Properties of Metal Nanoparticles with Arbitrary Shapes. *J. Phys. Chem. B* **107:** 6269-6275.
57. Kelly, K.L., Coronado, E., Zhao, L.L., and Schatz, G.C. (2003). The Optical Properties of Metal Nanoparticles: The Influence of Size, Shape, and Dielectric Environment. *J. Phys. Chem. B* **107:** 668-677.
58. Martiradonna, L., Stomeo, T., De Giorgi, M., Cingolani, R., De Vittorio, M. (2006). Nanopatterning of colloidal nanocrystals emitters dispersed in a PMMA matrix by e-beam lithography. *Microel. Engin.* **83,** 1478-1481.
59. Carraro, C., Maboudian, R., and Magagnin, L. (2007). Metallization and nanostructuring of semiconductor surfaces by galvanic displacement processes. *Surf. Sci. Rep.* **62**: 499-525.
60. Bansal, V., Jani, H., Plessis, D. J., Coloe, P. J., and Bhargava, S. K. (2008). Galvanic Replacement Reaction on Metal Films: A One-Step Approach to Create Nanoporous Surfaces for Catalysis. *Advanced Materials* **20**: 717-723.
61. Porter, L. A., Jr., Choi, H. C., Ribbe, A. E., and Buriak, J. M. (2002). Controlled Electroless Deposition of Noble Metal Nanoparticle Films on Germanium Surfaces *Nano Lett.* **2**: 1067-1071.
62. Schlesinger, M., and Paunovic, M. (2000) Modern Electroplating, 4th ed., John Wiley & Sons, Inc, New York.
63. Okinaka, Y., and Hoshino, M. (1998). Some Recent Topics in Gold Plating for Electronics Applications *Gold Bull.* **31**: 3-13.
64. Lee, D.-J., and Lee, H.S. (2006). Major factors to the solder joint strength of ENIG layer in FC BGA package. *Microelectron. Reliab.* **46**: 1119-1127.
65. Vasilica, R., and Dimitrov, N. (2005). Epitaxial Growth by Monolayer-Restricted Galvanic Displacement. *Electrochem. Solid-State Lett.,* **8**: C173-C176.
66. Lon, A., Porter, Jr., Choi, H. C., Schmeltzer, J. M., Ribbe, A. E., Elliott, L. C. C., and Buriak, J. M. (2002). Electroless Nanoparticle Film Deposition Compatible with Photolithography, Microcontact Printing, and Dip-Pen Nanolithography Patterning Technologies. *Nano Lett.* **2**: 1369-1372.
67. Aizawa, M., and Buriak, J. M. (2006). Nanoscale patterning of two metals on silicon surfaces using an ABC triblock copolymer template. *J. Am. Chem. Soc.* **128**: 5877-5886.
68. Liu, Y.-C., Yu, C.-C. and Hsu, T.-C. (2007) Trace molecules detectable by surface-enhanced Raman scattering based on newly developed Ag and Au nanoparticles-containing substrates. *Electrochem. Commun.* **9**: 639–644.

# 15

# Metal Enhanced Chemiluminescence

Yongxia Zhang, Kadir Aslan, Chris D. Geddes*
Institute of Fluorescence, laboratory for Advanced Medical Plasmonics and Laboratory for Advanced Fluorescence Spectroscopy
Medical Biotechnology Center, University of Maryland Biotechnology Institute, 701East Pratt St., Baltimore, MD, 21202, USA.

## 15.1 INTRODUCTION OF CHEMILUMINESCENCE

There is an increasing demand for the use of chemiluminescence as an analytical tool for quantitative detection in biotechnology[1, 2], such as in protein detection assays, immunoassays and in small molecule detection in more than 20 % of clinical laboratories in the USA today[2]. The attractiveness of chemiluminescence lies in the fact that no unwanted background luminescence, no excitation source and no optical filters are required, as compared to fluorescence [3-5]. However, chemiluminescence based detection is limited by the quantum efficiency of the chemiluminescence reaction or probe, i.e. poor signal-to–noise ratios at low analyte concentration and the long time before depletion of the reactants (i.e. slow glow). In this regard, an increased chemiluminescence yield and accelerated chemiluminescence reaction would clearly be beneficial for the sensitivity and rapidity of chemiluminescence based bioassays and technologies.

In this chapter, we summarize the Metal-Enhanced Chemiluminescence (MEC) phenomena [6, 7] in which sub-wavelength sized metallic nanoparticles amplify chemiluminescence emission, and other techniques subsequently derived from MEC, such as Microwave-triggered Metal–Enhanced Chemiluminescence (MT-MEC), which combines increasing chemiluminescence intensity with kinetic acceleration, and their subsequent applications in clinic assays and diagnostics[8, 9]. In addition, we also show how the use of Finite Different Time Domain simulations (FDTD), to visualize electric field distributions to optimize model geometries, can be used to enhance chemiluminescence[10]. Finally we discuss surface plasmon coupled chemiluminescence for creating directional chemiluminescence signatures as compared to the traditional isotropic emission. All of these chemiluminescence technologies provide for either significantly enhanced intensities and /or rapid chemiluminescence signatures.

## 15.2 METAL ENHANCED CHEMILUMINESCENCE (MEC)

Over the last several years, many groups have described the use of metallic surfaces and sub-wavelength sized metallic nanoparticles to modify both the far and near-field emissive properties of optically excited fluorophores, a technology named

*Metal-Enhanced Chemiluminescence.*
Edited by Chris D. Geddes.

Metal Enhanced Fluorescence (MEF) by Geddes [11-13]. The emission of fluorophores in close-proximity to metallic nanostructures was originally thought to originate solely from the fluorophore, the excited plasmons interacting with fluorophores and changing their free-space spectral characteristics (figure 15.1-Top). However, only recently the interpretation of metal-enhanced fluorescence has changed to one whereby excited fluorophores can non-radiatively transfer energy and couple to surface plasmons which in turn, radiate the fluorophores photophysical characteristics, in essence the fluorophore-metal system radiates (Figure 15.1-middle). It has also been reported that the interactions of silver nanoparticles with chemiluminescent species, results in an increase in the detectability of chemiluminescent reactions / species, with an approximately 20-fold increase in signal intensity, attributed to a plasmon-based luminescent enhancement. It was shown that surface plasmons can be directly excited by chemically induced electronically excited molecules (Figure 15.1-bottom). This phenomenon has been named metal-enhanced chemiluminescence (MEC)[6, 7].

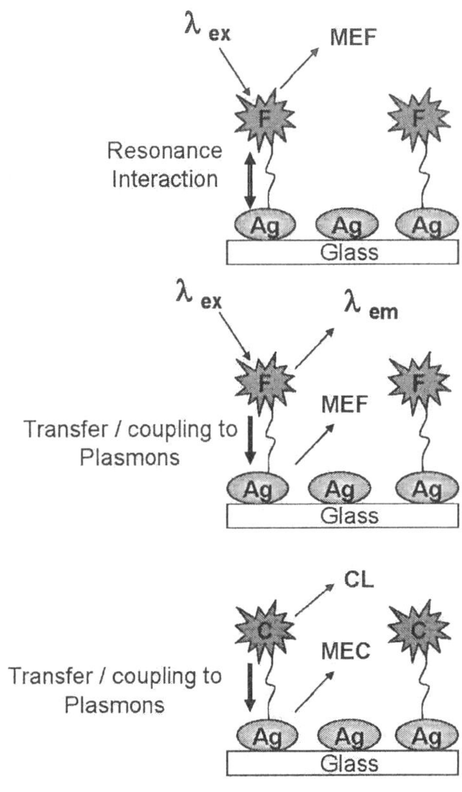


**Figure 15.1:** Graphical representations of Metal-Enhanced Fluorescence (MEF) (Top and Middle) and for Metal-Enhanced Chemiluminescence (MEC) (Bottom). F – Fluorophore, C – Chemiluminescence species / probe and CL – Chemiluminescence.

The MEC phenomenon was first demonstrated using a commercial light glow stick in the following experimental procedure. The glow stick contains the activating agent (hydrogen peroxide) and reaction chemicals. Activation of the chemicals is accomplished with a bend, snap and shake of the glass tube which breaks the glass capsule containing hydrogen peroxide. The hydrogen peroxide oxidizes the phenyl oxalate ester to a peroxyacid ester and phenol, the process chemically inducing an electronic excited state. After chemiluminescence initiation, approximately 70 μl of the glow stick fluid was placed between two glass microscope slides, clamped together. The glass slides contained silver island films on one end and were bare glass on the other end (a control sample). The bare end of the glass served as the control sample by which to compare the benefits of using the metal to plasmon-enhance chemiluminescence (Figure 15.2-bottom).

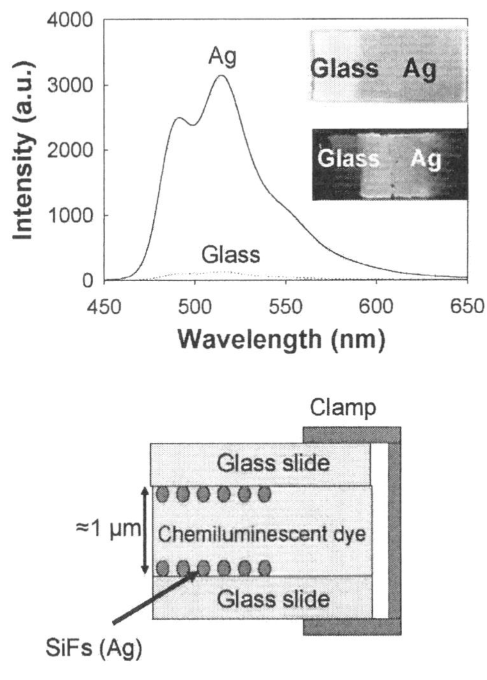


**Figure 15.2:** Chemiluminescence emission intensity from both the glass and the silvered surface (Ag) (Top). Insert – photographs of the silvered and glass surfaces, with (insert - Bottom) and without (insert - Top) chemiluminescence material in the experimental sandwich. The enhancement factor was > 20, i.e. intensity on Ag / intensity on glass. Experimental sample sandwich (Bottom). Reproduced from *Applied Physics Letters* 88: 173104. (2006).

The chemiluminescence emission spectra from between the silvered glass and glass plates are shown in figure 15.2 top. The emission from the silvered portion of the slide was spatially averaged to be about 4-5 times greater than the glass control side of the sample. In addition, the volume between both the sandwiched glass and silver slides was identical. Figure 15.2 –insert shows the photographs of the slides, both before and after the addition of the chemiluminescent material. Approximately 70 μL of fluid was enough to form a thin coating across both portions of the slide, held by capillary action as the slides were sandwiched. The enhanced chemiluminescence is clearly visible on the silvered portion and is very weak in the glass region of the slide, in fact almost undetectable.

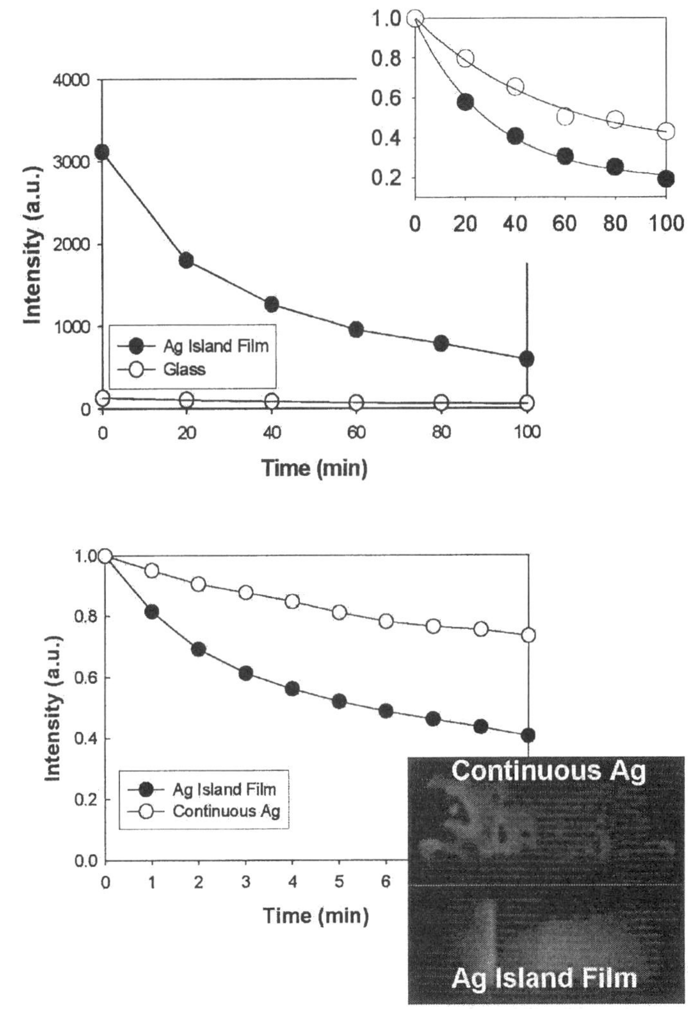


**Figure 15.3:** Chemiluminescence intensity decay measured on both SiFs and glass as a function of time (Top) and the data normalized (Top-insert). Normalized chemiluminescence intensity on both SiFs and a continuous silver film (Bottom). Photograph of the emission from both the continuous silver film and the

SiFs (Bottom – insert). Ag – Silver. SiFs – Silver Island Film. Reproduced from *Applied Physics Letters* 88: 173104 (2006).

The luminescence intensity as a function of time (chemiluminescence decay) is shown in figure 15.3. Clearly the enhanced luminescence from the SiFs is visible, with the initial intensity on silver ≈ 3100 a.u. (at t = 0) as compared to < 150 on glass. The rates of loss of chemiluminescence were compared after the curves were normalized, Figure 15.3 – top insert. From Figure 15.3 – Top insert, the rate of depletion on silver was found to be 1.7 times faster than on glass, 0.034 vs 0.019 $s^{-1}$ respectively. Two explanations could initially describe this observation: Firstly, silver catalysis of the chemiluminescence reaction, or secondly, the high rate of transfer / coupling of the chemiluminescence to surface plasmons, rapidly reducing the excited state lifetime of the chemiluminescence species. To eliminate silver based catalysis of the chemiluminescence reaction as an explanation for the enhanced signals, the luminescence rates on both SiFs and a continuous silver strip was measured. Interestingly, the rate of loss of luminescence was still found to be greater on the SiFs as compared to the continuous silver strip, Figure 15.3 – bottom. In addition, the emission intensity was very low indeed from the continuous strip of silver, Figure 15.3 – bottom insert. Given that the continuous strip is indeed *darker* and that the rate is slower than on SiFs, then silver based catalysis can be eliminated as a possible explanation of the observation of increased signal intensities on the SiFs. Subsequently, these observations suggest that chemically induced electronic excited states (chemiluminescence species) can readily induce/couple to surface plasmons, facilitating metal-enhanced chemiluminescence. Further, the reduced luminescence half-time and increased emission intensity observed, is very consistent with the findings for nanosecond decay time fluorophores sandwiched between identical silver nanostructures, similarly suggesting that the radiating plasmon model [14]is most likely also applicable to chemically induced electronic excited states.

## 15.3. MICROWAVE-TRIGGERED METAL-ENHANCED CHEMILUMINESCENCE (MT-MEC)

In addition to their utility in increasing chemiluminescence intensity, silver nanoparticles, in combination with low power microwaves, have also been shown to kinetically accelerate the chemical reactions that produce chemiluminescence[14].

In order to demonstrate the "on-demand" nature of the MT-MEC process, time-dependent chemiluminescent emission of a blue Acridan-based Chemiluminescence (CL) reagent on SiFs and glass surfaces, with multiple microwave exposures (Figure 15.4) and without any microwave exposures (Figure 15.4 inset) was recorded and compared for 2000 seconds. The exposure of the blue CL reagent to microwaves (multiple exposures, all 20 % power setting) results in an increase in the CL emission, which is observed as "triggered spikes" consistent with the rising edge of the microwave pulses in the graph (microwave pulses not shown). The largest increase (120 A.U. to 3300 A.U. on average) in CL intensity was observed during the first five microwave exposures and diminished upon further exposures, as chemiluminescent material is depleted. In all the experiments performed with low power microwaves, using both SiFs and glass, there was no evidence of surface drying. This is attributed to the previously made observations that the temperature increase of the aqueous solution on the surfaces due to microwave heating is only ~ $8^0$C (to ≈ 28 $^0$C) for a 30 μl of aqueous sample. The initial intensities at t = 0 seconds for both graphs are ≈ 120 A.U. The "triggered spikes" in the intensity, indicate the individual microwave exposure (10 seconds, 20% power). The inset shows the time-dependent emission (No microwave exposure) and the real-color photographs of the blue chemiluminescent reagent (before and after Mw exposure) on SiFs and glass surfaces.

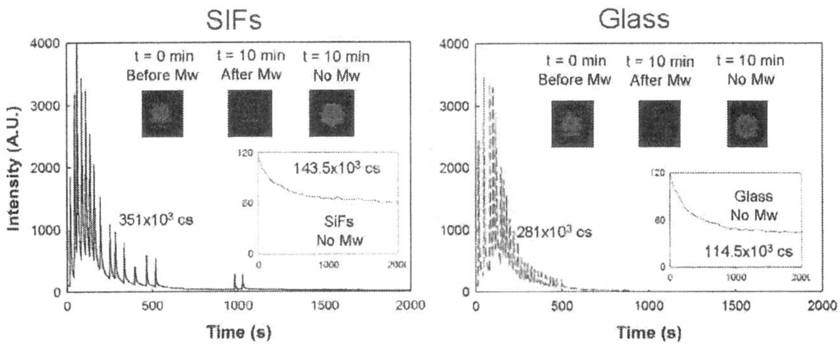


**Figure 15.4:** Time-dependent microwave-triggered chemiluminescence emission (intensity: arbitrary units) for a blue chemiluminescent reagent (10 ml) on silver island films (SiFs) (Left), and glass surfaces (Right) before, during and after low power Mw exposure. The initial intensities at t = 0 seconds for both graphs are 120 A.U. The "triggered spikes" in the intensity, indicate the individual Mw pulses (10 seconds, 20% power). The inset shows the time-dependent emission (No Mw exposure) and the real-color photographs of the blue chemiluminescent reagent (before and after Mw exposure) on SiFs and glass surfaces. The area under each curve, i.e. total photon flux, is given in terms of photon counts (cs). The final intensities at t = 2000 seconds for Figure 15.4-top and bottom are 30 and 25 A.U., respectively. Reproduced from *Journal of the American Chemical Society* 128: 13372-13373 (2006).

The number of photons detected from the blue CL reagent on the SiFs and glass surfaces after microwave exposures in 2000 seconds is 351 x $10^3$ and 281 x $10^3$ counts, respectively, which is significantly higher than those obtained *without* microwave exposures, 143.5 x $10^3$ and 114.5 x $10^3$ counts for SiFs and glass, respectively. This corresponds to a 2.45-fold increase in photon flux on both SiFs and glass surfaces. It is important to note that a 2.45-fold increase in photon flux represents the *average* increase in the overall photon flux from the ensemble of chemiluminescent species for 2000 seconds. One can complete the CL reaction with a single microwave exposure for 10 seconds that will yield a similar final photon flux for the CL reactions without microwave heating (i.e. 10 sec Vs 2000 sec). Considering the fact that the CL reactions currently in use are usually completed within 5 hours, the MT-MEC technique [14] provides researchers with an increased detectability (Plasmon enhanced and microwave accelerated) and a significant reduction in CL detection time, to as low as 10 seconds.

### 15.3 1. *Application of MT-MEC to Ultra-fast and Ultra-Sensitive Clinical Assays*

To demonstrate MT-MEC as a useful platform for protein quantification, a simple surface biotin-avidin assay was constructed[15, 16]. In the assay, biotinylated-BSA is incubated on both silvered and glass substrates (Figure 15.5). HRP-streptavidin is then added to the surface, localizing the enzyme catalyst in close proximity to the silver for MT-MEC. The peroxide and Acridan (lumophore) are then added to initiate the chemiluminescence reaction. While this assay in essence determines BSA concentration, this *model assay* could indeed be fashioned to both localize and sense other proteins / DNAs of interest.

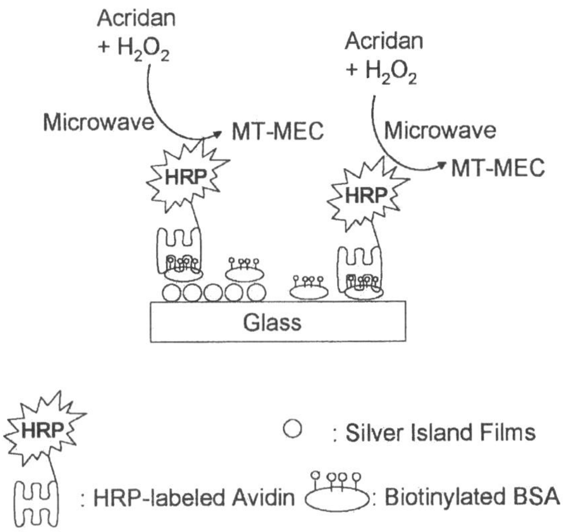


**Figure 15.5:** HRP-acridan chemiluminescence assay on both glass and silvered slides. Reproduced from *Anal Chem* 78: 8020-8027, (2006).

Figures 15.6 and 15.7 demonstrates the corresponding signal enhancement on glass and Ag substrates in the presence of low power microwave pulses. The Bottom insets in Figures 15.6 and 15.7 show the real-color photographs of the chemiluminescent reagents (before and after Mw exposure) on glass and the SiF surfaces. The high photon flux seen upon delivery of microwave pulses to the metal surface is attributed to localized heating around and above the metal surfaces. The local temperature increase not only accelerates the rate of the chemiluminescence reactions, but the proximity to the silver *additionally allows* for metal-enhanced chemiluminescence (MEC). These results clearly demonstrate the 'on-demand' nature of the chemiluminescent reactions in the presence of low power microwave pulses. Thus, chemiluminescence emission can be induced for discrete time intervals and subsequently 'on demand' photon flux generated. Figure 15.8 not only shows the sensitivity of the MT-MEC assay with respect to the integrated background counts on silver and glass substrates (dashed lines), but also that the triggered signal intensity can be used for quantitive protein detection [9, 17]. From these observations, we can see that the implementation of low power microwaves increases the detectability in protein based assays, and could equally be applied to the detection of DNA and RNA's as well.

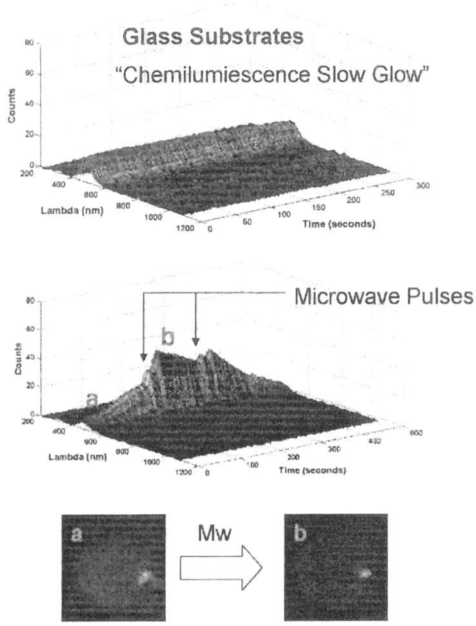


**Figure 15.6:** 3D plots of Acridan assay emission as a function of time from glass slides without (Top) and with low power microwave exposure / pulses (Middle). Bottom – photographs showing the Acridan emission both before (a) and after a low power microwave pulse (b). Mw – Microwave pulse. The concentration of BSA-Biotin was 1.56 pM. Adopted from *Anal Chem* 78: 8020-8027 (2006).

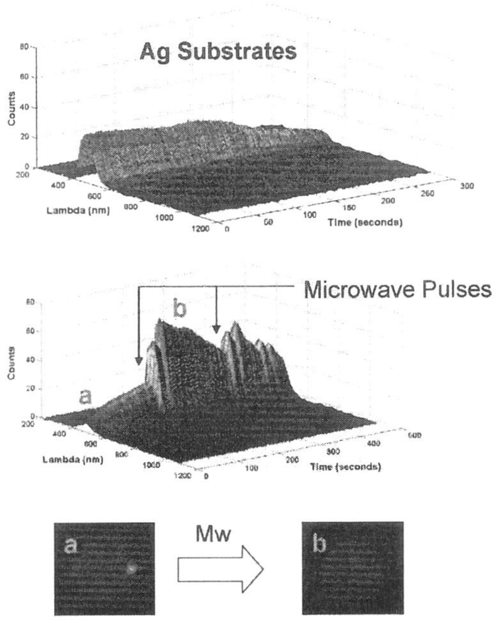


**Figure 15.7:** 3D plots of the Acridan assay chemiluminescence emission as a function of time from silvered glass slides (Ag) without (Top) and with low power microwave exposure / pulses (Middle). Bottom – photographs showing the Acradan emission both before (a) and after a low power microwave pulse (b). Mw – Microwave pulse. The concentration of BSA-Biotin was 1.56 pM. Reproduced from *Anal Chem* 78: 8020-8027, (2006).

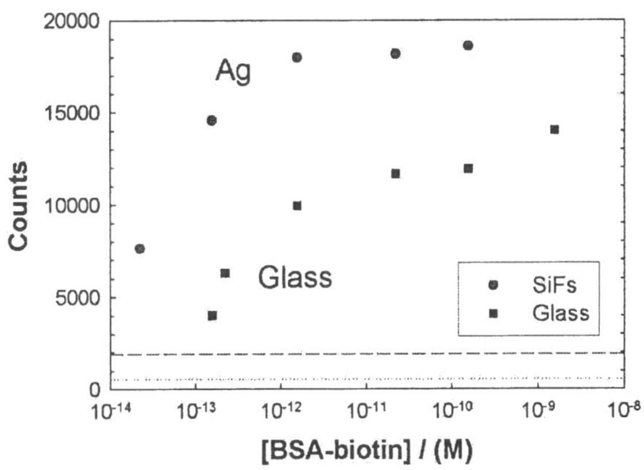

**Figure 15.8:** Integrated photon flux of the assay shown in Figure 15.5, for different concentrations of BSA-Biotin from both glass (G) and silvered surfaces (Ag). Reproduced from *Anal Chem* 78: 8020-8027, (2006).

## 15.3 2. *Application of MT-MEC to Blotting technologies*

Blots are still not considered reliable methods for accurate quantification of low protein concentrations, they are traditionally limited by antigen-antibody recognition steps that are generally kinetically very slow, which require long incubation times; e.g., western blots typically require processing times in excess of 4 hr. However, more recently, a western blotting methodology (One-Step Western Blot Analysis Kit, GenScript Corp.) has been made available that claims protein detection can be performed in less than 1 h, but this method still does not address the need for reliable protein quantification at low protein concentrations. Thus, both the rapidity and sensitivity of blots assays are still critical issues to be addressed to improve protein detection Figure 15.9. MT-MEC offers protein quantification for blots with ultrafast assay times, i.e., <2-min total versus 80 min traditionally. Using low power microwaves to induce the combined effect of kinetically accelerated binding, increased binding specificity and accelerated and 'triggered' chemiluminescent reactions, it is anticipated that Microwave Triggered Metal-Enhanced Chemiluminescence (MT-MEC) can be implemented to not only accurately quantify protein concentrations in an assay format, but also alleviate the current bottlenecks in blotting approaches.

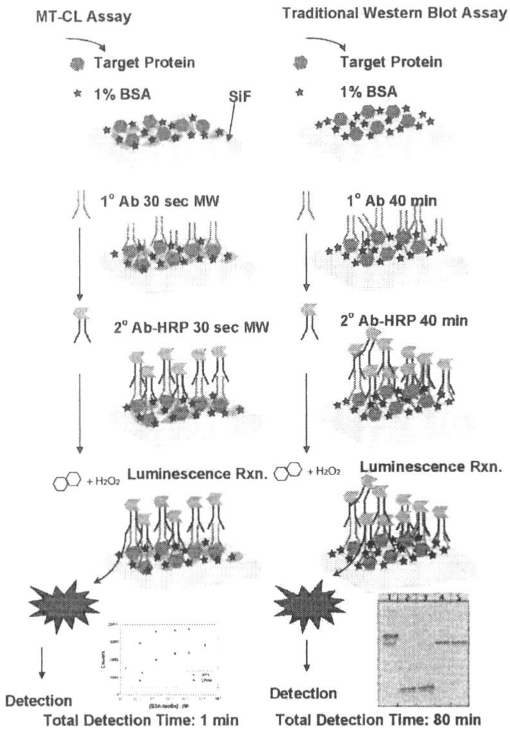


**Figure 15.9:** Procedure for the MT-MEC immunoassay (Mw, low-power microwave heating). Adapted from *Anal Chem* 78: 8020-8027, (2006).

## 15.3 3. *Microwave-Triggered Chemiluminescence with Aluminium Planar Geometrical Shapes*

### 15.3.3 1. *Finite Different Time Domain Simulations (FDTD)*

In the last 2 sections we saw how silver deposits, combined with microwave acceleration, provide for both rapid and intense chemiluminescence signatures. In those demonstrations the random nature of the deposits, provided for a near-uniform heating across the substrates. However, as we will now describe, planar geometric shapes, fabricated from virtually any conducting metal, can be used to focus microwaves, for the spatial triggering of chemiluminescence. Therefore, in an attempt to design small inexpensive microwave structures that locally accelerate chemical reactions, FDTD simulations have been used to visualize electric field distributions for aluminum structures in a microwave field [10]. For a single triangle geometry, intensity field enhancements are observed proximal to the tips of the triangles with the maximum enhancement at the triangle's apex when simulated with a 2.45 GHz total field scattered field (TFSF) that propagates from left to right. Subsequently, we designed two triangles, separated by a gap distance of 1 mm

(Figure 15.10). With this configuration, the propagation of the 2.45-GHz microwave field (TFSF, TE-polarized) flows across the metal surface, such that charge builds up on the triangle tips of the bow-tie, the maximum field enhancement localized in essence at the gap between the two triangles. While additional gap sizes (from 1 nm to 18 nm) were simulated, 1-mm gap sizes were found to provide convergent solutions and substantial field enhancements, which diminish for larger gap sizes and which lead to dielectric breakdown for smaller gaps (Figure 15.10 Bottom). In order to experimentally demonstrate that regions of maximum field enhancements via FDTD, for triangle and disjointed bow-tie geometries spatially correlate with regions of maximum chemiluminescent enhancements, 12.3-mm aluminum triangles 75 nm thick were deposited onto silanized glass microscope slides. Glass and aluminum triangle-modified substrates were cut into equal sized samples to minimize variations of the convective microwave heating that may arise from variations in the size of the glass substrates. For bow-tie and triangle structures, image wells were affixed to the substrates, such that the tip of single triangle and junction of the bow-tie geometry were exposed to the solution in the well (Figure 15.11D). Wells on the respective sample geometries were subsequently filled with 6 μL of blue chemiluminescence material (Figures 15.11A-D, blue circles). Photographs of each of the sample geometries before the application of low-power microwave pulses were taken and show that the pre-microwave chemiluminescence intensities for each of the samples are approximately equivalent (Figures 15.11E-H). Samples were subsequently exposed to a short, low-power microwave pulse, and photographs of each of the sample geometrieres after the application of low-power microwave pulses were subsequently taken (Figures 15.11M-P). The spatial profile of the resultant chemiluminescent signal enhancements for the glass substrates modified with Al geometries correlate very well with regions of maximum field enhancements for simulated structures. For glass surfaces modified with the aluminum triangle substrates, we observed >100-fold enhancement in "on-demand" photon flux for the single triangle geometry and >1000- fold enhancement for the bow-tie geometry. For the chemiluminescence solution placed at the center of the triangle, there is almost no enhancement in on-demand photon flux observation, which is consistent with the absence of any electric field distribution in the simulated images at the center of the one triangle geometry (Figure 15.11 B). In all of the geometrics studied, no wavelength spectrum shifts were observed, only changes in the intensity as shown in Figure 15.11, I-P.

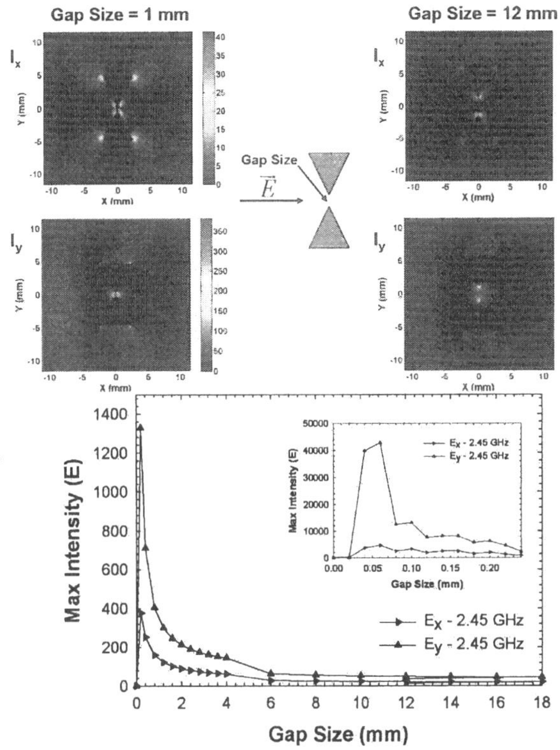


**Figure 15.10:** Simulated intensity images, Ix (top) and Iy (bottom) of the electromagnetic field distribution for 2.45 GHz microwave frequencies incident upon (2) 2-D equilateral triangles with 12.3 mm length and oriented with the sample geometry shown (middle). The incident field is held constant and the gap size is varied in subsequent simulations, 1 mm (left) and 12 mm (right) gap size examples are shown. Bottom: Maximum I (Ix (Ey2) + Iy (Ey2) ) pixel intensity versus gap (Top, Inset), expanded view of maximum pixel intensity versus gap size.

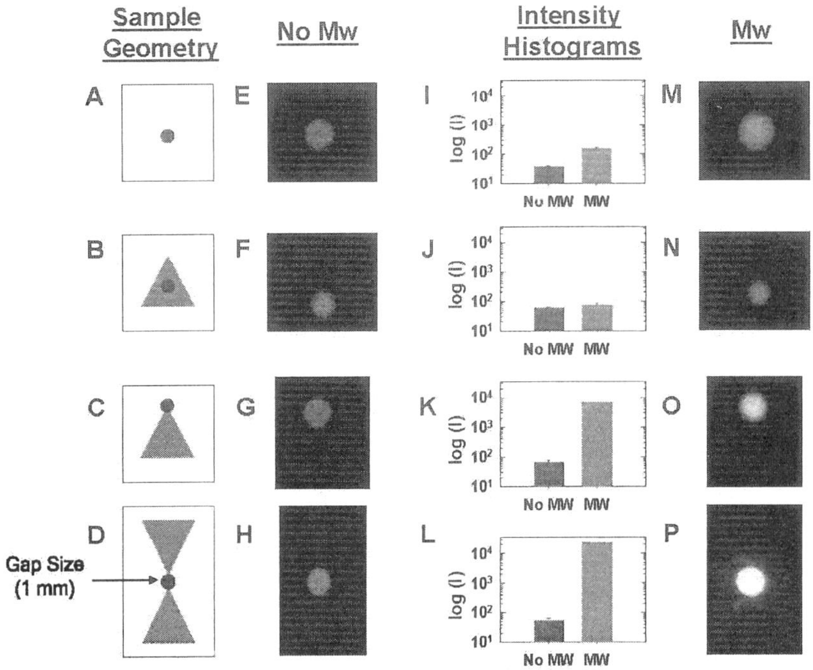


**Figure 15.11:** (A-D) The sample geometry depicts the chemiluminescence sample (blue circle), glass substrate (white square) and glass substrates modified with vapor deposited aluminum triangles 75 nm thick (12.3 mm length; 1 mm gap size for 'bow-tie' geometry). (E-H) Without an incident low power microwave pulse, the chemiluminescence signal is approximately equivalent for all sample geometries. (I-L) The relative chemiluminescent signals for all the sample geometries are measured both before (blue bars) and after (red bars) the application of the low power microwaves. (M-P) Upon application of low power short (5 seconds) microwave pulses, the clear differences in signal enhancement are represented visually in the Mw images. Adapted from *Anal Chem* 79: 7042-7052 (2007).

### 15.3.3 2. *Transferable Aluminum Substrates for Disposable Surface Assay Applications*

In order to access the feasibility of creating transferable aluminum structures for disposable sensing applications, sample coverslips were employed with deposited triangles on a base glass substrate. Since three-dimensional FDTD simulation data of the triangle geometries show that the enhanced field *also persists* in the z dimension, (vertically into the sample) imaging chambers were affixed to No. 1 coverslips (Corning Labware & Equipment) and 6 μL of chemiluminescent solution added. The imaging chamber and solution on top of the coverslip are placed on the top of the glass substrates, whereby the center of the imaging chamber is positioned proximal to triangle structures (Figure 15.12 A). The samples were again exposed to short, low-power microwave pulses. With the coverslip sample geometries (Figure 15.12A), the ratio of the signal was calculated after microwave pulsing to the signal before microwave pulsing and compared (Figure 15.12B). It was observed that the presence of the coverslip *did result* in minor signal enhancement loss for the bow-tie geometry (≈ 2-fold), but not substantially. The significance of this result lies in the fact that for spatially triggering chemiluminescence signatures, the sample and solution, can be immobilized onto a different substrate, the bow-tie geometries not fouled and therefore reuseable.

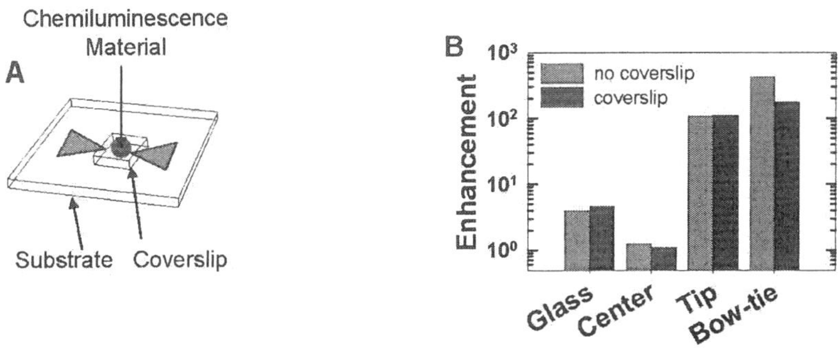


**Figure 15.12:** (A) Coverslip chemiluminescence reaction geometry scheme. Imaging chambers are affixed to No. 1 coverslips and filled with 6 uL of chemiluminescent material (blue circle). Coverslips are positioned on plain glass substrates and glass substrates modified with an aluminium triangle (12.3-mm length; 75 nm thick; 1 mm –gap size for two triangles geometries (insets, middle) (B) Enhancement is calculated as the ratio of chemiluminescence, with and without the coverslip. Adapted from *Anal Chem* 79: 7042-7052 (2007).

### 15.3.3 3. *Transferable Triangle Structures to Trigger Chemiluminescence from Commonly Used Substrates*

In addition to the possibility of creating transferable aluminum structures for disposable sensing applications with glass, it was further demonstrated that common substrates used in biosensing applications could also be used[10]. 12.3 mm triangle geometries from aluminum sheets (100 μm thick) were cut and affixed to different substrates (Figure 15.13), such that the gap size was 1 mm for the disjointed bow tie geometry. Image wells were again placed at the tip of a single aluminum triangle, between two aluminum triangles, at the center of the aluminum triangle, and on plain substrates (Figure 15.13). Wells on the respective sample geometries were filled with 6 μL of blue chemiluminescence material and subsequently exposed to a 5-second microwave pulse. From the recorded intensity data, almost equivalent enhancement trends were observed from the common protein detection sensing substrates (Figure 15.13), suggesting the substrate has little effect on the triggered chemiluminescence.

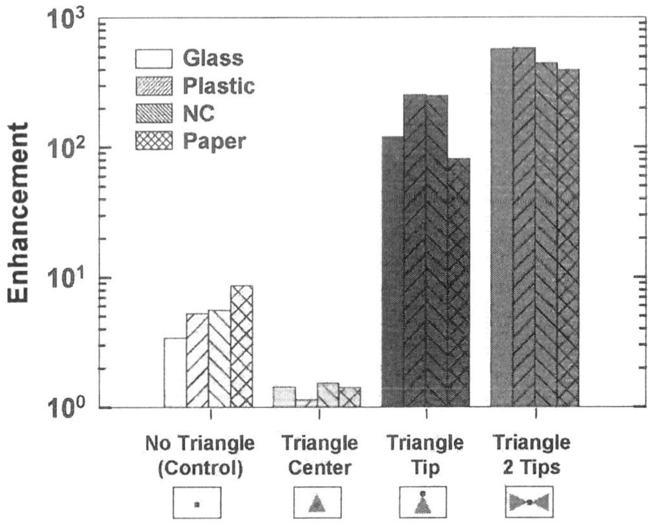


**Figure 15.13:** Chemiluminescent microwave (Mw) enhancement ratios (Mw/no Mw) upon application of low power microwave pulses (Mw) for different sample geometries on various dielectric substrates NC- nitrocellulose.

## 15.3.3 4. *Multiplexed Chemiluminescent Assay Format*

It has also been demonstrated that aluminum structures can be implemented to create highly sensitive multiplexed chemiluminescent assays for high-throughput screening with readily available and inexpensive materials[10]. At the corners of an 8-mm square structure 100 μm thick, the microwave enhancement of chemiluminescence emission is >300-fold (Figure 15.14A). While on a plain glass substrate in the absence of aluminum foil structures, there was only a 3-fold enhancement. Thus, the on-demand photon flux at the corners of the aluminum square geometries is 100-fold greater than the on-demand photon flux achieved with conventional microwave heating. For the solution placed on the center of the aluminum structure, negligible enhancement (1.4-fold) was observed upon exposure to a low-power microwave pulse. Images of four chemiluminescence solutions of different colors at the corners of 8-mm square aluminum structures are shown in Figures 15.14 B,C (before and after the application of microwave pulses), to demonstrate the ability of how to easily adapt this technology to a multiplexed or high-throughput chemiluminescent assay platform. One can readily imagine different multicolor assays, located at the tips of thin-film geometrical shapes.

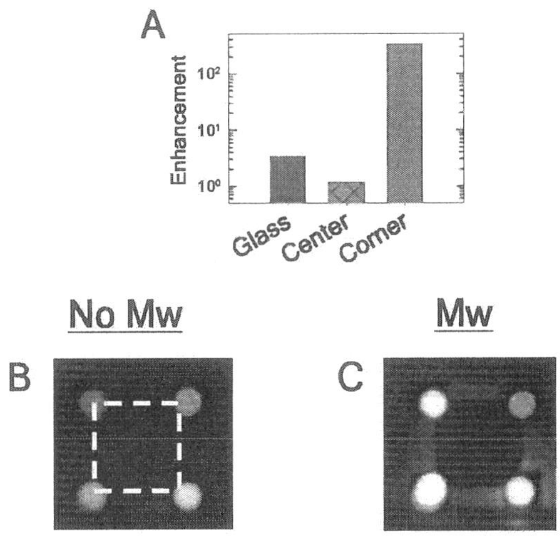


**Figure 15.14:** (A) The ratio of chemiluminescent intensities after Mw pulses to intensity before Mw pulses is shown for glass (blue bar), the center of the square geometry (hatched red bar), and the corner (solid red bar). (B) Chemiluminescence signal in a multiplexed format is approximately equivalent from all positions on the 8 mm square aluminum foil structure (dashed box) before the application of low power pulses and (C) significantly enhanced after the application of low power microwave pulses. Reproduced from *J. Fluorescence* 17: 279-287 (2007).

## 15.3.3 5. *Application of Microwave Triggered Chemiluminescence to Biological Assays and Western Blots*

As a further extension of the reusable triangle concept, it has been shown that disposable planar aluminum structures can be affixed to common sensing substrates, i.e., nitrocellulose membranes, to locally trigger enzyme (HRP)-catalyzed chemiluminescent reactions i.e. actual assays (Figure 15.15). With this approach we can readily see a greater 'triggered' chemiluminescence from the bow-tie geometries, as compared to the other structures. This observation suggests that bow-tie structures can be used to enhance and 'lift out' weak bands on western blots, Figure 15.16. Interestingly, the MW field exists in the z-direction as shown in Figure 15.12, suggesting that the bow tie geometrics can be supplied on separate glass cover slips, which are simply placed over the weak gel band, before microwave exposure.

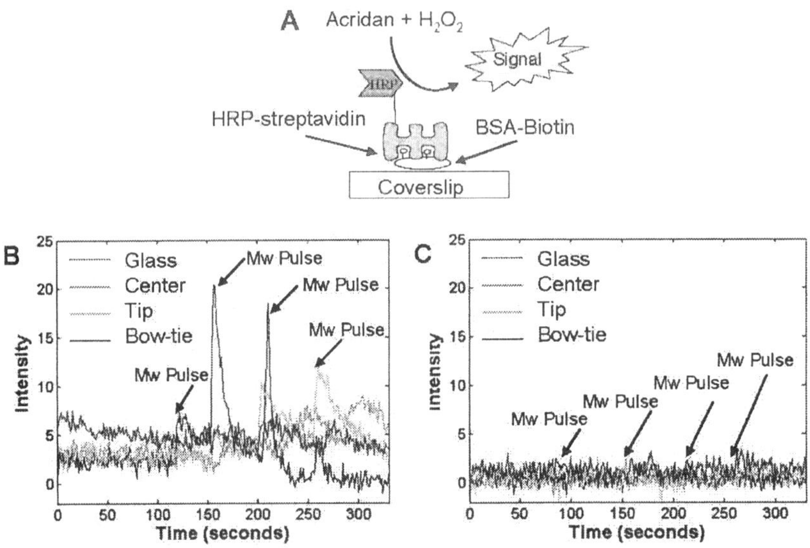


**Figure 15.15:** A) Model BSA-biotin, HRP-streptavidin chemiluminescent assay scheme. B) Acridan chemiluminescence emission as a function of time from HRP modified glass coverslips coated with 1 mM BSA-biotin and 1 mM HRP-streptavidin and positioned glass substrate geometries with and without 12.3 cm Al triangle 75 nm thick (left). C) Acridan chemiluminescence background emission as a function of time for glass coverslips incubated with 1.5% BSA and 1 mM HRP-streptavidin (control) positioned on glass substrate geometries with and without 12.3 cm Al triangle 75 nm thick shapes (right). All samples were exposed to four 10 second microwave pulses (Mw pulse) at 10% power. Adapted from *Anal Chem* 79: 7042-7052 (2007).

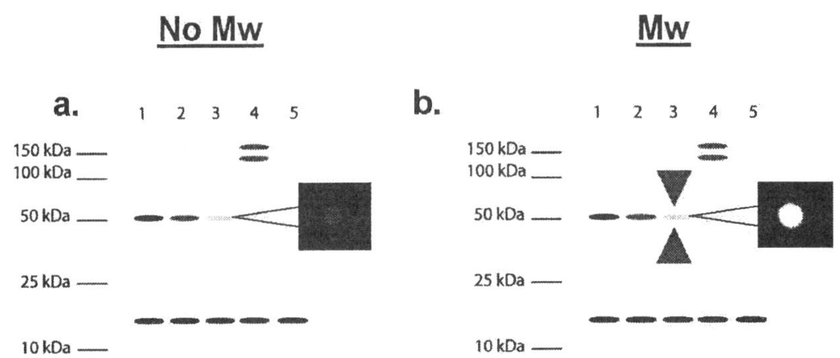


**Figure 15.16:** Cartoon illustrating a *potential* application of the technology: amplifying dim bands from a Western blot scheme, whereby weak chemiluminescence signal from a Western blot (A) without affixed triangle

geometries (left, inset) can be amplified (B) with a disjointed 'bow-tie' geometry attached. Adapted from *Anal Chem* 79: 7042-7052 (2007).

## 15.4 1. *Surface Plasmon Coupled Chemiluminescence (SPCC)*

The concept of surface plasmon coupled cheluminescence has recently been reported by Geddes and co-workers [18]. The observation of surface plasmon-coupled chemiluminescence (SPCC)[18], where the luminescence from chemically induced electronic excited states couples to surface plasmons in a thin continuous metal film has been demonstrated for numerous metals [18]. This technology results in highly directional and polarized emission of the chemiluminescence from the prism side of the thin film in the SPCC geometry, as compared to traditional chemiluminescence isotropic slow-glow.

The experimental geometry and instrument set-up used for the SPCC studies is shown in Figure 15.17.

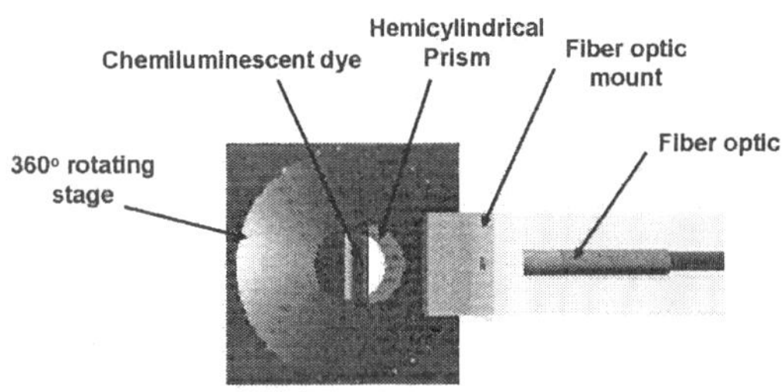


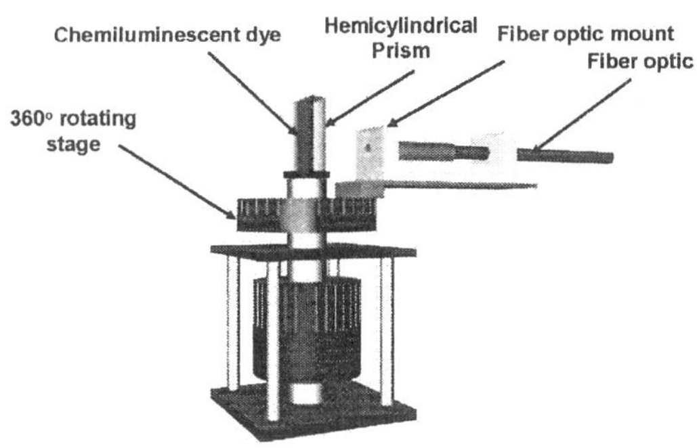


**Figure 15.17:** Experimental geometry used for surface plasmon coupled chemiluminescence (SPCC). Top – View from the top, Bottom – side view. Reproduced from *Journal of Physical Chemistry B* 110: 22644-22651, 2006.

Figure 15.18 (Top Left) shows the surface plasmon-coupled chemiluminescence (SPCC) and the free-space emission from a blue chemiluminescent dye on a 20 nm aluminum thin-film layer. It can be seen that the free-space emission is of much higher magnitude than the SPCC signal. This is because the sample chamber is ≈1 mm thick and only chemiluminescent species within approximately 250 nm of the surface of silver (or other metals) are known to excite and couple-to surface plasmons. Hence, the majority of the chemiluminescence material in the chamber does not couple to plasmons and so radiates as free-space isotropic emission. Figure 15.18 (Top Right) is an enlarged figure showing the highly directional and predominantly *p*-polarized SPCC emission. This is in contrast to the free-space emission which does not show any polarization or directional preference and is indeed isotropic. It is however worth noting that the signal at the SPCC peak angle is not entirely *p*-polarized. Figure 15.18 (Bottom) is the normalized SPCC and free-space emission spectra showing a high degree of overlap between the spectra. This suggests the plasmon-coupled chemiluminescence has not undergone any changes in its photophysical properties due to the interaction between the chemiluminescent species and the metal surface.

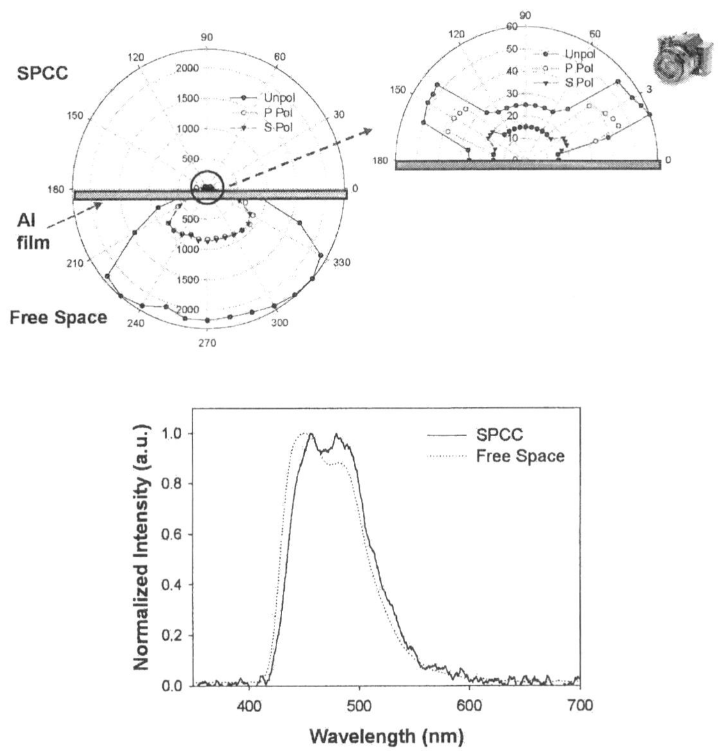


**Figure 15.18:** Surface Plasmon coupled chemiluminescence from 20 nm thick aluminium films. Top Right – enlarged directional SPCC, Top Left – Free space chemiluminescence and SPCC, Bottom – Emission spectra of both the free space chemiluminescence and SPCC. Reproduced from: *Journal of Physical Chemistry B* 110: 22644-22651, 2006. SPCC – Surface Plasmon Coupled Chemiluminescence.

Figure 15.19 shows real color photographs of the coupled emission (from the prism side) at the respective SPCC peak angle from various chemiluminescent dyes at both *s* and *p* polarizations as well as with no polarization. This figure clearly shows that the emission at the SPCC peak angle is predominantly *p*-polarized for all three dyes (on all three metals), thus suggesting that *surface plasmons* are responsible for the SPCC signal, as has been demonstrated for fluorophores [19, 20]. It can be seen that the *p*-polarized signal intensity at the SPCC peak angle is lower in magnitude than the unpolarized signal. This occurs because the entire SPCC signal consists of both *p* and to a lesser degree *s*-polarized light, and also because the sheet polarizers used in the experiment have only 30 - 40 % peak transmission efficiency for both polarizations.

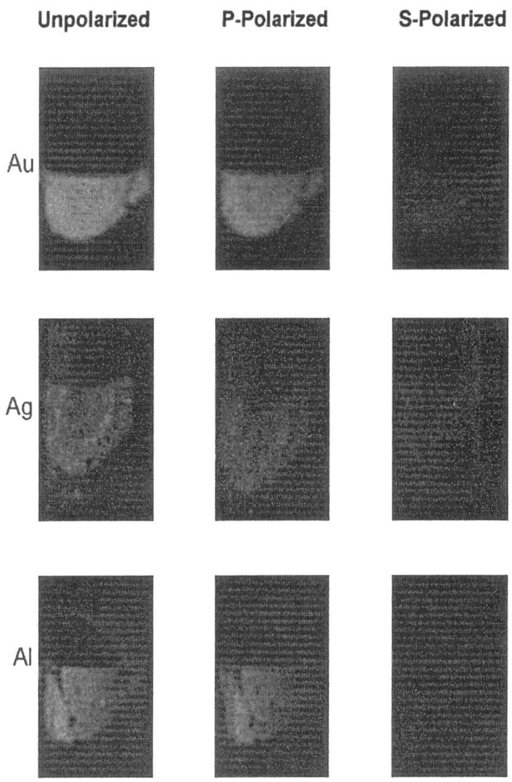


**Figure 15.19:** Photographs of the coupled Chemilumiescence emission at various polarizations for Gold, Silver and Aluminum films, top to bottom respectively, taken at their respective SPCC peak angles. Reproduced from *Journal of Physical Chemistry B* 110: 22644-22651, 2006

Experiments were also performed to determine the rate of decay of luminescence for blue and green chemiluminescent dyes as a function of time for both the free-space emission as well as the SPCC emission (with *p*-polarizers so that only plasmon coupled emission was measured). Figure 15.20 (Top) shows the decay of free-space and SPCC emission as a function of time for the blue CL dye on aluminium, and Figure 15.20 (Bottom) showing both the decay intensities normalized to their respective values at t = 0. The chemiluminescence decay, which is a due to the depletion of solution reactants and therefore a depletion over time of excited states, was found to follow first-order decay kinetics and can be modeled to a simple exponential function. The rate of depletion of the SPCC signal for the blue dye on aluminium was found to be only minimally greater than the free-space emission, 0.0003 versus 0.0002 $s^{-1}$, respectively. Since both the SPCC signal and the free-space emission signal decay is highly dependent on the rate of depletion of the same reactants (depletion of excited states) in the sample chamber over time, the measured decay rate for both the signals as shown in Figure 15.20 are almost identical. However, this finding *does indicate* that there are no localized catalytic effects of the aluminium on the chemiluminescence reaction, as this would be expected to manifest in a larger difference in the SPCC luminescence decay rate (from the Free Space decay rate) than is currently observed. Similar results of the decay of free-space and SPCC emission as a function of time for the green chemiluminescent dye on silver has been reported and are shown in Figure 15.21. This again indicates no localized catalytic or chemical effects of the silver on the chemiluminescence reaction studied.

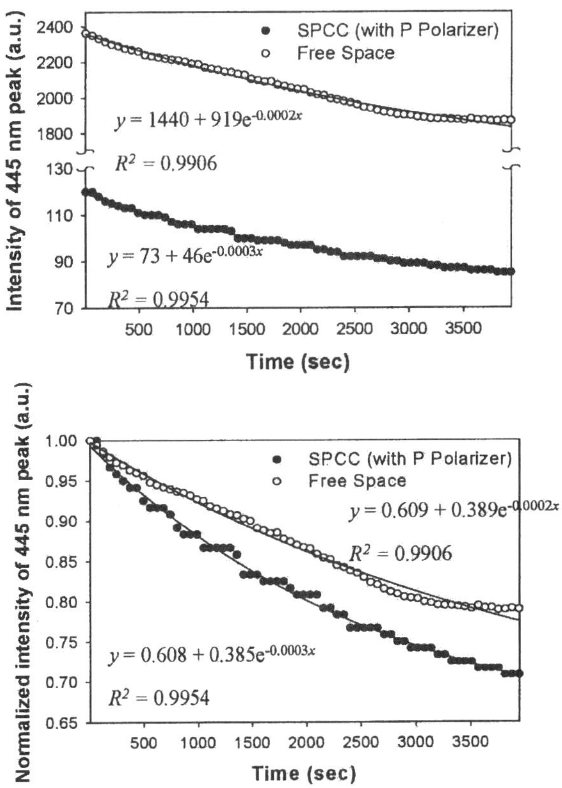

**Figure 15.20:** Chemiluminescence intensity decays from Aluminum films for both free space and coupled (Top), and normalized to the same initial intensity (Bottom). Reproduced from *Journal of Physical Chemistry B* 110: 22644-22651, 2006

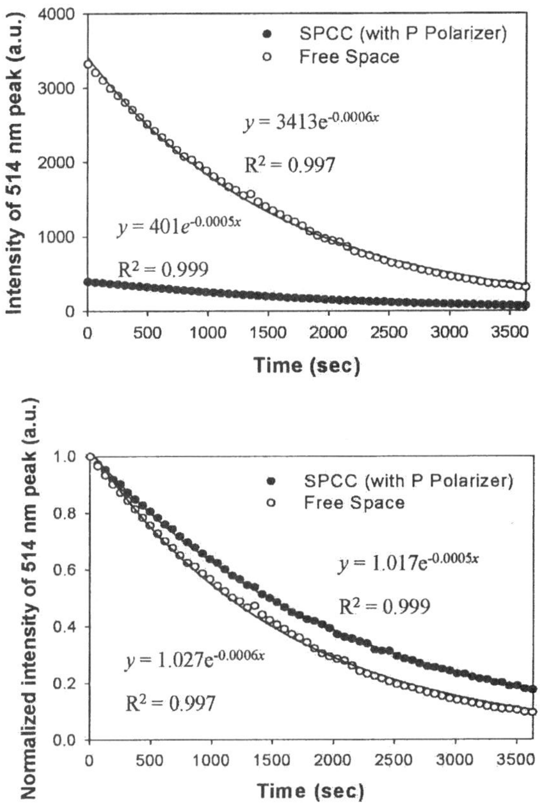


**Figure 15.21:** Chemiluminescence intensity decays from Silver films for both free space and coupled (Top), and normalized to the same initial intensity (Bottom). Reproduced from *Journal of Physical Chemistry B* 110: 22644-22651, 2006

## 15.5 CLOSING REMARKS

In this review chapter we have demonstrated the favorable effects of both Plasmon resonant particles and / or microwaves on both the intensity and glow time of chemiluminescence. In addition, thin metallic films, only a few 10's of nanometers thick, are able to directionally radiate chemiluminescence, as compared to the commonly observed isotropic emission. These technologies allow rapid and high sensitivity chemiluminescence practices, such as for protein or even DNA/RNA assays to be realized. While chemiluminescence receives favorable attention from researchers due to the simplistic nature of instrumentation required, as compared to fluorescence, its major disadvantage has been its long glow times, with blots taking traditionally several hours to develop using either film or digital scanners. However, by exposure to short low power microwave pulses, chemiluminescence photon fluxes can be rapidly increased with chemiluminescence kinetics over within several minutes. Subsequently, the amalgamation of plasmonics and microwave energies with chemiluminescence looks likely to catalyze a new realm of high sensitivity ultra fast chemiluminescence-based technologies and diagnostic tools.

## 15.6 REFERENCES


1. Bronstein, I., Martin, C. S., Fortin, J. J., Olesen, C. E., and Voyta, J. C. (1996). Chemiluminescence: sensitive detection technology for reporter gene assays *Clinical Chemistry* **42**: 1542-1546.
2. Kricka, L. J. (2000) in (Ed.), *Bioluminescence and Chemiluminescence, Pt C*, pp. 333-345.
3. Kindzelskii, A. and Petty, H. R. (2004). Fluorescence spectroscopic detection of mitochondrial flavoprotein redox oscillations and transient reduction of the NADPH oxidase-associated flavoprotein in leukocytes *European Biophysics Journal with Biophysics Letters* **33**: 291-299.
4. Knobloch, H., Brunner, H., Leitner, A., Aussenegg, F., and Knoll, W. (1993). Probing the Evanescent Field of Propagating Plasmon Surface-Polaritons by Fluorescence and Raman Spectroscopies *Journal of Chemical Physics* **98**: 10093-10095.
5. Kreiter, M., Neumann, T., Mittler, S., Knoll, N., and Sambles, J. R. (2001). Fluorescent dyes as a probe for the localized field of coupled surface plasmon-related resonances *Physical Review B* **64**.
6. Chowdhury, M. H., Aslan, K., Malyn, S. N., Lakowicz, J. R., and Geddes, C. D. (2006). Metal-enhanced chemiluminescence *Journal of Fluorescence* **16**: 295-299.
7. Chowdhury, M. H., Aslan, K., Malyn, S. N., Lakowicz, J. R., and Geddes, C. D. (2006). Metal-enhanced chemiluminescence: Radiating plasmons generated from chemically induced electronic excited states *Applied Physics Letters* **88**: 173104.
8. Aslan, K., Previte, M. J., Zhang, Y., and Geddes, C. D. (2008). Microwave-accelerated surface plasmon-coupled directional luminescence 2: A platform technology for ultra fast and sensitive target DNA detection in whole blood *J Immunol Methods* **331**: 103-113.
9. Previte, M. J., Aslan, K., Malyn, S. N., and Geddes, C. D. (2006). Microwave triggered metal enhanced chemiluminescence: quantitative protein determination *Anal Chem* **78**: 8020-8027.
10. Previte, M. J., Aslan, K., and Geddes, C. D. (2007). Spatial and temporal control of microwave triggered chemiluminescence: a protein detection platform *Anal Chem* **79**: 7042-7052.

11. Geddes, C. D. and Lakowicz, J. R. (2002). Metal-enhanced fluorescence *Journal of Fluorescence* **12**: 121-129.
12. Johansson, P., Xu, H. X., and Kall, M. (2005). Surface-enhanced Raman scattering and fluorescence near metal nanoparticles *Physical Review B* **72**.
13. Gersten, J. I., Theory of fluorophore-metallic surface interactions, in *Topics in Fluorescence Spectroscopy*, vol. 8, C. D. Geddes and J. R. Lakowicz, Eds. New York: Springer, 2004, pp. 197-221.
14. Aslan, K., Malyn, S. N., and Geddes, C. D. (2006). Multicolor microwave-triggered metal-enhanced chemiluminescence *Journal of the American Chemical Society* **128**: 13372-13373.
15. Wilchek, M. and Bayer, E. A. (1988). The avidin-biotin complex in bioanalytical applications *Analytical Biochemistry* **171**: 1-32.
16. Wilchek, M. and Bayer, E. A. (1990). Applications of avidin-biotin technology: literature survey *Methods Enzymol* **184**: 14-45.
17. Previte, M. J., Aslan, K., Malyn, S., and Geddes, C. D. (2006). Microwave-Triggered Metal-Enhanced Chemiluminescence (MT-MEC): Application to Ultra-fast and Ultra-sensitive Clinical Assays *J Fluoresc* **16**: 641-647.
18. Chowdhury, M. H., Malyn, S. N., Aslan, K., Lakowicz, J. R., and Geddes, C. D. (2006). Multicolor directional surface plasmon-coupled chemiluminescence *Journal of Physical Chemistry B* **110**: 22644-22651.
19. Previte, M. J. R., Zhang, Y. X., Aslan, K., and Geddes, C. D. (2007). Surface plasmon coupled fluorescence from copper substrates *Applied Physics Letters* **91**.
20. Geddes, C. D., Gryczynski, I., Malicka, J., Gryczynski, Z., and Lakowicz, J. R. (2004). Directional surface plasmon coupled emission *Journal of Fluorescence* **14**: 119-123.

# 16

# Enhanced Fluorescence From Gratings

Chii-Wann Lin*, Nan-Fu Chiu, Jiun-Haw Lee,
Chih-Kung Lee
National Taiwan University, No. 1, Sec. 4, Roosevelt Road, Taipei, 10617 Taiwan, Republic of China.

## 16.1 INTRODUCTION

In semiconductor-based optoelectronics devices made with organic materials have the potential advantages of low cost, structural flexibility, and simple fabrication processes. They have been shown to possess both electrical and optical properties that are associated with metals and semiconductors. Increasing attention is being paid to the electro-excitation or optical-excitation process between the metal electrodes and the active organic layers, where metal / organic interfaces are present. They are now thought to be one of the device parameters that significantly influence the device performance, for instance in organic light-emitting diodes (OLED) [1], organic photovoltaic cells [2], organic field-effect transistors [3] or new organic bio-sensors [4-7].

This chapter is intended to elucidate the effects of coupled active surface plasmon polaritons (SPPs) on a metallic lamellar grating nanostructure with organic material on the surface. It demonstrated how an energy gap for SPPs propagating on such a grating nanostructure can be used to modify the emission properties of an adjacent thin layer of organic semiconductor material ($Alq_3$, tris-(8-hydroxy-quinoline)-aluminum). We have fabricated several grating devices with differences in pitch size and coupled organic / metal nanostructure with symmetric and asymmetric dielectric SP band gap materials. It is found that emission is significantly inhibited in the vicinity of the band gap and that the modified emission spectrum is determined by the wavelength dependence of the density of SPP states. We present recent experimental results and discuss potential applications of such an active plasmonics for biosensor with enhanced resonance energy and highly directional emission due to local interactions on the organic / metal nano-grating.

### *The Effect Of Metal / Organic Interface On Radiative Decay Properties*

The radiative decay in the form of light emission near metal / organic interface following excitation can enhance the luminance efficiency. The pathways for the excited states to undergo various types of radiative decay are depicted by the Jablonski diagram as shown in Fig. 16.1(a) and (b) for electroluminance (EL) and photoluminance (PL). For EL type, light of well-defined color can be produced upon applying potential across the electrodes with organic layer(s) sandwiched in-between [8]. This happens because molecular excitons that decay radatively are formed once

*Enhanced Fluorescence from Gratings*
Edited by Chris D. Geddes.

electrons and holes are injected into the organic layer. In the case of photoluminance, a fluorophore is usually excited to some higher vibrational level of either $S_1$ or $S_2$. With a few rare exceptions, molecules in condensed phases rapidly relax to the lowest vibrational level of $S_1$. This process is called internal conversion (non-radiative) and generally occurs in $10^{-12}$ seconds or less [9]. These excited molecules then return to the highest vibrational level in the ground-state and quickly reach thermal equilibrium. An interesting consequence of the emission from the highest vibrational level in the ground-state is that the emission spectrum is typically a mirror image of the absorption spectrum for the $S_0 \rightarrow S_1$ transition. Molecules in the $S_1$ state can also undergo a spin conversion to the first triplet state, $T_1$. Emission from $T_1$ is termed phosphorescence and is generally shifted to lower energy relative to the fluorescence. Conversion of $S_1$ to $T_1$ is called intersystem crossing. In general, fluorescent is resulted from the lowest singlet excited state and phosphorescent is resulted from the lowest triplet excited state, whereas the ground state is always a singlet.

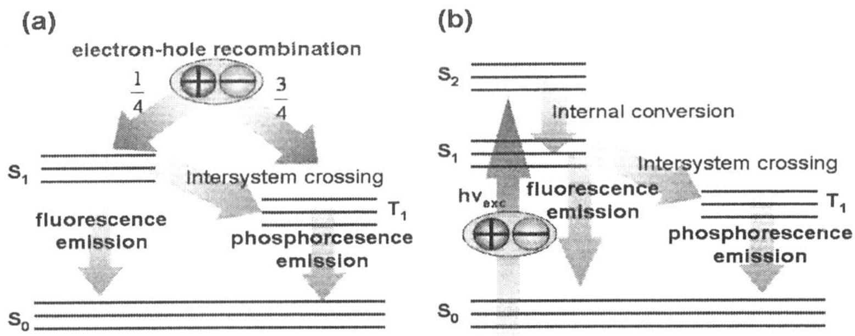


**Figure 16.1:** Schematic diagrams show the pathways of radiative decays for (a) electroluminescence (EL), and (b) photoluminescence (PL).

The energy transfer shown is equivalent to the near-field coupling of the dipole to SP modes in the organic / metal interface. Such a dipole can form exciton and contribute to its optical properties which subjected to quenching process near the metal / organic interface. This irreversible energy transfer from the excited organic molecules to the metal (Förster transfer) is due to the dipole-dipole interaction [10-12]. In addition, the diffusion of excitons to the metal / organic interface is possible, which makes the non-radiation energy transfer higher as compared to localized excitons. The interface dipole field is coupled to the surface modes causing excitons to oscillate and thus absorb energy. Normally, surface plasmons are not coupled to the radiation field. It does not violate the rule in this case because the short-range interface dipole field is not regarded as a radiation field without proper matching condition in the energy transfer processes [10]. The excitation of SPPs is one of the possible non-radiation routes by which the energy transfer of the organic semiconductor molecules can be absorbed by the metal. Quenching by the surface plasmon absorption of the metal due to Förster energy transfer occurs as shown in Fig. 16.2, where the dipole layer interacts between metal and organic thin film.

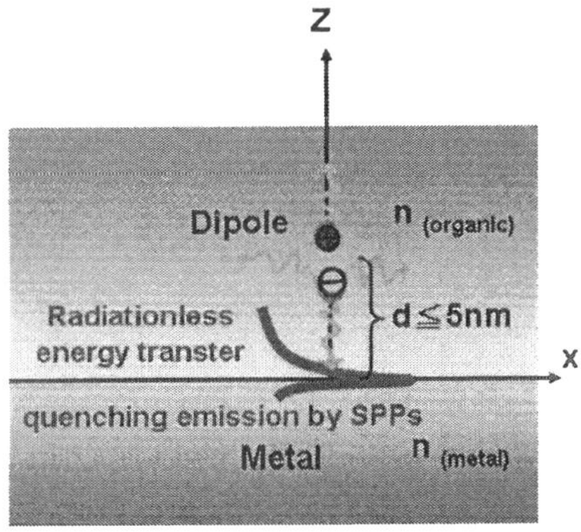


**Figure 16.2:** Schematic representation showing metal / organic interface energy transfer which is equivalent to the near-field coupling of the interface dipole to surface plasmons in the plane metal. The energy transfer across the interface could cause quenching of the interface dipole at metal / organic interface.

Furthermore, appropriate organic / metal interface can be used to manipulate the radiative decay properties. This approach is sometimes referred to as radiative decay engineering (RDE) [13-15]. Recently, enhanced transmission and emission of organic semiconductor on metallic thin film mediated by energy transfer of coupled SPPs has been used in the design of various opto-electronics. Initially, coupled SPPs transmission through a thin metal film was investigated on surface with few hundred nanometers in roughness [16] or corrugated microstructure [17-20]. Others have made SPPs coupled emissions from excited fluorescent molecules by evanescent field near the metallic surface [21-23]. The coupled emission around SPR angle has been observed by using attenuated total reflection (ATR) prism in the Kretschmann and reverse Kretschmann configuration [24-27], grating coupler [28-30] and nanoparticles [31]. One can thus achieve enhanced quantum yield or increased radiative rate of fluorescent emission. The existence of non-radiative decay near metallic surface can have effects on fluorescent intensity and lifetimes. The further interactions of coupled resonance energy by the so-called surface plasmon coupled emission (SPCE) can have interesting optical properties for many applications [32-34].

In Figure 16.3, the excitation / relaxation electric fields that can interact with organic semiconductor are from the incident light field (E) and metal field ($E_m$) for excitation and near metal surface quenching of non-radiative decay rates ($k_{ny}$), the radiative decay rate ($\Gamma$), metal radiative rate ($\Gamma_m$), and metal radiative decay rate ($k_m$) for relaxation. The emission of organic semiconductor layer with a thin metal layer greater than 5 nm in thickness is almost completely quenched [10, 13-15]. Additionally, the dipole of organic semiconductor can induce a field in the metal, which consequently increases the intrinsic radiative decay rates and quantum yields by the nearby metal surfaces; as the result decreasing the lifetime of the organic semiconductor. In summary, the quantum yields, intensity, resonance energy transfer and lifetimes of the emission light are governed by the magnitudes of the radiative decay rate and the sum of the non-radiative decay rates.

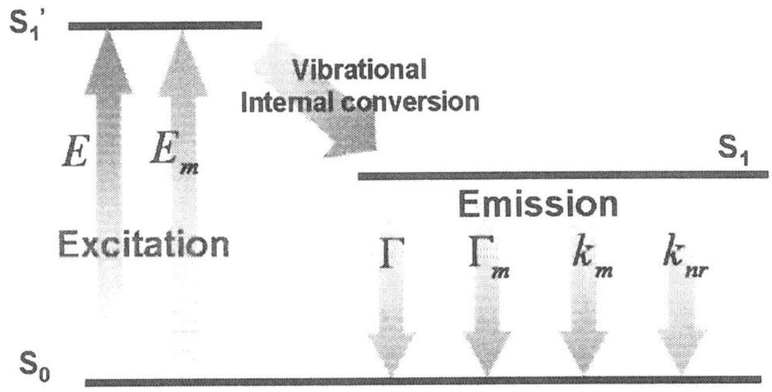


**Figure 16.3:** Modified Jablonski diagram shows the energy absorption effects of near metal surface enhanced fluorescence. The process involves creating an excited electronic singlet state by optical absorption and subsequent emission of fluorescence with different decay paths.

The effects of transferred energy on a plane occurred over distances are comparable to the excitation and emission wavelengths. For a planar metallic film system, all the energy non-radiatively transferred to the metal is ultimately lost as heat and the emission from the device is strongly quenched. For a grating metallic system, the energy transferred to SPPs can be recovered as photons as shown in Fig. 16.4. A grating is then incorporated into this basic sample structure. The combined effects of the multi-layered substrate and the grating-induced directional emission further increase the effective enhancement in a narrow angular range. Enhanced energy transfer can be combined with directional emission to result in higher sensitivity. It is known that fluorescence can be excited by the evanescent wave due to surface plasmons [13-15, 21-23]. In this case, the metal-induced increase in the transfer rate will result in transfer over longer distances and the emission light will become detectable with an increase in the transfer efficiency.

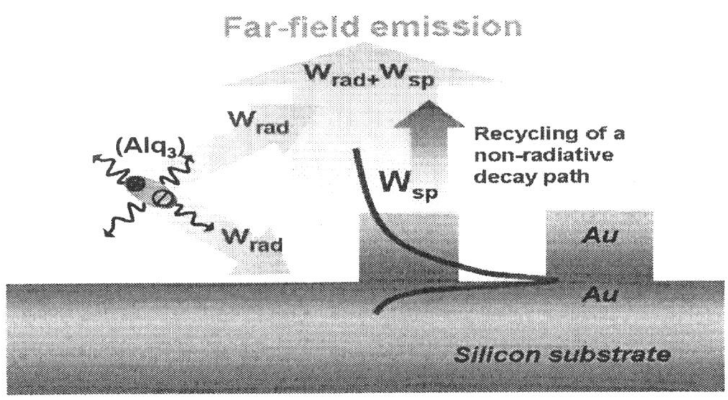


**Figure 16.4:** SPs radiates into the far field using a grating to resolve the momentum mismatch, which can turn the "non-radiative" plasmon into a far-field photon: Slow emitter becomes a fast emitter: $W=W_{rad}+W_{sp}$.

Figure 16.5 shows the possible relaxation pathways of excited organic molecules following absorption of photons. Elementary excitation dynamics, including generation, relaxation and deactivation are considered here, which are similar to those occurring in most optoelectronic devices.

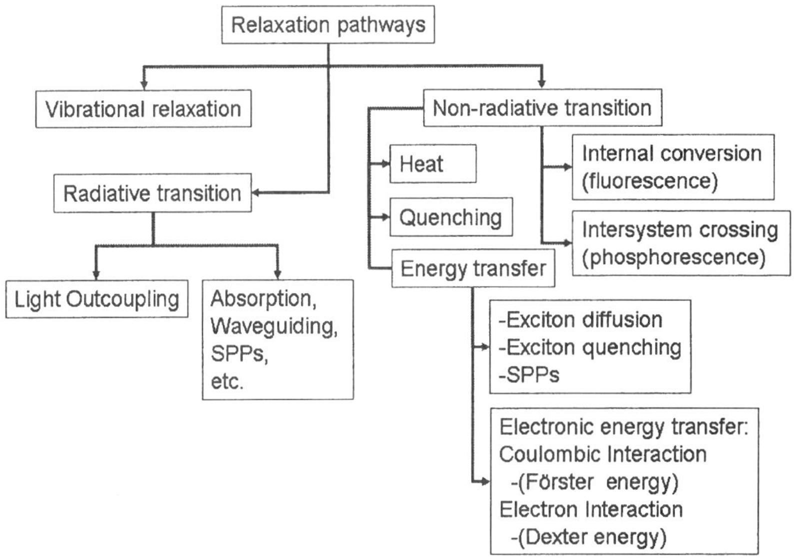


**Figure 16.5:** Processes involve in the fluorescence relaxation pathways of metal / organic optoelectronic devices.

## 16.3 EXPERIMENTAL SETUPS

There are a lot of research interests in the interactions between electromagnetic radiation and material structure in periodicity on the scale of the wavelength of light. In much the same way that energy band gaps occur for electrons propagating in periodic crystalline structures, the interaction between light and the material can lead to energy band gaps for propagating modes in such media [35-38]. We have designed a series of different pitch size gratings geometry with a thin layer of organic semiconductor ($Alq_3$) deposited onto the surface of such a grating. Excited $Alq_3$ molecules can relax by generating SPPs that propagate at the metal / $Alq_3$ interface.

### 16.3 1. *Fabricated Devices*

The samples are prepared by Electron Beam Lithography system (EBL). The EBL ELS-7500EX is a scanning electronic microscope (SEM) equipped with a lithographic system. The shape of the individual grating was a slightly elongated line of one-dimensional patterns of nanostructure gratings, including 400 nm, 500 nm,

600 nm and 800 nm, were adopted for the devices, exposure area 1.2×1.2 mm and beam current 300 pA with dose time 0.9 μs. Figure 16.6 shows the PR grating topography by atomic force microscope (AFM).

We have set up a photoluminescence (PL) and electroluminescent (EL) measurement systems as shown in Fig. 16.7 for the angular emission spectra produced by the designated surface plasmon grating coupled emission (SPGCE) from nanostructure. In brief, a 405 nm light source (Spectral Luminator 69050, Newport Oriel Inc., USA) or 405 nm diode laser (BWB-405-20E (20mW), B&W TEK Inc, USA) is used to excite $Alq_3$ molecules on the nano-grating device. The device was placed at the center of a high-resolution rotary stage with computer controllable incident angle, $\theta_i$, emission angle, $\theta_e$, and azimuthal angle, $\varphi$. The SPGCE output light was collected and measured by a 2-inch lens with focal distance of 5 cm and a 12-bit spectrometer (USB2000, Ocean Optics Co., USA). Two motorized rotary stages (SGSP-120YAW-W, Sigma Koki, Japan) and its controller (SHOT-204MS, Sigma Koki, Japan) are used to control $\theta_i$ and $\theta_e$, between sample and detector stages. The nominal angular resolution is 0.0025 degree. The system uses temperature, humidity sensor (Galltec TFG80J) and data acquisition card (PCI-6070E, National Instrument Inc, USA) to monitor the environmental operating condition for higher accuracy and reproducibility.

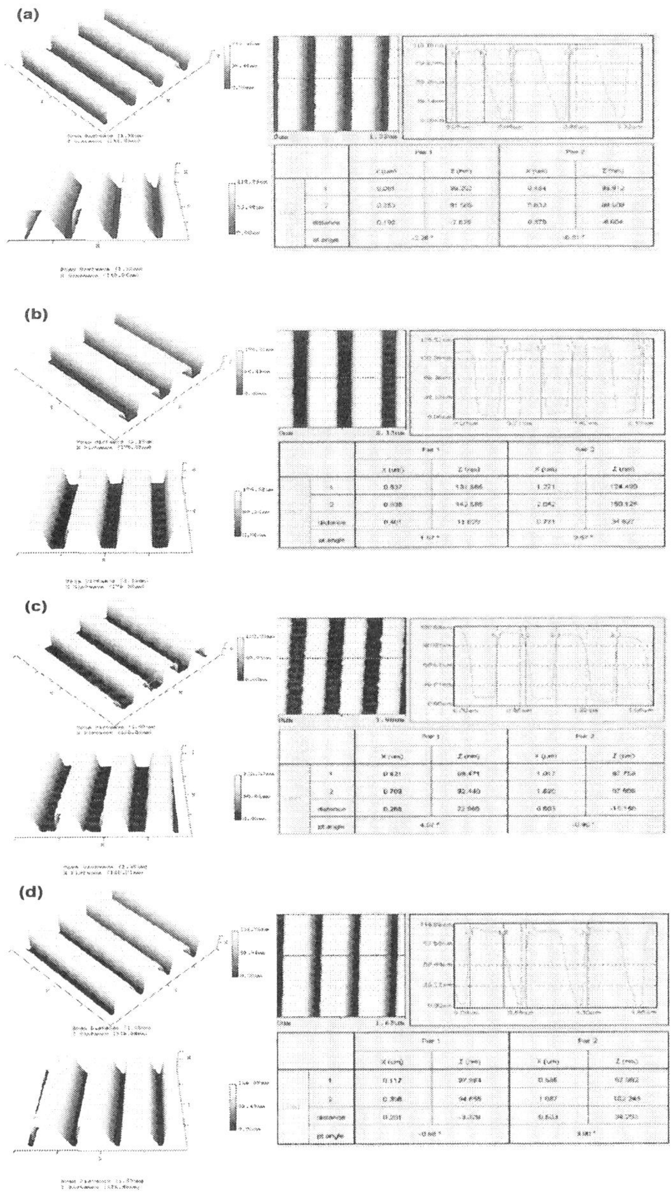

**Figure 16.6:** The PR grating profiles of (a) pitch 400 nm (b) pitch 500 nm (c) pitch 600 nm (d) pitch 800 nm are used to check the shape, lateral size, and height of the structure by AFM.

### 16.3. 2 *Instrumentation Set-Up*

The system experimental setup is shown as in Fig. 16.7. The chip was mounted on an azimuthal (local rotary 360°) stage and sample stage to allow scanning of the angle of incidence. The detector arm of the sample stage could move either a CCD or spectrometer for measuring the reflected intensity or emission light intensity respectively. Set-up 1: the grating was rotated and moved through the resonance while the spectrometer rotated by the same angle to maintain the same sample perspective. Set-up 2: the grating was kept at a fixed angle of incidence to excite organic semiconductors by grating coupled surface plasmons emission. The spermatic moved to measure the fluorescence intensity at different emission angles relative to the surface normal of the grating. The set-up 3 mode was used to record azimuthal-angle-dependent reflectivity data from a diffraction grating. The grating was placed on the rotating table with the incident light at a fixed stage and detector angle to measure the fluorescence intensity.

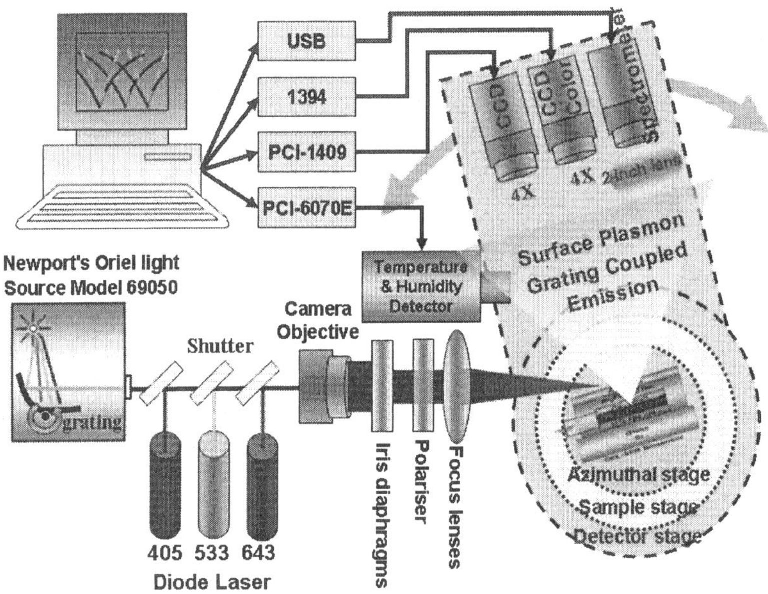


**Figure 16.7:** Schematic diagram of a surface plasmon spectroscopy setup for the grating coupled configuration. The chip was mounted on an azimuthal stage (local rotary 360°) and sample stage to allow scanning of the angle of incidence. The detector arm of the sample stage could move either a CCD or spectrometer for measuring the reflected intensity or emission light intensity respectively.

## 16.4 MODEL OF ACTIVE PLASMONIC VIA SPGCE

The SPGCE is based on the fluorescent emission from the organic $Alq_3$ molecules in the active layer and its diffraction in a periodically modulated surface with possible SPPs enhancement due to multilayer structure of metallic thin film.

Using 2-layer device as a simplified model, the possible interaction mechanisms are shown in Fig. 16.8. The $Alq_3$ molecules in the active layer can be electrical- or photo-pumping to provide non-oriented internal light source to generate SPPs on the metal / dielectric interfaces and de-couple through grating structure for detectable radiative emission or non-radiative waveguide mode. Therefore, there are four possible wave vectors, i.e., $k_{rad}$ ($k_{//}$), $k_{nrad}$ ($k_w$), $k_{SP(Au/air)}$, and $k_{SP(Au/Alq3)}$, that can exist between different interfaces of lamellar grating nanostructure, i.e. air, organic-layer, and two metal / dielectric interfaces. The two SP modes are associated with metal/organic and metal / air interfaces and their corresponding wave vectors are $k_{SP(Au/air)}$ and $k_{SP(Au/Alq3)}$. Through these possible mechanisms, the emitted light from $Alq_3$ could couple with surface plasmons under matching conditions while propagating along the grating surface. It can become radiative light again through the decoupling nanostructure for specific directional emission and enhancement in its optical properties.

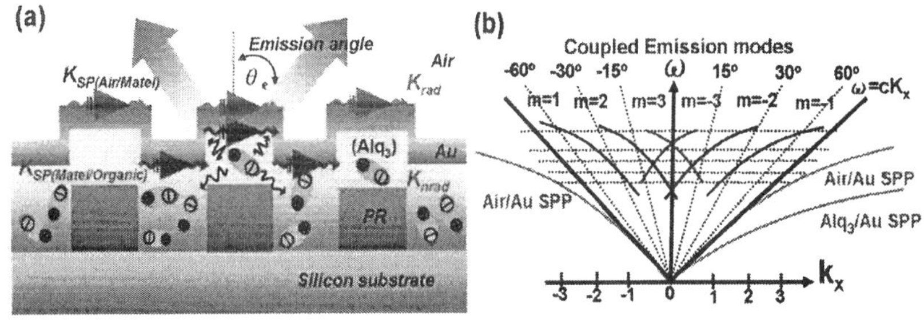


**Figure 16.8:** With a simplified 2-layer active plasmonic model device, the Alq3 molecules in the active layer can provide non-oriented internal light source to generate SPPs on the metal / dielectric interfaces and de-couple through grating structure for detectable radiative emission or non-radiative internal propagation. (a) The cross-section view of a simplified active plasmonic model device with Si / PR / Alq3 / Au layer structure. (b) The schematic representation of dispersion curves for SPPs propagating along metal grating surface of an Au / air and Alq3 / Au boundaries by SPGCE cross-coupling become radiative light at different angles.

The SP grating coupled emission, $k_{//}$, at certain wavelength of λ can result from the matching momentums of the internal $Alq_3$ emission ($\theta_e$) through grating coupler condition (dielectric constant of $Alq_3$ and pitch size of Λ) to the metal / air ($ksp_{(Au/air)}$) dispersion curve as shown in the Eq. (1). The SP modes for Au / air interface on the grating layer can be de-coupled into air if its wave vector is smaller than that of the air. Equation (2) gives such a matching condition for the guided mode of the organic layer and the de-coupling of light emission. According to Fresnel's transmission of optics, the guided mode in the $Alq_3$ layer propagates inside the organic film. The wave vectors for the organic layer mode are represented by Eqs. (2) and (3) shown below.

$$k_{//} = \sqrt{\varepsilon_d}\left(k_0 \sin(\theta_{emission}) \pm m\frac{2\pi}{\Lambda}\right) = k_0\sqrt{\frac{\varepsilon_{Au}\cdot\varepsilon_{air}}{\varepsilon_{Au}+\varepsilon_{air}}} = k_{SP(Au/air)} \quad (1)$$

$$k_w = k_{SP(Au/Alq3)} \pm m\frac{2\pi}{\Lambda} \quad (2)$$

$$k_{SP(Au/Alq3)} = k_0\sqrt{\frac{\varepsilon_{Au}\cdot\varepsilon_{Alq3}}{\varepsilon_{Au}+\varepsilon_{Alq3}}} \quad (3)$$

where $k_{//}$ and $k_w$ are the wave vectors parallel to the surface of the emitted light and the guided mode, respectively. Ksp is the SP wave vector parallel to the surface of the interfaces between metal and dielectric, such as the organic / metal and metal / air interface. The emission photon with wave-vector component $k_0\sqrt{\varepsilon_d}\sin(\theta)$ at an azimuthal angle (φ) may now couple directly with SPPs [39].

## 16.5 RESULTS AND DISCUSSION

### 16.5 1. *Enhancement And Tunability Of Active Plasmonic By Multilayer Grating Coupled Emission*

The effect of coupled mode surface plasmon polaritons (SPPs) on the active emission of a nanostructure grating with organic semiconductor material, $Alq_3$, on the surface was investigated in this study. We report surface plasmon grating coupled emission (SPGCE) from excited organic layer on metal grating in both organic / metal (2-Layer) and organic / metal / organic / metal (4-Layer) structures. The dispersion relation was obtained from angle-resolved photoluminescence measurement. The resultant emission intensity can have up to 6 times enhancement on the 4-Layer device and the Full-Width Half-Maximum (FWHM) is less than 50 nm. The combination of SPPs on organic / metal interface allows specific directional emission and color appearance of $Alq_3$ fluorophores. Potential applications of such an active plasmonics with enhanced resonant energy emission due to interactions on the organic / metal nano-grating as biosensor were presented and discussed.

We prepared the nanostructure with 1-D grating patterns by electron-beam lithography (EBL). We have designed, fabricated, and characterized two SPR configurations, a 2-layer one of [Si / grating (PR) / $Alq_3$(50 nm) / Au(20 nm)] and a 4-layer device of [Si / grating (PR) / $Alq_3$(50 nm) / Au(20 nm) / $Alq_3$(50 nm) / Au(20 nm)] symmetrical organic dielectric films, with grating line width and pitch size of 400 nm and 800 nm, respectively. We used the emission of $Alq_3$ organic molecules to excite SPPs on multilayer grating coupled emission. The emissions correspond to the resonant condition of SP modes on the $Alq_3$ / Au interface and grating couple to the Au / air interface for the emission of light. This technique has surface plasmon grating coupled emission (SPGCE) of light passing through metal and is a multilayer grating approach for the excitation of SPPs. Our experimental results show that these devices can have specific directional emission, enhanced emission intensity, and reduced Full-Width Half-Maximum (FWHM). We also best fit the measurement results to a revised theoretical model of grating coupler on a thin layer of metal which confirms the existence of SPP mode with the momentum-matching condition of a surface plasmon resonance in our experimental configuration. Further investigations will facilitate the development of novel bio-sensing device having multilayer organic / metal nanostructure for grating coupler active plasmonic biosensor and the use of admittance loci design method for such a purpose [40-42].

Cross sections of the 2-layer and 4-layer devices are shown in Fig. 16.9 (a) and (b), respectively. Our samples were prepared by an Electron Beam Lithography system. The exposure was carried out for 2 μsec by a pixel map of 60000×60000 dots to give a total exposure area of 1.2×1.2 $mm^2$. Next, 50 nm of organic $Alq_3$ and subsequent 20 nm of gold were deposited on the grating by a thermal evaporator with vacuum level and evaporation rate of approximately $2*10^{-6}$ Torr and 0.2 Å/s, respectively.

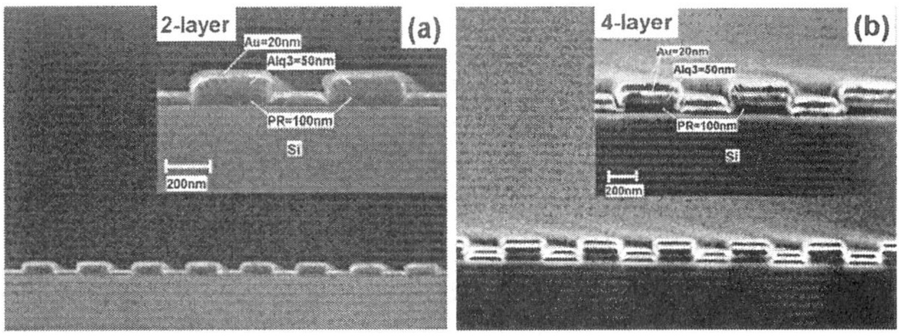


**Figure 16.9:** SEM images of the gratings cross section, which show the arrangement of a periodically lamellar layer a) 2-layer structure of Alq3 / Au and b) 4-layer structure of (Alq3 / Au / Alq3 / Au) on top of a 100 nm PR structure. Reprinted with permission from [Chiu et al., OPTICS EXPRESS. 15, 11608 (2007)]. Copyright 2007, Optical Society of America.

We have set up a photoluminescence (PL) measurement system for the angular emission spectra produced by the designated SPGCE from grating structure as shown in Fig. 16.7. The phenomena of plasmon enhancement effects on the fabricated devices are mainly from the top side of the device through ultra-thin Au film. The spectrometer moved at $\theta_e$ angle to measure the PL intensity at each specific emission angle to study the changes of emitted light due to interactions of SPPs. The emission spectra from both 2-layer and 4-layer devices are obtained by optical pumping $Alq_3$ layer with a 405 nm light source at fixed incident angle of $\theta_i$= 45°. At this angle, it is not corresponding to SPR resonance angle, which is greater than 65° in grating pitch [43], and thus would not be able to excite $Alq_3$ molecules through evanescent field. The emission of organic $Alq_3$ layer will be the source to excite the SPPs from thin grating metal film and cross couple for far-field measurement. We have measured SPGCE spectra for each of our fabricated samples, as shown in Fig. 16.9, by using PL measurement system to collect the emitted light intensity from -10 to 15 degree with 1 degree per step. We can then also examine the effect of SPPs excited SPGCE on the angular dispersion of emission spectra.

The PL spectra and corresponding luminescence changes at different emission angles are measured and shown as in Fig. 16.10(a) and (b). The enhanced PL can be due to the SPR effect as mentioned above.

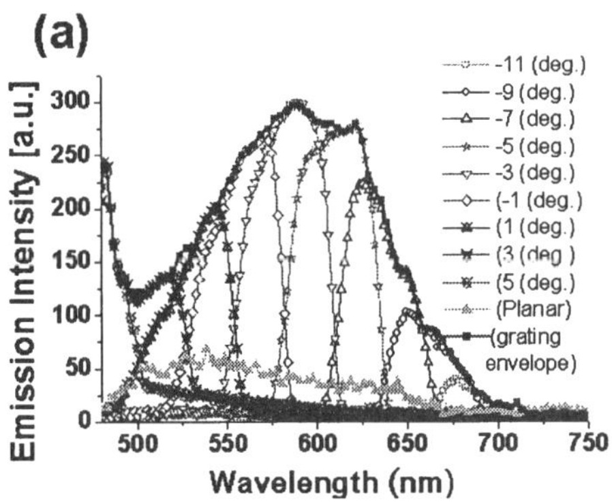


**Figure 16.10a:** The PL emission obtained from a grating of 4-layer sample, grating size (line 400nm, pitch 800nm, area size 1.2×1.2mm). Fig. 16.10(a) shows the PL of grating and non-grating (planar) samples as well as the integration of overall emission angles (envelope). The PL is measured on organic / metal grating. Reprinted with permission from [Chiu et al., Appl. Phys. Lett. 91, 083114 (2007)]. Copyright 2007, American Institute of Physics.

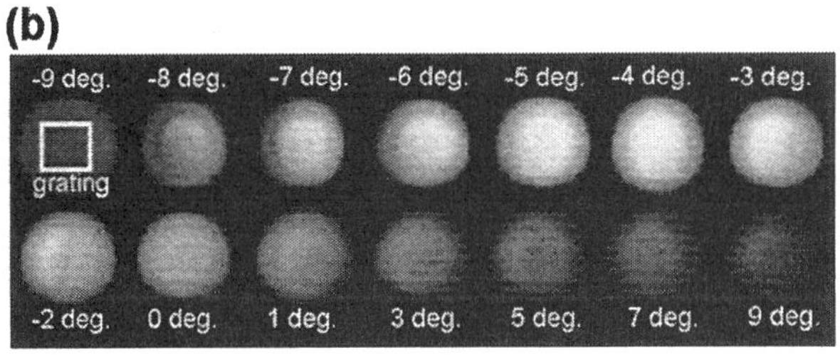


**Figure 16.10b:** The changes in SPR were measured from the colour of different angular spectra is shown here. Reprinted with permission from [Chiu et al., Appl. Phys. Lett. 91, 083114 (2007)]. Copyright 2007, American Institute of Physics.

The resultant angular emission spectra of enhanced luminescence from metal / organic grating from multiple emission angles are composed into colour coded three dimensional spectrogram as shown in Figs. 16.11(a) and 16.11(b) for 2-layer and 4-layer device, respectively. It is quite obvious from these two figures that 4-layer one does have higher intensity and smaller FWHM. The average shift in peak wavelength is 14 nm / degree for the grating with pitch size of 800 nm. The emission spectra can shift from 750 nm to 480 nm by changing the emission angle for measurement. It results in an angular dependent tuneable colour device with specific structural parameters, e.g., pitch constant ($\Lambda$), the thickness of each layer of the grating and the optical indices of used materials, to satisfy the SPP resonant conditions.

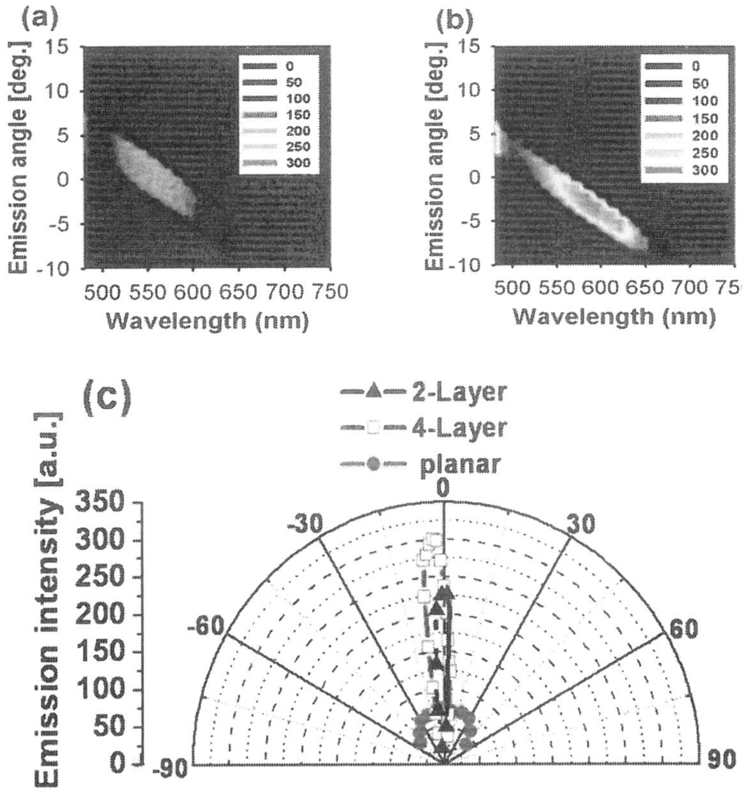


**Figure 16.11:** The PL emission obtained from a grating sample having 2-layer and 4-layer structure (grating size: line 400 nm, pitch 800 nm, area size 1.2×1.2 mm$^2$). The (a) and (b) are shown PL 3-D emission image obtained from a grating sample. The dependence of the emission spectra on observation angle (θ) is shown in (a) and (b) for 2-layer and 4-layer structure, respectively. The (c) shows the planar, 2-layer and 4-layer. The emission maximum was about 0° and -3° for 2-layer, 4-layer devices, respectively. Reprinted with permission from [Chiu et al., Appl. Phys. Lett. 91, 083114 (2007)]. Copyright 2007, American Institute of Physics.

The enhanced emission spectra measured with the coupling at different angles are shown in the characteristic diagram of PL emission in polar coordinates (Fig. 16.11(c)) for planar devices, 2-layer, and 4-layer, respectively. The ratio of the maximal intensity of these three devices is 1:4:6. The intensity from the 4-layer structure can be strongly enhanced by recovering from three possible mechanisms, i.e., coupled SPP from $Alq_3$ scattering emission, non-radiative mode, and the long-range surface plasmon polaritons (LR-SPPs) with symmetrical dielectric structure. The LR-SPPs are associated with the interactions of symmetric or antisymmetric magnetic fields on both sides of metal interface [44, 45]. The fields can constructively interact inside the thin metal film (20 nm) and then result in the LR-SPPs, which can extend into both $Alq_3$ layer for excitation.

Figure 16.12 shows the effect of pitch size on the directional emission spectra of SPGCE for three different grating structures. The resultant angular dispersion is about -35°~-5° for device with pitch size of 500 nm while for pitch size of 800 nm is limited to -10°~10°.

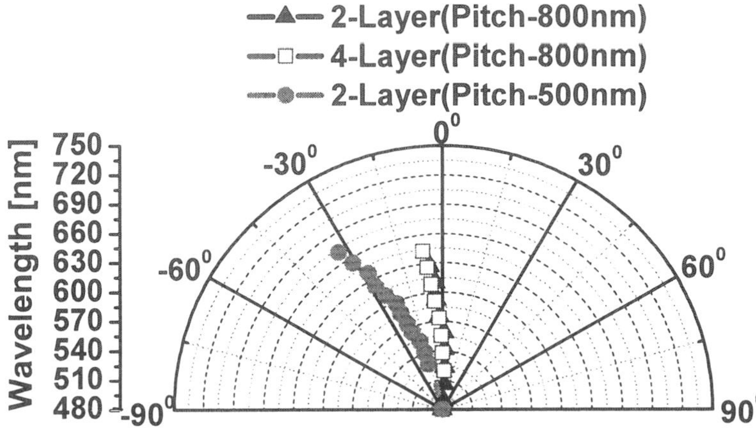


**Figure 16.12:** This shows experimental PL SPGCE diagram in directional emission spectra of pitch 500 nm and 800 nm structures. Reprinted with permission from [Chiu et al., OPTICS EXPRESS. 15, 11608 (2007)]. Copyright 2007, Optical Society of America.

It can be concluded that by using 4-layer structure, the SPGCE efficiency was improved greatly. Figure 16.13(a) shows the CIE-1931 chromaticity coordinates of the coupled emission in the emission angle and color changes from both 2-layer (blue line) and 4-layer (red line) devices. The colors of SP coupled emission are calculated and plotted on a CIE-1931 diagram, which shows angular dependent spectral changes from red to blue. We also measured the FWHM from the coupled emission spectra at different angles as shown in the Fig. 16.13(b). For the 4-layer structure, the FWHM is in the range of 40 to 50 nm, whereas for the 2-layer structure is from 57 to 70 nm, which may come from the stronger intensity of the 4-layer devices.

From the 4-layer measurement data, we can use peak emission wavelength at each emission angle to calculate its dispersion curve with m=-1 and Λ=800 nm by Eq. (1) and show as blue circle in Fig. 16.14(a). This result is related to the theoretical dispersion relation of SP-grating coupled emission at Au / air interface ($K_{SP}$(Au / air)) with m=1, dielectric constants of Alq3 ($\varepsilon_r = 2.973$, $\varepsilon_i = 0.012$) and Au ($\varepsilon_r = 7.393$, $\varepsilon_i = 1.918$) and wavelength at 550 nm by a matching momentum, ΔK of 24.55 $\mu m^{-1}$. It can be explained by using grating coupler dispersion relation as shown in Fig. 16.14(b), which shows the SPPs propagating along a grating surface has to reduce its wave vector by ΔK in order to have SPGCE. This light emission is a consequence of decoupling via grating structure from the photoluminance of Alq3 molecules. Such scheme allows a momentum-match between Alq3 emission photons and SPPs; consequently, tuneable SPP mediated light emission was achieved.

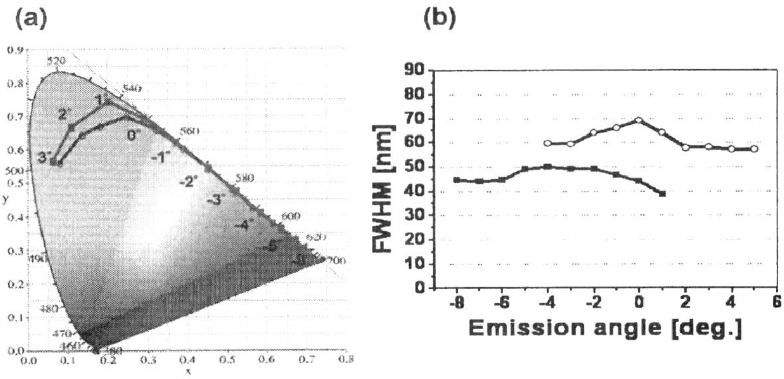


**Figure 16.13:** The SP-coupled emission is via a grating mediated by energy transfer of SP tuneable-color and FWHM at 2-layer (—o—) and 4-layer (—■—) grating structure. The (a) shows the coupled emission spectrum to 1931 CIE chromaticity diagram, with the coordinates of the spectra and angle. The (b) shows the FWHM shifts at different angle. Reprinted with permission from [Chiu et al., OPTICS EXPRESS. 15, 11608 (2007)]. Copyright 2007, Optical Society of America.

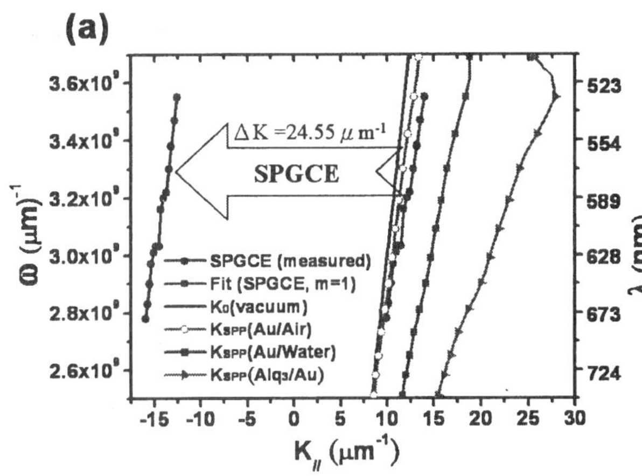

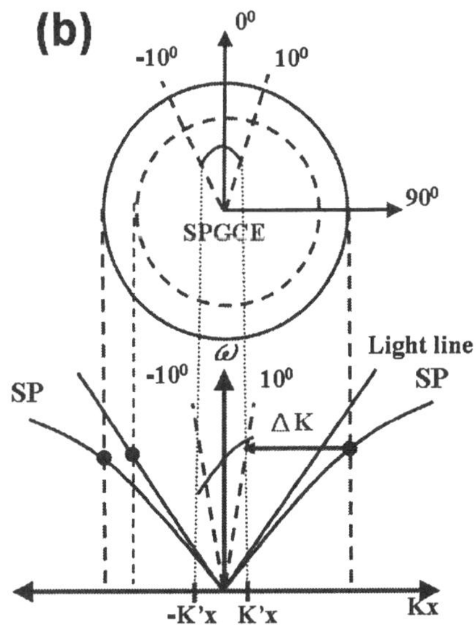


**Figure 16.14:** The figure give fitting results and theoretical interpretation. (a) is Frequency vs. wave vector for the measured data (—•—) and fitting data (—•—), the theoretical dispersion relation on interface surface Plasmon dispersion relation Au / air (—○—), $Alq_3$ / Au (—▸—), Au / water (—▪—), and the light in vacuum (——). The data were taken from the sample with 800 nm pitch. The explained that intrinsic $Alq_3$ emission and excitation into the Au / air coupler SP emission angle as shown in (b). Reprinted with permission from [Chiu et al., OPTICS EXPRESS. 15, 11608 (2007)]. Copyright 2007, Optical Society of America.

## 16.5 2. *Pitch Size Effect On Quantitative Active Surface Plasmon Grating Coupled Emission*

We prepared the nanostructure with 1-D grating patterns by electron-beam lithography (EBL). We demonstrate the surface plasmon grating coupled emission (SPGCE) from excited organic layer on different metal grating in organic / metal structure as shown in Fig. 16.15. The emissions correspond to the resonant condition of SP modes on the $Alq_3$ / Au interface and grating couple to the Au / air interface for the emission of light. In our experiments, we used different pitch sizes to control plasmonics band-gap which produced highly directional SPGCE with enhanced intensity. The experimental and theoretical results showed that SPGCE at different pitch can match a linear shifting of momentum (△K) of about 4.79 $\mu m^{-1}$ per 100 nm pitch size. Bio-SPGCE is proposed for the development of novel devices, which is expected to improve the capability of electroluminescent bio-plasmonic resonance measurement devices in the future.

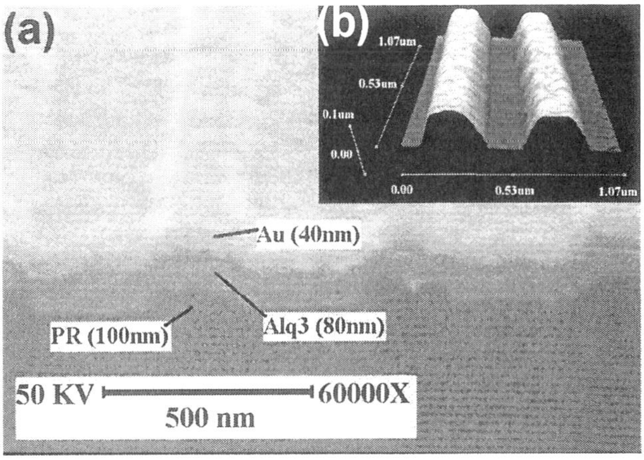


**Figure 16.15:** Excitation of SPGCE on periodically modulated lamellar grating nanostructure with SPPs propagation. (a) The SEM images of Au (40 nm), Alq3 (80 nm) and photoresist, PR, (100 nm) fabricated on silicon substrate. (b) The 3-D AFM images of PR-grating profiles with 500 nm pitch size.

We have set up a photoluminescence (PL) measurement system for the angular emission spectra produced by the designated SPGCE from grating structure as shown in Fig. 16.7. The use of grating pitch to control plasmonics band-gap might result in different electronic / photonic energy redistribution pathways in organic molecules on a metal grating surface. The resultant angular emission spectra of enhanced luminescence from fabricated devices at different emission angles were turned into color with 500 nm pitch size and different angles ($\theta_e$= -35°~ -10°). The emission plasmon model provides a rational approach for the use of metal grating nanostructures to collect, manipulate, and enhance the energy from excited luminescence. The measurement of the PL emission in a plane perpendicular to the grating pitch size of 400 nm, 500 nm, 600 nm, and 800 nm are shown in Fig. 16.16. The nano-grating coupled emission is very directional. It behaves like a visible light antenna, and has many SPPs with different wave vectors at different angles of emission $\theta_e$.

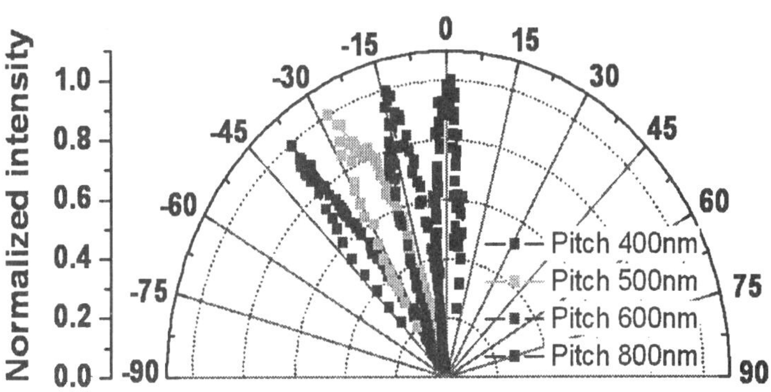

**Figure 16.16:** The experimental PL emission obtained from a sample with grating structure shows SPGCE diagram in directional emission spectra.

From the measured data of four different pitch samples, we can use peak emission wavelength at each emission angle to calculate its dispersion curve by Eq. (1). Figure 16.17 shows the plots with solid symbols for m = -1 and Λ= 400 nm, 500 nm, 600 nm and 800 nm on the left, respectively. These results are compared to the theoretical dispersion relation of SP-grating coupler at Au / air interface ($K_{SP}$(Au / air)) with m = 1, which are the denser lines on the right-hand side of the light line in vacuum. It results in a linear shifting of momentum (ΔK) of about 4.79 $\mu m^{-1}$ per 100 nm pitch size.

The measurement and fitting results of SPGCE light coupled emission device used to verify our theoretical calculations are summarized in Table 16.1. We demonstrate coupled mode in different pitch structure, the m = -1 can coupled emission at specific angle range and momentum (ΔK).

Figure 16.18 shows the theoretical simulation of SPGCE with different refractive index of biochemical and DNA molecules in contact with the surface of metallic thin films. The results are based on the calculations with grating pitch of 500 nm in air medium with $n_{Au}$ = 0.355+i2.695 at wavelength 550 nm, $n_{alcohol}$ = 1.329, $n_{ethanol}$ = 1.363, and $n_{DNA}$ = 1.405, respectively. It shows that the proposed grating nanostructure can result in the shifting of plasmon resonance toward higher wave number upon the adsorption of molecules.

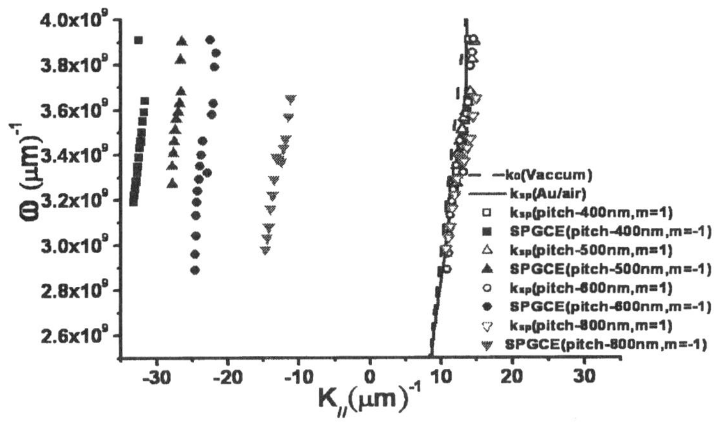


**Figure 16.17:** The figure gives four different pitch samples of fitting by theoretical calculation and experimental results. Angular frequency vs. wave number for the coupled emission measured data (m = -1), pitch 400 nm (solid square), 500 nm (solid up-triangle), 600 nm (solid circle) and 800 nm (solid down-triangle), the theoretical dispersion relation SP in Au / air (solid line), and the light in vacuum (dash line). The coupled emission at m = 1 propagating along Au / air interface match to the theoretical dispersion relation.

**Table 16.1:** The SPGCE relation between the coupling order m and emission angle of the match momentum.

| Pitch(nm) | ($\theta_e$) | ($MP\theta_e$) | Order (m) | $\triangle K(\mu m)^{-1}$ |
|---|---|---|---|---|
| 400 | -40° ~ -30° | -37° | -1 | 45.187 |
| 500 | -35° ~ -10° | -28° | -1 | 40.389 |
| 600 | -25° ~ -5° | -20° | -1 | 35.923 |
| 800 | -10° ~ 10° | 0° | -1 | 25.696 |

($\theta_e$) is Coupled emission angle.
($MP\theta_e$) is Max peak of emission angle.

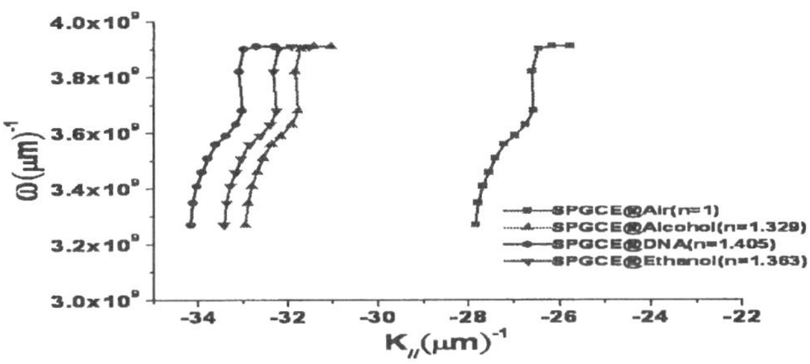


**Figure 16.18:** SPGCE theoretical simulation shows different biochemical and DNA molecules in metallic thin films surface. This means that the concentration can be detected as a change in the refractive index of the solution.

## 16.6 SUMMARY AND CONCLUSIONS

We have demonstrated the phenomenon of 2-layer and 4-layer modulation grating structure for active surface plasmon polaritons propagating along 1-D rectangular lamellar grating in an organic / metal interface via SPGCE for enhancing and tuning far-field light emission. Our results showed that strong coupling resonances in SP-coupled emission from the interactions of $Alq_3$ / Au and Au / air symmetric mode leads to the enhanced optical properties of directional emission, intensity and FWHM for active plasmon devices. The resultant emission intensity can have up to 6 times enhancement on the 4-layer device and the spectral bandwidth (FWHM) is less than 50 nm.

These pitch modulation experiments have demonstrated the effect of four different grating nanostructure devices to control plasmonics band-gap on the cross coupling of SPPs and $Alq_3$ for enhancing directional emission of light. The grating coupled emission of active $Alq_3$ molecules through the metal showed different wavelengths at different angles. The experimental and theoretical results showed that SPGCE at different pitch can match a linear shifting of momentum (△K) of about 4.79 $\mu m^{-1}$ per 100 nm pitch size.

We have demonstrated that the changes in SPR were measured from the colour of different angular spectra on metal / organic grating.

- Pitch modulation:
  −(emission angle): $$\frac{\Delta k}{\Delta Pitch} = \frac{4.79\mu m^{-1}}{100nm}, \quad \frac{\Delta \theta_e}{\Delta Pitch} = \frac{11^{\circ}}{100nm}$$
- Layer modulation:
  −(intensity): Planar < 2 Layer < 4 Layer. (Enhance up to 6 times)
  −(FWHM): 2 Layer > 4 Layer

The combination of SPPs on organic / metal interface allows specific directional emission and color appearance of $Alq_3$ fluorophores. Such scattering taking place through a metal film has an important bearing on the generation of useful light. Further investigations will be performed on SPPs with the integration of optimized organic electroluminescent plasmonic for active biosensor devices in biochemical analysis and immunoassay.

## 16.7 ACKNOLEDGEMENTS

This project is supported in part by National Science and Technology Program in Pharmaceuticals and Biotechnology, National Science Council, Taiwan, R.O.C., NSC 96-2323-B002-021, NSC 96-2120-M-002-004, NSC 96-2218-E-002-026 and MOEA 96-EC-17-A-05-S1-0017, NTU-96R0061-01.

## 16.8 REFERENCES

1. Tang, C.W., and VanSlyke, S.A. (1989). Organic electroluminescent diodes. Appl. Phys. Lett., **51**:913-915.
2. Tang, C.W. (1986). Two-layer organic photovoltaic cell. Appl. Phys. Lett. **48**: 183-185.
3. Horowitx, G. (1998). Organic field-effect transistors. Advn. Mater.**10**: 365-377.
4. Torsi, L. (2006). Organic thin-film transistors as analytical and bioanalytical sensors. Anal. Bioanal. Chem. **384**: 309.
5. Bartic, C. and Borghs, G. (2006). Organic thin-film transistors as transducers for (bio)analytical applications. Anal. Bioanal. Chem. **384**: 354-365.
6. Chiu, N.-F., Lin, C.-W., Lee, J.-H., Kuan, C.-H., Wu, K.-C., Lee, C.-K. (2007). Enhanced luminescence of organic / metal nanostructure for

grating coupler active long-range surface plasmonic device. Appl. Phys. Lett. **91**: 083114.

7. Chiu, N.-F., Yu, C., Nien, S.-Y. Lee, J.-H., Kuan, C.-H. Wu, K.-C. Lee, C.-K. and Lin, C.-W. (2007). Enhancement and tunability of active plasmonic by multilayer grating coupled emission. Opt. Express **15**: 11608-11615.
8. Kalinowski, J. (2005). Organic light-emitting diodes :principles, characteristics, and processes, Marcel Dekker, New York , pp. 1-45.
9. Lakowicz, J. R. (1999). Principles of Fluorescence Spectroscopy, 2nd Ed. Kluwer Academic / Plenum Publishers, New York.
10. Chance, R.R., Prock, A., and Silbey, R. (1975). Comments on the classical theory of energy transfer. J. Chem. Phys. **62**: 2245-2253.
11. Burin, A.L., and Ratner, M.A. (2000). Exciton migration and cathode ouenching in organic light emitting diodes. J. Phys. Chem. A. **104**: 4704-4710.
12. Choong, V.-E., Park, Y., Gao, Y., Mason, M.G., and Tang, C.W. (1998). Photoluminescence qunching of Alq3 by metal deposition: a surfaceanalytical investigation. J. Vac. Sci. Technol. A. **16**: 1838-1841.
13. Lakowicz, J.R., Malicka, J., Gryczynski, I., Gryczynski, Z., and Geddes, C.D. (2003). Radiative decay engineering; the role of photonic mode density in biotechnology. J. Phys. D; Appl. Phys. **36**: R240-R249.
14. Lakowicz, J.R. (2001). Radiative decay engineering; biophysical and biomedical applications. Anal. Biochem. **298**: 1-24.
15. Lakowicz, J.R., Shen, Y., D'Auria, S., Malicka, J., Fang, J., Gryczynski, Z., and Gryzynski, I. (2002). Radiative decay engineering 2. effects of silver island films on fluorescence intensity, lifetimes, and resonance energy transfer. Anal. Biochem. **301**: 261-277.
16. Fang, N., Liu, Z., Yen, T.-J. and Zhang, X. (2003). Regenerating evanescent waves from a silver superlens. Opt. Express **11**: 682-687.
17. Gruhlke, R. W., Holland, W. R., and Hall, D. G. (1986). Surface-plasmon cross coupling in molecular fluorescence near a corrugated thin metal film. Phys. Revi. Lett. **30**: 2838-2841.
18. Gifford, D. K., and Hall, D. G. (2002). Emission through one of two metal electrodes of an organic light-emitting diode via surface-plasmon cross coupling. Appl. Phys. Lett. **81**: 4315-4317.
19. Wedge, S., Barnes, W. L. (2004). Surface plasmon-polariton mediated light emission through thin metal films. Opt. Express **12**: 3673-3685.
20. Feng, J., Okamoto, T., Simonen, J., and Kawata, S. (2007). Color-tunable electroluminescence from white organic light-emitting devices through coupled surface plasmons. Appl. Phys. Lett. **90**: 081106.
21. Pockrand and A. Brillante, "Nonradiative decay of excited molecles near a metal surface." Chem. Phys. Lett. **69**, 499-504 (1980).
22. S. C. Kitson, W. L. Barnes, J. R. Sambles, "Photoluminescence from dye molecules on silver gratings," Opt. Commun. **122**, 147-154 (1996).
23. Y. -J. Hung, I. I. Smolyaninov, C. C. Davis, and H.-C. Wu, "Fluorescence enhancement by surface gratings." Opt. Express **14**, 10825-10830 (2006).
24. Nakano, T., Kobayashi, H., Shinbo, K., Kato, K., Kaneko, F., Kawakami, T., and Akamatsu, T. (2001). Emission light properties from

Adrhodamine-B LB films due to surface plasmon excitations in the Kretschmann and reverse configurations. Mat. Res. Soc. Symp. **660**: 1-6.

25. Shinbo, K., Toyoshima, S., Ohdaira, Y., Kato, K., and Kaneko, F. (2005). Surface plasmon emission light property due to molecular luminescence and molecular interaction. J. J. Appl. Phys. **44**: 599-603.
26. Winter, G., and Barnes, W. L. (2006). Emission of light through thin silver films via near-field coupling to surface plasmon polaritons. Appl. Phys. Lett. **88**: 051109.
27. Enderlein, J., and Ruckstuhl, T. (2005). The efficiency of surface-plasmon coupled emission for sensitive fluorescence detection. Opt. Express **13**: 8855-8865.
28. Kitson, S. C., Barnes, W. L., and Sambles, J. R. (1996). Photoluminescence from dye molecules on silver gratings. Opt. Commun. **122**: 147-154.
29. Kalkman, J., Strohhofer, C., Gralak, B., and Polman, A. (2003). Surface plasmon polariton modified emission of erbium in a metallodielectric grating. Appl. Phys. Lett. **83**: 30-32.
30. Hung, Y.-J., Smolyaninov, I. I., Davis, C. C., and Wu, H.-C. (2006). Fluorescence enhancement by surface gratings. Opt. Express **14**: 10825-10830.
31. Maier,S. A., Kik, P. G., Atwater, H. A., Meltzer, S., Harel, E., Koel, B. E., and. Requicha, A. A. G. (2003). Local detection of electromagnetic energy transport below the diffraction limit in metal nanoparticle plasmon waveguides. Nat. Mater. **2**: 229-232.
32. Lakowicz, J. R., Malicka, J., Gryczynski, I. and Gryczynski, Z. (2003). Directional surface plasmon-coupled emission: a new method for high sensitivity detection. Biochem. Biophys. Res. Commun. **307**: 435-439.
33. Lakowicz, J. R. (2004). Radiative decay engineering 3. Surface Plasmon-coupled direction emission. Anal. Biochem. **324**: 153-169.
34. Gryzynski, I., Malicka, J., Gryczynski, Z., and Lakowicz, J.R. (2004). Radiative decay engineering 4. Experimental studies of surface plasmon-coupled directional emission. Anal. Biochem. **324**: 170-182.
35. Raether, H. (*1988*). Surface plasmons on Smooth and Rough Surfaces and on Gratings, Springer Tracts Mod. Phys., Verlag, Berlin., **111**.
36. Kitson, S. C., Barnes, W. L., and Samblesc, J. R. (1995). Surface plasmon energy gaps and photoluminescence. Phys. Rev. B **52**: 11441-11445.
37. Kittson, S. C., Barnes, W. L., Sambles J. R., and N. Cotter, P. K. (1996). Excitation of molecular fluorescence via surface plasmon polaritons. J. of modern optics **43**: 573-582.
38. Amos,R. M., and Barnes, W. L. (1999). Modification of spontaneous emission lifetimes in the presence of corrugated metallic surfaces. Phys. Rev. B **59**: 7708-7714.
39. Hibbins, A. P., Sambles, J. R., and Lawrence, C. R. (1998). Azimuth-angle-dependent reflectivity data from metallic gratings. J. of modern optics **45:** 1019-1028.
40. Lin, C. -W., Chen, K. -P., Lin, S. -M., Lee, C. -K. (2006). Design and fabrication of an alternating dielectric multi-layer device for surface plasmon resonance sensor. Sens. Actuators B: Chem. **113**: 169-176.

41. Lin, C. -W., Chen, K. -P., Su, M. -C., Hsiao, T. -C., Lee, S. –S., Lin, S. -M., Shi, X. -J., and Lee, C. -K. (2006). Admittance loci design method for multilayer surface plasmon resonance devices. Sens. Actuators B: Chem. **117**: 219-229.
42. Lin, C. -W., Chen, K. -P., Su, M. -C., Lee, C. -K., and Yang, C. -C. (2005). Bio-plasmonics: Nano / micro structure of surface plasmon resonance devices for biomedicine. Opt. Quantum Electron **37**: 1423-1437.
43. Homola, J., Koudela, I., and Yee, S. S. (1999). Surface Plasmon resonance sensor based on diffraction gratings and prism couplers: sensitivity comparison. Sens. Actuators B **54**: 16-24.
44. Andrew, P.m and Barnes, W. L. (2004). Energy transfer across a metal film mediated by surface plasmon polaritons. Science **306**: 1002-1005.
45. Sarid, D. (1981). Long-Range Surface-Plasma Waves on Very Thin Metal Films. Phys. Rev. Lett. **47**: 1927-1930.

# 17

# Enhancing Fluorescence with Sub-Wavelength Metallic Apertures

Steve Blair[1] and Jérôme Wenger[2]
[1] Department of Electrical and Computer Engineering, University of Utah, Salt Lake City, UT, USA.
[2] Institut Fresnel, CNRS, Domaine Universitaire de Saint-Jérôme, Marseille, France.

## 17.1 INTRODUCTION

The ability to reliably produce nanometric structures with a resolution down to a few nanometers opens the way for promising photonics applications (1, 2). With these nanodevices, light can be locally confined, inducing large absorption and diffusion cross sections and enhancing the local electromagnetic fields. Such effects hold strong relevance for molecular detection methods based on photonic emission (fluorescence and Raman scattering) that have recently gained a wide popularity in various fields such as biology, medicine, chemistry and materials sciences. The sensitivity of most applications relies directly on the intrinsic molecular brightness. Therefore, increasing the optical emission of molecules using metallic nanostructured substrates has become a key issue (3, 4, 5).

The fluorescence emission of a single molecule can be greatly enhanced by properly tailoring its photonic environment, opening new opportunities for molecular detection. Since the founding works of Purcell (6), Drexhage (7) and Kleppner (8), it is well recognized that the spontaneous de-excitation of a quantum emitter can be controlled by its environment, leading to modifications of the total de-excitation rate and spatial emission distribution. Following Fermi's golden rule, the spontaneous de-excitation rate is proportional to the local density of states (LDOS) (9, 3), which can be altered by a wide range of structures, such as planar interfaces (9), nanoparticles (10, 11), or nanoantennas (12, 13).

Up to now, a large part of the international scientific attention was devoted to the study of metallic nanoparticles (either single particles or in a colloid ensemble) (3, 14). Milling nanometric apertures in a metallic film is an intuitive way to manufacture new nanophotonics devices that are robust and highly reproducible. Although this concept appears very simple, such apertures exhibit attractive physical properties, such as localization of excitation light, strong isolation from emission produced by unbound species, and an increase in apparent absorption and emission yield. The simplicity of the structures and their ease of use should further expand their application towards the real-time detection and identification of a small number of molecules.

The purpose of this chapter is to overview some of the key results in fluorescence emission from sub-wavelength metal apertures, both single apertures and in array arrangements. While a comprehensive theory is currently lacking, sufficient understanding can be obtained from computational models to qualitatively describe experimental results.

*MEF from Sub-Wavelength Apertures.*
Edited by Chris D. Geddes.

## 17.2 BACKGROUND

The publication of the unexpected optical properties of arrays of sub-wavelength metal apertures by Ebbesen and co-workers in 1998 (15) sparked tremendous interest in the light transmission properties of ordered and disordered arrangements of metal apertures. When normalized to the transmission of an isolated aperture (16), transmission enhancements of about 3 were obtained, where very little transmission would be expected from an isolated aperture (the original series of publications normalized total transmission to the Bethe-Bouwkamp theory (17, 18) for transmission through a single aperture $T \propto (d/\square)^4$, where $d$ is the diameter, which produced enhancement factors greater than $10^3$). This enhancement was attributed to coupling into surface plasmon polariton (SPP) modes of the metal film, mediated by the periodic arrangement of apertures.

Significant attention has also been directed towards the properties of an individual circular (19, 20, 21, 22, 23, 24) or rectangular (25, 26) aperture. Metal apertures support propagating modes with extended cutoff wavelength (25, 27, 26), along with a surface plasmon mode propagating down the walls that has no cutoff (27). The situation for a finite-height aperture is somewhat more complex in that localized plasmon (LSP) modes exist at the entrance and exit (21, 28, 29); the effects of LSP modes are more pronounced for noble metals such as silver and gold (30). These modes are analogous to the LSP modes of metallic nanoparticles via Babinet's principle (31) and can couple to the propagating modes of the nanocavity, resulting in surprisingly high transmission (32, 29).

A paper in 1986 suggested that fluorescence enhancement and reduced observation volumes could be obtained from random arrangements of small metal apertures (33). In 2003, single molecule detection in an epifluorescence arrangement from a single sub-wavelength aperture was demonstrated in a high-concentration environment (20) (in which no fluorescence enhancement was reported), followed by a report of enhancement from a fluorescing monolayer self-assembled within an ordered array of apertures (34) under trans-illumination. Subsequently, there have been reports of enhanced Raman scattering (35, 36), real-time monitoring of molecular capture (37), electroluminescence of an organic LED (38), and fluorescence from a polymer layer (39, 40). Computational modeling of dipole emission (41) provided a partial explanation for the enhancement that can be attributed to a single aperture. The fact that significant enhancement in excitation also occurs within a single nanoaperture was first demonstrated in a series of papers on single-molecule epifluorescence in round (42, 36) and rectangular (43) apertures. Excitation enhancement is further supported by computational models (44, 45, 29, 46). Additional studies demonstrated single molecule detection under trans-illumination (47), with comparatively lower enhancement levels, and in C-shaped apertures (48).

## 17.3 FLUORESCENCE ENHANCEMENT BY A SINGLE APERTURE

Nanometric apertures milled in an opaque metallic film allow for a simple and elegant way of generating a reduced analysis volume. As illustrated on Figure 17.1, the nanostructure itself acts as a pinhole filter directly located at the object plane of a standard epi-illumination microscope, similar to the original idea proposed by E. H. Synge in 1928 (49). When the aperture diameter is reduced sufficiently below the cut-off diameter of the fundamental excitation mode that may propagate through the (waveguiding) hole, the light inside the aperture is primarily confined to a rapidly decaying evanescent mode, with a decay length of a few tens of nanometers (such devices have thus been named zero-mode waveguides (20)).

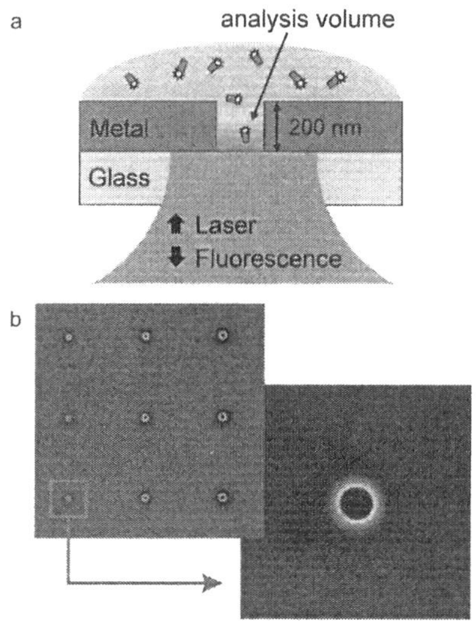


**Figure 17.1:** (a) Arrangement for single nanoaperture enhanced fluorescence. A fluorescent molecule located at the bottom of the aperture emits significantly more photons as the same molecule put in the diffraction limited spot of a high-numerical aperture microscope objective. (b) SEM image of 150, 220, 360 nm diameter apertures milled by focused ion beam on a 200 nm thick aluminum film deposited over a standard glass coverslip, and close-up view of a 150 nm aperture [image courtesy of J. Dintinger].

Beyond the optical confinement to isolate a single molecule in a highly concentrated (micromolar) solution, nanometric apertures can significantly enhance the fluorescence emission of a single molecule by properly tailoring its photonic environment. A nanometric aperture can affect the fluorescence emission in three ways: (i) by locally enhancing the excitation intensity, (ii) by increasing the emitter's radiative rate and quantum efficiency, and (iii) by modifying its radiation pattern, towards a higher emission directionality to the detectors. Determining the specific influence of these processes is a crucial issue to characterize nanodevices for enhanced fluorescence, and has been a topic of great interest for the last decade (9, 50). The inherent challenge in this task originates from the fact that the detected signal results from a product of excitation and emission processes. Excitation depends on the interaction between the driving field and the nanostructure, while at moderate optical intensity, the emission efficiency is set by the balance of radiative and non-radiative decays and the modification of the radiation pattern.

To our knowledge, the first implementation of nanoaperture-enhanced fluorescence dates back to 1986 (33), where apertures of randomly-distributed diameters between 180 and 480 nm in thin films of silver or gold supported by glass slides were used as probes to detect the properties of an adjacent liquid containing fluorescent molecules. Even in the absence of efficient detectors and laser sources, sensitive measurements of refractive-index differences of the order of $10^{-4}$ were demonstrated, and a global enhancement of the fluorescence brightness was reported. Yet, the lack of efficient detectors and laser excitation together with a rather involved experimental configuration and distribution of aperture sizes complicated the analysis of the measured data.

More recently, a single sub-wavelength nanoaperture was demonstrated to significantly enhance the fluorescence rate emitted by a single fluorescent molecule. Using single rhodamine 6G molecules in isolated 150nm diameter apertures milled in an aluminum film, a 6.5 fold enhancement of the fluorescence rate per molecule was reported as compared to free solution (42). This phenomenon was further investigated with a combination of FCS and fluorescence lifetime measurements to resolve the photokinetic rates of rhodamine 6G molecules in a water-glycerol mixture (51). The fluorescence lifetime appeared to be dramatically reduced inside the aperture, indicating that the molecular energy levels' branching ratios were strongly affected. For a properly tailored nanoaperture diameter of 150 nm, the local density of states alteration brought by the aperture allowed for a higher radiative rate without decreasing the fluorophore quantum efficiency (fluorescence quenching). The combination of this effect together with an increase in the local excitation intensity led to the overall fluorescence enhancement.

## 17.3 1. *Background*

The fluorescence enhancement □$_F$ in a nanoaperture is defined as the ratio of the detected fluorescence count rate per molecule in the aperture $CRM_{aper}$ and in open solution $CRM_{sol}$ for the same (fixed) excitation power, that is □$_F$ = $CRM_{aper}$ / $CRM_{sol}$. To a good approximation, most of the fluorescent molecules can be treated as a system of three energy levels in steady state (singlet ground state $S_0$, first excited singlet state $S_1$ and dark non-fluorescing state D). Under steady-state conditions, the fluorescence rate per molecule CRM is given by (52)

$$\mathrm{CRM} = \kappa\,\phi\,\frac{\sigma I_e}{1 + I_e / I_s} \qquad (1)$$

where □ is the light collection efficiency, □ = $k_{rad}$ / $k_{tot}$ is the quantum yield, and $I_s$ = $k_{tot}$ / □ * 1/(1+$k_{isc}$ / $k_d$) is the saturation intensity. □ $I_e$ stands for the excitation rate, where □ denotes the excitation cross-section and $I_e$ the excitation intensity. $k_{rad}$ and $k_{nrad}$ are the rate constants for radiative emission and non-radiative deexcitation from $S_1$ to the ground state $S_0$. $k_{isc}$ and $k_d$ are the rate constants for inter-system crossing to the dark state D and relaxation to the ground state $S_0$ respectively. The total de-excitation rate from the excited singlet state $S_1$ is noted $k_{tot}$, and is the inverse of the excited state lifetime.

In the low excitation regime $I_e << I_s$, Eq. (1) reduces to

$$CRM_{low} = \square\ \square\ \square\ I_e, \qquad (2)$$

which indicates that the fluorescence rate per molecule is proportional to the fluorescence collection efficiency, the dye quantum yield and the local excitation intensity. The fluorescence enhancement $\square_F$ can therefore be expressed as

$$\eta_F = \frac{CRM_{aper}}{CRM_{sol}} = \eta_\kappa \eta_\phi \eta_{I_e} = \eta_\kappa \frac{\eta_{krad}}{\eta_{ktot}} \eta_{I_e} \qquad (3)$$

where $\square_{\square}$, $\square_{\square}$, $\square_{Ie}$ are the enhancements in the collection efficiency, quantum yield and excitation rate respectively (the free solution serves as a reference for enhancements calibration). $\square_{krad}$ and $\square_{ktot}$ stand for the enhancements in radiative and total decay rates. This expression highlights the complex balance between the collection, excitation and emission efficiencies needed to obtain a significant total fluorescence enhancement.

## 17.3 2. *Experimental studies on fluorescence from nanoapertures*

Two distinct sets of experimental studies have been performed, both providing information on the enhancement of fluorescence from metal nanoapertures. The first set of experiments has been performed using self-assembled monolayers of fluorescing molecules while the second set of experiments has been performed on single molecules diffusing from solution. Despite the very different measurement methodologies, both sets of experiments provide corroborating data on enhanced fluorescence in nanoapertures.

### 17.3.2 1. *Studies with self-assembled monolayers*

Disordered arrangements of nanoapertures can be used to approximate the properties of an individual aperture while allowing measurements to be made across a large area. When the reciprocal space of the arrangement lacks the strong isolated peaks of a periodic lattice, SPP excitation on the surface is not coherently reinforced, thereby reducing inter-aperture interactions, and allowing the interpretation of the results as an ensemble across multiple, independent, apertures.

Similar to the early work by Fischer (33), fluorescence enhancement has been studied in disordered arrangements of apertures, but with the important difference of having well-defined sizes (37, 41, 46). In these experiments, apertures of ~200nm diameter and average spacing of 1□m were defined in PMMA using electron beam lithography (EBL) and dry etched into a 70nm layer of gold on a quartz substrate. The nanostructured metal films were coated on the top surfaces with a fluorescing monolayer, which consisted of avidin labeled with Cy-5 dye. Cy-5 has peak absorption near 649nm and peak emission near 670nm. A solution of 1□M avidin in phosphate buffered saline (PBS, pH 7.5, containing 0.02% sodlium azide as a preservative) was labeled with Cy-5, with a ratio of Cy-5 to avidin of 0.95 (37). This solution was then diluted to the desired Cy-5 concentration (10nM for 1%

dye concentration) in 1□M pure avidin in PBS. The diluted solution was then used to coat the sample surface, and left for 3 hours at room temperature to allow formation of the labeled monolayer. Then, the unadsorbed species were removed by washing the samples in TE buffer, which contained 10mM Tris buffer (pH7.4) and 1 mM EDTA. Finally, the samples were dried in vacuum for 1 hour and immediately used. Given the size of an individual avidin molecule, the average distance between the fluorophore and the surface was estimated at about 5nm.

Figure 17.2 shows the experimental setup for measurement of light transmission and Cy-5 fluorescence. Detection was performed from the backside to reduce collection of fluorescence emission from the top surface. Incident light from a narrow-line tunable diode laser set to 653nm was p-polarized and attenuated by an OD 3 filter to reduce photobleaching effects during long measurements. The light was spectrally filtered by a 1.5nm width band-pass filter to suppress broad-band spontaneous emission, and passed through a chopper before being focused on the sample. A photomultiplier tube (PMT) was mounted on an adjustable-length arm attached to a computer controlled rotation stage, while the sample was mounted at the center of rotation of this stage on a manual rotation stage. With this configuration, the laser incidence angle and the transmission detection angle could be adjusted either collectively or independently. Transmission of incident laser light was measured in the zeroth-order by a photodetector. For fluorescence transmission measurements, the detector was removed and the fluorescence signal imaged through a 670nm band-pass filter onto the PMT. The PMT was connected to a lock-in amplifier, the output of which was read by a computer.

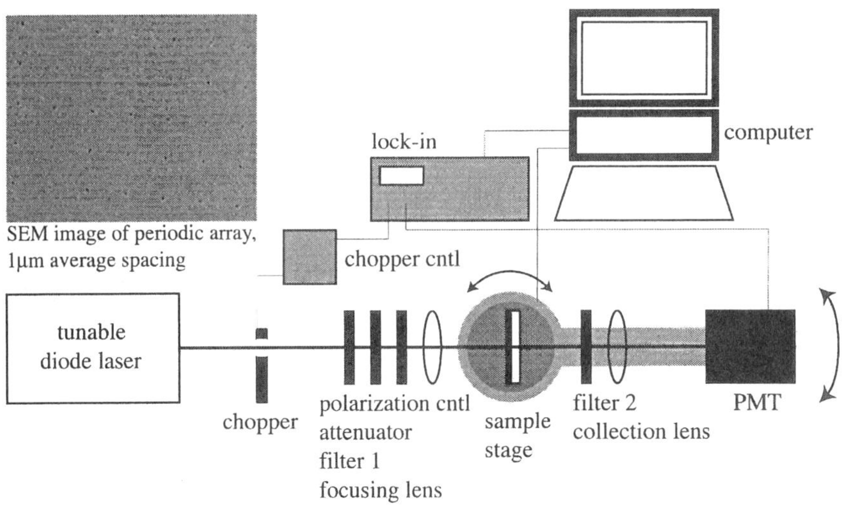


**Figure 17.2:** Experimental setup for measuring fluorescence from nanoaperture arrays. Narrow-line incident light (635nm) illuminates the sample from the top side, while detection occurs through the back side, using a standard fluorescence emission filter. The sample and detector can be independently rotated.

Figure 17.3 shows the transmission and fluorescence enhancement from a disordered nanoaperture array versus incidence angle. Both quantities are nearly constant with angle, indicating minimal influence from SPP coupling as expected. In these measurements, a reference fluorescence level was obtained from a bare substrate that underwent the same chemical modifications as the nanoaperture samples. Since the illumination area and collection optics were the same, a direct ratio of fluorescence signal could be used to determine the enhancement factor. This ratio must be corrected for the fill fraction of the nanoapertures, since with backside detection, fluorescence from the top surface was negligible (as determined by measuring fluorescence from the backside of a uniform gold film with the same surface treatment), and fluorescence emission originated from within the apertures. Since the avidin monolayer covered the entire interior surface area of the apertures, the fill fraction was 7.5% (rather than 3.1% based upon the nanoaperture diameter), leading to a fluorescence enhancement factor of 7, averaged across angle all angles of incidence.

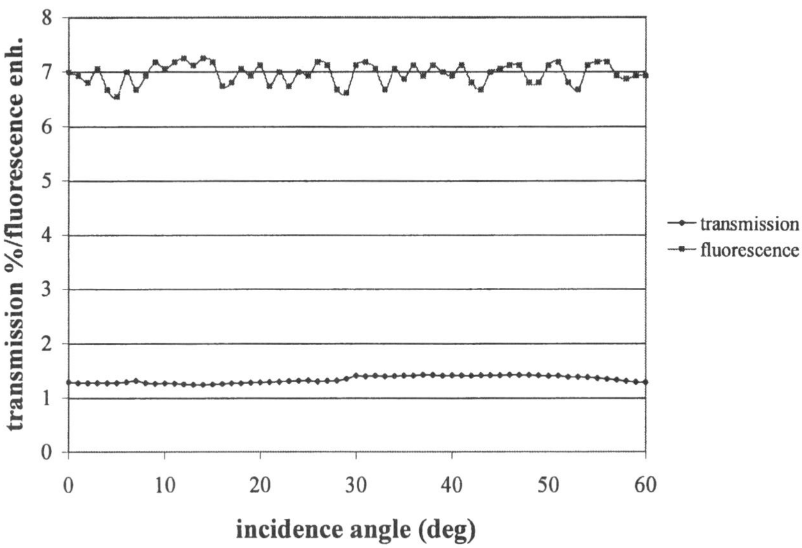


**Figure 17.3:** Transmission of excitation light and fluorescence emission from Cy-5 from a disordered arrangement of 200nm diameter nanoapertures in 70nm thick gold film, with average spacing of 1□m. In these measurements, the sample angle was varied with respect to the incident light, while detection occurred in a direction collinear with the incident light. Fluorescence is normalized to a quartz substrate with the same monolayer and corrected for fill fraction.

Performing the same measurements on samples where the gold surfaces had been passivated with mPEG-thiol (see section 4.1), leaving the subsequently self-assembled monolayers only at the bottom of the nanocavities, average fluorescence enhancement (across multiple samples) of 6.9±0.7 for Cy-5 has been obtained (46). In this case, the fill fraction for fluorescence emission was 3.1%. What the comparison of these measures shows is that the enhancements with and without coverage of the aperture sidewalls are about the same, a conclusion that is supported by simulation results (46).

## 17.3.2 2. *Single molecule studies*

In order to understand the physical phenomena behind nanoaperture-enhanced fluorescence, investigating an isolated nanoaperture is of major interest, which eliminates any effects of aperture coupling. Further, data from a single molecule may reveal information hidden by ensemble measurements, such as variances in kinetic rates, memory effects, or transient states (52).

In order to get an accurate estimation of the CRM and fluorescence enhancement factor of a single nanoaperture, it is of crucial importance to simultaneously quantify the total fluorescence intensity and the number of emitters from the aperture. Fluorescence Correlation Spectroscopy (FCS) is a robust method capable of both (53, 52). FCS can in principle provide information about any molecular dynamic process on the nanosecond and longer time scale that induces a change in fluorescence intensity. For instance, fluctuations occur when molecules diffuse in and out of an observation volume, or when reaction kinetics or conformational changes induce a change in the fluorescence brightness. Statistical analysis of the fluctuations is generally achieved by computing the fluorescence intensity correlation function, that is, the temporal fluctuations of the fluorescence intensity are recorded, and the autocorrelation of this signal is computed. Numerical analysis of the FCS data then provides the average number of detected molecules which is used to compute the fluorescence count rate per molecule CRM.

The FCS setup for working with single nanoapertures is shown in Figure 17.4. Initial measurements were performed with round aperture in 300nm thick aluminum films, with aperture diameters ranging from 110 to 420nm (42). These apertures were prepared by focused ion beam milling. Total fluorescence enhancement factors of up to nearly 7 were obtained for Rh6G dye, excited at 488nm, as reproduced in Figure 17.5b. Following the first demonstration by Cornell University (20), it was shown that nanoapertures could result in significant reduction in observation volumes, leading to the ability to perform single molecule measurements a high solution concentrations, as shown in Figure 17.6.

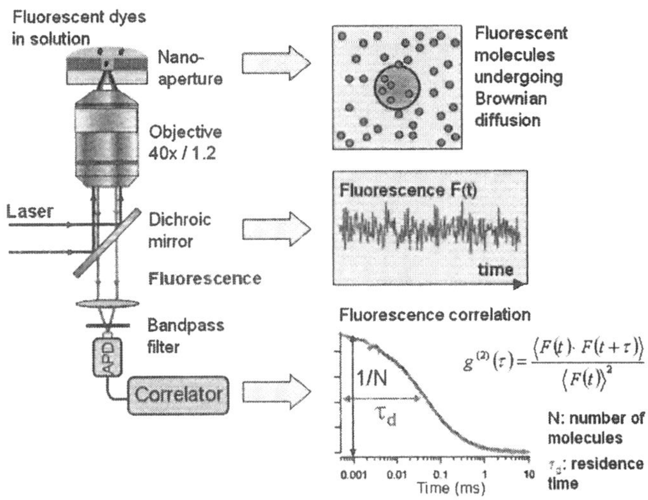


**Figure 17.4:** Principle of fluorescence correlation spectroscopy: the fluorescence intensity temporal fluctuations originating from a well-defined volume are recorded and correlated to estimate the average number of molecules observed and the characteristic fluctuation time. This data is used to compute the average detected fluorescence rate per molecule in the observation volume.

Experiments have also been conducted on rectangular sub-wavelength apertures milled in aluminum films (43). This shape allows the activation of two distinct modes by changing the angle between the linearly-polarized incoming electric field and the longitudinal axis of the aperture. Since the waveguide's cut-off is different for each axis, the penetration depth of the excitation light within the aperture can thus be significantly tuned, offering new possibilities to dynamically tailor the observation volume in FCS. Whereas for both polarization directions the molecular environment remained almost the same (as shown by the same fluorescence lifetime), the fluorescence rate enhancement varied strongly with the polarization direction: a high transmission of the excitation field did not yield any fluorescence enhancement whereas an evanescent coupling of the pump field induced a significant increase in the detected count rate per molecule. These results confirmed that the fluorescence enhancement found for rhodamine 6G was mainly related to the excitation near field intensity within the sub-wavelength aperture.

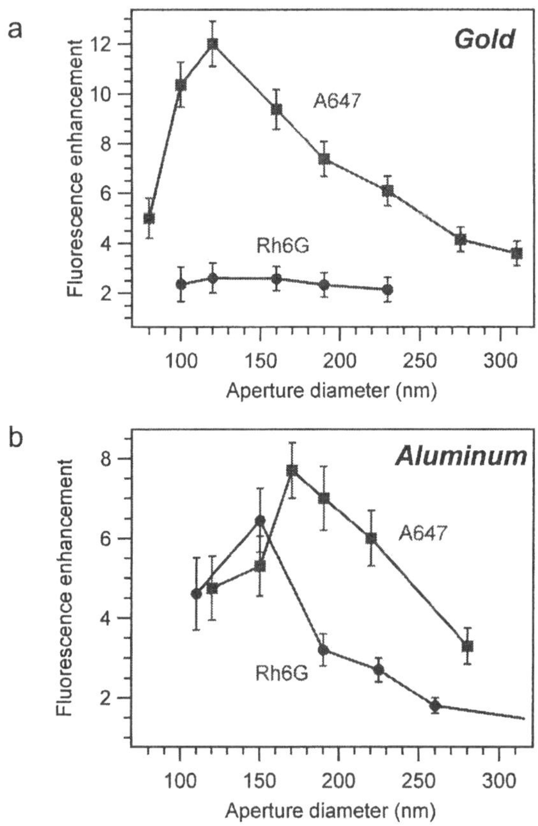


**Figure 17.5:** Experimental single molecule fluorescence enhancement versus aperture diameter, for apertures milled in gold (a) and aluminum (b) films. Rhodamine 6G (Rh6G) dyes were excited at 488 nm, and their fluorescence was collected in the 510-560 nm range. Alexa Fluor 647 (A647) dyes were excited at 633 nm, and their fluorescence was collected in the 650-690 nm range. These curves show significant differences between apertures milled in gold or in aluminum, and highlight the need for a layer exhibiting a strong metallic character at the excitation wavelength to obtain enhanced fluorescence.

All the above-mentioned experiments have been conducted with a high quantum yield dye and aluminum films, which exhibits a reduced plasmonic response in the visible range. A metal supporting strong surface plasmon resonances in the visible range, such as gold, allows for even larger fluorescence enhancement factors. Two recent studies have focused on the case of Alexa 647 dye (quantum yield in water solution 30%) and nanoapertures milled in gold films (30, 54). Nanometric apertures milled in gold exhibit significantly higher (50%) fluorescence enhancement factors than apertures in aluminium, with a maximum enhancement of 12 for a 120 nm diameter aperture in gold, as shown in Figure 17.5. This effect is related to a larger enhancement of the excitation intensity and radiative rate. Due to the intrinsic permittivity of gold and its larger skin depth than aluminum, the maximum enhancement for gold also occurs at a smaller aperture diameter, which is beneficial for single-molecule detection at high concentrations. Comparison with numerical simulations reveals that the enhancement factor is maximum when the group velocity of the guided mode inside the aperture is close to zero, that is, when the photonic density of modes is maximum. This diameter is significantly shifted (up to 50 nm) as compared to the intuitive position of the cut-off of the fundamental guided mode due in part to the skin depth of the metal. Such a significant shift has to be considered carefully while designing nanoapertures for high-efficiency single-molecule analysis.

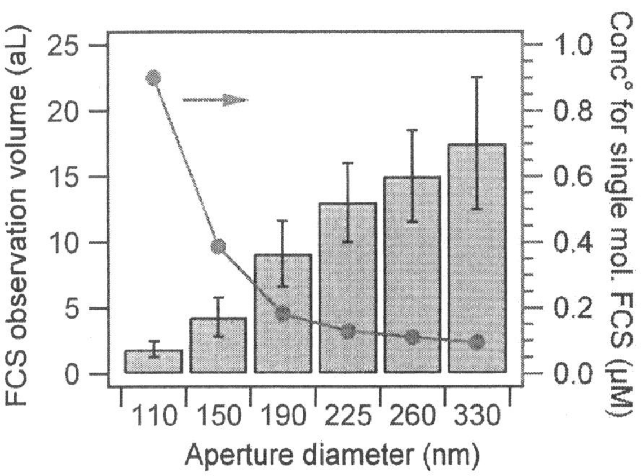


**Figure 17.6:** Observation volume measured by FCS for nanoapertures milled in aluminum (rhodamine 6G dyes, 488 nm excitation) and the corresponding concentration to ensure an average number of one molecule in the observation volume (right scale).

In a second contribution (54), the respective roles of excitation and emission in the enhanced fluorescence have been determined for a broad range of nanoapertures with diameters from 80 to 310 nm. For this purpose, a specific fluorescence characterization procedure combining FCS with fluorescence lifetime measurements has been developed, with results presented in Figure 17.7. This data allows discrimination among different physical effects and relation of photokinetic enhancements to the local photonic density of states; for example, the increase in radiative rate at saturation excitation power, as shown in Figure 17.8. For aperture diameters above 175 nm (at 633 nm excitation with gold films), the photokinetic rates increase as the aperture diameter is decreased. This region corresponds to a propagative excitation field. As the aperture diameter is decreased, more electromagnetic confinement is obtained at the aperture entrance, which translates into an increase in the emission enhancement along with the excitation enhancement. On the other hand, for aperture diameters below 100 nm, a decrease in the fluorescence enhancement is observed as the aperture diameter is reduced. Diameters below 100 nm lead to strongly evanescent fields inside the aperture and large propagative losses, which contributes to the reduction in emission and excitation enhancement. Lastly, diameters between 100 and 175 nm appear as a trade-off between the two extreme cases. This region is close to the position where the group velocity is minimum, leading to a maximum of the photonic density of states. Moreover, the metal-dielectric interface set by the aperture may allow fluorescence energy transferred to a localized surface plasmon to be coupled out into the radiated field at the aperture rim, contributing to the emission.

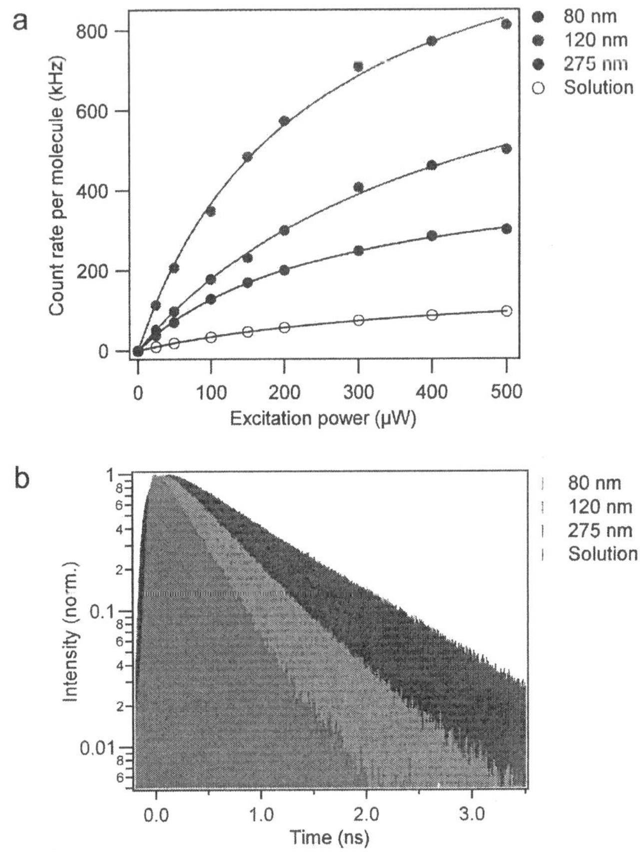


**Figure 17.7:** (a) Fluorescence rates per molecule versus excitation power in free solution and in single gold nanoapertures of 80, 120 and 275 nm diameter. Circles are experimental data, lines are numerical fits according to three energy states model. Nanoapertures not only allow for large fluorescence enhancements at low excitation powers, but also provide count rates per molecule largely above a few hundred thousands events per second. (b) Corresponding normalized fluorescence decay traces measured in free solution and in single nanoapertures, showing a clear lifetime reduction as the aperture diameter is decreased.

Combining FCS with fluorescence lifetime measurements turns out to be very efficient in resolving the contributions of excitation and emission to the overall fluorescence process. This procedure can be straightforwardly extended to other types of plasmonic nanostructures, as long as the observation volume is well defined. The limit of this method is that all the presented results account for spatial averaging over all the possible molecular orientations and positions inside the analyzed volume.

There is no sensitivity to individual molecular trajectories or dipole orientations, but one ends up directly with global figures to characterize the emitted fluorescence. Besides, distinguishing between the contributions of the radiative rate and the collection efficiency remains a challenge, mainly because of the intrinsic difficulty to reliably measure collection efficiency. Lastly, the fluorescence enhancement factors are spectrally averaged within the fluorescence bandpass detection window. However, further investigations can provide some additional knowledge on these last two points, as we will discuss hereafter.

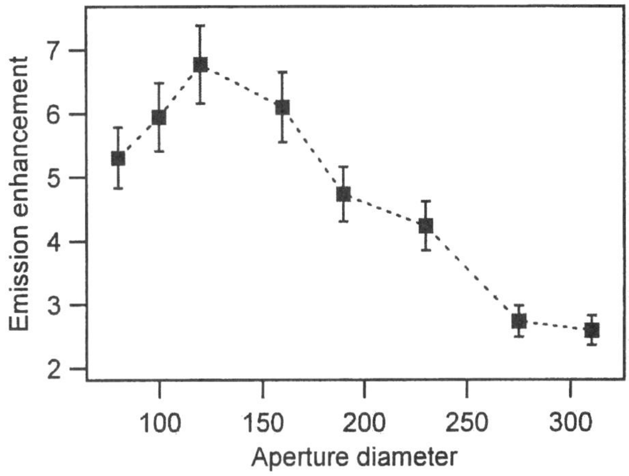


**Figure 17.8:** Fluorescence enhancement at saturation of the absorption / emission cycle, for A647 molecules in single nanoapertures milled in gold. This factor represents the product of the gain in collection efficiency □$_{□}$ by the enhancement in radiative rate □$_{krad}$. This figure highlights the contribution of the enhanced emission independently of the altered excitation intensity.

To address the question of spectral averaging within the fluorescence bandpass filters, fluorescence spectra were taken from a single gold nanoaperture (30) using a spectrograph with nitrogen-cooled CCD detector. The parallel use of FCS quantifies the number of detected molecules per experimental run, which allows the *a posteriori* normalization of the raw fluorescence spectrum, and the calculation of the fluorescence spectrum per single molecule. Experiments performed on Alexa 647 molecules excited at 633 nm revealed that the detected count rate per molecule was significantly enhanced over the whole emission spectral range, with a maximum enhancement for 120 nm diameter apertures, as observed in FCS. The shapes of the spectra do not change noticeably between the different apertures and the free solution. The same behavior has been observed for apertures in aluminium, and for Alexa 680 dyes. In the case of single nanoapertures, we point out that white light transmission measurements through a single aperture of diameter 120–300 nm yielded transmission curves almost flat over the 650–710 nm spectral range covered by the dye emission. This range, therefore, seems too narrow to monitor any clear spectral link between white light transmission and fluorescence.

## 17.3 3. *Simulation results*

Rigorous electromagnetic computations of light diffraction in cylindrical geometry confirmed the excitation intensity enhancement observed experimentally (44). At low excitation powers (well below saturation), the fluorescence enhancement was linked to the appearance of the fundamental mode propagating inside the aperture. The cutoff condition led to modes with a low group velocity, and to an increased local density of states allowing a higher excitation intensity as compared to a diffraction-limited beam. These results are verified by the computational studies presented in this section.

A generalized computational geometry for a single aperture is illustrated in Figure 17.9; the size of the computational space is 1.0×1.0×1.1μm, with scattering boundary conditions on all six faces. Electromagnetic computations were performed using COMSOL Multiphysics version 3.4. A quartz substrate is assumed, on top of which a 100-nm or 150-nm thick layer of gold or aluminum is placed; the upper region is air or water. Dielectric properties of gold and aluminum are incorporated via the complex dielectric constant, as measured by spectroscopic ellipsometry from 300 to 1,600nm. A single aperture is placed in the metal layer.

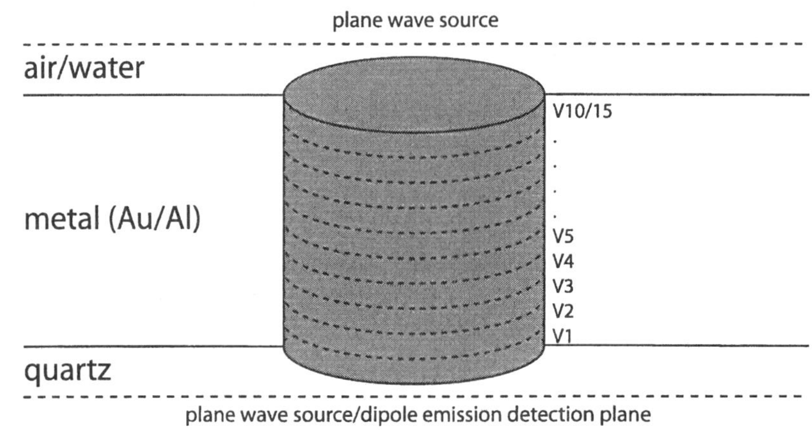


**Figure 17.9:** Illustration of general computational geometry for a single nanoaperture.

As an illustration of the fluorescence excitation and emission enhancement mechanisms within an aperture, Figure 17.10 plots the intensity distribution (sum of the intensities of all three field components) in the *xz* cross-section of an gold nanoaperture of 125nm diameter in water and 100nm thickness upon excitation by an *x*-polarized plane wave at 635nm wavelength originating from the substrate side (left half) and the enhancement of emission from a dipole (at 670nm wavelength, averaged across all three dipole orientations) located within the nanoaperture, as measured at a plane just below the metal/substrate interface (right half). In the excitation map, there are two contributions. The first is a component of relatively uniform lateral intensity, but exponential decay through the depth of the aperture; this is the fundamental aperture mode in cutoff. The second component is the localized surface plasmon (LSP) resonances, which are most prominent at the terminal regions of the aperture (29) and along the direction of the incident polarization. These resonances are analogous to LSP resonance modes of metal nanoparticles, and are most pronounced at the incidence side. They also bear strong connections with the electromagnetic concentration at the apex of a metallic tip.

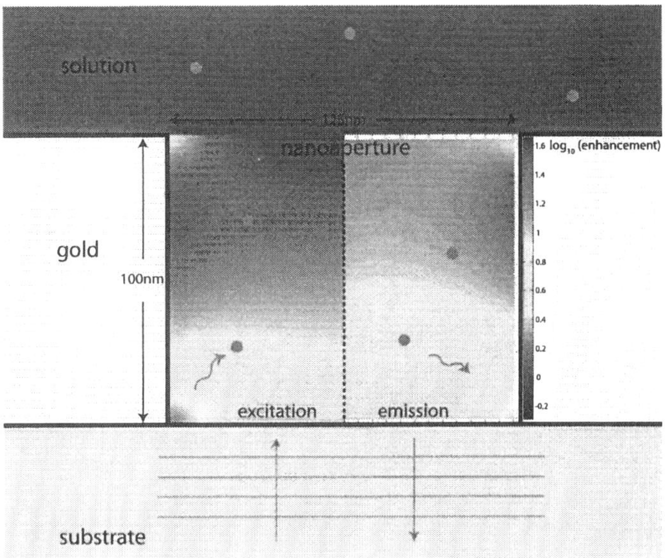


**Figure 17.10:** Image of calculated excitation (left half) and radiative emission (right half) of a molecule within a single nanoaperture. Illumination at 635nm wavelength is from below, where the left half image indicates the intensity distribution within the aperture. Emission on the right half is plotted as a function of dipole position, where the color map indicates the radiative enhancement at 670nm wavelength as detected on a plane just below the aperture/substrate interface. The white space around the emission map results from the lack of data points for dipole positions at the boundaries (a finite dipole size, 1nm, was assumed).

Similar contributions are present for dipole emission. Emission to the far field is enhanced the most when the dipole is located near the measurement plane, and especially, near the aperture rims where LSP resonances occur. As the dipole moves away from the substrate, the emission decays exponentially. The excitation and emission enhancement characteristics of nanoapertures will next be discussed in more detail.

### 17.3.3 1. *Excitation enhancement*

In the calculations of this section, a plane wave (at a specific wavelength) is launched 500nm from below the aperture. Average excitation enhancement is calculated by integrating the total intensity within a defined volume of an aperture (a 10nm thick cross-section slice, as illustrated in Figure 17.9) and dividing by the integrated intensity within the same volume, but in the absence of the metal layer. Figure 17.11 plots the averaged intensity within these cross-section volumes for aluminum and gold nanoapertures of 150nm thickness and varying diameters. In these simulations, the excitation light (at 488, 532, or 635nm) originates from the substrate side, and the upper region is air.

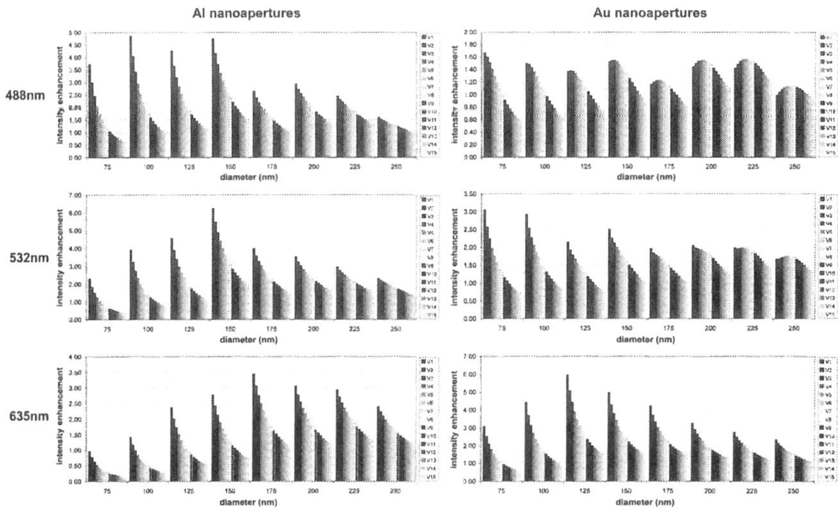


**Figure 17.11:** Calculated excitation enhancement at three different wavelengths for nanoaperture diameters ranging from 75nm to 250nm in both aluminum (left) and gold (right) films of 150nm thickness. The intensity enhancement is averaged within 10nm thick volumes within the apertures, with V1 representing the volume adjacent to the substrate. Plane wave excitation occurs from the substrate side. The upper region is air.

Even though the plasmonic resonance dominates the cross-section intensity map, as shown in Figure 17.10, it only occupies a small region of the aperture, and does not significantly affect the enhancement averaged within the 10nm thick volumes. This is shown in *xy* cross-section for a 125nm diameter gold nanoaperture in Figure 17.12, where the cross-section is taken 5nm above the substrate.

Numerous observations, consistent with experimental results, can be made from the calculations. For example, as the wavelength increases, in general, the aperture diameter providing the maximum enhancement increases. Further, the intensity decreases exponentially with distance away from the substrate; for the larger ratios of diameter to wavelength, the decay constant is clearly reduced, indicating non-cutoff conditions associated with a dramatic reduction in intensity enhancement.

Aluminum apertures provide enhancement over the visible spectrum, whereas gold drops dramatically in the green and blue (compare to Figure 17.5). Nevertheless, in the red, gold provides nearly twice the enhancement of aluminum, as expected (30). For aluminum apertures, the optimal aperture diameter for 488nm excitation is in the 100-150nm range, and for 635nm illumination, the optimal aperture diameter is near 175nm; these results are consistent with Figure 17.5. It is interesting that for small aluminum apertures (75nm diameter) and red excitation, no enhancement is expected. This is consistent with the original Cornell results (20). Similarly, for the gold aperture calculations, the enhancement is relatively flat with diameter for 488nm excitation, as expected from Figure 17.5, and less than a factor of 2, while a slight peak in enhancement is obtained for smaller aperture diameters under 532nm illumination. Under 635nm excitation, the optimal aperture diameter is near 125nm, also expected from the experimental results.

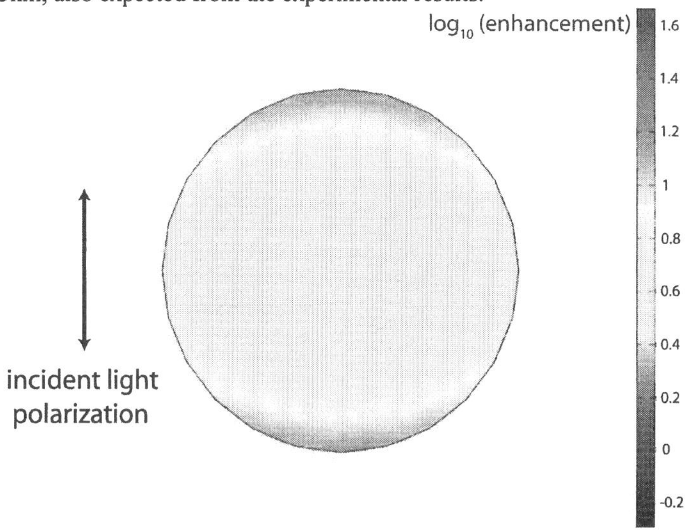


**Figure 17.12 :** Calculated intensity map in the *xy* plane of a 125nm diameter gold nanoaperture at a position 5nm above the substrate. Excitation is at 635nm wavelength from the substrate.

For high quantum efficiency dyes, such as Rh6G, excitation enhancement will be the dominant mechanism of increased fluorescence count rate in the non-saturating regime according to Eq. 3. The calculated results of Figure 17.11 have good quantitative agreement with measurements on enhanced fluorescence from Rh6G under 488nm illumination. Lower quantum efficiency dyes, such as Cy-5/Alexa647 for example, can also experience significant increase in quantum yield.

### 17.3.3 2. *Emission*

The emission properties of a fluorophore in a single nanoaperture have been theoretically studied (41, 46). These calculations do not depend on the excitation mechanism, and are therefore valid for chemiluminescence, electroluminescence, and photoluminescence, making the conclusions relevant to a number of biotechnology applications.

Dipole emission is a rather more complicated process to simulate than excitation, so some simplifications will be made here. The calculations presented here follow that of reference (46), in which the increase in radiative output of a dipole lying within a nanoaperture, as measured in a plane just below the interface with the substrate, is determined. The enhancement is calculated by taking the ratio of the power flow through this plane divided by the power flow through this same plane, but in the absence of the metal, for all three dipole orientations. This calculated enhancement corresponds approximately to the measured fluorescence enhancement in the saturation regime under strong excitation due to modification of the radiative rate. The more general case of estimating total fluorescence enhancement in the non-saturating regime, where the quantum efficiency of a given fluorophore needs to be determined, is deferred to a later publication.

Figure 17.13 plots the emission enhancement averaged over the 15 10nm thick volumes of 150nm tall apertures. Within each volume, dipoles are placed uniformly along the radius in 10nm increments; smaller increments are used near the edge. The averaged enhancement within each volume is a weighted average of the enhancement from each of these dipole locations, weighted by $r^2$, where $r$ is the distance from the aperture axis. There are some similarities between these results and the excitation enhancement results. In particular, the enhancement for gold is significantly greater than for aluminum, even at 570nm. Further, the diameter for maximum enhancement increases with the emission wavelength, with the diameters roughly corresponding to the optimal diameters in the excitation cases at the same wavelengths.

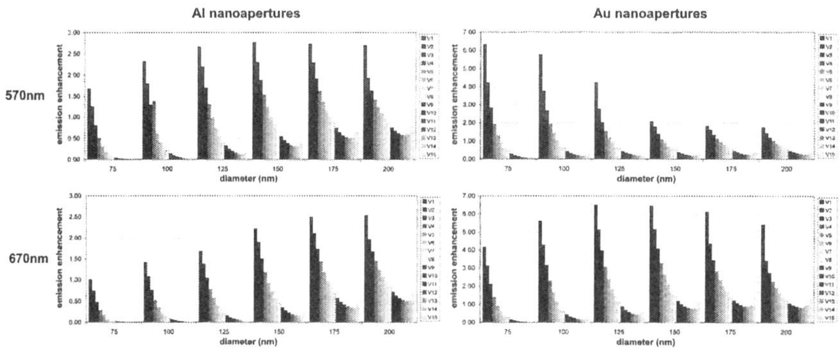


**Figure 17.13:** Calculated radiative emission enhancement at two different wavelengths for a dipole placed with nanoapertures of diameters ranging from 75nm to 200nm in both aluminum (left) and gold (right) films of 150nm thickness. The intensity enhancement is averaged for dipoles located within 10nm thick volumes within the apertures, with V1 representing the volume adjacent to the substrate. Emitted power is measured through a plane on the substrate side. The upper region is air.

The LSP modes also play a role in the emission process. This is evidenced in the plots by the peaks in emission enhancement in V1 and V15, leading to non-exponential decay throughout the aperture. The effect at V1 is strongest for the larger aperture diameters because of the reduced attenuation in propagation through the apertures. The LSP resonances are more pronounced in emission than in excitation. The difference is that any dipole positioned near the rims of the aperture interacts with the LSP resonance. In other words, the LSP enhancement occurs at any location near the rims, and therefore provides a greater contribution to the weighted-average enhancement.

Comparing to Figure 17.8, good agreement is obtained with the calculations. Calculations predict that the greatest change in radiative rate for a red dye (such as A647) occurs for a gold nanoaperture of 125-150nm diameter. Looking at the enhancement values calculated for V1, enhancements in the range 6-7 are anticipated, which is also agrees well with the experimental data.

### 17.3.3 3. *Radiation pattern*

An open question is whether a nanoaperture modifies the radiation pattern of emitters placed inside it. Any beaming effect towards the detectors would increase the molecular detection efficiency and, thus, the enhancement factors. According to recent calculations (55), the far-field directivity of the beam transmitted through a single sub-wavelength aperture is highly dependent on the hole size and on the permittivity of the material surrounding the aperture. From these calculations, the directivity of the beam tends to increase when the modulus of the permittivity decreases and when the aperture size increases. For instance, gold has a lower permittivity than aluminum, and should therefore provide a better directivity. Yet, to yield similar fluorescence enhancements, apertures in gold have sizes much lower than with aluminum. Altogether, the competition between these two effects leads to a similar far-field directivity for both metals. Experimentally, no beaming effect could be detected within the collection half-angle of 64° with neither gold nor aluminum from single aperture studies (30). Further, experimental results of the next section, where far-field radiation patterns from aperture arrays are shown, do not provide conclusive evidence of beaming effects either. Therefore, in the absence of detailed computational or experimental study of dipole radiation patterns from nanoapertures, any effect of directionality on measured fluorescence enhancement cannot be quantified at this time.

## 17.4 APERTURE ARRAYS AND STRUCTURED APERTURES

The extraordinary optical transmission (EOT) phenomenon discussed in Section 2 is generally attributed to coupling into surface plasmon polariton (SPP) modes of the metal film, mediated by the periodic arrangement of nanoapertures. The explanation of the extraordinary transmission through an array of nanoapertures can be built upon the properties of individual nanoapertures - the entire system can be viewed as an array of coupled resonators, where the coupling in general occurs through surface waves, which, under the proper phase matching conditions, are in fact the SPP waves.

The periodic arrangement adds two additional fluorescence enhancement mechanisms - increased excitation intensity due to SPP excitation (34) and modification of the radiation emission pattern. The latter effect has been referred to as ``beaming" (56), or more appropriately for fluorescence, surface plasmon coupled emission (SPCE) (57), and results from the coupling of multiple nanocavities via the interstitial surface regions.

In the 2-D periodic metallic structure, coupling between radiation modes and surface plasmons is described by

$$\vec{k}_t + nK\hat{x} + mK\hat{y} = \vec{k}_{sp} \qquad (4)$$

where $n$ and $m$ are integers. Because we use a square lattice, the coupling angles along the $x$ and $y$ axes are identical. The dispersion relation for a surface plasmon polariton (SPP) at a planar interface between a metal and a dielectric is given by

$$k_{sp} = \frac{\omega}{c}\sqrt{\frac{\varepsilon_m \varepsilon_d}{\varepsilon_m + \varepsilon_d}} \quad (5)$$

where □$_m$ and □$_d$ are the dielectric constants of the metal and dielectric material, respectively. Note that, due to the periodicity of an aperture array, the dispersion relation of equation 5 is modified by the details of the aperture size, shape, and spacing (58, 59, 60).

## 17.4 1. *Experiments*

In these experiments, periodic 2-D arrays of nanoapertures were produced in gold films using e-beam lithography and dry etching, as shown in the inset image of Figure 17.2. The underlying substrate for these structures is quartz. The films were coated on the top surfaces with a fluorescing monolayer, consisting of avidin labeled with Cy-5 dye, as before. The experimental setup of Figure 17.2 was used, which allowed measurements of fluorescence versus incidence angle and the radiation emission pattern.

As has already been established for fluorescence emission from individual nanoapertures, it is advantageous to perform detection from the backside of the substrate in order to minimize the contribution of background fluorescence. For experimental simplicity, all measurements were made in the trans-illumination geometry.

### 17.4.1 1. *Effects of EOT*

The EOT phenomenon in arrays of nanoapertures is the result of light coupling through the apertures. That being the case, it can reasonably be expected that increased light intensity within the apertures results. The first evidence that light transmitted directly through the apertures was obtained from near-field scanning optical microscopy across the backside of a sample (61), which clearly indicated localization of light at the apertures. Evidence of intensity enhancement within the apertures was provided in 2003 (34) using fluorophores as local intensity probes. In these measurements, aperture diameters of 100nm and 150nm were used, with spacings of 1□m and 750nm in 70nm thick gold films. In all cases, fluorescence intensity reached its maximum value under the conditions of EOT. Furthermore, even under non-EOT conditions, enhanced fluorescence was obtained, presumably owing to the enhancement effects of the individual apertures.

Figure 17.14 plots transmission and fluorescence emission versus incidence angle for an array of ~200nm diameter apertures with 1□m spacing in 70nm thick gold (fluorescence from a disordered array, Figure 17.3, is also plotted for comparison) (37). The fluorescence curves have been normalized against fluorescence from a bare glass substrate and corrected for the fill fraction of the apertures, which is 7.5% (including the bottom of the aperture and the sidewalls). It is clear that the fluorescence curve follows transmission. The first transmission peak occurs at an incidence angle of 25°, corresponding to the excitation of the (1,0) surface-plasmon mode at the metal-air interface; the location of this peak can be shifted (for a given wavelength) by changing the aperture spacing. The fluorescence also peaks at this angle, suggesting a corresponding increase in excitation intensity. On either side of this peak is evidence of suppression in the excitation light, where the fluorescence levels drop below that of a disordered array (which approximately represents the fluorescence enhancement from an individual aperture).

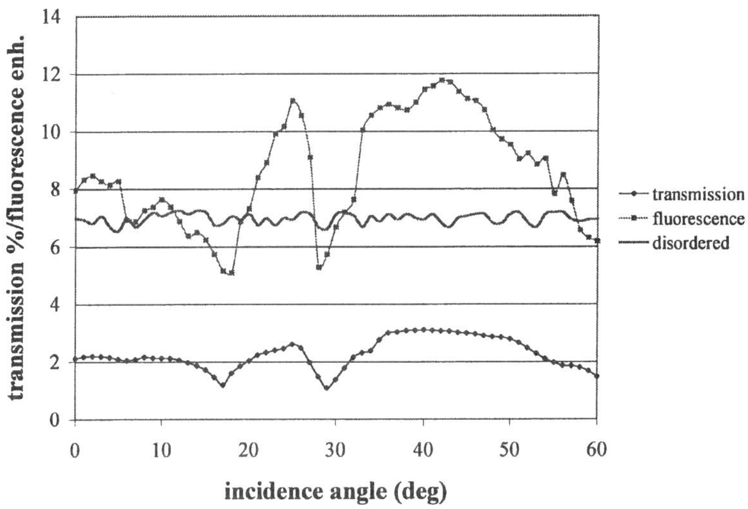


**Figure 17.14:** Transmission of excitation light and fluorescence emission from Cy-5 from a periodic arrangement of 200nm diameter nanoapertures in 70nm thick gold film, with spacing of 1□m. In these measurements, the sample angle was varied with respect to the incident light, while detection occurred from the backside in a direction collinear with the incident light. Fluorescence from a disordered array is also plotted for reference. Fluorescence is normalized to a quartz slide with the same monolayer, and corrected for fill fraction.

In these experiments presented so far, and in related experiments by others (39, 62, 40), all of the samples had fluorophores coating the entire surface, not just within the nanoapertures. The question then arises as to what components of fluorescence emission are detected through the backside.

### 17.4.1 2 *SPCE*

It is well established that fluorophores located near a metal surface can directly transfer energy from their excited state to surface plasmon modes (63). When the metal has a surface texture, these surface plasmon modes can directionally scatter to the far field (64). This effect has been most widely studied in reflection, but it is known that it can occur in transmission as well (65). SPCE in transmission has been studied in arrays of sub-wavelength apertures such that the magnitude of this component can be compared to the direct emission from the apertures (66). The structures used in the experimental results reported here were 2-D periodic array of apertures in 70nm thick gold film, with aperture spacing of 1□m and aperture diameter of 200nm (37).

The measurement of fluorescence emission from the 2-D patterns is shown in Figure 17.15. In this measurement, the sample is fixed at normal incidence to the excitation light and the detector is scanned. Because of the symmetry of diffraction from the 2-D structure, the fluorescence emission pattern was only measured for angles along the $x$ axis. The emission pattern exhibits a broad background with isolated peaks at angles corresponding to the calculated scattering angles using equation (4). When measured through an $s$-oriented analyzer, the emission pattern consists only of the broad background, as the $p$-polarized SPCE is blocked (66).

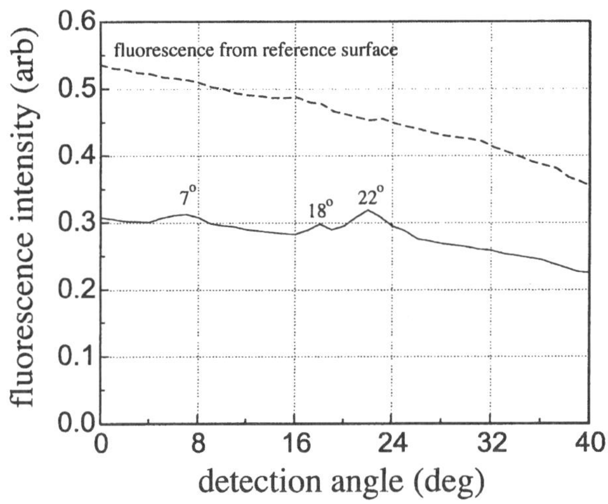


**Figure 17.15:** Unpolarized angular emission profile from a Cy-5/avidin fluorescing monolayer covering the entire surface of a nanoaperture array of 200nm diameter in 1□m spacing in 70nm thick gold. In this measurement, the sample was fixed at normal incidence and the detector on the back side was scanned (zero degrees is normal to the sample). Angular fluorescence emission from a glass substrate is shown for reference.

The broad background pattern in the measurements is therefore due to direct fluorescence emission from the apertures (in addition to a much weaker contribution due to the direct transmission of fluorescence through the metal), which is the dominant transmission mechanism. Emission from the apertures is incoherent in that there is no correlation from one aperture to another; the collective emission pattern is therefore representative of the emission from a single aperture under far-field measurement. For reference, the fluorescence emission pattern from a glass substrate is also plotted in Figure 17.15. This pattern is similar to that obtained from the nanoapertures. One might be tempted to conclude that the emission pattern is slightly more directional from the glass surface than from the nanoapertures. However, more thorough experimental study should be performed before drawing any such conclusions.

### 17.4.1 3. *Eliminating SPCE*

Even though the SPCE component is weak upon backside detection, it can be reduced by using backside illumination and/or controlling the fluorescence collection angle (37). For biotechnology applications, it is desirable to prevent molecular capture on the surface since transduction of these molecules would be less efficient than the transduction of molecules bound within the nanoapertures. This can be accomplished by chemical passivation of the top surface, which minimizes subsequent binding (37). An illustration of passivated versus non-passivated structures is shown in Figure 17.16.

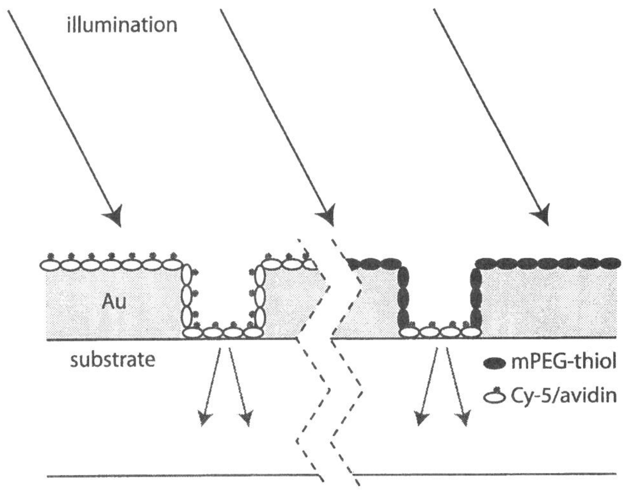


**Figure 17.16:** Illustration of non-passivated (left) and passivated (right) nanoaperture samples. For the non-passivated samples, a fluorescing monolayer covers the entire exposed surface, while only the bottom surfaces of the nanoapertures are covered in the passivated samples.

In the passivation procedure for gold (note that a passivation procedure for aluminum has recently been demonstrated (67)), mPEG-thiol in powdered form is dissolved in ethanol to a concentration of 1μM. The solution is applied to the samples for 24h in a nitrogen glovebox at room temperature and atmospheric pressure, then rinsed in ethanol and dried under a nitrogen gun. The labeled avidin solution is then applied as before. After this procedure, a reference gold surface did not produce any measurable fluorescence, while a reference quartz surface produced a fluorescence signal roughly 85% of the level of a coated quartz surface without passivation (which is probably the result of a reduction in bound surface concentration).

The unpolarized fluorescence emission pattern from the passivated nanoaperture array is shown in Figure 17.17. The absence of the distinct fluorescence peaks (as compared to Figure 17.15) demonstrates that the SPCE component is eliminated through prevention of immobilization of Cy-5/avidin on the top surface.

Again, the emission pattern from a glass substrate is also shown for reference. Figure 17.18 shows the fluorescence output as a function of incidence angle for the passivated sample. At the surface-plasmon incidence angle, the total fluorescence enhancement compared to the reference is 12 (normalized to the 3.1% fill-fraction of the bottom surface of the nanoapertures), which is comparable to the enhancement obtained under full interior surface coverage (Figure 17.14). Therefore, the fluorescence enhancement (per unit area) is comparable for fluorophores on the bottom as for fluorophores on the sidewalls with backside detection, as before with individual apertures.

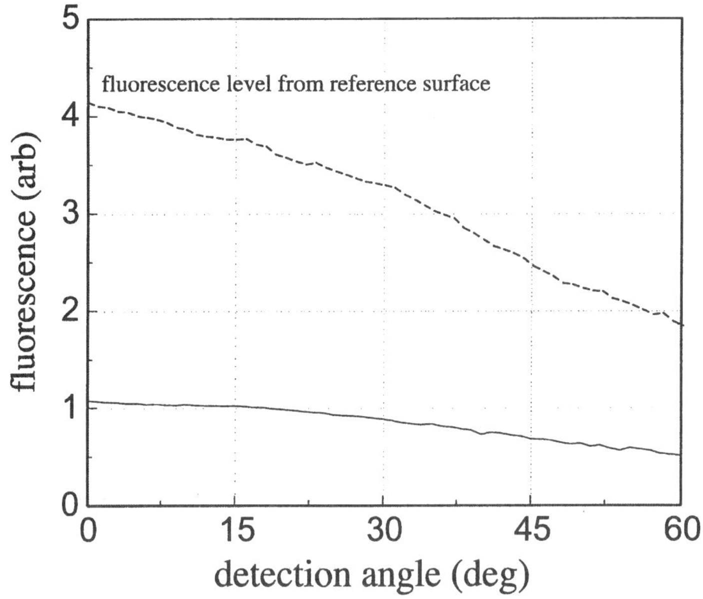


**Figure 17.17:** Unpolarized angular emission profile from a Cy-5/avidin fluorescing monolayer covering the exposed glass surface regions of a passivated nanoaperture array of 200nm diameter in 1□m spacing in 70nm thick gold. In this measurement, the sample was fixed at normal incidence and the detector on the back side was scanned (zero degrees is normal to the sample). Angular fluorescence emission from a glass substrate is shown for reference.

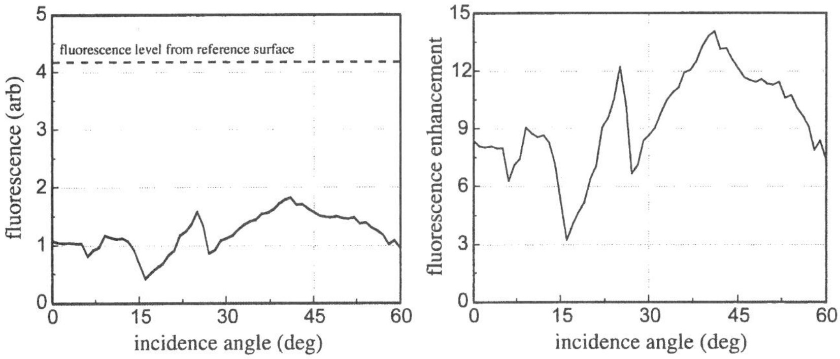


**Figure 17.18:** Fluorescence emission from a Cy-5/avidin fluorescing monolayer covering the exposed glass surface regions of a passivated nanoaperture array of 200nm diameter in 1□m spacing in 70nm thick gold. In this measurement, the sample was rotated with respect to normal incidence. The average fluorescence level from a glass substrate is indicated for reference.

As a conclusion to the experimental results, a more complete study of fluorescence enhancement from nanoaperture arrays was recently reported (46), where measurements were made across multiple samples. From these results, an enhancement factor of 12±1 was obtained using the same sample geometries described above.

## 17.4 2. *Simulations*

Considerable effort has been devoted to modeling the optical transmission through arrays of nanoapertures in order to maximize transmission; this work will not be further reviewed here. What is of interest is how the array modifies the enhancement mechanisms obtained from a single aperture, a topic which has received comparably little attention (58). As a result, only a brief overview of array effects will be presented here, in anticipation that more comprehensive studies will be performed in the future.

We presently make the assumption that the emission of a molecule within an aperture is not significantly affected by the presence of neighboring apertures in an array. Two conditions must hold in order for this assumption to be valid: 1) the apertures are sufficiently separated such that the dipole-like LSP modes, into which molecular emission can be coupled, do not overlap, and 2) coupling of molecular emission from the apertures into surface waves (19) is weak; clearly, both effects will have an exponential-like decay with aperture separation. There is some experimental evidence that supports these assumptions. For example, if efficient aperture coupling occurred in emission (where the top surface is passivated and molecules are localized within the apertures), then distinct scattering peaks would be observed in Figure 17.18 at angles corresponding to the dispersion relation of SPP waves at the metal-substrate interface. No such peaks are apparent. It is possible that the aperture spacing in these experiments was too large for efficient coupling to occur. Further, a recent study suggested the fluorescence emission from an aperture array is dominated by the aperture (40). Nevertheless, more investigation needs to be performed.

We therefore assume that the major effect of the array is in increased collection of light within the apertures. In the simulation results of this section, we only consider briefly the excitation component to enhanced fluorescence emission. As already established by experiments, the light intensity within periodic arrangements of nanoapertures increases in association with EOT and is a result of aperture coupling effects via surface waves under external excitation. This coupling therefore modulates the response obtained for an isolated aperture. Simulations were performed in the same manner as described before, except for the fact that periodic boundary conditions were used along the sides of the simulation space boundary.

Figure 17.19 plots the intensity enhancement averaged over the lower 10nm volume of the nanoaperture (i.e. V1) for the cases of a single aperture and an array, with illumination from the substrate side. For these calculations, 150nm apertures in a 100nm thick gold layer were considered. For the array, the aperture spacings are 400nm, 500nm and 600nm. It is clear from the calculation that the array produces strong modulation of the intensity enhancement of a single aperture, where significantly greater enhancement can be obtained. Further, the locations of the enhancement peaks can be shifted based upon the aperture spacing, with greater enhancements occurring in the red part of the spectrum (where the enhancement line shapes resemble Fano-type resonances). In the blue/green part of the spectrum, the enhancement drops dramatically owing to the dielectric properties of gold. To our knowledge, detailed optimizations of excitation enhancement with aperture arrays have not yet been performed, so this is an area of further study.

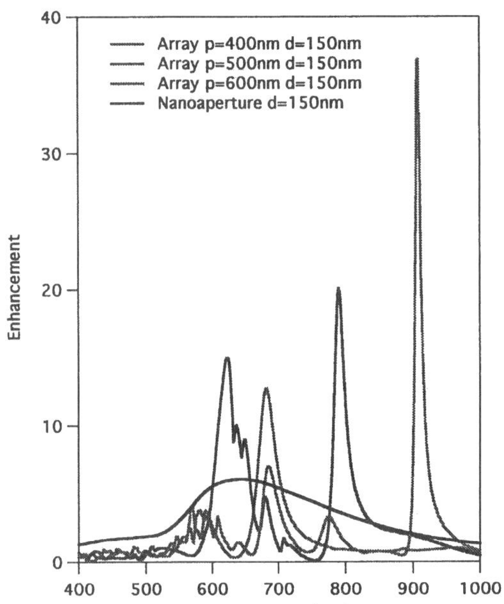


**Figure 17.19:** Calculated excitation intensity enhancement as a function of wavelength at the bottom 10nm slice (V1) of a single 150nm diameter nanoaperture

in 100nm thick gold compared to an array of nanoapertures of spacings 400nm, 500nm and 600nm. Excitation occurs from the substrate side and the upper region is air.

## 17.4 3. *Structured apertures*

In addition to nanoaperture arrays, structured apertures could provide higher fluorescence rate enhancement than single apertures, while still enabling efficient signal-to-background discrimination. Examples include the bull's eye structures (56, 23), nanopockets (68) or combined slit and aperture (69).

As already mentioned for aperture arrays, the periodicity of the structure can lead to significant local field enhancement by resonantly exciting surface plasmons. To translate this resonance effect to an isolated aperture, the metal surface surrounding the aperture can be structured in a periodic manner in order to efficiency excite the SPP. Most designs use concentric grooves around a central nanoaperture, which is called "bull's eye aperture" (56, 70, 71).

When the excitation light hits the grooves at resonance, extra focusing is obtained, which gives rise to an enhanced electromagnetic field (45, 29). This phenomenon has been first demonstrated for enhanced optical transmission trough a single (structured) nanoaperture (56). More recently, enhanced second harmonic generation (72) was obtained and ultra-compact silicon detectors have been demonstrated (73). Very elegantly, the bull's eye antenna structure compensates for the reduced active nanoaperture area. This opens promising opportunities for advanced optoelectronics devices assisted by plasmonics.

When the output surface surrounding the aperture is corrugated, a narrow beam can be generated, with a divergence of a few degrees (56). The backside grooves are used to scatter the surface waves into freely propagating light, which contributes to the total detected intensity. In addition, interference between the scattered light and the light emerging from the hole generates a low-divergence beam, which can be more efficiently collected. However, this process has not been studied in the context of dipole emission from a structured aperture.

From the point of view of nanoaperture-enhanced fluorescence, molecules are only sensitive to the local excitation intensity. Therefore, the effective excitation intensity is the critical parameter that has to be considered for practical applications. With the bull's eye aperture surrounded by concentric grooves, both the surface structure and the incident wave have to cover a diameter of about 4 times the wavelength. This dramatically limits the incoming intensity (i.e. constant power distributed across a larger surface area). Therefore, even with large enhancements, the excitation intensity in the central aperture may not be significantly higher that the intensity reached with a tightly focused laser beam without the nanostructure.

In the case of a bull's eye structure, the relative electromagnetic enhancement is about 100 (computed for an incoming plane wave and consistent with experiment (72)), but the structure radius and thus the laser beam focus spot are about 4 times the diffraction limit (minimum focus size). If we compare the relative intensity enhancement to the case of a diffraction-limited spot, the intensity with the bull's eye is increased by $100 \mathrm{x}(1/4)^2 = 6.25$, which is about the enhancement with a single (bare) aperture with a tightly focused beam. Of course, this argument has to be supplemented with the influence the emission rate and the directivity of the fluorescence beam, which contribute to the overall enhanced fluorescence. However, the situation is better with arrays of structures, where uniform illumination would be used.

Recently, the use of a single nanoaperture surrounded by a single channel groove was suggested (74). Such a single channel is much simpler to fabricate than the larger-area bull's eye, and is shown in Figure 17.20. Numerical analysis of diffraction shows the possibility of a 50-fold increase of the electric field intensity inside the central aperture, when compared to the incident field. The simultaneous excitation of cavity modes in the central aperture and in the surrounding coaxial channel can lead to almost 50-fold increase in the electric field intensity in the central aperture, compared to the intensity of the incident field. The coupling between the two cavity modes in the central aperture and in the surrounding coaxial channel is made through the surface plasmon wave that propagates along the metal-cladding interface. This coupling is the strongest when the field of the channel cavity mode matches the field of the plasmon surface wave, i.e, when the channel modes are below their cut-off.

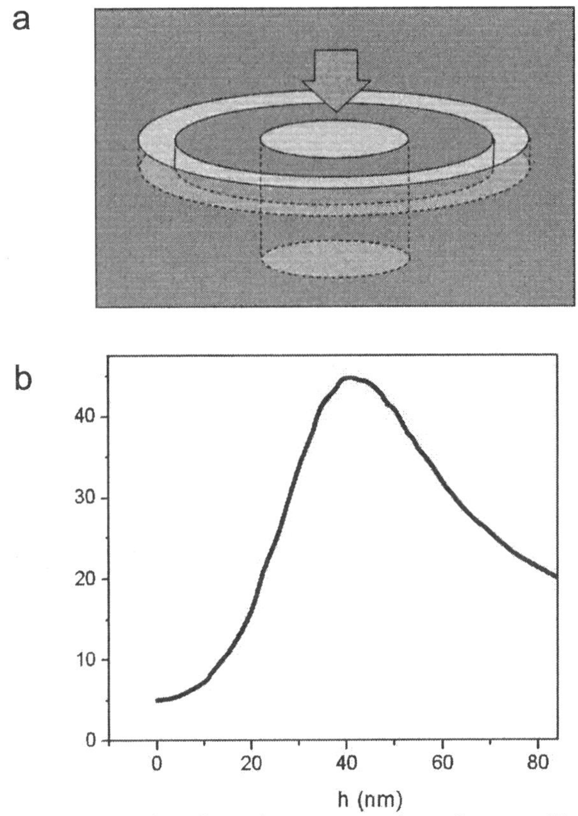


**Figure 17.20:** (a) Schematic representation of a metallic film with a circular aperture and a circular channel around it. The excitation beam impinges of the corrugated surface. (b) Field enhancement averaged at the entrance plane of the aperture as a function of the channel depth. The aperture is milled in aluminum, with a diameter of 150 nm (488 nm excitation wavelength). The groove has 400nm and

560nm inner and outer diameters respectively. [Image courtesy of N. Bonod and E. Popov].

## 17.5 APPLICATIONS

This review of nanoaperture enhanced fluorescence reveals some unique properties of nanoapertures useful for biotechnology applications. The primary advantage is in a highly localized observation volume, resulting in the reduction of background signal from sources lying outside the apertures. Related to this is a well-defined analysis volume, where localization of chemical reactions can be obtained.

### 17.5 1. *Enhanced single molecule analysis in solution*

A central issue in single-molecule analysis is to discriminate the fluctuating signal to the background noise, which mainly originates from laser-induced Rayleigh and Raman scattering, and from the electronic noise of the photon-counting detectors. To increase the signal-to-background ratio, a first strategy is to maximize the collected signal by using a high numerical aperture (NA) objective and naturally bright fluorophores. A second strategy is to reduce the background by rejecting the scattered light with high-quality optical filters (spectral filtering) and by restricting the detection volume (spatial filtering). Therefore, the common approach uses a confocal microscope with a high NA objective, which provides detection volumes of about 0.5 fL (= 0.5 $\mu m^3$). This limits the concentration for single-molecule analysis to a few nM. For artificial environments, this low concentration may not be a problem; however, it may be a crucial limitation for biological and biochemical studies which typically involves concentrations in the □M to the mM range (20). Many enzymatic reactions are also naturally effective at ligand concentrations in this range. Arbitrarily reducing the ligand concentration may lead to chemical pathway alteration and artifacts. Performing FCS at biologically more relevant concentrations requires the development of analysis volumes with nanometric dimensions, yielding a volume reduction by at least three orders of magnitude as compared to standard confocal microscopy.

To further improve the FCS technique and go beyond confocal microscopy, the use of micro- and nanostructures is of major interest. By tailoring the molecules environment, photonic structures can enhance the fluorescence emission and collection efficiencies. This simultaneously increases the molecular brightness and reduces the detection volume, leading to a lower amount of background scattered light, and to a higher concentration available for single-molecule studies, as shown in Figure 17.6. Several methods allowing a simultaneous decrease of the detection volume and increase of the fluorescence signal have been proposed so far, and are reviewed in (75).

In spite of their conceptual simplicity, nanoapertures possess appealing properties to increase the effectiveness of fluorescence-based single-molecule detection, such as localization of excitation light, strong isolation from emission produced by unbound species, and an increase in apparent absorption and emission yield (76, 77, 78). The significant fluorescence increase obtained for small-radius apertures appears especially interesting because it allows for the possibility to significantly reduce the observation volume while still detecting a sufficient signal. This yields an efficient signal-to-background discrimination, even with attoliter volumes and single molecule resolution. Using nanoapertures, a large range of biologic processes have been efficiently monitored with single molecule resolution at micromolar concentrations (20, 79, 47, 80).

Moreover, the studies on nanoaperture-enhanced fluorescence point out that for a properly tailored aperture, count rates per molecule greater than a few hundred thousands photons per second were readily obtained, whereas for a single molecule in open solution, fluorescence saturation prevents the count rate from exceeding a few tens of kilocounts per second. This allows for fast and reliable screening for single molecules.

We point out that for experiments conducted on ultra-small structures (diameters typically below 70 nm), the signal to noise ratio comes close to one, as a consequence of different effects. First, the fluorescence emission is strongly affected by the metal in the close vicinity of the structure, leading to a decrease in fluorescence enhancement (42, 46). Second, the background noise becomes then significant, due to the metal auto-fluorescence and the large pool of dyes above the aperture that induce some fluorescence leakage back to the detector. It should be noted that the weakly-fluctuating nature of the background still enables a discrimination against the signal, even if it tends to decrease the amplitude of the FCS correlation curve. Therefore, detecting the presence of a millisecond-scale enzymatic reaction event or a microsecond-scale diffusion event is always possible even for a low signal-to-noise ratio, which is a general property of FCS.

## 17.5 2. *Real time single molecule DNA sequencing*

Performing high-throughput, high-accuracy DNA sequencing at low costs has become a major issue, largely attracted by the growing potential of quantitative genomics. For this cutting edge application, nanometric apertures offer specific advantages as they enable the observation of single fluorophores against a dense background by maintaining a high signal-to-noise ratio. Moreover, nanoapertures can be operated in massive parallelism, allowing for high-speed DNA analysis.

This concept has been developed since 2004 by Pacific Biosciences, a private company formed by researchers at Cornell University. Each nanoaperture forms a nano-observation chamber for watching DNA polymerase as it performs sequencing by synthesis (Figure 17.21). Within each chamber, a single DNA polymerase molecule is attached to the bottom surface so that it persists in the detection volume. Fluorescently labeled nucleotides diffuse into the reaction solution at high concentrations to promote enzyme speed and accuracy. When the DNA polymerase encounters the nucleotide complementary to the next base in the template, it is incorporated into the duplicated DNA chain. During the incorporation process, the enzyme holds the nucleotide in the detection volume for a few tens of milliseconds, much longer than the average residence time of the diffusing nucleotides. This causes a millisecond fluorescence flash to occur, which is detected by the collection system. As each nucleotide (A, T, G, C) bears a different color marker, the spectral analysis of the fluorescence flash allows to discriminate the incorporated nucleotide, which constructs the DNA sequence step by steps. Then, as part of the natural incorporation cycle, the polymerase cleaves the bond holding the fluorophore in place and the dye diffuses out of the detection volume. This ensures that the duplicated DNA strand is purely natural, i.e. it doesn't bear any fluorescing material. Following incorporation, the signal quickly returns to the baseline. Then, the polymerase advances to the next base and the process continues to repeat, adding nucleotides one by one.

DNA sequencing is observed as it occurs in real-time across thousands of nanoapertures that can be operated simultaneously. Acting uninterrupted, the DNA polymerase keeps on incorporating bases at a speed of tens per second. Researchers at Pacific Biosciences demonstrated that DNA strands thousands of nucleotides in length could be accurately sequenced.

This sequencing has a wide range of applications, yet many technological issues have to be addressed. These mostly deal with polymerase enzymatic activity and large scale nanoaperture detection. Selective aluminum passivation for targeted immobilization of single DNA polymerase molecules in nanoapertures is described in (67). Improved fabrication techniques of nanoapertures on a large scale wafer are described in (81).

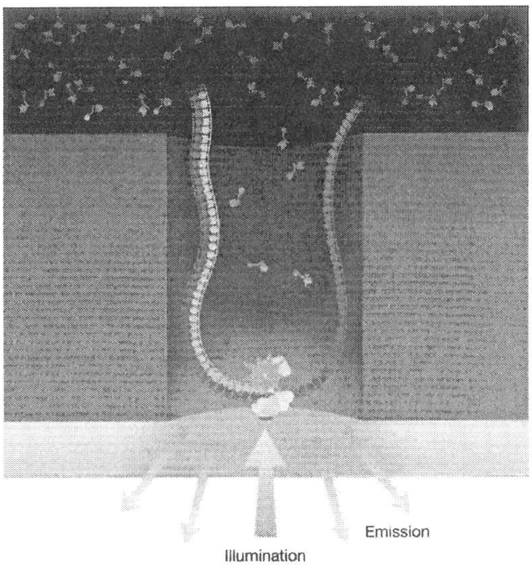


**Figure 17.21:** Single nanometric aperture for real time single molecule DNA sequencing. A DNA polymerase molecule attached to the bottom surface is used to successively incorporate fluorescent nucleotides complementary to the DNA strand, causing fluorescence bursts for each incorporation process (see text for details). Image copyright of Pacific Biosciences Inc., reprinted with permission.

## 17.5 3. *Sub-diffraction diffusion analysis within lipidic membranes*

Revealing the dynamic organization of the cell plasma membrane at the submicron level is a challenging task. On one hand, electron microscopy resolves nanometric features, and yet it cannot be easily applied to live cells under physiological conditions. On the other hand, standard optical microscopy unravels details of live cell membranes, but is limited to a resolution of about 200 nm.

Therefore, new methods providing dynamic information on lipids and proteins in live cell membranes with nanometric resolution are required.

The first proof-of-principle towards the use of nanometric apertures to probe the plasma membrane of a live cell on a sub-diffraction area was reported in 2005 (82). A single aperture acts as a pinhole directly located under the cell membrane and restricts the observation area below the diffraction limit. Recently, cell membrane invagination within the nanostructure was shown to be highly dependent upon actin filaments but not on microtubules (83). The same concept also works for lipid bilayers, which are especially relevant to study systems involving ligand-receptor interactions (84). Moreover, much information on the molecular diffusion is obtained by performing measurements with increasing nanoaperture diameters (85). Since the diffusion of molecules in the cell plasma membrane is highly sensitive to heterogeneities, monitoring the lateral diffusion of lipids and membrane proteins is highly relevant in order to reveal the structure and the role of the membrane. Experiments reported in (86) showed that the aperture limited the observed membrane area, but did not alter the diffusion process. Extensive FCS analysis on fluorescent-labelled ganglioside in cell membranes revealed membrane structures of 30 nm radius, well below the optical resolution of confocal microscopes.

Nanoapertures thus appear as valuable tools to monitor the diffusion of molecular components in the plasma membrane at high spatial and temporal resolutions, with the cells being kept under physiological conditions and with ultra-low amounts of incident laser light. Combined with FCS, this yields a technique having both high spatial and temporal resolution together with a direct statistical analysis.

## 17.5 4. *Biosensing applications of sub-wavelength apertures*

Detecting molecules in real-time with high sensitivity and molecular specificity is of great interest in many fields of bioscience. Intensive world-wide research on new biosensing techniques is motivated by numerous applications in clinical diagnostics, genetic screenings, proteomics, and single-molecule detection. In this context, the combination of molecules and subwavelength apertures is a promising area of application. The electromagnetic field enhancement, the sensitivity to the dielectric medium in contact with the surface and the simplicity of integrating the arrays have motivated efforts to use them to detect molecules and enhance spectroscopic contrasts (41).

A first application of the fluorescence enhancement was devoted to the detection of DNA binding events (37). To perform this affinity sensing, nanoapertures in gold were spotted with probe molecules specific to a target of interest. In order to immobilize probe molecules directly on the bottom (quartz) surface of the apertures, exposed gold surfaces were first passivated with mPEG-thiol, leaving only the exposed quartz surface available for avidin monolayer formation. Biotinylated oligonucleotide probes of 20-base length were then immobilized. During hybridization, 20mer target oligos accumulate within the nanoapertures due to specific recognition and binding to the probes. Bound targets produced enhanced fluorescence in transmission, while (non-enhanced) fluorescence emitted from unbound molecules in solution lying outside did not couple efficiently through the structure. Performing detection through the backside of the substrate therefore yields high signal-to-background rejection, enabling real-time detection, as demonstrated in these proof-of-principle experiments (37).

Nanoaperture arrays are particularly relevant for integration with high-throughout screening (HTS) methodologies, for applications in drug screening or evolutionary biotechnologies (87). Combined with FCS or coincidence analysis (88), nanostructures appear very well suited to design miniaturized HTS fluorescence sensors (80).

## 17.6 CONCLUSIONS AND FUTURE DIRECTIONS

In summary, we have tried to overview the state of the field of nanoaperture enhanced fluorescence. While much is known currently about the photophysics of an isolated aperture, it is clear that much more work needs to be performed in order to understand, and maximize, fluorescence enhancement effects from arrangements of apertures and structured apertures. In particular, the relative contributions of radiative and non-radiative processes have only been studied in a couple of experiments, and limited computational work has been performed. In addition, the role of emission directionality is not clear at present and requires further study.

In addition to the directions for further study mentioned above, there are quite a few additional avenues of exploration. For example, the blue/UV range has not been explored. EOT in the blue has been demonstrated with aluminum aperture arrays (89); one might reasonably expect fluorescence enhancement to also occur in this regime (in particular, Figure 17.11 shows significant excitation enhancement in the blue/green). In addition, the infrared region is promising since surface plasmon dissipation losses are significantly reduced as compared to the visible range, and very large enhancement would be expected.

Despite some of the limitations of the current state of knowledge, nanoapertures are enabling significant research and even commercial applications, largely due to their unique properties of simultaneous enhancement, background isolation properties, and molecular localization properties.

## 17.7 ACKNOWLEDGEMENTS

S. B. acknowledges Y. Liu for early experimental and computation work, and F. Mahdavi and M. Diwekar for new contributions to this chapter. Financial support was provided by the National Institutes of Health, the National Science Foundation, and the State of Utah Centers of Excellence Program.

J. W. acknowledges H. Rigneault, E. Popov, T.W. Ebbesen, N. Bonod, D. Gerard and financial support from the Agence Nationale de la Recherche ANR and Centre National de la Recherche Scientifique CNRS.

## 17.8 REFERENCES


1. Barnes, W. L., Dereux, A., and Ebbesen, T. W. (2003). Surface plasmon subwavelength optics. *Nature* 424:824-830.
2. Ozbay, E. (2006). Plasmonics: Merging Photonics and Electronics at Nanoscale
Dimensions. *Science* 311:189-193.
3. Lakowicz, J. R. (2005). Radiative decay engineering 5: metal-enhanced fluorescence and plasmon emission. *Anal. Biochem.* 337:171-194.
4. Aslan, K., Gryczynski, I., Malicka, J., Matveeva, E., Lakowicz, J. R., and Geddes, C. D. (2005). Metal-enhanced fluorescence: an emerging tool in biotechnology. *Curr. Op. Biotech.* 16:55-62.
5. Lakowicz, J. R. (2006). Plasmonics in biology and plasmon-controlled fluorescence," *Plasmonics* 1:5-33.
6. Purcell, E. M. (1946). Spontaneous emission probabilities at radio frequencies. *Phys. Rev.* 69:681.
7. Drexhage, K. H. (1974). Interaction of light with monomolecular dye layers. *Prog. Opt.* 12:163.
8. Kleppner, D. (1981). Inhibited spontaneous emission. *Phys. Rev. Lett.* 47, 233-236.
9. Barnes, W. L. (1998). Fluorescence near interfaces: the role of photonic mode density. J. Mod. Opt. 45:661-699.
10. Anger, P., Bharadwaj, P., and Novotny, L. (2006). Enhancement and quenching of single molecule fluorescence. *Phys. Rev. Lett.* 96:113002.
11. Kuhn, S., Hakanson, U., Rogobete, L., and Sandoghdar, V. (2006). Enhancement of single-molecule fluorescence using a gold nanoparticle as an optical nanoantenna. *Phys. Rev. Lett.* 97:017402.
12. Farahani, J. N., Pohl, D. W., Eisler, H.-J., and Hecht, B. (2005). A single quantum dot coupled to a scanning optical antenna: A tunable superemitter. *Phys. Rev. Lett.* 95:017402.
13. Muskens, O. L., Giannini, V., Sanchez-Gil, J. A., and Rivast, J. G. (2007). Strong enhancement of the radiative decay rate of emitters by single plasmonic nanoantennas. *Nano Letters* 7:2871-2875.
14. Lakowicz, J. R., Malicka, J., Gryczynski, I., Gryczynski, Z., and Geddes, C. D. (2003). Radiative decay engineering: the role of photonic mode density in biotechnology. *J. Phys. D: Appl. Phys.* 36:R240–R249.
15. Ebbeson T. W., Lezec H. J., Ghaemi H. F., Thio T., and Wolff, P. A. (1998). Extraordinary optical transmission through subwavelength hole arrays. *Nature* 391:667–669.
16. Lezec H. J., and Thio T. (2004). Diffracted evanescent wave model for enhanced and suppressed optical transmission through subwavelength hole arrays. *Opt. Express* 12:3629–3651.
17. Bethe H. A. (1944). Theory of diffraction by small holes. *Phys. Rev.* 66:163–182.
18. Bouwkamp C. J. (1950). On Bethe's theory of diffraction by small holes. *Philips Res. Rep.* 5:321–332.
19. Wannemacher, R. (2001). Plasmon-supported transmission of light through nanometric holes in metallic thin films. *Optics Communications* 195:107-118.

20. Levene, M. J., Korlach J., Turner S. W., Foquet M., Craighead H. G., and Webb W. W. (2003). Zero-mode waveguides for single-molecule analysis at high concentrations. *Science* 299:682–686.
21. Degiron A., Lezec H. J., Yamamoto N.,and Ebbesen T. W. (2004). Optical ransmission properties of a single subwavelength aperture in a real metal. *Opt. Commun.* 239:61–66.
22. Yin L., Vlasko-Vlasov V. K., Rydh A., Pearson J., Welp U., Chang S.-H., Gray S. K., Schatz G. C., Brown D. B., and Kimball C. W. (2004). Surface plasmons at single nanoholes in Au films. *Appl. Phys. Lett.* 85:467–469.
23. Popov E., Bonod N., Neviere M., Rigneault H., Lenne P., Chaumet P. (2005). Surface plasmon excitation on a single subwavelength hole in a metal sheet. *Appl. Opt.* 44:2332–2337.
24. Catrysse P. B., Shin H., and Fan S. (2005). Propagating modes in subwavelength cylindrical holes. *J. Vac. Sci. Technol. B* 23:2675–2678.
25. Gordon R., and Brolo A. G. (2005). Increased cut-off wavelength for a subwavelength hole in a real metal. *Opt. Express* 13:1933–1938.
26. Collin S., Pardo F., and Pelouard J.-L. (2007). Waveguiding in nanoscale metallic apertures. *Opt. Express* 15:4310–4320.
27. Shin H., Catrysse P. B., and Fan S. (2005). Effect of the plasmonic dispersion relation on the transmission properties of subwavelength cylindrical holes. *Phys. Rev. B* 72:085436.
28. Chang S.-H., Gray S. K., and Schatz G. C. (2005). Surface plasmon generation and light transmission through isolated nanoholes and arrays of nanoholes in thin metal films. *Opt. Express* 13:3150–3165.
29. Shuford K. L., Ratner M. A., Gray S. K., and Schatz G. C. (2007). Electric field enhancement and light transmission in cylindrical nanoholes. *J. Comput. Theor. Nanosci.* 4:239–246.
30. Gérard, D., Wenger, J., Bonod, N., Popov, E., Rigneault, H., Mahdavi, F., Blair, S., Dintinger, J., and Ebbesen, T. W. (2008). Nanoaperture-enhanced fluorescence: Towards higher detection rates with plasmonic metals. *Phys. Rev. B* 77:045413.
31. Prikulis J., Hanarp P., Olofsson L., Sutherland D., and Kall M. (2004). Optical spetroscopy of nanometric holes in thin gold films. *Nano Lett.* 4:1003–1007.
32. Heng X., Cui X., Knapp D. W.,Wu J., Yaqoob Z., McDowell E. J., Psaltis D., Yang C . (2006). Characterization of light collection through a subwavelength aperture from a point source. *Opt. Express* 14:10410–10425.
33. Fischer U. C. (1986). Submicrometer aperture in a thin metal film as a probe of its microenvironment through enhanced light scattering and fluorescence. *J. Opt. Soc. Am. B* 3:1239–1244.
34. Liu Y., and Blair S. (2003). Fluorescence enhancement from an array of sub-wavelength metal apertures. *Opt. Lett.* 28:507–509.
35. Brolo A. G., Arctander E., Gordon R., Leathem B., and Kavanagh K. L. (2004). Nanohole-enhanced Raman scattering. *Nano Lett.* 4:2015–2018.
36. Wenger J., Dintinger J., Bonod N., Popov E., Lenne P.-F., Ebbesen T. W., and Rigneault H. (2006). Raman scattering and fluorescence emission in a single nanoaperture: optimizing the local intensity enhancement. *Opt. Commun.* 267:224–228.

37. Liu Y., Bishop J., Williams L., Blair S., and Herron J. N. (2004). Biosensing based upon molecular confinement in metallic nanocavity arrays. *Nanotechnology* 15:1368–1374.
38. Liu C., Kamaev V., and Vardeny Z. V. (2005). Efficiency enhancement of an organic light-emitting diode with a cathode forming two-dimensional periodic hole array. *Appl. Phys. Lett.* 86:143501.
39. Brolo A. G., Kwok S. C, Moffitt M. G., Gordon R., Riordon J., and Kavanagh K. L. (2005). Enhanced fluorescence from arrays of nanoholes in a gold film. *J. Am. Chem. Soc.* 127:14936–14941.
40. Kim, J. H., and Moyer, P. J. (2007). Laser-induced fluorescence within subwavelength metallic arrays of nanoholes indicating minimal dependence on hole periodicity. *Applied Physics Letters* 90:131111.
41. Liu Y., Mahdavi F., and Blair S. (2005). Enhanced fluorescence transduction properties of metallic nanocavity arrays. *IEEE J. Sel. Top. Quantum Electron.* 11:778–784.
42. Rigneault H., Capoulade J., Dintinger J., Wenger J., Bonod N., Popov E., Ebbesen T. W., and Lenne P.-F. (2005). Enhancement of single-molecule fluorescence detection in subwavelength apertures. *Phys. Rev. Lett.* 95:117401.
43. Wenger, J., Lenne, P.-F., Popov, E., Rigneault, H., Dintinger, J., and Ebbesen, T. W. (2005). Single molecule fluorescence in rectangular nano-apertures. *Opt. Express* 13:7035-7044.
44. Popov, E., Nevière, M., Wenger, J., Lenne, P.-F., Rigneault, H., Chaumet, P., Bonod, N., Dintinger, J., and Ebbesen, T. W. (2006). Field enhancement in single subwavelength apertures. *J. Opt. Soc. Am. A* 23:2342-2348.
45. Yuan H. X., Xu B. X., Wang H. F., Chong T. C. (2006). Field enhancement of nano-sized metal aperture with and without surrounding corrugations through resonant surface plasmons. *Jpn. J. Appl. Phys.* 45:787–791.
46. Mahdavi, F., Liu, Y., and Blair, S. (2007). Modeling fluorescence enhancement from metallic nanocavities. *Plasmonics* 2:129-141.
47. Leutenegger, M., Gösch, M., Perentes, A., Hoffmann, P., Martin, O. J. F., Lasser, T. (2006). Confining the sampling volume for Fluorescence Correlation Spectroscopy using a sub-wavelength sized aperture. *Opt. Express* 14:956-969.
48. Fore, S., Yuen, Y., Hesselink, L., and Huser, T. (2007). Pulsed-Interleaved Excitation FRET Measurements on Single Duplex DNA Molecules Inside C-Shaped Nanoapertures. *Nano Lett.* 7, 1749-1756.
49. Synge, E. H. (1928). A suggested method for extending the microscopic resolution into the ultramicroscopic region. *Philos. Mag.* 6:356-362.
50. Fort, E., and Grésillon, S. (2008). Surface enhanced fluorescence. *J. Phys. D: Appl. Phys.* 41:013001.
51. Wenger, J., Cluzel, B., Dintinger, J., Bonod, N., Fehrembach, A.-L., Popov, E. Lenne, P.-F., Ebbesen, T. W., and Rigneault, H. (2007). Radiative and nonradiative photokinetics alteration inside a single metallic nanometric aperture. *J. Phys. Chem. C* 111:11469-11474.
52. Single-Molecule Detection in Solution - Methods and Applications, edited by C. Zander, J. Enderlein and R.A. Keller (VCH-Wiley, Berlin/New York, 2002).

53. Rigler, R., and Elson, E. S. (2001). Fluorescence correlation spectroscopy, Theory and Applications (Springer, Berlin).
54. Wenger, J., Gérard, D., Dintinger, J., Mahboub, O., Bonod, N., Popov, E., Ebbesen, T. W., and Rigneault, H. (2008). Emission and excitation contributions to enhanced single molecule fluorescence by gold nanometric apertures. *Opt. Express* 16:3008-3020.
55. Popov, E., Nevière, M., Sentenac, A., Bonod, N., Fehrembach, A.-L., Wenger, J., Lenne, P.-F., and Rigneault, H. (2007). Single-scattering theory of light diffraction by a circular subwavelength aperture in a finitely conducting screen. *J. Opt. Soc. Am. A* 24:339-358.
56. Lezec, H. J., Degiron, A., Devaux, E., Linke, R. A., Martin-Moreno, L., Garcia-Vidal, F. J., and Ebbesen, T. W. (2002). Beaming light from a subwavelength aperture. *Science* 297:820-822.
57. Lakowicz, J. R. (2004). Radiative decay engineering 3: Surface plasmon-coupled directional emission. *Analytical Biochemistry* 324:153-169.
58. Salomon, L., Grillot, F., Zayats, A. V., and de Fornel, F. (2001). Distribution of optical transmission of periodic subwavelength holes in a metal film. *Physical Review Letters* 86:1110-1113.
59. Kretschmann, M., and Maradudin, A. A. (2002). Band structures of two-dimensional surface-plasmon polaritonic crystals. *Physical Review B* 66:245408.
60. Baudrion, A. L., Weeber, J. C., Dereux, A., Lecamp, G., Lalanne, P. and Bozhevolnyi, S. I. (2006). Influence of the filling factor on the spectral properties of plasmonic crystals. *Physical Review B* 74:125406.
61. Thio, T., Ghaemi, H. F., Lezec, H. J., Wolff, P. A., and Ebbesen, T. W. (1999). Plasmon-enhanced transmission through hole arrays in Cr films. *Journal of the Optical Society of America B* 16:1743-1748.
62. Brolo, A. G., Kwok, S. C., Cooper, M. D., Moffitt, M. G., Wang, C.-W.,Gordon, R., Riordon, J., and Kavanagh, K. L. Surface plasmon-quantum dot coupling from arrays of nanoholes. *Journal of Physical Chemistry* 110:8307–8313.
63. Weber, W. H., and Eagen, C. F. (1979). Energy transfer from an excited dye molecule to the surface plasmons of an adjacent metal. *Opt. Lett.* 4:236–238.
64. Kitson, S. C., Barnes, W. L., and Sambles, J. R. (1995). Surface plasmon energy gaps and photoluminescence. *Phys. Rev. B* 52:11441–11445.
65. Gruhlke, R. H., Holland, W. R., and Hall, D. G. (1986). Surface-plasmon cross coupling in molecular fluorescence near a corrugated thin metal film. *Phys. Rev. Lett.* 56:2838–2841.
66. Liu, Y., and Blair, S. (2004). Fluorescence transmission through 1-D and 2-D periodic metal films. *Optics Express* 12:3686-3693.
67. Korlach, J., Marks, P.J., Cicero, R.L., Gray, J.J., Murphy, D.L., Roitman, D.B., Pham, T.T., Otto, G.A., Foquet, M., and Turner, S.W (2008). Selective aluminum passivation for targeted immobilization of single DNA polymerase molecules in zero-mode waveguide nanostructures. *Proc. Natl. Acad. Sci. USA* 105:1176-1181.
68. Liu, G. L., Kim, J., Lu, Y. and Lee, L. P. (2006). Fluorescence enhancement of quantum dots enclosed in Au nanopockets with subwavelength aperture. *Appl. Phys. Lett.* 89, 241118.

69. Liu, C., Chen, N., and Sheppard, C. (2007). Nanoillumination based on self focus and field enhancement inside a subwavelength metallic structure. *Appl. Phys. Lett.* 90, 011501.
70. Garcia-Vidal, F. J., Martin-Moreno, L., Lezec, H. J., and Ebbesen, T. W. Focusing light with a single subwavelength aperture flanked by surface corrugations. *Appl. Phys. Lett.* 83:4500-4502.
71. Martin-Moreno, L., Garcia-Vidal, F. J., Lezec, H. J., Degiron, A., and Ebbesen, T. W. (2003). Theory of highly directional emission from a single subwavelength aperture surrounded by surface corrugations. *Phys. Rev. Lett.* 90:167401.
72. Nahata, A., Linke, R. A., Ishi, T., and Ohashi, K. (2003). Enhanced nonlinear optical conversion from a periodically nanostructured metal film. *Opt. Lett.* 28:423-425.
73. Ishi, T., Fujikata, J., Makita, K., Baba, T., and Ohashi, K. (2005). Si nano-photodiode with a surface plasmon antenna. *Jpn. J. Appl. Phys.* 44:L364–L366.
74. Bonod, N., Popov, E., Gérard, D., Wenger, J., and Rigneault, H. (2008). Field enhancement in a circular aperture surrounded by a single channel groove. *Opt. Express* 16:2276-2287.
75. Blom, H., Kastrup, L., and Eggeling, C. (2006). Fluorescence fluctuation spectroscopy in reduced detection volumes. *Curr. Pharm. Biotechnol.* 7:51-66.
76. Craighead, H. G. (2006). Future lab-on-a-chip technologies for interrogating individual molecules. *Nature* 442, 387-393.
77. Mannion, J. T., and Craighead, H. G. (2007). Nanofluidic structures for single biomolecule fluorescent detection. *Biopolymers* 85:131-143.
78. Genet, C., and Ebbesen, T. W. (2007). Light in tiny holes. *Nature* 445:39-46.
79. Samiee, K. T., Foquet, M., Guo, L., Cox, E. C., and Craighead, H. G. (2005). Lambda repressor oligomerization kinetics at high concentrations using fluorescence correlation spectroscopy in zero-mode waveguides. *Biophys. J.* 88:2145-2153.
80. Wenger, J., Gérard, D., Lenne, P.-F., Rigneault, H., Dintinger, J., Ebbesen, T. W., Boned, A., Conchonaud, F., and Marguet, D. (2006). Dual-color fluorescence cross-correlation spectroscopy in a single nanoaperture : towards rapid multicomponent screening at high concentrations. *Opt. Express* 14:12206-12216.
81. Foquet, M., Samiee, K. T., Kong, X., Chaudhuri , B. P., Lundquist, P. M., Turner, S. W., Freudenthal, J., and Roitman, D. B. (2008). Improved fabrication of zero-mode waveguides for single-molecule detection. *J. Appl. Phys.* 103:034301.
82. Edel, J. B., Wu, M., Baird, B., and Craighead, H. G. (2005). High spatial resolution observation of single molecule dynamics in living cell membranes. *Biophys. J.* 88:L43-L45.
83. Moran-Mirabal, J. M., Torres, A. J., Samiee, K. T., Baird, B., and Craighead, H. G. (2007). Cell investigation of nanostructures: zero-mode waveguides for plasma membrane studies with single molecule resolution. *Nanotechnology* 18:195101.

84. Samiee, K. T., Moran-Mirabal, J. M., Cheung, Y. K., and Craighead, H. G. (2006). Zero Mode Waveguides for Single-Molecule Spectroscopy on Lipid Membranes. *Biophys. J.* 90:3288-3299.
85. Wenger, J., Rigneault, H., Dintinger, J., Marguet, D., Lenne, P. F. (2006). Single-fluorophore diffusion in a lipid membrane over a subwavelength aperture. *J. Biol. Phys.* 32:SN1-SN4.
86. Wenger, J., Conchonaud, F., Dintinger, J., Wawrezinieck, L., Ebbesen, T. W., Rigneault, H., Marguet, D., and Lenne, P. F. (2007). Diffusion analysis within single nanometric apertures reveals the ultrafine cell membrane organization. *Biophys. J.* 92:913-919.
87. Koltermann, A., Kettling, U., Bieschke, J., Winkler, T., and Eigen, M. (1998). Rapid assay processing by integration of dual-color fluorescence cross-correlation spectroscopy: high throughput screening for enzyme activity. *Proc. Natl. Acad. Sci. USA* 95:1421-1426.
88. Winkler, T., Kettling, U., Koltermann, A., and Eigen, M. (1999). Confocal fluorescence coincidence analysis: an approach to ultra high throughput screening. *Proc. Natl. Acad. Sci. USA* 96:1375-1378.
89. Ekinci, Y., Solak, H. H., and David, C. (2006). Extraordinary optical transmission in the ultraviolet region through aluminum hole arrays. *Optics Letters* 32:172-174.

# 18

# Enhanced Multi-Photon Excitation of Tryptophan-Silver Colloid

Renato E. de Araujo[1], Diego Rativa[1] and Anderson S. L. Gomes[2]

[1] Department of Electronic and Systems, Universidade Federal de Pernambuco, Recife, PE, Brazil, renato.earaujo@ufpe.br

[2] Department of Physics, Universidade Federal de Pernambuco, Recife, PE, Brazil

## 18.1 INTRODUCTION

Metal nanoparticles have attracted considerable interest due to their properties and applications related to size effects, which can be appropriately studied in the framework of nanophotonics [1]. Metal nanoparticles such as silver, gold and copper can scatter light elastically with remarkable efficiency because of a collective resonance of the conduction electrons in the metal (i.e., the Dipole Plasmon Resonance or Localized Surface Plasmon Resonance). Plasmonics is quickly becoming a dominant science-based technology for the twenty-first century, with enormous potential in the fields of optical computing, novel optical devices, and more recently, biological and medical research [2]. In particular, silver nanoparticles have attracted particular interest due to their applications in fluorescence enhancement [3–5].

Metal nanoparticles (NPs) do not fluoresce, but they can rather enhance the emission of nearby fluorophores, owing to the influence of local field effects. This property can be explored in optical microscopy to analyze cell structure and functions. More specifically, nonlinear microscopy (by two-photon autofluorescence and by third-harmonic generation) of both live and fixed cells was demonstrated exploring plasmon resonance of metal nanoparticles [4]. In particular, for bioimaging, multi-photon fluorescence microscopy has important advantages over conventional epifluorescence or confocal microscopy, especially for the imaging of thick biological specimens. The most important of these is that photobleaching occurs only in the immediate focal region, rather than over the complete volume as in confocal fluorescence. In nonlinear microscopy usually an infrared source is used, also resulting in less scattering and greater penetration through thick tissue. The low phototoxicity produced by two and three-photon imaging compared to wide-field or confocal imaging offers exciting prospects for long-term *in vivo* imaging. Moreover, two and three-photon absorption can be used to excite levels corresponding to wavelengths deep into the UV, without the necessity of UV optics. Three-photon excitation has been predicted to yield greater resolution and deeper excitation into the UV absorption levels than two-photon excitation using the same excitation wavelength. Multi-photon cell microscopy was demonstrated exploring cell autofluorescence [6]. The cell autofluorescence chromophores are those which occur naturally, these includes the aromatic amino acids (Tryptophan, Phenylalanine and

*Multiphoton Excited Tryptophan-Colloid*
Edited by Chris D. Geddes.

Tyrosine), NADH and Flavins. The Tryptophan ($C_{11}H_{12}N_2O_2$) is the dominant source of UV absorption and emission in proteins. Most of the intrinsic fluorescence emission of a cell is due to excitation of Tryptophan residues, with additional contributions of Tyrosine and Phenylalanine. In addition, Tryptophan is a relatively rare amino acid, therefore many proteins contain only one or a few Tryptophan residues. Tryptophan is a fluorescent molecule which has been subject of a recent detailed study [7] aiming at providing a new insight into the interpretation of the fluorescence origin. Furthermore, the Tryptophan fluorescence is a useful tool for spectral analysis, especially to studies of protein conformational changes [8]. For many organisms, including humans, it is one of the essential amino acids (building blocks of proteins), which cannot be synthesized by the human body and therefore must be part of its diet. Tryptophan is a precursor for serotonin, one of the key brain chemicals involved in mood regulation [9]. It is also necessary for the production of Niacin, a water-soluble vitamin whose derivatives such as NADH play essential roles in energy metabolism in the living cell [10]. Furthermore, Tryptophan is involved in the body regulation of sleep. It has also been found that people suffering from migraine headaches have abnormal levels of Tryptophan, and its monitoring and control may be helpful. It is thus important to understand Tryptophan not only from the basic point of view but also due to its applications and implications in health sciences.

Several articles and reviews on different aspects of multi-photon excitation of biomolecule system are available. For example, Birch [11] considerations concentrate mainly on the impact of multi-photon techniques to the time-resolved fluorescence spectroscopy. Lakowicz and Gryczynski [12] have discussed examples of three-photon excited fluorescence. Rehms and Callis studied the two-photon excited fluorescence emission of aromatic amino acids [13]. Kierdasz et al analyzed emission spectra of Tyrosine- and Tryptophan-containing proteins using one-photon (270-3 10 nm) and two-photon (565-6 10 nm) excitation [14].

This chapter is devoted to describe the impact of metallic nanosphere to the multi-photon excitation fluorescence of Tryptophan, and little further consideration to multi-photon absorption process will be given, as the reader can find several studies in [11-14]. In section II, the nonlinear light-matter interaction in composite materials is discussed through the mechanism of nonlinear susceptibilities. In section III, experimental results of fluorescence induced by multi-photon absorption in Tryptophan are reported and analyzed. Section IV described the main results of this chapter, which is the effect of metallic nanoparticles on the fluorescent emission of the Tryptophan excited by a multi-photon process. Influence of nanoparticle concentration on the Tryptophan-silver colloids is observed and discussed based on a nonlinear generalization of the Maxwell Garnett model, introduced in section II. The main conclusion of the chapter is given in section IV.

## 18.2 NONLINEAR LIGHT-MATTER INTERACTION IN COMPOSITE MATERIALS

When light propagates trough a colloid, two basic processes need to be considered: scattering and absorption.

Scattering of light results from the spatial changes on the refraction index of the light path [15]. A scattering process can be characterized by the scattering cross-section or scattering coefficient. A special kind of scattering is called Rayleigh scattering. It refers to processes where the scattering particles dimensions are smaller than the wavelength of the incident light. In this case, the scattered intensity is inversely proportional to the fourth power of the wavelength [15]. More complex scattering theory, known as Mie theory, describe the wave diffraction on a dielectric sphere, regardless scatter dimensions [15]. There are several techniques to measure scattering properties of different materials. A well known one uses double-integrating sphere geometry and it was first applied by Derbyshire et al to measure optical tissue properties [16]. Wavelength-dependence of scattering processes is an important issue and, in general, the scattering coefficient is smaller for longer wavelength.

Light absorption by a material is a fundamental process in nature. This interaction can induce chemical reactions, thermal effects, or fluorescent emission. Light absorption can be characterized by an absorption coefficient. Fluorescence excitation cross-section is another parameter used to describe an absorption process. It quantifies the photon probability of being absorbed by a fluorophore, generally one-photon absorption process. However, higher order processes, such as two and three-photons, lead to excitation cross-section which quantify the excitation probability by multi-photons nonresonant mechanism. One and two-photon absorption are governed by different selections rules [17]. Forbidden one-photon transition can be allowed by two-photon process. Electronic, vibrational and rotational states, which cannot be reached by one-photon absorption, may be excited by a two-photon process. The low probability of multi-photon transition requires the interaction of fluorophores with high intensity fields, which can be readily obtained with short-pulses infrared lasers. Multi-photon excitation cross-section can be strongly enhanced by resonance with intermediate states [17]. One and multi-photon absorption in a three level system are represented in figure 18.1.

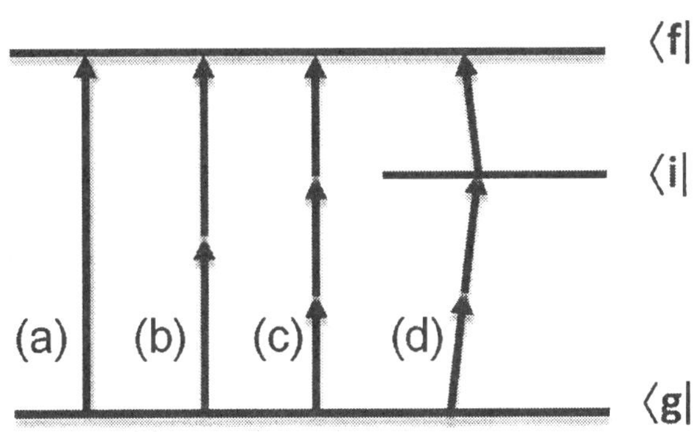


**Figure 18.1:** One and multi-photon transitions: one-photon process (a), two-photon single step process (b), three-photon single step process (c); three-photon two steps process (d)

Nonlinear optical process, like multi-photon absorption, can be understood through analyses of the induced electric polarization. When an electric field is applied to a medium, charges bound in each molecule will react to the applied field and will execute perturbed motions, changing the molecular charge density of the media, polarizing it. The relation of the induced electric polarization, $\vec{P}$, and the applied electric field, $\vec{E}$, can be written as $\vec{P}(\omega) = \chi^{(1)} \cdot \vec{E}(\omega)$ [18], where $\omega$ is the frequency of the electrical field and $\chi^{(1)}$ is the electric susceptibility of the medium. The real part of the electric susceptibility can be related to the material refractive index, $n_0$, while its imaginary portion is related to the absorption coefficient, $a_0$, of the medium. High intensity electric fields interacting with a medium is better described by an expansion of $\vec{P}$ into power series of the electric field[17];

$$\vec{P}(\omega) = \chi^{(1)}(\omega) \cdot \vec{E}(\omega) + \chi^{(2)}(\omega) \cdot \vec{E}(\omega)\vec{E}(\omega) + \chi^{(3)}(\omega) \cdot \vec{E}(\omega)\vec{E}(\omega)\vec{E}(\omega) + \ldots$$

The first term of this expression is the linear response of the media to the incident field, as the one-photon absorption (1PA) process. The other terms describe the nonlinear reaction of the material. The coefficient $\chi^{(2)}$ is related to processes like second harmonic generation, sum and difference-frequency generation and electro-optics effects. The coefficient $\chi^{(3)}$ is associated to many others phenomena as third harmonic generation, two-photon absorption (2PA), Stimulated Raman Scattering [16]. Moreover, the simultaneous absorption of three-photons (3PA) is related to an even higher order nonlinear susceptibility, $\chi^{(5)}$ [19].

In particular, the third-order nonlinear susceptibility is a complex quantity, $\chi^{(3)}(\omega) = \mathrm{Re}[\chi^{(3)}(\omega)] + i \cdot \mathrm{Im}[\chi^{(3)}(\omega)]$. The real part of $\chi^{(3)}$ is related to the to the nonlinear refraction index, $n_2$ [20], and its imaginary component – which is the component of interest to our work - is associated to the material nonlinear absorption coefficient, $a_2$ [20], through

$$\mathrm{Im}[\chi^{(3)}] = \frac{n_o \varepsilon_o c^2}{\omega} a_2 \tag{1}$$

where $n_o$ is the linear index of refraction, $\varepsilon_o$ is the permittivity of vacuum, and $c$ speed of light. The absorption coefficient of a material, $a(I)$, can be written as

$$a(I) = a_o + a_2(I) \tag{2}$$

where $I$ denotes the irradiance within the sample [20].

In a composite material where small particles (inclusions) are distributed in a host material, an overall third-order nonlinear susceptibility should be a function of both host and inclusions third-order nonlinear susceptibilities. Here we consider a composite topology where the inclusions are sphere of radius $a$, and we define a characteristic distance between inclusions to be $b$, with $a<<b<<\lambda$, where $\lambda$ is the wavelength of light in vacuum (see Figure 18.2).

The referred topology, known as Maxwell Garnett geometry [21], is considered to be isotropic macroscopically, and an effective-medium dielectric constant can be written as

$$\varepsilon_{eff} = \varepsilon_h \left( 1 + \frac{3\beta f}{1 - \beta f} \right) \tag{3}$$

where $\varepsilon_h$ is the host dielectric constant, $f$ is volume filling factor (defined as the ratio of the total volume occupied by the nanoparticles divided by the total volume of the solution), and $\beta$ is given by:

$$\beta = \frac{\varepsilon_i - \varepsilon_h}{\varepsilon_i + 2\varepsilon_h} \tag{4}$$

Here, $\varepsilon_i$ is the inclusion dielectric constant. Equation 3 is known as the Maxwell Garnett result [21].

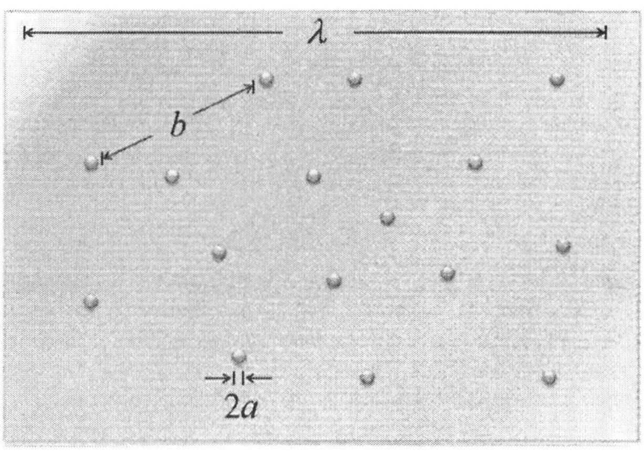


**Figure 18.2:** Topology of a composite material

An effective third-order nonlinear susceptibility of a composite optical material was determined by Sipe and Boyd [21]. Accordingly, the effective nonlinear susceptibility of a composite optical material comprised of spherical inclusion particles contained within a host material is given by Eq. (5):

$$\chi_{eff}^{(3)} = f \left| \frac{3\varepsilon_h}{\varepsilon_i + 2\varepsilon_h} \right|^2 \left[ \frac{3\varepsilon_h}{\varepsilon_i + 2\varepsilon_h} \right]^2 \chi_i^{(3)} + \chi_h^{(3)} \tag{5}$$

In Eq. (5), $\chi_i^{(3)}$ and $\chi_h^{(3)}$ are the third-order nonlinear optical susceptibilities of the inclusions and the host, respectively.

In a composite material, as described here, the effective third-order nonlinear susceptibility should depend linearly with the concentration of the inclusions in a low filling fraction regime. In that way, the nonlinear absorption coefficient of the medium, associated to the $\mathrm{Im}[\chi^{(3)}(\omega)]$ and consequently to the two-photon absorption processes, should also be a function of the inclusions' concentration.

## 18.3 FLUORESCENCE INDUCED BY MULTI-PHOTON EXCITATION

Fluorescence emission is just one, but probably the most convenient, of several methods available for observing multi-photon absorption. In a multi-photon excited fluorescence process, the fluorescence intensity $I_{fl}$ does not increase linearly with increasing of the excitation intensity, $I_{exc}$. Instead, $I_{fl}$ and $I_{exc}$ are related by

$$I_{fl} \propto \sigma_i I_{exc}^i \tag{6}$$

where $\sigma_i$ is the absorption cross-section for *i* photons. The value of the absorption cross-section $\sigma_i$ decreases dramatically, as the number *i* of photons increases. For example, for the Tryptophan molecule, $\sigma_i$ = 6.3 x10$^{-18}$ cm$^2$ (1PA at 290nm) [22], $\sigma_2$ = 4.9 x10$^{-50}$ cm$^4$s$^{-1}$ photon$^{-1}$ (2PA at 580nm) [22], and $\sigma_3$ = 1x10$^{-84}$ cm$^6$ s$^2$ photon$^{-2}$ for a three-photon absorption (at 710nm) [23]. This fact alone demands the use of high power pulsed laser excitation. Commercial femtosecond laser systems, when focused down to ~10 *u*m spot size, can provide an excitation power density up to GW cm$^{-2}$, sufficient to generate both two, three and even greater numbers of photon excited fluorescence in most fluorophores.

Due to its relevance to the next section, we observed and analyzed the fluorescent emission of Tryptophan in water solution excited by one, two, and three-photon absorption. For that, three different light sources were used: a UV (180–375 nm) lamp, the second harmonic of a Q-switched Nd:YAG laser (with 8 ns pulse duration at 532 nm) and a Ti-Sapphire laser delivering pulses at 76 MHz, with 150 fs pulse duration and 500 mW average power at 800 nm.

The fluorescence emission obtained by the UV lamp excitation was measured using a model DV-Z500 spectrophotometer (Beckman, USA). For both laser excitation setups the light beam was focused into the sample with a 10cm focal length lens (L1). The emitted light was collected along a direction perpendicular to the incident beam and optical filters (F) were used remove the pump scattered light. The fluorescence was sent to a monochromator attached to a GaAs photomultiplier tube connected to a lock-in amplifier (SR530 Stamford Research Dual Phase), as shown in figure 18.3. The lock-in amplifier was triggered by an electrical signal obtained from the laser control electronics. All experiments were performed with the samples contained in a 10 mm long quartz cuvette at room temperature.

The fluorescence spectrum of the Tryptophan solution is shown in figure 18.4. The Tryptophan fluorescence is characterized by a broad band (70nm) emission with its peak at around 360nm. The Tryptophan emission spectra dependence with the excitation intensity is present in the inset of figure 18.4. Particularly, the inset on figure 18.4(b) shows that the fluorescence intensity of the Tryptophan solution has a square dependence with the laser pump intensity (at 532nm), consistent with simultaneous absorption of two photons. As expected, a cubic dependence of the Tryptophan emission intensity with the intensity of the 800nm pump beam was observed, see inset of figure 18.4(c).

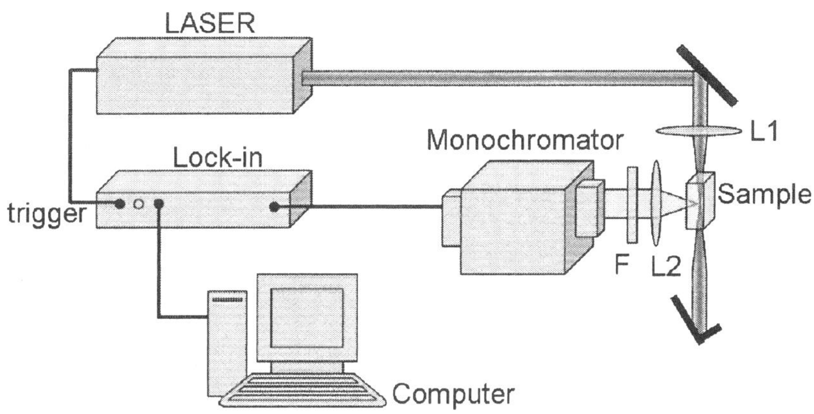


**Figure 18.3:** Experimental setup for nonlinear spectroscopy.

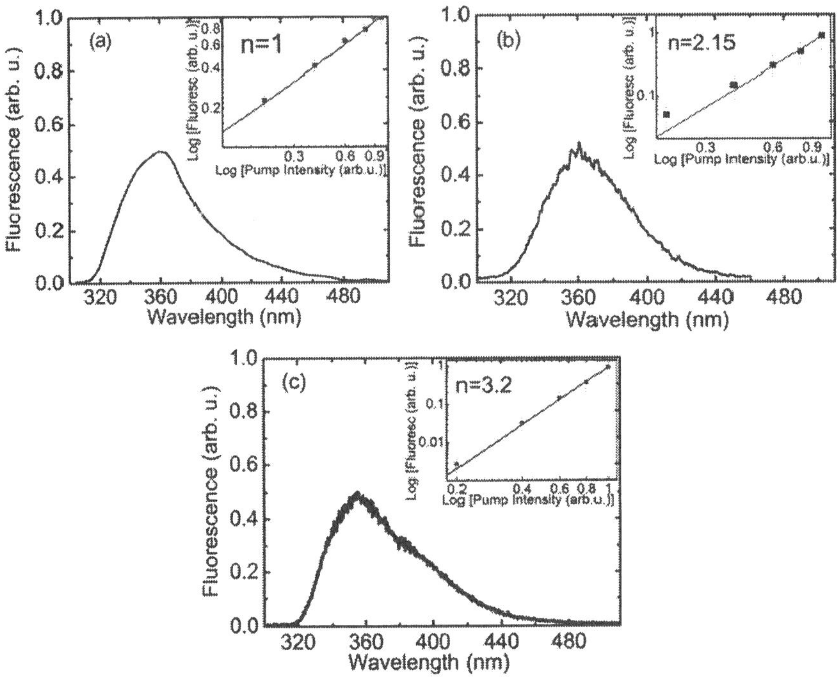


**Figure 18.4:** Fluorescence spectrum of the Tryptophan solution excited by a) one, b) two and c) three-photon absorption process. Insets show the excitation intensity dependence of the fluorescence.

## 18.4. MULTI-PHOTON EXCITATION OF TRYPTOPHAN-SILVER COLLOID

By definition, a colloid is a mechanical mixture of small particles (<1000 nm) in a dispersion medium. In this work, silver colloid suspensions were added to the Tryptophan water solution (amino acid concentration of 0.1 μM and pH = 6±1). The objective of the introduction of the silver nanoparticles is to enhance the emitted fluorescence, which in turn can be useful on multi-photon microscopy studies. The silver colloid was synthesized according to the method described in [24], then was laser ablated using the second harmonic of a Q-switched Nd:YAG laser (8ns, 10Hz), for about one hour (depending upon the prepared volume) leaving as a results spherical nanoparticles of 9nm mean diameter. The characterization of the prepared colloids were performed by measuring their UV-visible absorption spectra, over the 200–700 nm range using a model DV-Z500 spectrophotometer (Beckman, USA), and by analyzing the nanoparticle size distribution through electron microscopy imaging.

When metallic nanoparticles are added to a biological medium, its linear and nonlinear optical properties are changed. In particular, the analysis of the linear optical absorption is one of the methods most commonly used to study colloids optical properties. It is known that aggregation of Tryptophan to metal clusters can lead to remarkable changes of the optical absorption as compared to the bare molecule [25]. Localized Surface Plasmon Resonance is sensitive to changes in the local dielectric environment. Therefore, NP-molecules interaction can also lead to displacement of the Plasmon peak and to the narrowing of the Plasmon bandwidth [26]. Figure 18.5 shows the Plasmon band spectra of 9nm silver spheres in pure water and in the Tryptophan solution. A 6nm displacement and bandwidth reduction of 3nm indicates that some adsorption of molecules on the surface of the metal spheres may have taken place. As Ag NPs are added to the Tryptophan solution, an increase on the Plasmon peak intensity was observed but no other significant displacement of the Plasmon band was detected. Differently from other metal nanospheres, Ag NPs present a relatively narrow Plasmon band (~44 nm), and no strong absorption is observed in the UV spectral region ($\lambda$<340 nm). Silver nanoparticles have an advantage over other metal nanoparticles (i.e., gold and copper) since the Surface Plasmon Resonance energy of Ag is located far from the interband transition energy. As a consequence, in a silver nanoparticle-contained compound, one can investigate the nonlinear optical effects solely based on plasmon contributions.

With the one-photon excitation, the increase of NPs concentration was followed by a quenching in the amino acid emission. The quenching process can be understood by damping of the dipole oscillators by the nearby metal particles [27, 28]. Similar quenching was also observed when the colloid was excited by a two-photon absorption process. The decrease in the fluorescent peak intensity as a function of the colloid filling factor is shown in figure 18.6, for one and two-photon excitation. The fluorescence intensity was normalized to the case of no NPs in the solution. Each solid line in figure 18.6 is a two exponential decay fitting. Same decay parameters were used to fit the experimental data, indicating that the NP-Tryptophan interaction is similar for one and two-photon excitation.

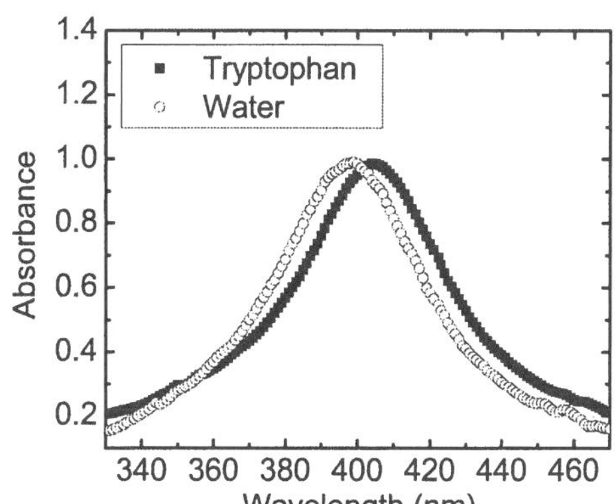


**Figure 18.5:** Plasmon band spectra of 9nm silver spheres in pure water and in the Tryptophan solution.

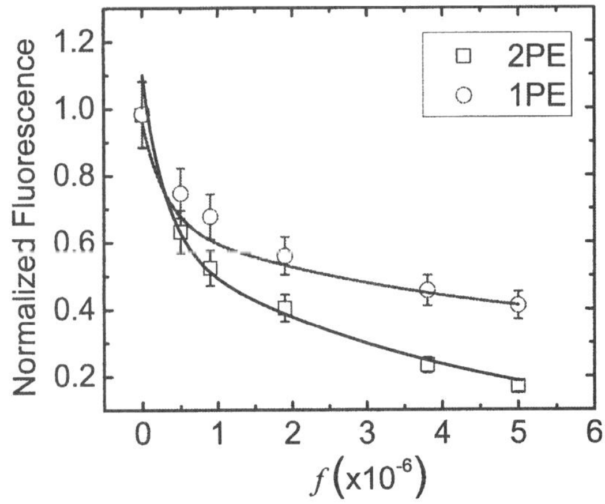


**Figure 18.6:** Fluorescent peak intensity of Tryptophan-Ag colloids excited by one and two-photon absorption process

Excitation and the relaxation (radiative and non radiative) processes of the Tryptophan solution and the colloids are represented in figure 18.7. The new relaxation pathways introduced by the metal nanoparticles (nonradiative decay rate, $\kappa_{np}$) are shown in figure 18.7(b). Although, one photon at 270nm and two-photons at 532nm are resonant with the excited states of the molecule, these wavelengths are not in resonance with the Plasmon energy level.

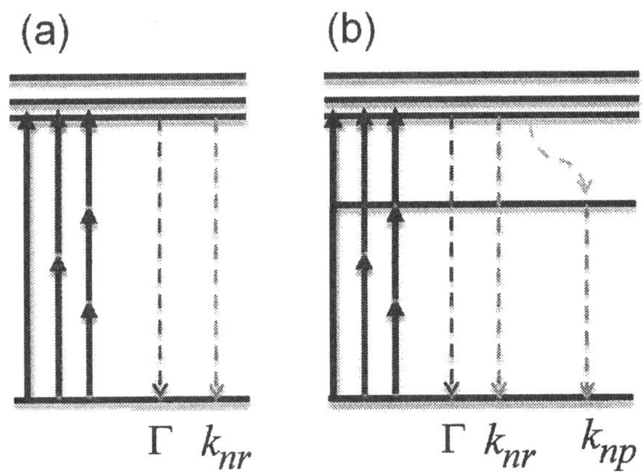


**Figure 18.7:** The excitation and the relaxation processes of a) the Tryptophan solution and b) the Tryptophan-Ag colloid.

When the Tryptophan-Ag colloid is excited by three-photons, an increase of the fluorescence is observed as the NPs concentration increase. The presence of nanoparticles introduces a new energy level in the solution which matches with the energy of two photons at 800nm (figure 18.7b), providing an enhancement for absorption of a third photon by the amino acid. Figure 18.8 shows the increase of the fluorescent peak intensity as a function of the colloid filling factor. One can notice that, for low values of the filling factor the fluorescence intensity increase linearly with *f*. A saturation behaviour is also observed for *f* higher than 3 x $10^{-6}$. To confirm whether the excitation of the Plasmon energy state is the responsible for the fluorescence enhancement of the Tryptophan-silver colloid a time-resolved fluorescence analysis of the emission and a dual pump resonant experiment were performed [5]. Both independently performed experiments confirmed the role played by the Plasmon resonance.

By pumping the Tryptophan-Ag colloid with 800nm, the excitation of the Tryptophan molecule is assisted by silver particles' two-photon absorption. Resonant absorption of two-photon by the NPs is related to the imaginary part of the effective third-order optical susceptibility, $\chi^{(3)}$. According to the generalized Maxwell-Garnett model (equation 5), the effective third-order nonlinear susceptibility should depend linearly with the concentration of the inclusions (in a low filling fraction regime).

To have a better insight into the origin and importance of the nonlinearity in the process, we also determined experimentally the behaviour of the imaginary part of the nonlinear susceptibility of the Tryptophan-Ag colloid as a function of the NP concentration. There are several techniques to measure $\chi^{(3)}$ of a biosample, such as four wave mixing, Kerr gate, Z-scan and we explored a modification of the well known Z-scan, named thermally managed eclipse Z-scan technique (TM EZ-scan), recently introduced by us [29, 30]. The TM-EZ scan technique presents the sensitivity of eclipse Z-scan [31] and allows for the simultaneous measurements of the nonthermal and thermal nonlinearities of the material under study. The use of lasers with high repetition rate allows measurements with large sensitivity due to the better signal-to-noise ratio obtained. In short, the TM-EZ scan method consists in acquiring the time evolution of the EZ-scan signal, for the sample placed along the pre- and post-focal positions of the laser focal plane. The nonlinear absorption is measured detecting the whole beam transmitted by the sample. A Ti-sapphire laser (800 nm, 150 fs, 76 MHz) was used as the pump source. The maximum average power reaching the sample was about 400mW. The technique has been described in detail in ref [29].

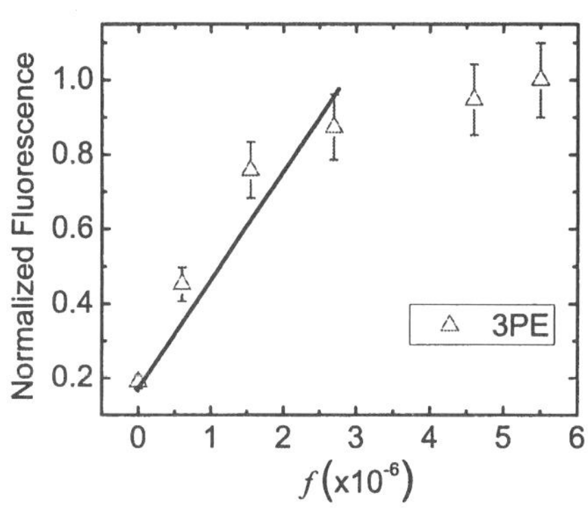


**Figure 18.8:** Fluorescent peak intensity of Tryptophan-Ag colloid excited by three-photon absorption process

The TM-EZ scan technique was applied to the Tryptophan-Ag colloid with different NPs concentration. The values obtained for the nonlinear susceptibilities are summarized in figure 18.9. One can also observe that, in a low filling fraction regime, the imaginary part of the third-order nonlinear susceptibility increase linearly with the concentration of NPs, and a saturation behaviour is observed for $f > 3\text{x}10^{-6}$, in accordance with the results obtained from the fluorescence spectroscopy (figure 18.8).

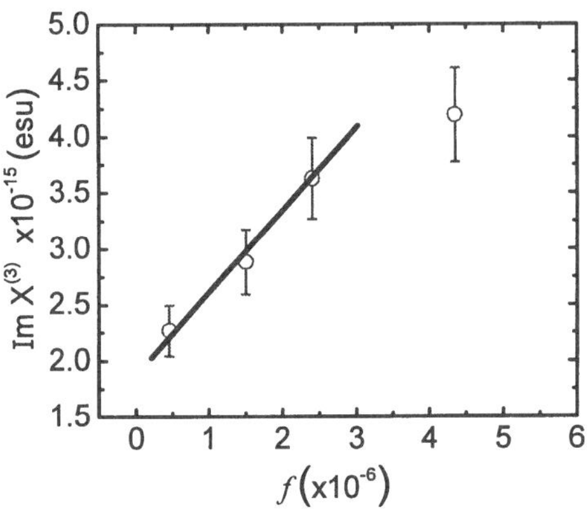


**Figure 18.9:** Imaginary part of the third-order nonlinear susceptibility of Tryptophan-Ag colloid, with different NPs concentration.

## 18.5. CONCLUSIONS

Multi-photon excitation is an important tool for diagnostics of biological systems. In this chapter, we considered the impact of metallic nanosphere to the multi-photon excitation fluorescence of Tryptophan. Metallic nanoparticles can affect molecular chromophores in a close proximity to their surface. We showed that the excitation of Tryptophan can be assisted by silver nanoparticles' two-photon absorption, due to the Plasmon resonances. This process can overcome fluorescence quenching due to metal NP- fluorophore interaction. The multi-photon excitation of colloid was described trough the analysis of the imaginary part of the third-order nonlinear optical susceptibility of the composite medium. Moreover, the influence of nanoparticle concentration on the Tryptophan-silver colloid fluorescence was observed and discussed based on a nonlinear generalization of the Maxwell Garnett model. The three photon absorption process assisted by NPs can be explored as a new tool in fluorescence microscopy and possibly diagnostic procedures.

## 18.6 ACKNOWLEDGEMENTS

This work has been supported by the Brazilian Agencies CNPq and CAPES, under the Nanophotonics Network Program and the Millenium Institute of Nonlinear Optics, Photonics and Biophotonics. It has also been part of the PhD thesis of D. Rativa.

## 18.7 REFERENCES

1. Prasad, P. N. (2004). *Nanophotonics*, John Wiley & Sons, New York.
2. Lakowicz, J. R. (2006). Plasmonics in biology and plasmon-controlled fluorescence. *Plasmonics* 1:5-33.
3 Lakowicz, J. R., Geddes, C. D., Gryczynski, I., Malicka, J., Gryczynski, Z., Aslan, K., Lukomska, J., Matveeva, E., Zhang, J., Badugu, R., Huang., J. (2004) Advances in Surface-Enhanced Fluorescence. *J. of Fluores.* 14: 425-441.
4 Yelin, D., Oron, D., Thiberge, *S.*, Moses, E., and Silberberg, Y. (2003). Multiphoton plasmon-resonance microscopy. Opt. Express 11:1385-1391.
[5] Rativa, D., Gomes, A. S. L., Wachsmann-Hogiu, S., Farkas, D. L., and de Araujo, R. E. (2008). Nonlinear Excitation of Tryptophan Emission Enhanced by Silver Nanoparticles. *J. Fluoresc*, Online.
6 Zipfel, W. R., Williams, R. M., Christie, R., Yu Nikitin, A., Hyman, B.T. and Webb, W. W. (2003). Live tissue intrinsic emission microscopy using multiphoton-excited native fluorescence and second harmonic generation. *PNAS* 100:7075-7080.
7. Albani, J. R. (2007). New insights in the interpretation of tryptophan fluorescence, J Fluoresc 17:406–417
8. Lakowicz, J. R. (1999) Principles of Fluorescence Spectroscopy. 2 Edition. Springer, New York.
9. Schaechter, J.D., Wurtman, R.J. (1990). Serotonin release varies with brain tryptophan levels. *Brain Res.* 532: 203-210.
10. Ikeda, M., Tsuji, H., Nakamura, S., Ichiyama, A., Nishizuka, Y., and Hayaishi, O. (1965) Studies on the biosynthesis of nicotinamide adenine dinucleotide. II. A role of picolinic carboxylase in the biosynthesis of nicotinamide adenine dinucleotide from tryptophan in mammals. *J. Biol. Chem.* 240:1395-401.
11 Birch D. J. S. (2001) Multiphoton excited fluorescence spectroscopy of biomolecular systems. *Spectrochimica Acta* 57: 2313-2336.
12 Lakowicz, J. R., and Gryczynski, I. (1999). Three-photon excitation of fluorescence. In *Applied Fluorescence in Chemistry, Biology and Medicine*, Rettig, W., Strehmel, B., Schrader, S., and Seifert, H. (Eds), Springer-Verlag, Berlin Heidelberg, pp. 137.
13 Rehms, A. A., and Callis, P. R. (1993). Two-photon fluorescence excitation spectra of aromatic amino acids. *Chemical Physics Letters* 208: 276-282.

14 Kierdaszuk, B., Gryczynski, I:, Modrak Wojcik, A., Bzowska, A., Shugar, D:, and Lakowicz, J. R. (1995) fluorescence of tyrosine and tryptophan in proteins using one- and two-photon excitation. *Photochemistry and Photobioiogy.* 61: 319-324.

15 Ishimaru., A. (1978) *Wave Propagation and Scattering in Random Media,* Academic Press, New York.

16 Derbyshire, G. J., Bogen, D. K., and Unger, M. (1990) Thermally induced optical property changes in myocardium at 1.06 mm. *Lasers Surg. Med.* 10: 28-34.

17 Shen, Y. R. (2003) *The principles of Nonlinear Optics,* John Wile & Sons, Inc. New Jersey.

18. Jackson, J. D. (1998) *Classical Electrodynamics*, Wiley, New York, 1998.

19. Boyd, R. W. (2002) *Nonlinear* Optics, Academic Press.

20. Sheik-Bahae, M., Said, A. A., Wei, T. H., Hagan, D. J., and Van Stryland, E. W. (1989) Sensitive Measurement of Optical Nonlinearities Using a Single Beam. *IEEE Journal of Quantum Electronics.* QE-26: 760-769.

21. Sipe, J. E., and Boyd, R. W. (1992) Nonlinear susceptibility of composite optical materials in the Maxwell Garnett model. *Phys. Rev. A* 46: 1614-1629.

22. Xu, C., and Webb, W. W. (1997) Multiphoton excitation of molecular fluorophores and nonlinear laser microscopy. in *Topics in Fluorescence Spectroscopy.* J. R. Lakowicz (Ed.), Plenum Press, New York, pp. 475-540.

23. Maiti, S., Shear, J. B., Williams, R. M., Zipfel, W. R., and Webb, W. W. (1997) Measuring Serotonin Distribution in Live Cells with Three-Photon Excitation. *Science* 275: 530 – 532.

24. Lee, P. C., and Meisel D. (1982) Adsorption and surface-enhanced Raman of dyes on silver and gold sols. *J. Phys. Chem.* 86: 3391-3395.

25. Compagnon, I., Tabarin, T., Antoine, R., Broyer, M., Dugourd, P., Mitric, R., Petersen, J., and Koutecký, V. B. (2006) Spectroscopy of isolated, mass-selected tryptophan-Ag3 complexes: A model for photoabsorption enhancement in nanoparticle-biomolecule hybrid systems. *J. Chem. Phys.* 125: 164326.

26. Gómez, L. A., de Araújo, C. B., Brito Silva, A. M, and Galembeck, A. (2007) Influence of stabilizing agents on the nonlinear susceptibility of silver nanoparticles. *J. Opt. Soc. Am. B* 24:2136-2140.

27. Lakowicz, J. R. (2001) Radiative Decay Engineering: Biophysical and Biomedical Applications. *Anal. Biochem*, 298: 1-24

28. Campion, A., Gallo, A., Harris, C., Robota, H. J., and Whitmore, P. (1980) Electronic energy transfer to metal surfaces: A test of classical image dipole theory at short distances. *Chem. Phys. Letts.* 73: 447-450.

29. Gomes, A. S. L., Falcão-Filho, E. L., de Araújo, C. B., Rativa, D., and de Araujo, R. E. (2007) Thermally managed eclipse Z-scan. *Opt. Express* 15: 1712-1717.

30. Rativa, D., de Araujo, R. E., de Araújo, C. B., Gomes, A. S. L., and Kassab, L. R. P. (2007) Femtosecond nonlinear optical properties of lead-germanium oxide amorphous films. *Appl. Phys. Lett.* 90: 231906.

31. Xia, T., Hagan, D. J., Sheik-Bahae, M., and Van Stryland, E. W. (1994) Eclipsing Z-scan Measurements of □/$10^4$ Wavefront Distortion. *Opt. Lett.* 19, 317-319.

# 19

# Plasmon-enhanced radiative rates and applications to organic electronics

Lewis Rothberg[1] and Shanlin Pan[2]

[1] University of Rochester, Department of Chemistry, Rochester, NY 14627.

[2] The University of Alabama, Department of Chemistry, Tuscaloosa, AL 35487.

## 19.1 INTRODUCTION

Molecular spectroscopy and photophysics can be dramatically altered when molecules are near nanotextured metal surfaces. An early example comes from the work of Van Duyne et al (1-2) who showed that electrochemically roughened silver electrodes could enhance Raman scattering from adsorbed pyridine by more than $10^6$ times. Surface enhancement of Raman spectra remains a subject of practical and fundamental interest (3-4). In spite of the widespread utilization and acceptance of surface-enhanced Raman scattering, it was nevertheless surprising when the discovery of Raman scattering from single molecules was made by several groups in the late 1990s (5-8). Since Raman cross-sections are typically of order $10^{-30}$ cm$^2$, this meant that processes near the metal surfaces including near-field enhancement of the optical fields and chemical enhancement effects associated with charge transfer resonance between the molecule and metal must increase the cross-section by nearly 14 orders of magnitude to account for those observations.

Analogous effects of metal surfaces on interactions between electromagnetic fields and molecules have also been observed to enhance molecular absorption and emission (9-12). Surface plasmon coupled luminescence is a well-documented phenomenon resulting in increased emission rates for molecules near gold and silver films (9-12, 13-17). The physical origin of radiative enhancement can be understood in terms of concentration of the electromagnetic field by interaction with the electron plasma in the metal to volumes small compared to its usual size scale of $\sim \lambda^3$ where $\lambda$ is the wavelength of the radiation. Strong field localization is predicted and observed at wavelengths near the plasma resonance for the metal particle or particle assembly. Even larger effects are possible by coupling to surface plasmons in nanotextured metal assemblies given proper control over metal morphology, molecule-particle distance and orientation, and excitation wavelength. It is easy to understand how this picture applies to the interaction of incident radiation with molecules near metal nanoparticles. While it is less clear how this applies to enhancement of spontaneous emission rates where there is no photon prior to excited state decay, one can still understand emissive rate enhancement in terms of spatial concentration of the density of vacuum states responsible for spontaneous emission.

There are many potential applications for plasmon enhancement of absorption, emission and higher order interactions of the electromagnetic field with molecules. These include biomolecular sensing (18-20), enhancement of nonlinear spectroscopies (21), up-conversion of infrared radiation (22) and enhanced Raman

*Plasmon-Enhanced Radiative Rates in Organic Electronics*
Edited by Chris D. Geddes.

spectroscopy (5-8). In the present chapter, we will consider applications of plasmon enhanced radiative processes in organic electronics. Organic electronics are potentially attractive due to the processability of organic materials making them compatible with inexpensive manufacturing processes such as roll-to-roll coating and ink-jet printing. Of particular interest are organic light-emitting diode (OLED) technology for high information content displays (23-24) and organic photovoltaic (OPV) technology (25) for solar energy conversion. Great strides have been made in both efficiency and stability but further improvement is needed to make these technologies commercially successful. Our particular focus here will be to investigate how plasmonic enhancement of the optical processes intrinsic to OLED and OPV devices can be used to ameliorate some of the shortcomings of the devices. We will provide a rationale for why plasmon-enhancement could be useful, document some of its properties, review recent attempts to apply plasmonics to organic electronics and speculate on the challenges and prospects for this strategy.

In the case of OLEDs, device efficiencies are now adequate to support commercial applications and the primary issue for the technology is stability. Since emissive quantum yields of many chromophores used in OLEDs verges on 100 %, there is little to be gained by fluorescent enhancement. However, to the extent that excited state processes leading to material degradation limit OLED longevity, increasing radiative rates may have the benefit of reducing degradation by reducing excited state lifetime. Processes known to degrade fluorophores such as intersystem crossing to long-lived triplets susceptible to chemical oxidation can be reduced by more than an order of magnitude and, indeed, photobleaching suppression has been documented by Lakowicz and coworkers (26-27). A related strategy to improve OLED stability that would be enabled by plasmonic enhancement of luminescence would be to choose very stable materials with lower emissive yields and use plasmon enhancement to improve their luminescence. This could be very important in niches like the viability of blue phosphorescent devices (28). Increases in phosphorescent rates could also help to remediate problems like triplet-triplet annihilation that can occur in OLEDs operating at high brightness.

The general issues pertinent to organic solar cells are quite different but these devices may also be amenable to substantial improvements by exploiting plasmon enhancement of light absorption. The poor charge transport properties of organic materials dictate the use of very thin layers of active materials (~ 100 nm) and it is difficult to absorb all of the incident radiation effectively. This is especially true at wavelengths where organic materials are marginal absorbers such as the near infrared region of the spectrum that carries a substantial fraction of the energy emitted by the sun. In spite of these facts, organic photovoltaics have achieved efficiencies near 10 % in the case of liquid polyelectrolyte based Gratzel-type cells (29-30) and about 5 % in the case of solid state cells based on blended conjugated polymers (31-33). Plasmon enhancement of the absorption and modification of the absorber's spectrum can be used to good effect and, given long term material stability, perhaps improvements of as little as a factor of two in efficiency would make these approaches commercially important.

## 19.2 OUTLINE OF THE CHAPTER

In order to forecast the potential impact of plasmonic enhancement on OLED and OPV technologies, we will review the physics of plasmonic enhancement of radiative rates and a series of experiments designed to address some of the questions relevant to how to integrate nanotextured metallic assemblies capable of supporting plasmon resonance into organic optoelectronic devices. We review a simple and computationally useful picture of electromagnetic field concentration using finite element modeling and then present empirical data for what we have learned about how to design nanoparticle morphologies that exhibit very large field concentration and why these are effective. Next, we will develop a model for how molecular photophysics is altered near metal nanoparticles. In particular, we lay the groundwork for an approach to determining how much observed photoluminescence enhancement can be attributed to emissive rate enhancement by the excited state and how much can be ascribed to increases in molecular absorption by the ground state. This distinction is central for applications to organic electronics since emissive rate enhancement is of no value for OPV applications since charge transfer quenching of the excited donor occurs very rapidly. Similarly, absorption enhancement is of no value for OLED applications since the excited state is generated electrically through electron and hole recombination. After showing how to differentiate between these contributions, we present a series of experiments that address a number of practical issues such as the optimal spacing of molecular chromophores from metallic surfaces to avoid charge transfer quenching, the effects of plasmon resonance on the molecular emission and absorption spectra, and an estimate of how large plasmon effects on emissive rates could be with optimal engineering. Finally, we present some early results of trying to adapt these ideas to improvement of OLED and OPV technologies and provide some perspective for the future.

## 19.3 GENERAL PICTURE OF ABSORPTION AND LUMINESCENCE ENHANCEMENT

A great deal of insight into the magnitude and origin of the plasmon enhancement of radiative rates can be derived from straightforward computation of the electromagnetic field in fixed geometries of metal nanoparticles. Two basic computational approaches are available, the finite difference time domain (FDTD) method (34-36) and the discrete dipole approximation (DDA) (37-39). For the purposes of qualitative understanding of the phenomenon, we present a simple DDA calculation of the intensity distribution near a silver nanoparticle dimer with particle sizes of 15 nm and separation 2 nm (Figure 19.1A). The DDA approach has the advantage of conceptual simplicity and a demonstrated ability to calculate scattering coefficients for arbitrarily shaped particles. This method is based on representing the particles by a periodic array of sites, usually cubic, lying within the surface of the particle. Each site is assigned a polarizability determined from the Clausius-Mossotti equation that relates applied fields and dielectric constants to material polarization. The pairwise interactions between all sites are described by a system of simultaneous linear equations that account for fields created by the incident oscillating field together with the oscillating field from all the other induced dipoles. For illustrative purposes, we present the electromagnetic field distribution at a wavelength near the plasmon resonance around the nanoparticle dimer relative to what it would be in free space. For simplicity, to model the structure of Figure 19.1A using the DDA, we have taken the nanoparticle pair to be 2-dimensional and created a mesh of uniform 1 nm squares neglecting the effects of sharp corners in fitting the circular geometry. The qualitative features of the field distribution and rough magnitude of enhancement are nonetheless approximately correct.

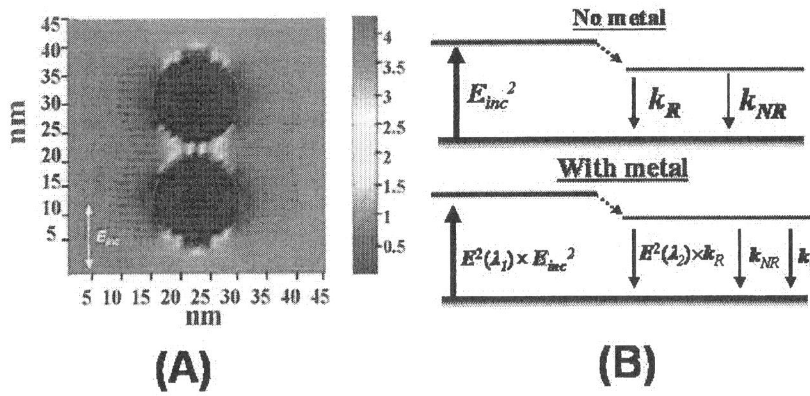


**Figure 19.1:** (A) 2D projection of the calculated local field intensity distribution around a pair of 15 nm diameter silver nanoparticles excited with $\lambda_1$ = 400 nm light polarized along the interparticle axis. The edge-to-edge particle separation is 2 nm and the free space incident light intensity $|E_{inc}|^2$ is taken to be unity. The local field intensity near the pair is shown in false color. The calculation was done using dipole-dipole approximation (DDA) method with each dipole unit being a square with sides of 0.2 nm. (B) Model of the photophysics of a molecule represented by a three level system and how the excitation and decay dynamics are affected by plasmon enhancement of radiative rates and the introduction of a rate for quenching $k_Q$ of the excited state due to proximity to the metal surface. $E^2(\lambda_1)$ and $E^2(\lambda_2)$ are the field enhancements at the position of the molecule for the excitation and emission wavelengths respectively. $k_R$ and $k_{NR}$ represent the radiative and non-radiative decay rates of the molecule in the absence of plasmon enhancement.

As a calculation of how much photoluminescence enhancement can be expected for a chromophore embedded in such a structure, this approach is too naïve since it fails to consider tradeoffs between enhancement at the excitation and emission wavelengths as well as non-radiative processes. Nevertheless, several salient features of the enhancement are evident that are useful in practical design of geometries for large radiative rate enhancement. In particular, note the concentration of the field in crevices between nanoparticles. Further computation on this simple model structure shows that the frequency of the plasmon resonance where the field is strongly enhanced depends sensitively on particle separation. It also suggests that separations somewhat less than particle radius and larger particle sizes are optimal. Of course, the details of whether there is a single chromophore in the center of the gap or whether there are many chromophores filling space are important considerations in what is "optimal". The fact that the chromophore is of molecular dimensions (a few nm at most) dictates that enhancements for single chromophores will benefit substantially less from larger nanoparticle size than would be the case for circumstances with many chromophores. The model also exhibits strong sensitivity of the enhancement to the polarization of the electromagnetic field with orientation along the interparticle axis being best. Obviously, there is an implicit correspondence with the preferred orientation of the chromophore transition dipoles relative to the nanoparticle assembly. More sophisticated multiple nanoparticle arrangements are considered in the work of the Stockman (40-41) and Shalaev groups (42) where the magnitudes of enhancement are so large that it is possible to rationalize the extreme enhancements characteristic of single molecule Raman scattering.

Understanding the field enhancement of radiative rates is insufficient to predict how molecular photophysical properties such as enhancement of fluorescence quantum yield will be affected by interactions of the molecule with plasmons. A more detailed model of the photophysics that accounts for non-radiative rates is necessary to deduce effects on photoluminescence (PL) yields. Such a model must include decay pathways present in the absence of metal nanoparticles as well as additional pathways such as charge transfer quenching that are associated with the introduction of the metal particles. Schematically, we depict the simplest conceivable model in Figure 19.1B. Note that both the contributions of radiative rate enhancement and the excited state quenching by proximity to the metal surface will depend on distance of the chromophore from the metal assembly. In most circumstances, one expects the optimal distance of the chromophores from the surface to be dictated by the competition between quenching when it is too close and reduction of enhancement when it is too far. The amount of PL will be increased both due to absorption enhancement and emissive rate enhancement. Hence, it is possible to increase PL substantially even for molecules with 100 % fluorescence yield in the absence of metal nanoparticles.

## 19.4 DEVELOPMENT OF NANOTEXTURED METAL ASSEMBLIES FOR STRONG PLASMONIC ENHANCEMENT OF PHOTOLUMINESCENCE

The principles in the electromagnetic field calculations of section 3 are useful in developing approaches to fabrication of nanotextured substrates exhibiting high plasmonic enhancement. Generally speaking, it is desirable to make many nanoscale crevices for chromophores to occupy and to tailor particle density according to the principle that having more strongly interacting particles will lead to redder plasmon resonance. We have found the Stockman nanolens (40-41) picture to be extremely useful in guiding our thinking. In that work, Li et.al. show that three carefully placed silver nanoparticles can produce sufficient photon localization to provide enhancements of greater than 3 orders of magnitude in electromagnetic field. The particles are each spherical and have widely different sizes and uneven separations. While we have chosen of necessity to optimize our nanotextured silver substrates empirically, thc idcas bchind that design correspond well to our conclusions about surface optimization.

There are many approaches to fabrication of silver-based plasmonic substrates and we have studied a representative set that produces widely varying morphologies. In particular, we performed a study comparing Raman enhancement on glass substrates treated by evaporation of silver, by self-assembly of colloidal silver nanoparticles on thiol-functionalized glass, by precipitation of colloidal silver onto glass and by silver particle formation from the Tollens ("silver mirror") reaction (43). Figure 19.2 presents a topological analysis using the atomic force microscope image of surfaces produced in these ways. Specifically, we plot the growth of the root mean square surface roughness versus selected area. In the limit of zero area (a single spot), the RMS roughness is zero and it increases with measured area until it converges to the long range surface average. The slope of the growth of the RMS roughness on a logarithmic plot reflects surface dimensionality. We consistently find that surfaces exhibiting two-dimensional behavior show poor Raman enhancement while surface that have fractional dimensionality have enhancement capable of supporting single molecule Raman scattering. These results are consistent with our investigations of photoluminescence enhancement as well. In that case, however, electronically inert spacers between the emitting molecule and the metal are necessary to prevent charge transfer quenching of the emissive excited state.

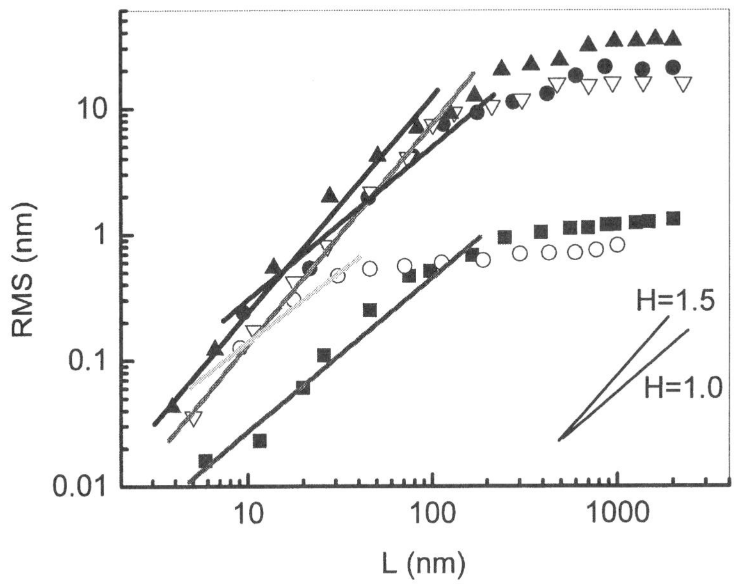


**Figure 19.2:** Self-similarity analysis for nanotextured silver surfaces prepared in different ways. The root mean square roughness inferred from atomic force microscopy is plotted versus measurement area. The various surfaces are 100 nm thick evaporated silver films (*solid squares, red line*); 5.2 nm thick evaporated silver films (*open circles, green line*); nanoparticle films assembled from colloid attachment to self-assembled monolayers (*solid circles, blue line*); films from deliberate precipitation of silver colloid (*solid up-triangles, black line*); Tollens reaction films (*open down-triangles, orange line*). Lines with slopes H = 1.0 and H = 1.5 representing two-dimensional and 1.5 dimensional surfaces respectively are

shown in the bottom right for reference. Reprinted from reference 43 with permission of *Proc. Natl. Acad. Sciences USA*.

One practical conclusion from our studies is that use of the Tollens reaction is a simple and practical way to create nanotextured substrates for plasmonic enhancement of radiative transitions in molecules. It produces much cleaner surfaces than colloidal precipitation, much higher enhancement than evaporated silver island films and surfaces with much smaller crevices than are achievable with electron beam lithography. Coverage is relatively easy to control. Unfortunately, wet processing methods, especially those utilizing water, are likely to be incompatible with many protocols for organic device fabrication and finding better approaches is an important challenge to address if plasmon-enhancement is to have impact in organic electronics.

At the time we performed our topological studies, we concluded that the fractal character of the surface might be important and, in fact, the sizes and separations in the Stockman nanolens calculation were taken to be self-similar. Further consideration has led us to believe, however, that the self-similarity is not intrinsically important but is highly correlated to properties we think are essential for optimum enhancement, namely size inhomogeneity and spatial disorder. In requiring both size distribution and spatial randomness, we avoid symmetric arrangements where field intensity is necessarily distributed. From a mechanistic point of view, size inhomogeneity allows large plasmonic particles with huge scattering cross-sections to "capture" the electromagnetic field and smaller particles nearby to "feel" incident fields scattered by the large particles that are enhanced by something on the order of the Q factor of the plasmon resonance (around 30 in the case of silver). The participation of relatively small particles allows very tight localization. Thus, confluences of as few as three or four particles can produce field enhancements of order $10^3$ corresponding to Raman enhancements as high as 12 orders of magnitude when considering intensity enhancement of the incident and scattered beams. Spatial disorder is also important since translationally invariant structures allow for propagation of energy and inhibit plasmonic focusing. This is akin to the physics of Anderson localization of charges in highly disordered semiconductors and is a phenomenon general to wave propagation in random media. A remarkable illustration of the need to prevent plasmon propagation is presented in Figure 19.3 where we have studied the silver coverage dependence of Raman scattering (44). As soon as the surfaces reach the percolation threshold and become conducting, the magnitude of the Raman enhancement drops dramatically.

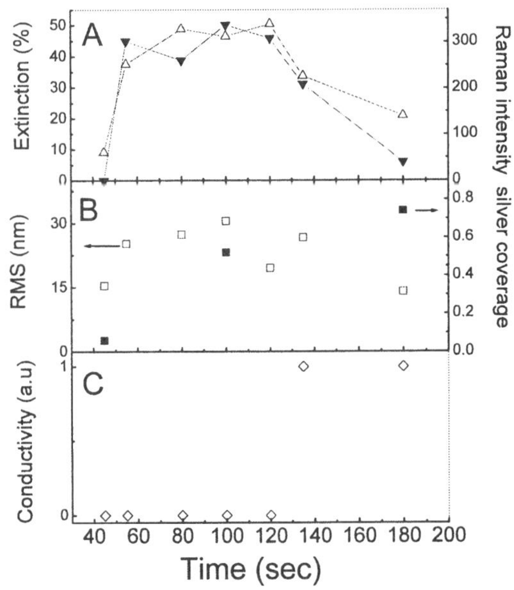


**Figure 19.3:** Correlation of Raman scattering intensity with optical, morphological and electrical properties of substrates fabricated by the Tollens reaction with different silver deposition times. (A) Raman intensity of the 1100 cm$^{-1}$ mode (*open triangles*) and extinction at the excitation wavelength of 633 nm (*filled triangles*) versus deposition time. (B) Surface RMS roughness from atomic force microscope images (*open squares*) and silver coverage from SEM images (*solid squares*) versus silver deposition time. (C) Steady state conductivity versus deposition time with 0 being insulating and 1 conducting. Reprinted with permission from reference 44.

## 19.5 MODELING OF PLASMONIC EFFECTS ON MOLECULAR PHOTOPHYSICS: SEPARATION OF ABSORPTION, EMISSIVE RATE ENHANCEMENT AND EXCITED STATE QUENCHING EFFECTS

As noted above, observations of large enhancements of the photoluminescence are insufficient to guarantee utility for application of plasmon-enhanced emission in OLEDs where the excited state is not photogenerated. In principle, increases in photoluminescence observed experimentally could be completely due to absorption enhancement. Even observation of reduced excited state lifetimes in conjunction with increased emission is insufficient to prove radiative rate enhancement since the lifetime reduction could be due to excited state quenching by the metallic surface and compensated by large absorption enhancements.

We have developed a quantitative procedure for approximately separating the various contributions to the photoluminescence enhancement based on a limited set of experimental data (45). The photophysics implicit in Figure 19.1B can be described in terms of 5 parameters, the local field enhancements at the absorption and emission wavelengths, $E^2(\lambda_1)$ and $E^2(\lambda_2)$, the excited state decay rates due to quenching through interactions with the metal $k_Q$, other nonradiative processes that relax the emissive state $k_{NR}$, and the radiative rate $k_R{}^0$ where the superscript denotes the absence of silver. These parameters can be measured or constrained with five experimental observations. In particular, one needs to measure the relative photoluminescence with silver, $PL/PL^0$, the excited state lifetimes with and without silver, $\tau$ and $\tau^0$, the luminescence quantum yield without silver $\Phi_F{}^0$, the extinction coefficient at the excitation wavelength with silver and the photoluminescence excitation spectrum with silver. The measurements $\Phi_F{}^0$ and $\tau^0$ specify the radiative and nonradiative rates in the absence of silver since $k_R{}^0 = \Phi_F{}^0\ \tau^0$ and $(k_{NR} + k^0{}_R)^{-1} = \tau^0$. Given that nonradiative decay rate $k_{NR}$ is unaffected by the presence of silver, the observed lifetime $\tau$ in the presence of silver must correspond to a limited set of possible emissive rate enhancements $E^2(\lambda_2)$ and quench rates $k_Q$ that could account for the observed reduction in lifetime. If we choose a particular $k_Q$, then the absorption enhancement $E^2(\lambda_1)$ is determined. We can set upper and lower limits on what the absorption enhancement can be through consideration of the extinction and excitation spectra. The maximum the absorption enhancement can be is the total extinction at the excitation wavelength in the presence of silver. The minimum possible absorption enhancement can be determined from the ratio of the photoluminescence excitation spectra with and without silver by assuming that the difference between these can only be accounted for by absorption enhancement. A more complete quantitative treatment can be found in reference (45). We apply the formalism to the case of enhanced phosphorescence of platinum octaethyl porphyrin (PtOEP) in the next section.

## 19.6 APPLICATION OF THE FORMALISM TO ENHANCED EMISSION BY AN OLED PHOSPHOR

For reasons clear from the introduction, enhancement of phosphorescence is a particularly attractive application of plasmonics to OLED technology. Since carriers are injected into an OLED from separate contracts, their spins are uncorrelated and spin statistics dictate preferential formation of triplet excited states. Since these are generally poor emitters at ambient temperature, metallic enhancement of the phosphorescent rate would be desirable. Moreover, triplet states are typically long-lived and prone to oxidation reactions so that reduction of the triplet lifetime could potentially improve stability of the phosphors. PtOEP is a model phosphor and its application to electroluminescence was pioneered by the groups of Forrest and Thomson (46-47). We have investigated plasmonic enhancement of the PtOEP phosphorescence on silver surfaces prepared using the Tollens reaction. Dilute PtOEP in a polymer binder was spin cast onto substrates with various densities of nanotextured silver and assumed to deposit conformally, the spin speed being used to control the approximate thickness of the overlayer.

The extinction spectra of the substrates with various silver densities are illustrated in Figure 19.4A. The bottom spectrum ("1") is for the glass slide in the absence of silver while the spectra labeled "2"-"5" are for successively increasing amounts of silver controlled by deposition time using the Tollens reaction. Note the evolution of the spectrum from one dominated by the isolated silver nanoparticle plasmon resonance at around 400 nm to that of a broadly distributed set of plasmon resonances due to nanoparticle interactions to that of a highly reflective metallic surface. The corresponding spectra of the sharp PtOEP phosphorescence at 650 nm as a function of excitation wavelength are presented in Figure 19.4B for films of PtOEP dilutely doped into polystyrene on the silver that are approximately 6 nm thick, this being the thickness found to exhibit optimum luminescence enhancement (45). Remarkably, phosphorescence increases by more than a factor of 200 relative to those without silver are observed at an excitation wavelength of 535 nm. Consistent with the observations of the enhancement of Raman scattering with silver coverage in Figure 19.3, a dramatic reduction in enhancement is observed as the silver becomes opaque and electrically continuous.

Figure 19.4C presents data on the phosphorescence decay dynamics on two of the samples in Figure 19.4B, the one without silver and the one having optimum enhancement. These exhibit the characteristic signature of plasmon-enhanced emission, namely that increases in luminescence are accompanied by *decreases* in excited state lifetime. Ordinarily, radiative rates are fixed by quantum mechanical matrix elements and variation in excited state lifetime is due to changes in nonradiative rates so that increases in lifetime correspond to lower non-radiative rates and increases in luminescence yield. Here, the lifetime is reduced by about a factor of 3 due to increased emissive rate even as the luminescence increases 215 times. These large enhancements cannot be accounted for simply by increase in phosphorescence yield since the yield is greater than 10 % in the absence of silver. It is evident that a substantial fraction of the increase must be accounted for by absorption enhancement.

Further analysis of our data enables us to apply the formalism of the previous section to make quantitative inferences about the contributions of absorption enhancement, increases in the fraction of excited states that emit due to emissive rate enhancement and quenching introduced by the proximity of PtOEP to metallic surfaces. The result of this analysis is summarized in Figure 19.5 and further details are supplied in reference 45. The lines generated represent the set of possible decompositions into absorption enhancement $E^2(\lambda_1)$ and emissive rate enhancement $E^2(\lambda_2)$ where the precise combination is specified by a choice of parameter $f_Q$ (= $k_Q\tau$) which is the fraction of excited states that are quenched by interactions with the metal. The shaded bar on the left represents the upper and lower limits for the absorption enhancement as inferred from extinction and excitation spectra respectively. These constraints imply that the actual fraction of PtOEP excited states quenched through interaction with the silver is at the low end of its range. This conclusion is confirmed by comparison of the possible set of solutions for 6 nm films with that for 2 nm films that exhibit less enhancement even though it seems likely that they experience at least as large absorption enhancement. A likely approximate decomposition to explain the ~ 200-fold increase in phosphorescence is about 40 times absorption enhancement and about 10 times emissive rate enhancement. The latter value corresponds to increases in true quantum yield for phosphorescence in our films, the fraction of PtOEP triplets created that emit, by about a factor of 5 to around 50%.

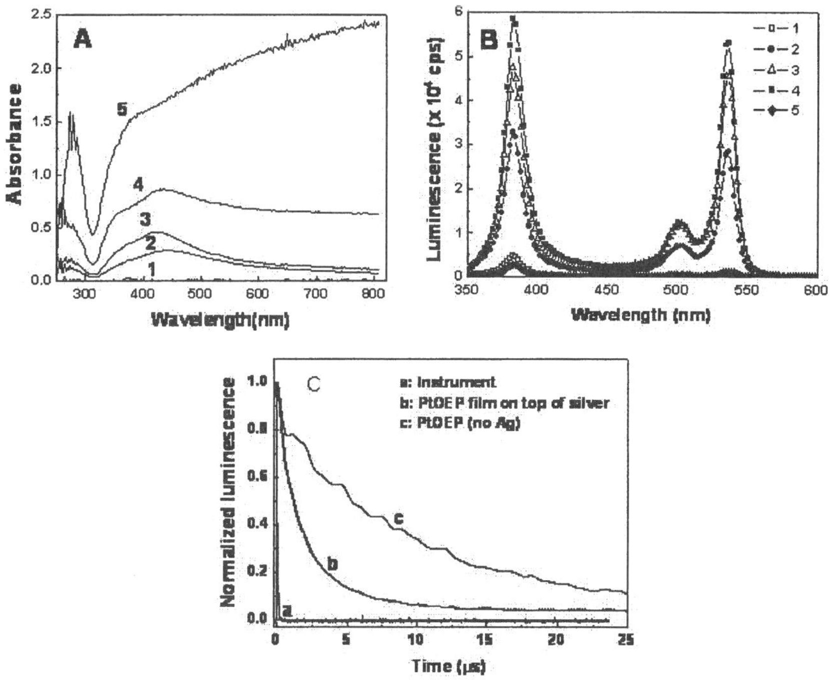


**Figure 19.4:** (A) Extinction spectra of silver nanoparticle films used for PtOEP emission enhancement. (B) Corresponding excitation spectra monitored at 650 mm of 6 nm films of PtOEP in a polystyrene binder spin cast onto the silver films. (C) Excited state decay dynamics of the PtOEP phosphorescence for 6 nm films excited by 5 ns pulses at 532 nm with no silver (c) and on substrates like number 4 from A with silver coverage to optimize enhancement (b). The instrument resolution when detecting scattering of the excitation pulse (a) is shown for reference. Reprinted from reference 45 with permission of the American Chemical Society.

It is interesting to note that the phosphorescence enhancement observed in trace 4 of Figure 19.4B is 7 times as great for excitation at 535 nm than for 380 nm excitation in spite of the silver extinction at these wavelengths being roughly equal. This is consistent with the idea the plasmon resonance in the red associated with multiple particles is much more effective than single particle plasmon resonance (~ 400 nm) in concentrating the electromagnetic fields. This observation is confirmed throughout our work and is consonant with our discussion of figure 19.1A and the importance of crevices in plasmonic enhancement.

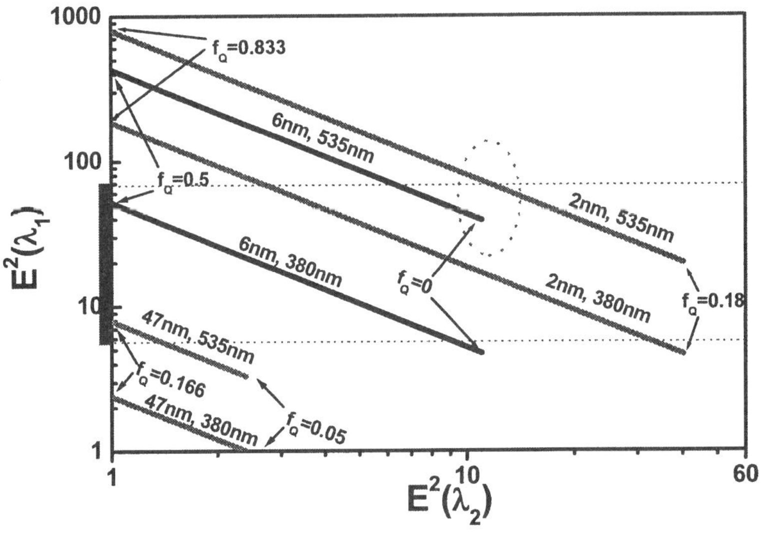


**Figure 19.5:** Schematic diagram showing decomposition of total phosphorescence enhancement of PtOEP on silver films into absorption enhancement $E^2(\lambda_1)$ and emissive rate enhancement $E^2(\lambda_2)$ based on the photophysical model described in the text and data from steady state and transient spectroscopy of PtOEP films with various thicknesses and excitation wavelengths as labeled. The lines represent the possible combinations that could explain the experimentally observed changes in photoluminescence where each position on the line represents a different choice of $f_Q$, the fraction of the excited states that are quenched nonradiatively by interactions between the molecule and the metallic surface. The blue shaded region on the vertical axis is the range of possibilities allowed by constraints from extinction and excitation spectra as explained in the text. The dotted oval is what we believe to be the most likely decomposition for the 6 nm films characterized in Figure 19.4 as discussed in the text. Reprinted from reference 45 with permission of the American Chemical Society.

The model of the photophysics we have advocated does not take into account spin-orbit coupling effects associated with silver as a heavy atom that might affect phosphorescent and nonradiative decay rates for the triplet state. The theoretical justification for this is that heavy atom effects are quite short range since they require wavefunction overlap. Effects of the silver are in any case likely to be much smaller than those of the Pt atom embedded in the porphyrin. Experimentally, we can rule out the importance of these effects since we do not observe phosphorescence enhancement on top of uniform evaporated silver films nor on films that become essentially continuous as for the thickest films in Figure 19.4.

The fact that the maximum luminescence is observed for 6 nm thick films suggests that the optimal spacing of a PtOEP molecule from the silver is on the order of 3 nm. One clear consequence of our photophysical model is that the ideal spacing of a chromophore from the silver will be different for molecules with higher or lower emissive quantum yield. For low quantum yield species, reducing distance from silver would increase the fraction of states that decay radiatively in the absence of quenching until $k_R = E^2(\lambda_2)\, k_R^0 >> k_{NR}$. Though quenching would naturally increase closer to the silver, this would be offset by increased emissive rates. The same reasoning would mean that ideal spacings will be larger for higher quantum yield emitters. Likewise, the theory predicts that the ideal spacing will depend on the excited state lifetime in the absence of silver since short-lived species will tend to be less affected by quenching processes. The benefits of proximity to the silver will therefore be more pronounced for molecules with already short excited state lifetimes and shorter spacers will be helpful in improving emission.

## 19.7 SYSTEMATIC STUDY OF DISTANCE DEPENDENCE OF ENHANCEMENT AND INCREASED COVERAGE ARTIFACTS ON NANOTEXTURED SURFACES

The study of distance dependence of photoluminescence enhancement in the previous section is very crude in that the experimental geometry is poorly defined and the PtOEP is distributed throughout the overlayer. Moreover, no attempt was made to account for changes in the total number of PtOEP molecules introduced by the increase in surface area on the silver surfaces relative to flat glass substrates. In the present section, we will report a more somewhat more controlled study of these effects. Having said that, however, the geometry of the randomly functionalized substrates we use and the placement of the chromophores is always poorly defined. From a practical point of view, however, the type of study reported here tells us a lot about how to optimize the thickness of inert spacers to maximize luminescence enhancement. This is of particular interest to us in the context of plasmon-enhanced OLEDs. Since the emissive region in a bilayer OLED is localized to the interface between electron and hole transport layers, knowledge of the spacer requirements tells us where we want to try to embed metal nanoparticles in order to most effectively increase emissive efficiency in OLEDs.

For this study, we chose to take advantage of layer-by-layer electrostatic assembly of polyelectrolytes as inert spacers where bilayers of sulfonated polystyrene (PSS) and polyallyl amine (PAH) can serve as conformal spacers with nm resolution. As illustrated in Figure 19.6, it is possible to effectively integrate these spacers with silver deposited from colloidal solution since the negatively charge coatings on the colloid effectively adhere to the PAH. We used polymeric chromophores based on water-soluble poly-paraphenylenevinylenes since we found that the self-assembled layers had enough porosity that small chromophores such as fluorescein could not be precisely placed in the structure. Comparing the two structures in Figure 19.6 also enables us in principle to get some handle on increases in total coverage of chromophores due to substrate texture introduced by the nanoscale metal particles. Further details of the fabrication are provided in reference (48).

Figure 19.7A reports the fluorescent enhancement of the conjugated polymer emitter for structure I when it is deposited on top of various thicknesses of polyelectrolyte spacers for silver treated versus untreated substrates. The fluorescence increases nearly 40 fold for 2 bilayers (roughly 4-6 nm) and then drops though it remains dramatically enhanced even for as much as 13 bilayers (25 nm).

This behavior illustrates the balance between quenching when the chromophores are too close to the silver and reduced enhancement as the distance from the silver nanoparticles is increased. The fact that the enhancement does not fall rapidly after a few tens of nm probably reflects the fact that the spacings associated with the large number of bilayers are not meaningful since the surface roughness is several tens of nm and chromophores can fall in between silver particles. A more careful analysis of the surface topology allows us to estimate the increase in surface area and hence number of chromophores on nanotextured substrates (48). We find this to bc about a factor of 2 and believe that the actual enhancement per chromophore is only about half of what is reported in Figure 19.7A. Figure 19.7B shows how the emission spectrum is changed by the plasmon resonance. The spectra on silver are pulled to the blue, towards the dominant features of the plasmon resonance. We present a more detailed study of this phenomenon in the next section.

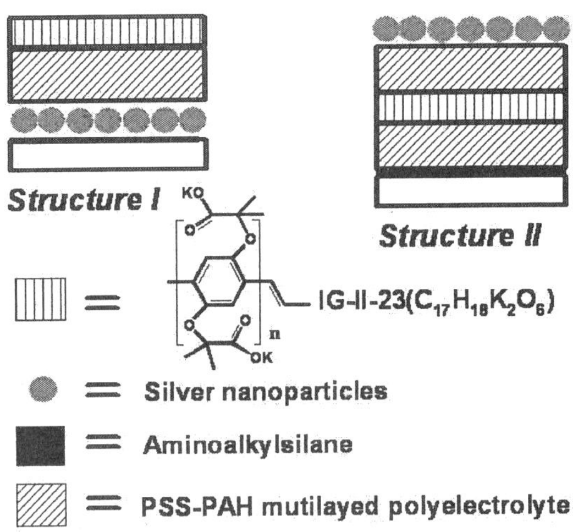


**Figure 19.6:** Experimental geometries for distance dependent control of fluorescence enhancement on top of silver nanoparticles (structure I) and underneath silver nanoparticles (structure II), respectively. Reprinted from reference 48 with permission of the SPIE.

The data for the same fluorophores embedded in structure II are illustrated in Figure 19.7C. We observe enhancement but very little dependence on spacer thickness. Since the silver layer has thickness of the same order as the largest spacer, the variation of average chromophore distance from the silver is relatively mild. The small enhancement is also easy to understand since the chromophores will not be placed in the interstices between silver particles when films are assembled with silver on top.

We have also studied how the fluorescence in structure II varies depending on the silver coverage with fixed spacer thickness (data not shown). As observed in sections 3 and 6, we find an optimal thickness (O.D. ~ 0.25) beyond which the enhancement is reduced. It is interesting to note that the optical density associated with maximum enhancement is smaller in the experiments of Figure 19.7 than those in Figure 19.4. We believe that this is due to the fact that the surfaces are produced by different deposition methods. Utilizing deposition from a uniform silver colloid as in Figure 19.7 is less effective in creating plasmon enhancement and peaks at lower extinction than surfaces treated with the Tollens reaction. In part, this is because surfaces with the same optical density at ~ 400 nm show much more extinction (plasmon resonance) in the red using the Tollens reaction and that is better matched to where these chromophores emit. We believe this also reflects relatively better size homogeneity and spatial order in the nanoparticle assemblies made from the colloid. As argued in section 4, we think that spatially disordered surfaces and inhomogenous size distributions allow for better electromagnetic field localization.

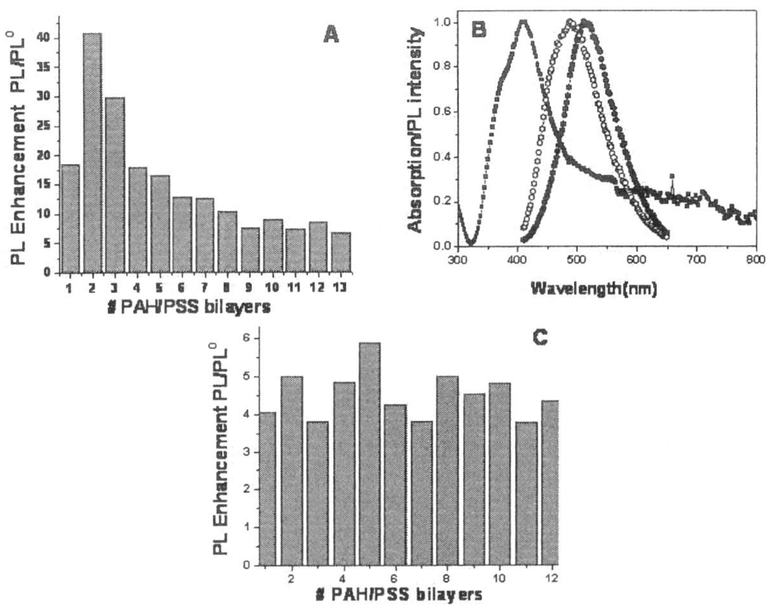


**Figure 19.7:** Fluorescence enhancement and spectral changes for conjugated polyelectrolyte emitters in structures like those depicted in Figure 19.6. (A) Enhancement of photoluminescence on top of nanotextured silver films and PAH/PSS polyelectrolyte bilayer spacers (structure I). (B) Normalized spectra of the silver extinction (solid squares) and emission spectra with no silver (solid circles) and with silver (open circles) using structure I and 13 bilayers. (C) Enhancement of photoluminescence for emitters in structure II. Reprinted from reference 48 with permission of the SPIE.

One common misunderstanding about the underlying physics of the plasmon-enhanced fluorescence is made clear by the distance dependence data. Researchers often cite Förster transfer from the molecular chromophore to the electron plasma in the metal as a quenching mechanism for the excited state. If this mechanism were operative, it would be impossible to see fluorescence enhancement at distances from the metal much shorter than the Förster radius (> 10 nm if we take the plasma as simply an acceptor) but, in fact, Figures 19.4 and 19.7 make it quite clear that this is what is observed. In our view, this reasoning is flawed since it is impossible to think about the dipoles associated with the plasmon and the molecule independently when they are strongly coupled.

## 19.8 EFFECTS OF PLASMON-ENHANCEMENT ON MOLECULAR SPECTROSCOPY

We pointed out in Figure 19.7B that the molecular emission can appear to be "pulled" by the plasmon resonance since different parts of the molecular spectrum, regardless of whether it is homogeneously or inhomogeneously broadened, will experience different enhancement. In this section, we will look at a case study of rhodamine red monolayers covalently attached to glass with silver deposited directly on top (49). Once again, we assemble silver nanoparticle assemblies on top of the immobilized dye from the colloidal suspension as in structure II of section 7, the purpose being to rule out PL changes due to changes in the fluorophore coverage. Figure 19.8 presents extinction spectra and emission spectra of the dye monolayers with varying amounts of silver assembled on top. The absorption of the dye is evident in the extinction spectra and we can use the literature extinction coefficient of Rhodamine Red™ ($\varepsilon \sim 129{,}000\ cm^{-1}M^{-1}$) with the observed optical density of about 0.01 at 580 nm to estimate surface density of dye molecules at around $10^{13}$ per $cm^2$, nearly a dense monolayer.

As silver density increases from sample b to e, we observe concomitant increases in the extinction due to the surface plasmon resonance characteristic of single isolated nanoparticles at wavelengths around 400 nm. A broad peak around 700 nm appears with increasing silver density that can be ascribed to the coupling between the localized plasmon modes of two or more particles that are closely spaced in the direction of the excitation polarization. Nearly sixfold increases in average fluorescent enhancement were obtained for excitation at 450 nm when the silver nanoparticle optical density at 400 nm was around 0.3 corresponding to surface coverages of silver nanoparticles around 50% as judged from scanning electron micrographs. Assuming that fluorescence from dye molecules directly contacted by the metallic surface is totally quenched, the measured emission underestimates the enhancement for the molecules whose emission is not completely quenched. Once again, at sufficiently high silver coverage, the total fluorescence intensity decreases (cf. sample e in Figure 8B) regardless of excitation or collection direction.

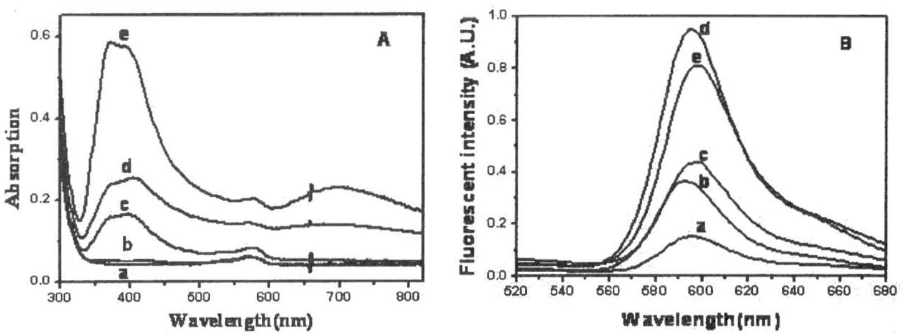


**Figure 19.8:** Photoluminescence enhancement of rhodamine monolayers covalently attached to glass with various amounts of nanotextured silver on top. (A) Extinction spectra of substrates with no silver (a) and increasing amounts of silver from (b) to (e). (B) Resulting photoluminescence from these subtrates for excitation at 450 nm. Reprinted from reference 49 with permission of the American Chemical Society.

Of particular interest is the behavior of the fluorescence spectrum with silver coverage. At low density of silver, the emission spectrum is pulled to the blue as in Figure 19.7B, consistent with coupling of the molecules to the isolated nanoparticle plasmon resonance that peaks near 400 nm. If scattering or reabsorption by the silver were at play, those would tend to produce a red shift since the extinction of the silver rises monotonically to the blue. With additional silver deposition, the dye's emission spectrum shifts back to the red and, in fact, sample e has greater emission than sample d for colors to the red of 640 nm. This is because the areas of the silver producing red plasmon resonance are aggregates that support much stronger field localization and therefore better enhancement of the red luminescence. In the case of sample e, effects of scattering or reabsorption of the emitted light by the silver would tend to blue shift the spectrum. The fact that this does not happen implicates interaction of the molecule with the local plasmon resonance as the root cause of the spectral shift. The upshot of these data is that there are definite effects of plasmon resonance on molecular spectra but these are difficult to predict from the extinction spectra of the metal particles. There is good reason to believe, however, that if we could engineer substrates to have small distances between metal nanoparticles or islands and could spectrally concentrate plasmon resonance, that both photoluminescence enhancements and spectral changes could be made much larger.

## 19.8 LIMITS TO PLASMON ENHANCEMENT OF PHOTOLUMINESCENCE AND OLEDS

It would be both interesting and useful to know what might be the largest enhancement of photoluminescence that one could observe if we could achieve optimal engineering of the metal nanoparticle morphology for a given luminophore.

Once again, the answer to this question would clearly depend on the whether we are discussing single molecules at the "hottest" spot of electromagnetic enhancement or an entire film of molecules. Moreover, it would depend on the quantum yield for luminescence in the absence of nanotextured metal since much larger enhancement is possible for systems with very low fluorescence. As a practical matter for OLED technology, we are primarily interested in reasonably good fluorophores with tens of percent luminescence yield in a relatively dense confined layer such as the rhodamine monolayer studied in the previous section. This problem is theoretically intractable so we resort to an empirical study based on the following idea. We assume that each resolution unit of area ~ $\lambda^2$ will exhibit an enhancement peculiar to the local silver nanoparticle morphology. If we can measure the distribution of enhancements by sampling many such small areas using confocal fluorescence microscopy, we can get an idea of what would be the maximum enhancement for such a layer. Such a study (49) leads us to conclude that a factor of at least 50 enhancement is possible in "ideal" spots on the rhodamine monolayer whose fluorescence quantum yield is about 12 %. When we account for absorption enhancement, we estimate true quantum yield increases (i.e. the fraction of singlets formed that emit) of over a factor of 4 to well over 50 %. Presumably, if we could implement this in an OLED, this is the improvement that could be realized.

## 19.10 APPLICATION OF PLASMON ENHANCEMENT OF EMISSIVE RATES TO ORGANIC ELECTROLUMINESCENT DEVICES

In order to study influence of silver nanostructures on electroluminescence efficiency, we selected a model OLED structure composed of an electron transport layer of tris (8-hydroxyquinoline aluminum) ($Alq_3$) and a hole transport layer of N, N'-diphenyl-N, N'-bis (3-methylphenyl) - [1, 1'-biphenyl] -4, 4'- diamine (TPD). There are a number of reasons to choose this common device. First, the majority of successful OLED systems are based on $Alq_3$ and devices based on $Alq_3$ are easy to make and robust. Second, we thought that, even though plasmon enhancement using evaporated is not optimal, that evaporation would be the most effective way to place nanotextured silver in a precise location without damaging the organics. Third, the quantum yield of $Alq_3$ is modest (~ 25%) so that modifying the radiative decay rate of $Alq_3$ using field enhancement should noticeably improve the device efficiency. Finally, we have shown that we can make the quantum yield for $Alq_3$ fluorescence close to 100 % in studies of surface enhanced photoluminescence (50). Figure 19.9 (bottom) shows the fluorescence spectra of 10 nm $Alq_3$ overlayers on surfaces created by assembling various densities of 20 nm diameter silver nanoparticles onto quartz substrates. The $Alq_3$ overlayers were thermally evaporated onto these substrates and have the same thickness for all the samples. Comparing samples having silver with the control, we observe dramatic enhancement of fluorescence while increasing the silver optical density. Maximum fluorescence at 520 nm was achieved when there are silver nanoparticles with an optical density of around 0.3 at 400 nm underneath the 10 nm emissive layer. Using the approach of sections 5 and 6 to decompose the observed enhancement into absorptive and emissive components, we infer nearly 100 % quantum yield for emission from the best of the samples in

Figure 19.9 so that four-fold increases in electroluminescence would be possible if we could nondestructively embed analogous silver morphologies into an OLED. The spectra of the $Alq_3$ emission in Figure 19.9 indicate that plasmon resonance pulls them to the blue, quite different from what we observe for the rhodamine monolayer (cf. section 8). This might be due to the green emission of $Alq_3$ which is closer to the wavelength for isolated silver nanoparticle plasmon resonance. There may also be differences due to the relative thickness of the $Alq_3$ layer or the fact that the silver resides underneath rather than above the $Alq_3$.

Figure 19.10 shows a schematic of the TPD/$Alq_3$ light emitting diodes we fabricated in order to study the effects of silver surface plasmons on the electroluminescence. The silver layer is thermally evaporated onto a 55 nm TPD hole transport layer and capped with an additional 5 nm of TPD to prevent direct charge transfer quenching of $Alq_3$ excited states. A set of current voltage characteristics and the corresponding electroluminescence is shown in Figures 19.10 A and B, respectively. Unfortunately, our results show that the electroluminescent efficiency of such devices was significantly decreased with addition of silver. Although the silver doped devices still emitted light, the diode turn-on thresholds increased dramatically. We attribute the decreased electroluminescence and higher turn-on voltages to deep trapping of charges by the silver particles. We have tried strategies to reduce trapping as will be discussed below but these have not yet succeeded. It may be that simply treating the ITO electrode with nanotextured silver results in some device enhancement (51), but it is nontrivial to prove that the origin of the improvements are plasmonic in nature since modification of contacts will also change injection barriers and charge balance.

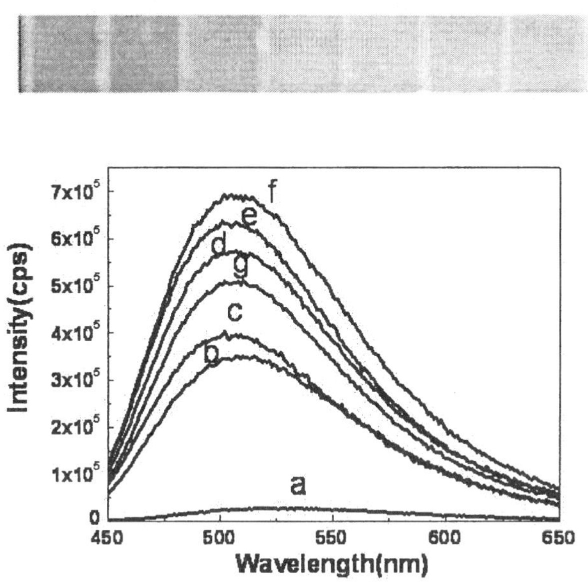


**Figure 19.9:** Top. Fluorescence images of thin $Alq_3$ films evaporated on top of silver nanoparticles with increasing density increased from left to right. Bottom.

Corresponding photoluminescence spectra of the segments of the same sample (a-g from left to right) excited at 400 nm.

We present more details on the charge trapping behavior of silver doped devices both to provide a flavor for the challenges to be overcome in embedding metal nanoparticles into OLEDs without degrading the electrical properties and because there may be practical applications in the form of organic flash memory devices. The silver doped TPD/$Alq_3$ devices show interesting hysteresis behaviors in their current voltage characteristics as shown Figure 19.11A. There have been a number of previous studies on two terminal memory devices using organic semiconductors and good progress has been made towards nonvolatile storage based on inexpensive organic materials doped with metal nanoparticles in the last several years as demonstrated by the groups of Yang (52-54) and Scott (55). The devices presented here may have some advantage in their bipolar construction since the memory can be erased by electron injection as well as hole extraction.

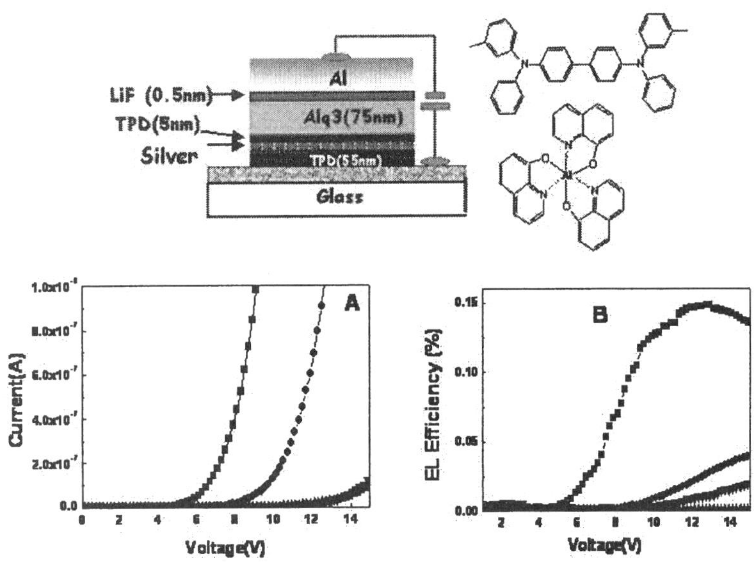


**Figure 19.10:** $Alq_3$/TPD based organic light emitting diode structure used to investigate effects of an evaporated silver layer embedded 5 nm from the emissive interface inside the hole transport layer TPD. The molecular structures of TPD and $Alq_3$ are illustrated in the upper right hand corner of the Figure. Panels A and B show I-V curves and corresponding electroluminescence efficiencies for various amounts of silver incorporated into the devices. Squares, circles, down triangles and up triangles are for 0 nm, 1 nm, 5 nm and 10 nm silver layer thicknesses respectively.

The charge traps result in a strong dependence of the current on silver density and on the direction and sweep rate of the voltage. Electron storage inside the silver nanoparticles and possibly hole storage in TPD can account for the memory behavior. In our efforts to make plasmon-enhanced OLEDs, we have observed analogous hysteresis when metal nanoparticles are doped inside either the hole transport layer TPD, inside the electron transport layer $Alq_3$, or placed at the interface between them. The resistance of the silver doped TPD/$Alq_3$ diode can be switched easily by applying a positive voltage and then switched back to the initial value by applying a negative voltage. A small probe voltage as low as 5 V can be applied to read the conductivity of the device without destroying the carrier status inside the device. Figure 19.11B shows the voltage pulses sequences we applied to the devices to test the memory quality of a TPD/$Alq_3$ diode doped with 10 nm average thickness silver layer. Pulses of 15 V amplitude were used to write or erase the storage, and 5 V was applied before and after switch. Figure 19.11C shows the corresponding current trace of a device without silver doping where there is no difference in current level at 5 V before and after operating the device at 15 V. With silver incorporation, two well-resolved current levels are obtained when the device is "read" at 5 V before and after erasure as shown in Figure 19.11D. We have applied thousands of pulses to test and durability of the memory behavior of the device and found no significant changes in memory behavior as shown in Figure 19.11E. Silver doped bilayer structures composed of conjugated polymers/$C_{60}$ or pentacene/$C_{60}$ incorporated with silver nanoparticles are found to have similar memory characteristics, an observation relevant to the discussion of organic photovoltaics in the next section.

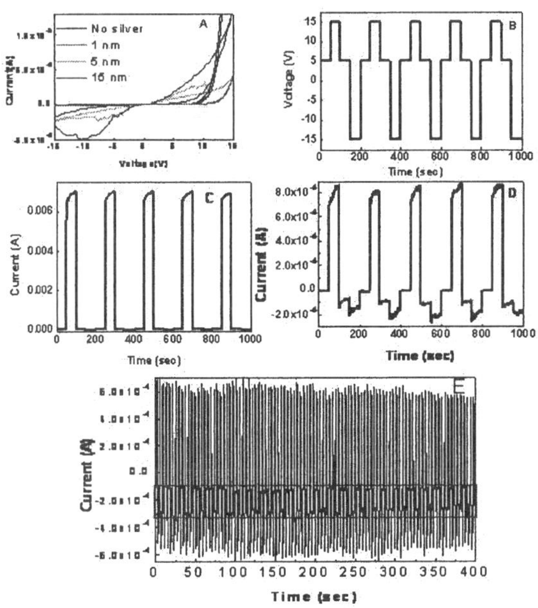

**Figure 19.11:** (A) I-V characteristics of silver doped organic light emitting diodes illustrating hysteresis. The structures are identical to those shown in Figure 19.10 except that the silver layer is embedded 10 nm inside the TPD. The amounts of silver are shown in the Figure legend. (B) Voltage applied to the 15 nm silver doped

OLED. The 15 V forward bias "writes" the memory state by introducing trapped charges onto the silver while the –15 V "erases" the device. The 5 V intermediate level is used to "read" the state of the device. (C) For the device without silver, little current is observed during the read period and there is no difference whether the write or erase pulse precedes the read. (D) With 15 nm silver, the current level after writes is significantly different than after erases. (E) is obtained under same conditions as (D) but illustrating the ability to write and erase reproducibly for hundreds cycles. The red rectangle highlights the area where there are two states probed by the low voltage with appreciable difference in their current level.

While we have not yet achieved metal particle enhanced EL in OLEDs due to charge trapping by the metallic nanoparticles, we remain cautiously optimistic about several strategies to remediate the problem. Silver particles capped by suitable electrically isolating organic/inorganic cladding layers could be use to prevent electron trapping if we can find a suitable fabrication approach to introduce them. We also think there is promise in plasmonic enhancement of other forms of organic light emitting devices whose mechanism does not depend on electron and hole transport such as light emitting electrochemical cells (LECs). In LECs, different redox states of mobile ions are formed under applied voltage and the redox species can diffuse toward each other and react to form an emissive excited state. Our preliminary attempts in this direction are encouraging and show that use of silver cathode can improve the light intensity and shorten the response time of Ruthenium (II) compound based cells relative to aluminum electrodes, but this may be due to the stability and the higher work function of the silver electrode. More detailed research has to be conducted to determine whether the brightness enhancement is due to the coupling between the dipoles of the dye molecules and the surface plasmons. Third, fabrication of nanoscale light emitting diodes could be done so that silver nanoparticles can be arranged peripheral to the carrier flow channel to avoid trapping but configured so that the emitters can nevertheless "feel" local field enhancement by silver nanoparticles.

## 19.11 PLASMON-ENHANCED ORGANIC PHOTOVOLTAICS

One of the most promising potential applications of organic electronics is fabrication of inexpensive solar cells using easily processed organic materials to make photovoltaics. When configured with donor and acceptor layers, devices with structures very similar to those of the OLEDs in Figure 19.10 can be used to absorb incident radiation and generate separated charges whose potential energy can be stored. The enhancement of photoluminescence observed in the experiments of sections 4-10 in the presence of silver nanoparticles was demonstrated to contain contributions from both increased radiative decay rate and improved light absorption by the fluorophores. Since the excited state dissociation leading to charge separation occurs on subpicosecond time scales at donor-acceptor interfaces in organic photovoltaic cells, enhanced radiative decay rates of the excited state do not significantly reduce charge photogeneration. However, the improvement of the light absorption cross-section can be used to good effect due to design constraints on organic photovoltaics. In particular, the poor charge mobilities in organic materials require that very thin films or very high voltages be used to effectively sweep out photogenerated charges. For reasons of power efficiency, voltages high compared to the photon energy are unacceptable and this limits practical organic film thicknesses to ~ 100 nm, thicknesses where absorbing all of the incident radiation is problematic. For bilayer device architectures, relatively short collection lengths for photogenerated excited states can limit the effective length even further since excitation must be near (within ~ 20 nm of) the donor acceptor junction. Therefore, incorporation of metallic nanostructures could be used to enhance absorption and hence efficiency in organic solar cells. Forrest and coworkers (56-58) have recently demonstrated the ability to exploit surface plasmon resonance of an ultrathin silver film in a tandem cell. Kamata et al (59-60) reported photocurrent amplification by metal nanoparticles incorporated into single heterojunction organic solar cells and dye-sensitized solar cells. Metallic nanoparticles (gold and silver) have also been investigated as novel sources for enhancing solar energy conversion by sensitizing the photoelectrochemical activity of a $TiO_2$ film (61-62).

We have recently demonstrated that silver interlayers as thin as 2 nm incorporated into a stacked heterojunction pentacene/$C_{60}$ cells improved both their photocurrent and open circuit voltage (63). A schematic of the solar cell we studied is shown in Figure 19.12A. The improved open circuit voltage is due to the formation of two separated pentacene/$C_{60}$ cells in series where the silver islands can act as carrier recombination centers. However, the photocurrent response is also increased by a higher efficiency of light absorption caused by surface plasmon resonance of the nanotextured silver interlayer. The evidence for that assertion is shown in Figure 19.12B where we compare the wavelength response of photovoltaics with and without silver incorporation. This allows us to approximately separate improvements due to the silver acting as a recombination center and improvements due to silver acting as a plasmonic antenna. Appreciable power efficiency enhancement is observed near the plasmon resonance of isolated silver nanoparticles (~ 400 nm) and we attribute this to absorption enhancement in the pentacene layer. Relative to the cell operating without silver, the power efficiency near the plasmon resonance is enhanced 5 times the amount that it is enhanced at wavelengths far from the plasmon resonance. Because 400 nm light is poorly represented in the solar spectrum, enhancement ar 400 nm does little to improve the device in Figure 19.12 from the point of view of solar collection efficiency. However, we anticipate substantial improvements will be achieved by optimizing the geometry and size of the silver or gold nanostructures in tandem photovoltaics since much greater effects on overall solar cell efficiencies will be obtained for enhancement of red absorption where solar energy is concentrated and organics tend to be poor absorbers. The use of silver in non-tandem geometries such as bulk heterojunction devices will be hindered by reasons similar to those we found problematic in OLEDs. Again, strategies such as surface protection of silver clusters to preserve strong absorption enhancement without disrupting charge transport are under investigation.

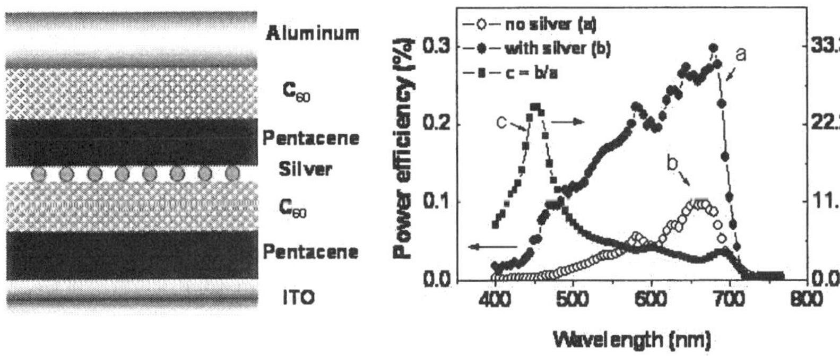


**Figure 19.12:** Left. Structure of the tandem solar cell based on pentacene and $C_{60}$ with a thin, discontinuous evaporated silver layer incorporated for efficiency improvement. Right. Power efficiency versus excitation wavelength for tandem solar cells with 2 nm silver middle layer (solid circles) and without the silver layer (open circles). The right hand scale is for the ratio of efficiency of the two cells to illustrate the significant difference in dependence on excitation wavelength. Reprinted from reference 63 with permission of the SPIE.

## 19.12 CONCLUSIONS AND PROSPECTS

Plasmon enhancement of the radiative properties of molecules near nanotextured metal raises exciting possibilities to improve organic optoelectronic devices such as light-emitting diodes and solar cells. We have documented substantial increases in photoluminescence near nanotextured metal surfaces and evaluated various ways of controlling surface morphology to maximize these effects. Of paramount importance to assess the utility of radiative rate enhancements for organic devices is a better understanding of the relative contributions of absorption and emissive rate enhancement in the observed metal-enhanced fluorescence. Specifically, we expect emissive rate enhancement to be useful in improving OLED technology and absorption enhancement in improving solar cell technology. To that end, we have developed a simple photophysical model that enables us to separate these contributions experimentally and applied it to several systems.

The ability to improve fluorescence quantum yield by competing with non-radiative processes could enable the use of more stable OLED materials with poorer intrinsic fluorescence efficiency. In the best case, it could enable the use of a wider spectrum of phosphors so that one can take advantage of the bias towards triplet formation when charges with uncorrelated spins recombine. Another benefit of increased radiative rate might be achievable even with very efficient emitters in that reduction of excited state lifetimes could improve material stability in a fashion analogous to photobleaching suppression observed in the presence of silver films (27). While we have been able to observe large radiative rate enhancements routinely across a wide variety of luminophores, our progress in translating these effects to OLEDs has been hampered to date by charge trapping effects that emerge when metal nanoparticles are incorporated into OLED structures. There may be strategies involving chemical protection of silver or changes in device architecture that allow for further progress.

The ability to increase absorption has potential to remedy one of the problems with organic photovoltaics of interest for solar energy conversion, namely the difficulty in efficient absorption of solar radiation by thin films. In that case, we and others have demonstrated promising results showing improvements in organic photovoltaic efficiency that can be traced back to plasmon-enhanced absorption. Prospects for further improvements appear excellent.

## 19.13 ACKNOWLEDGEMENTS

We would like to thank Eastman Kodak for partial support of this work and the NSF for grant DMR-0804960.

## 19.14 REFERENCES

1. Jeanmaire, D. L. and Van Duyne, R. P. (1977). Surface Raman spectroelectrochemistry: Part I. Heterocyclic, aromatic, and aliphatic amines adsorbed on the anodized silver electrode. *J. Electroanal. Chem.* **84**: 1-20.
2. Albrecht, M. G. and Creighton, J. A. (1977). Anomalously intense Raman spectra of pyridine at a silver electrode. *J. Am. Chem. Soc.* **99**: 5215-5217.
3. Van Duyne, R. P. (1979). In *Chemical and biochemical applications of lasers*, Moore, C. B. (Ed.), academia press, New York, pp101-185,
4. Chang, R. K. and Furtak, T. E. (1982). Surface enhanced Raman scattering, Plenum press, New York.
5. Nie, S., and Emory, S. R. (1997). Probing single molecules and single nanoparticles by surface-enhanced Raman scattering. *Science* **275**:1102-1106.
6. Zhang, W., Yeo, B. S., Schmid, T. and Zenobi, R., (2007). Single molecule tip-enhanced Raman spectroscopy with silver tips. *J. Phys. Chem. C.* **111**: 1733-1738.
7. Jiang, J., Bosnick, K., Maillard, M. and Brus, L. (2003). Single molecule Raman spectroscopy at the junctions of large Ag nanocrystals. *J. Phys. Chem. B.* **107**: 9964-9972.
8. Kneipp, K., Kneipp, H. and Kneipp, J. (2006). Surface-enhanced Raman scattering in local optical fields of silver and gold nanoaggregates-from single-molecule Raman spectroscopy to ultrasensitive probing in live cells. *Acc. Chem. Res.* **39**: 443-450.
9. Kümmerlen, J., Leitner, A., Brunner, H.; Aussenegg, F. R., Wokaun, A. (1993) Enhanced dye fluorescence over silver island films: analysis of the distance dependence *Mol. Phys.* **80**: 1031-1046.
10. Schalkhammer, T., Aussenegg, F. R., Leitner, A., Brunner, H., Hawa, G., Lobmaier, C. and Pittner, F. (1997). Detection of fluorophore-labeled

antibodies by surface-enhanced fluorescence on metal nanoislands. *Proc. SPIE: Int. Soc. Opt. Eng.* **2976**: 129-136.

11. Lakowicz, J. R. (2001). Radiative decay engineering: biophysical and biomedical applications. *Anal. Biochem.* **298**:1-24.
12. Lakowicz, J. R., Shen, Y., D'Auria, S., Malicka, J., Fang, J., Gryczynski, Z. and Gryczynksi, I. (2002). Radiative decay engineering. 2. effects of silverisland films on fluorescence intensity, lifetimes, and resonance energy transfer. *Anal. Biochem.*, **301**:261-277.
13. Tomokatsu H., S. Tamil S., and Masayuki N. (1999) Field enhancement effect of small Ag particles on the fluorescence from $Eu^{3+}$ -doped $SiO_2$ glass. *Appl. Phys. Lett.* **74**:1513-1515.
14. Kramer, A., Trabesinger, W., Hecht, B. and Wild, U. P. (2002). Optical near-field enhancement at a metal tip probed by a single fluorophore. *Appl. Phys. Lett.* **80**: 1652-1654.
15. Konstantin S., George C., and Therese M. C. (1998) Enhancement of molecular fluorescence near the surface of colloidal metal films. *Anal. Chem.*, **70**: 3898-3905.
16. Roy, S., Kim, J. H., Kellis, J. T., Jr.,Poulose, A. J., Robertson, C. R. and Gast, A. P. (2002) Surface plasmon resonance/surface plasmon enhanced fluorescence: an optical technique for the detection of multicomponent macromolecular adsorption at the solid/liquidinterface. *Langmuir*, **18**: 6319-6323.
17. Yu, F., Yao, D. and Knoll, W. (2003) Surface plasmon field-enhanced fluorescence spectroscopy studies of the interaction between an antibody and its surface-coupled antigen. *Anal. Chem.* **75**: 2610-2617.
18. Gryczynski I., Malicka J., Gryczynski Z., Lakowicz J.R. and Geddes C.D., (2002), Dramatic increases in resonance energy transfer have been observed between fluorofores bound to DNA above metalic silver islands: opportunities for long-range immunoassays and new DNA arrays. *J. Fluoresc.*, **12**: 131-133.
19. Geddes, C. D., Gryczynski, I., Malicka, J., Gryczynski, Z. and Lakowicz J. R. (2003). Metal-enhanced fluorescence: potential applications in HT. *combinatorial Chem. & High Throughput Screening* **6**:109-117.
20. Geddes, C. D., Cao, H., Gryczynski, I., Gryczynski, Z., Fang, J. and Lakowicz, J. R. (2003). Metal-enhanced fluorescence (MEF) due to silver colloids on a planar surface: Potential applications of indocyanine green to in vivo imaging, *J. Phys. Chem. A.*, **107**:3443-3449.
21. Danckwerts, M. and Novotny, L. (2007). Optical frequency mixing at coupled gold nanoparticles. *Phys. Rev. Lett.* **98**: 026104-1-4.
22. Da Silva, D. M., Kassab, L. R. P., Lüthi, S. R.; de Araújo, C. B., Gomes, A. S. L., Bell, M. José V. (2007). Frequency upconversion in $Er^{3+}$ doped PbO-$GeO_2$ glasses containing metallic nanoparticles. *Appl. Phys. Lett.* **90**: 081913-081915.
23. Tang, C.W. and Van Slyke, S. A. (1987). Organic electroluminescent diodes, *Appl. Phys. Lett.* **51**: 913-915.
24. Cleave, V., Yahioglu, G., Barny, P. L., Friend, R. H. and Tessler N. (1999). Harvesting singlet and triplet energy in polymer LEDs. *Adv. Mater.* 11: 285-288.

25. Tang, C. W. (1986). Two layer organic photovoltaic cell, *Appl. Phys. Lett.* 48: 183-185.
26. Fu, Y. and Lakowicz, J. R. (2006) Enhanced fluorescence of Cy5-labeled oligonucleotides near silver island films: a distance effect study using single molecule spectroscopy. *J. Phys. Chem. B* **110**: 22557-22562.
27. Malicka J., Gryczynski I., Fang J., Kusba J. and Lakowicz J.R. (2002). Photostability of Cy3 and Cy5-labeled DNA in the presence of metallic silver particles. *J. Fluoresc.*, 12: 439-437.
28. Holmes, R. J., Forrest, S. R., Sajoto, T., Tamayo, A., P. Djurovich, I., Thompson, M. E., Brooks, J., Tung,Y. J., D'Andrade, B. W., Weaver, M. S., Kwong, R. C. and Brown, J. J. (2005). Saturated deep blue organic electrophosphorescence using a fluorine-free emitter. *Appl. Phys. Lett.* **87**: 243507-1-3.
29. Wang, P., Zakeeruddin, S. M., Moser, J. E., Nazeeruddin, M. K., Sekiguchi, T. and Gratzel M. (2003). A stable quasi-solid-state dye-sensitized solar cell with an amphiphilic ruthenium sensitizer and polymer gel electrolyte. *Nat. Mater.* **2**: 498-498.
30. O'Regan, B. and Gratzel, M. (1991). A low-cost, high-efficiency solar cell based on dye-sensitized colloidal $TiO_2$ films. *Nature* 353:737 – 740.
31. Sariciftci, N.S. and Smilowitz, L. Heeger, A.J. and Wudl, F. (1992) Photoinduced electron transfer from conducting polymers onto buckminsterfullerene, *Science* **258**: 1474-1476.
32. Brabec, C. J., Sariciftci, N. S. and Hummelen J. C. (2001). Plastic solar cells, *Adv. Funct. Mater,* **11**: 15-26.
33. Chu, C. W., Yang, H., Hou, W., Huang, J., Li, G. and Yang, Y. (2008) Control of the nanoscale crystallinity and phase separation in polymer solar cells. *Appl. Phys. Lett.* **92**: 103306-1-3.
34. Ni, W., Kou, X., Yang, Z. and Wang, J. (2008) Tailoring longitudinal surface plasmon wavelengths, scattering and absorption cross sections of gold nanorods. *ACS Nano* **2**: 677-686.
35. Klar, T., Perner, M., Grosse, S., Plessen, G. von, Spirkl,W. and Feldmann, J. (1998). Surface-plasmon resonances in single metallic nanoparticles, *Phys. Rev. Lett.* **80:** 4249-4252.
36. Saj, W. (2005). FDTD simulations of 2D plasmon waveguide on silver nanorods in hexagonal lattice. *Opt. Express*, **13**: 4818-4827.
37. Draine, B.T., and Flatau, P.J. (1994). Discrete-dipole approximation for scattering calculations, *J. Opt. Soc. Am. A*, **11**: 1491-1499.
38. Draine, B.T., and Goodman, J.J. (1993) Beyond clausius-mossotti: wave propagation on a polarizable point lattice and the discrete dipole approximation, *Astrophysical J.*, **405**: 685-697.
39. Draine, B.T., and Weingartner, J.C. (1996). Radiative torques on interstellar grains: I. superthermal rotation, *Astrophysical J.*, **470**: 551-565.
40. Li, K., Li, X., Stockman, M. I. and Bergman, D. J. (2005) Surface plasmon amplification by stimulated emission in nanolenses, *Phys. Rev. B* **71**: 115409-1-5.
41. Li, K., Stockman, M. I. and Bergman, D. J. (2003) Self-similar chain of metal nanospheres as an efficient nanolens, *Phys. Rev. Lett.* **91**: 227402-1-4.

42. Shalaev, V.M. (2001). Fractal nano-composites: giant local-field enhancement of optical responses, In *Nanoscale linear and nonlinear optics*, Bertolotti, M. and Sibilia, C. (Ed.) AIP.
43. Wang, Z.J., Pan, S.L., Krauss, T.D., Du, H. and Rothberg, L. J. (2003) The structural basis for giant enhancement enabling single-molecule Raman scattering, *Proc. Natl. Acad. Sci. USA*, **100**: 8638-8643
44. Wang, Z.J. and Rothberg, L.J., (2006) Silver nanoparticle coverage dependence of surface-enhanced Raman scattering, *Appl. Phys. B*, **84**: 289-293
45. Pan, S.L. and Rothberg, L. J. (2005) Enhancement of Platinum Octaethyl Porphyrin Phosphorescence near Nanotextured Silver Surfaces. *J. Am. Chem. Soc.* **127**: 6087-6094
46. D'Andrade, B. W., Holmes, R. J. and Forrest, S. R. (2004) Efficient organic electrophosphorescent white-light-emitting device with a triple doped emissive layer *Adv. Mater.* **16**: 624 – 628.
47. Baldo, M. A.,O'Brien, D. F., You, Y., Shoustikov, A., Sibley, S., Thompson, M. E. and Forrest, S. R. (1998) Highly efficient phosphorescent emission from organic electroluminescent devices. *Nature,* **395**: 151-154
48. Pan, S. L., Rothberg, L.J., Nolte, A.J.,Rubner, M. F., Swager, T. M.(2005). Distance dependent investigation of surface plasmon resonance enhanced fluorescent emission using sequentially adsorbed polyelectrolyte mutilayer spacer. *Proc. SPIE* **5927**: 592705
49. Pan, S. L., Wang, Z.J. and Rothberg, L.R. (2006) Enhancement of Adsorbed Dye Monolayer Fluorescence by a Silver Nanoparticle Overlayer, J. Phys. Chem. B **110**: 17383-17387.
50. Li, W.M., Pan, S.L. and Rothberg, L. J. (2008). Fluorescence enhancement in tris(8-quinolinolato)-aluminum ($Alq_3$). *Proc. SPIE,* (to be published)
51. Park, J.W., Ullah, M. H., Park S.S. and Ha, C.S. (2007). Organic electroluminescent devices using quantum-size silver nanoparticles. *J Mater Sci: Mater. Electron.* **18**: S393–S397
52. L.P. Ma, J. Liu and Yang, Y. (2002) Organic electrical bistable devices and rewritable memory cells *Appl. Phys. Lett.* **80**:2997.
53. Ma, L.P., Pyo, S., Ouyang, J.Y., Xu, Q.F. and Yang Y. (2003). Nonvolatile electrical bistability of organic/metal-nanocluster/organic system . *Appl. Phys. lett.* **82**:1419.
54. Ma, L.P., Liu, J., Pyo, S. and Yang, Y. (2002) Organic bistable light-emitting devices. *Appl. Phys. Lett.* **80**, 362-364.
55. Bozano, L.D., Kean, B. W., Deline, V. R., Salem, J. R. and Scott, J. C. (2004) Mechanism for bistability in organic memory elements. *Appl. Phys. Lett.* **84**: 607.
56. Peumans P.; Yakimov A.; Forrest S. R. Small molecular weight organic thin-film photodetectors and solar cells (2003) *J. Appl. Phys.*, **93**: 3693-3723.
57. Yakimov; A.; Forrest; S. R. (2002) High photovoltage multiple-heterojunction organic solar cells incorporating interfacial metallic nanoclusters. *Appl. Phys. Lett.* **80**: 1667-1669.
58. Rand B. P., Peumans, P. and Forrest, S. R. (2004) Long-range absorption enhancement in organic tandem thin-film solar cells containing silver nanoclusters. *J. Appl. Phys.*, **96**: 7519-7526.

59. Uemura S.,Yoshida M., Kodzasa T. and Kamata, T. (2003) Surface plasmon resonance effect on photocurrent amplification. *Synth. Met.*, **137**: 1443-1444.

60. Wen C., Ishikawa K., Kishima M. and Yamada K. (2000) Effects of silver particles on the photovoltaic properties of dye-sensitized TiO2 thin films. *Sol. Energy Mater. Sol. Cells.* **61**: 339-351.

61. Chandrasekharan N. and Kamat, P. V. (2000). Improving the photoelectrochemical performance of nanostructured $tio_2$ films by adsorption of gold nanoparticles. *J. Phys. Chem. B*, **104**:10851-10857.

62. Tian, Y. and Tatsuma, T. Mechanisms and applications of plasmon-induced charge separation at $tio_2$ films loaded with gold nanoparticles (2005). *J. Am. Chem. Soc.* **127**: 7632-7637.

63. Pan, S.L. and Rothberg, L.J. (2007). Photovoltaic efficiency enhancement by silver metallic ultrathin film incorporated in pentacene/C60 tandem cell, *Proc. SPIE.* **6641**: 664109

# 20

# Fluorescent Quenching Gold Nanoparticles: Potential Biomedical Applications

Xiaohua Huang,[1] Ivan H. El-Sayed,[2] and Mostafa A. El-Sayed[1]

[1]Laser Dynamics Laboratory, School of Chemistry and Biochemistry, Georgia Institute of Technology, Atlanta, GA 30332-0400, USA.

[2]Department of Otolaryngology-Head and Neck Surgery, Comprehensive Cancer Center, University of California at San Francisco, San Francisco, CA 94143, USA.

## 20.1 SURFACE PLASMON RESONANCE OF GOLD NANOPARTICLES

Gold nanoparticles have generated widespread interests for centuries. They are optically active in multiple modalities useful for biomedical imaging including fluorescence, absorption, scattering and Raman spectroscopy. In fluorescence, gold can serve as a fluorescent probe under certain conditions, or modify the signal of an adjacent fluorophore. The effect of gold nanoparticles on fluorescence spectroscopy depends on the stimulating light source, the relation of gold nanoparticle to an adjacent fluorohpore and the surrounding medium. For instance, gold particles can enhance or quench fluorescence of adjacent fluorophores or can quench fluorophores to which they are covalently bound.

To better comprehend how gold nanoparticles can demonstrate multiple effects with a range of potential biomedical applications, an understanding of the particle-light interaction is necessary. When stimulated by the electromagnetic field of the light, the free electrons of the metal nanoparticle undergo a collective coherent oscillation on the surface of the ionic metallic lattice (Figure 20.1).

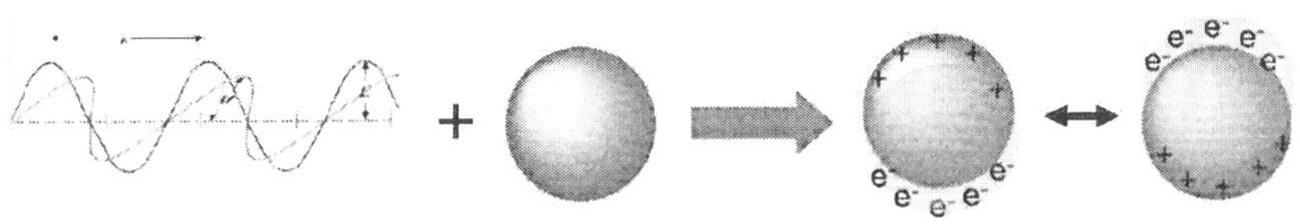

**Figure 20.1:** Surface plasmon resonance of gold nanoparticles.

This collective oscillations of the free electrons confined on the particle surface is called surface plasmon. The surface plasmon is resonant at a specific frequency of the incident light called surface plasmon resonance (SPR). This optical behavior of plasmonic nanoparticles can be explained by Mie theory [1] which involves the salvation of the maxwell's equation for an electromagnetic light wave interacting with a small sphere. For nanoparticles much smaller than the wavelength of light (< 20nm), only the dipole oscillation contributes significantly to the extinction cross section and thus Mie's theory is reduced to the following equation:

*Biomedical Applications of Gold Nanoparticles.*
Edited by Chris D. Geddes.

$$\sigma_{ext}(\omega) = 9\frac{\omega}{c}\varepsilon_m^{3/2} V \frac{\varepsilon_2(\omega)}{[\varepsilon_1(\omega)+2\varepsilon_m]^2+\varepsilon_2(\omega)^2} \quad (1)$$

Where $V$ is the particle volume, $\omega$ is the angular frequency of the exciting light, $c$ is the speed of light, and $\varepsilon_m$ and $\varepsilon(\omega) = \varepsilon_1(\omega) + i\,\varepsilon_2(\omega)$ are the dielectric functions of the surrounding medium and the material itself, respectively. The resonance condition is fulfilled when $\varepsilon_1(\omega) = -2\,\varepsilon_m$ if $\varepsilon_2$ is small or weakly dependent on $\omega$. The maximum SPR frequency depends on the metal composition, particle size, shape, structure and the surrounding medium. Gold nanoparticles exhibit a strong surface plasmon resonance in the visible spectra region around 520 nm while silver nanoparticles show a SPR around 400 nm.

The surface plasmon resonance induces a strong absorption in the visible region responsible for the intense red color of the particle solution. This surface plasmon absorption maximum is slightly red shifted with the increase of the particle size (Figure 20.2A). When the shape of the nanoparticle is changed from spheres to rods, the surface plasmon resonance band is split into two bands [2]. A stronger band in the near infrared (NIR) region corresponds to the electron oscillations along the longitudinal direction and a weaker band in the visible region corresponds to the electron oscillations along the transverse direction. The longitudinal band of gold rods is very sensitive to the aspect ratio (Length/Width) of the rods. Increasing the aspect ratio significantly red shifts the absorption maximum from the visible to the NIR region (Figure 20.2B).

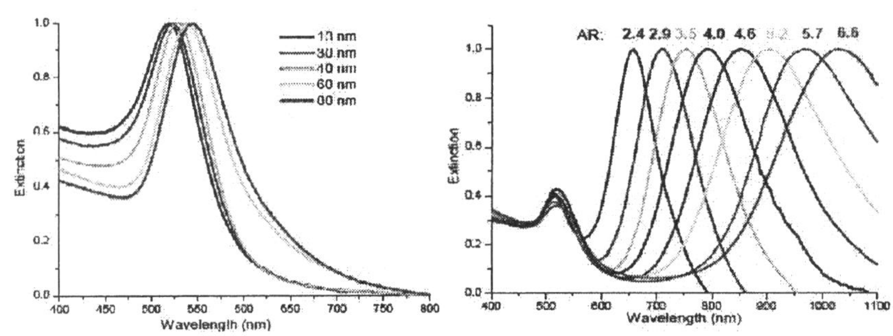


**Figure 20.2:** Size and shape dependence of the surface plasmon resonance absorption spectra of gold nanoparticles. (A) Spheres in different sizes; (B) Rods in different aspect ratios.

## 20.2 ENHANCED OPTICAL PROPERTIES OF GOLD NANOPARTICLES

The SPR phenomena strongly enhance multiple optical properties including fluorescence, absorption, scattering and surface enhanced Raman scattering of absorbed species [3] (Figure 20.3). The enhancement originates from the surface plasmon decays. The surface plasmon oscillation decays by two pathways: radiation decay and non-radiative decay [4]. In the radiative decay process, photons with the same energy as the incident light are emitted, which is called SPR light scattering. This SPR involved light scattering of gold nanoparticles is orders of magnitude stronger in intensity than that of polymer beads and the brightest dye molecules [5]. The nonradiative decay occurs via electron-hole transitions either within the conduction band (intraband) or between the d band and the conduction band (interband). These excited electrons interact with phonons in the metal over a few picoseconds resulting in a hot lattice [6]. The heat of the metallic lattice is dissipated to surrounding medium through phonon-phonon interactions on a time scale from tens to hundreds of picoseconds causing temperature increases of the species surrounding the particles. This photon-to-thermal energy conversion by the metallic nanoparticles is well suited for photothermal applications in diseases treatment.

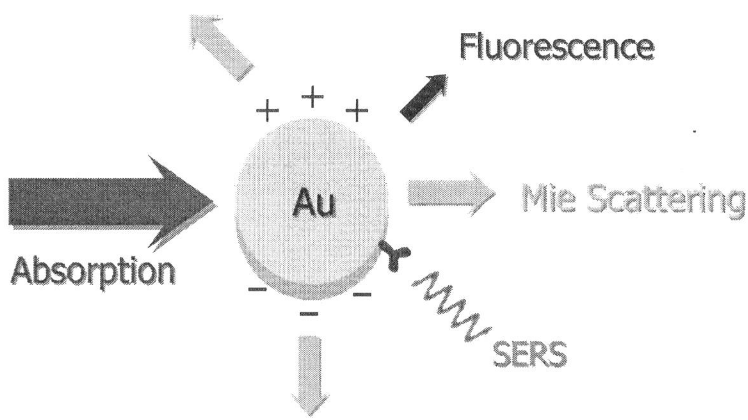


**Figure 20.3**: Enhanced optical properties of gold nanoparticles resulting from the interaction of light with a gold nanoparticle including absorption, Mie scattering, fluorescence and surface-enhanced Raman scattering of adsorbed molecules. Reprinted with permission from ref 3. Copyright 2007 Future Medicine.

The excitation of the surface plasmon effect also induces strongly enhanced fluorescence properties of gold nanoparticles due to the enhancement in the radiative rate of the inter-band electronic transitions relative to that in bulk metals. Metal nanoparticles, especially gold nanorods exhibit enhanced two-photon luminescence (TPL) and multi-photon luminescence (MPL) [7, 8]. Strongly-enhanced TPL has been observed from individual particles [9, 10] and particle solutions [11] under femtosecond NIR laser excitation. This observation raises the possibility of nonlinear optical imaging in the NIR region, where water and biomolecules have minimal absorption. In addition, TPL with the gold nanorods is preferable in biologic tissues demonstrating a significant background light scattering.

The strongly resonant surface plasmon oscillation can be very simply visualized as a photon confined to a small size particle. These confined photons constitute an intense electric field localized on the particle, known as the plasmonic near field. For a single silver nanoparticle, the field can range up to 180 times stronger than the electric field of the incoming light [12]. This near field enhancement depends on the particle composition, size and shape, as well as the distance to the particle surface. If two particles are proximate to each other, the field at the junction of the two particles is extremely high due to the field coupling [13]. As a result, adsorbed species on the surface of the particle demonstrate significant surface enhanced Raman scattering as the Raman intensity is proportional to the square of the electric field.

## 20.3 PARTICLE-FLUOROPHORE INTERACTIONS

### 20.3 1. *Mechanism of molecular fluorescence*

At room temperature molecules occupy on the lowest vibrational level of the ground electronic state ($S_0$). Upon light absorption, molecules are excited and electrons transit on a femotosecond time scale from a vibrational level in the electronic ground state to one of the many vibrational levels in the electronic excited state such as $S_1$ or $S_2$ depending on the incident laser wavelength (Figure 20.4).

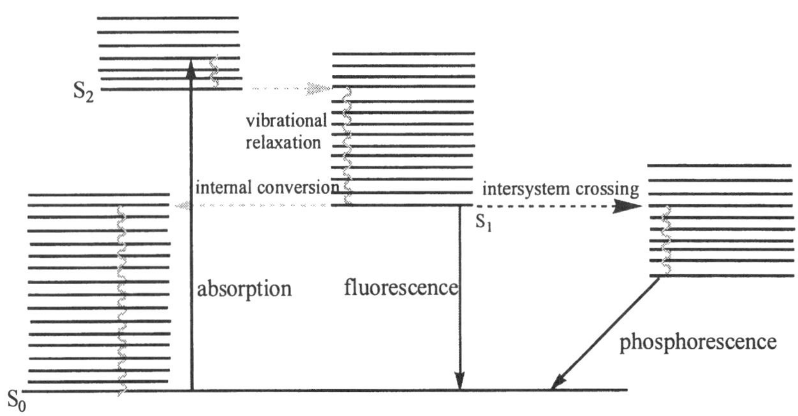


**Figure 20.4:** Scheme of the mechanism of molecular fluorescence. $S_0$: electronic ground state, $S_1$: electronic first excited singlet state, $S_2$: electronic second excited singlet state, $T_1$: electronic first excited triplet state.

If the electrons reach a higher vibrational level of an excited state, collisions with other molecules cause the excited molecule to lose vibration and electric energy and fall to the lowest vibrational level of this state and then loss energies via internal conversion to the same vibrational levels of a lower excited state until reaching the lowest vibrational level of the first excited state $S_1$. From the lowest vibrational level of the first excited state $S_1$, the molecule can return to any of the vibrational levels of the ground state, emitting its energy in the form of fluorescence or crossing to the triplet state by the process known as intersystem crossing. This emission process occurs on a nano- to millisecond time scale. The molecule will also distribute the excess energy to other possible modes via nonradiative processes especially internal conversion to the same energy level in the ground electronic state $S_0$ where the molecule relaxes to the lowest vibrational level of this state, and/or intersystem crossing to the same vibratinal level of the triplet excited state $T_1$ where the molecules relaxes to the lowest vibrational level of this state and then return to electronic ground state $S_0$ by loss of vibration energy (heat) or if in a rigid medium and avoid oxygen collision quenching, it emits phosphorescence [14, 15].

Quantum yield of the fluorescence processes is defined by the ratio of emitted photons to absorbed photons which can be further expressed in the form of radiative and nonradiative decay rate:

$$QY = \text{Emitted photons} / \text{Absorbed photons} = k_r / (k_r + \sum k_{nr}) \quad (2)$$

Where $\sum k_{nr}$ includes all processes of internal conversions $k_{ic}$ as well as other possible nonradiative processes that participate in the energy dissipation of the excited electronic state. The fluorescence life time $\tau$, i.e. the time the molecule stays in its excited state before emitting a photon is defined as:

$$\tau = 1/ (k_r + \sum k_{nr}) \quad (3)$$

## 20.3 2. *Donor-acceptor probes*

Fluorescence spectroscopy is widely applied in biochemistry and molecular biology for analyte detection and cellular imaging. It is a well established analytical technique useful for chemical sensing in dilute nonscattering samples and providing biochemical, structural and functional information of probed biomolecules in vitro and in vivo. While single labeling probes are quite useful, dual-labeled probes containing a fluorescence donor and a receptor provide an additional tool with increased sensitivity and specificity based on the fluorescence quenching or enhancing mechanisms. Conventional fluorescent receptors are organic dye molecules capable of quenching the donor fluorescence based by either static or dynamic quenching processes including Förster energy transfer, collision quenching, and other mechanisms such as electron transfer or excited state complex formation [16] (Figure 20.5).

In static quenching, the donor and acceptor bind together as one complex. The strong dipole-dipole coupling causes the coupling of the excited electronic states of the two dye molecules. In the Förster resonant energy transfer (FRET) process between two chromophores, a donor chromophore in its excited state can transfer energy by a nonradiative, long-range dipole-dipole coupling mechanism to an acceptor in close proximity (typically $<10$nm). The efficiency of energy transfer is given by

$$E = k_{ET} / (k_r + k_{ET} + k_{nr}) = 1/[1+(R/R_0)^6] \quad (4)$$

Where R is the distance between the donor and acceptor and $R_0$ is distance between donor and acceptor at which the FRET efficiency is 50% which is given by

$$R_0^6 = 8.8 \times 10^{-28} \kappa^2 n^{-4} Q_0 J \qquad (5)$$

where $\kappa^2$ is the dipole orientation factor, $n$ is the refractive index of the medium, $Q_0$ is the fluorescence quantum yield of the donor in the absence of the acceptor and $J$ is the spectral overlap integral between donor emission and acceptor absorption spectra. Therefore, the most important two parameters that cause higher efficiency of energy transfer are the distances between the donor and acceptors and the overlap of the absrotpion spectrum of the acceptor with the emission spectrum of the donor.

Donor-acceptor probes are very useful as fluorescent lables in biochemical assay and sensing based on the fluorescence quenching mechanisms. In a closed state form where the donor and acceptor are close to each other, the fluorescence of the donor (reporter) is highly quenched. In an open state, where the donor and receptor are spaced away from each other due to biochemical reactions, the donor fluorescnece is restored. The changes of the fluorescence intensity of the reporter has been used for DNA detection, immunoassay, enzyme sensing and detection of many other bimolecules.

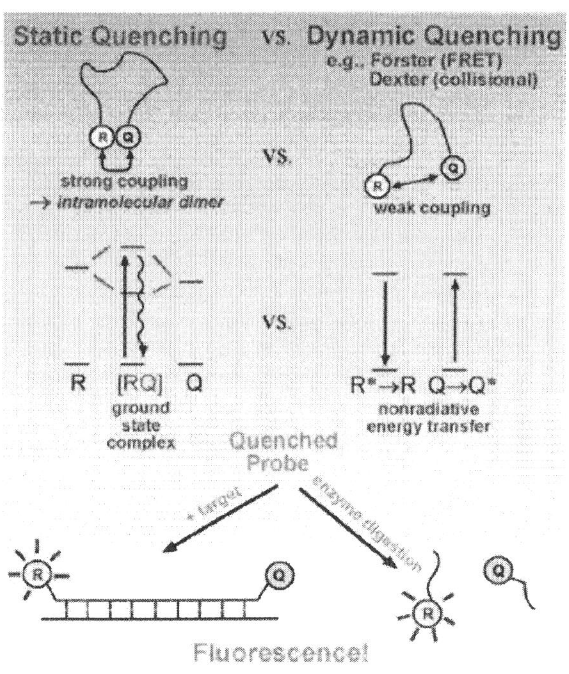


**Figure 20.5:** Static and dynamic quenching mechanisms. R: reporter (donor), Q: quencher (acceptor). Reprinted with permission from ref 16. Copyright 2003 WILEY-VCH Verlag GmbH & Co. KGaA.

## 20.3 3. *Donor-AuNP probes*

Gold nanoparticles can modify the emission of dipole fluorophores through dipole-dipole interactions that are dependent on the gold particle size, shape, and distance from the fluorophore. Further, the effect is dependent on the orientation of the molecular dipole with respect to the particle and the overlap of the emission of the molecule with the particle's absorption spectrum [17]. Either quenching or enhancement of the fluorescence may occur depending on the various parameters that determine the quantum yield of the emission of the fluorophore molecules. These parameters include the excitation electric field E, the radiative rate $k_r$ and nonradiative rate $k_{nr}$.

### 20.3.3 1. *Fluorescence enhancement*

The surface plasmon oscillation can be simply visualized as a photon confined to a small particle size. These confined photons constitute an intense oscillating electric field localized on the particle surface which can be described by Equation (6) for a spherical nanoparticle which has a radius of r, dielectric function of ε and distance to the molecule of d [18].

$$E_{induced} = E_0 r^3 (\varepsilon-\varepsilon_0)/(\varepsilon+2\varepsilon_0)(r+d)^3 \quad (6)$$

The field reaches its maximum at the surface plasmon resonance frequency when $\varepsilon = -2\varepsilon_0$ where $\varepsilon_0$ is the dielectric constant of the medium surrounding the particle surface. This induced field of the metallic nanoparticles provides an external field for the fluorescence excitation of the molecules in addition to the electric field of the incident light and thus increases the absorption rate which is responsible for the enhanced fluorescence intensity.

In addition to the field enhancement, the increases of the radiative decay rate of the molecule also lead to the fluorescence enhancement. This happens when molecules are 5 -20nm away from metal nanoparticles aggregated on surfaces [19-21]. Lakowicz and coworkers have characterized this phenomena by using silver island films deposited on the internal surface of two quartz plates which sandwich a bulk fluorophore solution [20]. The fluorophores are physically placed close to silver islands so that there are a range of distances between the fluorophore and metal. The fluorescence enhancement is accompanied by decreased lifetimes and increased photostability. This phenomenon shows that the silver island increases the radiative decay rate of the fluorophore and therefore induces the fluorescence enhancement.

### 20.3.3 2. *Fluorescence quenching*

In addition to the enhancing effects, gold nanoparticles can also quench adjacent fluorophores through nonradiative processes including FRET, electron transfer, quenching collisions, decreases in radiative decays, re-absorption of the emitted light by the nanoparticles and chemical reactions that changes the ground state of the molecules. The nonradiative processes decrease the quantum yield and correspondingly decreases the fluorescence lifetime. The decreases in the radiative rate will also lead to decreases in the fluorescence lifetime. The re-absorption of the emitted light and chemical reactions that change the concentration of the emitting molecules decreases the quantum yield without altering the fluorescence lifetime as these processes do not involve the excited electronic states.

In the presence of metal nanoparticles, the electronic excitation energy transfers from the dye molecules to the surface plasmon resonance of the metal nanoparticles and/or the electron-hole pairs resulting in fluorescence quenching. In the case of the excitations of surface plasmon, the small dipole in the dye molecues induces a large dioplc in the particle as of a collective electron excitation on the particle surface, refered to as surface-energy-transfer (SET). The surface plasmon resonance effect significantly enhances the quenching efficiency at a factor of $10^4$-$10^5$ compared to organic quenchers [22] and extends the energy distances as far as 70-100 nm which is 10 times longer than typical Förster distance [23]. The probability of the energy transfer depends on the overlap of the emission band of the donor fluoorophore and the absorption band of the metal nanoparticles. Experiments [24] using fluorophore tagged DNA have shown that the quenching efficiency decreases with the increasing particle size due to the lower surface area to volume ratio and higher rate of nonradiative life time incrementation. The rate of quenching efficiency is different for two different particle size ranges: particles smaller than 16 nm greatly quench the fluorescence of 1-methylaminopyrene while larger particles up to 70 nm show a smaller quenching rate [25]. When the dye molecues are statically adsorbed onto the particle, nearly 100% quenching efficiency is observed [26]. Theoretical calculations on account of both surface plasmon excitations and electron-hole pair excitations show a $d^{-6}$ dependence of the energy transfer rate on the distance d between the dye and a 1.4 nm nanoparticle [27]. However, some experiments show a $d^{-4}$ dependence [28] the explanation of which is still unclear.

Electron-transfer process has been observed between fluorophore and metal surface by photocurrent measurement [9, 29, 30]. Transient absorption studies have shown the charge separation between the dye molecule and gold nanoparticles upon pulse laser excitation [31]. In the collision quenching process, the nonradiative rate is proportional to the quencher concentration [19].

At distances less than 2 nm, the radiative rate $\Gamma$ also decreases as the donor dipole and the induced dipole on the particle radiate out of phase if the molecules are tangentially oriented to the gold nanoparticles [32]. When the distance is more than 2 nm, the decrease in quantum efficiency is primarily due to the decreases in the radiative rate [33]. The radiative rate increases monotonically when the distance d is increased from 2 to 16 nm, where the radiative rate reaches the intrinsic value. The radiative rate can drop more than one order of magnitude within 1 nm distance between the dye molecules and gold nanoparticles [34]. For instance, gold nanoparticles have a 99.8% quenching efficiency of lissamine molecular fluorophores 1 nm away from the surface of a 2 nm gold nanoparticle with 51 times longer radiative lifetime and 14 times shorter nonradiative lifetime [34].

## 20.4 BIOMEDICAL APPLICATIONS

### 20.4 1. *In vitro DNA detection*

Gold nanoparticles have potential as nano-quenchers for DNA detection [26, 35-39]. The detection method is based on the distance dependence of the fluorescence of a fluorophore that is bound to gold nanoparticles through a DNA linker which is also used for DNA sensing. Typically a single-stranded DNA (ssDNA) molecule is functionalized with a dye molecule at one end and a gold nanoparticle at the other end. When ssDNA is conjugated to gold nanoparticles, the fluorescence of the dye molecules is quenched due to either dye adsorption onto the nanoparticle particle surface or due to short distance between the dye and gold due to a curved conformation of ssDNA. DNA hybridization stiffens the DNA chain and separates the dye molecule away from gold surface resulting in an increase in the fluorescence intensity.

Figure 20.6 shows a schematic diagram of the DNA detection probe and principle by Maxwell et al. [26]. 2.5 nm gold nanoparticles are used both as a nano-scaffold to load sensing DNA sequence and a nano-quencher. The sensing oligonucleotide molecules 5'-F-T6-TAG GAA ACA CCA AAG ATG ATA TTT – T6-SH-3' are functionalized with thiols at one end for gold binding through Au-S bounds and fluorescein molecules at the other end for fluorescence detection. The fluorescein molecules are electrostatically adsorbed onto the particle surface causing a constrained conformation. In this conformation, the fluorescein is completely quenched by energy transfer process and the oligo sequence is exposed for specific hybridization. Upon targeted binding, the fluorescein is pulled away from the gold surface due to a dramatic increase in the DNA chain rigidity after hybridization. Since energy transfer is only efficient within 1-2nm for 2-3nm gold particles, in this opened state, no energy transfer occurs from the excited dye molecules to gold nanoparticles.

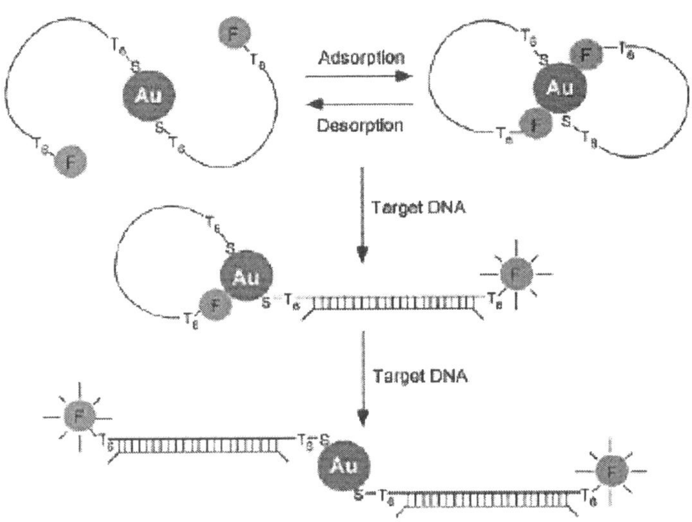


**Figure 20.6:** Schematic diagram of the DNA detection using fluorescence quenching by gold nanoparticles. Oligonucleotide molecules are conjugated to gold surface vial Au-S bonds and the fluorescein molecules at the other end of the oligo sequence is adsorbed onto gold nanoparticles to form a constrained conformation.

When the oligonuceotides are bound to complementary sequences, the constrained form opens and fluorescein leaves particle surface and the quenched fluorescence in the constrained form is restored. Reprinted with permission from ref 26. Copyright 2007 American Chemical Society.

The gold quenched fluorescence method can detect single-mismatched DNA sequences as the DNA hybridization could increases the dye fluorescence by a factor of as much as several thousand. Dubertret et al has demonstrated that a 1.4 nm gold nanoparticles show 100 times higher quenching efficiency for dyes emitting in the near infrared region [35]. A 25 base oligonucleotide modified with a primary amine at 3' end for dye conjugation such as fluorescein, Rhodamine 6G, Texas red and Cy5 and a disulfide at 5' for gold conjugation via gold-sulfur bonds. Due to the hybridization of the five nucleotides at the extremities, the oligonuceotide sequence adopts a hairpin-loop structure at room temperature. This loop structure opens easily upon hybridization of the nucleotide to its target with a resultant enhancement in the fluorescence intensity. Due to the significant fluorescence enhancement after DNA hybridization, the method is sensitive enough to detect single-mismatch DNA sequences. This single-mismatch sensing is eight fold greater than other molecular probes.

## 20.4 2. *In vitro immunoassay*

Immunoassay is the use of antibody-antigen specific interactions to measure the concentration of a substance in solution and is widely used for anylate in biologic fluids such as urine or serum. Immunoassay is usually performed using one of two methods: sandwich format and competitive format. In the sandwich format, the primary antibody is adsorbed onto a microtiter plate or microparticle solid phase. The unknown antigen is then bound to the antibodies on the solid phase. A detection antibody which is radio-labeled, enzyme labeled or fluorophore labeled secondary antibody is bound to the antigen, with the antigen sandwhiched between the two antibodies. The label activity is used for the quantitative measurement of analyte concentration. In the competitive immunoassay, known antigen or antibody is conjugated to the solid phase and labled antibody or labled antigen then bind to the antigen or antibody on the solid phase. The unknown antigen will compete with the known antigen for the reaction sites on antibodies. In such a case, the labled antibody or antigen is displaced from the solid phase for detection by the lables on the antibody or antigen.

By modifying the solid phase with gold nanoparticles, a novel, simple and sensitive homogenous immunoassay can be developed based on the fluorescence quenching property using gold nanoparticles. A simple competitive immunoassay can be designed by incorporating gold nanoparticles with the polymer bead solid phase [39-43]. As described by Kato et al., gold naoparticles are deposited via electrostatic interactions onto submicron-sized polystyrene latex spheres that are precoated with alternating multilayers of PSS polyanion and PEI polycation molecules (Figure 20.7).

The gold nanoparticle/polyelectrolyte coated latex particles are then conjugated with biotin molecules through a layer of biotinylated poly(allylamine hydrochloride) that is deposited on the particle surface before biotin binding. Fluorescein isothiocyanate labled anti-biotin immunoglobulin (FITC-anti-biotin IgG) is bound to the biotin on the surface of the gold-polymer complex particles. In this binding complex, the FITC is quenched. When analyte biotins are mixed with the complex, the biotin will bind to the dye labeled antibodies on the particle surface and forms biotin-FITC labeled antibodies which are suspended in solution. These analyte biotin-FITC labeled antibodies greatly enhance the dye fluorescence compared to the quenched fluorescence of the dye onto the particle surface. This biotin-functionalized gold/polymer coated particles show a dynamic sensing range of 1-50 nmol.

In the non-competitive heterogenous sandwiched immunoassay, secondary antibodies are labeled with gold nanoparticles. One example is the assay demonstrated by Peng et al. for the sensing of human IgG based on the dissociation of the immunocomplex and flureoscence quenching by gold nanoparticles [39]. Anti-human immunoglobulin G coated polystyrene microwells capture human IgG analyte on the microplate which then bind to antibody conjugated to gold nanoparticles. The sandwich-type immunocomplex is subsequently dissociated by the mixed solution of sodium hydroxide and trisodium citrate forming free gold labeled secondary antibodies. To this solution, fluorescein is added to be adsorbed onto the gold surface and the fluorescence intensity of the fluorescein at 517 nm is measured. The fluorescence intensity is inversely proportional to the logarithm of the concentration of human IgG as more IgG analyte bind more gold labeled secondary antibodies which then adsorb more fluorescein molecules. This method shows a dynamic range of 10–5000 $ngmL^{-1}$ with a detection limit of 4.7 $ngmL^{-1}$.

The immunoassay, using the quenching properties of gold nanoparticle, can be modified in a similar way as in immunoassay without gold nanoparticles. In recent years, magnetic nanoparticles are used as solid phase due to their easy of separation under a magnetic field. The magnetic nanoparticles are coated with primary antibodies which capture with analyte antigen in sample solution. Gold nanoparticle labeled secondary antibodies are then bound to the antigen. This sandwich complex is then separated under a magnetic field. In the method of Ao et al., fluorophores such as FITC are quenched by adsorbtion to gold nanoparticles and the IgG is quantitatively measured by the decrease of the fluorescence intensity of the FITC molecules.

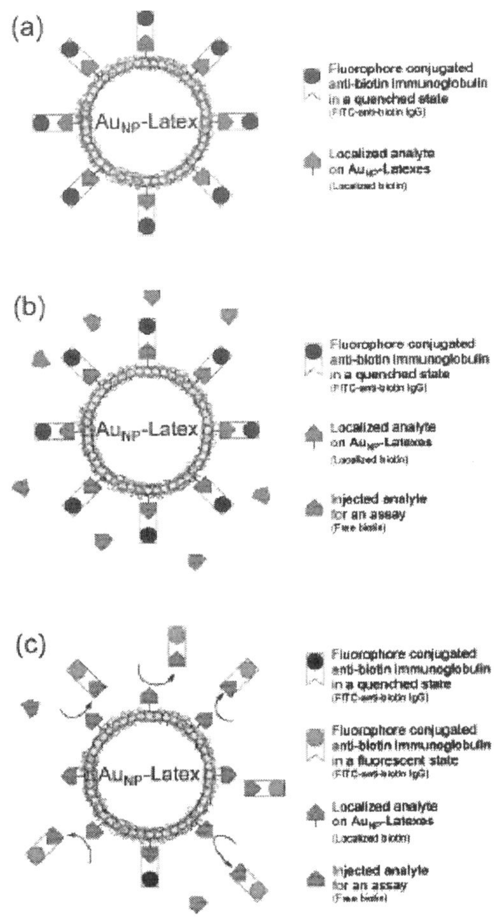


**Figure 20.7:** Schematic illustration of competitive FQIA using a suspension of AuNP/PE-coated latexes. (a) Fluorophore-labeled antibodies are quenched before injection of analyte (biotin). (b) Injection of analyte. (c) Fluorophore-labeled antibodies fluoresce after being competitively released by biotin. Reprinted with permission from ref 40 by Kato et al. Copyright 2007 American Chemical Society.

## 20.4 3. *In vivo tumor imaging*

Gold nanoparticles - chromophore complex can be used for fluorescence imaging based on the high quenching efficiency of the probe fluorophore. A near-infrared-fluorescence (NIRF)-quenched gold-nanoparticle imaging probe for in vivo drug screening and protease activity determination has been reported by Lee et al [44]. The probe is used to sense matrix metalloprotease (MMP) activity or their inhibitors which are mainly involved in cancer, inflammation and vascular disease. The probe is composed of 20 nm spherical gold nanoparticles loaded with Cy5.5 dye molecules that are conjugated to MMP substrate Gly-Pro-Leu-Gly-Val-Arg-Gly-Cys (Figure 20.8). In this conjugated state, the fluorescence of Cy5.5 dye is completely quenched. When the gold probe meets protease enzymes, the enzyme cleaves the substrate between gold and the dye molecule and the dye molecules are separated from gold nanoparticles. Due to the separation of the dye molecules from the surface of gold nanoparticles, the fluorescence of the dyes greatly increases. As tested by DTT cleavage of the Cy5.5-substrate from the nanoparticles, the fluorescence of thedye results in enhanced fluorescence from the dye molecules (Figure 20.8E).

In vivo tumor imaging of a squamous cell cancer (SCC) line with these NIRF-quenched gold probes have been demonstrated using a preclinical optical imaging system with an excitation wavelength of 670 nm and emission wavelength of 799 nm [44]. The SCC7 cell overexpresses MMP2- levels compared to immortalized benign cell lines. Figure 20.8F shows a comparison of the images after subcutaneous injection or intratumoral injection of the probes in three cases: normal mice, tumor bearing mice, tumor bearing mice and inhibitor which is administered into the tumors 3 minutes prior to probe injection. In the mice lacking tumor, there are only weak NIRF signals that do not change with time. In the mice with tumors, an increasing intensity of fluorescence is measurable over time. When inhibitor is injected, there is minimal fluorescence similar to the control group of mice. The high contrast imaging of tumors due to the cleavage of the MMP-2 enzyme from the substrate in the probes results in a much higher intensity of fluorescence compared to the quenched fluorescence in the probes. This method allows for a simple visual monitoring of the activities of both protease and the inhibitors.

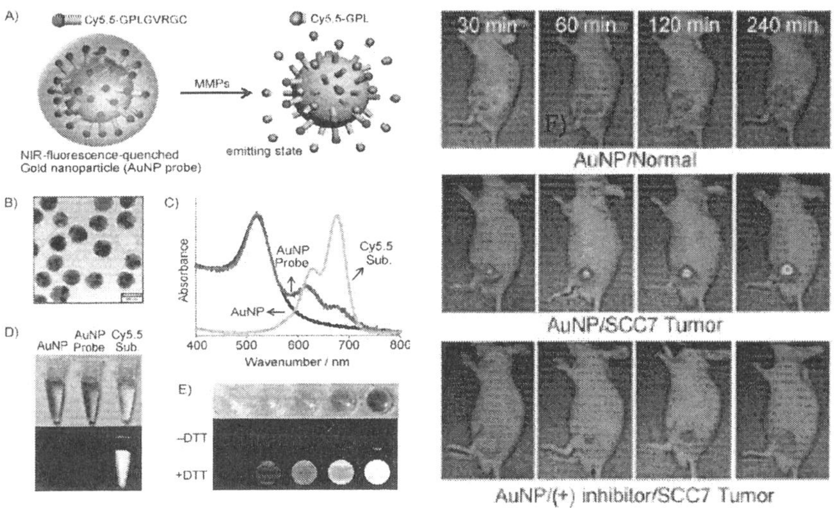


**Figure 20.8:** A) The MMP-sensitive AuNP probe. B) TEM image of the AuNP probe. C) UV/Vis spectra of AuNP, AuNP-probe, and Cy5.5- substrate solutions. D) Bright and NIRF image of vials containing AuNP, AuNP-probe, and Cy5.5-substrate solutions. E) Bright and NIRF image sections of a 96-well microplate containing 0, 2.7, 7, 14, and 27nm of the AuNP probes (left to right) with and without DTT, F): NIRF tomographic images of normal and subcutaneous- SCC7- tumor-bearing mice after injection of the AuNP probe with and without inhibitor.

## 20.4 4. *Cancer and cellular imaging of intrinsic fluorophores*

### 20.4.4 1. *Autofluorophore quenching*

Biologic tissues and extracellular matrix contain a variety of intrinsic fluorophores that can be characterized by their absorption and emission spectra. Figure 20.9 shows the absorption and emission spectra of some of the important fluorophores in tissue [45]. The endogenous fluorescence allows optical assessment of the tissue biochemical and metabolic status. Among many autofluorophores, nicotinamide adenine dinucleotide, collagen, elastin and flavins plays dominant roles due to their strong fluorescence upon excitation between 335 nm and 360 nm. Protein fluorescence is contributed by amino acids including tryptophan, tyrosine and phenylalaine.

(A)

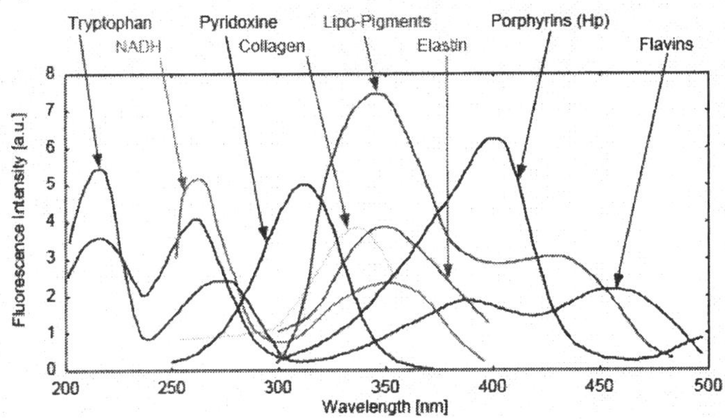


(B)

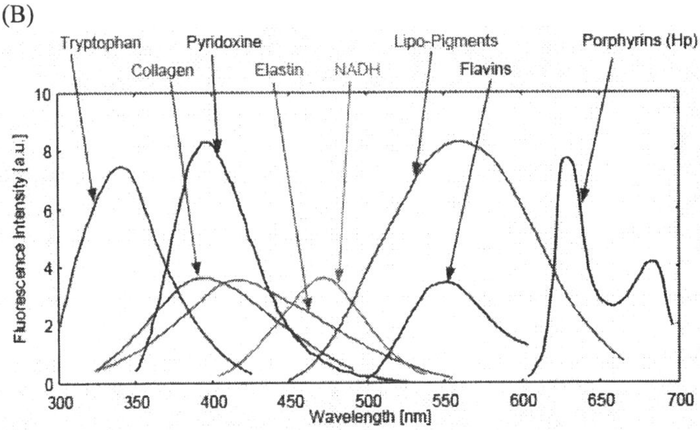

**Figure 20.9:** (A) Excitation spectra of some biomolecules; (B) Emission spectra of some biomolecules. Cited from ref 45.

Nicotinamide adenine dinucleotide is one of the most important coenzymes found in every cell. It is involved in the production of energy in all cells, and is necessary for oxidization of all foods including carbohydrate, fats, and amino acids to produce Adenosine Triphosphate (ATP), which is used by every cell in the body [46]. It serves as a hydrogen and electron carrier and alternates between the oxidized state nicotinamide adenine dinucleotide ($NAD^+$) and the reduced state – dihydronicotinamide adenine dinucleotide (NADH) in cellular respiratory and metabolic processes [47] (see structure in Figure 20.10). The extra hydrogen on position 4 of the planar pyridine ring in NADH results in the significant spectral differences between NADH and $NAD^+$. In contrast with oxidized $NAD^+$, the reduced coenzyme NADH absorbs strongly around 340 nm and fluoresces in the blue spectral region with a maximum around 460 nm [48]. This reduced fluorescent form is one of the major endogenous fluorophores on cellular fluorescence [49].

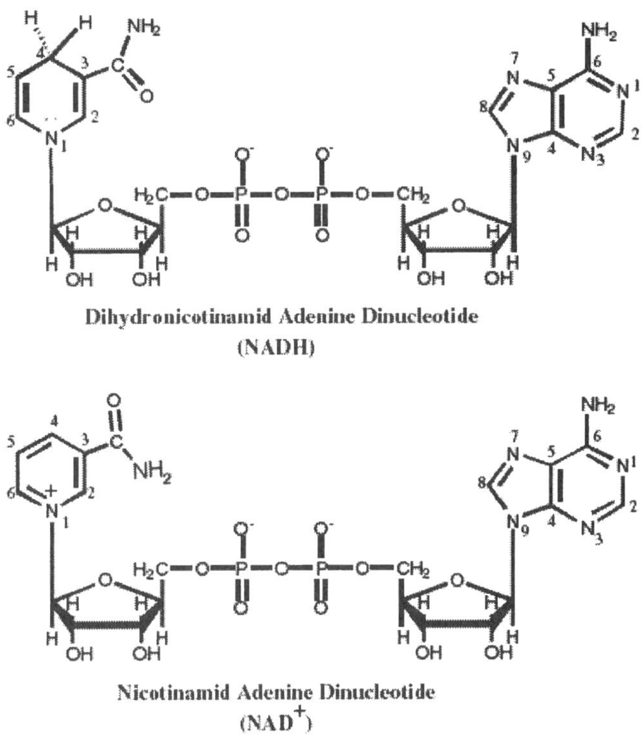


**Figure 20.10:** Structure of NADH and $NAD^+$. Cited from reference 52.

Collagen is a fibrous component of bone, teeth, cartilidge, tendons, blood vessels and matrix of skin. It contains 3 protein chains forming triple helix [50]. All the three helix contains GLY-X-Y-repeating sequence (X often Pro, Y often Hyp).

Collagen fluorescence is associated with two types of cross-links. One is hydroxylysyl pyridinoline (HP) and another is lysyl pyridinoline (LP) [51] (Figure 20.11). Both are fluorescent with emission maximum at 400 nm when excited at 325 nm.

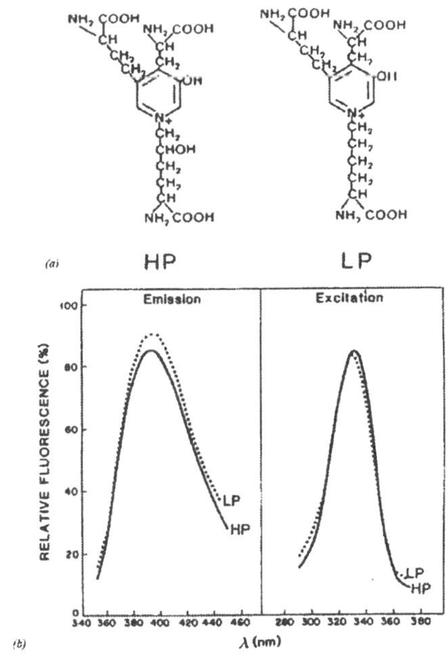


**Figure 20.11:** Fluorescence components in collagen. Reprinted with permission from ref 51. Copyright 1984 Annual Reviews.

### 20.4.4.1 1. *NADH quenching*

Gold nanoparticles have been found to quench the fluorescence of NADH solution by using the optical fluorescence and absorption spectroscopy [52]. When gold nanoparticles are added into NADH solution, the emission band of NADH with a maximum at 460 nm ($\lambda_{exci}$ = 325 nm) is quenched (Figure 20.12A). Increasing gold concentrations lead to increases in the quenching efficiency up to 95%. The normalized fluorescence decay (Inset in Figure 20.12(A)) shows that the fluorescence lifetimes in the presence of gold nanoparticles are not changed compared to pure NADH solution, which suggests that the quenching effects are not due to the coupling of Au nanoparticles with the excited states of NADH, but due to the change of their ground states. Figure 20.12 (B) and the inset spectrum, after subtracting gold absorption, shows that the intensity of absorption band of NADH at 340 nm, originating from the n-π* transition of the dihydronicotinamide in NADH, greatly decreases while the intensity of the 260 nm band originating from the π-π* transition of the adenine ring slightly increase.

This observed decrease of 340 nm absorption bands suggests that the structure of NADH is changed due to the addition of gold nanoparticles. Therefore, NADH must be converted to some other species. The concurrent intensity increase of the 260 nm band indicates $NAD^+$ production, since $NAD^+$ has larger extinction coefficient at 260 nm ($18\times10^{-3}$) than that of NADH ($14.4 \times10^{-3}$) due to the contribution of oxidized nicotinamide [53]. The effect of the addition of NADH is found to slightly red shift and increase the intensity of the surface plasmon absorption band of gold nanoparticles at 520 nm. These results show that gold nanoparticles catalyze the oxidization of NADH to $NAD^+$ resulting in the NADH fluorescence quenching. This chemical effect only changes the amount of ground state NADH while not affecting the fluorescence lifetime.

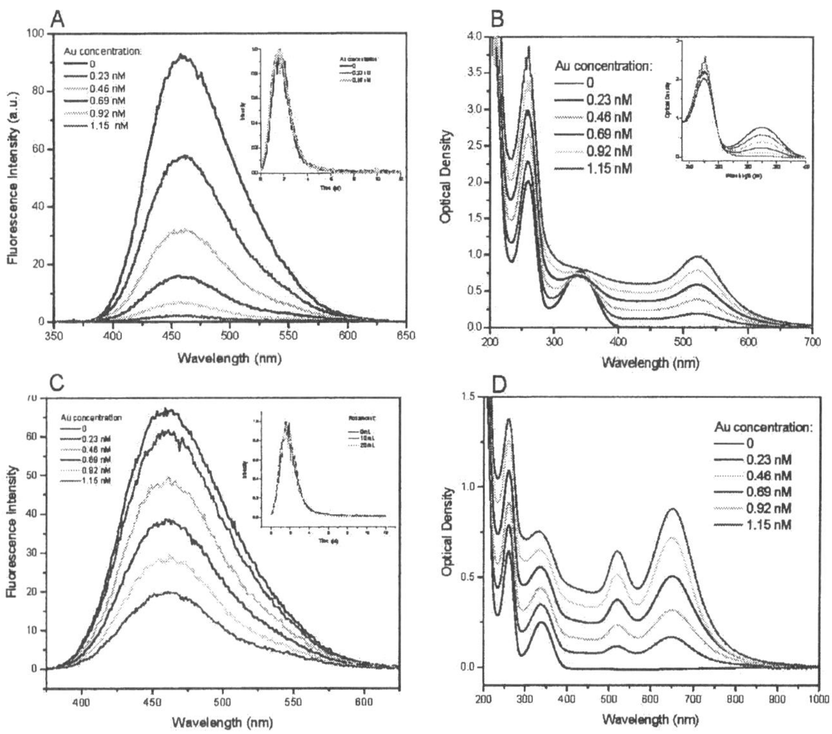


**Figure 20.12:** (A) Fluorescence spectra of NADH in the presence of different concentrations of gold nanospheres (OD=1.0). Inset: the fluorescence decay of NADH at 460 nm in the presence of different concentrations of gold nanospheres (λexci = 335 nm); (B) Absorption spectra of NADH (0.08mM) in the presence of different concentrations of gold nanospheres (OD=1.0); Inset: absorption spectra of NADH after subtraction of absorptions from gold nanospheres; (C) Fluorescence spectra of NADH in the presence of different concentrations of gold nanorods (OD=1.0). λexci = 325 nm; Inset: the fluorescence decay of NADH at 460 nm in the presence of different concentrations of gold nanorods (λexci = 335 nm); (D)Absorption spectra of NADH (0.08mM) in the presence of different concentrations of gold nanorods (OD=1.0). Cited from reference 52.

Gold nanorods also quench NADH fluorescence, but in a different mechanism from the spheres. With increases in the nanorods concentration, the fluorescence of NADH decreases greatly (Figure 20.12C). The results from time resolved fluorescence experiment (Figure 20.12C inset) indicates that no energy transfer processes or radiative decay changes are involved. Unlike spheres, the intensity of 340 band from NADH molecules and 260 nm from both NADH and $NAD^+$ do not change after the addition of gold nanoparticles. This suggests that there are no oxidization reactions. The quenching is only due to surface plasmon absorption of the NADH fluorescence light by gold nanoparticles.

#### 4.4.1 2. *Collagen quenching*

Gold nanoparticles, both spheres or rods, quench collagen fluorescence by photonic absorption of the emission light of the fluorophore by the nanoparticles. The addition of Au nanospheres quenches the collagen fluorescence (Figure 20.13A) but it does not affect the absorption spectra of collagen (Figure 20.13B).

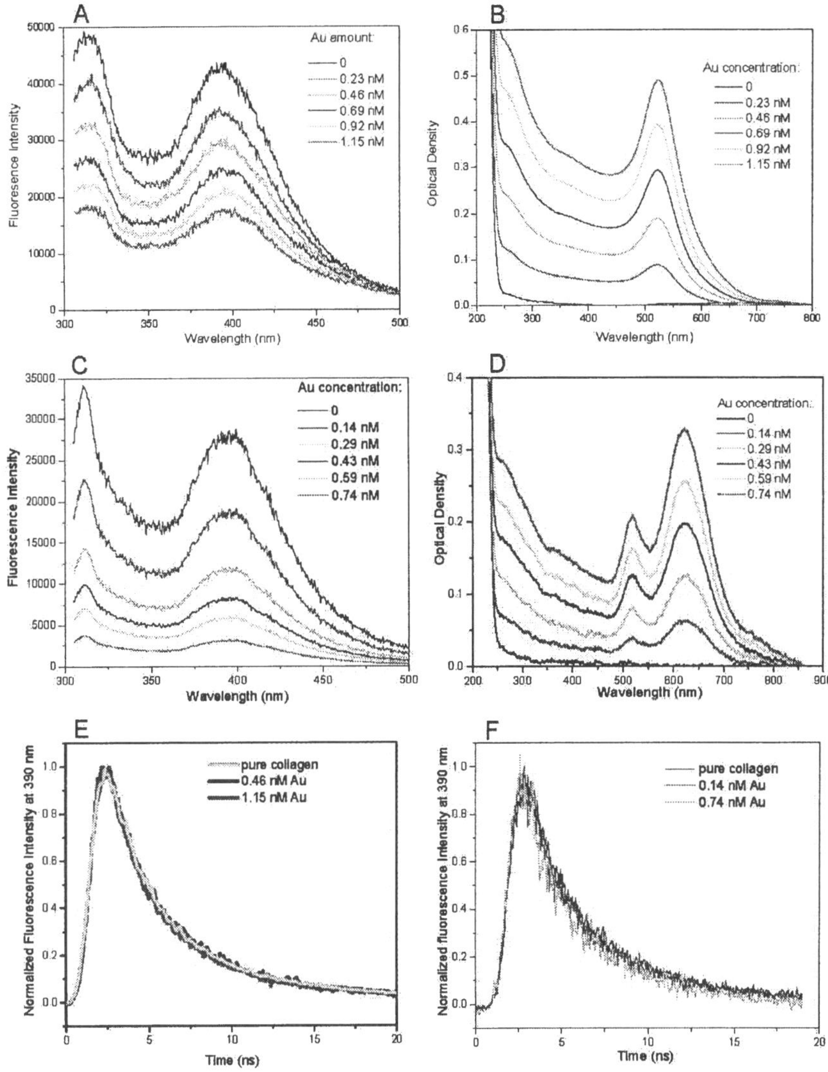


**Figure 20.13:** Fluorescence (A) and absorption (B) spectra of collagen in the presence of different concentrations of gold nanospheres (OD=1.0); Fluorescence (C) and absorption (D) spectra of collagen in the presence of different concentrations of gold nanorods (OD=1.6); the fluorescence decay of collagen at 390 nm in the presence of different concentrations of gold nanospheres (E) and gold nanorods (F). (λexci = 335 nm). Cited from reference 52.

Similar to nanospheres, the nanorods quench collagen fluorescence (Figure 20.13C) but do not affect the collagen absorption spectra (Figure 20.13D). Time resolved fluorescence shows that both spheres (Figure 20.13E) and rods added to collagen solution (Figure 20.13F) show the similar lifetime of collagen emission at 400 nm as that of pure collagen solution which is 2.4 ns. These results suggest that either the collagen ground state or excited states are not affected by the nanoparticles. So the quenching of the collagen emission is due to the absorption of the emission light at 400 nm by the interband absorption of the plasmonic nanoparticles. The stronger quenching efficiency by nanorods compared to spheres is due to higher absorption efficiency of nanorods than that of spheres.

### 20.4.4 2. *Whole cell quenching*

Live cells show one strong absorption band around 280 nm derived from complex set of cellular proteins (Figure 20.14A). When excited at the absorption maximum of 280 nm, the cells show a very strong and similar fluorescence band around 336 nm (Figure 20.14B). When excited at NADH absorption maximum around 340 nm, they show a much weaker band around 440 nm to 450 nm (Figure 20.14C). Figure 20.11C shows that the normalized profile for each cell type is different. The normal HaCat cells have higher fluorescence at wavelength shorter than 400 nm and higher wavelength over 480 nm regions. HSC cancer cells have higher fluorescence intensity from 420nm to 450 nm. Collagen fluorescence (380nm region) and flavin (520nm region) in HaCaT noncancerous cells are stronger than that in HOC and HSC cancerous cells. Elastin fluorescence (420nm region) is stronger in HSC cancerous cells than that in HaCaT noncancerous cells and HOC cancerous cells.

It is known that gold nanoparticles exert different effects on fluorophores emission intensity depending on their distance from the nanoparticle surface. If a fluorophore is bound to the particle surface, then quenching can occur by one of two methods: 1) chemical, which does not cause a decrease in lifetime of emission, or 2) physical, due to energy transfer to the particle. If there is energy transfer to the particle, there is a decrease in the observed lifetime of the emission. If the fluorophore is not directly on the NP surface, but is located within its strong plasmon field, then enhancement of the fluorescence is observed. However, at longer distances and outside the nanoparticle's plasmon field, quenching can again be observed due to simple reabsorption and scattering of light by the Au NPs. In this case no decrease in emission lifetime occurs. To elucidate these affects on whole cell fluorescence nanoparticles are either extracelluarly placed around cells by simple mixing or intracellularly located inside cells by incubation live cells with particle suspension over 12 hours.

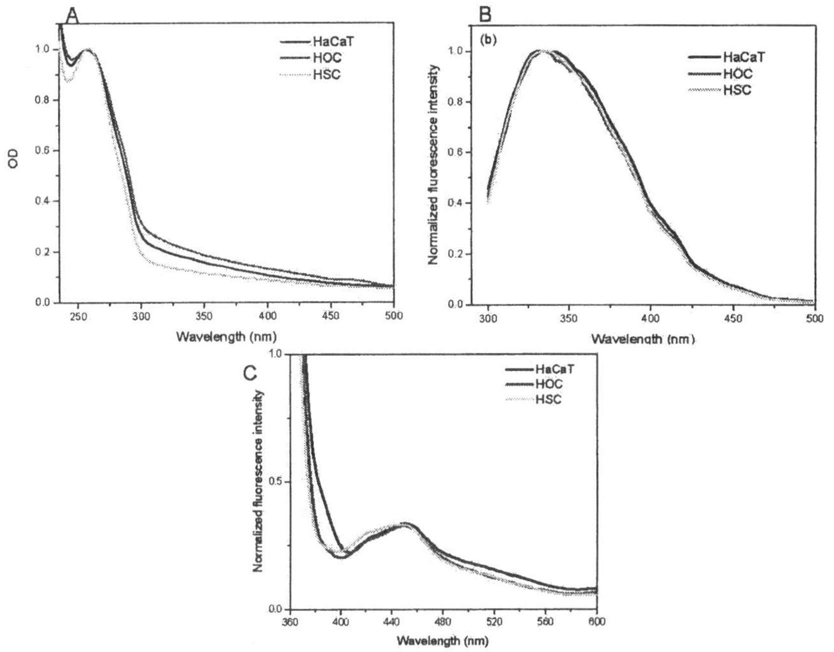


**Figure 20.14:** Absorption (A) and fluorescence spectra (B,C) of Hacat normal cells, HSC-3 cancer cells and HOC 313 cancer cells without nanoparticles. (B: $\lambda_{exci}$ = 280 nm, (C): $\lambda_{exci}$ = 340 nm. Cited from reference 52.

When cells are mixed with gold nanoparticle suspension, the nanoparticles are located in the vicinity of cells in the extracellular region. Figure 20.15 shows a comparison of the protein fluorescence spectra of whole cells with and without spherical gold nanoparticles. Obviously gold nanoparticles quench the fluorescence of cells at their protein bands when excited at protein absorption wavelength of 280 nm. There are no obvious differences between normal cells and cancer cells.

The nanoparticles can be incorporated into cells by incubation the nanoparticles suspended in DMEM medium with the cells for 12 to 48 hours. From dark field light scattering imaging [54], it has been established that gold nanoparticles are taken intracellularly into these cell lines. The nanoparticles are accumulated inside cytoplasm of live cells by receptor-mediated endocytosis processes. Figure 20.16 shows that for most of samples gold nanospheres quench the protein fluorescence at 12 h incubation time. However, in some samples fluorescence enhancement is observed.

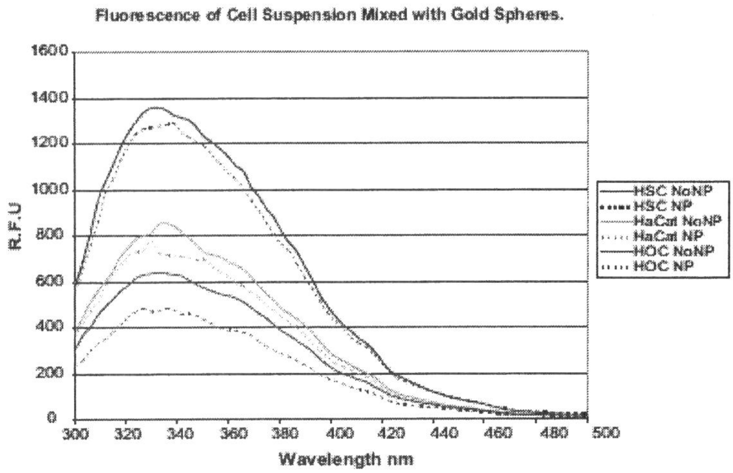


**Figure 20.15:**. Fluorescence spectra of whole cells with and without gold nanospheres. Each cell type demonstrates characteristic decrease in fluorescence for all samples tested.$\lambda_{exci}$ = 280 nm. RFU: relative fluorescence unites. NP: nanoparticles. Cited from reference 52.

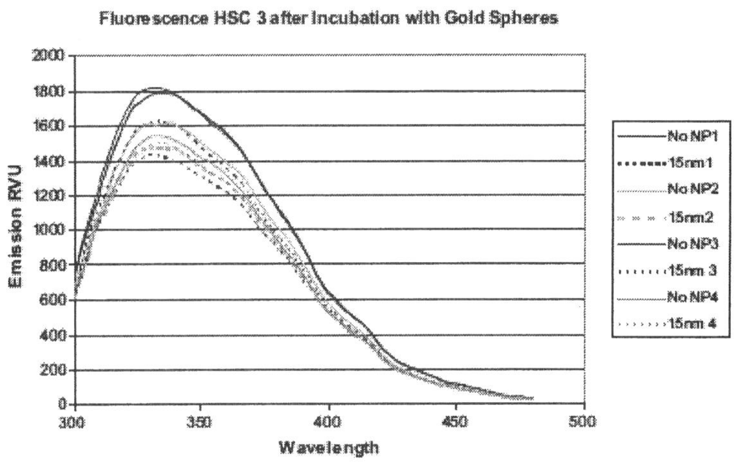


**Figure 20.16:** Fluorescence spectra of HSC 3 cells incubated with nanosphere solution. $\lambda_{exci}$ = 280 nm. Overall, gold nanospheres decrease the fluorescence of cells; however, in some samples an increase in fluorescence is observed. The data is similar for all three cell lines (HaCat, HOC 313, and HSC 3) when the nanoparticles were incubated with cells. Cited from reference 52.

Figure 20.17 shows a proportion comparison of the average amount of quenching for the normal and two cancer cell lines in the case of nanoparticles inside cells. Proportion is determined using the maximum peak value at 335 nm for each sample. 26 of 32 samples are quenched. Comparison of the mean fluorescence value with excitation at 280 nm and the peak maximum of 333nm revealed a mean relative value unit of 1244 (no NPs) and 1059 (NPs). The difference between means reveals an average quenching of 12.8% ($p<.00005$). No difference is observed for samples incubated with particles and mixed with particles in solution just prior to measurements with sample size. Some samples demonstrate an increase in fluorescence intensity. With this small sample size, no difference could be determined between cell types or incubation times with nanoparticles.

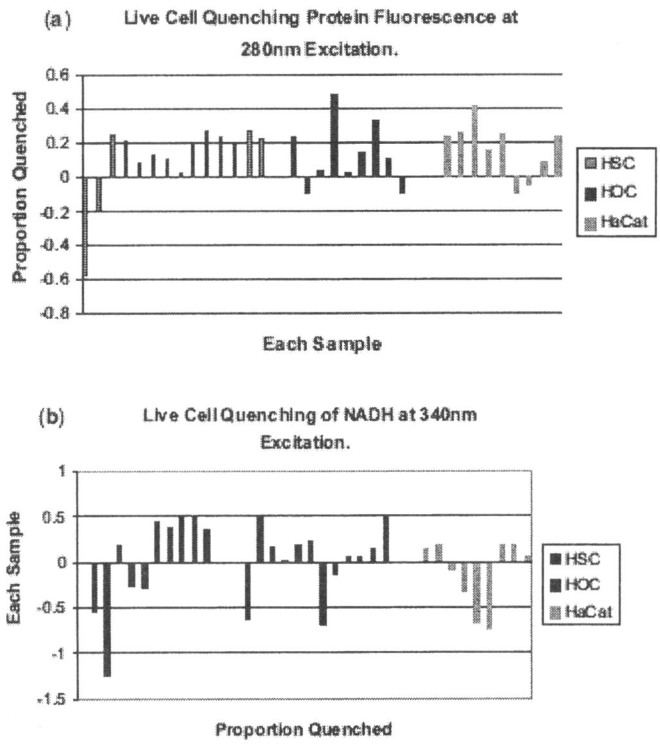


**Figure 20.17:** (**a**) Proportion of fluorescence intensity decreased by 15nm nanospheres for all samples. $\lambda_{exci}$ = 280 nm and $\lambda_{emiss}$ =335 nm corresponding to protein fluorescence. (**b**) Proportion of total NADH fluorescence quenched by nanospheres. $\lambda_{exci}$ = 340 nm and $\lambda_{emiss}$ = 440 nm corresponding with NADH fluorescence. Cited from reference 52.

## 20.5 SUMMARY

Gold nanoparticles have several potential applications in fluorescent spectroscopy due to the surface plasmon resonance that occurs on the surface of the particle. The SPR provides a highly charged electric field that can interact with photonic processes to enhance or quench fluorescent processes. Several applications exist either through direct conjugation of fluorophores to gold nanoparticles, or through proximity of the particle to the fluorophore. In the case of whole cell imaging, intrinsic fluorophores provide a source of information that may be useful for imaging and possible diagnostic applications. The mechanism of quenching of whole cell autofluorescence appears to be due to photonic absorption of light as it passes through the sample. However, through improved targeting and adjustment of the relation of the gold probe to fluorophores, it may be possible in the future to invoke other mechanisms of quenching or even enhancement of intrinsic fluorophores.

## 20.6 ACKNOWLEDGEMENTS

We like to thank the support of the Chemical Science, Geosciences and Bioscience Division of the Department of Energy (Grant DE-FG02-97ER14799), the Hearing Research Institute (Pilot Grant), San Francisco, Ca, and Randal Kramer, Department of Cell and Tissue Biology and Anatomy, University of California San Francisco.

## 20.7 REFERENCES

1. Mie, G. (1908). Contribution to the optics of turbid media, especially colloidal metal suspensions. *Ann. Phys.* **25**:377-445.
2. Gans, R. (1915). Form of ultramicroscopic particles of silver. *Ann. Phys.* **47**:270-284.
3. Huang, X. H., P. Jain, I.H. El-Sayed, M.A. El-Sayed. (2007). *Nanomed.* **2**:
4. Sönnichsen, C., Franzl T, Wilk T, Plessen GV, Feldmann J (2002). Drastic reduction of plasmon damping in gold nanorods. *Phys. Rev. Lett.* **88**:
5. Jain, P. K., Lee KS, El-Sayed IH, El-Sayed MA (2006). Calculated absorption and scattering properties of gold nanoparticles of different size, shape, and composition: applications in biological imaging and biomedicine. *J. Phys. Chem. B* **110**:
6. Link, S., El-Sayed MA (2000). Shape and size dependence of radiative, non-radiative and photothermal properties of gold nanocrystals. *Int. Rev. Phys. Chem.* **19**:
7. Boyd, G. T., Yu ZH, Shen YR (1986). Photoinnduced induced luminescennce from the nobel metals and its enhancement on roughened surfaces. *Phys. Rev. B: Condens. Matter* **33**:
8. Bouhelier, A., Beversluis MR, Novotny L (2003). Characterization of nanoplasmonic structures by locally excited photoluminescence. *Appl. Phys. Lett.* **83**:

9. Imura, K., Nagahara, T., and Okamoto, H. (2004). Plasmon mode imaging of single gold nanorods. *J. Am. Chem. Soc.* **126**:12730-12731.
10. Wang, H., Huff TB, Zweifel DA, He W, Low PS, Wei, A, Cheng JX (2005). In vitro and in vivo two-photon luminescence imaging of single gold nanorods. *Proc. Natl. Acad. Sci. U.S.A.* **102**:
11. Durr, N. J., Larson T, Smith DK, Korgel BA, Sokolov K, Ben-Yakar A (2007). Two-photon luminescence imaging of cancer cells using molecularly targeted gold nanorods. *Nano lett.* **7**:
12. Hao, E., Schatz G, Hupp J (2004). Synthesis and optical properties of anisotropic metal nanoparticles. *J. Fluor.* **14**:
13. Hao, G., Schatz GC (2004). Electromagnetic fields around silver nanoparticles and dimers. *J. Chem. Phys* **120**
14. Avouris, P., W. M. Gelbart, M. A. El-Sayed. (1977). Non-Radiative Electronic Relaxation Under Collision-Free Conditions. *Chem. Rev.* **77**:
15. Lower, S. K., M. A. El-Sayed (1966). The Triplet State and Molecular Electronic Processes in Organic Molecules. *Chem. Reviews* **66**:
16. Johansson, M. K., R.M. Cook. (2003). *Chem. Eur. J.* **9**:
17. Gersten, J., and Nitzan, A. (1981). Spectroscopic properties of molecules interacting with small dielectric particles. *J. Chem. Phys.* **75**:1139-1152.
18. Kneipp, K., Kneipp, H., Itzkan, I., Dasari, R. R., and Feld, M. S. (2002). Surface-enhanced Raman scattering and biophysics. *J. Phys. Condens. Matter* **14**:R597-R624.
19. Lakowicz, J. R. (2001). Radiative decay engineering: biophysical and biomedical applications. *Anal. Biochem.* **298**:1-24.
20. Lakowicz, J. R., Shen, Y., D'Auria, S., Malicka, J., Fang, J., Gryczynski, Z., and Gryczynski, I. (2002). Radiative Decay Engineering. 2. Effects of silver island films on fluorescence intensity, lifetimes, and resonance energy transfer. *Anal. Biochem.* **301**:261-277.
21. Lakowicz, J. R. (2005). Radiative decay engineering 5: metal-enhanced fluorescence and plasmon emission. *Anal. Biochem.* **337**:171-194.
22. Gersten, J., and Nitzan, A. (1985). Photophysics and photochemistry near surfaces and small particles. *Surf. Sci.* **158**:165-189.
23. Stryer, L. (1978). Fluorescence energy transfer as a spectroscopic ruler. *Annu. ReV. Biochem.* **47**:819-846.
24. Kim, J. H., Kim, J. S., Choi, H., Lee, S. M., Jun, B. H., Yu, K. N., Kuk, E., Kim, Y. K., Jeong, D. H., Cho, M. H., and Lee, Y. S. (2006). Nanoparticle probes with surface enhanced Raman spectroscopic tags for cellular cancer targeting. *Anal. Chem.* **78**:6967-6973.
25. Ghosh, S. K., Pal, A., Kundu, S., Nath, S., and Pal, T. (2004). Fluorescence quenching of 1-methylaminopyrene near gold nanoparticles: size regime dependence of the small metallic particles. *Chem. Phys.* **395**:366-372.
26. Maxwell, D. J., Taylor, J. R., and Nie, S. (2002). Self-assembled nanoparticle probes for recognition and detection of biomolecules. *J. Am. Chem. Soc.* **124**:9606-9612.
27. Swathi, R. S., and Sebastian, K. L. (2007). Resonance energy transfer from a fluorescent dye molecule to plasmon and electron-hole excitations of a metal nanoparticle. *J. Chem. Phys.* **126**:234701-1-234701-5.
28. Yun, C. S., Javier, A., Jennings, T., Fisher, M., Hira, S., Peterson, S., Hopkins, B., Reich, N. O., and Strouse, G. F. (2005). Nanometal surface

energy transfer in optical rulers, breaking the FRET barrier. *J. Am. Chem. Soc.* **127**:3115-3119.

29. Enger, O., Nuesch, F., Fibbioli, M., Echegoyen, L., Pretsch, E., and Diederich, F. (2000). Photocurrent generation at a fullerene self-assembled monolayermodifed gold electrode cast with a polyurethane membrane. *J. Mater. Chem.* **10**:2231-2233.
30. Akiyama, T., M. Nakada, N. Terasaki, S. Yamada. (2006). Photocurrent enhancement in a porphyrin-gold nanoparticle nanostructure assisted by localized plasmon excitation. *Chem. Commun.* 395-397.
31. Ipe, B. I., Thomas, K. G., Barazzouk, S., Hotchandani, S., and Kamat, P. V. (2002). Photoinduced charge separation in a fluorophore-gold nanoassembly. *J. Phys. Chem. B* **106**:18-21.
32. Kim, C. K., Rajamohan, R. K., Singh, J. P., Fortner, A., Griffin, J., Darbha, G. K., and Ray, P. C. (2006). Gold-nanoparticle-based miniaturized laser-induced fluorescence probe for specific DNA hybridization detection: studies on size-dependent optical properties. *Nanotechnol.* **17**:3085-3093.
33. Dulkeith, E., Ringler, M., Klar, T. A., Feldmann, J., Javier, A. M., and Parak, W. J. (2005). Gold particles quench fluorescence by phase induced radiative rate suppression. *Nano Lett.* **5**:585-589.
34. Dulkeith, E., Morteani, A. C., Niedereichholz, T., Klar, T. A., and Feldmann, J. (2002). Fluorescence quenching of dye molecules near gold nanoparticles: radiative and nonradiative effects. *Phys. Rev. Lett.* **89**:203002-203004.
35. Dubertret, B., Calame, M., and Libchaber, A. J. (2001). Single-mismatch detection using gold-quenched fluorescent oligonucleotides. *Nat. Biotechnol.* **19**:365-370.
36. Wua, Z. S., Jianga, J. H., Fua, L., Shen, G. L., and Yua, R. Q. (2006). Optical detection of DNA hybridization based on fluorescence quenching of tagged oligonucleotide probes by gold nanoparticles. *Anal. BioChem.* **353**:22-29.
37. Kim, J. H., Estabrook, R. A., Braun, G., Lee, B. R., andReich, N. O. (2007). Specific and sensitive detection of nucleic acids and RNases using gold nanoparticle-RNA-fluorescent dye conjugates. *Chem. Commun.* 4342-4344.
38. Ray, P. C., Darbha, G. K., Ray, A., Walker, J., and Hardy, W. (2007). Gold nanoparticle based FRET for DNA detection *Plasmonics* **2**:173-183.
39. Peng, Z., Chen, Z., Jiang, J., Zhang, X., Shen, G., and Yu, R. (2007). A novel immunoassay based on the dissociation of immunocomplex and fluorescence quenching by gold nanoparticles. *Anal. Chim. Acta.* **583**:40-44.
40. Kato, N., and Caruso, F. (2005). Homogeneous, competitive fluorescence quenching immunoassay based on gold nanoparticle/polyelectrolyte coated latex particles. *J. Phys. Chem. B* **109**:19604-19612.
41. Ao, L., Gao, F., Pan, B., He, R., and Cui, D. (2006). Fluoroimmunoassay for antigen based on fluorescence quenching signal of gold nanoparticles. *Anal. Chem.* **78**:1104-1106.
42. Yang, W., Trau, D., Renneberg, R., Yu, N. T., and Caruso, F. (2001). Layer-by-Layer construction of novel biofunctional fluorescent microparticles for immunoassay applications. *J. Colloid & Interface Sci.* **234**:356-362.

43. Yu, F., Persson, B., Lofas, S., and Knoll, W. (2004). Surface plasmon fluorescence immunoassay of free prostate-specific antigen in human plasma at the femtomolar level. *Anal. Chem.* **76**:6765-6770.
44. Lee, S., Cha, E. J., Park, K., Lee, S. Y., Hong, J. K., Sun, I. C., Kim, S. Y., Choi, K., Kwon, I. C., Kim, K., and Ahn, C. H. (2008). A near-infrared-fluorescence-quenched gold-nanoparticle imaging probe for in vivo drug screening and protease activity determination. *Angew. Chem. Int. Ed.* **47**:2804-2807.
45. *www.gfmer.ch/Presentations_En/ Pdf/Instrumentational_setup.pdf*
46. Jeffrey, J. (1980). Dehydrogenases Requiring Nicotinamide Coenzymes. Birkhgoldser Verlag, Basel, Boston and Stuttgart.
47. Dolphin, D., Poulson, R., and Avramovic, O. (1987). Pyridine Nucleotide Coenzymes. Wiley, New York.
48. White, A., P. Handler, E.L. Smith, R.L. Hill, I.R. Lehman. (1978). Principle of Biochemistry. McGraw-hill, New York.
49. Köng, K., Bers, M. W., and Tromberg, B. J (1997). Time-resolved and steady-state fluorescence measurements of beta-nicotinamide adenine dinucleotide-alcohol dehydrogenase complex during UVA exposure. *Photochem. Photobio. B: Biol.* **37**:91-95.
50. *www.udel.edu/chem/robinson/ courses_doc/527_S05/lecture_4.ppt.*
51. Eyre, D., and Paz, M (1984). Cross-linking in collagen and elastin. *Annu. Rev.Biochem.* **53**:717-748.
52. El-Sayed, I. H., Huang, X., Macheret, F., Humstoe, J. O., Kramer, R., and El-Sayed, M. A. (2007). Effect of plasmonic gold nanoparticles on benign and malignant cellular autofluorescence: a novel probe for fluorescence based detection of cancer. *Technol. Cancer Res. & Treat.* **6**:403-412.
53. Campbell, I. D., R.A. Dwek. (1984). Biological Spectroscopy. Menlo Park, CA, Benjamin Cummings.
54. El-Sayed, I. H., Huang, X., and El-Sayed, M. A. (2005). Surface plasmon resonance scattering and absorption of anti-EGFR antibody conjugated gold nanoparticles in cancer diagnostics: applications in oral cancer. *Nano lett.* **5**:829-834.

# INDEX